T0142268

Advances in Intelligent Systems and Computing

Volume 1069

Series Editor

Janusz Kacprzyk, Systems Research Institute, Polish Academy of Sciences,
Warsaw, Poland

Advisory Editors

Nikhil R. Pal, Indian Statistical Institute, Kolkata, India
Rafael Bello Perez, Faculty of Mathematics, Physics and Computing,
Universidad Central de Las Villas, Santa Clara, Cuba
Emilio S. Corchado, University of Salamanca, Salamanca, Spain
Hani Hagras, School of Computer Science and Electronic Engineering,
University of Essex, Colchester, UK
László T. Kóczy, Department of Automation, Széchenyi István University,
Gyor, Hungary
Vladik Kreinovich, Department of Computer Science, University of Texas
at El Paso, El Paso, TX, USA
Chin-Teng Lin, Department of Electrical Engineering, National Chiao
Tung University, Hsinchu, Taiwan
Jie Lu, Faculty of Engineering and Information Technology,
University of Technology Sydney, Sydney, NSW, Australia
Patricia Melin, Graduate Program of Computer Science, Tijuana Institute
of Technology, Tijuana, Mexico
Nadia Nedjah, Department of Electronics Engineering, University of Rio de Janeiro,
Rio de Janeiro, Brazil
Ngoc Thanh Nguyen⬤, Faculty of Computer Science and Management,
Wrocław University of Technology, Wrocław, Poland
Jun Wang, Department of Mechanical and Automation Engineering,
The Chinese University of Hong Kong, Shatin, Hong Kong

The series "Advances in Intelligent Systems and Computing" contains publications on theory, applications, and design methods of Intelligent Systems and Intelligent Computing. Virtually all disciplines such as engineering, natural sciences, computer and information science, ICT, economics, business, e-commerce, environment, healthcare, life science are covered. The list of topics spans all the areas of modern intelligent systems and computing such as: computational intelligence, soft computing including neural networks, fuzzy systems, evolutionary computing and the fusion of these paradigms, social intelligence, ambient intelligence, computational neuroscience, artificial life, virtual worlds and society, cognitive science and systems, Perception and Vision, DNA and immune based systems, self-organizing and adaptive systems, e-Learning and teaching, human-centered and human-centric computing, recommender systems, intelligent control, robotics and mechatronics including human-machine teaming, knowledge-based paradigms, learning paradigms, machine ethics, intelligent data analysis, knowledge management, intelligent agents, intelligent decision making and support, intelligent network security, trust management, interactive entertainment, Web intelligence and multimedia.

The publications within "Advances in Intelligent Systems and Computing" are primarily proceedings of important conferences, symposia and congresses. They cover significant recent developments in the field, both of a foundational and applicable character. An important characteristic feature of the series is the short publication time and world-wide distribution. This permits a rapid and broad dissemination of research results.

** Indexing: The books of this series are submitted to ISI Proceedings, EI-Compendex, DBLP, SCOPUS, Google Scholar and Springerlink **

More information about this series at http://www.springer.com/series/11156

Kohei Arai · Rahul Bhatia · Supriya Kapoor
Editors

Proceedings of the Future Technologies Conference (FTC) 2019

Volume 1

 Springer

Editors
Kohei Arai
Faculty of Science and Engineering
Saga University
Saga, Japan

Rahul Bhatia
The Science and Information
(SAI) Organization
Bradford, West Yorkshire, UK

Supriya Kapoor
The Science and Information
(SAI) Organization
Bradford, West Yorkshire, UK

ISSN 2194-5357   ISSN 2194-5365   (electronic)
Advances in Intelligent Systems and Computing
ISBN 978-3-030-32519-0   ISBN 978-3-030-32520-6   (eBook)
https://doi.org/10.1007/978-3-030-32520-6

© Springer Nature Switzerland AG 2020
This work is subject to copyright. All rights are reserved by the Publisher, whether the whole or part of the material is concerned, specifically the rights of translation, reprinting, reuse of illustrations, recitation, broadcasting, reproduction on microfilms or in any other physical way, and transmission or information storage and retrieval, electronic adaptation, computer software, or by similar or dissimilar methodology now known or hereafter developed.
The use of general descriptive names, registered names, trademarks, service marks, etc. in this publication does not imply, even in the absence of a specific statement, that such names are exempt from the relevant protective laws and regulations and therefore free for general use.
The publisher, the authors and the editors are safe to assume that the advice and information in this book are believed to be true and accurate at the date of publication. Neither the publisher nor the authors or the editors give a warranty, expressed or implied, with respect to the material contained herein or for any errors or omissions that may have been made. The publisher remains neutral with regard to jurisdictional claims in published maps and institutional affiliations.

This Springer imprint is published by the registered company Springer Nature Switzerland AG
The registered company address is: Gewerbestrasse 11, 6330 Cham, Switzerland

Editor's Preface

After the success of three FTCs, the Future Technologies Conference (FTC) 2019 was held on October 24–25, 2019, in San Francisco, USA, a city of immense beauty—be it the charming neighborhoods such as Castro/Upper Market or landmarks such as the Golden Gate Bridge—and even denser fog. FTC 2019 focuses on technological breakthroughs in the areas of computing, electronics, AI, robotics, security, and communications.

The ever-changing scope and rapid development of computer technologies create new problems and questions, resulting in the real need for sharing brilliant ideas and stimulating good awareness of this important research field. The aim of this conference is to provide a worldwide forum, where the international participants can share their research knowledge and ideas on the recent and latest research and map out the directions for future researchers and collaborations.

For this conference proceedings, researchers, academics, and technologists from leading universities, research firms, government agencies, and companies from 50+ countries submitted their latest research at the forefront of technology and computing. After the double-blind review process, we finally selected 143 full papers including 6 poster papers to publish.

We would like to express our gratitude and appreciation to all of the reviewers who helped us maintain the high quality of manuscripts included in this conference proceedings. We would also like to extend our thanks to the members of the organizing team for their hard work. We are tremendously grateful for the contributions and support received from authors, participants, keynote speakers, program committee members, session chairs, steering committee members, and others in their various roles. Their valuable support, suggestions, dedicated commitment, and hard work have made FTC 2019 a success.

We hope that all the participants of FTC 2019 had a wonderful and fruitful time at the conference and that our overseas guests enjoyed their sojourn in San Francisco!

Kind Regards,
Kohei Arai

Contents

Project BUMP: Developing Communication Tools
Claudia B. Rebola and Shi He

VEDAR: Accountable Behavioural Change Detection

Amit Kumar[(⊠)], Tanya Ahuja, Rajesh Kumar Madabhattula, Murali Kante, and Srinivasa Rao Aravilli

Department of Innovation @Cisco Systems, Inc., San Jose, USA
{amitkku,tanahuja,rmadabha,mukante,saravill}@cisco.com

Abstract. With exponential increase in the availability of telemetry/streaming/real time data, understanding contextual behavior changes is a vital functionality in order to deliver unrivalled customer experience and build high performance and high availability systems. Real time behavior change detection finds a use case in number of domains such as social networks, network traffic monitoring, ad exchange metrics, etc. In streaming data, behavior change is an implausible observation that does not fit in with the distribution of rest of the data. A timely and precise revelation of such behavior changes can give substantial information about the system in critical situations which can be a driving factor for vital decisions. Detecting behavior changes in streaming fashion is a difficult task as the system needs to process high speed real time data and continuously learn from data along with detecting anomalies in a single pass of data. This paper illustrates a novel algorithm called Accountable Behavior Change Detection (VEDAR) which can detect and elucidate the behavior changes in real time and operates in a fashion similar to human perception. The algorithm is bench-marked by comparing its performance on open source anomaly data sets against industry standard algorithms like Numenta HTM and Twitter AdVec (SH-ESD). The proposed algorithm outperforms above mentioned algorithms for behaviour change detection, efficacy is given in Sect. 5.

Keywords: Telemetry · Real time · Streaming · Accountable · Behavioural change

1 Introduction

Accountable behavior change detection is a fundamental problem in data mining which has been well explored in the past few decades. But with an exponential increase in the availability of telemetry data, there is a need of a behavioral change detection system that can work well in dynamic and streaming environment. The expansion of Internet of Things (IOT) has added innumerable sources of Big Data into the Data Management landscape. Cloud-Servers, smart

© Springer Nature Switzerland AG 2020
K. Arai et al. (Eds.): FTC 2019, AISC 1069, pp. 1–17, 2020.
https://doi.org/10.1007/978-3-030-32520-6_1

phones, sensors on machines, all generate huge amount of real time and continuously changing data for IOT. This leaves us to reconsider the problem of behavior change detection in streaming fashion.

Real time accountable behavioral change detection is gaining practical and significant usage across many industries like analysis of network traffic, monitoring real time tweets, analyzing trends in online shopping, monitoring ad exchange data, monitoring system resource utilization in cloud servers and many more. The significance of behavior change detection is due to the fact that abnormalities in data often lead to significant, often critical, actionable information.

In this paper, change in behavior is referred as an observation or a group of observations that are significantly different from rest of the data or the patterns that do not conform to the notion of normal behavior. Abnormal behavior in data can either be spatial or temporal. In spatial behavior change, an individual data point is significantly different from rest of the data, independent of its location in the data stream. The spikes in Fig. 3 represent instances of spatial behavior change. Temporal anomalies occur when a data point is abnormal only in specific temporal context, e.g. the first anomalous point (marked in green) in Fig. 4. Such deviations may be caused by a positive factor like increased traffic to a site or a negative reason like CPU utilization exceeding the limits. But in either case such changes can lead to actionable intelligence.

Behavioral change detection has traditionally been handled using rule-based techniques applied to static data in batches. But with the number of scenarios out-growing in streaming data, such techniques are difficult to scale. Moreover, even with huge amount of streaming data available, the chances of such data being labelled is very rare which moves the traditional classification algorithms out of scope. The behavioral change detection systems face two major challenges in streaming environment. Firstly, streaming analytics requires models that can learn continuously in real time, without storing the entire data, have low operational complexity and are fully automated. Secondly, the definition of anomaly continuously changes as systems evolve and behaviors change. With streaming data being dynamic, the model needs to re-train quickly and adjust to the changing data distribution in real time.

This paper illustrates a novel algorithm that detects changes in data behavior in real time. The algorithm takes into consideration, the factors such as seasonal repetition of values and trend in data while determining behavior changes. This system can be applicable to streaming data with different scales like number of bytes written to disk/sec and CPU utilization percentage as it employs a data normalization module. It uses data-driven, dynamic rules to detect abnormalities that can quickly detect sudden changes and also adjust to long term changes in statistics of data. The resulting system is computationally efficient and does not require any prior parameter adjustment. This algorithm provides an edge over existing systems as it adds the feature of accountability. All these layers will be explained in greater depth in Sect. 3. An end to end deployment architecture of VEDAR is shown in Fig. 1.

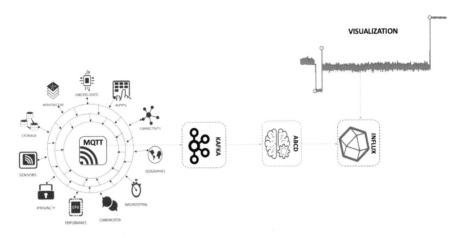

Fig. 1. The figure shows Accountable Behavior Change detection deployment architecture in Cloud, IOT, Networking, Operations, Management and Services.

2   Related Work

Behavioral change detection in telemetry/streaming data has been a well-explored research area in past few decades. Some classical statistical algorithms like setting threshold, moving average and moving median were extensively used for batch anomaly detection. Some other techniques like clustering and exponentially weighted moving average EWMA [12] have also been studied as a solution for behavioral change detection in streaming data. But these techniques can only be used to detect spatial abnormalities. These techniques are capable of detecting any abrupt change in the process behavior but they fail to adjust to gradual shift in distribution of data. Moreover, these techniques are incapable of taking factors like seasonality and trend into consideration.

A well-known technique for detecting spatial as well as temporal anomalies in streaming data is change point detection technique [1]. These techniques are fast and are also applicable to multivariate time series data. The only limitation of these algorithms is that their performance is sensitive to hyper parameters such as window size and thresholds. Due to this they might result in more false positives. These methods also do not take seasonality into account.

Another algorithm for detecting temporal behavior changes is ARIMA [5]. It is better than above mentioned algorithms as it is a combination of both auto-regressive and moving-average. Seasonal-ARIMA [2] can also detect seasonality patterns of fixed periods like weekly, daily seasonality. But it fails to re-adjust to shifts in the seasonal patterns in real time.

Deep learning algorithms like LSTM autoencoders [10] have also been explored for detecting behavior changes. It reconstructs models where some form of reconstruction error is used as a measure of a behaviour change. Deep learning models have to be re-trained frequently in order to stay updated with new data. Also, they require huge amount of data for training purpose.

Numenta introduced a machine learning algorithm derived by neuroscience called Hierarchical Temporal Memory [3] which models the spatial and temporal patterns in streaming data. HTM performs better than above mentioned algorithms as it comprehends the seasonal and trend factors in the streaming data and also adjusts with changing data statistics. But HTM does not provide any accountability for detected abnormalities and the cause and type of detected anomalies cannot be interpreted.

Twitter also released an open-source algorithm for detecting both spatial and temporal anomalies in streaming analytics called TwitterAdVec (SH-ESD) [8]. Although this algorithm models the seasonal patterns, just like HTM, it does not provide any accountability for the detected anomalies. Moreover, the precision of SH-ESD is less as compared to VEDAR because of large number of false positives. Comparison of results of proposed algorithm with HTM and Twitter Anomaly Detection Algorithm in the comparison section.

3   VEDAR for Real Time Accountable Behavioral Change Detection

For the purpose of this paper, behavior change is defined as an observation or a group of observations that deviate significantly from underlying distribution of rest of the data. This paper focuses primarily on three types of behavior changes: (i) Seasonality interruption change: when the observation deviates from expected seasonal value, illustrated in Fig. 2, (ii) Erratic change: an abrupt transient change that is short lived, as shown in Fig. 3, (iii) Linear change: when observations gradually proceed towards abnormal behavior, depicted in first anomaly point in Fig. 4. For the task of change detection, it is vital to detect erratic changes, initially alert in case of seasonality interruption and have the model must quickly re-adjust to such change and also alert in case the signal is gradually approaching towards abnormal behavior.

In order to identify change in signal behavior, VEDAR scales the input signal, redirects the normalized signal to analyze any non-stationary factors like trend and seasonality. If either of these are present, the signal is made stationary by subtracting out trend and seasonal components. The residual is further smoothened out using exponential and linear quadratic smoothening techniques. The smoothened signal is further utilized by Non-Parametric Estimation module in order to generate likelihood of the observation belonging to the underlying distribution. The system detects an event stream as an anomaly based on empirical rules. Figure 5 illustrates a block diagram of proposed algorithm.

Each layer of VEDAR is explained in following subsections:

3.1   Layer 1 - Data Comprehension

In the first layer, system understands the nature of streaming data by extracting features like trend, seasonality and scale of data. According to the observed

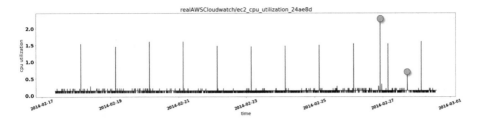

Fig. 2. The figure is a real-world CPU utilization data from AWS Cloud Watch collected over 5 min interval. Data contains periodic spikes occurring at a frequency of 1 day. The anomalous points are caused by interruption of seasonality.

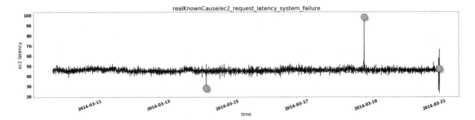

Fig. 3. The figure is a real-world ec2-request-latency-system-failure data collected from AWS Cloud Watch servers over 5 min interval. Anomalous points depict an erratic change in system behavior.

characteristics of data, the algorithm processes the input signal in order to perform equally well on datasets with different characteristics. This layer consists of three sub-modules namely, data scaling, trend and seasonality extraction. These sub-layers are explained in below subsections.

Data Scaling. Scaling of data is a feature in order to deal with data sets of varying scales and also to detect anomalies at desired granularity. For instance, if the input data is in bytes, algorithm is able to detect behavior changes at the scale of Gigabytes by scaling the entire data to Gigabytes. This provides the user with the flexibility to choose the granularity of detected behavior changes.

VEDAR is capable of detecting behavior changes in input signals with different scales like number of bytes written to disk/sec which can range from 0 to hundreds of Gigabytes as well as percentage CPU utilization which lies between 0 to 100. In order to operate equally well on such different scales, the system internally maintains a data-driven scaling/normalization factor. The input signal is divided by this normalization factor to scale it. The value of this scaling factor can be controlled in two ways: (1) Users can explicitly provide a value for the scaling factor based on the granularity of data and percentage of data expected as anomalies. For instance, if the streaming data denotes the number of bytes written to disk/sec and the user expects the fluctuations of Gigabytes/sec to be detected as behavior changes then a scaling factor of 1024 or 1024 * 1024

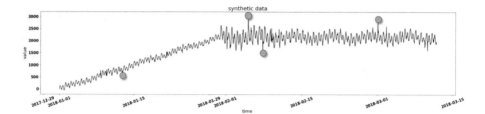

Fig. 4. The figure is an artificial data depicting observations linearly trending towards the upper limit followed by periodic data with random spikes. First anomaly point captures the linear trend and second and third anomalies depict erratic changes in signal value.

is appropriate. This converts entire data into Gigabytes scale and behavior changes are detected at this scale. (2) Alternatively, VEDAR automatically determines the scaling factor by observing set of data points. The normalization factor is dynamic and is computed frequently by the system, thus enabling us to detect any behavior change.

Seasonality and Trend Extraction. As already mentioned, some data streams are seasonal in nature i.e., the system behavior changes periodically resulting in periodic fluctuations. It is essential to detect any periodicity present in data in order to avoid false alarms in case of periodic behavior changes. In absence of seasonality and trend extraction module, the algorithm would raise an alarm at every seasonal fluctuation which leads to huge number of false positives.

Since the input signal might be non-stationary, VEDAR first extracts the trend and seasonal components from the signal in order to make it stationary. As already mentioned, in a streaming data some of the observations or behavior changes occur quasi-periodically. That is, values in same range occur at a similar time every hour/day/week/year. In order to calculate the periodicity of data, the proposed algorithm uses YIN [13], an auto-correlation [6] based method which is explained in greater depth in. In YIN, the difference function is calculated as follows:

$$D'(\rho, t_j) = \frac{D(\rho, t_j)}{\frac{1}{p}\sum_{k=1}^{\rho} D(k, t_j)} \tag{1}$$

where,

$$D(\rho, t_j) = \sum_{k=1}^{h-\rho}(C_\xi^k(t_j) - C_\xi^{k+\rho}(t_j))^2 \tag{2}$$

The terms of above function are explained clearly in section 3.4.2 of [13]. On calculating correlation values D' for each possible value of ρ, algorithm detects the lowest value (troughs) in the de-trended correlation values. The frequency of periodicity is the value ρ, whose correlation corresponds to the lowest trough. Let us denote this value by ρ'.

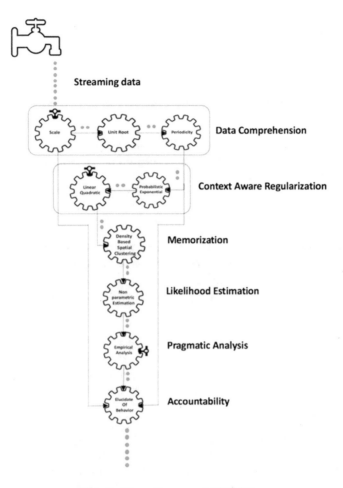

Fig. 5. Flow diagram of VEDAR.

The scaled input signal is first de-seasonalized if any periodicity is detected in the previous observations. In order to de-seasonalize the input signal, a window of size w is taken around the previous seasonal data point (observation occurring exactly ρ' points prior to current point in the data stream) and current data point is subtracted from the most similar observation in the seasonal window.

VEDAR outperforms a number of available algorithms by quickly adapting to changes in periodicity of data. As illustrated in Fig. 6, seasonality of data varies with time. VEDAR adapts to such change in periodicity by continuously updating the value of ρ'. At the end of each day, periodicity value ρ' is re-calculated based on past 2 weeks of data.

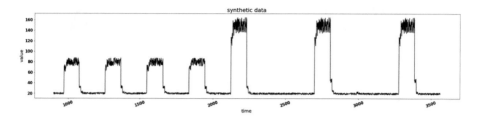

Fig. 6. The figure is an artificial data generated at a frequency of 5 min. The data illustrates periodic steps after every 24 h. At certain point, periodicity shift from 24 h to 48 h is evident in the graph.

3.2 Layer 2 - Context Aware Regularization

Despite of the presence of seasonality in data, the odds of such seasonality being perfect are very rare i.e., periodic fluctuations are seldom of equal magnitudes. The residual signal obtained after seasonality and trend extraction incorporates a lot of white noise which is caused as a consequence of disproportionate periodic behavior changes.

The presence of white noise in data leads to high number of false positives as any fluctuation caused as a consequence of this white noise is detected by the algorithm as behavior change. In order to eliminate white noise from data VEDAR performs a context aware smoothening of the residual signal. Figure 7(a) shows the synthetic signal data and Fig. 7(b) shows the de-seasonalized signal data containing white noise.

VEDAR performs two levels of smoothening in order to eliminate white noise from the signal while keeping the behavior changes intact. The primary smoothening step which evens out very small fluctuations is probabilistic exponential smoothening. Once the smaller variations are handled, second level of smoothening is applied which tackles any persisting fluctuations while leaving significant behavior changes unaffected. This layer is called linear quadratic smoothening. The necessity for two levels of smoothening arises in order to perform constrained smoothening. The following subsections explain the smoothening techniques used in VEDAR:

Probabilistic Exponential Smoothening. This is an advanced version of Exponential smoothening [12] for time series data. Exponential moving average computes local mean μ_t of a time series X_t by applying exponentially decreasing weight factors to the past observations. Weighting factor α determines the amount of weight given to historic values.

$$\mu_t = \alpha\mu_{t-1} + (1 - \alpha)X_t \tag{3}$$

Due to fixed weighting factor α, moving mean μ_t is highly susceptible to any abrupt behavior.

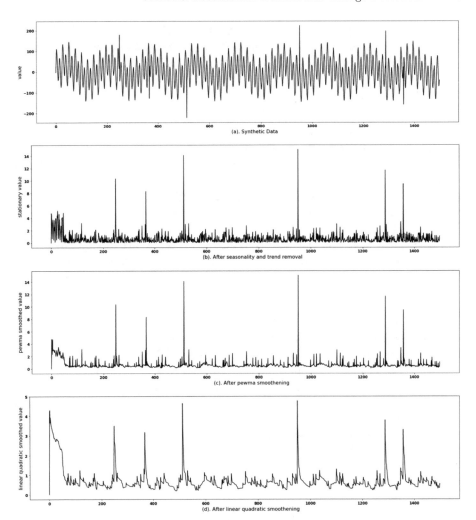

Fig. 7. The figure contains 4 graphs. First graph shows a synthetic data generated at a frequency of 5 min. Data contains periodic peaks and troughs after every 224 data points. Second graph displays the de-seasonalized and de-trended signal. Data in Graph 2 possesses a lot of white noise. Third graph shows smoothened signal obtained after applying probabilistic exponential smoothening. It levels out smaller variations. Fourth graph contains the residual signal obtained by applying Linear Quadratic smoothening on the previous signal.

Probabilistic Exponential Smoothening which is a modified version of PEWMA [9] adjusts the weighting parameter α based on probability of current observation X_t.

$$\mu_t = \alpha(1 - \beta P_t)\mu_{t-1} + (1 - \alpha(1 - \beta P_t))X_t \tag{4}$$

Where P_t is the probability of X_t based on underlying data distribution and β is the weightage given to P_t. For the purpose of smoothening, any observation with $|X_t - \mu_t| < 3\alpha_t$ is substituted with whereas the observations that are greater than $3\alpha_t$ away from μ_t are kept unaffected. PEWMA smoothened signal is displayed in Fig. 7(c).

Linear Quadratic Smoothening. Linear quadratic smoothening [4] is a widely used technique for signal smoothening. It begins by predicting the next value based on previous observations. Based on the deviation of this predicted value from the actual data point, the predicted value is either directly utilized as smoothened value or the system is trained with actual value to reduce the error term. The signal obtained after applying Linear Quadratic Smoothening is shown in Fig. 7(d).

Linear Quadratic Smoothening consists of two steps: (1) Prediction: Linear Quadratic Smoothening produces estimates of the current state variables, along with their uncertainties.

$$X^- = FX \tag{5}$$

$$P^- = FPF^T \tag{6}$$

where X^- is the predicted value, X is the value predicted in the previous step. F is state transition matrix which computes X^- given X (For this use case, value of F is taken as 1). P^- is the predicted process co-variance matrix and P is the process co-variance matrix predicted in the previous step. These are used to model the variance of the system. (2) Error correction: Once the outcome of the next measurement (necessarily corrupted with some amount of error, including random noise) is observed, these estimates are updated using a weighted average, with more weight being given to estimates with higher certainty.

$$S = HP^-H^T \tag{7}$$

$$K = P^-H^TS^{-1} \tag{8}$$

$$Y = Z - HX^- \tag{9}$$

$$X = X^- + KY \tag{10}$$

$$P = (I - KH)P^- \tag{11}$$

where H is measurement function used to transform state variables into measurement space and S is the transformation of Process co-variance matrix in measurement space. K is Kalman gain is the ratio between uncertainty in prediction and uncertainty in measurement. Y is the deviation of actual data point Z from the predicted value X^-. X is the updated estimate value which is weighed sum of the actual prediction X^- and the residual Y. Kalman gain K determines the weight given to the residual value Y in order to update the predicted value. State co-variance matrix P is also updated using Kalman gain and measurement function H.

Details of Linear quadratic smoothening technique are available in [4].

3.3 Layer 3 - Memorization

In order to distinguish behavior changes from rest of the observations, some technique is required to segregate and memorize the past behavior. The absence of context knowledge leads to high number of false alarms as the algorithm fails to learn from previously available data.

Distribution generated over the past behavior is used by the algorithm to determine that any given observation is anomalous or not. Generating a distribution over entire historic data is a tedious job as it would increase the memory requirement along with slowing down the system. In order to resolve this issue, VEDAR samples data points from historic data.

VEDAR uses Density based Spatial Clustering technique (DBSCAN) [7] to segregate normal and anomalous behaviors. Samples from these clusters are then used to generate distribution for the data points which is used by the subsequent layer. Since streaming data can be highly dynamic, it leads to the formation of unspecified number of clusters with different densities. As DBSCAN can perform well on such datasets, it seemed to fit the use case.

3.4 Layer 4 - Likelihood Estimation

In order to determine the probability of any given data point belonging to the underlying signal distribution, an underlying data distribution needs to be established in the first place. VEDAR employs KDE [11] for this purpose.

Kernel Density Estimation is a non-parametric estimation technique which models combinations of distributions. KDE estimates the probability density function of random variables. It is applied at two different steps in VEDAR. (1) Each cluster of historic data generated by DBSCAN is supposed to have an underlying distribution. These distributions are modeled using KDE. (2) Once the distributions are identified, samples are generated from them using random sampling. The sampled data points from historic data along with current data buffer are used to train KDE for determining the likelihood of any given observation.

Given an input data point, KDE determines its likelihood by applying Kernel function. For the purpose of behavior change detection VEDAR uses Gaussian kernel.

$$P_k(y) = \sum_{i=1}^{N} K((y - x_i)/h) \tag{12}$$

where K is Gaussian kernel and h is the bandwidth which is a hyper-parameter controlling the spread of distribution over data. N is number of data points. y is data point for which likelihood is to be computed.

3.5 Layer 5 - Pragmatic Analysis

Empirical rule is often used in statistics for the purpose of forecasting. VEDAR uses empirical rule on the likelihood of data points to determine if given data point is anomalous or not. The algorithm applies two set of rules for analyzing behavior of streaming data.

(1) Empirical rule or Three Sigma Rule - If the likelihood of current data point deviates from the previous data point's likelihood by more than three sigma value, then the current observation is considered to be a significant behavior change. This rule fails to address the issue of linearly dropping probabilities.
(2) In case of observations gradually deviating from the normal behavior, the likelihood of each upcoming observation approaches more and more towards 0. In this case, the behavior change can not be identified with empirical rule as the likelihoods of both of the consecutive observations is exponentially small (10^{-10} scale and below) and so is the likelihood difference. In order to handle the case where the observations are linearly deviating from the underlying data distribution, VEDAR monitor the scale change in the likelihood of a set of data points rather than monitoring the actual difference of likelihood.

3.6   Layer 6 - Accountability

Accountability is the layer of VEDAR which makes it superior to rest of the algorithms available for behavior change detection. In this layer, the algorithm rationalizes the cause of such behavior change based on the type of anomaly occurred. Whenever a behavior change happens, the algorithms raises an alarm to the user which contains the following information: actual value, expected value and the type of behavior change detected. Based on the type of behavior change, user can take appropriate actions.

The three types of behavior changes detected by VEDAR have been explained earlier in Sect. 3. These are seasonality interruption change, erratic change and linear changes. Please refer to Figs. 2, 3 and 4 for visualization of these three categories.

Any behavior changes detected in signal containing seasonality are identified as seasonality interruption changes as the system behaves differently from the expected seasonal behavior.

In the datasets which do not contain periodicity, any sudden probability drop of an observation is denoted as erratic change whereas the linearly dropping scale of likelihood is designated as linear change.

Table 1. Performance results of HTM, Twitter anomaly detection and VEDAR on NAB datasets.

DataSet	Algorithm	True positive	False positive	False negative	Precision	Recall	F1-Score
realKnownCause/ec2_request_latency _system_failure	HTM	3	9	0	0.25	1	0.4
	TwitterAdVec	3	5	0	0.38	1	0.55
	VEDAR	3	1	0	0.75	1	0.86
realAWSCloudwatch/rds_cpu _utilization_e47b3b	HTM	2	2	0	0.5	1	0.67
	TwitterAdVec	1	7	1	0.13	0.5	0.2
	VEDAR	2	0	0	1	1	1
realTraffic/occupancy_6005	HTM	1	1	0	0.5	1	0.67
	TwitterAdVec	1	3	0	0.25	1	0.4
	VEDAR	1	0	0	1	1	1
realTweets/Twitter_volume_FB	HTM	2	4	0	0.33	1	0.5
	TwitterAdVec	2	26	0	0.07	1	0.13
	VEDAR	2	1	0	0.67	1	0.8
realAdExchange/exchange- 4_cpm_results	HTM	3	4	0	0.43	1	0.6
	TwitterAdVec	2	1	1	0.67	0.67	0.67
	VEDAR	3	1	0	0.75	1	0.86

4 Results

The proposed algorithm has been tested on the Numenta Anomaly Benchmark Dataset. The dataset comprises of real time anomaly detection data collected over five different domains. This section contains the results of VEDAR on one dataset from each category. Each of the Figs. 8, 9, 10, 11 and 12 represent a real-world telemetry data belonging to each of the categories. In each of the following graphs, the actual behavior changes, as provided by Numenta are marked by green dots while the behavior changes detected by VEDAR algorithm are marked by yellow dots. The overlapped green and yellow dots represent the correctly detected behavior changes/true positives while the individual green and yellow dots represent the false negatives and false positives, respectively.

The results of VEDAR have been compared with Numenta HTM and TwitterAdVec (SH-ESD) algorithm as well. The accuracy metrics used for this comparison are number of True Positives, False Positives, False Negatives, precision, recall and F1-Score.

The final comparison of all the three algorithms over these five datasets is shown in Table 1 in the comparison section.

Figure 8 shows the real CPU utilization data collected over AWS servers by Amazon Cloud Watch Service. The observations collected over a frequency of 5 min range between 0 to 100 as the data is percentage of CPU utilized. Other data sets in this category contain AWS server metrics like Network bytes In, Disk write bytes, CPU utilization etc. The data snapshot exhibits 2 significant behavior changes, first one being an erratic spike and the second one is a sudden drift in the distribution of data.

Figure 9 illustrates a traffic occupancy dataset from the category of Real Traffic over a frequency of 5 min. This category contains real time traffic data from

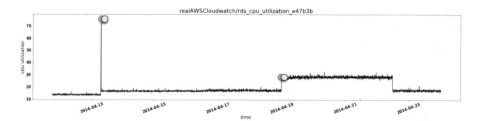

Fig. 8. The figure shows real CPU Utilization data collected by Amazon Cloud Watch Service over a frequency of 5 min. The data contains two actual significant behavior changes, both of which are detected accurately by VEDAR algorithm as represented by overlapped yellow and green dots.

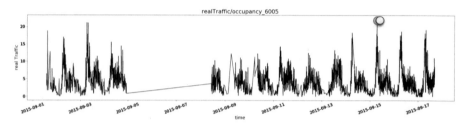

Fig. 9. The figure is a real traffic occupancy data collected over a frequency of 5 min. The data contains 1 significant behavior change marked by yellow and green overlapped dots. The change in behavior belongs to the category of periodicity interruption as the observed values are significantly larger than expected.

the twin Cities Metro area in Minnesota, collected by the Minnesota Department of Transportation. Metrics captured by the sensors in this domain include occupancy, speed and travel time. The dataset contains periodic repetition of values after every 14 h. Straight line in the figure indicates missing data. Behavior of the data deviates from the expected seasonal behavior once in the given snapshot which is accurately detected by VEDAR algorithm as shown in the figure.

Figure 10 shows a dataset from real Ad-Exchange category. This dataset captures cost per thousand impressions (CPM) which is a metric for online advertisement clicking rates. The dataset is collected over frequency of 1 h. The data contains rare erratic spikes which are detected by VEDAR accurately. While VEDAR detected all the actual behavior changes correctly, it also detected 1 behavior change (false positive) which is not mentioned by NAB.

Another category of datasets provided by Numenta is Real Tweets which is a collection of Twitter mentions of large publicly traded companies like Google, Facebook etc. Dataset used for demonstration in this paper contains the number of mentions of Facebook in tweets every 5 min. Results of VEDAR on Facebook Real tweets is illustrated in Fig. 11. The snapshot of data contains 2 significant

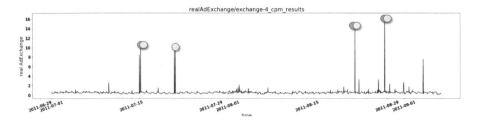

Fig. 10. The figure is a real ad exchange dataset which measures cost per thousand impressions. Data contain 4 significant spikes as illustrated in the figure.

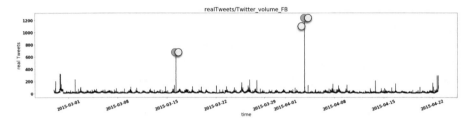

Fig. 11. The figure shows tweet mentions of Facebook in 5 min intervals. Dataset contains two sudden spikes detected by VEDAR. These behavior changes are displayed by overlapping green and yellow dots.

behavior changes which are correctly detected by VEDAR, while giving one false positive. Both of the behavior changes are sudden short-lived spikes.

The ec2-Request Latency System Failure dataset illustrated in Fig. 12 belongs to real Known Cause category. This dataset presents CPU utilization data from a server from Amazon's east coast data center. The data snapshot contains 2 erratic spikes and ends with a complete system failure as shown in Fig. 12. No periodicity is present in this dataset.

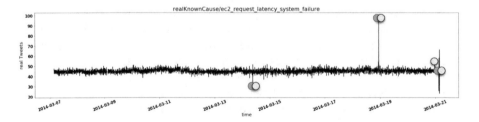

Fig. 12. The figure represents the CPU utilization data from one of Amazon's AWS server collected over a frequency of 5 min. As evident in the figure, data contains 4 behavior changes/spikes detected by VEDAR. The last behavior change represents complete system failure. VEDAR detected 3 true behavior changes along with one false alarm depicted in overlapped dots and an individual yellow dot, respectively.

5   Comparison

This section contains the consolidated results of three algorithms: VEDAR, HTM and Twitter Anomaly Detection on one dataset from each of the 5 domains from the Anomaly Benchmark datasets provided by *Numenta*. Metrics used for comparison of accuracy are number of True positives, false positives, false negatives, precision, recall and F1 score. HTM and Twitter Anomaly Detection results have been taken from *numenta* and *twitterAdVec*, respectively. The below formula is used for HTM and TwitterAdVec to determine anomaly:

$$anomaly\_score >= 1 - \epsilon \tag{13}$$

where, $\epsilon = 0.01$.

As it is evident from the results in Table 1, VEDAR gives the least number of false positives on all the datasets while not missing any of the actual behavior changes at the same time (0 false negatives). Precision and recall of VEDAR is higher than both HTM and TwitterAdVec.

where,

$$precision = \frac{truepositives}{truepositives + falsepositives} \tag{14}$$

$$recall = \frac{truepositives}{truepositives + falsenegatives} \tag{15}$$

$$F1 - score = 2 * \frac{precision * recall}{precision + recall} \tag{16}$$

6   Conclusion

This paper illustrates a novel algorithm for behaviour change detection that can work well in dynamic real time environment. With the exponential increase in availability of connected real time sensors, behaviour change detection is gaining much importance as an application of Machine learning in IOT.

As the results describe, VEDAR produces best in class results for behaviour change detection on NAB datasets with least number of false positives. While VEDAR is robust to both spatial and temporal anomalies, it also proves to be capable of adapting to changes in generative model of data. It detects both abrupt behaviour changes and slowly growing abnormal behaviour equally well. It is computationally efficient and needs no prior tuning of parameters.

The future extensions for VEDAR include the application of algorithm for multivariate version. VEDAR works as an ensemble of multiple layers where each layer performs a specific operation. In order to further improve the accuracy of the system, exploring other models for individual layers could potentially emerge useful.

References

1. Adams, R.P., Mackay, D.J.C.: Bayesian Online Changepoint Detection, vol. 1, p. 7 (2007)
2. Bustoni, I.A., Permanasari, A.E., Hidayah, I.: SARIMA(Seasonal ARIMA) implementation on time series to forecast the number of malaria incidence, vol. 1 (2013)
3. Ahmad, S., Purdy, S.: Real-time anomaly detection for streaming analytics (2016)
4. Aravkin, A.Y., Burke, J.V., Pillonetto, G.: Optimization viewpoint on Kalman smoothing, with applications to robust and sparse estimation (2013)
5. Bianco, A.M., Garcia Ben, M., Martinez, E.J., Yohai, V.J.: Outlier detection in regression models with ARIMA errors using robust estimates, vol. 1 (2001)
6. De Cheveigné, A., Kawahara, H.: YIN, a fundamental frequency estimator for speech and music, vol. 1 (2002)
7. Ester, M., Kriegel, H.P., Sander, J., Xu, X.: A density-based algorithm for discovering clusters in large spatial databases with noise, pp. 226–231 (1996)
8. Vallis, O.S., Hochenbaum, J., Kejariwal, A.: Automatic anomaly detection in the cloud via statistical learning, vol. 1 (2017)
9. Streilein, W.W., Carter, K.M.: Probabilistic reasoning for streaming anomaly detection (2012)
10. Malhotra, P., Ramakrishnan, A., Anand, G., Vig, L., Agarwal, P., Shroff, G.: LSTM-based encoder-decoder for multi-sensor anomaly detection, vol. 2 (2016)
11. Henderson, D.A., Jonesa, M.C.: Maximum likelihood kernel density estimation: on the potential of convolution sieves
12. Roberts, S.W.: Control chart tests based on geometric moving averages. Technometrics $1(3)$, 239–250 (1959)
13. Zhang, S., Liu, Y., Meng, W., Zhang, Y.: PreFix: switch failure prediction in datacenter networks, vol. 1 (2018)

AI Embedded Transparent Health and Medicine System

Yichi Gu$^{(\boxtimes)}$

Cultigene Medical Technology (Beijing) Ltd.,
Beijing, People's Republic of China
guyichi@cultigene.com

Abstract. Along with the rapid exploration of intelligent system, AI methods have been applied in health and medicine fields to manage the fast growing medical data in recent years. Scientists and engineers started their first attempts. Medical history is collected and studied for automated diagnose, skin cancer images are deeply learned for auto judgment and classification [1], lung nodules are segmented automatically [2], and the future of medicine is debated in [3]. In this paper, we set up the transparent health and medicine AI system for the future, based on etiological analysis, construct total resolution network, provide accurate orientation and precise treatment, applying intelligent algorithms.

Keywords: Artificial intelligence · Health law · Genetics · Gene · Medicine · Data analysis · Accurate and precise medicine · Machine learning · Deep learning

1 Representation of Health

Personal physical condition can be detected before coming to the world. How to represent the state of health? What elements are related with health state and how do they co-work? The definition of sickness and what causes sickness? How to treat and cure illness? A bunch of questions come up about physical condition.

Health is a traditional topic coming along with the human history and one of the most important things in the life. The old emperors paid great attention to and started the research of human health including the development of culture. Figure 1 shows the ancient blood-letting therapy. Every country or ethnic group sets up their own theory of health [4]. To uncover the mystery of fitness problem, people keeps exploring on the scientific fields of astronomy, geography, ecology, physics, chemistry, biology, physiology, psychology, etc. Beyond all doubt, health is closely related with environment; contains physiological and psychological aspects; presents both physical and chemical situations.

In this paper, we explore the artificial intelligence application on the health and medical system. We will study the history, principles, expression and complexities of health in Sect. 1 and construct the health and medical data system in Sect. 2. The AI application on the system analysis is discussed in Sect. 3 and the construction methods will be reported in the conclusion of Sect. 4.

© Springer Nature Switzerland AG 2020
K. Arai et al. (Eds.): FTC 2019, AISC 1069, pp. 18–26, 2020.
https://doi.org/10.1007/978-3-030-32520-6_2

Fig. 1. Blood-letting therapy

1.1   Ancient Law of Health

Among the famous medical classics, let's start with traditional Chinese Medicine [5]. Build upon the ancient science, Huang Di's Cannon of Internal Medicine [5] applies the format of conversation between the emperor and the doctor to show the general principles of health and medicine, not only including biological growth and development rules, but also studying the human internal organs and their connections with the environment.

In Chinese Medicine, plants and living things have their own characters: shape, color, frequency, scale, temperature, humidity, hardness, density, sound, smell etc. The ancients created Ying/Yang-Theory and Five-Phases-Law to present the relationships behind everything in the world. For example, strong/weakness, heaven/ground, sun/moon are normal samples of Yang/Ying, while the five internal organs: heart, liver, spleen, lungs and kidneys, correspond to five materials, five colors, five smells, five sounds, five emotions, etc. Fig. 2. The Health Law proposes that more then reduced, weak then stronger and the mutual promotion and resistant between five phases. Conversely, the physical condition can be tuned by seasons, life habit, music, pictures/videos, tools, etc.

1.2   Medicine Evolution

The Chinese Medicine is based on visceral manifestation, channels and collaterals, while the western Medicine studies anatomical physiology, histological embryology, biochemistry and molecular biology [6]. Western Medicine renovates the traditional medicines with advanced technologies and operation tools. In the hundreds of years' development of western medicine, modern accurate detection methods were invented to show the exact illness location and the characteristics, which break through the rough approximation approach in traditional medicine and create the new diagnose and treatment principles to follow.

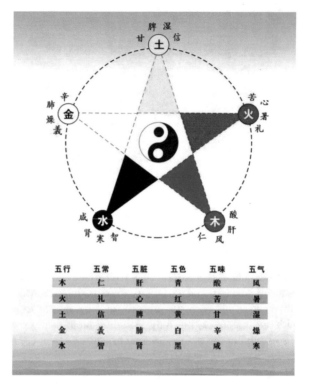

Fig. 2. Five-Phases-Law

Electropulsegraphs automatically detects the blood pulse volume which substitutes the pulse-taking. Pulse performance dominantly shows the functional information of vital energy, blood and internal organs while electrocardiogram detects the blood voltage along with the heart activity around the body. Routine pulse graph is one-to-one corresponded with nine kinds of personality health [7], while instant pulse is closely related to emotion [8]. The cardiac and vessel problems can be detected by the voltage graph [9].

With the development of genetics and genes, scientists have invented gene detection machines, gene comparison algorithms etc. The mystery between microworld and macroworld is being discovered in several directions. The supermicrometer shows the structure of molecule, DNA/RNA, virus and their function mechanism. Gene target drugs have been applied to patients [10]. CRSIPR/CASE9 technology has been applied to gene editing baby in 2018 [11].

1.3 Physical Expression

Human physical situation is a complicated high-dimensional system represented by the mapping.

$$HPS(x_1, x_2, \ldots, x_n) = F(x_1, x_2, \ldots, x_n).$$

The variables x_i include the scaled values of genetic and gene data, physical data like age, height, weight, etc., pulse, ECG, time, environment, blood analysis, mental condition data, internal organ data, symptom data, etc., the HPS contains the diagnose and treatment result.

Given the constraint of part of the variants, HPS mapping has conditional probability formation

$$HPS(x_1, x_2, \ldots, x_k) = F(x_1, x_2, \ldots, x_k | x_{k+1}, \ldots, x_n).$$

For a specific person in continuous time, HPS is a sequence of states on physical condition data. For a group of patient, HPS surveys the medical diagnose and treatment. The conditional HPS presents the global variation of physical situation which is applied to assist the medical and medicine research and drug tests.

1.4 Medical Network

In HPS mapping, the variants are correlated such as the physical variants: age, height, weight are not mutually independent and the joint distributions shows the general rule of the variants. There are several ways to explore the connection among the variants and covariants of HPS mapping, e.g. function fitting (Fig. 3), physical exploration, and Brownian simulation. The deep learning algorithm constructs the network of medical variants with input and output simulated by several formations, e.g. polynomials, wavelets, PDE, etc. The AI medical network not only makes linkage between large physical elements, but also controls the network by the coefficients of the simulated mapping. AI method is the most efficient way to cover the most situations and provides best solutions with probability. AI has been successfully applied to single problem-solving [12] and the multifunctional direction is under exploration.

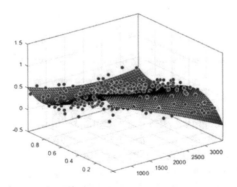

Fig. 3. Two dimensional fitting

2   Transparent Health and Medicine System

One direction of constructing Health and Medicine system is to make it transparent so that patients and doctors feel comfortable to handle the physical situation. The transparency has the following meanings:

A. Provide both general and specific facets;
B. Provide multiple solutions with probability analysis;
C. Individualized health and medical treatment;
D. Self-learning process;
E. Etiological analysis;
F. Health prediction.

2.1   Multiple Variables

There are several ways to manage the large dimensional physical variables: Classification according to their features, relativity, scale, recursion, relevance, graph, etc. Practically, the variables can be classified to the types of general physical data, environmental data, medical data, dietary plan, life/work schedule, health and medical rules, etc. Figure 4 recursively sub-classified to micro-groups; linked with relevance; and connected with graphs.

Fig. 4. Example of multi-classified variables

2.2   Total Resolution Network

The simulated network mapping HPS is a complicated multi-dimensional system. How to study such a system? Similar to games, we first set up the general rules and then start evaluation and gamble. Adapting multi-resolution method, the health and medical problem is studied step by step, from fitness judgment to diagnose with illness location to pathogenesis study to treatment; the multi-dimensional variables are recursively classified and linked in the graph. Figure 5 shows an example of multi-resolution decomposition of signal, where x is the original signal decomposed into the sum of approximation a6 and leveled details d1, ..., d6.

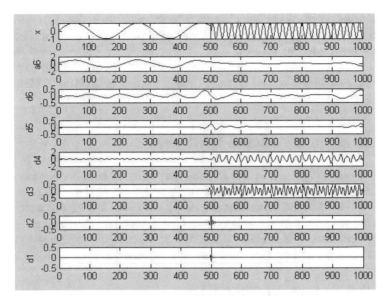

Fig. 5. Example of multi-resolution decomposition

Based on the multiple variables in Sect. 2.1, we build the total resolution network with multiple solutions to realize the transparent health and medicine system. Not only the input data and output result are multi-resolution, but also the simulated HPS mapping is layered with different resolutions. Therefore we obtain the total resolution health and medicine network which considers all physical elements, provides reasonable diagnose and treatment in several ways applying the AI algorithm. The following is an example.

Based on the organ images with nidus annotation, AI algorithm can judge and segment nidus and moreover detect the characteristics of nidus simultaneously or in separate processes. Both the data and prediction have different levels and will be designed to fulfill the task completely.

2.3   Accurate Orientation and Precise Treatment

The main theme of the system is attributable to the field of data analysis. The collected data contains Classic/Modern medicine principles and therapies, individual data, official data, medical detection, diagnose and treatment data, etc. According to the medical and medicine principles and history medical documentation, for most normal people, the system judges their fitness and suggests for the future health; for patients, the system provides the personalized medical treatment. So according to the individual data, life/work situation and history, the system provides the accurate directions of health judgment and suggestion, or medical detection and treatment process.

Whether a person is healthy or has discomfort, the system provides scientific health/medical orientation based on Classic/Modern medical law, evidence-based medicine, genetics and genes study, personalized health and medicine. The aim of the system is to help people easily handle their health problem and stay in good health. Although medical and medicine technologies have been developed for thousands of years, new approaches of regimen, diagnose and treatment keeps emerging one and another. The system will apply the law of new technology revolution, new regimen approaches, new diagnose and treatment methods, and collects new ideas through internet.

3 Quantization of Artificial Intelligence

AI methods especially machine learning and deep learning have being widely applied in most industry fields, not only substituting hard handwork, but also showing high efficiency and accuracy. AI quantizes the mapping between known and unknown with scaled noise level, so that AI works better than traditional algorithms such as characteristics extraction. For complex problems, the mixture of traditional and modern learning algorithms will provide more powerful mechanism.

3.1 Statistics of Health and Medical Data

Statistics is an excellent tool for discovering the general principle of nature. The distribution function shows the correspondence between variables' values and the probability with characters of average and variance. Bayesian analysis computes conditional probability under different specific cases. Inverse problem is one theme of statistics. Given some samples, we can approximate the exact distribution or conditional distribution, also the relationships between variables. The accuracy of the approximation depends on the quantity and quality of samples.

For health and medical data, statistics is applied to approximate the variables' characters, distributions, and relationships. Precisely, we'd like to approximate the HPS mapping or partial HPS mapping. We can pursue the degree of health, future health's hidden trouble, possible diagnoses and treatments with probability, drug's efficiency, and culture problem in medicine. It is of important sense to show the features of health and medicine variables and format their principles.

3.2 Intelligent Medical and Etiological Analysis

Etiological analysis plays the most import role in health and medicine system, with which one can diagnose and make appropriate treatment. As discussed in Sect. 2.1, they are all multi-resolution data. The specialty of etiological analysis is its connections with several branches of nature, environment, culture and medicine. Besides the collection of these health/medicine laws, AI is to find all the fitted rules for the situation. Artificial intelligent multi-classification is one way, while we can also study the cases, extract elements and simplify the correspondence.

3.3 Intelligent Case Diagnose and Treatment

AI provides the approximation mappings to construct this statistical correspondence. After we set up the network of variants and covariants, the noisy AI mapping is complete. The learning approaches are various, e.g. comparison learning, reinforcement learning, against and generated learning etc. and the learning function has various formations, e.g. convolutions, polynomials, PDEs, etc. Applying to the sequence of data consequently is the key to obtain the desired AI diagnose and treatment process.

The multi-resolution AI work has several applications such as case history study, imaging study, etiologic study, synthesis analysis. AI is designed to composite all the information and composes the best solutions.

4 Management and Conclusion

The health and medicine system discussed above is mainly framed as the HPS mapping to be resolved and approximated as total resolution network. AI algorithms will be firstly applied to obtain partial approximations and later be synthesized together.

4.1 System to Be Transparent

Generally, the HPS mapping performs as the kernel function of the health and medicine system. At the same time, we construct the health and medicine data base with public health and medical information and private client's data. The kernel of the system is recursively making approximations of conditional HPS mappings for medical detection, treatment, statistics and judgment, etc. These functional approximations are shared with the clients who can obtain the health level detection result and suggestions, medical diagnose and treatment processes with etiological analysis and transparent medical and medicine information. The statistics information and medicine principles are always public to all the clients.

4.2 AI Magic

Section 3 gives the quantization plan for artificial intelligent algorithms' application on the health and medicine system. It will be a magic total resolution network, approach and solve the problem recursively, record chronological individual data, renew the approximation upon the new data and technology periodically, open self-treatment section and set up virtual doctor, etc.

References

1. Esteva, A., et al.: Dermatologist-level classification of skin cancer with deep neural networks. Nature 542.7639, 115–118 (2017)
2. Gu, Y., et al.: Gaussian Filter in CRF based Semantic Segmentation. https://arxiv.org/abs/1709.00516, September 2017

3. Topol, E.: The Future of Medicine is in Your Hands. ZheJiang People's Publication, Translated by Jie Zheng (2016)
4. Climate variability and linguistic diversity, Tage S. Rai, Science, 24 May 2019, vol. 364, Issue 6442, pp. 747–748. https://doi.org/10.1126/science.364.6442.747-f
5. Huang Di's cannon of internal medicine
6. Baidu Research: https://baike.baidu.com/item/%E8%A5%BF%E5%8C%BB/6014741?fr=aladdin
7. Peng, Q., Xie, M.: Chemical industry publisher, Chinese medicine pulse diagnose clinic diagram (2018)
8. Zhang, J.: Pulse signal quality evaluation and application to emotion recognition. Master dissertation (2018)
9. ECG signal processing: Classification and interpretation. Springer, Adam Gacek (2012)
10. Li, A., Ma, S., Wu, M.: Application of molecular targeted drug to the treatment of liver cancer. J. Hepatopancreatobiliary Surv., May 2015
11. Baidu Research. https://baike.baidu.com/item/%E5%9F%BA%E5%9B%A0%E7%BC%96%E8%BE%91/18830535?fr=aladdin
12. https://deepmind.com/research/alphago/

Image Segmentation Based on Cumulative Residual Entropy

Z. A. Abo-Eleneen[1]([⊠]), Bader Almohaimeed[2],
and Gamil Abdel-Azim[3]

[1] Zagazig University, Zagazig 44519, Egypt
zaher_aboeleneen@yahoo.com
[2] Qassim University, P. Box 6644, Buraydah 51452, Al Qassim, Saudi Arabia
[3] Canal Suez University, Suez, Egypt

Abstract. Cumulative residual entropy (CRE) is an essential concept in information theory and have more general mathematical properties in contrast to entropy. However, it is observed that research on CRE has relatively little consideration in image processing. Image thresholding technique plays a crucial role in several of the tasks needed for pattern recognition and computer vision. In this paper, we study, implement, and apply the CRE measure for image thresholding. Firstly, we have defined a thresholding criterion, which is based on the CRE measure that related and based on the image. Secondly, the optimal solution of CRE function found. Finally, the proposed method is applied over data set of image such as nondestructive testing (NDT) images. Moreover, we compare this with several classic segmentation techniques on the same data set.

Keywords: Image segmentation · Histogram · Cumulative residual entropy · Information theory

1 Introduction

Segmentation of images into homogeneous region is found to be the one of the most important and latest research area in computer vision. Image thresholding is extensively applied in medical applications. It plays a necessary role in the image process. In particular, for handling, develop observation and anomaly detection [1, 2].

Shannon [3] introduced a measure of uncertainty of statistical variables, which known as the Shannon entropy. Different entropy image measurements can be defined during the construction of a statistical model for the imaging process. Recently, thresholding techniques based entropy and its related information are studied and discussed by many researchers. The interested reader is refer to [4–15].

Several techniques and algorithms are inbuilt the entropy of image and extensively employed in image processing precisely, e.g., medical imaging processing. The entropy use uncertainty as a measure to explain the information contained in a data, we obtain maximum information. In case of unavailability of prior knowledge, in this instance, the result is considered as maximum uncertainty in this case. This concept generalized to Renyi's entropy [16].

© Springer Nature Switzerland AG 2020
K. Arai et al. (Eds.): FTC 2019, AISC 1069, pp. 27–35, 2020.
https://doi.org/10.1007/978-3-030-32520-6_3

Entropy-based thresholding assumes that there are two probability distributions. First distribution is for the object class and other for the background. If the total of the entropies of the two classes is greatest, the segmentation of the image is carried out.

Sezgin and Sankur [17] stated that the thresholding strategies classified in six groups in line with the theory of the information exploit. Overall, these classes use entropy and its connected information as optimum criteria.

Generally, these strategies divided as: optimal criteria based entropy, optimal criteria based cross-entropy and optimal criteria based fuzzy entropy.

For example, the maximum entropy for image thresholding technique is discussed by Pun [18, 19]. Kapur et al. [20] presented entropy-based method, which is improved Pun criterion. Abutaleb [21] proposed a criterion, based on 2D entropy, which is similar to Pun's method. Brink's [22] developed an entropy criterion method, which is based on autocorrelation functions of the threshold histograms. Li and Lee [23] proposed entropy criterion that is based on relative cross entropy. By using an information-theoretic approach Kitlerl and Illingworth [24] developed a thresholding method that minimizes the segmentation errors. Chenge et al. [25] suggested threshold criterion, which is based on fuzzy entropy. Rao et al. [26] presented CRE as a new and robust measure of information.

This measure supported the accumulative distribution function instead of density function. CRE has some of the necessary properties of entropy, as well as overcoming some of the limitations of entropy. Rao et al. [26] obtained several properties of the CRE information and discussed some of its applications in image alignment and reliability. Zhang and Li [27] developed a registration algorithm based on combination ratio gradient and cross- CRE. Here, we propose a new algorithm to segment images. The technique, which we propose is based on the CRE information measure. The obtained algorithm tests over some nondestructive testing (NDT) and laser cladding (LC) images. It is shown through experiments that the objects extracted successfully. Regardless of the complexity of the background and the difference in class size, the obtained result is good.

2   Cumulative Residual Entropy Segmentation

Let I denotes a gray-scale image with L gray levels, $[0, 1, \ldots, L-1]$. Let for each gray level i, the number of pixels is given by m_i. Therefore, the total number of pixels is denoted by $M = m_0 + m_1 + \ldots + m_{L-1}$. Then each grey level i have the probability mass function of gray level i as:

$$f_i = \frac{m_i}{M}, \quad f_i \geq 0, \quad \sum_{i=0}^{L-1} f_i = 1. \tag{1}$$

Suppose that the pixels within the image are divided into two categories A and B by a gray level cutoff t. The set of pixels with levels $[0, 1, \ldots, t]$ represent A, and the set of pixels with levels $[t+1, t+2, \ldots, L-0]$ belongs to B. A and B are typically correspond to the object and the background, or the other way around. Then the mass probability functions of the two classes are given by the following:

$$f_A = \frac{f_1}{w_1}, \frac{f_2}{w_1}, \ldots, \frac{f_t}{w_1}, \tag{2}$$

$$f_B = \frac{f_{t+1}}{w_2}, \frac{f_{t+2}}{w_2}, \ldots, \frac{f_{L-1}}{w_2}. \tag{3}$$

Where

$$w_1(t) = \sum_{i=0}^{t} f_i, \, w_2(t) = 1 - w_1(t). \tag{4}$$

The priori CRE, for each class defined as:

$$CRE_A(t) = \sum_{i=1}^{t} \frac{(1 - F_i)}{w_1} \log \frac{(1 - F_i)}{w_1}, \tag{5}$$

$$CRE_B(t) = \sum_{i=t+1}^{L} \frac{(1 - F_i)}{w_2} \log \frac{(1 - F_i)}{w_2}, \tag{6}$$

Where F_i is the cumulative distribution function of ith pixel.

The CRE depends parametrically on threshold t for class A and class B. Therefore, the CRE measure within the two classes is defined as.

$$CRE(t) = w_1 CRE_A(t) + (1 - w_1) CRE_B(t). \tag{7}$$

We maximize the $CRE(t)$ inside the two classes. Once it is maximized, the brightness level t is taken into account to be the threshold. This criterion may achieve with a small computational effort.

$$t_{opt} = \arg \max [w_1 CRE_A(t) + (1 - w_1) CRE_B(t)]. \tag{8}$$

2.1 Algorithm

The proposed algorithm defines an objective function, which is based on CRE and corresponds to two threshold classes. The threshold value determines to maximize the objective function. The algorithm is described as follows:

---

1. Input Max=0, output= the maximum of CRE criterion. This is the threshold value.

2. For t =I to L.   L is Maximum of gray levels.

3. Compute the $CRE(t)$ function that corresponds to the gray level t

If $CRE(t)$ > Max, assign $CRE(t)$ to Max   and   Thopt = t.

Finally the threshold= Thopt.

End.

---

3   Experimental Results

In this experiment, we implement the developed similarity function, which is based on the CRE measure. The performance of the proposed criterion is verified, by examine it on NDT and LC images. This type of images has massive category variance distinction within the object and background. For example, objects have slight gray scale variations and small class variances, while backgrounds have large gray scale variations and large class variances. Therefore, most of the thresolding methods fail to segment this kind of images [28]. All the used images are 256 × 256, 8-bit types. The results obtain by the proposed strategies are compared to Otsu's and Kaptur's methods. These methods are the most common methods in literature. Moreover, the proposed strategies give more accurate results than Otsu and Kaptur's methods.

3.1   Experiments on NDT Images

First, we consider the problem of NDT image analysis. NDT means to detect objects by special instruments and strategies and quantify their potential defects without harmful effects. NDT is employed in an abroad kind of applications in astronautics physical science, and nuclear industry [28].

Two NDT real images are examined. The first and second images represent a lightweight research image of a material structure. Lightweight research often used to inspect the microstructure of a material to obtain information about its properties such as: consistency, particle size, distribution uniformity and so on. The third one is a printed circuit board (PCB) image.

Obviously, the histogram of gray scale information is essentially non-Gaussian. The two images are in Figs. 1, 2 and 3, respectively.

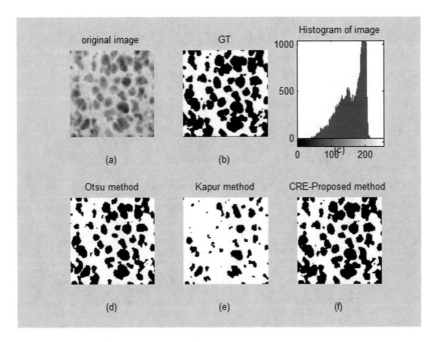

Fig. 1. Material image1: (a) original, (b) ground truth image, (c) histogram, (d) method of Otsu (T = 154), (e) Kapur (T = 119), (f) proposed method (T = 165).

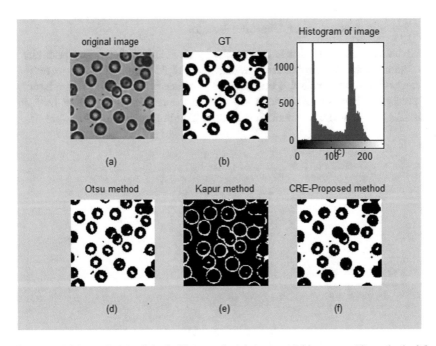

Fig. 2. Material image2: (a) original, (b) ground truth image (c) histogram, (d) method of Otsu (T = 111), (e) Kapur (T = 174) [28], (f) the proposed method (T = 135).

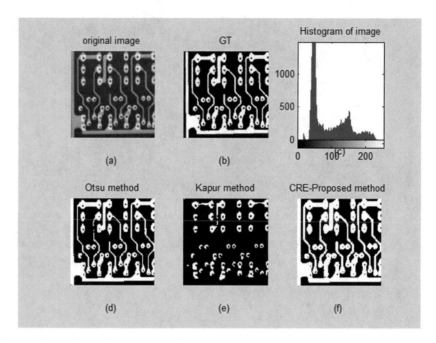

Fig. 3. Material image3: (a) original, (b) ground truth image (c) histogram, (d) method of Otsu (T = 104), (e) Kapur (T = 158) [28], (f) the proposed method (T = 61).

3.2 Experiments on Laser Cladding Images

The advanced material treatment technology by powder injection called laser cladding. It has applications in manufacturing, repair parts, rapid prototyping and metal plating. Whenever the powder melted, a thin layer of substrate is fabricated using a laser beam. The precise division of LC plays a key role in the rear feeding system [28]. In this subsection, we consider segmentation of laser cladding images (Figs. 4 and 5).

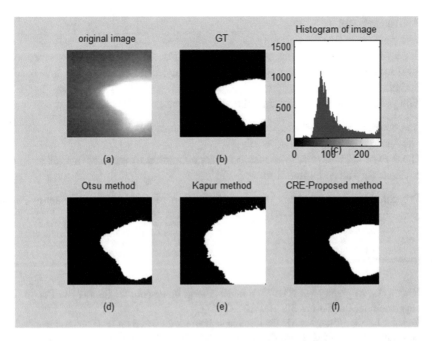

Fig. 4. Laser cladding image1: (a) original, (b) ground truth image, (c) histogram, (d) method of Otsu (T = 166), (e) Kapur (T = 109) [28], (f) the proposed method (T = 193).

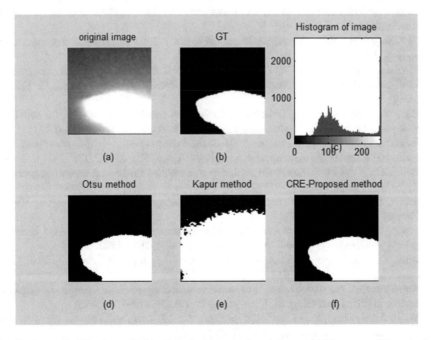

Fig. 5. Laser cladding image2: (a) original, (b) ground truth image (c) histogram, (d) method of Otsu (T = 175), (e) Kapur (T = 100) [28], (f) the proposed method (T = 194)

4   Conclusions

CRE plays a crucial role in statistical mechanics. In this article, we propose a straightforward, but effective methodology of image segmentation that builds on the CRE similarity measure.

The projected methodology has the advantages:

1. It is automatic threshold method and nonparametric.
2. The strategy is implemented easily.
3. The 2 D of CRE similarity measure for segmentation of medical images is an open problem for future exploration.

We conclude by remarking that the application of CRE measure in segmentation analysis ought to receive renewed impetus of this consideration.

References

1. Khelifi, L., Mignotte, M.: BMFM: a multi-criteria framework for the fusion of color image segmentation. Inf. Fusion **38**, 104–121 (2017)
2. Gui, L., Li, C., Yang, X.: Medical image segmentation based on level set and isoperimetric constraint. Eur. J. Med. Phys. **42**, 162–173 (2017)
3. Shannon, C.E.: The mathematical theory of communication. Bell Syst. Tech. J. **27**, 423–467 (1948)
4. Abdel Azim, G., Abo-Eleneen, Z.A.: A novel algorithm for image thresholding using non-parametric Fisher information. In: 1st International Electronic Conference on Entropy and its Application, pp. 1–15 (2014)
5. de Albuquerque, M.P., Esquef, I.A., Gesualdi Mello, A.R.: Image thresholding using tsallis entropy. Pattern Recogn. Lett. **25**, 1059–1065 (2004)
6. Duraisamy, P.S., Kayalvizhi, R.: A new multilevel thresholding method using swarm intelligence algorithm for image segmentation. J. Intell. Learn. Syst. Appl. **2**, 126–138 (2010)
7. Guo, W., Wang, X., Zhang, T.: Entropic thresholding based on gray-level spatial correlation histogram. In: International Conference on Pattern Recognition, pp. 1–4 (2008)
8. Ben Ishak, A.: Choosing parameters for Reni and Tsallis entropies within a two-dimensional multilevel image segmentation framework. Phys. A **466**, 521–536 (2017)
9. Ben Ishak, A.: A two-dimensional multilevel thresholding method for image segmentation. Appl. Soft Comput. **52**, 306–322 (2017)
10. Sahoo, P.K., Wilkins, C., Yeager, J.: Threshold selection using Reni's entropy. Pattern Recogn. **30**(1), 71–84 (1997)
11. Sahoo, P.K., Slaaf, D.W., Albert, T.A.: Threshold selection using a minimal histogram entropy difference. Soc. Photo-Opt. Instrum. Eng. **36**(7), 1976–1981 (1997)
12. Xiao, Y., Cao, Z., Zhong, S.: New entropic thresholding approach using gray-level spatial correlation histogram. Opt. Eng. **49**(12), 1–13 (2010)
13. Abo-Eleneen, Z.A., Abdel-Azim, G.: A novel statistical approach for detection of suspicious regions in digital mammogram. J. Egypt. Math. Soc. **21**, 162–168 (2013)
14. Yimit, A., Hagihara, Y., Miyoshi, T., Hagihara, Y.: 2-D direction histogram based entropic thresholding. Neurocomputing **120**, 287–297 (2013)

15. Xiao, Y., Cao, Z., Yuan, J.: Entropic image thresholding based on GLGM histogram. Pattern Recogn. Lett. **40**, 47–55 (2014)
16. Rényi, A.: On measures of entropy and information. In: Proceedings of the 4th Berkeley Symposium on Mathematical Statistics and Probability, vol. I, pp. 547–561. University California Press, Berkeley, California (1961)
17. Sezgin, M., Sankur, B.: Survey over image thresholding techniques and quantitative performance evaluation. J. Electron. Imaging **13**, 146–165 (2004)
18. Pun, T.: A new method for grey-level picture thresholding using the entropy of the histogram. Signal Process. **2**, 223–237 (1980)
19. Pun, T.: Entropic thresholding: a new approach. Comput. Graph. Image Process. **16**, 210–239 (1981)
20. Kapur, J.N., Sahoo, P.K., Wong, A.K.C.: A new method for gray level picture thresholding using the entropy of the histogram. Comput. Vis. Graph. Image Process. **29**, 273–285 (1985)
21. Abutaleb, A.S.: Automatic thresholding of gray-level pictures using two-dimensional entropies. Pattern Recogn. **47**, 22–32 (1989)
22. Brink, A.D.: Thresholding of digital images using two-dimensional entropies. Pattern Recogn. **25**, 803–808 (1992)
23. Li, C.H., Lee, C.K.: Minimum cross entropy thresholding. Pattern Recogn. **26**, 617–625 (1993)
24. Kittler, J., Illingworth, J.: Minimum cross error thresholding. Pattern Recogn. **19**, 41–47 (1986)
25. Cheng, H.D., Chen, J.R., Li, J.G.: Threshold selection based on fuzzy c-partition entropy approach. Pattern Recogn. **31**, 857–870 (1998)
26. Rao, M., Chen, Y., Vemuri, B.C., Wang, F.: Cumulative residual entropy: a new measure of information. IEEE Trans. Inform. Theory **50**, 1220–1228 (2004)
27. Zhang, Y.M., Li, J.Q.: Registration for SAR and optical image via cross-cumulative residual entropy and ratio operator. Adv. Mater. Res. **452–453**, 954–958 (2012)
28. Li, Z., Liu, C., Liu, G., Cheng, Y., Yang, X., Zhao, C.: A novel statistical image thresholding. Int. J. Electron. Commun. **64**, 1137–1147 (2010)

Knowledge-Based Adaptive Hypermedia with HAries

Yira Muñoz[1]($\boxtimes$), María de los Angeles Alonso[2], Iliana Castillo[2],
and Verónica Martínez[2]

[1] Higher Education School Ciudad Sahagun, Autonomous University
of Hidalgo State, Carretera Cd. Sahagún – Otumba s/n, Zona Industrial,
43990 Sahagun, Hidalgo, Mexico
`yira@uaeh.edu.mx`
[2] Computing and Electronic Academic Area, Autonomous University
of Hidalgo State, Carretera Pachuca-Tulancingo Km 4.5,
42184 Pachuca, Hidalgo, Mexico
`{marial,ilianac,vlazcano}@uaeh.edu.mx`

Abstract. The theory developed for the construction of an adaptive hypermedia is presented, which has the ability to make decisions and be adjusted to the user's needs. It is considered that besides the elements conforming any Hypermedia (nodes, links, multimedia, etc.), the specialized knowledge on how to handle information is included and too a working memory for each user which allows remember all the interaction that this one executes with the system. This type of knowledge is included in the knowledge representation scheme of hybrid language HAries, by means of three structures: Hypermedia HAries, Execution Variable and State Variable of the Hypermedia which were created for such purpose. The first stores all the hypermedia information related to the data and the knowledge of themselves in a database. This information is then used during the execution of the hypermedia to decide what to show to the user to see in every single moment. The last two structures have the function to start the execution of the hypermedia and the navigation control of the user through itself. All of them are then stored in a knowledge base created with the HAries language.

Keywords: Adaptive Hypermedia · Knowledge representation · Knowledge base · Dynamic links

1 Introduction

On broaden the concept of Hypertext using Multimedia techniques, the hypermedia has become a very useful tool to create systems that offer greater possibilities to the user by presenting the information in a more organized, intuitive and comprehensible way. As an example of this technique we could mention the helping systems, dictionaries, encyclopedias, manuals, among others.

The structure of a classic Hypermedia is made up from a group of nodes which are pages previously constructed with information (texts, images, sounds, videos, animations, etc.) on a specific topic and a group of links among the nodes to grant access to

© Springer Nature Switzerland AG 2020
K. Arai et al. (Eds.): FTC 2019, AISC 1069, pp. 36–45, 2020.
https://doi.org/10.1007/978-3-030-32520-6_4

such pages. The functioning of this type of hypermedia is based on the navigation of these pages (nodes) through the corresponding links, previously defined; so, this process presenting a static behavior.

It is commonly known that typical Hypertexts [1–3] given its static behavior, present several difficulties, which indeed have been transmitted to the systems that work on the Web, among the most common: the disorientation problem and the information overload mainly.

With the only purpose of solving the problems mentioned above, the researchers of these areas got interested in developing techniques which enable to customize the users' navigation and minimize these negative aspects. These works are commonly known as "Adaptative Multimedia Systems" or "Adaptative Hypermedias" [4–20]. The term "adaptative" refers to the incorporation of knowledge on the conventional components.

To incorporate knowledge to Adaptative Hypermedia, there are used several alternatives, both for information about user or node as to the strategies to decision making or link. These proposals have been focused to specific [5–9] or to general purpose [4, 11, 19]. In these proposals, the information about user is diverse, being some examples: Knowledge level through pre-test [5] or forms [17], case studies [11], learning styles [4, 12, 13, 18, 20] the students' needs [6], review's aspects and sentiments [14], psychological, physiological and/or user's behavioral state [16], student's cognitive and affective [9]. On the other hand, the strategies to give them the adaptive capacity, there are been based on adaptative rules [5, 6, 10, 12, 16, 20], red multi–entidad Bayesiana [15], graphs [7], asymmetrical affinity measure [19] y fuzzy logic system and a backward propagation neural network [9], among others, being the first the most used in this research context.

1.1   Adaptative Hypermedia HAries

The adjective adaptative, focuses on providing reasoning capabilities to the hypermedia so that the activation and use of the multimedia can be controlled by the automatic decision of the system and that the results shown be variable and can be adapted the user's needs.

Unlike the works that precede our proposal, the elements within this work which make the information adaptation possible are two:

The node's sensibility is achieved by sensible zones functionality. The sensitive zones (links) represent nodes with variable behavior and their activation will depend on the fulfillment of condition denominated Sensibility Condition. Based on this condition and the context, the execution process will present a specific text or multimedia. This condition represents a structure that is incorporated in knowledge base.

Activation of the multimedia elements to generate dynamically pages. The activation of a sensitive zone may cause one or several groups of diverse actions to take place. These groups of action are attached to such zones that represent nodes and their execution is also conditioned to the behavior of the user during the system interaction. One specific action involves an analytic, reasoning, or inference process. This allows the pages to generate themselves in a dynamic way and also that can be showed different multimedia in activation of a sensitive zone.

The nodes sensibility, through the generation of sensitive zones that are presented to the user as well as the activation of the diverse multimedia elements should be controlled to allow the hypermedia to act in a variable way according to the user and the actions executed by him. Such control is realized in HAries (Hybrid Artificial Intelligent Expert System), by means of two structures of knowledge representation generated for this whose structure and functioning will be further explained.

2 Hybrid Knowledge Representation Language HAries

HAries is a computing tool that is used to build knowledge-based systems. Its hybrid characteristic is due to its Knowledge Representation Scheme (KRS) combines the characteristics of the production systems with the object-oriented programming and procedures. For this, it is possible to achieve the modeling of adaptative behaviors based on inference mechanisms and uncertainty manipulation as well as the execution of common procedures of conventional programming among the ones we can mention are the skip and interrupt operations, assigning values, comparing and displaying the elaborated results by the self-programmer in developing time [21–23]. Its KRS is constituted by three basic components that are also known as: concept, relationship among the concepts and hybrid structures among the previous ones.

One concept can be considered as a structure with diverse components according to the knowledge being present. Its goal is to represent knowledge of very diverse kind, for example, objects taking values (uncertainty, numerical, qualitative, texts) and objects that perform actions (display a screen, a menu, different multimedia, make a comparison, organize a system, etc.) this type of representation is set up by means of a class with a group of components to define its attributes and methods or functioning.

A relationship is the kind of representation that allows establishing a dependency among the concepts which causes the activation of a series of mechanisms for the assessing of the related concepts with it. It is implemented as a conditional sentence If-Then-Else. The production rule is one of the different types of relationship that can be established in this language.

Lastly, the hybrid structures are representations of the concept type, previously described, which include within some of their components some relationship, providing a greater richness of expression to model the systems. As an example of this type of structure is the HAries propositions that represent facts or affirmations to which a value of uncertainty is associated to. These structures differ from the propositions of logic in that not only do they represent a text and an associated value, but they model other behaviors such as common sense and the execution of multimedia actions in the moment of their evaluation among others.

The propositions in their simple form can be combined with others through logic connectives to form expressions called compound propositions which can be used in the preceding of the relations to condition such relation.

3   Hybrid Knowledge Representation Language HAries

To include in the HAries language the possibility to create an adaptive Hypermedia was necessary to add a series of elements to the KRS that would allow the representation of this computing means. Next, we will describe every component corresponding to the behavior of them.

3.1   Adaptive Hypermedia

For language a hypermedia is a hybrid structure that contains not only the hypermedia information represented by texts, images, sounds, videos, animations but also the knowledge that defines how and when to use such elements.

The construction of the hypermedia is possible through a tool developed with such purpose whereas the execution and the navigation control are performed through HAries language.

The **Adaptive Hypermedia (IH)** in the KRS is represented as follows:

$$< EHi, NHi, GAi >   \tag{1}$$

In this structure, EHi represents a group of elements of the hypermedia, NHi represents a group of hypermedia nodes and GAi represents a group of actions.

According to what was mentioned above, if you wish to build a hypermedia HAries for a specific domain, it is necessary to define the elements: EHi, NHi and GAi. The definition of each of these elements involves an acquisition process of its data collection and the associated knowledge to them, and consequently its storing. To store data and knowledge in a logical way to facilitate its access a database was designed which will be automatically created for each hypermedia being constructed or will be updated for existing hypermedia.

3.2   Element of Hypermedia (EHi)

It is known as *Hypermedia Element* (EHi) every single one of the elements or computing objects that can be included in a hypermedia. The valid means that can be used within a hypermedia HAries are video, sound, images, texts, windows hypertexts, hypermedia pages and executable programs.

3.3   Construction of Hypermedia Pages

The interaction of the hypermedia with the user, including navigation, is executed by means of the information screens or pages constructed by the system.

A page is understood as the union of one or more hypermedia elements (EHi) presented as a unit on the monitor. The pages can be dynamic or static. In the first case, the developer previously constructs the information just as he wishes it to appear and will always be presented in the same way, as in the second case, they can constitute the result of the assessment of a group of actions (AGi) that can be sequential or parallel. An example of sequential actions could be several sounds which can be connected to each other to create a new one and an example of parallel actions can be the fusion of several texts or images to construct a page.

It is important to mention that the EHi that will form every page will be extracted from the Hypermedia Database in which they were previously defined.

3.4 Group of Actions (GAi)

A group of actions within the Hypermedia HAries definition represents a set of actions with a specific goal. A *Hypermedia Action* (AcHi) is defined as a relation that is expressed as follows:

$$C\&@ \rightarrow EhA1, EhA2, \ldots, EhAa, EhN1, EhN2, \ldots, EhNn \qquad (2)$$

In this relation, C is a condition of analysis that can be a proposition HAries simple or compound, @ represents the context or state, in which lies the execution of the hypermedia in the moment of assessing the relation. The elements EhA1, EhA2, ..., EhAa are the elements of hypermedia which will be visualized in case of the fulfillment of C & @, whereas EhN1, EhN2, ..., EhNn are the elements of hypermedia which will be visualized in case of the completing fulfillment of C & @. The symbol $\rightarrow$ represents the Hypermedia relation action type.

The expression "C & @" is known as the preceding of the action: EhA1, EhA2, ..., EhAa happening for the affirmation and EhN1, EhN2, ..., EhNn happening for the negation. It may be that one of the groups of the happening be empty but not both at once.

A group of action (GAi), is created by integrating two or more actions under the analysis of a same condition, expressed as follows:

$$< Cgi, AC1, AC2, \ldots, ACn > \qquad (3)$$

In this new structure, Cgi represents the analysis condition of i group, which can be expressed through a simple or compound proposition. So, every group of action can be associated a condition that allows assessing when it will make sense an analysis and when not. It is important to mention that one action can belong to more than one group of action at the same time. The elements AC1, AC2, even ACn, represent the group of action which makes up the group.

3.5 Hypermedia Nodes (HNi)

A Hypermedia Node (HNi) is an image or text zone which is sensibilized whose selection causes the execution of one or more hypermedia groups of action, according to the analysis of the condition Cgi of each group of action.

As we can see, in this definition a difference is introduced with the typical structure since in that case a node represents a complete page that can have several zones with links each to a specific place, whereas in this theory, the node is defined as possible actions to execute, even separating the concept of sensitive zone. This enables the self-node to be able to be linked to different sensitive zones.

Every hypermedia node will have one or more groups of these actions associated according to the needs of each problem. This means that the activation of one node will trigger an analysis to decide which one must be executed, which will be able to be solved by the system or the user accordingly. It is also possible that a node is linked to a condition that allows assessing if the same will be presented to the user as active or not.

3.6   Condition of Sensibility

It is known as a *Sensibility Condition of a Node* to the condition established for a node to serve as an active sensitive zone. This is expressed as follows:

$$C \& @ \tag{4}$$

This condition presents the same sense that the preceding of relation (2).

Its definition permits the hypermedia have information to decide under what circumstances and under what kind of context makes sense for a sensitive zone to appear as active on displaying in the page where this node is located, in other words, the sensitive zones acquire certain dynamism since these could be active under specific conditions and not active under other different circumstances.

3.7   Activation of Hypermedia Nodes

The analysis of nodes starts before they are presented to the user as part of an information page, this happens because first it must analyze the condition of sensibility (4) associated to each node, to determine if it will stay active or not.

If a node does not have an associated condition of sensibility or this happens, the analysis conditions associated to the different groups of action will be evaluated (condition Cgi of structure (3)), which permits the system to recognize beforehand the groups of action that will be evaluated. If the group's condition is given, the node will become active and it will be possible for the user to navigate through this link just with a "click" on it.

However, what happens if a node remains associated with several groups of action whose conditions of analysis are given? well, we will be in a situation where the user has several ways to continue the navigation and so he will have to choose which way he wants to use.

As we can see the activation of a node can cause a process of interaction user-system since the analysis conditions can be asked.

3.8   Hypermedia Execution

The hypermedia execution, as previously mentioned, is realized from the HAries language, through a structure of the concept type that allows defining the conditions of starting the hypermedia execution. This structure is called Hypermedia Execution Variable and is expressed as follows:

$$<T, Pa, A>$$ (5)

Where the T parameter represents a text that defines the concept of variable and the Pa and A parameters are defined as how the Hypermedia HAries will be executed. Pa is simple HAries proposition associated to the variable to cause assessing and so the execution of the hypermedia. Through the A parameter it will be defined the type of activation or way in which the hypermedia will begin.

So, the hypermedia can be activated in two ways: from a static page or by assessing of one or more groups of action, from which a starting page will result compound of several multimedia elements; considering that the information of the static pages and multimedia are located in the Hypermedia Database.

3.9   Navigation Control

At all times will the system store the interaction process with the user, as well as his navigation, and this information is not lost from one consultation to the other. This allows the system to have a continuous memory on the work the user has been doing on the information and also to be able to make decisions on base of the navigation executed.

To be able to use this information a structure has been defined, it is called *State of the Hypermedia Variable* that permits analyze numerous ways the process of executed navigation at any time.

3.10   State Variable of Hypermedia (Vj)

The navigation control as well as the hypermedia execution was set up in the HAries language through the representation of the knowledge of the concept type called *State Variable of the Hypermedia*, which is expressed as follows:

$$<T, Pa, \Re, C>$$ (6)

Represented through four parameters, the state variable of the hypermedia models the navigation control of the user. The Pa parameter represents a simple proposition that activates its assessing. $\Re$ represents a group of relations that can be established to follow up the user's behavior. The right and left members of the relations can be for example the number of times the Ni node activated with or without results, the number of times that the GHi group of action activated with or without results, the number of times that the Ni node or a constant has been shown. The relation can be any element of the group $\{=, <>, <=, <, >=, >\}$ finally, the C parameter constitutes the logical connective which will be established among the relations of group $\Re$ that can be "&" (conjunction), "V" (inclusive disjunction) or "I" (excluding disjunction).

The last three parameters conform an expression that will be used by the inference machine to control navigation of the user through the hypermedia.

4   General Conception of Adaptative Hypermedia HAries

Once explained all the elements of Hypermedia HAries, we are missing is to present the general idea for its construction and use, which it's described in the steps below:

1. Using the name tool, you may capture: the elements of hypermedia, the nodes, the sensibility conditions of actions and group of actions. All this backed up in a data base.
2. A knowledge base is created with the HAries language where a variable will be included: an execution variable of the hypermedia with the starting way of this and a state variable of the hypermedia with the conditions which must be supervised in the navigation by itself.
3. The knowledge base is executed from the HAries language and on requiring the value of the proposition associated to the execution variable of the hypermedia, this will be executed.
4. In the same way, the need of assessing the associated proposition of the state variable of the hypermedia, it will activate the navigation control of the user from the conditions that have been defined by means of the relations and logical connectives.
5. The generated navigation information is stored in a data base in such a way as it is not lost from one session to another in the interaction with the hypermedia.

5   Conclusion

In this work, with the providing knowledge to nodes and links of a hypermedia it's possible to define the form and the elements of the hypermedia behavior showed to user according to their necessities.

We take advantages of the possibilities to the hybrid language HAries, because of its KRS and the inference mechanisms had created three structures that represent the knowledge: Hypermedia Structure, Execution Structure and State Structure that let to analyze and to make decisions allowing the adaptative behavior in the hypermedia.

The possibilities that present the adaptative hypermedia in this theory can be summarize in four basic aspects, which are the hyper-space can be adapted during the execution according the user´s necessities, the hypermedia is designed for the decision making, the nodes could be form, for more than one hypermedia element and the links are dynamics for the users.

References

1. Nielsen, J.: Hypertext and Hypermedia. Academic Press Professional, California (1993)
2. Amaya, G., Gualdrón, E., Fernández, C.: Hipertexto, Influencia en la estructuración del conocimiento. Horizontes Pedagógicos **19**(1), 1–46 (2017)

3. Barbero, M.: Jóvenes entre el Palimpsesto y el Hipertexto. Nuevos Emprendimientos Editoriales, Barcelona (2017)
4. Gligora, M., Kadoić, N., Kovačić, B.: Selection and prioritization of adaptivity criteria in intelligent and adaptive hypermedia e-Learning systems. TEM J. Technol. Educ. Manage. Inform. **7**(4), 137–146 (2018)
5. Sfenrianto, S., Hartarto, Y., Akbar, H., Mukhtar, M., Efriadi, E., Wahyudi, M.: An adaptive learning system based on knowledge level for english learning. Int. J. Emerg. Technol. Learn. (iJET) **13**(12), 191–200 (2019)
6. Zhao, X.: Mobile english teaching system based on adaptive algorithm. Int. J. Emerg. Technol. Learn. (iJET) **13**(8), 64–77 (2018)
7. Tosheva, S., Stojkovikj, N., Stojanova, A., Zlatanovska, B., Martinovski Bande, C.: Implementation of adaptative "E-School" system. TEM J. Technol. Educ. Manag. Inform. **6**(2), 349–357 (2017)
8. Prado, T.R., Moro, M.M.: "Review recommendation for points of interest's owners" de In: HT '17 Proceedings of the 28th ACM Conference on Hypertext and Social Media (2017)
9. Zataraín, R., Barrón, M.L., González, F., Oramas, R.: Ambiente inteligente de aprendizaje con manejo afectivo para Java. Res. Comput. Sci. **92**, 111–121 (2015)
10. Mohd, J.K., Khurram, M.: Modelling adaptive hypermedia instructional system: a framework. Multimed. Tools Appl. **78**, 14397–14424 (2018)
11. Isaias, P., Lima, S.: Collaborative design of case studies applying an adaptive digital learning tool. In: Proceedings of EdMedia: World Conference on Educational Media and Technology, pp. 1473–1482 (2018)
12. El Guabassi, M., Al Achhab, I., Jellouli, B., Mohajir, E.L.: Personalized ubiquitous learning via an adaptive engine. Int. J. Emerg. Technol. Learn. (iJET) **13**(12), 177–190 (2018)
13. Hamza, L., Tlili, G.: The optimization by using the learning styles in the adaptive hypermedia applications. Int. J. Web-Based Learn. Teach. Technol. **13**(2), 16–31 (2018)
14. Mutlu, B., Veas, E., Trattnero, T.: Tags, titles or Q&As? choosing content descriptors for visual recommender systems. In: HT '17 Proceedings of the 28th ACM Conference on Hypertext and Social Media, pp. 262–274 (2017)
15. Tadlaoui, M.A., Carvalho, R.N., Khaldi, M.: A learner model based on multi-entity Bayesian networks and artificial intelligence in adaptive hypermedia educational systems. Int. J. Adv. Comput. Res. **8**(37), 148–160 (2018)
16. Hou, M., Fidopiastis, C.: A generic framework of intelligent adaptive learning systems: from learning effectiveness to training transfer. Theor. Issues Ergon. Sci. **18**, 167–183 (2017)
17. Benigni, G., Marcano, I.: Qué herramientas utilizar para diseñar sistemas hipermedia educativos adaptativos? Revista Espacios **35**(6), 13 (2014)
18. Yang, T.-C., Hwang, G.-J., Yang, S.J.-H.: Development of an adaptive learning system with multiple perspectives based on students' learning styles and cognitive styles. Educ. Technol. Soc. **16**(4), 185–200 (2013)
19. Messina, M., Di Montagnuolo, R., Massa, R.: Borgotallo: hyper Media News: a fully automated platform for large scale analysis, production and distribution of multimodal news content. Multimed. Tools Appl. **63**(2), 427–460 (2013)
20. Tsortanidou, X., Karagiannidis, C., Koumpis, A.: Adaptive educational hypermedia systems based on learning styles: the case of adaptation rules. iJET Int. J. Emerg. Technol. Learn. **12**(5), 150 (2017)
21. de los Angeles Alonso Lavernia, M., De la Cruz Rivera, A.V., Gutiérrez, A.: Knowledge representation language: HAries. In: Memorias de la 8th World Multiconference on Systemics, Cybernetics and Informatics (SCI 2004), Orlando, Florida (2004)

22. de los Angeles Alonso Lavernia, M., De la Cruz Rivera, A.V., Gutierrez, A.: HAries: Un lenguaje para la programación del conocimiento con facilidades para la construcción de material educativo. In: Memorias de la 3ª Conferencia Iberoamericana en Sistemas, Cibernética e Informática (CISCI 2004), Orlando (2004)
23. de los Angeles Alonso Lavernia, M.: Representación y manejo de información semática y heterogénea en interacción hombre-máquina. Tesis (Doctorado en Ciencias de la Computación) (2006).   https://tesis.ipn.mx/handle/123456789/21037?show=full.   Último acceso: 8 01 2019

Group Sales Forecasting, Polls vs. Swarms

Gregg Willcox, Louis Rosenberg$^{(\boxtimes)}$, and Hans Schumann

Unanimous AI, San Luis Obispo, CA, USA
{gregg, Louis}@unanimous.ai

Abstract. Sales forecasts are critical to businesses of all sizes, enabling teams to project revenue, prioritize marketing, plan distribution, and scale inventory levels. Research shows, however, that sales forecasts of new products are highly inaccurate due to scarcity of historical data and weakness in subjective judgements required to compensate for lack of data. The present study explores sales forecasting performed by human groups and compares the accuracy of group forecasts generated by traditional polling to those made using real-time Artificial Swarm Intelligence (ASI), a technique which has been shown to amplify the forecasting accuracy of human teams in a wide range of fields. In collaboration with a major fashion retailer and a major fashion publisher, three groups of fashion-conscious millennial women (each with 15 participants) were asked to predict the relative sales volumes of eight clothing products (sweaters) during the 2018 holiday season, first by ranking each sweater's sales in an online poll, and then using an online software platform called Swarm to form an ASI system. The Swarm-based forecasts were significantly more accurate than polling such that the top four sweaters ranked using Swarm sold **23.7%** more units than the top four sweaters as ranked by survey, (p = 0.0497). These results suggest that ASI swarms of small groups can be used to forecast sales with significantly higher accuracy than a traditional polling.

Keywords: Swarm intelligence · Artificial intelligence · Collective intelligence · Sales forecasting · Product forecasting · Customer research · Market research · Customer intelligence · Marketing · Business insights

1 Background

Accurate sales forecasting is critical to businesses of all sizes, enabling teams to project revenue, prioritize marketing, plan distribution, and scale inventory levels. In recent years, AI has been used to assist in sales forecasting, but traditional AI tools are heavily reliant on historical sales data. Unfortunately, there are many situations where little to no historical sales data exists to support forecasting. This is especially true when launching new products or features, or entering new markets [1]. When historical data is sparse, human judgement methods such as large-scale polling, focus groups, customer interviews, expert opinions and customer intention surveys are often used in place of the missing data to forecast the demand of new products [2]. In one study of the prevalence of new product forecasting techniques, customer and market research was found to be the most widely used technique for forecasting the sales of new products, and that it was used in the majority (57%) of all new product forecasts [3].

© Springer Nature Switzerland AG 2020
K. Arai et al. (Eds.): FTC 2019, AISC 1069, pp. 46–55, 2020.
https://doi.org/10.1007/978-3-030-32520-6_5

While human judgement methods such as customer surveys are widely used in new product forecasting, the overall forecasting accuracy of these techniques is low: a new product's forecasted sales have been found to be on average 42% off from the true unit sales, and the more novel the product, the higher this error rate [3]. More accurate forecasting techniques are therefore necessary to improve the efficiency of bringing new products to market.

The present study proposes a novel methodology for sales forecasting using an emerging form of AI called Artificial Swarm Intelligence (ASI). If effective, ASI offers a major benefit, as it does not require historical data but instead uses real-time customer input. The study then compares the accuracy of traditional polling, using consumer surveys, to real-time "customer swarms" using ASI when making forecasts.

Previous research has shown that human groups can significantly amplify group forecasting accuracy by using ASI technology [4–7]. This wide body of research has reliably shown that real-time "human swarms" produce forecasts of significantly higher accuracy than "Wisdom of Crowds" methods such as surveys, votes, and polls [8]. One study tasked human groups with predicting a set of 50 professional soccer matches in the English Premier League. The forecasts were 31% more accurate when participants collaborated as ASI swarms, as compared to when the group voted to predict the same games [9]. The ASI swarms also outperformed the BBC's machine-model known as "SAM" over those 50 games [10]. Other studies find similar increases in forecast accuracy, including a study at Stanford Medical School that showed small groups of radiologists were able to diagnose pneumonia with significantly higher accuracy when using Swarm software than when taking votes or diagnosing as individuals [11].

While a wide body of research has demonstrated that ASI can enable groups to more accurately forecast than existing individual and crowd-based methods, no formal study has been conducted to validate ASI in the domain of corporate sales forecasting.

1.1 From Crowds to Swarms

The phrase "Wisdom of the Crowd" is frequently used to refer to the process of collecting individual input from groups of human forecasters and averaging that input into a single forecast that is more accurate than the individual forecasts alone [12–14]. The difference between "crowds" and "swarms" is that crowd-based systems rely on participants each providing isolated input (i.e. separate from one another), which is then aggregated using a statistical model, whereas swarm-based systems rely on participants interacting in real-time, "thinking together" as a unified system. Thus, while crowds can be thought of as statistical constructs, swarms are real-time, closed-loop systems that converge on solutions through the interactions and reactions of participants.

ASI systems often mimic biological systems such as schools of fish, flocks of birds, and swarms of bees. The ASI technology used in this study was Swarm®, a software platform developed by Unanimous AI. The core algorithms behind this software employs a similar decision-making process to honeybee swarms, which have been shown to reach highly optimized decisions by forming real-time systems [9, 15]. In fact, at a structural level, the decision-making processes observed in honeybee swarms are very similar to the decision-making processes observed in human brains [16, 17].

Both swarms and brains employ large numbers of independent decision-making units (i.e., neurons and bees) that work together in real-time to (a) accumulate evidence about the world, (b) weigh competing alternatives when a decision needs to be made, and (c) converge on preferred decisions as a unified system. In both swarms and brains, decisions are made through a process of competition of support among groups of the independent decision-making units. When one group exceeds a threshold level of support, that group's alternative is selected by the system. Honeybees use this decision-making structure to mediate many hive-wide decisions, such as selecting the best hive location from a large set of candidate locations. Researchers have shown that honeybees converge upon the optimal solution to this life-or-death decision approximately 80% of the time [18, 19].

1.2   Creating Human Swarms

While fish, bees, and birds use natural swarming mechanisms to connect members into real-time systems, humans have not evolved such natural abilities. Birds, for example, rely on the positions of other birds around them, while schooling fish measure the fluctuations in water pressure caused by other fish around them. Honeybees have evolved one of the most advanced methods, communicating the position of hive sites using body vibrations called a "waggle dance". To enable human groups to form similar systems, specialized software is required to establish a feedback loop across all members. A software platform (**Swarm**®) was created to allow online groups to form real-time systems from anywhere in the world [6, 20]. Modeled on the decision-making process of honeybees, the Swarm platform enables groups of online users to work in parallel to (a) accumulate evidence, (b) weigh competing alternatives, and (c) converge on decisions together as a real-time closed-loop system.

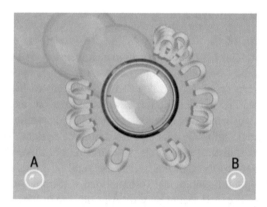

Fig. 1. Real-time ASI choosing between options

These human swarms answer questions as a group by moving a graphical puck to select among a set of options, as shown in Fig. 1 above. Each participant in the swarm influences the swarm by moving their on-screen cursor, shown in Fig. 1 as a magnet.

By positioning their magnet to pull towards an answer, participants express their intent to the rest of the swarm. Each participant provides a real-time, continually varying stream of input in the form of their magnet position, rather than a single vote for an answer or forecast. All swarm participants vary their intent in real-time, so the group collectively explores the answer choices, not based on one user's intent, but rather based on the emergent dynamics of the whole group. This system allows complex deliberations to emerge in real-time, empowering the group to converge on the answer that best represents their combined knowledge, wisdom, and insights.

It is important to note that participants do not vary only the orientation of their intent but also the magnitude of that intent by changing the distance between their magnet and the puck: the closer their magnet is to the puck, the more influence they have on the puck's movement. Participants try to keep their magnets as close to the puck's outer rim as possible to have a maximal impact on the swarm, even as the puck moves across the screen. The real-time positional control that participants have requires that participants remain engaged throughout the swarm's decision-making process, continuously evaluating and re-evaluating their individual thoughts and feelings with respect to the question at hand.

2 Method

To compare the predictive accuracy of ASI to traditional polling in sales forecasting, the relative sales performance of eight sweaters was forecast using both surveys and swarms. The sweaters were all part of a new line designed for the 2018 holiday season by a major fashion retailer, so there was little historical data that could be used to accurately forecast the relative sales performance of each sweater (see Fig. 2). In addition, each sweater was a different color or pattern, or featured a printed graphic, so forecasting the relative sales of each sweater was highly subjective.

Fig. 2. Example sweaters used for sales forecasting

The intended market for the sweaters were millennial women, so three groups of 15 millennial women were convened as a representative group of 'experts' to forecast the sales of each sweater. All of the participants in this study self-identified as interested in fashion with no previous sales forecasting experience. Participants were also not co-located for this experiment.

To compare the forecasting accuracy of traditional surveys to ASI swarms, each group was asked to predict relative unit sales (i.e., the ordered rank by sales) of the eight sweaters, first as individuals using an online survey, then by "thinking together" as an AI-optimized system using Swarm. The online survey consisted of one task: ranking the sweaters in order of total unit sales over the holiday period, from most (1st) to least (8th). The mean rank of each sweater over all surveys was used to generate an ordered list of sweaters, from most (1st) to least (8th) forecasted unit sales.

Next, each group's participants logged in remotely to the *swarm.ai* website, and used the Swarm software to collectively rate the sweaters on two critical metrics, Trendiness (from 1 to 5) and Breadth of Appeal (from 1 to 5). After rating the sweaters on those metrics, each group used a "process of elimination" method to sequentially rank the eight sweater styles on projected sales units, from least to most. The mean rank of each sweater over all three swarms was used to generate an ordered list of sweaters, from most (1st) to least (8th) forecasted unit sales.

Approximately three months later, the actual unit sales of each item over the holiday period was reported to the researchers by the fashion retailer, and the perfor-mance of each Sales Forecasting method was compared.

3   Results

The actual unit sales of each sweater, as percent of total sales, are reported in Table 1 on the next page along with the rankings generated by the average of all surveys and the average rank of all swarms. Sweater names and sales numbers have been anon-ymized for confidentiality.

Table 1. Actual sales performance along with forecasted sales rankings

Sweater	Percent of total sales	Survey average forecasted sales rank	Swarm average forecasted sales rank
A	23.4%	3	2
B	21.7%	7	3
C	16.8%	1	1
D	15.4%	2	5
E	12.0%	8	7
F	8.4%	5	4
G	1.2%	4	8
H	1.1%	6	6

4 Analysis

The results shown in Table 1 reveal that the swarms generated a far superior ranking than the survey, correctly identifying the top three sweaters by sales volume, as well as correctly identifying the bottom two. That said, a rigorous analysis is required to compare the rankings methods with statistical significance. To perform this comparison, the forecasted rank and real-world performance (in unit sales) of each item was used as a measure of the quality of the ranking: the more the unit sales of each item reflected the forecasted rank, the better the ranking.

Specifically, the quality of the rankings was compared by calculating the percentage of total unit sales accounted for by the top n items in each ranking (Table 2). The cumulative unit sales generated using the swarm to select sweaters is greater than or equal to the cumulative unit sales of the survey for all cutoffs **n**, indicating that participants were regularly able to predict the sales performance of each sweater more accurately as a swarm rather than by survey. In other words, if this group was asked to predict the top **n** sweaters by unit sales performance, they would have selected a far better-performing set of sweaters by swarming than by providing input through a traditional poll.

Table 2. Cumulative proportion of total unit sales and p-value of difference

Number "n" of sweaters selected	Top N survey unit sales	Top N swarm unit sales	p-value
1	16.8%	16.8%	0.9954
2	32.2%	40.2%	0.4140
3	55.6%	61.9%	0.0001
4	56.8%	70.3%	0.0497
5	65.2%	85.7%	0.0394
6	66.3%	86.8%	0.1573
7	88.0%	98.8%	0.0046
8	100.0%	100.0%	1.0000

Statistical Significance: to calculate the probability that the swarm outperformed the survey by chance, the average survey ranking of each sweater was bootstrapped 10,000 times by randomly resampling participant surveys with replacement from the original pool of surveys. An ordered list of forecasted sweaters was generated for each bootstrapped survey using the average ranking of each sweater in that bootstrap.

The cumulative sales for each bootstrap was recorded, and the proportion of bootstraps was calculated where the survey matched or outperformed the swarm's cumulative sales performance for each of the top N-ranked sweaters (a p-value). The p-values for each rank are listed in Table 2. We find that the swarm significantly outperforms the survey ($p < 0.05$) for the selection of the top 3, 4, 5, or 7 sweaters.

To visually compare swarm and survey performance, the cumulative unit sales of each method is plotted in Fig. 3, including a black dotted line that marks the cumulative sales that would have been achieved by perfectly ranking the sweaters.

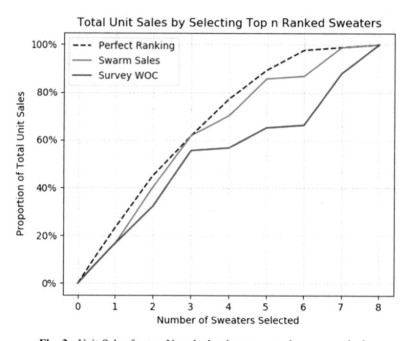

Fig. 3. Unit Sales for top N ranks by the swarm and survey methods

Looking at Fig. 3 above, we see that if a business wanted to predict the top selling products from a set of offerings, for example to plan inventory levels, they would get more accurate insights by using the swarm over the survey. In fact, we can compare the actual sales of the top four sweaters selected by the swarm to the top four sweaters selected by the survey. We find that the total sales generated by the swarm-based selections was $3.1 million **(91%** of maximum possible) compared to only $2.5 million **(74%** of maximum possible) for the survey-based picks, as illustrated in Fig. 4.

To more directly compare the quality of overall rankings between the swarm and the survey, the sum of the unit sales at each cutoff was calculated as the area under the curve (AUC) in Fig. 3. In fact, the swarm achieves an AUC of 94.4% of the best possible cumulative unit sales, while the survey achieves an AUC of only 81.0%. A histogram of the bootstrapped survey vs. swarm proportion of ideal sales AUC is shown in Fig. 5: we find that the swarm forecasts significantly outperform the survey forecasts using this aggregate metric (p < 0.001).

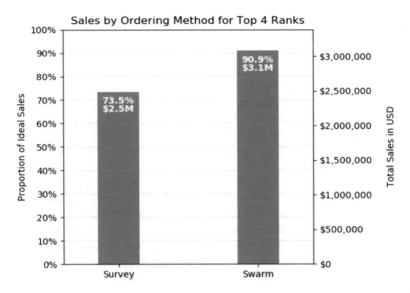

Fig. 4. Total sales (USD) and proportion of ideal sales for the top four sweaters selected from each forecasting method

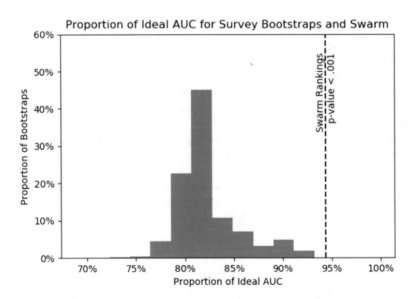

Fig. 5. Cumulative unit sales of all ranks from survey bootstraps and the swarm

5 Conclusions

Sales forecasting of new products is inherently challenging. By definition, there is little historical data on the market perception of the product. Similarly, when new versions of existing products are launched with substantially updated features or when new competitive products enter a market, historical data may provide weak or misleading guidance. Prior research shows that the inaccuracy in contemporary sales forecasts is often in excess of 42% of true unit sales, which significantly impairs the efficiency of business processes and strategic plans, and suggests that new methods are needed to improve the accuracy of sales forecasts.

In this study, we introduced Artificial Swarm Intelligence systems as a tool for sales forecasting using an online software platform called Swarm. We compared the forecasting accuracy using Swarm with the accuracy using traditional polling via online survey. Each method forecasted the sales of eight new sweaters distributed by a large fashion retailer during the 2018 holiday season. The ASI systems forecast the sales rankings of the sweaters significantly more accurately ($p < 0.001$) than the survey, achieving 91% of the optimal forecast for the top sellers as compared to only 74% for the survey method.

While the results from this study are extremely promising, the study was limited by sample size (i.e. number of groups, number of products forecasted), sample diversity (the type of products forecasted), and forecast timespan (i.e. 1 month, 6-month, 1 year performance). It would be valuable in future studies to test the use of ASI technology for sales forecasting across larger product sets, larger population pools, and in variety of different methods, regions, time-frames, and channels.

Future research may also investigate using the deep behavioral data from Human Swarms in combination with, and as a data source for, machine learning efforts that seek to forecast future events with higher accuracy. The combination of deeper human insights and machine learning has in the past been shown to provide exceptionally accurate results [21], so it would be valuable to investigate if this trend holds up for product forecasting and other challenging forecasting problems.

Acknowledgments. Thanks to Bustle Digital Group for supporting this project by sourcing participants and coordinating with the retail partner. Also, thanks to Unanimous AI for the use of the swarm.ai platform for this ongoing work.

References

1. Kahn, K.B.: Solving the problems of new product forecasting. Bus. Horiz. **57**(5), 607–615 (2014)
2. Lynn, G.S., Schnaars, S.P., Skov, R.B.: A survey of new product forecasting practices in industrial high technology and how technology businesses. Ind. Mark. Manag. **28**(6), 565–571 (1999)
3. Kahn, K.B.: An exploratory Investigation of new product forecasting practices. J. Prod. Innov. Manag. **19**, 133–143 (2002)

4. Rosenberg, L.: Human Swarms, a real-time method for collective intelligence. In: Proceedings of the European Conference on Artificial Life 2015, pp. 658–659 (2015)

5. Rosenberg, L.: Artificial swarm intelligence vs human experts. In: 2016 International Joint Conference on Neural Networks (IJCNN). IEEE (2016)

6. Metcalf, L., Askay, D.A., Rosenberg, L.B.: Keeping humans in the loop: pooling knowledge through artificial swarm intelligence to improve business decision making. Calif. Manag. Rev. (2019). https://doi.org/10.1177/0008125619862256

7. Rosenberg, L., Willcox, G.: Artificial swarm intelligence. In: Intelligent Systems Conference (IntelliSys), London, UK (2019)

8. Rosenberg, L., Baltaxe, D., Pescetelli, N.: Crowds vs swarms, a comparison of intelligence. In: IEEE 2016 Swarm/Human Blended Intelligence (SHBI), Cleveland, OH (2016)

9. Baltaxe, D., Rosenberg, L., Pescetelli, N.: Amplifying prediction accuracy using human swarms. In: Collective Intelligence 2017, New York, NY (2017)

10. McHale, I.: Sports Analytics Machine (SAM) as reported by BBC. http://blogs.salford.ac.uk/business-school/sports-analytics-machine/

11. Rosenberg, L., et al.: Artificial swarm intelligence employed to amplify diagnostic accuracy in radiology. In: IEMCON 2018, Vancouver, CA (2018)

12. Bonabeau, E.: Decisions 2.0: the power of collective intelligence. MIT Sloan Manag. Rev. **50**(2), 45 (2009)

13. Woolley, A.W., Chabris, C.F., Pentland, A., Hashmi, N., Malone, T.W.: Evidence for a collective intelligence factor in the performance of human groups. Science **330**(6004), 686–688 (2010)

14. Surowiecki, J.: The wisdom of crowds. Anchor (2005)

15. Seeley, T.D., Buhrman, S.C.: Nest-site selection in honey bees: how well do swarms implement the 'best-of-N' decision rule? Behav. Ecol. Sociobiol. **49**, 416–427 (2001)

16. Marshall, J., Bogacz, R., Dornhaus, A., Planqué, R., Kovacs, T., Franks, N.: On optimal decision-making in brains and social insect colonies. Soc. Interface (2009)

17. Seeley, T.D., et al.: Stop signals provide cross inhibition in collective decision-making by honeybee swarms. Science **335**(6064), 108–111 (2012)

18. Seeley, T.D.: Honeybee Democracy. Princeton University Press, Princeton (2010)

19. Seeley, T.D., Visscher, P.K.: Choosing a home: How the scouts in a honey bee swarm perceive the completion of their group decision making. Behav. Ecol. Sociobiol. **54**(5), 511–520 (2003)

20. Rosenberg, L., Willcox, G.: Artificial swarms find social optima. In: 2018 IEEE Conference on Cognitive and Computational Aspects of Situation Management (CogSIMA 2018), Boston, MA (2018)

21. Willcox, G., Rosenberg, L., Donovan, R., Schumann, H.: Dense neural network used to amplify the forecasting accuracy of real-time human swarms. In: IEEE International Conference on Computational Intelligence and Communication Networks (CICN), (2019)

Machine Translation from Natural Language to Code Using Long-Short Term Memory

K. M. Tahsin Hassan Rahit[1,2P(✉)], Rashidul Hasan Nabil[3,4],
and Md Hasibul Huq[5]

[1] Institute of Computer Science,
Bangladesh Atomic Energy Commission, Dhaka, Bangladesh
[2] Department of Bio-chemistry and Molecular Biology,
University of Calgary, Calgary, AB, Canada
kmtahsinhassan.rahit@ucalgary.ca
[3] Department of Computer Science,
American International University-Bangladesh, Dhaka, Bangladesh
merhnabil@gmail.com
[4] Department of Computer Science and Engineering,
City University, Dhaka, Bangladesh
[5] Department of Computer Science and Software Engineering,
Concordia University, Montreal, QC, Canada
mdhasibul.huq@mail.concordia.ca

Abstract. Making computer programming language more understandable and easy for the human is a longstanding problem. From assembly language to present day's object-oriented programming, concepts came to make programming easier so that a programmer can focus on the logic and the architecture rather than the code and language itself. To go a step further in this journey of removing human-computer language barrier, this paper proposes machine learning approach using Recurrent Neural Network (RNN) and Long-Short Term Memory (LSTM) to convert human language into programming language code. The programmer will write expressions for codes in layman's language, and the machine learning model will translate it to the targeted programming language. The proposed approach yields result with 74.40% accuracy. This can be further improved by incorporating additional techniques, which are also discussed in this paper.

Keywords: Text to code · Machine learning · Machine translation · NLP · RNN · LSTM

1 Introduction

Removing computer-human language barrier is an inevitable advancement researchers are thriving to achieve for decades. One of the stages of this advancement will be coding through natural human language instead of traditional

© Springer Nature Switzerland AG 2020
K. Arai et al. (Eds.): FTC 2019, AISC 1069, pp. 56–63, 2020.
https://doi.org/10.1007/978-3-030-32520-6_6

programming language. On naturalness of computer programming Knuth said, *"Let us change our traditional attitude to the construction of programs: Instead of imagining that our main task is to instruct a computer what to do, let us concentrate rather on explaining to human beings what we want a computer to do."* [6]. Unfortunately, learning programming language is still necessary to instruct it. Researchers and developers are working to overcome this human-machine language barrier. Multiple branches exists to solve this challenge (i.e. inter-conversion of different programming language to have universally connected programming languages). Automatic code generation through natural language is not a new concept in computer science studies. However, it is difficult to create such tool due to these following three reasons:

1. Programming languages are diverse.
2. An individual person expresses logical statements differently than other.
3. Natural Language Processing (NLP) of programming statements is challenging since both human and programming language evolve over time.

In this paper, a neural approach to translate pseudo code or algorithm like, human language expression into programming language code is proposed.

2   Problem Description

Code repositories (i.e. Git, SVN) flourished in the last decade producing big data of code allowing data scientists to perform machine learning on these data. In 2017, Allamanis *et al.* published a survey in which they presented the state-of-the-art of the research areas where machine learning is changing the way programmers code during software engineering and development process [1]. This paper discusses what are the restricting factors of developing such text-to-code conversion method and what problems need to be solved.

2.1   Programming Language Diversity

According to the sources, there are more than a thousand actively maintained programming languages, which signifies the diversity of these language[1], [2]. These languages were created to achieve different purpose and use different syntaxes. Low-level languages such as assembly languages are easier to express in human language because of the low or no abstraction at all whereas high-level, or Object-Oriented Programing (OOP) languages are more diversified in syntax and expression, which is challenging to bring into a unified human language structure. Nonetheless, portability and transparency between different programming languages also remains a challenge and an open research area. George *et al.* tried to overcome this problem through XML mapping [3]. They tried to convert codes from C++ to Java using XML mapping as an intermediate language. However, the authors encountered challenges to support different features of both languages.

---

[1] https://en.m.wikipedia.org/wiki/List_of_programming_languages.

[2] http://www.99-bottles-of-beer.net.

2.2 Human Language Factor

One of the motivations behind this paper is - as long as it is about programming, there is a finite and small set of expression which is used in human vocabulary. For instance, programmers express a *for-loop* in a very few specific ways [8]. Variable declaration and value assignment expressions are also limited in nature. Although all codes are executable, human representation through text may not due to the semantic brittleness of code. Since high-level languages have a wide range of syntax, programmers use different linguistic expressions to explain those. For instance, small changes like swapping function arguments can significantly change the meaning of the code. Hence, the challenge remains in processing human language to understand it properly which brings us to the next problem.

2.3 NLP of Statements

Although there is a finite set of expressions for each programming statements, it is a challenge to extract information from the statements of the code accurately. Semantic analysis of linguistic expression plays an important role in this information extraction. For instance, in case of a loop, what is the initial value? What is the step value? When will the loop terminate?

Mihalcea *et al.* has achieved a variable success rate of 70–80% in producing code just from the problem statement expressed in human natural language [8]. They focused solely on the detection of step and loops in their research. Another research group from MIT, Lei *et al.* use a semantic learning model for text to detect the inputs. The model produces a parser in C++ which can successfully parse more than 70% of the textual description of input [7]. The test dataset and model was initially tested and targeted against ACM-ICPC participantsínputs which contains diverse and sometimes complex input instructions.

A recent survey from Allamanis *et al.* presented the state-of-the-art on the area of naturalness of programming [1]. A number of research works have been conducted on text-to-code or code-to-text area in recent years. In 2015, Oda *et al.* proposed a way to translate each line of Python code into natural language pseudo code using Statistical Machine Learning Technique (SMT) framework [10] was used. This translation framework was able to - it can successfully translate the code to natural language pseudo coded text in both English and Japanese. In the same year, Chris *et al.* mapped natural language with simple if-this-then-that logical rules [11]. Tihomir and Viktor developed an Integrated Development Environment (IDE) integrated code assistant tool *anyCode* for Java which can search, import and call function just by typing desired functionality through text [4]. They have used model and mapping framework between function signatures and utilized resources like WordNet, Java Corpus, relational mapping to process text online and offline.

Recently in 2017, Yin and Neubig proposed a semantic parser which generates code through its neural model [12]. They formulated a grammatical model which works as a skeleton for neural network training. The grammatical rules are defined based on the various generalized structure of the statements in the programming language.

```
1   define function dummy with an argument x    def dummy(x):
2      if x is less than 0                          if x < 0:
3         return the string "foo"                      return "foo"
4      if x is divisible by 2                       elif x % 2:
5         return the string "bar"                      return "bar"
6      if not                                       else:
7         return the string "foobar"                   return "foobaar"

        source data in human language            target data in programming language
```

Fig. 1. Text-Code bi-lingual corpus

3 Proposed Methodology

The use of machine learning techniques such as SMT proved to be at most 75% successful in converting human text to executable code [2]. A programming language is just like a language with less vocabulary compared to a typical human language. For instance, the code vocabulary of the training dataset was 8814 (including variable, function, class names), whereas the English vocabulary to express the same code was 13659 in total. Here, programming language is considered just like another human language and widely used SMT techniques have been applied.

3.1 Statistical Machine Translation

SMT techniques are widely used in Natural Language Processing (NLP). SMT plays a significant role in translation from one language to another, especially in lexical and grammatical rule extraction. In SMT, bilingual grammatical structures are automatically formed by statistical approaches instead of explicitly providing a grammatical model. This reduces months and years of work which requires significant collaboration between bi-lingual linguistics. Here, a neural network based machine translation model is used to translate regular text into programming code.

Data Preparation: SMT techniques require a parallel corpus in the source and thr target language. A text-code parallel corpus similar to Fig. 1 is used in training. This parallel corpus has 18805 aligned data in it.[3] In source data, the expression of each line code is written in the English language. In target data, the code is written in Python programming language.

Vocabulary Generation: To train the neural model, the texts should be converted to a computational entity. To do that, two separate vocabulary files are created - one for the source texts and another for the code. Vocabulary generation is done by tokenization of words. Afterwards, the words are put into their contextual vector space using the popular *word2vec* [9] method to make the words computational.

---

[3] Dataset: https://ahclab.naist.jp/pseudogen/ [10].

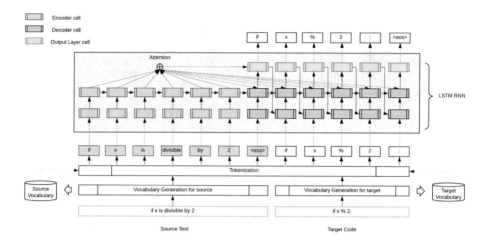

Fig. 2. Neural training model architecture of text-to-code

Neural Model Training: In order to train the translation model between text-to-code an open source Neural Machine Translation (NMT) - *OpenNMT* implementation is utilized [5]. *PyTorch*[4] is used as Neural Network coding framework. For training, three types of Recurrent Neural Network (RNN) layers are used – an encoder layer, a decoder layer and an output layer. These layers together form a LSTM model. LSTM is typically used in *seq2seq* translation.

In Fig. 2, the neural model architecture is demonstrated. The diagram shows how it takes the source and target text as input and uses it for training. Vector representation of tokenized source and target text are fed into the model. Each token of the source text is passed into an encoder cell. Target text tokens are passed into a decoder cell. Encoder cells are part of the encoder RNN layer and decoder cells are part of the decoder RNN layer. End of the input sequence is marked by a *<eos>* token. Upon getting the *<eos>* token, the final cell state of encoder layer initiate the output layer sequence. At each target cell state, *attention* is applied with the encoder RNN state and combined with the current hidden state to produce the prediction of next target token. This predictions are then fed back to the target RNN. *Attention* mechanism helps us to overcome the fixed length restriction of encoder-decoder sequence and allows us to process variable length between input and output sequence. *Attention* uses encoder state and pass it to the decoder cell to give particular attention to the start of an output layer sequence. The encoder uses an initial state to tell the decoder what it is supposed to generate. Effectively, the decoder learns to generate target tokens, conditioned on the input sequence. Sigmoidal optimization is used to optimize the prediction.

---

[4] https://pytorch.org/.

4   Result Analysis

Training parallel corpus had 18805 lines of annotated code in it. The training model is executed several times with different training parameters. During the final training process, 500 validation data is used to generate the recurrent neural model, which is 3% of the training data. We run the training with epoch value of 10 with a batch size of 64. After finishing the training, the accuracy of the generated model using validation data from the source corpus was 74.40% (Fig. 3).

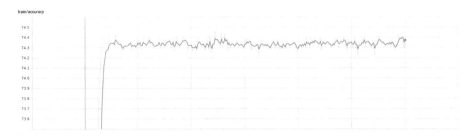

Fig. 3. Accuracy gain in progress of training the RNN

Although the generated code is incoherent and often predict wrong code token, this is expected because of the limited amount of training data. LSTM generally requires a more extensive set of data (100k+ in such scenario) to build a more accurate model. The incoherence can be resolved by incorporating coding syntax tree model in future. For instance,

"define the method tzname with 2 arguments: self and dt."

is translated into–

```
def __init__ ( self, regex) :.
```

The translator is successfully generating the whole code line automatically but missing the noun part (parameter and function name) part of the syntax.

5   Conclusion and Future Works

The main advantage of translating to a programming language is, it has a concrete and strict lexical and grammatical structure which human languages lack. The aim of this paper was to make the text-to-code framework work for general purpose programming language, primarily Python. In later phase, phrase-based word embedding can be incorporated for improved vocabulary mapping. To get more accurate target code for each line, Abstract Syntax Tree (AST) can be beneficial.

The contribution of this research is a machine learning model which can turn the human expression into coding expressions. This paper also discusses available methods which convert natural language to programming language successfully in fixed or tightly bounded linguistic paradigm. Approaching this problem using machine learning will give us the opportunity to explore the possibility of unified programming interface as well in the future.

Acknowledgment. We would like to thank Dr. Khandaker Tabin Hasan, Head of the Department of Computer Science, American International University-Bangladesh for his inspiration and encouragement in all of our research works. Also, thanks to Future Technology Conference - 2019 committee for partially supporting us to join the conference and one of our colleague - Faheem Abrar, Software Developer for his thorough review and comments on this research work and supporting us by providing fund.

References

1. Allamanis, M., Barr, E.T., Devanbu, P., Sutton, C.: A survey of machine learning for big code and naturalness. CoRR abs/1709.0 (2017). http://arxiv.org/abs/1709.06182

2. Birch, A., Osborne, M., Koehn, P.: Predicting success in machine translation. In: Proceedings of the Conference on Empirical Methods in Natural Language Processing - EMNLP 2008, pp. 745–754, October 2008. https://doi.org/10.3115/1613715.1613809

3. George, D., Priyanka Girase, N., Mahesh Gupta, N., Prachi Gupta, N., Aakanksha Sharma, N., Navi, V.: Programming language inter-conversion. Int. J. Comput. Appl. **1**(20), 975–8887 (2010). https://doi.org/10.5120/419-619

4. Gvero, T., Kuncak, V., Gvero, T., Kuncak, V.: Synthesizing Java expressions from free-form queries. In: Proceedings of the 2015 ACM SIGPLAN International Conference on Object-Oriented Programming, Systems, Languages, and Applications - OOPSLA 2015, vol. 50, pp. 416–432. ACM Press, New York (2015). http://dl.acm.org/citation.cfm?doid=2814270.2814295

5. Klein, G., Kim, Y., Deng, Y., Senellart, J., Rush, A.M., Seas, H.: OpenNMT: open-source toolkit for neural machine translation. In: Proceedings of the ACL, pp. 67–72 (2017). https://doi.org/10.18653/v1/P17-4012

6. Knuth, D.E.: Literate programming. Comput. J. **27**(2), 97–112 (1984). http://www.literateprogramming.com/knuthweb.pdf

7. Lei, T., Long, F., Barzilay, R., Rinard, M.C.: From natural language specifications to program input parsers. In: Proceedings of the 51st Annual Meeting of the Association for Computational Linguistics 1, pp. 1294–1303 (2013). http://www.aclweb.org/anthology/P13-1127

8. Mihalcea, R., Liu, H., Lieberman, H.: NLP (natural language processing) for NLP (natural language programming). In: Linguistics and Intelligent Text Processing, pp. 319–330 (2006). https://doi.org/10.1007/11671299_34

9. Mikolov, T., Chen, K., Corrado, G., Dean, J.: Distributed representations of words and phrases and their compositionality. CrossRef Listing of Deleted DOIs 1, pp. 1–9 (2000). https://arxiv.org/abs/1310.4546

10. Oda, Y., Fudaba, H., Neubig, G., Hata, H., Sakti, S., Toda, T., Nakamura, S.: Learning to generate pseudo-code from source code using statistical machine translation. In: 2015 30th IEEE/ACM International Conference on Automated Software Engineering (ASE), pp. 574–584. IEEE, November 2015. http://ieeexplore.ieee.org/document/7372045/

11. Quirk, C., Mooney, R., Galley, M.: Language to code: learning semantic parsers for if-this-then-that recipes. In: Proceedings of the 53rd Annual Meeting of the Association for Computational Linguistics and the 7th International Joint Conference on Natural Language Processing (Volume 1: Long Papers), pp. 878–888. Association for Computational Linguistics, Stroudsburg (2015). http://aclweb.org/anthology/P15-1085
12. Yin, P., Neubig, G.: A syntactic neural model for general-purpose code generation. In: Proceedings of the 55th Annual Meeting of the Association for Computational Linguistics (2017). http://arxiv.org/abs/1704.01696

ML Supported Predictions for SAT Solvers Performance

A. M. Leventi-Peetz[1](✉), Jörg-Volker Peetz[2], and Martina Rohde[1]

[1] Federal Office for Information Security, Godesberger Allee 185–189,
53175 Bonn, Germany
leventi-peetz@bsi.bund.de
[2] Bonn, Germany

Abstract. In order to classify the indeterministic termination behavior of the open source SAT solver CryptoMiniSat in multi-threading mode while processing hard to solve Boolean satisfiability problem instances, internal solver runtime parameters have been collected and analyzed. A subset of these parameters have been selected and employed as features vector to successfully create a machine learning model for the binary classification of the solver's termination behavior with any single new solving run of a not yet solved instance. The model can be used for the early estimation of a solving attempt as belonging or not belonging to the class of candidates with good chances for a fast termination. In this context, a combination of active profiles of runtime characteristics appear to mirror the influence of the solver's momentary heuristics on the immediate quality of the solver's resolution process. Because runtime parameters of already the first two solving iterations are enough to forecast termination of the attempt with good success scores, the results of the present work deliver a promising basis which can be further developed in order to enrich CryptoMiniSat or generally any modern SAT solver with AI abilities.

Keywords: AI (Artificial Intelligence) · ML (Machine Learning) ·
SAT solver · Security

1 Introduction

The significance of SAT[1] solvers as a core technology for the analysis, synthesis, verification and testing of security properties of hardware and software products is established and well known. Automated Reasoning techniques for finding bugs and flaws examine nowadays billion lines of computer code with the use of Boolean and Constraint Solvers on domains of interest. However, due to the continuous improvement of SAT solver efficiency during the last decades and the dramatic scalability of these solvers against large real-world formulas, many new application cases arise in which SAT solvers get deployed for tackling hard problems, which were believed to be in general intractable but yet get solved [5].

---

[1] SAT – *satisfiablity.*

© Springer Nature Switzerland AG 2020
K. Arai et al. (Eds.): FTC 2019, AISC 1069, pp. 64–78, 2020.
https://doi.org/10.1007/978-3-030-32520-6_7

SAT solvers are also known to be used for security protocol analysis [1, 21] as well as for tasks like the automated verification of access control policies, automatic *Anomaly Detection* in network configuration policies [15] and verification of general Access Control Systems where access rules are first encoded in SAT representation [6, 16]. Furthermore, SAT-based cryptanalysis methods are in advance and report increasing successes, like for example cryptographic key recovery by solving instances which encode diverse cipher attacks [9, 10, 17, 19], etc.

Also the solution of constraint optimization problems for real-world applications on the basis of already very competitive MaxSAT solvers, the great majority of which are core-guided, heavily relying on the power of SAT solvers, is the best way of proving unsatisfiability of subsets of soft constraints, or unsat cores, in an iterative way towards an optimal solution [8]. New algorithmic solutions instantiating innovative approaches to solve various data analysis problems in ML,[2] like correlation clustering, causal discovery, inference, etc. are based on SAT and Boolean optimization solvers. Decision and optimization problems in artificial intelligence profit by the application of SAT solvers [18] but also the opposite direction is pursued, namely the improvement of SAT solving using ML [4, 12, 13]. The intrinsic connection of the two subjects, SAT solvers and AI seems to be growing especially also under the scope of global trends intending to incorporate AI and ML technologies in the majority of the next generation cybersecurity solutions. The improvement of SAT solver performance by means of AI is a topic of intense and general interest and motivated this work.

1.1   Organization of This Paper and Contributions

First, a brief account is given here of previously gained experience with CryptoMiniSat [20], gathered while studying the solver's behavior when engaged for the solution of hard CNF instances representing KPA in cryptanalysis.[3] Similar instances whose solution is far from being trivial, are used also for the results produced in this work [11]. However these instances are here taken as an example of especially hard instances to solve while the achieved results should be relevant to the solution of arbitrary hard instances having similar features to those of the here employed ones. In what follows the motivation of the present work is substantiated and then the procedure followed to produce our results as well as the results themselves will be presented. In the last part we summarize about this work and discuss about further investigations planned. In the past the CMS[4] solver's performance has been studied by carrying out runtime tests both with the solver in default configuration and with various solver switches set. The tests showed that the solver runtime until the solution is found, in case the job doesn't previously stop because of some time-limit setting, is subjected to distinct statistical variations. This is due to the indeterministic behavior of the solver in multi-thread operation mode. The complexity of the here discussed problems

---

[2] ML – *machine learning.*

[3] CNF – *conjunctive normal form*; KPA – *known-plaintext attacks.*

[4] CMS – *CryptoMiniSat.*

though excludes one-thread operation from being an option. The performed runtime tests were in average highly time consuming both when a termination was reached and when the job had to be interrupted because it reached some previously defined upper run time limit. The job interruption practice was motivated by an amplitude of experience showing that if some long runtime limit has been surpassed without a solution found, the majority of test runs do not terminate at all. As a matter of fact, even under identical solver parameter configuration when running several tests to solve one and the same instance, one cannot avoid diverging solving times or finding no solution at all for instances which are solvable by construction. All cryptanalytic instances used in experiments possess by construction one single solution, which was in this case the sought-for cryptographic key. The performance of a large number of tests had allowed a statistical analysis of the non-deterministic solver runtimes to empirically define command line parameter combinations for CMS which yield best runtimes medians for the type of instances taken under examination. The application of an AAC[5] tool which followed [14], allowed a systematic exploration of the configuration parameter space reaching beyond the empirical tests to discover even better configuration parameter combinations. The results of those efforts have been quite encouraging, demonstrating a lowering of the median of runtimes by 30% to finally 90% with the application of the AAC [11]. The limitation we see however in further pursuing this approach is that when a semi-automatic tuning of the solver's configuration parameters is to be performed, this has to be carried out on the basis of previously cumulated values of best achieved runtimes. Those best runtimes are of course obtainable at a high computational and time cost which should probably have to be repeatedly afforded, in case the expensively discovered effective configuration parameter settings are not globally valid but rather problem specific.

1.2 Contributions of the Present Work

In the present work we show how to create a prognosis concerning the successful termination of a solving process not by properly adjusting some solver configuration parameters but by using internal runtime parameters of the beginning of the process during the process. Instead of analyzing effectiveness of the solver's configuration in a post-hoc manner, that is following the event of numerous successful terminations, we have here chosen to observe and analyze the joint evolution of solver internal parameters that are dynamically changing during the automatic state transformation of the instance while the solver is searching for a solution. These parameters which are issued by the solver when running in verbose mode, have been at first separately investigated in order to see if they demonstrate any correlation to the duration of job-runtimes and final successful termination. This question could not get uniquely answered. This circumstance motivated the tryout of a subset of the solver's dynamically changing runtime parameters as model features for building an ML model with the purpose to

---

[5] AAC – *automatic algorithm configuration*.

detect if the solver's active state evolution follows a direction with good chances to terminate timely or not. Taking into consideration the fact that the joint observation of parameters that belong to the two or three first iterations of the solving process showed to be sufficient to construct the ML model so as to get some decent classification results, this approach can be seen as a workable basis to later devise and incorporate an internal mechanism in the solver to trigger early changes of search strategy on the basis of internal short-termed *collective* parameter changes at a minimal time cost. Given the fact that the time length of the solver's iterations normally essentially grows with the iteration's number order, the limitation of the needed for the model runtime parameters on those of the first iterations, definitely helps avoid waste of resources on hopeless solving efforts involving many iterations. The innovation of the effort described here lies also in the fact that a combination of parameter profiles originating from both terminating and non-terminating test runs equally contributes to the model building. The ultimate task is building an ML model to be used for an almost real-time quality estimation of the solver's momentary search direction.

To our knowledge the here presented way of joint employment of solver's internal statistics in order to create solver-forecasts has not been attempted before.

2   Solver Runtime Parameters: Description, Analysis and Illustration

We have selected runtime parameters which we consider to be in general significant for characterizing the state of the solving process, and this not only in relation to the here regarded CNF instances. The parameter charts plotted in the graphics below are representative for the processing of instances of similar features. As features of the here employed CNF instances we observe the number of variables L, the number of CNF clauses N, the length of clauses (minimum, lower quartile, median, upper quartile, and maximum length), the sum of all occurrences of all variables, the fraction of clauses by length in the CNF instance, the occurrence of literals in clauses, and their mean value [7]. The three instances whose runtime parameters are plotted here belong to three different variations of the same mathematical problem (known-plaintext attack on the round-reduced AES-64 model cipher) with instance densities N/L assuming the values 303.4 (18-vs), 304.2 (20-vs) and 306.6 (30-vs). We could prove that similar parameter graphics are produced for instances created on the basis of the same mathematical problem but with different problem parameters (different number of plaintexts, for our instances 18, 20, and 30, respectively and/or different key).

The parameter names are almost self-explanatory in the context of the functionality of a CDCL[6] solver. We have experimented with two slightly different sets of parameters for the creation of the ML model and these sets are displayed in the first and second columns of Table 1, where the abbreviation props stands for propagations.

---

[6] CDCL – *conflict driven clause learning*.

Table 1. Runtime parameters as model features

Set 1	Set 2
all-threads	all-threads
conflicts/second	conflicts/second
blocked-restarts	blocked-restarts
restarts	restarts
props/decision	props/decision
props/conflict	literals/conflict
	decisions/conflict

The parameter `all-threads` expresses the total time consumed by a solver iteration calculated by adding the solving times contributed by the various threads during this iteration. All here discussed parameters are regularly delivered by the solver at the end of each and every iteration, with the solver running in verbose mode. In Fig. 1 the `conflicts/second` parameters for six different runs of the same instance that terminated at different times with the solver identically configured for all runs, is exemplarily depicted. There exist differences in the structure, especially the heights and in the positions of the peaks of the curves which seem to associate to the length of the runtime until a successful termination of the corresponding job (see legends in Fig. 1) has been reached.

It seems plausible to attribute to the steep rising number of conflicts at the beginning of the solving-process (that means during the first and/or the second iterations) an early generation of a lot of additional useful information for the solver. This event accelerates the learning effect which would justify the assumption of a direct association of a shorter solution runtime to the occurrence of many conflicts at the beginning of the solving process. Many conflicts per second along the search path naturally results in a shorter iteration time, as there is a limit of conflicts per iteration, allowed by the solver, which in this case gets earlier reached. Advantageous are therefore short iteration times, arising out of the occurrence of high conflict rates during short iterations.

The markers on the curves correspond to the begin/end of an iteration. A short first iteration time sets the begin of the corresponding plot closer to the y-axis and it is notable that curves starting far to the left describe evolutions of solutions with mostly shorter therefore better runtimes.

Curves which despite their starting far to the left do evolve to describe long and therefore bad termination runtimes, usually demonstrate also early negative or very flat gradients as regards the `conflicts/second` parameter. To this, one can compare the two relatively flat curves with termination times 215,108 and 123,359 s, respectively (see legend of Fig. 1). These two curves could also be categorized as not successfully terminating runs, if an upper time-limit had been set for the allowed runtime of the corresponding jobs.

Comparable results are depicted in Fig. 2 (Fig. 3) where the evolution of nine (seven) different runs of the instance 18-vs (30-vs) are shown. The instance in

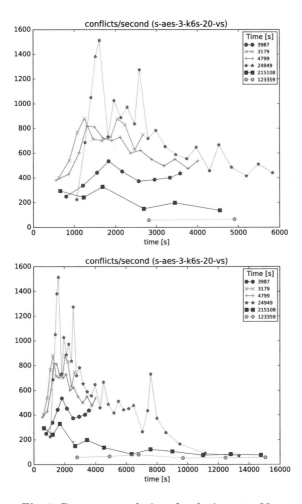

Fig. 1. Parameter evolutions for the instance 20-vs

Fig. 2 is the most difficult to solve in comparison to the other two instances of Figs. 1 and 3, respectively. In this more difficult case one observes several intensive or less intensive learning phases in a sequence, represented by more than one distinct peaks in the corresponding curve plots. Fine differences in the start values seem to play also a considerable role for the consequent runtime evolution. Not only the duration of the first iteration but also the initial value of the parameter `conflicts/second` is in this respect significant. Striking is again the appearance of the late starting and thoroughly flat curve corresponding to the very late termination time of 352,324 s (see legends of Fig. 2).

Considering the fact that no continuous correlation connecting any of the runtime parameters with the corresponding solver termination time is to be found, the question arises if one can use statistical traits of the joint evolution

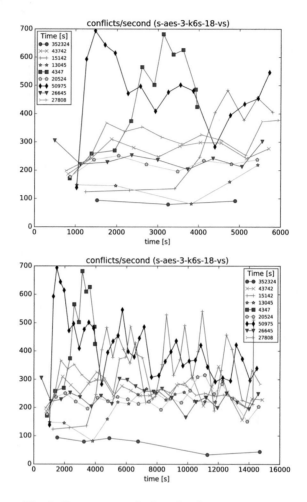

Fig. 2. Parameter evolutions for the instance 18-vs

of parameters in a combination, not in order to discover conditions for some optimal solution time but in order to distinguish between a solver path leading to a termination and one which does not.

The second runtime parameter here checked is that of the `restarts`. Restarts is a critically important heuristic according to many experts in the field of CDCL SAT-solver research. Restart schemes have been evaluated in detail and a particular benefit could be identified in frequent restarts when combined with *phase saving* [2]. Restarts are considered to *compact the assignment stack* and frequent restarts enhance the quality of the learnt clauses thus shorten the solution time. In a recent work an ML-based restart policy has been introduced to trigger a restart every time an unfavorable forecast regarding the quality of the expected new to be created learned clauses arises [13].

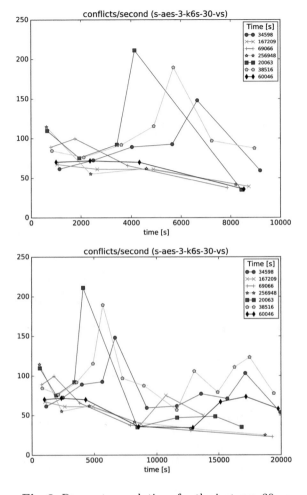

Fig. 3. Parameter evolutions for the instance 30-vs

In Figs. 4, 5 and 6 there are depicted the courses of the corresponding
`restarts` parameters for the same runs whose `conflicts/second` parameters have
been plotted above.

A comparison between these new three plots reveals as a common indicator of unpromising (not timely expected to terminate) runs, the corresponding
poorly structured, and at parts continuously evolving course of a curve. This
adverse curve-shape feature, if observed alone, becomes especially obvious only
when watched over a time period which is longer than two or three iterations.
In combination with the iteration length, given by the `all-threads` parameter
though, the `restarts` parameter can indeed count as a criterium to forecast a
timely termination during the first two iterations alone.

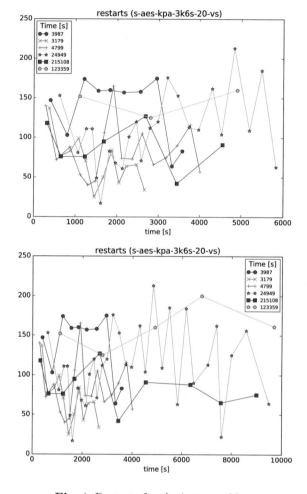

Fig. 4. Restarts for the instance 20-vs

Similar remarks apply well for `blocked-restarts`, the complementary parameter, and the rest of the parameters taken for the ML model building, whose
plots are not displayed here out of space considerations. In conclusion, one can
ascertain that a missing *agility* in the timely evolution of runtime parameters
and especially in combination with a low conflict rate at the beginning of the
solving process, plausibly signalize little chance for a timely termination of a
solving process.

Figure 7 shows pairwise scatterplots of the six (top) and seven (bottom) runtime parameters (compare first and second columns respectively of Table 1. Also
different CNF instances were used for each plot). These parameters correspond
to the first of the two iterations taken for building the ML model. The plots
help to reveal pairwise correlations between the model features. They indicate

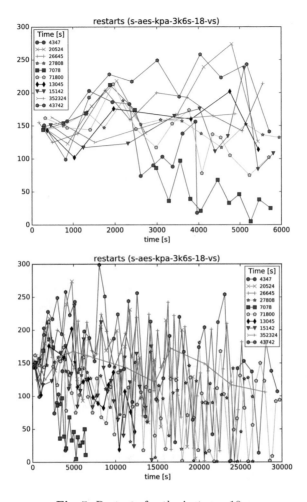

Fig. 5. Restarts for the instance 18-vs

also the intricacy implied in finding an analytical function for the classification. Points with different colors symbolize different classification tags.

3   ML Models with CMS-Runtime Parameters as Model Features

In order to investigate the feasibility of correctly and fast predicting the finding of a solution with the CMS solver on the basis of an ML model, we used CMS-runtime parameters of previous test-run cases to construct NN[7] classifiers. For the practical implementation we have employed the high-level Keras framework [3]. Keras is an open source neural-network library written in Python. It is

---

[7] NN – *neural network*.

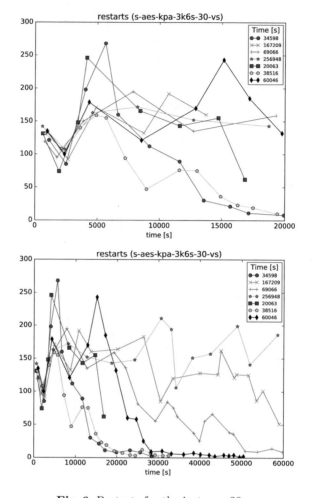

Fig. 6. Restarts for the instance 30-vs

capable of running on top of TensorFlow, Cognitive Toolkit (CNTK), or Theano. It was developed with a focus on enabling fast experimentation.

The training datasets for the models were compiled by extraction of the runtime parameters out of log-data originating in a set of test run cases, incorporating two equal numbers of terminating and not terminating processes (150 all together for each CNF instance) and which were then accordingly tagged with 1 or 0. We explored three different NN models for binary classification, each of them has been trained with the same training datasets. The test datasets had in each case half the magnitude of the corresponding training dataset (75 test cases).

The first NN model we tried was also the simplest one consisting of mainly two layers of neurons. For the first layer, the input layer, there was taken a

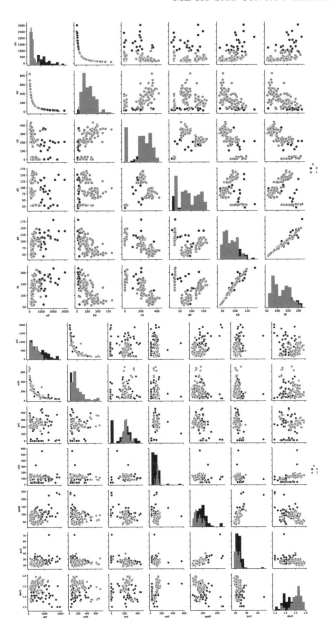

Fig. 7. Pairwise scatterplots of six (top) and of seven (bottom) CMS runtime parameters. Each point indicates a CMS run. Colors indicate the classification of the test run.

number of neurons equal to the runtime parameters of the CMS, planned to serve as features for the model. See Table 1. One and the same runtime parameter was considered as a different model feature if belonged to a different iteration, so that we had 12 (14) input parameters for the model. The second layer (output layer) had just one neuron and delivered the binary classification.

The second NN model had an intermediate layer added with half as much neurons as the first layer, here six. The additional, hidden layer helps an NN-model to better approximate the classification of non-linearly separable data.

The third NN model was derived from the second model by dropping a percentage of the input values for the second and the third layer which should help prevent overfitting. This model is called a multilayer perceptron for binary classification. In all NN-models non-linear activation functions were used.

Independent of model, when employing the 6 features of the first column of Table 1 we had for all three instance-cases an equally good hit ratio near 90% while the formal model accuracy was 100%. With the seven parameters as model features of the second column of Table 1 these simple models deliver less success with a hit score of about 70%. This shows that parameter (feature) selection is very important. Also, a more developed NN model might be necessary for a reliable classification prediction. These first results encourage the building of further more sophisticated models in the future.

4   Conclusions and Future Work

The SAT solver CryptoMiniSat reenacting produces a series of runtime parameter values while processing a CNF instance until solving or failing to solve it within a time limit. We showed that it is possible to select and employ subsets of these runtime parameters, the same for all runs, and use them as features to build an ML model for solver runtime classification. For any future attempt to solve a new CNF instance of features similar to those of the instance used to create the training data for the model, this model enables the early classification of the attempt as a fast or a late terminating run. By early is meant that the parameter profiles released during the first two solver iterations are already sufficient for a classification of this solving attempt.

In a future work, the model should be extended so as to become capable to generate forecasts for attempts to solve problems bigger than the ones which it has been trained for. Also the possibility to extend the model so as to generate forecasts for CNF instances of any features and especially for hard random instances of any parametrization lies in the focus of our plans.

References

1. Armando, A., Compagna, L.: SAT-based model-checking for security protocols analysis. Int. J. Inf. Sec. **7**(1), 3–32 (2006)
2. Biere, A., Fröhlich, A.: Evaluating CDCL restart schemes. In: Proceedings of the International Workshop on Pragmatics of SAT (POS 2015), 16 p. (2015). http://fmv.jku.at/papers/BiereFroehlich-POS15.pdf

3. Chollet, F., et al.: Keras (2015). https://keras.io
4. Devlin, D., O'Sullivan, B.: Satisfiability as a classification problem. In: Proceedings of the 19th Irish Conference on Artificial Intelligence and Cognitive Science (2008). http://www.cs.ucc.ie/~osullb/pubs/classification.pdf
5. Ganesh, V.: Machine Learning for SAT Solvers. Abstract. (2018) https://uwaterloo.ca/artificial-intelligence-institute/events/machine-learning-sat-solvers. Invited talk at Waterloo AI Institute, University of Waterloo, Canada
6. Hughes, G., Bultan, T.: Automated verification of access control policies using a SAT solver. Int. J. Softw. Tools Technol. Transf. **10**(6), 503–520 (2008)
7. Hutter, F., Xu, L., Hoos, H.H., Leyton-Brown, K.: Algorithm runtime prediction: methods & evaluation. Artif. Intell. **206**, 79–111 (2014)
8. Järvisalo, M.: Boolean satisfiability and beyond: algorithms, analysis, and AI applications. In: Proceedings of the 25th International Joint Conference on Artificial Intelligence (IJCAI 2016), pp. 4066–4069. AAAI Press (2016). http://www.ijcai.org/Proceedings/16/Papers/602.pdf
9. Kehui, W., Tao, W., Xinjie, Z., Huiying, L.: CryptoMiniSAT solver based algebraic side-channel attack on PRESENT. In: 2011 First International Conference on Instrumentation, Measurement, Computer, Communication and Control, pp. 561–565 (2011)
10. Lafitte, F., Nakahara Jr., J., Van Heule, D.: Applications of SAT solvers in cryptanalysis: finding weak keys and preimages. J. Satisfiability Boolean Model. Comput. **9**, 1–25 (2014)
11. Leventi-Peetz, A., Zendel, O., Lennartz, W., Weber, K.: CryptoMiniSat switches-optimization for solving cryptographic instances. In: Proceedings of the International Workshop on Pragmatics of SAT (POS 2018) (2018). https://easychair.org/publications/preprint/Q4kv
12. Liang, J.H., Ganesh, V., Poupart, P., Czarnecki, K.: Learning rate based branching heuristic for SAT solvers. In: Theory and Applications of Satisfiability Testing - SAT 2016, Bordeaux, France, Proceedings, pp. 123–140 (2016)
13. Liang, J.H., Oh, C., Mathews, M., Thomas, C., Li, C., Ganesh, V.: A machine learning based restart policy for CDCL SAT solvers. In: Proceedings of the 21st International Conference on Theory and Applications of Satisfiability Testing (SAT 2018), pp. 94–110. Springer (2018)
14. Lindauer, M., Eggensperger, K., Feurer, M., Falkner, S., Biedenkapp, A., Hutter, F.: SMAC v3: algorithm configuration in python (2017). https://github.com/automl/SMAC3
15. Lorenz, C., Schnor, B.: Policy anomaly detection for distributed IPv6 firewalls. In: Proceedings of the 12th International Conference on Security and Cryptography (SECRYPT-2015), pp. 210–219. Science and Technology Publications (2015)
16. Mauro, J., Nieke, M., Seidl, C., Yu, I.C.: Anomaly detection and explanation in context-aware software product lines. In: Proceedings of the 21st International Systems and Software Product Line Conference, SPLC 2017, Volume B, Sevilla, Spain, pp. 18–21 (2017). https://www.isf.cs.tu-bs.de/cms/team/nieke/papers/2017-SPLC.pdf
17. Otpuschennikov, I., Semenov, A., Gribanova, I., Zaikin, O., Kochemazov, S.: Encoding Cryptographic Functions to SAT Using Transalg System. arXiv:1607.00888 (2016)
18. Rintanen, J.: Solving AI Planning Problems with SAT (2013). http://www.epcl-study.eu/content/downloads/slides/rintanen_2013.pdf. Research talk at the EPCL Basic Training Camp

19. Semenov, A.A., Zaikin, O., Otpuschennikov, I.V., Kochemazov, S., Ignatiev, A.: On cryptographic attacks using backdoors for SAT. In: Proceedings of the Thirty-Second AAAI Conference on Artificial Intelligence, (AAAI-18), New Orleans, Louisiana, USA, 2018, pp. 6641–6648 (2018)
20. Soos, M., Nohl, K., Castelluccia, C.: Extending SAT solvers to cryptographic problems. In: Theory and Applications of Satisfiability Testing - SAT2009, 12th International Conference, SAT 2009, Swansea, UK, 30 June–3 July, 2009. Proceedings, pp. 244–257 (2009).https://github.com/msoos/cryptominisat
21. Vanhoef, M., Piessens, F.: Symbolic execution of security protocol implementations: handling cryptographic primitives. In: USENIX Workshop on Offensive Technology (USENIX WOOT) (2018). https://papers.mathyvanhoef.com/woot2018.pdf

The Application of Artificial Neural Networks to Facilitate the Architectural Design Process

Pao-Kuan Wu$^{(\boxtimes)}$ and Shih-Yuan Liu

Asia University, Wufeng, Taichung 41354, Taiwan
`paokuanwu0412@gmail.com`

Abstract. In this paper, the main purpose is how to apply Artificial Neural Networks (ANN) to facilitate the process of architectural site analysis which is one of the crucial steps in architectural design process. The experiment is based on a student design project which can demonstrate the way of AI application in this paper. The goal of this student design project is to arrange several studios including public studios and private studios on a particular parcel. In the experiment, the ANN models were trained by several environmental factors which can help students perceive better environmental features for the purposes of this design project; then the trained ANN models can be used to determine where the appropriate locations are for the arrangement of the public and private studios. Furthermore, by analyzing the ANN weight values can reveal more information about which environmental factors are more important than others.

Keywords: Artificial Neural Networks · Computer-aided design · Architectural site analysis · Geographic Information System

1 Introduction

Design a building is a complicated process. The decision-making process is a kind of obscure expert knowledge that only depends on designers' experiences [1]. In order to advance the efficiency of architectural design, using computational techniques to better improve the decision-making process is an important trend nowadays [2, 3].

Computer-aided design (CAD) has been developing for a long time. However, most of the CAD applications is mainly on improving design drawing by computer devices. There is not so much work related to the decision making of architectural design with AI techniques [2]. Artificial Neural Networks (ANN) is one of proper Artificial Intelligence (AI) techniques for the application of decision making with its pattern recognition ability [4–6]. ANN has been applied in lots of urban planning research [6–11], so we can see this as a reliable AI technique. And the process of architectural design thinking is a series of pattern recognition process which has been discussed by an important book: A Pattern Language: Towns, Buildings, Construction published by Christopher Alexander in 1977 [12]. Therefore, it is a feasible idea to apply ANN to facilitate the architectural design process.

© Springer Nature Switzerland AG 2020
K. Arai et al. (Eds.): FTC 2019, AISC 1069, pp. 79–89, 2020.
https://doi.org/10.1007/978-3-030-32520-6_8

This paper wants to demonstrate the beginning phase of applying ANN in the architectural design process. Therefore, applying ANN for architectural site analysis is the main purpose of our experiment. The architectural site analysis is a crucial and preliminary step to affect the decision making of architectural design. And it's rather simple to test the application of pattern recognition logic for improving the architectural design process.

After reviewing some relevant literature and pointing out the purpose of this paper in this section, the following Sect. 2 is about the experimental methodology and process. The discussion of experimental results is described in Sect. 3, followed by the conclusion and the discussion of further work in Sect. 4.

2   The Experimental Methodology and Material

2.1   Artificial Neural Networks

ANN is one of the pattern classification techniques derived from the imitation of human perception process. The mechanism of ANN is like the function of our brain that can receive outside incentives by neurons, and then recognize the situations according to our experience. The development of ANN has come a long way and had many resources to describe the details [4, 5, 13, 14]. The application of ANN for urban planning or design projects can date back to Fisher and Opershaw around 1998 [15, 16], and this AI technique is still popular to apply nowadays [6, 7, 10]. ANN has many advantages, such as solving the problems of optimization and approximation, the endurance of data noise and the great ability of prediction [14]. Furthermore, ANN is a kind of 'data-driven' techniques which means the ANN results are derived from the tendency of data distribution based on statistical regression theory. In that way, the experimental results can be more objective [13].

Feedforward ANN is applied in this paper and Fig. 1 is the data processing structure of this type of ANN. Generally, feedforward ANN consists of three components: input layer, hidden layer and output layer. The hidden layer is an awkward part for the application of ANN because of the notoriety of "black box" [17]. The application of ANN with the hidden layer means using non-linear statistic skills which can easily have high performance of ANN result, but blur the correlations between the target patterns and the input factors. In order to simplify the experimental operations and easily understand which factors are more effective to the target pattern, the ANN hidden layer doesn't be considered in the experiment.

The architectural design process is like the pattern recognition process [12, 18, 19]. A seasoned designer is like a well-trained ANN model with lots of empirical data and has the ability to recognize best design solution (i.e., to recognize the target pattern). Applying trained ANN to recognize optimal architectural patterns can be described by Eq. (1), and Fig. 2 is the illustration for this process.

$$Y_j = \sigma\left(\sum_{i=1}^{n} W_{ij}X_i + \beta\right) \tag{1}$$

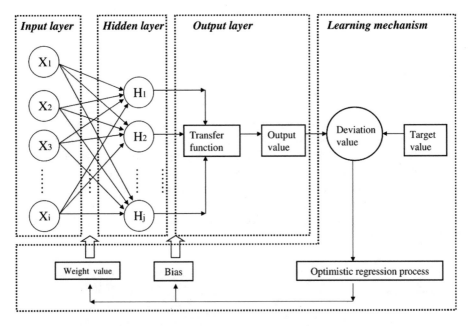

Fig. 1. The data processing structure of feedforward ANN.

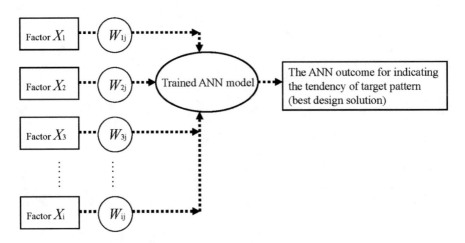

Fig. 2. The generation of ANN outcome from the trained ANN model

where Xi is i-th factor; where Yj is the ANN outcome for indicating the tendency of target pattern; where Wij is the ANN weight value; where σ is the transfer function; where β is a bias value.

2.2   Experimental Process and Material

The experiment a part of a student architectural design project. The goal of this design project is to design a group of buildings that consists of public and private studios. The demonstration of how to pick the appropriate building sites by the application of ANN is the main part of this paper.

The architectural site is set within our University. The architectural site is demonstrated in Fig. 3 and includes two main environmental features. The northeastern side is the complex of classrooms which is adjacent to the architectural site by the road. The southwestern side is the lake which is considered as a quieter side. In order to record the environmental factors, we use Arcgis software as a GIS (Geographic Information System) data platform to handle not only graphical but also numerical data. Additionally, we format the architectural site into grid formation like Fig. 4 shows. Each grid point is a potential studio location that can be recorded by several environmental factors.

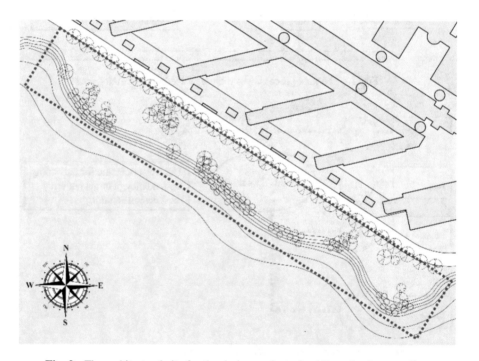

Fig. 3. The architectural site for the design project of public and private studios.

The number of locations is 8505. The environmental factors are listed in Table 1. There are 9 environmental factors separated by three categories: the distance from the specific environmental feature which has 4 factors, the specific zone which has 3 factors, and the limited condition which has 2 factors. All the 8505 locations can be described by these factors with the GIS data platform. All the ANN computing work is programmed by MATLAB. The next chapter is the experimental results and discussion.

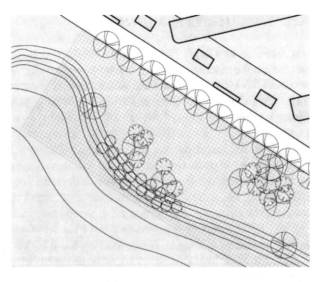

Fig. 4. The planning site with the grid formation; each grid point is the potential studio location.

Table 1. The environmental factors.

Category	The environmental factor	Denotation	Description	Numerical setting
The distance from the specific environmental feature	The road side	D1	The distance from the specific feature to the target location	Normalization in between of 0–1
	The lake side	D2		
	The eastern side	D3		
	The western side	D4		
The specific zone	The center zone	S1	Whether the target location is within the specific zone	If the target location is within the specific zone, then the factor value is 1; otherwise is 0
	The edge zone	S2		
	The tree protection zone	S3		
The limited condition	The building zone can't be out of the site boundary	L1	Whether the target location has the limited conditions	If the target location has the limited conditions, then the factor value is 0; otherwise is 1
	The building zone can't cover the tree protection zone	L2		

3   The Experimental Results and Discussions

In the experiment, there are two target patterns needed to be recognized, those are the locations proper for public studios and the locations proper for private studios. Therefore, we set two ANN models to handle the pattern recognition work respectively. Both ANN models have the same conditions during the training step: 500 random locations for training the ANN models and another 150 random locations for validating the efficiency of the trained ANN models. In the ANN training step, if one selected location is considered as the target pattern, then the target pattern value is set as 1 otherwise is 0. In addition, logistic sigmoid transfer function is used as the ANN transfer function in order to restrict the ANN outcome in the range of 0–1.

After the ANN training steps, the accuracy performance of pattern classification for public studio locations is 88%, and the accuracy for the private studio locations is 89.33%. Then all the 8505 locations can generate the final ANN outcomes by the trained ANN models according to Eq. (1). If one location has a higher value of ANN outcome, which means it has a higher tendency to judge as the target pattern. We can use the trained ANN models to denote which locations are appropriate for public studios or private studios.

The GIS platform can demonstrate the ANN experimental results clearly. All the locations are classified in five levels depending on their ANN outcomes: 0–0.2, 0.21–0.4, 0.41–0.6, 0.61–0.8, and 0.81–1 which listed in Table 2. Figures 5 and 6 are the GIS demonstrations for displaying where the locations are more appropriate to set public or private studios respectively. The darker color means that this location is better to set the target studio.

Table 2.   The final ANN outcomes classified in five levels.

Target pattern	ANN outcome level	The number of locations
For public studio	L1: 0–0.2	5253
	L2: 0.21–0.4	1610
	L3: 0.41–0.6	1457
	L4: 0.61–0.8	170
	L5: 0.81–1	15
		Total: 8505
For private studio	L1: 0–0.2	397
	L2: 0.21–0.4	4413
	L3: 0.41–0.6	3166
	L4: 0.61–0.8	490
	L5: 0.81–1	39
		Total: 8505

Since the hidden layer was not considered in the ANN experimental process, the further information about which environmental factors from Table 1 are more important for ANN to recognize the target patterns can be evaluated by the analysis of ANN

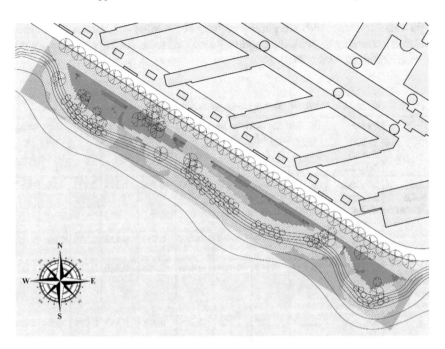

Fig. 5. The GIS demonstration for displaying the appropriate locations of the public studio; the darker color is more suitable to set public studio.

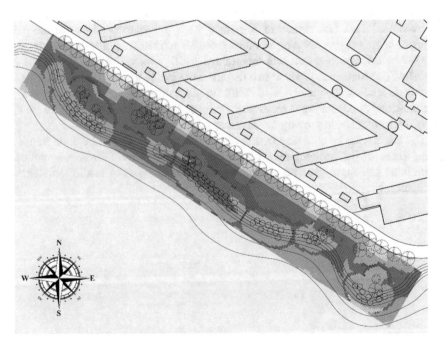

Fig. 6. The GIS demonstration for displaying the appropriate locations of the private studio; the darker color is more suitable to set private studio.

weight values. All the ANN weight values from both trained ANN models are listed in Table 3. By checking these ANN weight values in absolute values can figure out which factors are more important [13].

Table 3. The analysis of ANN weight values.

	Factor category	The distance from the specific environmental feature				The specific zone			The limited condition	
	Factor denotation	D1	D2	D3	D4	S1	S2	S3	L1	L2
Trained ANN for public studio	Original ANN weight value	0.739	−0.440	−0.669	0.435	−0.792	−0.174	0.275	0.36	−0.293
	Absolute value	**0.739**	0.440	0.669	0.435	**0.792**	0.174	0.275	0.36	0.293
Trained ANN for private studio	Original ANN weight value	−0.353	3.517	1.709	−2.456	−0.305	0.859	0.889	0.284	0.183
	Absolute value	0.353	**3.517**	1.709	**2.456**	0.305	0.859	0.889	0.284	0.183

The first two ANN weight values from the trained ANN for the public studio are S1 and D1, which means that this trained ANN is more depending on 'the center zone' and 'the roadside' these two factors to determine the appropriate locations for the public studios. The first two ANN weight values from the trained ANN for the private studio are D2 and D4, which means that this trained ANN is more depending on 'The lake side' and 'The western side' these two factors to determine the appropriate locations for the private studios. The analysis of ANN weight values can help us review that the environmental factor setting is reasonable or not and give us more information to adjust the factor setting for improving the next experiment.

The automation of choosing both types of studio locations can be done by further setting some of the programming conditions. Figure 7 is the GIS demonstration of determining the appropriate construction sites for both types of studio automatically and randomly.

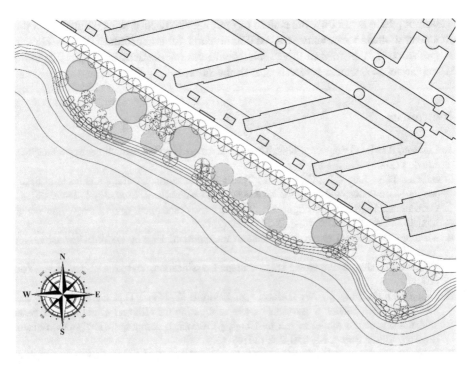

Fig. 7. The demonstration of determining the appropriate construction sites automatically and randomly; the large circles are the locations for the public studios, the small circles are the locations for the private studios.

4 Conclusions

The expertise of architectural design is often seen as obscure knowledge. How to apply AI technique to clarify this expertise as much as possible is the motivation of this paper. In this paper, we demonstrate how to apply ANN to facilitate an important step of architectural design- determining the appropriate building site, or we can call this step as 'the architectural site analysis'. According to the experimental results, the application of ANN can assist us in analyzing and determining the appropriate locations to set the different types of studios. Furthermore, by analyzing the ANN weight values can reveal more information about which environmental factors are more important to decide the appropriate building sites. That is the important information to guide us in adjusting the factor setting in order to be more precise to our decision judgment.

This paper is the beginning step to try to apply AI techniques in the architectural design process. There are lots of problems needed to be improved, like the setting of grid resolution. The higher quality of demonstrating the results of architectural site analysis depends on higher grid resolution, but that also means more demanding work on computing. The balance between the acceptable analytic demonstration and the setting of grid resolution is the issue for different types of architectural design project.

In addition, there is not only the pattern of building site needed to be defined, but other architectural design elements (e.g., different parts of architectural forms). How to comprehensively consider all these architectural design elements by the application of AI techniques is our next research goal in the future.

References

1. Lawson, B.: How Designers Think: The Design Process Demystified, 4th edn. Architectural, London (2006)
2. Salman, H.S., Laing, R., Conniff, A.: The impact of computer aided architectural design programs on conceptual design in an educational context. Des. Stud. **35**, 412–439 (2014)
3. Oxman, R.: Digital architecture as a challenge for design pedagogy: theory, knowledge, models and medium. Des. Stud. **29**, 99–120 (2008)
4. Bishop, C.M.: Neural Networks for Pattern Recognition, 1st edn. Oxford University, New York (1995)
5. Duda, R.O., Hart, P.E., Stork, D.G.: Pattern Classification, 2nd edn. Wiley, New York (2001)
6. Pijanowski, B., Tayyebi, C.A., Doucette, J., Pekin, B.K., Braun, D., Plourde, J.: A big data urban growth simulation at a national scale: configuring the GIS and neural network based land transformation model to run in a high performance computing (HPC) environment. Environ. Model Softw. **51**, 250–268 (2014)
7. Basse, R.M., Omrani, H., Charif, O., Gerber, P., Bódis, K.: Land use changes modelling using advanced methods: Cellular automata and artificial neural networks. The spatial and explicit representation of land cover dynamics at the cross-border region scale. Appl. Geogr. **53**, 160–171 (2014)
8. Patuelli, R.S., Reggiani, L.A., Nijkamp, P.: Neural networks and genetic algorithms as forecasting tools: a case study on German regions. Environ. Plan. B Plan. Des. **35**, 701–722 (2008)
9. Heppenstall, A.J., Evans, A.J., Birkin, M.H.: Genetic algorithm optimisation of an agent-based model for simulating a retail market. Environ. Plan. B Plan. Des. **34**, 1051–1071 (2007)
10. Grekousis, G., Manetos, P., Photis, Y.N.: Modeling urban evolution using neural networks, fuzzy logic and GIS: the case of the Athens metropolitan area. Cities **30**, 193–203 (2013)
11. Alajmi, A., Wright, J.: Selecting the most efficient genetic algorithm sets in solving unconstrained building optimization problem. Int. J. Sustain. Built Environ. **3**, 18–26 (2014)
12. Alexander, C., Ishikawa, S., Silverstein, M.: A Pattern Language: Towns, Buildings, Construction. Oxford University Press, New York (1977)
13. Marzban, C.: Basic statistics and basic AI: neural networks. In: Haupt, S.E., Pasini, A., Marzban, C. (eds.) Artificial Intelligence Methods in the Environmental Sciences, 1st edn., pp. 15–47. Springer, Heidelberg (2009)
14. Haupt, S.E., Lakshmanan, V., Marzban, C., Pasini, A., and Williams, J.K.: Environmental science models and artificial intelligence. In: Haupt, S.E., Pasini, A., Marzban, C. (eds.) Artificial Intelligence Methods in the Environmental Sciences, 1st edn., pp. 3–13. Springer, Heidelberg (2009)
15. Fischer, M.M.: Computational neural networks: a new paradigm for spatial analysis. Environ. Plan. A **30**, 1873–1891 (1997)
16. Openshaw, S.: Neural network, genetic, and fuzzy logic models of spatial interaction. Environ. Plan. A **30**, 1857–1872 (1998)

17. Zhang, G.P.: An investigation of neural networks for linear time-series forecasting. Comput. Oper. Res. **28**, 1183–1202 (2001)
18. Beirão, J.N., Nourian, P., Mashhoodi, B.: Parametric urban design: an interactive sketching system for shaping neighborhoods. Presented at the the 29th International Conference on Education and research in Computer Aided Architectural Design in Europe, Ljubljana, Slovenia: University of Ljubljana (2011)
19. Beirão, J.N., Duarte, J.P., Stouffs, R.: Creating specific grammars with generic grammars: towards flexible urban design. Nexus Network J. **13**, 73–111 (2011)

An Agent-Based Simulation of Corporate Gender Biases

Chibin Zhang[1] and Paolo Gaudiano[1,2(✉)]

[1] Aleria PBC, New York, USA
paolo@aleria.tech
[2] Quantitative Studies of Diversity and Inclusion, City College of New York, New York, USA

Abstract. Diversity & Inclusion (D&I) is a topic of increasing relevance across virtually all sectors of our society, with the potential for significant impact on corporations and more broadly on our economy and our society. In spite of the value of human capital, Human Resources in general and D&I in particular are dominated by qualitative approaches. We introduce an agent-based simulation that can quantify the impact of D&I on corporate performance. We show that the simulation provides a compelling explanation of the impact of hiring and promotion biases on corporate gender balance, and it replicates the patterns of gender imbalance found in various industry sectors. These results suggest that agent-based simulations are a promising approach to managing the complexity of D&I in corporate settings.

Keywords: Diversity & inclusion · Agent-based simulation · Workforce analytics

1 Introduction

In spite of the growing body of evidence showing that companies with greater gender representation in leadership roles tend to outperform companies with fewer women [1–3], many industries continue to exhibit a sharp gender imbalance at senior and executive levels. These imbalances contribute to some of the severe problems we see across a variety of industries, ranging from gender pay gaps [4, 5] and high churn rates [6, 7] to discrimination lawsuits [8, 9]. In turn, these problems lead to high costs and internal instabilities, and expose companies to significant reputational risk. Beyond the private sector, gender imbalances also impact academia and the public sector [10, 11].

Given the extensive studies showing that greater gender inclusion can lead to corporate, economic and societal benefits, given the tangible negative implications of gender imbalances, and given the ongoing efforts ranging from individual activism to legislation, why are there still such significant gender imbalances?

We believe that the relative lack of progress is due primarily to the sheer complexity of the problem, and the lack of tools that can deal with this degree of complexity. Today, Human Resource management is considered a "soft" skill; workforce analytics platforms rely on measurements and statistical analysis, but are unable to

© Springer Nature Switzerland AG 2020
K. Arai et al. (Eds.): FTC 2019, AISC 1069, pp. 90–106, 2020.
https://doi.org/10.1007/978-3-030-32520-6_9

capture the myriad events, interactions, attitudes and subjective preferences of employees.

Further, each organization is a unique "ecosystem", and what works for one organization is unlikely to work the same way for another. When you also consider that the impact of personnel initiatives can take months or years to be observed, and that missteps can be extremely costly, it's no wonder that leaders are reluctant to take decisive action.

In this paper we use an agent-based simulation, one of the primary tools of *Complexity Science*, to study a particular aspect of corporate gender imbalance: by simulating the career advancement of employees at typical companies, we can analyze the impact of introducing gender bias in the promotion process.

Under reasonable assumptions, we find that gender biases in promotion can yield the kinds of gender imbalances that are typical of many companies, with decreased representation of women at higher corporate ranks. We also find that by adjusting gender biases in hiring as well as promotions, it is possible to develop gender imbalances that match the patterns observed in different industries.

Our findings, in line with other studies, show that agent-based simulations are a powerful tool for workforce analytics. More importantly because the simulations capture detailed aspects of individual behaviors and interactions, our approach holds great promise for theoretical and applied research into Diversity & Inclusion (D&I) – a topic of great current interest with significant economic and societal implications.

After introducing some background materials, the remainder of this paper describes the simulation we have developed, and then presents some of the results we obtained with the simulation, including matching to published data about gender imbalances across multiple industries. The paper is brought to a close with some conclusions and suggestions for future opportunities to expand this line of work.

2   Background

2.1   Gender Biases in the Workplace

Although female labor force participation is increasing, women still are severely underrepresented at the top level of organizations. According to the U.S. Department of Labor, women account for almost half of the total labor force in the U.S., and more than 40 percent of those have college degrees; however, in 2018, women only held 5% of chief executive officer positions in S&P 500 companies, 21.2% of board seats and 26.5% of the executive and senior-level management positions [12].

Bielby and Baron [13] categorized the causes of the gender gap in upper management positions into *supply-side* and *demand-side* explanations. According to supply-side explanations, the divergence in employment outcomes between women and men are mainly due to differences in gender-specific preferences and productivity [14], therefore individual attributes determine the gender inequalities in the workplace [15]. For example, some believe that balancing family life and work lowers women's promotion rate as women need to take on a larger share of domestic and parental responsibilities [16]; others hypothesize that women, in general, are less competitive than men so they may be reluctant to compete for promotion [17].

In contrast, demand-side explanations suggest that gender stratification in the workplace is primarily due to gender-specific barriers; demand-side explanations focus on the institutional constraints and managerial biases faced by women in climbing the career ladder. For example, a male-dominated board of directors may prefer to hire male executives [14]; women must meet higher performance standards for promotion than their male colleagues [18, 19]; stereotypes of leadership style differences favor men in advancing to leadership roles [20].

In this paper we present a simulation that provides indirect but compelling evidence for a demand-side explanation. Specifically, we simulate the typical career advancement of employees in an organization without incorporating gender-specific preferences and characteristics, such as education level and career interruption. Rather, we focus on demand-side barriers, specifically focusing on the impact of introducing gender biases in the promotion process, which we believe is a significant institutional barrier that prevents women from attaining senior roles. We find that under some very simple assumptions about the presence of gender biases in promotion, it is possible to replicate the types of gender disparities that are observed in typical companies across a variety of industries. Hence, while we cannot conclusively prove that gender imbalances are due to gender-specific barriers, we demonstrate that the existence of gender-specific barriers would yield the kinds of imbalances that are observed empirically.

Another gender-specific barrier that we examined is bias in hiring. As suggested in a recent report by McKinsey & Company and the nonprofit organization *Lean In* [21], women may remain underrepresented at manager levels and above because they are less likely to be hired into entry-level jobs, in addition to being less likely to be hired or promoted into manager-level positions.

We extended our simulation to capture a simple form of hiring bias, and found that by adjusting the hiring and promotion biases simultaneously we are able to replicate industry-specific gender balance patterns as reported in the McKinsey study.

Before introducing our simulation, we provide some additional background information on agent-based simulation and its application to corporate D&I.

2.2   Agent-Based Simulation

The issue of diversity in companies – from early-stage startups to global corporations – is a highly complex problem, with interconnections and ripple effects that range from individuals to entire corporations and even society as a whole.

Our work is rooted in the theory and application of *Complexity Science*, a discipline that first took shape in the late 1960s with the seminal work of Thomas Schelling on the emergence of segregation [22], and became a full-fledged area of academic inquiry in 1984 with the creation of the Santa Fe Institute.

Complexity Science is a broad field that encompasses a variety of technologies for studying complex systems, *i.e.*, systems whose behavior depends in complex and often unpredictable ways on the behaviors of many individuals who interact with one another and with their environment [23]. One of the primary tools for the analysis of complex systems is *agent-based simulation*, a methodology that combines behavioral science and computer modeling [24]. Agent-based simulations capture the behaviors of individuals, and their interactions with other individuals and with their environment, to simulate the way in which the overall behavior of a system emerges through these complex chains of interactions.

Our team has previously developed dozens of agent-based simulations to solve complex problems across many sectors and many types of organizations, including corporations, government agencies and foundations. The majority of these applications involved simulating and analyzing the behavior of human systems, including consumer marketplaces [25, 26], energy consumption in commercial buildings [27], manpower and personnel management for the U.S. Navy [28], healthcare [29] and computer security [30]. In this paper we focus specifically on the application of agent-based simulations to corporate D&I.

2.3 Simulating Corporate D&I

A majority of the research on gender inequality in the workplace is conducted by applying statistical methodology on collected data. These statistical approaches have significant limitations, including the fact that they hide any dynamic information as well as details about individuals.

In contrast, agent-based simulations are ideally suited to analyze and predict the performance of human organizations by capturing the mutual relationships between individual employees and their organization: the performance of an employee influences the success of an organization and, conversely, the environment created by the organization influences the success of the individual employee. This sort of "feedback loop" is part of what makes workforce management so complex, and it is exactly the type of problem that lends itself to analysis using agent-based simulation.

In this light, agent-based simulation promises to be a valuable tool to capture the impact of D&I on corporate environments: to the extent that a company influences people's experiences differently based on personal traits, the company's performance will in turn be impacted. In fact, the connection between D&I and complexity has been proposed by others [31]. More specifically, agent-based simulation has already been used to analyze issues related to corporate D&I. For example, the simulation built by Bullinaria [32] shows how ability differences and gender-based discriminations can lead to gender inequality at different hierarchical levels within an organization; Takács et al. [33] found that discrimination can emerge due to asymmetric information between employer and job applicants even without hiring biases; Robison-Cox et al. [34] used agent-based simulation to test the possible explanations of gender inequality at the top level of corporations, and found that giving men favorable performance evaluations significantly contributes to the gender stratification of top-level management.

Our agent-based simulation takes a step further and simulates ongoing activities and transitions, to show the dynamics of gender imbalance at each level of the hierarchy that result from imposing promotion biases or hiring biases. The simulation replicates the week-by-week operations of a typical company with men and women distributed across four levels: entry-level employees, managers, vice presidents and executives. In a pilot project, we were able to simulate the impact of gender biases in the promotion process, which leads to the kinds of gender imbalances seen in real companies, with increasing representation of men in higher levels of the company [35]. We also reported some preliminary results showing that removing biases in a company that is already imbalanced is not a very effective strategy, as gender inequalities can persist for significant periods of time. Our findings are in line with those of Kalev et al. [36], who found that programs targeting lower levels of management, such as diversity training and performance evaluations do not help to increase diversity at higher corporate levels, and that imbalances persist after organizations adopt these diversity management programs.

In this paper, we focus on the establishment of gender imbalances as a result of promotion biases and hiring biases, comparing our results to recent reports.

3 The Simulation

One of the most powerful aspects of agent-based simulation is that it captures the way real-world systems work in an intuitive, human-centric fashion. This means that anyone who has familiarity with the problem can contribute to the design of the simulation. In a sense, agent-based simulation democratizes analytics, because it does not require knowledge of advanced mathematical or computational techniques. Compared to more traditional approaches to analytics – in which a data scientist analyzes large amounts of data using analytical tools to look for patterns, and the domain expert is relegated to making sense of the identified patterns – agent-based simulation allows the domain experts to be more closely involved both with the model design and with the analysis of the results [25].

The simulation we developed for this paper is a good example, as it is based on simple assumptions about the operations of a typical company. One of us has significant experience developing simulations of organizations, but the simulation we now introduce should be intuitive to virtually anyone who has even a basic understanding of the functioning of companies.

3.1 Core Elements of the Simulated Company

In our simulated company, there are two types of employees, which we associate with men and women. Notice that because we are only interested in studying the impact of institutional barriers, our simulation assumes that men and women have identical abilities and that their performance is also identical. Similarly, gender-specific preferences and characteristics like education level and career interruption are not incorporated in the simulation. All of these details could easily be added if one were interested in studying related phenomena. However, for the present study we wanted to focus exclusively on the impact of biases in promotions and hiring.

Employees fall into one of four increasing levels: entry level, managers, VPs and executives. The company starts with a user-defined number of employees, distributed across levels in a way that matches a typical company, with smaller numbers of employees in higher levels.

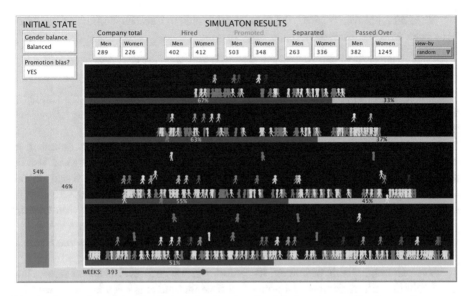

Fig. 1. Screenshot of the simulation, showing four levels of employees. Employees that appear between levels are in the process of being promoted, while employees that hover just above a level are in the process of separating. Men are shown in blue (dark), women in yellow (light).

The company is assumed to grow over time, creating vacancies at each level. In addition, employees may separate from the company, creating additional vacancies. Vacancies at the entry level are filled through hiring from a hypothetical external pool of candidates, while vacancies at all other levels are filled by promoting individuals from the level immediately below, which, in turn, create additional vacancies at the level from which someone was promoted.

For this version of the simulation we do not simulate direct hiring into higher levels, nor do we allow promotions to skip levels. However, these and other details could be added if we were interested in exploring the effects of such modifications.

Figure 1 is a screenshot of the simulation, which was developed with the NetLogo simulation platform [37]. Each "floor" acts as a simple histogram, showing the percentage of men and women at that level. The two vertical bars on the left show the overall gender balance across the entire organization.

3.2 Simulating the Hiring Process

Vacancies at the entry level can occur as a result of overall company growth, or from entry-level employee separations, or from entry-level employees being promoted to the manager level.

When a vacancy occurs at the entry level, the simulation assumes that a new employee will be hired from a potentially infinite pool of candidates. By default, there is an equal chance of hiring a man or a woman. Gender biases in hiring are simulated by setting a *hiring-bias* parameter that changes the probability of men (or women) being hired.

When an employee is hired, the simulation begins to track the amount of time they have been with the company.

3.3   Simulating Promotions

When a vacancy appears at any level above the entry level, promotions are simulated by identifying a pool of "promotion candidates" from the level immediately below, and then choosing randomly one employee from that pool.

By default, the promotion pool is set as a percentage of the total number of employees at that level, and is based on the promotion score of each employee. Also by default, the promotion score of each employee is a normalized value based on the amount of time the employee has spent at the current level, relative to the amount of time spent at that level by the most senior employee at that level. In other words, in the absence of a bias, promotions are based strictly on seniority within each level.

Gender biases in the promotion process are simulated by adding a *promotion-bias* value to the promotion scores of men[1]. This bias is applied uniformly at all levels.

When an employee is promoted, its seniority at the new level is set to zero, while the simulation still tracks the total amount of time that the employee has been with the organization.

3.4   Simulating Separations

We simulate employee separations without regard as to whether the employee left the company voluntarily or by being terminated. We simulate separations at each level as a random event, whose frequency is proportional to the number of people at that level.

Employees are chosen for separation through a process similar to promotions: a pool of separation candidates is formed by selecting a percentage of employees at that level that have the lowest promotion scores. The actual employee to separate from the company is chosen randomly from that pool.

If the promotion-bias is set to a non-zero value, the same bias that influences promotions will also influence separations.

3.5   Tracking Individual and Aggregate Results

One advantage of agent-based simulations is that they can simultaneously provide aggregate, population-level metrics to match those commonly used for workforce analytics, while preserving all of the details about individuals. For example, we can track traditional metrics such as overall company size, gender balance and churn rates, but we can also track how many times any given individual was passed over for a promotion (*i.e.*, when someone less senior at the same level was promoted to the next higher level).

We use the aggregate metrics to ensure that the simulation is behaving "reasonably" by matching performance metrics to what is seen in real companies, and then we can

---

[1] The promotion-bias parameter can be positive or negative to simulate biases that favor men or women, respectively.

drill down to perform a much finer level of analysis, for instance studying the impact of promotion biases on the likelihood that any given individual will be passed over for promotion.

We now turn to a description of the experiments we ran to test the impact of promotion and hiring biases.

4   Simulation Setup and Results

4.1   Parameter Settings

For all results reported here, the company begins with a total of 300 employees, 150 women and 150 men. The employees are distributed across the four levels in a way that roughly simulates a typical company: 40% at the entry level, 30% at the manager level, 20% at the VP level and 10% at the executive level. Note that we have tested the simulation with other settings and found that as long as there are at least 100 or so employees at the start, the overall results are consistent.

During the simulation, each time tick is set to correspond to a seven-day period. At each time step, random numbers are drawn to determine whether separations, hires or promotions need to take place. The frequencies of each of these occurrences are set so that, over the course of a simulated year, the overall growth and churn rates are within a range that would be consistent with a "generic" company: the company grows by 10% each year, while we target an annual churn rate of 20%, which is in accordance with national average churn rates [38].

For promotions we set a candidate pool size of 15%: in other words, when someone needs to be promoted to a higher level, we select the 15% of employees with the highest scores, and then randomly choose one of them for promotion. For separations, we set the candidate pool size to 50%, meaning that someone is selected randomly from half of the employees at each level with the lowest promotion scores. We have tested different size candidate pools and found that the results do not change significantly[2].

The simulation includes several steps that invoke a random number generator. This results in variance across simulations[3]. Unless otherwise specified, each of the results shown below was obtained by averaging the results from five *Monte Carlo* simulations with each parameter setting. We found that results do not vary significantly as we increase the number of simulations.

4.2   Experiment 1: The Impact of Gender Biases in Promotion

In the first set of experiments we wanted to establish a baseline and then test the impact of systematically increasing the degree of gender bias in the promotion process. Figure 2 shows the gender balance at each level during a 30-year simulation when there is no promotion bias or hiring bias. The figure shows that, in the absence of

---

[2] Smaller pool sizes coupled with non-zero promotion biases tend to create cyclical behaviors which do not influence the overall results but create some unnatural dynamics.

[3] To ensure reproducibility, we have the ability to select the seed for the random number generator.

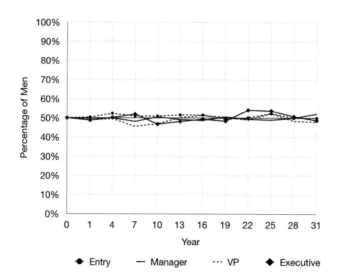

Fig. 2. Timeline showing the fluctuations in gender balance across all four levels during a 30-year simulation. For this figure the promotion and hiring biases are set to zero.

biases, the gender balance stays at 50–50 throughout the simulation, with only small fluctuations due to the inherent randomness of the simulations.

Next, we tested the impact of increasing the promotion bias to 0.1, 0.3 and 0.5. As mentioned earlier, in the absence of biases, each simulated employee's promotion score is simply its seniority relative to the most senior employee at that level. Hence all the promotion scores prior to the application of a gender bias are between 0 and 1.

Adding a promotion bias of 0.1 thus means that while women's scores will still be in the range [0, 1], men's promotion scores will be between 0.1 and 1.1. Similarly, at the highest level of bias reported here (0.5), men's promotion scores will be between 0.5 and 1.5, while women's promotion scores will stay in the range [0, 1]. In all three sets of simulations, the hiring bias is set to zero.

As can be seen in Fig. 3, a bias of 0.1 in promotions begins to show an interesting pattern: while the entry and manager levels continue to stay roughly at 50-50, the VP level (dashed line with no symbol) is starting to show an imbalance in favor of men, while men now make up roughly 60% of the executive level.

The pattern becomes much more evident in Fig. 4, which shows the gender balance at each level when the promotion bias is set to 0.3 (top) and 0.5 (bottom). Several interesting phenomena are worth pointing out.

First, we see that, even though the promotion bias is a single parameter that works uniformly at each level, the successive promotions compound the effect, so that the gender imbalance is greatest at the executive level: after several simulated years, the executive level reaches a make-up of approximately 80% men when the bias is at 0.3, and exceeds 90% men when the bias is at 0.5.

Second, increasing the bias has the effect of increasing the degree of imbalance, but also the speed with which the imbalance spreads through the organization: notice that with a bias of 0.3, the imbalance at the executive level builds gradually over a span of

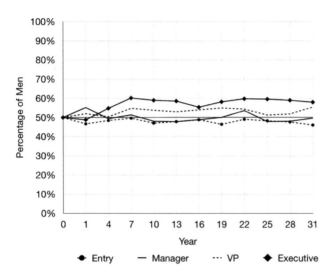

Fig. 3. Timeline showing the fluctuations in gender balance across all four levels during a 30-year simulation when the promotion bias is 0.1 and hiring bias is zero.

nearly 20 years; but when the bias is 0.5, the executive level crosses the 80%-male mark in less than five years, and has essentially leveled off by year 10.

Third, there is a surprising effect at the entry level: even though there is no hiring bias (we confirm in the simulation that the same number of men and women are hired), and even though women are being terminated more often than men (because the promotion score influences separations) we see that women make up an increasing percentage of the entry-level population, reaching roughly 60% when the bias is 0.3 and 70% when the bias is 0.5. The reason for this "reverse imbalance" is that men are being promoted at a much higher rate than women, so that women are being left behind. In reality, this is not uncommon in the real world: in many industries you find greater numbers of women in entry-level positions, and women often describe the negative experience of being "stuck" while their male colleagues get promoted.

This last observation illustrates another great aspect of agent-based simulations: unlike typical "black-box" models, with an agent-based simulation it is possible to dig into the detailed activities to understand the origin of observed macroscopic phenomena, *i.e.*, emergent behaviors. We will come back to this point in the closing section.

Overall, the results of these experiments show that, starting with a very simple assumption, we can capture some qualitative phenomena that match our observations of real-world companies: increasing gender imbalance at higher levels, and women being stuck in lower levels.

In the next section, we show how, modifying an additional parameter, we can start to customize the simulation to capture data observed from specific industries.

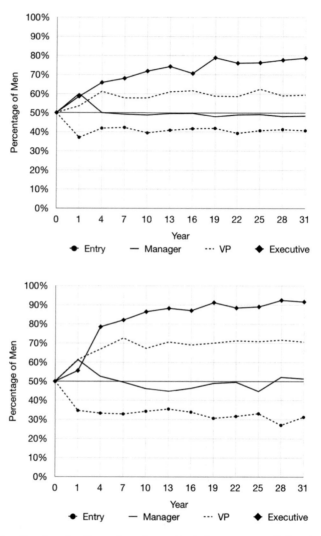

Fig. 4. Timeline showing the fluctuations in gender balance across all four levels during a 30-year simulation when the promotion bias is 0.3 (top) and 0.5 (bottom). See text.

4.3   Experiment 2: Combining Promotion and Hiring Biases to Match Industry-Specific Imbalances

While the patterns shown in Figs. 2, 3 and 4 already look remarkably like those we observe in real companies, we wanted to see if, using a minimal set of assumptions, we could match real-world data on gender imbalances for more specific cases. To this end, we used our simulation to match data from McKinsey's and LeanIn's *Women in the Workplace* report [21].

In the figures that follow we ran simulations for ten years, and measured the gender (im)balance at each level. In all simulations we modified two parameters from the baseline case: the promotion bias and the hiring bias. In general, as mentioned earlier, higher promotion biases create larger imbalances at higher levels, and can lead to reverse-imbalance at the entry level. In other words, if we think of the company's gender balance as a funnel, that funnel is somewhat shallow at low levels of promotion bias, and steeper at high levels of promotion bias.

In contrast, the hiring bias has an immediate impact only on the entry levels. Hence we expect that increasing the hiring bias will make the overall funnel narrower, while the promotion bias level will influence the steepness of the funnel.

In Fig. 5, we show the ten-year gender balance data from our simulation using a format that is meant to mimic the format used in the McKinsey report, which show gender balance as a horizontal funnel, with entry level on the left, and executive level

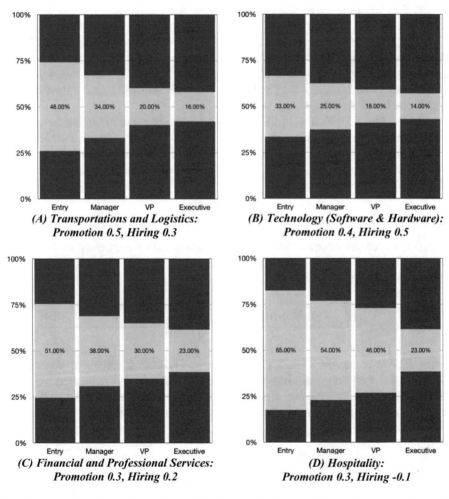

Fig. 5. Simulating the gender make-up of different industries by adjusting both promotion and hiring biases. See text.

on the right. The McKinsey report uses six levels (entry, manager, director, VP, SVP and C-Suite), so we selected the four levels that match the levels used in our simulation: entry, manager, VP and Executive (C-Suite).

Starting with Fig. 5(A), we see that setting the promotion bias to 0.5 and hiring bias to 0.3 results in a relatively steep funnel, with just under half women at the entry level, but only 16% women in the top ranks – a threefold reduction in representation. This shape matches closely the gender imbalances observed in the *Transportations and Logistics* sector in the McKinsey study.

In Fig. 5(B), the promotion bias is lowered to 0.4, but the hiring bias is raised to 0.5. As expected, the overall funnel becomes much narrower, with only 33% women at the entry level, and it is not as steep: while the top rank, at 14% women, is slightly lower than in Fig. 5(A), this means that the representation of women in the top ranks represent a 2.4x reduction relative to the entry level. This graph matches closely the gender data from the *Technology (Software and Hardware)* sector in the McKinsey study.

In Fig. 5(C), the promotion bias is further lowered to 0.3, and the hiring bias is only 0.2. The lower hiring bias results in women making up almost exactly half of the entry level, while the lower promotion bias leaves nearly 25% of women in the top ranks. These results match closely the gender data reported by McKinsey for the *Financial and Professional Services* sector.

Finally, Fig. 5(D) shows a pattern that resembles the gender imbalances observed in the *Hospitality* sector, which tends to be dominated by women at the entry level, but with only a modest female representation at the top ranks. We obtained this graph by keeping the promotion bias to 0.3, but setting the hiring bias to −0.1, *i.e.*, the hiring bias actually favors women.

4.4   Simulation Accuracy

To test the accuracy of our simulation, we calculated the root-mean-squared deviation (RMSD) between each simulation and the data provided from the McKinsey study, given by the formula:

$$RMSD_i = \sqrt{\frac{(L0_D - L0_S)^2 + (L1_D - L1_S)^2 + (L2_D - L2_S)^2 + (L3_D - L3_S)^2}{4}} \qquad (1)$$

Where $RMSD_i$ is the RMSD for a given industry i, and each squared term captures the difference between the gender balance data (subscript D) and the simulation (subscript S) at a given level, and the division by 4 represents the fact that we are averaging the result across the four levels. In all cases shown in Fig. 5, the RMSD was below 6%.

5   Discussion

We have introduced an agent-based simulation that captures, in a simplified form, some of the gender imbalances that are observed across a variety of industries. What is perhaps most surprising about our findings is that we are able to capture several phenomena through some very simple assumptions, and by varying a small number of parameters.

Of course, the fact that our model is able to reproduce some of the observed phenomena does not mean that we are accurately capturing the true causes of these phenomena: it is possible that the mechanisms we hypothesized are not representative of real-world corporate functions, and that the similarity between our results and real-world observations are purely coincidental.

However, what we have been able to show is that by making some very simple assumptions about the functioning of a company, and introducing a minimalistic notion of bias, the company dynamics result in gender imbalances that are very similar to those observed in the real world. Because our model is capturing the causal links between the behaviors of individuals and the emergent behaviors of a company, and because our model is very parsimonious in its assumptions, we are confident that our model, while certainly simplistic, is capturing some fundamental aspects of corporate function that reflect real-world contexts.

It is worth noting that most of the research on gender inequality in the workplace is conducted by applying statistical methods to observed data. This more common approach has a number of limitations, including the fact that it is only capturing present conditions, it removes any information about dynamics, it hides details about individual interactions, and, most importantly, it tends to identify correlations that may or may not be due to causal relationships. But the most significant advantage of our agent-based approach relative to statistical simulations is that, because we are capturing causal links, the same tool that is used to match observed data can then be used to test the likely impact of different initiatives. In other words, we believe that the simulation we presented can be used to help corporations understand the likely outcomes of different D&I initiatives.

We see this project as the beginning of a systematic study of the impact of D&I on corporate performance. This highlights one more advantage of the agent-based approach: it is possible to add details to a model to increase its predictive power without having to throw away the previous model. For instance, we could explore the impact of the candidate pool not being infinite and perfectly balanced; in fact, a former student of ours developed an additional agent-based simulation that shows how job candidates may be influenced by the perceived level of inclusion and diversity of a company, and how this will impact the talent pool available to any company [39].

Even within the promotion model itself, there are many ways in which we could increase the fidelity of the model to explore the impact of different assumptions and of different initiatives. For example, we could add the ability to hire people directly into higher levels. We could add the notion of employee satisfaction, and tie it to individual experiences in a way that influences retention rates: in the current model, as bias increases, women tend to get passed over for promotion much more frequently than men, and end up being stuck for a long time; in the real world, these factors undoubtedly lead women to quit, resulting in lower retention rates for women – a phenomenon that is common across many male-dominated industries. We could also simulate the impact of having managers of a different gender on satisfaction and career advancement. In other words, this model can serve as the basis to explore a large number of hypotheses about the sources of gender disparities, and to test the likely impact of different interventions.

Finally, although in this paper we have focused on gender, it is possible to represent other personal characteristics that impact an employee's experience, such as ethnicity, race, religious beliefs, sexual orientation, physical and cognitive abilities, and so on. We have actually begun to develop some agent-based simulations that include other facets of diversity, and have already encountered some complex, fascinating issues that suggest entirely different ways of thinking about diversity and inclusion. In all, we are optimistic that our work and the work of other complexity scientists can lead to a dramatic shift in how people think about diversity and inclusion, and, more importantly, what corporate leaders can do about it.

References

1. Hunt, V., Yee, L., Prince, S., Dixon-Fyle, S.: Delivering through diversity (2018). https://www.mckinsey.com/business-functions/organization/our-insights/delivering-through-diversity. Accessed 17 Feb 2019
2. Flabbi, L., Macis, M., Moro, A., Schivardi, F.: Do female executives make a difference? The impact of female leadership on gender gaps and firm performance. IZA Discussion Paper No. 8602 (2014). SSRN: https://ssrn.com/abstract=2520777
3. Dezsö, C.L., Ross, D.G.: Does female representation in top management improve firm performance? A panel data investigation. Strateg. Manag. J. **33**(9), 1072–1089 (2012)
4. Blau, F.D., Kahn, L.M.: The gender pay gap: have women gone as far as they can? Acad. Manag. Perspect. **21**(1), 7–23 (2007)
5. Biagetti, M., Scicchitano, S.: A note on the gender wage gap among managerial positions using a counterfactual decomposition approach: sticky floor or glass ceiling? Appl. Econ. Lett. **18**(10), 939–943 (2011)
6. Becker-Blease, J., Elkinawy, S., Hoag, C., Stater, M.: The effects of executive, firm, and board characteristics on executive exit. Financ. Rev. **51**(4), 527–557 (2016)
7. Shaffer, M.A., Joplin, J.R., Bell, M.P., Lau, T., Oguz, C.: Gender discrimination and job-related outcomes: a cross-cultural comparison of working women in the United States and China. J. Vocat. Behav. **57**(3), 395–427 (2000)
8. Hirsh, E., Cha, Y.: For law and markets: employment discrimination lawsuits, market performance, and managerial diversity. Am. J. Sociol. **123**(4), 1117–1160 (2018)
9. Murphy, T.: Morgan stanley settles sex-discrimination suit, 12 July 2004. https://www.forbes.com/2004/07/12/cx_tm_0712video3.html
10. Jokinen, J., Pehkonen, J.: Promotions and earnings – gender or merit? Evidence from longitudinal personnel data. J. Labor Res. **38**(3), 306–334 (2017)
11. Bain, O., Cummings, W.: Academes glass ceiling: societal, professional-organizational, and institutional barriers to the career advancement of academic women. Comp. Educ. Rev. **44**(4), 493–514 (2000)

12. Catalyst, Pyramid: Women in S&P 500 Companies, 1 January 2019. https://www.catalyst. org/knowledge/women-sp-500-companies. Accessed 3 Mar 2019
13. Bielby, W.T., Baron, J.N.: Men and women at work: sex segregation and statistical discrimination. Am. J. Sociol. **91**, 759–799 (1986)
14. Matsa, D.A., Miller, A.R.: Chipping away at the glass ceiling: gender spillovers in corporate leadership. Am. Econ. Assoc. **101**(3), 635–639 (2011). 0002-8282
15. Olsen, C., Becker, B.E.: Sex discrimination in the promotion process. Ind. Labor Relat. Rev. **36**, 624–641 (1983)
16. Kossek, E.E., Su, R., Wu, L.: "Opting Out" or "Pushed Out"? Integrating perspectives on women's career equality for gender inclusion and interventions. J. Manag. **43**(1), 228–254 (2016)
17. Niederle, M., Vesterlund, L.: Do women shy away from competition? Do men compete too much? Quart. J. Econ. **122**(3), 1067–1101 (2007)
18. Gjerde, K.A.: The existence of gender-specific promotion standards in the U.S. Manag. Decis. Econ. **23**(8), 447–459 (2002)
19. Lyness, K.S., Heilman, M.E.: When fit is fundamental: performance evaluations and promotions of upper-level female and male managers. J. Appl. Psychol. **91**(4), 777–785 (2006)
20. Vinkenburg, C.J., Engen, M.L., Eagly, A.H., Johannesen-Schmidt, M.C.: An exploration of stereotypical beliefs about leadership styles: is transformational leadership a route to women's promotion? Leadersh. Quart. **22**(1), 10–21 (2011)
21. McKinsey & Company: Women in the workplace (2015)
22. Schelling, T.: Models of segregation. Am. Econ. Rev. **59**(2), 488–493 (1969)
23. Waldrop, M.M.: Complexity: The Emerging Science at the Edge of Order and Chaos. Simon and Schuster, New York (1992)
24. Bonabeau, E.: Agent-based modeling: methods and techniques for simulating human systems. Proc. Natl. Acad. Sci. **99**, 7280–7287 (2002)
25. Gaudiano, P.: Understanding attribution from the inside out (2016). https://www. exchangewire.com/blog/2016/05/16/understanding-attribution-from-the-inside-out/. Accessed 5 Apr 2018
26. Duzevik, D., Anev, A., Funes, P., Gaudiano, P.: The effects of word-of-mouth: an agent-based simulation of interpersonal influence in social networks. In: 2007 Word of Mouth Research Symposium. Word of Mouth Marketing Association, Las Vegas (2007)
27. Gaudiano, P.: Agent-based simulation as a tool for the built environment. Ann. N. Y. Acad. Sci. **1295**, 26–33 (2013)
28. Garagic, D., Trifonov, I., Gaudiano, P., Dickason, D.: An agent-based modeling approach for studying manpower and personnel management behaviors. In: Proceedings of the 2007 Winter Simulation Conference, Washington (2007)
29. Gaudiano, P., Bandte, O., Duzevik, D., Anev, A.: How word-of-mouth impacts medicare product launch and product design. In: 2007 Word of Mouth Research Symposium. Word of Mouth Marketing Association, Las Vegas (2007)
30. Shargel, B., Bonabeau, E., Budynek, J., Buchsbaum, D., Gaudiano, P.: An evolutionary, agent-based model to aid in computer intrusion detection and prevention. In: Proceedings of the 10th International Command and Control Research and Technology Symposium, MacLean, VA (2005)
31. Page, S.E.: The Diversity Bonus. Princeton University Press, Princeton (2017)
32. Bullinaria, J.: Agent-based models of gender inequalities in career progression. J. Artif. Soc. Soc. Simul. **21**(3), 1–7 (2018)

33. Takács, K., Squazzoni, F., Bravo, G.: The network antidote: an agent-based model of discrimination in labor markets. Presented at MKE 2012 Conference, Budapest, Hungary, 20–21 December 2012 (2012)
34. Robison-Cox, J.F., Martell, R.F., Emrich, C.G.: Simulating gender stratification. J. Artif. Soc. Soc. Simul. **10**(3) (2007). http://jasss.soc.surrey.ac.uk/10/3/8.html
35. Gaudiano, P., Hunt, E.: Equal opportunity or affirmative action? A computer program shows which is better for diversity (2016). Forbes.com. http://www.forbes.com/sites/gaudianohunt/2016/08/22/equal-opportunity-or-affirmative-action-a-computer-program-shows-which-is-better-for-diversity/
36. Kalev, A., Dobbin, F., Kelly, E.: Best practices or best guesses? Assessing the efficacy of corporate affirmative action and diversity policies. Am. Sociol. Rev. **71**(4), 589–617 (2006)
37. Wilensky, U.: An Introduction to Agent-Based Modeling: Modeling Natural, Social, and Engineered Complex Systems with NetLogo. MIT Press, Cambridge (2015)
38. North American Employee Turnover: Trends and Effects. https://www.imercer.com/ecommerce/articleinsights/North-American-Employee-Turnover-Trends-and-Effects. Accessed 20 Mar 2019
39. Naghdi Tam, A.: Agent-based simulation as a tool for examining the impact of a company's reputation on attracting diverse talent. Unpublished Master's thesis, City College of New York (2017)

Machine Learning Analysis of Mortgage Credit Risk

Sivakumar G. Pillai[✉], Jennifer Woodbury, Nikhil Dikshit, Avery Leider, and Charles C. Tappert

Department of Computer Science, Pace University, Pleasantville, NY 10570, USA
{sp57299w,jw82676n,nd20961n,aleider,ctappert}@pace.edu

Abstract. In 2008, the US experienced the worst financial crisis since the Great Depression of the 1930s. The 2008 recession was fueled by poorly underwritten mortgages in which a high percentage of less-credit-worthy borrowers defaulted on their mortgage payments. Although the market has recovered from that collapse, we must avoid the pitfalls of another market meltdown. Greed and overzealous assumptions fueled that crisis and it is imperative that bank underwriters properly assess risks with the assistance of the latest technologies. In this paper, machine learning techniques are utilized to predict the approval or denial of mortgage applicants using predicted risks due to external factors. The mortgage decision is determined by a two-tier machine learning model that examines micro and macro risk exposures. In addition a comparative analysis on approved and declined credit decisions was performed using logistic regression, random forest, adaboost, and deep learning. Throughout this paper multiple models are tested with different machine learning algorithms, but time is the key driver for the final candidate model decision. The results of this study are fascinating and we believe that this technology will offer a unique perspective and add value to banking risk models to reduce mortgage default percentages.

Keywords: Machine Learning Model · Mortgages · Credit risk · Logistic Regression · Random Forest Classifier · Deep Neural Network · Classification and regression trees

1 Introduction

In September 2018, the Board of Governors of the Federal Reserve published the total US mortgage debt outstanding totaling $15.131 trillion across all holders, an increase of 5% over the previous 12 months. The jump sends a strong indication that a recovery has continued in the housing market.

Noticeably the debt levels are also closer to the levels of 2008 US housing market melt down. In order to prevent another crisis it is imperative that market participants avoid the pitfalls of the 2008 US housing market melt down again. Mortgage originators and financial institutions practice stricter underwriting guidelines in comparison to the pre-crisis era. Regulators, along with Congress,

© Springer Nature Switzerland AG 2020
K. Arai et al. (Eds.): FTC 2019, AISC 1069, pp. 107–123, 2020.
https://doi.org/10.1007/978-3-030-32520-6_10

Fig. 1. Source: NAHB, Haver Analytics, Deutsche Bank Global Markets Research, Indicates that the US Housing market remains strong with interested participants. Also Consumer interest in home buying analysis indicates it is at 2006 level.

implemented the Dodd-Frank Wall Street Reform and Consumer Protection Act [13] to assist in preventing a recurrence of the market meltdown from a macro perspective. Regulations to ensure appropriate consumer practices is just the beginning. Additional work is necessary to ensure that borrowers have the ability and commitment to pay their mortgage.

The objective of this study is to build a model that has the ability to assess the credit risk of mortgage related exposures to financial institutions [11]. The model factors in borrower-level (micro) and market-level stresses (macro) derived while utilizing machine learning technologies. This paper is designed to provide a background of the mortgage market industry, micro and macro level risk exposures, technologies and methodologies used to design and implement the credit risk model. The results and findings of the study provides an analysis of approved and declined mortgages using Machine Learning models, with market-level risks applied [8].

2   Literature Review

2.1   Business Risks and Key Mortgage Elements

Current US Housing Market. The current state of the US Housing Market reflects a slowdown in popular US regions - Seattle, Silicon Valley and Austin, Texas. Historically, trends in popular US regions set the tone for the market. With rising mortgage rates and prices climbing at a faster rate than income, buyers are getting squeezed and will hit a limit [3]. But market participants continue to view the housing sector as strong due to a healthy labor market and steady economic growth. This indicates price stabilization and not crisis-level conditions. The US Housing market remains strong with interested participants (see Fig. 1).

Risk Assessment Models. Financial institutions rely on proprietary underwriting models to assess their risk exposure to mortgage loans. Underwriters carefully examine personal information and credit profiles to ensure borrower eligibility. Key elements that play a role in the decision making process include: FICO (Fair Isaac Corporation) score, occupation, total household income, DTI (debt to income) ratio, property location, loan amount, loan to value ratio (LTV), full documentation availability, property type and occupancy status [7].

These elements are used as cohorts to determine if a borrower qualifies for a mortgage. Researchers conclude that buyers with low LTV ratios and high FICO scores typically qualify for lower mortgage rates [3]. A borrower with less than pristine credit does not automatically disqualify a borrower. But other factors together may reduce the risk associated with the borrower. For example, a borrower with a low LTV (less than 70%), high FICO (700+), proof of income/high income, and low DTI (less than 20%) is viewed as a "less risky" profile [2]. The loan amount and LTV ratio combined can be used to determine whether or not private mortgage insurance (PMI) is required.

The first objective of our study was to utilize key mortgage variables to construct a borrower profile who has less chances of defaulting on their mortgage loan.

2.2 Current Market Utilization of Machine Learning

Analysis of Feature Selection Techniques in Credit Risk. Utilizing the best prediction features in credit analysis is crucial in assessing risk. We looked at credit risk assessment to get a better understanding of variables used to assess mortgage credit risk. In "Analysis of feature selection techniques in credit risk assessment", Rama and Kumasaresan [12] found that the most important features in credit risk assessments as illustrated in Fig. 2 are checking account status, credit history, duration in months, saving account balances, purpose, credit score, property type, present employment, occupancy, age, installment plans, personal status, and sex. The feature selection was done using information gain, gain ratio, and chi square correlation. Data used in this research was public data of German credit that consists of 1000 instances in which 700 of them are creditworthy applicants and 300 of bad credit applicants.

A Machine Learning Approach for Predicting Bank Credit Worthiness Analysis of UCI machine learning dataset revealed that there is a relationship between the customer's age and their level of account balances. Customers who are between 20 and 40 years of age and have small bank account balance are most prone to become defaulters. The dataset contained 23 variables from which researchers picked 5 most important features. Using the 5 selected features they performed multiple classifications. Each of these algorithms achieved an accuracy rate between 76% to over 80% [14] (see Fig. 3).

Attribute No	Name of the attribute	Value of gain ratio
1	Status of checking account	123.7209
3	Credit history	61.6914
2	Duration in month	46.8311
6	Savings account	36.0989
4	Purpose	33.3564
5	Credit amount	26.9528
12	Property	23.7196
7	Present employment since	18.3683
15	Housing	18.1998
13	Age in year	16.3681
14	Other installment plans	12.8392
9	Personal status and sex	9.6052
20	Foreign worker	6.737
10	Others debtors	6.6454
17	Job	1.8852
19	Telephone	1.3298
18	Number of people being liable to provide maintenance	0
8	Installment rate in percentage	0
11	Present residence since	0
16	Number of existing credits at this bank	0

Fig. 2. Source: R.S. Ramya and S. Kumaresan, "Analysis of feature selection techniques in credit risk assessment," 2015 International Conference on Advanced Computing and Communication Systems, Coimbatore, 2015, pp. 1–6. Variables affecting Credit risk assessment and assess mortgage credit risk.

2.3   Market Risk Factors

GDP. Gross domestic product (GDP) measured quarterly and annually, provides insight into the growth rate of a nation's economy. GDP is measured as

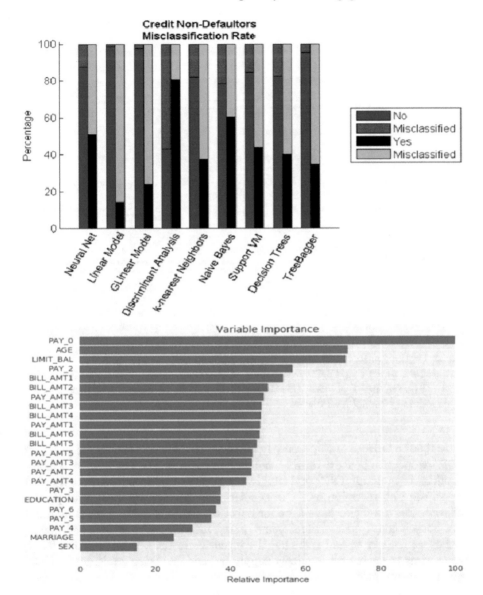

Fig. 3. Source: R.E. Turkson, E.Y. Baagyere and G.E. Wenya, "A machine learning approach for predicting bank credit worthiness," 2016 Third International Conference on Artifcial Intelligence and Pattern Recognition (AIPR), Lodz, 2016, pp. 1–7. Provides analysis of selected features they performed multiple classifcations. Each of these algorithms achieved an accuracy rate between 76% to over 80%.

nominal GDP (inflation included) and real GDP (excludes inflation) [9]. Two major factors that affect GDP include: inflation and recession. US GDP over the

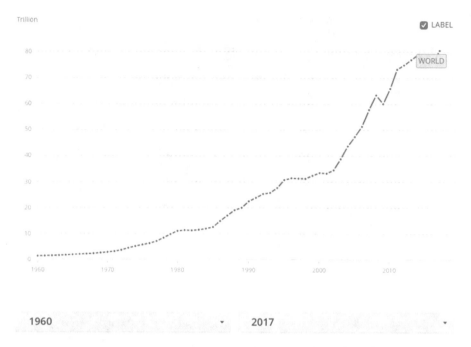

Fig. 4. United States Gross Domestic Product 1960–2018 clearly indicates that there is consistent growth in GDP irrespective of short term economic fluctuations. The Federal Reserve uses 2 policies to maintain GDP and the Economy Expansionary and Contractionary Monetary Policies.

years 1960 to 2018 as in Fig. 4 clearly indicates that there is consistent growth in GDP irrespective of short term economic fluctuations. The Federal Reserve uses 2 policies to maintain GDP and the Economy: **Expansionary Monetary Policy** to ward off recession and **Contractionary Monetary Policy** to prevent inflation. Both policies have a major effect on the disposable income of American households. Under expansionary policy, interest rates are lowered making it cheaper to borrow and reducing the incentive to save. While contractionary policy aims to decrease the money supply by reducing price levels, and increase private consumption.

The US Federal Reserve utilizes certain tools to maintain GDP and the US Economy: open market operations, discount rate, and reserve requirements.

- Open Market Operations: Central Banks buy and sell securities in the open market.
- Discount Rate: The rate that Central Banks charge its members to borrow at its discount window.
- Reserve Requirement: The required money that banks are mandated to hold overnight.

Unemployment. Definition: Unemployment occurs when a person who is actively searching for employment is unable to find work. Unemployment is often used as a measure of the health of the economy. The most frequent measure of unemployment is the unemployment rate, which is the number of unemployed people divided by the number of people in the labor force [10].

Effect of Unemployment to the Economy:

- Unemployment causes an individual to be deprived of resources that trickle down to the benefit of society.
- The economy produces 70% which contributes to direct consumption. If people begin losing their jobs, the whole cycle will be hampered. As a result GDP is reduced and the country drifts away from making efficient allocation of the resources.
- Unemployment triggers inflationary conditions causing a rise in the general price of goods and services, while the purchasing power of currency decreases.

3   Methodology

3.1   Exploratory Data

The Housing and Economic Act of 2008 (HERA) [1] stipulates that certain mortgage information must be made publicly available and stored in a public use database. FHLB adheres to this rule by storing census-level data relating to mortgages purchased by the organization.

For the purpose of this study, we extracted data from FHLB's Public Use Database (PUDB) [6] for mortgage loans acquired from 2010 to 2017 to perform exploratory analysis for creating a base line credit profile. Key fields extracted from the database include: Year, FIPSStateCode, FIPSCountyCode, Income, IncomeRatio, UPB, LTV, MortDate, Purpose, Product, Term, AmortTerm, Front, Back, BoCredScor. Additional fields were derived: State, County (State and County were mapped from the US Census Bureau) [15] and PMT (derived from Rate, AmortTerm, and Amount). See Table 1 for key fields extracted from FHLB PUDB.

3.2   Data Input for ML Model

In order to improve the accuracy results of the machine learning model, additional data was needed that displayed both approved and declined mortgage applicants. As a primary source of data, we extracted 2009–2017 annual data reported by financial institutions required by the Home Mortgage Disclosure Act (HMDA) [5]. In 1975, the United States Congress enacted a regulation that required financial institutions to track and ensure fair lending practices throughout the United States.

The focal fields extracted from this data set include: Action Taken Type, Year, Loan Type, Loan Purpose, Property Type, Occupancy, Amount, State Code, County Code, Income, Denial Reason, Purchaser Type.

Table 1. Extracted data from FHLB PUDB used for exploratory analysis

FHLB - HMDA Loan Application Register Code Sheet	
Year	
State	Two-digit FIPS state identifier
County	Three-digit FIPS county identifier
Loan Type	1 – Conventional (any loan other than FHA, VA, FSA, or RHS loans)
Property Type	1–One to four-family (other than manufactured housing)
Loan Purpose	1–Home purchase
Owner-Occupancy Loan Amount	In thousands of dollars
Action Taken	
Approved	1–Loan originated
Approved	2–Application approved but not accepted
Declined	3–Application denied by financial institution
Approved	6–Loan purchased by the institution
Declined	7–Preapproval request denied by financial institution
Income	Gross Annual Income
Reasons for Denial	
	1–Debt-to-income ratio
	2–Employment history
	3–Credit history
	4–Collateral
	5–Insufficient cash (downpayment, closing costs)
	6–Unverifiable information
	7–Credit application incomplete
	8–Mortgage insurance denied
	9–Other

3.3   Machine Learning Techniques

The initial phase of our study was dedicated to deriving a baseline profile, stratified by state and application year. The extracted FHLB data is distributed in multiple panda dataframes to cleanse NaN values, and remove unwanted data. The next steps was data exploration, in which we have identified critical variables and plotted on a graph (Matplotlib, Seaborn) to determine correlation and highlight key variables that are most impactful to the outcome. Final exploratory steps include normalizing and preprocessing the remaining data. The outcome serves as the basis for baseline assumptions. High level process flow our study is depicted in Fig. 6.

The second data set extracted from HMDA serves as primary input for the Classification Model. The categorical data details approved and declined mortgage transactions from 2009–2017. A logistic regression is then performed on the data to determine the binary outcome of the model.

Datasets before and after pre-processing were stored in buckets for analysis purposes.

Accuracy of the model was calculated using cross entropy and according to the loss obtained, the weights and bias were adjusted to obtain a higher accuracy. Defined Success criteria as: The loan was approved after risk assessment.

And Failure criteria as:

- The loan was approved by the bank, but was declined by the applicant due to a better alternative.
- The application was rejected by the bank because the applicant did not meet the criteria for approval.

4 Results

4.1 Data Processing - Preprocessor 1.5

As an initial step the Raw Data is cleaned and normalized in a Jupyter Notebook termed "Preprocessor 1.5". The preprocessor utilizes python script to stratify and analyze the mortgage data extracted from FHLB public database [5].

The cleaning process extracts all loans that fall outside the set criteria. For example loan balances less than \$10 000, Income Ratios less than 0.01 and equal to 1. There were approximately 1500 loans deleted from the raw data set. After the loan extractions the data is ready to be analyzed.

Key fields were selected and statistical analysis performed to determine the mean, median, mode, distribution and standard deviation. These same fields were grouped by distribution and the mean or mode is used to determine the base case model. Critical field distributions are provided in Fig. 5.

4.2 Preprocessor for Machine Learning Model

Data was extracted from Home Mortgage Disclosure Act (HMDA) website. We performed feature engineering on the raw data in a separate Jupyter Notebook termed "Preprocessor_ML.ipynb". The data was processed as follows:

- Loan_Type filtered for 30 year Conventional Loans.
- Property_Type filtered for one-to-four family dwellings.
- Statecode mapped to State's abbreviation.
- Rate mapped to the state's 30 year Fixed Rate Mortgage (FRM) for the year of mortgage application else used interest rates provided.
- PMT - monthly mortgage payment derived with pmt pandas function; parameters include: loan amount, rate, and loan amortization term.
- GDP - GDP rate mapped for each state by year.

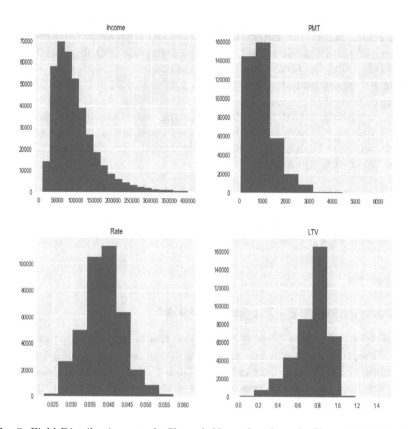

Fig. 5. Field Distributions +vely Skewed, Normal and −vely Skewed Distribution

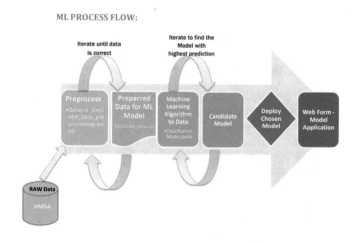

Fig. 6. Machine learning process flow

- IncomeRatio - Derived from dividing monthly payment by monthly gross income.
- Credit Score - assumed failing Credit Score if reason for denial included credit history, else credit score assumed to be passing.
- LTV - LTV is provided a pass or fail based on loan denial reason. If "mortgage insurance denied" was reason for the loan's denial, the LTV was assigned 100% else LTV was assigned 75%.

All fields were extracted and saved as "ML_Processed_Data.csv". This csv file serves as the primary input for the Classification Model - Machine Learning Model.

4.3   Cart Model - Baseline Model Determination

The model was built on a balanced set of training and testing data. Classification and Regression Tree (CART) algorithm [4] is used to predict the approval or denial outcome. Each root node of the decision tree represents a variable input and split point on the variable. The leaf nodes are the output variables that were tabulated to form a final prediction score. Our study used Decision tree algorithm to calculate the applicant's prediction score. The prediction score obtained is based on economic indicators for each state. For the purpose of this model, state unemployment rates, GDP and real state growth% are considered indicators of each state's economic condition. In an optimal scenario, if real state growth increases, GDP increases and unemployment% decreases, the prediction score threshold would be lowered resulting in greater mortgage application approval decisions. At the opposite end of the spectrum, if unemployment% increases, GDP decreases and real state growth% decreases, the prediction score threshold would increase, resulting in lower mortgage application approval decisions.

Variable inputs and splits are:

- Is your income greater than $10,000?
- Is your Borrower Credit Score greater than 700?
- Is your Income Ratio less than 20%?
- Is the loan amount greater than $100,000?
- Is the Loan-To-Value (LTV) less than 75%?
- Has the unemployment rate increased or decreased from the previous year?
- Has the real state growth increased or decreased from the previous year?
- Has GDP increased or decreased from the previous year?

4.4   Classification Model_Machine Learning

ML_Processed_Data.csv is initially read and stored as a dataframe. CSV file generated from Preprocessor_Machine Leaning Model.

Label or Target Creation. Action Type drives the approved or declined labeling process as follows:

- Approved: Action Type 1, 2, 6 – Label = 1
- Declined: Action Type 3, 7 – Label = 0

Filtering and Selection Fields. The following interested fields were selected to train the model:

- interested = AgencyCode, LoanType, PropertyType, LoanPurpose, Occupancy, Amount, ActionType, StateCode, CountyCode, Income, PurchaserType, ApplicationDateIndicator, PropertyLocation, USPSCode, GDP, RealState-Growth%, Rate, PMT, IncRat, Unemployemnt, AmorTerm, BoCreditScor, LTV'

Extracting Categorical and Continuous Features. Interested fields were classified as categorical or continuous categories as below:

- categorical = Agency_Code, Loan_Type, Property_Type, Loan_Purpose, Occupancy, USPS_Code, County_Code, BoCreditScor, LTV
- continuous = Amount, Income, GDP, RealStateGrowth%, Rate, PMT, IncRat, Unemployment

One Hot Encoding. Applied One Hot encoding on the categorical features

Normalize Continuous Features. Normalized the continuous features to return an outcome in the range of 0 to 1.

Balance Dataset. For optimal results the dataset needs to be balance, preferably 50%–50% split.

Train Validate Test Split. Total dataset contains 14,000 records and split:
Train Data = 2014, 2015, 2016 (288,000 Sample)
　　Validate Data = 2017 (40,000 Sample)
　　Test Data = 2017 (40,855 Sample)

Train Machine Learning Model. After performing data pre-processing, a classification algorithm is used to classify the elements into two groups (Approved = 1 and Declined = 0).

Logistic Regression (LR) is used as an efficient machine learning model. LR model is trained on Train data set and the trained model is tested with Test data set and an accuracy of 85.85% is achieved. Figure 7 represents the Confusion Matrix which displays the performance of the algorithm.

Test Data Size: n = 80855		Predicted		
		Declined	Approved	
Actual	Declined	**10370**	3619	13989
	Approved	7825	**59041**	66866
		18195	62660	

Accuracy:	85.85%

Fig. 7. Confusion matrix logistic regression

Candidate Models. In total, nine supervised classification models were tested on the data set. The top four models with the high performance results include: Logistic Regression, Random Forest Classifier, AdaBoost Classifier and Deep Neural Network.

All of the chosen parameters were selected after tuning and testing, performing hyper parameter tuning, cross-validation, etc. and the featured parameters are the best for our data selection. Each model parameters are presented below

Model 1 Logistic Regression Hyper Parameters:

- C =1000,
- max_iter = 100,
- solver = liblinear,
- penalty = l2

Model 2 Random Forest Classifier Hyper Parameters:

- n_estimators = 100,
- criterion = 'entropy'

Model 3 AdaBoost Classifier Hyper Parameters:

- algorithm = 'SAMME.R',
- base_estimator = None,
- learning_rate = 1.0,
- n_estimators = 50

Model 4 Deep Neural Network

- Layer 1: 32 units Dense Layer with activation 'relu'
- Layer 2: 64 units Dense Layer with activation 'relu'
- Layer 3: 1 units Dense Layer with activation 'sigmoid'

– Loss Function: binary_crossentropy
– Optimizer: ADAM
– Epochs: 5
– Batch Size: 64
– Validation split: 0.20

Candidate Model Results: Model Selection. The accuracy levels of each model range between 84%–85% and all F1-Scores are above 0.85. Accuracy results appear to be similar across models but training time was ultimately the differentiating factor.

In an attempt to improve on the model's accuracy results we re-evaluated the Random Forest Candidate Model. Large datasets are slow to use, making it difficult to handle overfitting, by combining the predictions of many different large neural networks at test time. Dropout is a technique for addressing this problem, significantly reducing overfitting with significant improvements over other regularization methods.

After selecting the important features and training the model using the random forest, our accuracy is less than what we have obtained using the logistic regression approach. Figures 8 and 9 shows Candidate model outcome and Time taken for Training.

	Logistic Regression	Random Forest Classifier	AdaBoost Classifier	Deep Neural Network
Accuracy	85.84%	85.35%	84.15%	84.82%
Precision	0.88	0.89	0.87	0.87
Recall	0.86	0.85	0.84	0.84
F1- Score	0.87	0.86	0.85	0.85

Fig. 8. Candidate model results

	Logistic Regression	Random Forest Classifier	AdaBoost Classifier	Deep Neural Network
Training Time (Seconds)	64.01	174.85	61.19	204.06

Fig. 9. Training time model results

5   Project Approach

5.1   Front End

The baseline CART Model has been deployed to a web interface. Users have the ability to input their income, debt to income ration, FICO score, loan balance, loan to value ratio, property state, etc. and a mortgage decision (approved or denied) will output.

Form Handling:

- If the request uses a get method, the input form is displayed.
- If the request uses a post method, but an input value is not valid, the input form is displayed with an error message.
- If the request uses a post method, and all input fields are valid, the prediction decision is displayed.

Software Used:

- Python 3.6+
- python packages:
- Flask
- Pandas
- Sci-kit Learn

5.2   Programming Language

Python will be the programming language utilized throughout the project. All project code is stored in Jupyter Notebooks.

5.3   Machine Learning Framework and Libraries

- Scikit Learn Library
- Keras (with Tensorflow as Backend)

5.4   Google Cloud Storage

The data in .csv format is uploaded to a Google cloud storage buckets.

5.5   Data Processing Framework and Libraries

- Pandas
- Numpy

5.6   Visualization

- Matplotlib
- Seaborn
- Ploty
- Tableau

6 Recommendations

The model takes a two-prong approach in applying machine learning techniques to the mortgage decision process. The first layer applies a standard base credit profile for each state. We will implement a scoring system that will determine an individual's score compared to their state's baseline profile. Individual scores are derived from the baseline and recalibrated based on micro and macro market elements. From this section, the model has the ability to predict future approval/denial decisions based on economic conditions.

After training and testing the ML model to obtain high accuracy we concluded that a single baseline across all states would not be optimal. Instead we implemented a single variable selection process to create base line models for each state and vintage year of mortgage origination.

In addition we discovered HMDA data was the best source of public data that provided approved and declined decisions. After preprocessing and feature engineering, a higher accuracy level was achieved. The model has been trained on 2014–2016 mortgage data. For test purposes we utilized 2017 data, which was not used to train the machine model. Projections based on Market risk factors were analyzed with approved and declined mortgages to show necessity to apply External factors on current Mortgage underwriting and Credit Risk analysis.

As part of the study, multiple models were tested with different ML Algorithms and observed roughly similar results in all the models, while Logistic Regression was slightly better with the training set we used. It took 64.01 seconds to train the model with the accuracy of 85.85% and F1 score of 0.87.

References

1. Arthur, B.: Housing and economic recovery act of 2008. Harv. J. Legis. **46**, 585 (2009)
2. Calcagnini, G., Cole, R., Giombini, G., Grandicelli, G.: Hierarchy of bank loan approval and loan performance. Economia Politica, 1–20 (2018)
3. Critchfield, T., Dey, J., Mota, N., Patrabansh, S., et al.: Mortgage experiences of rural borrowers in the united states: insights from the national survey of mortgage originations. Technical report, Federal Housing Finance Agency (2018)
4. Denison, D.G.T., Mallick, B.K., Smith, A.F.M.: A Bayesian cart algorithm. Biometrika **85**(2), 363–377 (1998)
5. Federal Financial Institutions Examination of Council: Home Mortgage Disclosure Act (2018)
6. Federal Home Loan Bank: Fhfa.gov, Federal Home Loan Bank Member Data — Federal Housing Finance Agency (2018). https://www.fhfa.gov/DataTools/Downloads/Pages/Federal-Home-Loan-Bank-Member-Data.aspx
7. Guiso, L., Pozzi, A., Tsoy, A., Gambacorta, L., Mistrulli, P.E.: The cost of distorted financial advice: evidence from the mortgage market (2018)
8. Hippler, W.J., Hossain, S., Kabir Hassan, M.: Financial crisis spillover from wall street to main street: further evidence. Empirical Econ., 1–46 (2018)

9. Inekwe, J.N., Jin, Y., Valenzuela, M.R.: The effects of financial distress: evidence from US GDP growth. Econ. Model. **72**, 8–21 (2018)
10. Kaplan, R.S., et al.: Discussion of economic conditions and key challenges facing the us economy. Technical report, Federal Reserve Bank of Dallas (2018)
11. Fieldhouse, A.J., Mertens, K., Ravn, M.O.: The macroeconomic effects of government asset purchases: evidence from postwar US housing credit policy. Q. J. Econ. **1**, 58 (2018)
12. Ramya, R.S., Kumaresan, S.: Analysis of feature selection techniques in credit risk assessment. In: 2015 International Conference on Advanced Computing and Communication Systems, pp. 1–6. IEEE (2015)
13. SEC Emblem: Implementing the Dodd-Frank Wall Street Reform and Consumer Protection Act (2018). https://www.sec.gov/spotlight/dodd-frank.shtml
14. Turkson, R.E., Baagyere, E.Y., Wenya, G.E.: A machine learning approach for predicting bank credit worthiness. In: International Conference on Artificial Intelligence and Pattern Recognition (AIPR), pp. 1–7. IEEE (2016)
15. United States Census Bureau: FIPS Codes for the States and the District of Columbia (2018)

Efficient Bayesian Expert Models for Fever in Neutropenia and Fever in Neutropenia with Bacteremia

Bekzhan Darmeshov[1] and Vasilios Zarikas[2,3(✉)]

[1] Mechanical and Aerospace Engineering Department, School of Engineering, Nazarbayev University, Nur-Sultan (Astana) 010000, Kazakhstan
[2] School of Engineering, Nazarbayev University, Nur-Sultan (Astana) 010000, Kazakhstan
vasileios.zarikas@nu.edu.kz
[3] Theory Division, General Department, University of Thessaly, Volos, Greece

Abstract. Bayesian expert models are very efficient solutions since they can encapsulate in a mathematical consistent way, certain and uncertain knowledge, as well as preferences strategies and policies. Furthermore, the Bayesian modelling framework is the only one that can inference about causal connections and suggest the structure of a reasonable probabilistic model from historic data. Two novel expert models have been developed for a medical issue concerning diagnosis of fever in neutropenia or fever in neutropenia with bacteremia. Supervised and unsupervised learning was used to construct these two the expert models. The best one of them exhibited 93% precision of prediction.

Keywords: Bayesian networks · Expert model · Cancer · Neutropenia · Bacteraemia

1 Introduction

The identification of patients' reliable diagnosis is prioritized in medicine to prevent disease and its complications. To improve diagnostic systems, Bayesian reasoning can be used to build an expert system [38–40]. Improvements originate from the high accuracy identification structures, which allow machine learning for decision making. For example, Bayesian machine learning algorithms can drive reliable decisions about the stage of malignancy of a tumor.

Bayesian Networks can be used in any medical related problem that requires decision making based on historic data or based on expert fuzzy rules [14, 40]. For example, Bayesian sequential segmentation was used to build an expert system [14, 33]. The system was able to identify patients with high risk of Myocardial Infarction. Hence, the period of stay in hospital for patients under coronary care can be safely reduced. With the help of Bayesian Network, the correlation between physical activities, stress, sleep, mass of the objects and diabetes existence throughout the population of black and white people was found [15, 31]. According to this study, compound of sedentary lifestyle, bad sleeping regimes, high stress existence and extra

© Springer Nature Switzerland AG 2020
K. Arai et al. (Eds.): FTC 2019, AISC 1069, pp. 124–143, 2020.
https://doi.org/10.1007/978-3-030-32520-6_11

body weight factor resulted in high risk of diabetes. Moreover, with the help of Bayesian Belief Network, the relation between short sleep (6>) and long sleep (8<) with stroke and cardiovascular diseases was found [30–32]. The validation of the research has shown approximately the same results, which states the position of unhealthy lifestyle as leading reason of diseases. Furthermore, method of the finite variables was used to predict possibility of radiation pneumonia [36, 37, 41]. Also, Bayesian Networks was used to identify low glycaemic index and hypoglycaemia risk of patients suffering from diabetes type 1 [21, 34]. Another application is relying on the theory of Bayesian networks managed to calculated risks of public health incidents [33, 41]. At the same time, blood samples and enhanced detection accuracy of mean corpuscular haemoglobin concentration were analysed using a decision tree.

Application of Bayesian statistical methods has also spread to the gene expression analysis as a solution of early detection of cancer. Another model with Bayesian Networks allows to choose correct systematic analysis method for a predefined microarray [12, 28, 29]. The main goal of such analysis of microarrays is to emphasize the panel of genes that causes diseases. The method of comparison of microarrays could also be used in the data of microarrays of DNA that affects development of breast cancer [7, 11, 42]. In addition, to predict prostate cancer possibility in patients an expert system with the help of Markov Blanket model (implemented with the BayesiaLab software) was constructed, Renard Penna et al. 2016. This model performed 80% of the preciseness for surgical Gleason score [8, 9, 26, 27]. Similarly, a Bayesian Network was constructed for the diagnosis of fine needle aspiration cytology of the breast [21, 38]. Furthermore, an application of Bayesian Networks can be found in diagnostics of atypical adenomatous hyperplasia of the prostate or to improve pathological staging of prostate cancer [10, 11, 20]. It is also used in the systemization of endometrial hyperplasia. Moreover, using PCCSM model in Bayesian Networks, risk assessment of development of cervical cancer or cervical precancer was made [13, 21]. Finally, a model of Naïve Bayesian network assessing breast cancer or ductal carcinoma in situ and its development possibility was proposed [21–25].

The main goal of the present study is to demonstrate two medical expert systems created with BayesiaLab software. The models organize a system of identification of risks related to fever in neutropenia for patients with diagnosed cancer. Annually, 2300 young patients up to seventeen age, die from cancer development in US and around 1800 paediatric cancer patients perished due to the same reason [1, 3–6, 12, 16, 17]. As cancer is a severe illness, frequently, patients suffer from complications during treatment [1, 3–6, 12, 18, 19]. One of such complications for non-adolescent diseased patients is fever in neutropenia or fever in neutropenia with bacteremia. Fever in neutropenia can lead to lethal result. Following the main goal of the study, a prediction solution, expert system, constructed in BayesiaLab. Hopefully, such systems can reduce lethal cases during chemotherapy due to overtemperature of patients. Thus, the project is related with diagnostic methods for cancer therapy consequences.

Our system was developed using the presented data [1–6, 19]. The development of expert systems of pre-detection of patients with risk of development serious complication of chemotherapy is profoundly important, since it is related with lethal cases during cancer treatment. According to the research, 583 out of 800 patients were diseased cancer, which is 73% of total [2]. In our study all of the cases were tested using eleven factors. As a result, a table of 800 patient categorized by eleven characteristics was created. BayesiaLab was used to identify the most important factors that affect the complication of chemotherapy treatment, and then to create algorithm of identification of patients with predisposition to FN. The novelty of the project is related to the machine learning contraction of Bayesian expert systems, where predictive decisions are made with accuracy up to 95% and even more.

In the following of the paper, Sect. 2 describes the design and statistical methods that were used to formulate the expert systems. Section 3 gives an extensive discussion of the results of the validation of the two expert models and finally Sect. 4 provides a summary and the final overall conclusions.

2   Design and Methods

To create the expert systems, BayesiaLab software was used. Data obtained from the paper "Pediatric Patients at Risk for Fever in Chemotherapy-Induced Neutropenia in Bern, Switzerland, 1993–2012." was divided into two parts: (a) fever in neutropenia (FN); (b) fever in neutropenia with bacteraemia (FN with bacteraemia) [1–6]. The reason is to understand accuracy of the system in diagnosis in both cases separately. In the first case (a), factor of FN with bacteraemia is excluded, while for the case (b), FN data is included. In addition, some unnecessary data was deleted (year of observations and random ID). Year of observation is considered as unnecessary data, because the calendar time of observations is not efficient factor in creation of diagnostic system of cancer and its complications. Data consists of 583 patients with cancer and several data lines for each patient at different observation periods (2113 samples in total), each line roughly can be defined as separate case for BayesiaLab.

2.1   Statistical Methods

In this subsection of the paper, steps to obtain an expert diagnostic system of risk identification are presented. The original data source can be reviewed in given references [2]. All the calculations were done in last version of BayesiaLab [35]. As always, the first step of the creation of the expert system is always the preparation of data and a descriptive analysis of them to check the probability density functions. Subsequent steps are important in the data processing, because they direct relation to efficiency of the system.

For both cases (FN and FN with bacteremia) random test sampling with test samples of 30% was set, (30% percentage of the data for test set, and 70% as a learning set). The determination of variables to use appropriate discretization method is very important too. The factors "observation period", "age group", "number of malignancies", "length of observation period" and "chemotherapy intensity" were defined as continuous type of data, as they give increasing intensity values in a cardinal scale. Meanwhile, "relapsed malignancy", "bone marrow involvement", "central venous access device", "earlier FN" and "earlier FN with bacteremia" were set discrete type because of binary nature of data. For the test of FN, two subsets were analyzed: with FN bacteremia values and without FN bacteremia test. FN bacteremia values were defined as continuous type. For the case of FN bacteremia, FN was defined as continuous. "Diagnostic group of malignancy" and "sex" parameters were chosen to be not distributed type of the data.

Next step is the identification of the discretization type for continuous values. Equal distances discretization method was used for both cases. After choosing the discretization type to the model variables, unsupervised and supervised learning of the system was implemented. Two cases of FN were tested, and the procedure of learning implementations was made similarly. In both cases, unsupervised learning was accomplished initially to check results. EQ algorithm was used in the case of unsupervised learning. The reason is that unsupervised learning is easy and convenient model for initial observations of the data that guarantees workability of the system. However, the main tool used to create an expert system was the supervised learning.

Unsupervised learning creates arrow type relation on the graph, which will be demonstrated in results section of the paper. The discovery of the correlation between variables was done using the Pearson's correlation method. As a result, strength of linear relation between variables can be obtained [30].

Finally, implementation of supervised learning was performed, considering the FN or FN with bacteremia as target variable. Tree Augmented Naïve Bayes model was used and learning of the system is implemented. After, analysis of the network performance was accomplished. To check correctness of the results, K-folds analysis was executed. Tables with results are presented in results section of the paper.

Description of Key Algorithms Used

EQ Algorithm
The algorithm is to search equivalent classes in any Bayesian Network. The method is to understand the correlation of factors, avoiding a set of local minima and it helps to decrease learning time.

Tree Augmented Naïve Bayes
Predefined target node is the "father" of all other nodes. This is the specificity of the model, which is similar to Naïve architecture. The difference is the consideration of "child" nodes of the network to create more robust system. However, learning time is bigger than in "Naïve Bayes" type of learning.

Equal Distances
Discretization type is adequate even for big data, where the algorithm searches for maximum and minimum of each variable, consequently dividing the range to predefined range width.

Pearson's Correlation
Model demonstrates strength of correlation between variables. Formula to find the number is:

$$r_{XY} = \frac{\sum_{i=1}^{n}\left(X_i - \overline{X}\right)\left(Y_i - \overline{Y}\right)}{\sqrt{\sum_{i=1}^{n}\left(X_i - \overline{X}\right)^2}\sqrt{\sum_{i=1}^{n}\left(Y_i - \overline{Y}\right)^2}}$$

Where, X and Y are two variables correlation of which is required to find.

3   Results and Discussions of Research

In this section, we describe two Bayesian expert models that are able to predict FN status and FN with bacteraemia status. In the two following subsections details of results are presented for each one of the two models. For both these expert models, we have used 30% percentage of the data for test set, and 70% as a learning set.

3.1   Results Received in FN Test

In this subsection, we study two different cases. The first one is FN test without the effect of FN bacteraemia values. The factors of the first case are "observation period", "age group", "number of malignancies", "length of observation period", "chemotherapy intensity", "relapsed malignancy", "bone marrow involvement", "central venous access device", "earlier FN" and "earlier FN with bacteremia". The second case is the model for FN that includes the factor of FN with bacteremia too.

FN Test without the Effect of FN Bacteraemia

Our analysis begins using an unsupervised learning algorithm, see Figs. 1 and 2 that concern the model without considering the factor of FN bacteraemia. We have used EQ algorithm for this case of unsupervised learning. Furthermore, we have selected discretization of continuous variables as equal distances method. And to see performance of the network, Pearson's correlation was used.

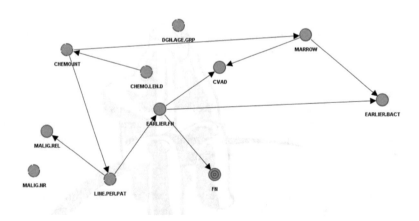

Fig. 1. EQ unsupervised learning.

From Figs. 1 and 2, we understand that the number of malignancy and age groups do not affect the model. In addition, the strongest correlation is between variables "earlier FN" and "earlier FN bacteraemia", and the weakest is between "bone marrow involvement" and "earlier FN bacteraemia". Strong correlation can be observed between "relapsed malignancy" variable and "observation period".

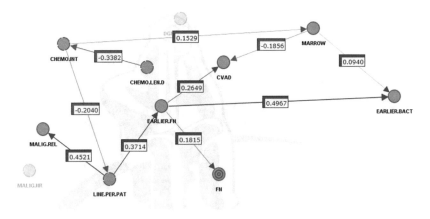

Fig. 2. EQ unsupervised learning with Pearson's correlation.

A second analysis has been carried out with supervised learning algorithm "Tree Augmented Naïve Bayes", see Table 1. Figure 3, demonstrates correlation between all variables and target variable. Therefore, it can be noted that the supervised modelling is more reliable and accurate. The precision of the test model of FN can be observed in Table 2. For the expert system of FN risks identification, the precision value is approximately 75.7%. The accuracy is considered as acceptable but requires further improvements.

Table 1. Tree Augmented Naïve Bayes learning results for FN.

Target: FN						
Value	0	1	2	3	4	5
Gini Index	11.0284%	25.9385%	57.0300%	67.2534%	5.3712%	74.5656%
Relative Gini Index	39.8915%	34.1353%	58.2258%	68.0055%	5.3883%	74.6835%
Lift Index	1.1594	1.5782	2.1396	2.3452	1.7608	2.0630
Relative Lift Index	87.6163%	65.1051%	44.1299%	43.1448%	26.9707%	29.3520%
ROC Index	69.9464%	67.0683%	79.1191%	84.0027%	52.6941%	87.3418%
Calibration Index	69.6574%	88.2996%	29.7505%	19.0471%	86.7409%	100.0000%
Binary Log-Loss	0.4941	0.4934	0.1715	0.2614	0.0632	0.0546

K-folds test results in Tables 3 and 4 demonstrate the correctness of the previously obtained results of "Tree Augmented Naïve Bayes" learning. In general, prediction is not very high, although it is acceptable. The reason lays on the diversity of data. Considering precision values for predicted stages of FN, the number allocation is imbalanced among states of FN. Most of data are associated with state "0" of FN, less data concern state "1" and much less for other states. This fact reduces overall precision of the model. The "0" case is predicted with the percentage of precision of 97.1%.

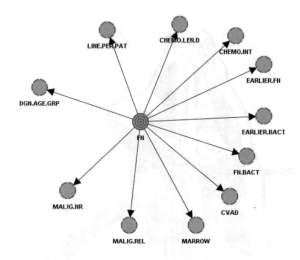

Fig. 3. Supervised learning of FN without the effect of FN bacteraemia.

Table 2. Tree Augmented Naïve Bayes precision results for FN.

Overall Precision	75.6714%
Mean Precision	20.8010%
Overall Reliability	76.1537%
Mean Reliability	27.4118%
Overall Relative Gini Index	39.1426%
Mean Relative Gini Index	46.7216%
Overall Relative Lift Index	80.5422%
Mean Relative Lift Index	49.3865%
Overall ROC Index	69.5721%
Mean ROC Index	73.3621%
Overall Calibration Index	72.8565%
Mean Calibration Index	65.5826%
Overall Log-Loss	0.8377
Mean Binary Log-Loss	0.2564

While for other categories, the percentage of precision is less than 22%. Of course, the obvious solution of this problem is to utilize data from larger set of patients (Table 5).

FN Test with the Effect of FN Bacteraemia Values
This time, similar test of FN was performed with the concern of FN bacteraemia effect. The factors were "observation period", "age group", "number of malignancies", "length of observation period", "chemotherapy intensity", "relapsed malignancy", "bone marrow involvement", "central venous access device", "earlier FN", "earlier FN with bacteremia" and "FN bacteraemia". In Fig. 4, the network constructed by "Tree

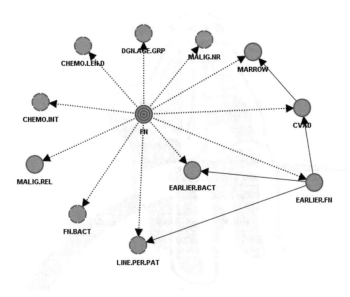

Fig. 4. Supervised learning of FN with the effect of FN bacteraemia

Table 3. K-fold test results for FN.

K-Fold (10)						
Tree Augmented Naive Bayes						
Target: FN						
Value	0	1	2	3	4	5
Gini Index	12.5292%	29.9848%	66.8091%	57.6014%	−11.8919%	0.0000%
Relative Gini Index	44.76x51%	39.3979%	68.9576%	58.2888%	−11.9728%	0.0000%
Lift Index	1.1607	1.6450	2.3055	2.1232	0.1949	0.0000
Relative Lift Index	87.4124%	67.8786%	52.3543%	41.0480%	3.4948%	0.0000%
ROC Index	72.3929%	69.7116%	84.5018%	69.1444%	14.0136%	0.0000%
Calibration Index	62.9797%	76.5845%	49.3585%	83.5206%	89.2016%	100.0000%
Binary Log-Loss	0.5216	0.5317	0.1441	0.1045	0.0962	0.0004

Augmented Naïve Bayes" learning model can be seen. As it can observe, FN factor depends on all other factors. It was shown, that the difference from the Bayesian network of FN test without the effect of FN bacteraemia values is in the extra correlations appeared in the network. As a result, overall precision was increased to 76.3%. It can be reviewed in Table 6. The overall performance of the model was increased; however, not always data about FN bacteraemia are available. Thus, under these

Table 4. Occurrences, reliability and precision values for K-fold test of FN.

Occurrences

Value	0 (1065)	1 (353)	2 (43)	3 (15)	4 (4)	5 (0)
0 (1332)	1034	260	23	12	3	0
1 (100)	10	77	11	2	0	0
2 (45)	19	15	9	1	1	0
3 (2)	2	0	0	0	0	0
4 (1)	0	1	0	0	0	0
5 (0)	0	0	0	0	0	0

Reliability

Value	0 (1065)	1 (353)	2 (43)	3 (15)	4 (4)	5 (0)
0 (1332)	77.6276%	19.5195%	1.7267%	0.9009%	0.2252%	0.0000%
1 (100)	10.0000%	77.0000%	11.0000%	2.0000%	0.0000%	0.0000%
2 (45)	42.2222%	33.3333%	20.0000%	2.2222%	2.2222%	0.0000%
3 (2)	100.0000%	0.0000%	0.0000%	0.0000%	0.0000%	0.0000%
4 (1)	0.0000%	100.0000%	0.0000%	0.0000%	0.0000%	0.0000%
5 (0)	0.0000%	0.0000%	0.0000%	0.0000%	0.0000%	0.0000%

Precision

Value	0 (1065)	1 (353)	2 (43)	3 (15)	4 (4)	5 (0)
0 (1332)	97.0892%	73.6544%	53.4884%	80.0000%	75.0000%	0.0000%
1 (100)	0.9390%	21.8130%	25.5814%	13.3333%	0.0000%	0.0000%
2 (45)	1.7840%	4.2493%	20.9302%	6.6667%	25.0000%	0.0000%
3 (2)	0.1878%	0.0000%	0.0000%	0.0000%	0.0000%	0.0000%
4 (1)	0.0000%	0.2833%	0.0000%	0.0000%	0.0000%	0.0000%
5 (0)	0.0000%	0.0000%	0.0000%	0.0000%	0.0000%	0.0000%

circumstances, mean performance is the precision value of FN test without the effect of FN bacteraemia.

As it can be observed from the Table 8 of K-fold test, Tree Augmented Naïve Bayes test suggests that the expert model is valid about its in performance. Table 7 also shows similar results in overall performance of the model because of the effect of FN bacteraemia. Moreover, highest value of precision is again in state of "0" of FN, with 98.5%, which supports the idea of performance reduction in precision due to the imbalanced data population.

3.2   Results of FN Bacteraemia Test

This time, the expert system under construction concerns the case of FN bacteraemia considering all relevant factors. The list of factors is "observation period", "age group", "number of malignancies", "length of observation period", "chemotherapy intensity", "relapsed malignancy", "bone marrow involvement", "central venous access device", "earlier FN", "earlier FN with bacteremia" and "FN".

We begin our analysis by implanting unsupervised learning of the model. Figures 5 and 6 concerns Bayesian Network of the unsupervised learning of the FN bacteraemia test. As it can be observed from Figs. 5 and 6, number of malignancy and age group of patients are two non-affecting variables in this case. For the unsupervised learning of FN bacteraemia case, EQ algorithm is used as well. Moreover, discretization of continuous variables as equal distances method has been selected while for the performance of the network, Pearson's correlation was used. In Figs. 7 and 8 one can observe that considering Pearson's correlations, the strongest correlations are between "observation period" and "relapsed malignancy", and between "FN" and "FN with bacteremia". Meanwhile, the weakest correlation is between "bone marrow involvement" parameter and "chemotherapy intensity".

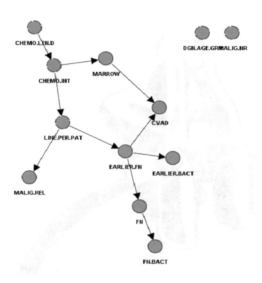

Fig. 5. Unsupervised learning of FN bacteraemia.

Seeing the overall performance of the test of FN bacteraemia, Table 9, precision is increased significantly to the value of 92.9%, as can be observed in the Table 10. Consequently, the identification of risks of FN bacteraemia can be done more accurate. Chance of making incorrect decision is lower than 8%, which is relatively good result on this stage of the research. Further improvements can be accomplished increasing the size of historic data. Like the first test results, imbalanced number allocation affected overall performance. Most of data are associated with state "0" of FN, less data concern state "1" and much less for other states. As a solution, utilization of data from larger set of patients can be taken.

Fig. 6. Pearson's correlation of unsupervised learning of FN bacteraemia.

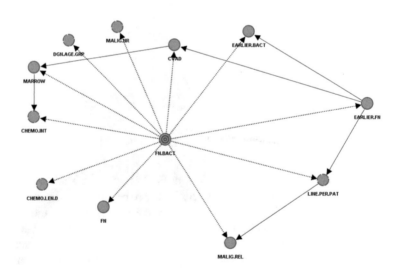

Fig. 7. Tree Augmented Naïve Bayes learning of FN bacteraemia.

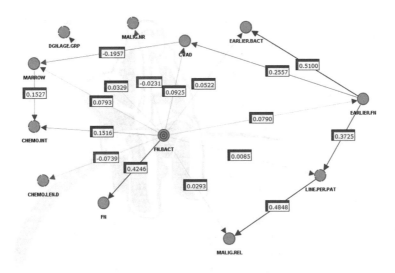

Fig. 8. Pearson's correlation of Tree Augmented Naïve Bayes learning of FN bacteraemia.

Table 5. Tree Augmented Naïve Bayes learning results for FN with the effect of FN bacteraemia.

Target: FN						
Value	0	1	2	3	4	5
Gini Index	12.0963%	29.5928%	57.2974%	66.5764%	6.1611%	74.5656%
Relative Gini Index	43.7542%	38.9444%	58.4988%	67.3209%	6.1807%	74.6835%
Lift Index	1.1722	1.6318	2.2111	2.4552	1.9005	2.0630
Relative Lift Index	88.5809%	67.3155%	45.6061%	45.1691%	29.1115%	29.3520%
ROC Index	71.8777%	69.4729%	79.2556%	83.6604%	53.0903%	87.3418%
Calibration Index	65.9042%	89.6500%	54.4270%	52.2059%	89.8878%	100.0000%
Binary Log-Loss	0.4822	0.4855	0.1424	0.2597	0.0628	0.0546

In addition, Table 13, K-fold test has shown that FN bacteremia learning with Tree Augmented Naïve Byes model is accurate enough. However, precision values for separate cases are still deviating significantly, see Tables 11 and 12. The precision values are 97.8% and 18.8% for "0" and "1" categories, respectively.

Table 6. Tree Augmented Naïve Bayes precision results for FN with the effect of FN bacteraemia.

Overall Precision	76.3033%
Mean Precision	21.0199%
Overall Reliability	74.0085%
Mean Reliability	27.8946%
Overall Relative Gini Index	43.0928%
Mean Relative Gini Index	48.2304%
Overall Relative Lift Index	81.8304%
Mean Relative Lift Index	50.8558%
Overall ROC Index	71.5471%
Mean ROC Index	74.1165%
Overall Calibration Index	71.3487%
Mean Calibration Index	75.3458%
Overall Log-Loss	0.8078
Mean Binary Log-Loss	0.2479

Table 7. K-fold test result for Tree Augmented Naïve Bayes learning with the effect of FN bacteraemia.

K-Fold (10)						
Tree Augmented Naive Bayes						
Target: FN						
Value	0	1	2	3	4	5
Gini Index	12.4814%	30.1615%	68.2259%	63.8851%	−11.8919%	0.0000%
Relative Gini Index	44.5935%	39.6324%	70.4553%	64.6380%	−11.9728%	0.0000%
Lift Index	1.1642	1.6605	2.4356	2.3972	0.1949	0.0000
Relative Lift Index	87.6787%	68.5196%	55.5939%	45.9123%	3.4948%	0.0000%
ROC Index	72.3071%	69.8288%	85.2627%	72.3305%	14.0136%	0.0000%
Calibration Index	66.0269%	74.2443%	66.0119%	84.6483%	86.5370%	100.0000%
Binary Log-Loss	0.5091	0.5255	0.1202	0.1021	0.0961	0.0005

Table 8. Occurrences, reliability and precision values of K-fold test for Tree Augmented Naïve Bayes learning with the effect of FN bacteraemia.

Occurrences

Value	0 (1065)	1 (353)	2 (43)	3 (15)	4 (4)	5 (0)
0 (1356)	1049	265	27	12	3	0
1 (114)	15	84	12	2	1	0
2 (8)	1	3	4	0	0	0
3 (1)	0	0	0	1	0	0
4 (1)	0	1	0	0	0	0
5 (0)	0	0	0	0	0	0

Reliability

Value	0 (1065)	1 (353)	2 (43)	3 (15)	4 (4)	5 (0)
0 (1356)	77.3599%	19.5428%	1.9912%	0.8850%	0.2212%	0.0000%
1 (114)	13.1579%	73.6842%	10.5263%	1.7544%	0.8772%	0.0000%
2 (8)	12.5000%	37.5000%	50.0000%	0.0000%	0.0000%	0.0000%
3 (1)	0.0000%	0.0000%	0.0000%	100.0000%	0.0000%	0.0000%
4 (1)	0.0000%	100.0000%	0.0000%	0.0000%	0.0000%	0.0000%
5 (0)	0.0000%	0.0000%	0.0000%	0.0000%	0.0000%	0.0000%

Precision

Value	0 (1065)	1 (353)	2 (43)	3 (15)	4 (4)	5 (0)
0 (1356)	98.4977%	75.0708%	62.7907%	80.0000%	75.0000%	0.0000%
1 (114)	1.4085%	23.7960%	27.9070%	13.3333%	25.0000%	0.0000%
2 (8)	0.0939%	0.8499%	9.3023%	0.0000%	0.0000%	0.0000%
3 (1)	0.0000%	0.0000%	0.0000%	6.6667%	0.0000%	0.0000%
4 (1)	0.0000%	0.2833%	0.0000%	0.0000%	0.0000%	0.0000%
5 (0)	0.0000%	0.0000%	0.0000%	0.0000%	0.0000%	0.0000%

Table 9. Tree Augmented Naïve Bayes learning results for FN bacteraemia.

Target: FN.BACT				
Value	0	1	2	3
Gini Index	5.7899%	78.5420%	−55.9242%	99.5261%
Relative Gini Index	83.2960%	83.9815%	−56.1014%	99.6835%
Lift Index	1.0652	2.7402	0.2686	6.0285
Relative Lift Index	99.3655%	73.5510%	4.1146%	85.7722%
ROC Index	91.6499%	91.9908%	21.9493%	99.8418%
Calibration Index	76.5048%	80.9714%	87.8170%	62.5000%
Binary Log-Loss	0.3075	0.2476	0.1100	0.0026

Table 10. Overall performance of Tree Augmented Naïve Bayes learning of FN bacteraemia.

Overall Precision	92.8910%
Mean Precision	28.3614%
Overall Reliability	90.3354%
Mean Reliability	35.0062%
Overall Relative Gini Index	82.9259%
Mean Relative Gini Index	52.7149%
Overall Relative Lift Index	97.3710%
Mean Relative Lift Index	65.7008%
Overall ROC Index	91.4647%
Mean ROC Index	76.3579%
Overall Calibration Index	76.8077%
Mean Calibration Index	76.9483%
Overall Log-Loss	0.3601
Mean Binary Log-Loss	0.1670

Table 11. K-fold test of Tree Augmented Naïve Bayes learning of FN bacteraemia.

K-Fold (10)				
Tree Augmented Naive Bayes				
Target: FN.BACT				
Value	0	1	2	3
Gini Index	5.6947%	76.1080%	−5.0676%	1.5541%
Relative Gini Index	82.0214%	81.6418%	−5.1020%	1.5646%
Lift Index	1.0636	2.6291	0.0287	0.0865
Relative Lift Index	99.2096%	71.3445%	0.5137%	1.5515%
ROC Index	91.0485%	90.8411%	2.4490%	5.7823%
Calibration Index	63.9899%	65.8907%	100.0000%	100.0000%
Binary Log-Loss	0.2010	0.2002	0.0239	0.0239

Table 12. Occurrences, reliability and precision values of K-fold test of Tree Augmented Naïve Bayes learning of FN bacteraemia.

Occurrences				
Value	0 (1377)	1 (101)	2 (1)	3 (1)
0 (1429)	1347	81	1	0
1 (48)	28	19	0	1
2 (3)	2	1	0	0
3 (0)	0	0	0	0
Reliability				
Value	0 (1377)	1 (101)	2 (1)	3 (1)
0 (1429)	94.2617%	5.6683%	0.0700%	0.0000%
1 (48)	58.3333%	39.5833%	0.0000%	2.0833%
2 (3)	66.6667%	33.3333%	0.0000%	0.0000%
3 (0)	0.0000%	0.0000%	0.0000%	0.0000%
Precision				
Value	0 (1377)	1 (101)	2 (1)	3 (1)
0 (1429)	97.8214%	80.1980%	100.0000%	0.0000%
1 (48)	2.0334%	18.8119%	0.0000%	100.0000%
2 (3)	0.1452%	0.9901%	0.0000%	0.0000%
3 (0)	0.0000%	0.0000%	0.0000%	0.0000%

Table 13. Statistics of K-fold test of Tree Naïve Bayes learning of FN bacteraemia.

Statistics						
Measure	Overall	Mean	Min	Max	Standard Deviation	Best Network#
R	0.4305	0.4392	0.2719	0.5337	0.0886	4
R2	0.1853	0.2008	0.0739	0.2849	0.0697	4
RMSE	0.2471	0.2452	0.1700	0.2704	0.0304	4
NRMSE	8.2374%	21.3737%	8.9931%	26.9485%	5.8585%	2
Overall Precision	92.2973%	92.2973%	87.8378%	96.6216%	2.2040%	4
Mean Precision	29.1583%	29.4445%	23.7226%	33.1823%	3.0875%	7
Overall Reliability	90.4029%	90.6125%	85.3460%	96.0019%	2.8959%	4
Mean Reliability	33.4613%	33.9202%	23.0496%	40.9770%	4.9411%	4
Overall Relative Gini Index	81.8367%	81.8367%	59.2736%	93.6474%	9.7768%	2
Mean Relative Gini Index	40.0315%	40.0315%	24.0968%	50.7366%	7.2331%	2
Overall Relative Lift Index	97.1617%	97.1617%	95.5694%	99.1890%	1.2340%	4
Mean Relative Lift Index	43.1548%	43.1548%	38.1467%	49.6626%	3.2139%	2

(*continued*)

Table 13. (*continued*)

Statistics

Measure	Overall	Mean	Min	Max	Standard Deviation	Best Network#
Overall ROC Index	90.9549%	90.9549%	79.6791%	96.8743%	4.8913%	2
Mean ROC Index	47.5302%	47.5302%	41.0937%	62.8956%	5.6208%	2
Overall Calibration Index	64.2029%	64.2029%	31.7433%	84.4683%	14.6900%	4
Mean Calibration Index	82.4701%	82.4701%	65.9170%	91.8425%	6.9163%	4
Overall Log-Loss	0.2474	0.2474	0.0961	0.6215	0.1500	4
Mean Binary Log-Loss	0.1123	0.1123	0.0481	0.2548	0.0597	4

4   Conclusions

Applications of Bayesian network have spread to a variety of fields. However, the most popular applications of Bayesian Network are about building expert systems in medical cases. In this field significant improvement of diagnostics can be accomplished.

To sum up, this work concerns treatment of cancer related complications. Based on BayesiaLab Networks, two expert systems of fever in neutropenia and fever in neutropenia with bacteraemia were created. For the case of paediatric patients suffering from cancer, these complications during treatment of chemotherapy lead to lethal cases. Thus, this problem required a concrete solution to identify risks of individual patients to develop FN or FN with bacteremia. Hence, we have proposed the usage of novel expert systems regarding the risks identification based on the previously collected data of FN and FN with bacteremia.

In the paper, data collected from 800 patients of ages from zero to 17 were considered, through several factors such as "observation period", "age group", "number of malignancies", "length of observation period", "chemotherapy intensity", "relapsed malignancy", "bone marrow involvement", "central venous access device", "earlier FN", "earlier FN with bacteremia", and "FN". Furthermore, two methods of learning were used. The first method was the unsupervised EQ algorithm, while the second method was the supervised Tree Augmented Naïve Bayes learning.

As a result, it became clear, that risk of fever in neutropenia can be identified with accuracy of 75.7%, while risk of development of fever in neutropenia with bacteremia is detected by the accuracy of 92.9%. Obtained results were supported by K-folds method, which demonstrated approximately the results and did not found weaknesses in the models.

The achieved accuracy is limited by the imbalanced distribution of the data. In future the present study will be extended and improved using a larger set of data. The number of patients with and without the fever in neutropenia should be increased. In addition, several other statistical tests will be applied to allow comparison and increase on the performance of the expert system.

References

1. Alexander, S.W., Wade, K.C., Hibberd, P.L., Parsons, S.K.: Evaluation of risk prediction criteria for episodes of febrile neutropenia in children with cancer. J. Pediatr. Hematol. Oncol. **24**, 38–42 (2002)
2. von Allmen, A.N., Zermatten, M.G., Leibundgut, K., Agyeman, P., Ammann, R.A.: Pediatric patients at risk for fever in chemotherapy-induced neutropenia in Bern, Switzerland, 1993–2012. Sci. Data **5** (2018). https://doi.org/10.1038/sdata.2018.38
3. Ammann, R.A.: Predicting adverse events in children with fever and chemotherapy-induced neutropenia: the prospective multicenter SPOG 2003 FN study. J. Clin. Oncol. **28**, 2008–2014 (2010)
4. Ammann, R.A., Aebi, C., Hirt, A., Ridolfi Lüthy, A.: Fever in neutropenia in children and adolescents: evolution over time of main characteristics in a single center, 1993–2001. Support. Care Cancer **12**, 826–832 (2004)
5. Ammann, R.A., Teuffel, O., Agyeman, P., Amport, N., Leibundgut, K.: The influence of different fever definitions on the rate of fever in neutropenia diagnosed in children with cancer. PLoS ONE **10**, e0117528 (2015)
6. Binz, P.: Different fever definitions and the rate of fever and neutropenia diagnosed in children with cancer: a retrospective two-center cohort study. Pediatr. Blood Cancer **60**, 799–805 (2013)
7. Bodey, G.P., Buckley, M., Sathe, Y.S., Freireich, E.J.: Quantitative relationships between circulating leukocytes and infection in patients with acute leukemia. Ann. Intern. Med. **64**, 328–340 (1966)
8. Boragina, A., Patel, H., Reiter, S., Dougherty, G.: Management of febrile neutropenia in pediatric oncology patients: a canadian survey. Pediatr. Blood Cancer **48**, 521–526 (2007)
9. Walsh, T.J., Finberg, R.W., Arndt, C., Hiemenz, J., Schwartz, C., Bodensteiner, D., Pappas, P., Seibel, N., Greenberg, R.N., Dummer, S., Schuster, M.: Empirical therapy in patients with persistent fever and neutropenia. New England J. Med. **340**(10), 764–771 (1999). https://doi.org/10.1056/NEJM199903113401004
10. Gafter-Gvili, A.: Antibiotic prophylaxis for bacterial infections in afebrile neutropenic patients following chemotherapy. Cochrane Database Syst. Rev. CD004386 (2012)
11. Hann, I., Viscoli, C., Paesmans, M., Gaya, H., Glauser, M.: A comparison of outcome from febrile neutropenic episodes in children compared with adults: results from four EORTC Studies. International Antimicrobial Therapy Cooperative Group (IATCG) of the European Organization for Research and Treatment of Cancer (EORTC). Br. J. Haematol. **99**, 580–588 (1997)
12. Hinds, P.S., Drew, D., Oakes, L.L., Fouladi, M., Spunt, S.L., Church, C., Furman, W.L.: End-of-life care preferences of pediatric patients with cancer. J. Clin. Oncol. **23**(36), 9146–9154 (2005). https://doi.org/10.1200/JCO.2005.10.538
13. Kern, W.V.: Oral versus intravenous empirical antimicrobial therapy for fever in patients with granulocytopenia who are receiving cancer chemotherapy. International Antimicrobial Therapy Cooperative Group of the European Organization for Research and Treatment of Cancer. N. Engl. J. Med. **341**, 312–318 (1999)
14. Lauritzen, S.L., Spiegelhalter, D.J.: Local computations with probabilities on graphical structures and their application to expert systems. J. Roy. Stat. Soc.: Ser. B (Methodol.) **50**(2), 157–224 (1988)
15. Lehrnbecher, T.: Guideline for the management of fever and neutropenia in children with cancer and hematopoietic stem cell transplantation recipients: 2017 update. J. Clin. Oncol. **35**, 2082–2094 (2017)

16. Lehrnbecher, T., Sung, L.: Anti-infective prophylaxis in pediatric patients with acute myeloid leukemia. Expert Rev. Hematol **7**, 819–830 (2014)
17. Macher, E.: Predicting the risk of severe bacterial infection in children with chemotherapy-induced febrile neutropenia. Pediatr. Blood Cancer **55**, 662–667 (2010)
18. Michel, G.: Incidence of childhood cancer in Switzerland: the Swiss childhood cancer registry. Pediatr. Blood Cancer **50**, 46–51 (2008)
19. Morgan, J.E., Cleminson, J., Atkin, K., Stewart, L.A., Phillips, R.S.: Systematic review of reduced therapy regimens for children with low risk febrile neutropenia. Support. Care Cancer **24**, 2651–2660 (2016)
20. Nimah, M.M., Bshesh, K., Callahan, J.D., Jacobs, B.R.: Infrared tympanic thermometry in comparison with other temperature measurement techniques in febrile children. Pediatr. Crit. Care. Med. **7**, 48–55 (2006)
21. Onisko, A., Druzdzel, M.J., Marshall Austin, R.: Application of Bayesian network modeling to pathology informatics. Diagn. Cytopathol. **47**(1), 41–47 (2019). https://doi.org/10.1002/dc.23993
22. Phillips, R.S.: Predicting microbiologically defined infection in febrile neutropenic episodes in children: global individual participant data multivariable meta-analysis. Brit. J. Cancer **114**, 623–630 (2016)
23. Phillips, R.S., Bhuller, K., Sung, L., Ammann, R.A.: Risk stratification in febrile neutropenic episodes in adolescent/young adult patients with cancer. Eur. J. Cancer **64**, 101–106 (2016)
24. Pizzo, P.A., Robichaud, K.J., Wesley, R., Commers, J.R.: Fever in the pediatric and young adult patient with cancer. A prospective study of 1001 episodes. Med. (Baltimore) **6**, 153–165 (1982)
25. ProceedingsVol2.Pdf: n.d. Accessed 28 Mar 2019. https://iris.unipa.it/retrieve/handle/10447/221666/444325/proceedingsVol2.pdf#page=27
26. Rackoff, W.R., Gonin, R., Robinson, C., Kreissman, S.G., Breitfeld, P.B.: Predicting the risk of bacteremia in children with fever and neutropenia. J. Clin. Oncol. **14**(3), 919–924 (1996). https://doi.org/10.1200/JCO.1996.14.3.919
27. Penna, R., Raphaele, G.C.-T., Comperat, E., Mozer, P., Léon, P., Varinot, J., Roupret, M., Bitker, M.-O., Lucidarme, O., Cussenot, O.: Apparent diffusion coefficient value is a strong predictor of unsuspected aggressiveness of prostate cancer before radical prostatectomy. World J. Urol. **34**(10), 1389–1395 (2016). https://doi.org/10.1007/s00345-016-1789-3
28. Robinson, P.D., Lehrnbecher, T., Phillips, R., Dupuis, L.L., Sung, L.: Strategies for empiric management of pediatric fever and neutropenia in patients with cancer and hematopoietic stem-cell transplantation recipients: a systematic review of randomized trials. J. Clin. Oncol. **34**, 2054–2060 (2016)
29. Schlapbach, L.J.: Serum levels of mannose-binding lectin and the risk of fever in neutropenia pediatric cancer patients. Pediatr. Blood Cancer **49**, 11–16 (2007)
30. Sedgwick, P.: Pearson's correlation coefficient. BMJ **345**, e4483 (2012). https://doi.org/10.1136/bmj.e4483
31. Seixas, A.A., Henclewood, D.A., Langford, A.T., McFarlane, S.I., Zizi, F., Jean-Louis, G.: Differential and combined effects of physical activity profiles and prohealth behaviors on diabetes prevalence among blacks and whites in the US population: a novel Bayesian belief network machine learning analysis. Research Article. J. Diab. Res. (2017). https://doi.org/10.1155/2017/5906034
32. Azizi, S., Dwayne, H., Stephen, W., Olajide, W., April, R., Gbenga, O., Girardin, J.-L.: Abstract WP171 long sleep is a stronger predictor of stroke than short sleep: comparative analysis of multiple linear regression model and Bayesian belief network model. Stroke **47** (suppl_1), AWP171 (2016). https://doi.org/10.1161/str.47.suppl_1.wp171

33. Singer, D.E., Mulley, A.G., Octo Barnett, G., Thibault, G.E., Morgan, M.M., Skinner, E.R.: The course of patients with suspected myocardial infarction - prediction of complications using a computer-based databank. In: Proceedings of the Annual Symposium on Computer Application in Medical Care, vol. 3, pp. 1590–1593, November 1980

34. Tramsen, L.: Lack of effectiveness of neutropenic diet and social restrictions as anti-infective measures in children with acute myeloid leukemia: an analysis of the AML-BFM 2004 trial. J. Clin. Oncol. **34**, 2776–2783 (2016)

35. USA, Bayesia. n.d.: BayesiaLab 8 - Bayesian Networks for Research, Analytics, and Reasoning. Accessed 18 Feb 2019. https://www.bayesialab.com

36. Walsh, T.J., Finberg, R.W., Arndt, C., Hiemenz, J., Schwartz, C., Bodensteiner, D., Pappas, P., et al.: Liposomal amphotericin B for empirical therapy in patients with persistent fever and neutropenia (1999)

37. Wicki, S.: Risk prediction of fever in neutropenia in children with cancer: a step towards individually tailored supportive therapy? Pediatr. Blood Cancer **51**, 778–783 (2008)

38. Bapin, Y., Zarikas, V.: Smart building's elevator with intelligent control algorithm based on bayesian networks. Int. J. Adv. Comput. Sci. Appl. (IJACSA) **10**(2), 16–24 (2019)

39. Zarikas, V.: Modeling decisions under uncertainty in adaptive user interfaces. Univ. Access Inf. Soc. **6**(1), 87–101 (2007)

40. Zarikas, V., Papageorgiou, E., Regner, P.: Bayesian network construction using a fuzzy rule based approach for medical decision support. Expert Syst. **32**(3), 344–369 (2015)

41. Zhang, Y., Guo, S.-L., Han, L.-N., Li, T.-L.: Application and exploration of big data mining in clinical medicine. Chin. Med. J. **129**(6), 731–738 (2016). https://doi.org/10.4103/0366-6999.178019

42. Knudsen, S.: Cancer Diagnostics with DNA Microarrays. Wiley-Liss, Hoboken (2006)

Preventing Overfitting by Training Derivatives

V. I. Avrutskiy[(✉)]

Moscow Institute of Physics and Technology,
Institutsky lane 9, Dolgoprudny, Moscow Region 141700, Russia
avrutsky@phystech.edu

Abstract. A seamless data-driven method of eliminating overfitting is proposed. The technique is based on an extended cost function which includes the deviation of network derivatives from the target function derivatives up to the 4th order. When gradient descent is run for this cost function overfitting becomes nearly non existent. For the most common applications of neural networks high order derivatives of a target are difficult to obtain, so model cases are considered: training a network to approximate an analytical expression inside 2D and 5D domains and solving a Poisson equation inside a 2D circle. To investigate overfitting, fully connected perceptrons of different sizes are trained using sets of points with various density until the cost is stabilized. The extended cost allows to train a network with $5 \cdot 10^6$ weights to represent a 2D expression inside $[-1, 1]^2$ square using only 10 training points with the test set error on average being only 1.5 times higher than the train error. Using the classical cost in comparable conditions results in the test set error being $2 \cdot 10^4$ times higher than the train error. In contrast with the common techniques of combating overfitting like regularization or dropout the proposed method is entirely data-driven therefore it introduces no tunable parameters. It also does not restrict weights in any way unlike regularization that can hinder the quality of approximation if its parameters are poorly chosen. Using the extended cost also increases the overall precision by one order of magnitude.

Keywords: Neural networks · Overfitting · Partial differential equations · High order derivatives · Function approximation

1 Introduction

Overfitting is a common fault of neural networks. It occurs in many different applications [11,15,16,27]. Consider the case of supervised learning: a network $\mathcal{N}$ is trained to represent a vector function $F : X \to Y$ known only by a finite set of its arguments $\mathbf{x}_a \subset X$ and the corresponding outputs $\mathbf{y}_a \subset Y$, $1 \le a \le n$. It is a common practice to subdivide the total number of n vector pairs $\mathbf{x}_a$, $\mathbf{y}_a$ into at least two sets: $n - m$ training pairs $\mathbf{x}_b$, $\mathbf{y}_b$, $1 \le b \le n - m$ and m test pairs $\mathbf{x}_c$, $\mathbf{y}_c$, $n - m < c \le n$. Whatever algorithm is used for minimization, it

© Springer Nature Switzerland AG 2020
K. Arai et al. (Eds.): FTC 2019, AISC 1069, pp. 144–163, 2020.
https://doi.org/10.1007/978-3-030-32520-6_12

is not run for pairs $\mathbf{x}_c$, $\mathbf{y}_c$. Instead they are used later to observe the network's performance on previously unknown data. Overfitting can be broadly described as a different behavior of the cost function $E = ||\mathcal{N}(\mathbf{x}) - \mathbf{y}||$ for $\mathbf{x}_b$, $\mathbf{y}_b$ and $\mathbf{x}_c$, $\mathbf{y}_c$. For example, its average values on those two sets can be orders of magnitude away. A ratio $\overline{E_c}/\overline{E_b}$ between the average cost for the test and train sets can show how much attention the training procedure pays to the actual input rather than to the function behind it. However, this relation between averages is not important all by itself, since the absolute values of E_c on the test set $\mathbf{x}_c$, $\mathbf{y}_c$ are the ultimate goal of the training procedure. For example some techniques for avoiding overfitting can equalize the cost on the test and train sets but produce a lower absolute precision due to the impairment of approximation ability.

There are numerous ways to avoid overfitting. It is possible to directly address negative effects for E_c by introducing another subset of patterns called cross validation set. Its elements are not fed to minimization algorithm as well, but during the training the cost function for them is observed. When it starts to increase, while the cost on the training set continues to lower, one can draw a conclusion that the negative effects of overfitting started to overpower positive effects of the training itself. If one chooses to stop, the decision effectively puts a lower bond on the approximation error. One might also say that before E_c started to increase, any presence of overfitting was not practically significant. Many ways to handle such situations were developed [6,17,22]. For example, the network can be pruned [20], or some special statistical rule for updating weights can be used [8,25,29].

Generally, the more approximation abilities a neural network has the more it is prone to overfitting so methods for avoiding it often force weights into a smaller subspace. Such approaches usually introduce an additional tunable parameter which must be chosen carefully not to restrict the network too much (L2 regularization) or not to affect the training too much (dropout). The presence of extra parameters is not very desirable, and many papers were dedicated to creating algorithms that could determine them on the fly [2,3,9,19,26]. From this point of view an absence of a parameter is more beneficial than an auto-tuning procedure. In this paper the overfitting is avoided by adding extra terms to the cost function. Along with a norm $||\mathcal{N}(\mathbf{x}) - \mathbf{y}||$ the cost function includes terms like $||\frac{\partial}{\partial x_i}\mathcal{N}(\mathbf{x}) - \frac{\partial}{\partial x_i}\mathbf{y}||$ where x_i is a certain component of an input vector $\mathbf{x}$. Derivatives of the network $\mathcal{N}$ with respect to x_i can be obtained via automatic differentiation and derivatives of the target mapping $F : X \rightarrow Y$ can be calculated using a finite difference stencil. The stencil requires to evaluate the target function multiple times for different but close arguments [28] so the derivative can be approximated by their linear combination. The information that prevents overfitting is derived directly from the target itself and does not depend on the internal structure of the network.

Suppose a random noise is present and the network is overtrained, i.e. the algorithm was run for long enough to start paying too much attention to the training set. This almost certainly leads to the worsening of performance on the test set, since the noise on the training set is generally not correlated with the noise on the test set. However, if the data is noiseless, then even having an extremely low error on the training points can bring no adverse effects to the cost on the test set. This paper is using various derivatives of the target function, therefore, it is not possible to include the noise directly, since even the slightest variations of neighboring points can have profound consequences for numerically calculated derivatives. Vice versa, if derivatives up to say the 4th order are known for one point, then the nearby points can be calculated via Taylor approximation virtually without any noise. Therefore, in the scope of this paper the overfitting is defined not by the worsening of performance on the test set, which was not observed, but rather by the ability of a network to pay more attention to the training set. An absence of overfitting can then be expressed as an inability to distinguish between the statistical properties of the cost function on the test and training sets. Since the real world data almost always include some noise, initial denoising would be required to apply the presented procedure.

The material is structured as follows: Sect. 2 briefly describes prior papers where derivatives of neural networks with respect to their inputs were used in the training process. Section 3 formulates the method of avoiding overfitting using such derivatives. Section 4 describes the results of this approach on simple model cases. Section 5 concludes the paper and contains remarks on limitations and the possible scope of a future work.

2 Background

The algorithms of supervised training with cost functions that include derivatives were described in a number of publications [5,13]. The most notable are double backpropagation [4] and tangent propagation [24]. Both consider an image classification problem and their cost functions in addition to the squared difference between the output and the target classes include the first order derivatives of that difference. The double backpropagation calculates those derivatives with respect to the values of individual input pixels. Such a derivative can be viewed as the slope of network's reaction to the subtle distortion of a pixel. The slope of a target for such distortion is zero since it does not change the class. Thus, the training process tries to minimize the derivatives of the outputs with respect to inputs for each meaningful pattern. The tangent propagation is more elaborate and calculates derivatives along special directions of the image space. For each image I a direction is another pseudo-image J that can be multiplied by a small scalar α and added to I so that $I + \alpha J$ simulates an infinitesimal geometrical distortion of I. For example it can be rotation through a small angle. Such distortions do not change class either so derivative of the output with respect to α is trained to be zero. Note that such derivatives depend on both - an initial image and the form of distortion. Those two methods are rarely used in practice which might be due to their modest benefits, an increase in training time and a tricky process of calculating derivatives.

This study does not consider complicated mappings like image classifiers and the examples will be limited to somewhat artificial: target functions are smooth and their derivatives are easily computable. This allows to determine the effect of high order derivatives on the training process. It allows to highlight the benefits and to establish a technique that might be generalized for more complex target functions later. Since any continuous function defined on a compact set can be represented by a smooth function with arbitrary precision, the approach is justified.

Another case when derivatives appear in a cost function is when neural networks are used for solving differential equations [12,14,18,23]. The established approach is to construct a neural network that maps each point of the space of independent variables to the value of solution. The technique can be best described by example. Consider a boundary value problem inside a region Γ for a function $u(x, y)$ written as

$$U(x, y, u, u_x, u_y, u_{xx}, ...) = 0$$

$$u|_{\partial\Gamma} = f$$

The first step is to make a substitution using the relation

$$u(x, y) = v(x, y) \cdot \phi(x, y) + f \tag{1}$$

where $\phi(x, y)$ is a known function that vanishes on a boundary $\partial\Gamma$ and is non zero inside Γ. Function $v(x, y)$ is to be constructed by a neural network [14]. The term f is now a smooth continuation of the boundary condition into Γ that is supposed to exist. The equation is written as

$$V(x, y, v, v_x, v_y, v_{xx}, ...) = 0$$

and provided the solution is smooth enough the boundary condition is simply

$$v|_{\partial\Gamma} < \infty$$

A network with two inputs, one output, and a suitable number of hidden layers and neurons is then created. Its weights are initialized and trained to minimize a cost function E - a measure of a local cost function $e = V^2$ on Γ calculated using a set of grid points. If E reaches small enough values, $v(x, y) \cdot \phi(x, y) + f$ can be considered as an approximate numerical solution. As long as the procedure converges to a finite v, the boundary condition is satisfied. An extension of the classical cost is required due to the presence of terms like v_{xx}, which are derivatives of the neural network's output with respect to the input. Various types of differential equations can be solved using this approach [13,30]. The smaller effect of overfitting is the fewer points are required to capture the behavior of the function e on Γ for a proper discretization of E and, therefore, to solve the equation in the entire region.

Author's recent study [1] showed significant improvements from using high order derivatives in the learning process. The paper was mostly focused on precision and it found that more capable networks were required to enhance the approximation quality but at the same time the number of training points could be reduced. This suggests that some anti-overfitting capabilities are present. They are investigated in this paper.

3   Method

3.1   Direct Function Approximation

A neural network is trained to represent a scalar function $f : \mathbb{R}^n \to \mathbb{R}^1$ defined inside a region $\Gamma \in \mathbb{R}^n$. Components of an input vector $\mathbf{x}$ are denoted by x_i. The network's output is denoted by $\mathcal{N} = \mathcal{N}(\mathbf{x}) = \mathcal{N}(x_1, ..., x_n)$. To gather the cost function the following terms with an index k are used:

$$e_k = \sum_i \left[\frac{\partial^k}{\partial x_i^k} \mathcal{N} - \frac{\partial^k}{\partial x_i^k} f \right]^2 \tag{2}$$

Index i can run through all or some of the input components as explained in the results section. If $k = 0$ all terms are the same and the sum can be omitted thus the expression is a regular cost $(\mathcal{N} - f)^2$. The next order $k = 1$ is used in the double backpropagation. The cost of the order s for one pattern (i.e. local) contains all derivatives up to the sth and is defined as

$$e_s^{\text{local}} = \sum_{k=0}^{s} e_k \tag{3}$$

If the network is trained on a set of M patterns, the total cost of the order s is

$$E_s = \frac{1}{M} \sum_{\alpha=1}^{M} e_s^{\text{local}}(\mathbf{x}_\alpha)$$

In the results section the cost with $s = 0$ will be compared to the one with $s = 4$. Since the extended cost uses more information about the function, the number of patterns for $s = 0$ and $s = 4$ should be different. Namely, it should equalize the total amount of data as will be explained in the results section.

3.2   Solving Differential Equation

A boundary value problem inside two-dimensional region Γ is considered:

$$U(x_1, x_2, u, u_{x_1}, u_{x_1 x_1}, ...) = 0$$

$$u|_{\partial \Gamma} = f,$$

After function substitution is made according to (1) the equation is written for v

$$V(x_1, x_2, v, v_{x_1}, v_{x_1 x_1} ...) = 0$$

$$v|_{\partial \Gamma} < \infty$$

which is to be represented by a neural network $v = \mathcal{N}$. The boundary condition is satisfied automatically. Similarly to (2) the cost terms with index k are written as

$$e_k = \sum_{i=1}^{2} \left[\frac{\partial^k}{\partial x_i^k} V \right]^2$$

The sum can be omitted for $k = 0$. The classical method of solving PDE involves only $e_0 = V^2$. The extended local cost function of the order s is

$$e_s^{\text{local}} = \sum_{k=0}^{s} e_k \tag{4}$$

If a grid is close to regular, the total cost function which is a measure μ of e_s^{local} on Γ can be discretized on its M points as

$$E_s = \underset{\mathbf{x} \in \Gamma}{\mu} \left(e_s^{\text{local}} \right) \simeq \frac{1}{M} \sum_{\alpha=1}^{M} e_s^{\text{local}}(\mathbf{x}_\alpha)$$

In the results section the cost with $s = 0$ will be compared to the one with $s = 4$. The number of grid points for $s = 0$ and $s = 4$ will equalize the total number of arithmetic operations per epoch.

4   Results

Analytical functions are approximated inside 2D box and 5D unit sphere. Boundary value problem is solved inside a 2D unit circle. Regions are denoted by Γ and their boundaries by $\partial \Gamma$. For all cases the cost is minimized for points of nearly regular grids. They are generated in a similar manner and comprised of two parts - internal and surface. The internal part is a Cartesian grid, which is generated in three steps. At first, points with spacing λ are generated inside a sufficient volume. Two random vectors in the input space are then chosen, and the grid is rotated from one to another. Finally, it is shifted along each direction by a random value from an interval $[-\frac{\lambda}{4}, \frac{\lambda}{4}]$, and after that all points outside of Γ are excluded. The surface part contains equidistant points from $\partial \Gamma$ with distance $\tau = \lambda$ unless otherwise stated. The training is performed for a number of various spacings. The initial value of λ is rather small so that the training set is dense enough to prevent overfitting for nearly all network configurations and both orders s. As λ gradually increases networks with greater capacity began to overfit for $s = 0$, up until only few points left and all orders and networks become severely overfitted. This process was designed to observe as the density of grid points becomes low enough to produce overfitting, and to determine the difference between $s = 0$ and $s = 4$.

Results for $s = 0$ and $s = 4$ are compared against each other not via the number of grid points, but rather via their effective number N. The value N coincides with the number of grid points M when only the values of function are used in the cost. It is equal to $(p + 1)M$, if p extra derivatives in each point are included. This can be justified since any derivative of the target can be found using a finite difference stencil which is a linear combination of function values in nearby points. A stencil for the first derivative requires two points, for the second one three, and so on, thus the factor $(p + 1)$. Therefore N represents the number of times the target function needs to be evaluated in order to construct the training set. Plotting the results against N also nearly equalizes the total number of arithmetic operations per one epoch provided networks have the same architecture. For each training grid, order s and network the process is repeated 5 times, and the average values are presented. The space above and below each curve is filled according to the maximum and minimum values obtained in those 5 runs.

Training is based on RProp [21] procedure with parameters $\eta_+ = 1.2$, $\eta_- = 0.5$. Weights are forced to stay in $[-20, 20]$ interval, no min/max bonds for steps are imposed. Initial step Δ_0 is set to $2 \cdot 10^{-4}$. Due to the absence of a minimal step a lot of them are reduced to zero after certain number of epochs. To tackle that, after each 1000 epochs all zero steps are restored to 10^{-6}. The original minimum step value suggested by [21] leads to oscillations. Single precision arithmetic is used. Weights are initialized [7] with random values from range $\pm 2/\sqrt{s}$, where s is a number of senders. Thresholds are initialized in range ± 0.1. All layers are fully connected and have non linear sigmoid activation function:

$$\sigma(x) = \frac{1}{1 + \exp(-x)}$$

except for the input and output layers which are linear. To make sure that backpropagation is run long enough to produce overfitting, the following rule for stopping is implemented: during both classical and extended training modes, the root mean of e_0 for grid points $\sqrt{\langle e_0 \rangle_{\text{train}}}$ is tracked. After its best value has changed by less than 10% for the last 1000 epochs, or when the 10000th epoch is reached, the training is stopped. The weights from the best iteration are saved, and for them $\sqrt{\langle e_0 \rangle_{\text{test}}}$ is calculated on a much finer grid.

4.1   Direct Function Approximation

2D Function. Networks are trained to represent the following expression inside $[-1, 1]^2$ square:

$$f = \frac{1}{2}\left(-x_1^2 - x_2^2 + 1\right) + 2\tanh\left(x_1 \cdot \sin\frac{x_2}{2}\right) - \sin x_1 \cdot \cos x_2$$

A series of grids with various point densities is generated. It starts with a grid containing 804 points ($\lambda = 0.073$). For each next grid λ is increased, so that the total number of points would be approximately 10% less. The last grid has 5 points, 4 on the boundary and 1 near the middle ($\lambda = 1.45$). In each point derivatives from the 1st up to the 4th order are calculated with respect to x_1 and x_2. The extended training is using $i = 1, 2$ in expression (2) and $s = 4$ in (3). Therefore, the total number of parameters for a grid of M points is $N = 9M$. For the classical training $s = 0$ and $N = M$. To conduct the tests within the same interval for N the extended training ($s = 4$) is run on grids from 88 to 5 points (λ from 0.24 to 1.45), and the classical one ($s = 0$) is run on grids from 804 to 43 points (λ from 0.073 to 0.37). Calculating the target function $9M$ times is exactly enough to evaluate derivatives up to the 4th order with respect to x_1

and x_2 in M points using finite difference stencils. To investigate how network's architecture affects the overfitting three different layer configurations are used (asterisks denote linear layers):

$$2^*, 64, 64, 64, 64, 64, 64, 1^*$$

$$2^*, 512, 512, 512, 512, 512, 512, 1^*$$

$$2^*, 1024, 1024, 1024, 1024, 1024, 1024, 1^*$$

They comprise $2 \cdot 10^4$, $1 \cdot 10^6$ and $5 \cdot 10^6$ weights respectively and are referred to on figures by those numbers. After the training is finished, the performance of each network is tested on a grid with 3400 points ($\lambda = 0.035$). Base 10 logarithm of the ratio $\sqrt{\langle e_0 \rangle_{\text{test}}} / \sqrt{\langle e_0 \rangle_{\text{train}}}$ is plotted in Fig. 1. The network with the most number of weights trained with $s = 4$ is compared with three networks trained with classical cost function ($s = 0$). The ratio itself and its variance, which is related to amount of filling above and below each curve, are much smaller for the extended method. Figure 2 shows the overfitting ratio for networks trained with the extended cost ($s = 4$). Plots for different architectures are very close to each other, unlike when the same networks are trained with the classical cost function. From about $N = 200$ and higher, which corresponds to 23 grid points, all curves lie mostly below the zero, which means that the root mean square error calculated on the dense test grid is a bit smaller than the one calculated on the training grid. In other words, all local maximums of e_0 lie very close to the training grid points. Figure 3 depicts the final root mean square error on the test set. It represents an inevitably more important aspect in a sense that the previous plots alone do not necessary mean that the training was successful. The precision of the proposed method is about one order of magnitude higher than the one obtained with the classical cost. Only one curve is shown for $s = 4$ since plots for other networks are very close to it.

5D Function. Networks are trained to represent the following expression inside a 5D unit sphere $\Gamma : r \leq 1$

$$f = \frac{1}{2}(-x_2^2 - x_5^2 + 1) + 2 \tanh \left(x_3 \cdot \sin \frac{x_4}{2} \right) - \sin x_1 \cos x_2$$

A similar series of grids is generated. Unlike 2D case, $\tau = 1.6\lambda$. The first grid has 1579 points ($\lambda = 0.336$). Each next grid has about 10% points less. The last grid has 11 points ($\lambda = 1.1$), 10 of them are on the boundary and 1 near the center. In each point derivatives from the 1st to the 4th order are calculated with respect to two randomly chosen variables x_p and x_q out of five possible. The extended training has $i = p, q$ in the expression (2) and $s = 4$ in (3), therefore, the total number of parameters for a grid of M points is $N = 9M$. To conduct the tests within the same interval for N the extended training is run on grids from 166 to 11 points (λ from 0.55 to 1.1) and the classical training on grids from 1579 to 98 points (λ from 0.336 to 0.62). Calculating the target function $9M$ times is exactly enough to evaluate derivatives with respect to two mentioned variables.

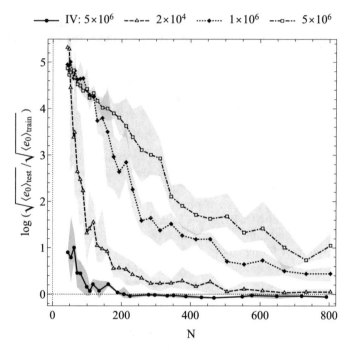

Fig. 1. Approximating 2D function. Base 10 logarithm of the ratio between root mean square errors on the test and train sets is plotted against the effective number of training points. The solid line represents the network with $5 \cdot 10^6$ weights trained with $s = 4$ (marked by IV), and the rest of the curves are for the networks trained by the classical cost ($s = 0$). As the effective number of points decreases, the overfitting increases steadily for classical training.

The test grid has 11000 points ($\lambda = 0.15$). Network configurations are the same as for 2D case, except the input layers have 5 neurons instead of 2.

Similarly to the previous case, Fig. 4 shows the overfitting ratio for the biggest network trained with the extended cost $s = 4$ and all networks trained with the regular cost. Unlike 2D case even on the most dense grids the overfitting is very strong for larger networks trained without derivatives. Figure 5 shows the same ratio for three architectures trained with the extended cost $s = 4$. It is small, barely different and has low variance between training attempts. Finally, Fig. 6 compares the performance of networks on the test set. It shows that the extended cost not only produces very close test and train errors, but also gives a better approximation. Different curves for the extended cost are not shown since they are very close to each other. One can notice that among the results for the classical cost, the $1 \cdot 10^4$ network (the line marked with triangles) shows both: a lower overfitting in Fig. 4 and a better test cost in Fig. 6. This could make the network a preferable choice for the task. However, when derivatives are included, any architecture demonstrates much smaller overfitting ratio and nearly one extra order in final precision. The only advantage left for smaller

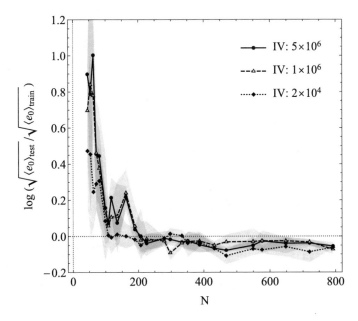

Fig. 2. Approximating 2D function. Base 10 logarithm of the ratio between root mean square errors on the test and train sets is plotted against the effective number of training points. Comparison between the networks with different capacities trained with the extended cost ($s = 4$ or IV). The solid line is the same as in Fig. 1.

network is training time. Despite this being a 5D function, for each input vector it was enough to train derivatives with respect to two variables, provided they are randomly chosen from point to point.

4.2 Solving Differential Equation

Poisson equation inside a 2D unit circle $\Gamma : x_1^2 + x_2^2 \leq 1$ with vanishing boundary condition is considered.

$$\Delta u = g$$
$$u|_{\partial \Gamma} = 0 \tag{5}$$

According to (1) the following substitution can be made:

$$u = v(x_1, x_2) \cdot (1 - x_1^2 - x_2^2)$$

After which v can be found using a neural network. To make results comparable to the previous case of function approximation, the analytical solution is chosen as

$$u_a = f(x_1, x_2) \cdot (1 - x_1^2 - x_2^2)$$

where $f(x_1, x_2)$ is the expression used for 2D approximation. The function u_a is substituted into (5) to calculate the source g that would produce it. In the

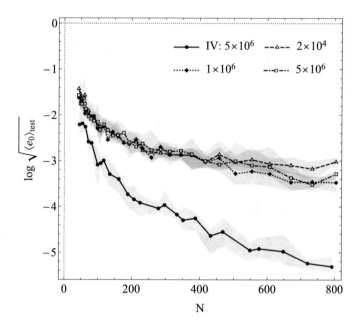

Fig. 3. Approximation of 2D function. Base 10 logarithm of root mean square error on the test set. The solid line is for the extended cost, which utilizes derivatives up to the IV order ($s = 4$).

process of minimizing the residual of the equation the neural network would have to fit exactly the same function as in the case of 2D approximation. The distinction from the direct function approximation is that the cost e_0 is not a squared difference between the output and target, but is the squared residual of the equation. A series of grids with various density is generated: the first one has 704 points ($\lambda = 0.07$) and the last one has 3 points ($\lambda = 1.62$). The classical training is using $s = 0$ in (4), therefore, in addition to values themselves, 4 derivatives of v encountered in the residual V must be calculated in each point:

$$\frac{\partial}{\partial x_1}, \frac{\partial}{\partial x_2}, \frac{\partial^2}{\partial x_1^2}, \frac{\partial^2}{\partial x_2^2}$$

The extended training casts 8 extra operators on V

$$\frac{\partial}{\partial x_i}, \frac{\partial^2}{\partial x_i^2}, \frac{\partial^3}{\partial x_i^3}, \frac{\partial^4}{\partial x_i^4}$$

where $i = 1, 2$. Due to overlapping this produces only 24 different derivatives of v. Therefore, an equivalent classical grid for M points in this mode is $N = 5M$. To conduct the tests within the same interval for N the extended training ($s = 4$) is run on grids from 139 to 3 points (λ from 0.16 to 1.62), and classical one is run on grids from 704 to 16 points (λ from 0.07 to 0.6). The test grid has 3000 points ($\lambda = 0.033$). Neural networks have the same configurations as for 2D function

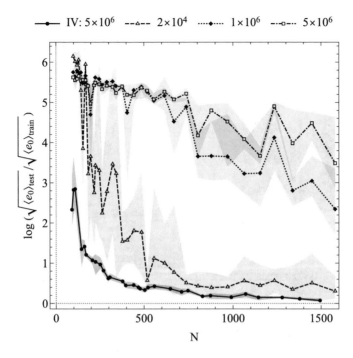

Fig. 4. Approximation of 5D function. Base 10 logarithm of the ratio between root mean square errors on the test and train sets is plotted against the effective number of training points. Solid line is for the extended cost function ($s = 4$, marked IV), and the rest are for regular one ($s = 0$).

approximation. Figure 7 compares overfitting ratio between one network trained using the extended cost and all three trained with the classical cost. Figure 8 compares the performance of the same networks on the test set. Results are similar to those in Figs. 1 and 3. Overfitting ratios for different architectures trained with the extended cost are nearly indistinguishable and are not shown. The extended training is more than one order ahead of the classical one.

A particular detail for boundary value problem is worth mentioning. As the grid spacing λ increases, the last "accurate" solution ($\max |u - u_a| < 3 \cdot 10^{-5}$) is obtained by the classical training on a grid with $\lambda = 0.09$ (441 points) and by the extended one on a grid with $\lambda = 0.29$ (52 points). The last "meaningful" solution ($\max |u - u_a| < 1 \cdot 10^{-3}$) is obtained by the classical training on a grid with $\lambda = 0.17$ (139 points) and by the extended one on a grid with $\lambda = 0.68$ (12 points). Those results are very close to what one can expect from the 4th and 6th order finite difference schemes [28]. Namely, the solution u is based on the equation for v, in which terms $4x_i \partial v / \partial x_i$, $i = 1, 2$ would be the biggest source of the stencil error ω. For 2D Poisson equation, the value of ω is very close to the maximum deviation from the exact solution. The leading term of the error for the first order derivative calculated with the 4th order central stencil

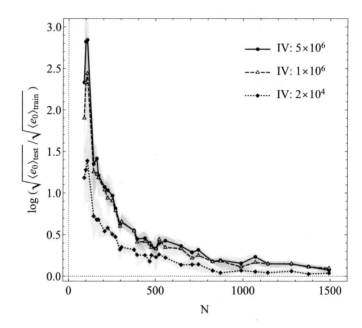

Fig. 5. Approximation of 5D function. Base 10 logarithm of the ratio between root mean square errors on the test and the train sets is plotted against the effective number of training points. Comparison between networks with different capacities trained with the extended cost ($s = 4$), the solid line is the same as in Fig. 4.

is $\lambda^4/30 \cdot \partial^5 v/\partial x_i^5$ and with the 6th order stencil is $\lambda^6/140 \cdot \partial^7 v/\partial x_i^7$. After substituting those expressions into the error source and noting that all derivatives of v are of the order of 1, the 4th order stencil has $\omega = 1 \cdot 10^{-5}$ for $\lambda = 0.09$ and $\omega = 1 \cdot 10^{-4}$ for $\lambda = 0.17$. The 6th order stencil has $\omega = 1.6 \cdot 10^{-5}$ for $\lambda = 0.29$ and $\omega = 3 \cdot 10^{-3}$ for $\lambda = 0.68$. However, the 6th order stencil requires at least 7 points in each direction, therefore, it can never be used on a grid with 12 points in total. On the 52-point grid its implementation is possible, but unlikely that precise, since in a lot of points derivatives will have to be calculated using not central but forward or backward stencil, which approximation error ω is about 20 times higher. This means that neural networks can use grid points more effectively than finite differences.

4.3   Elimination of a Few Trivial Explanations

The simplest explanation of lower overfitting for the extended cost could be an early stopping. To summarize the data the lower 3-quantiles for the number of epochs were calculated for each problem, training mode, and network size. For 2D classical training they are 7000, 6000 and 4000, and for the extended one 5000, 4000 and 4000 for $2 \cdot 10^4$, $1 \cdot 10^6$ and $5 \cdot 10^6$ networks respectively. For 5D case they are 6000, 4000 and 3000 for classical and 6000 for all networks for extended mode. Finally the boundary value problem has got 5000, 3000 and 2000

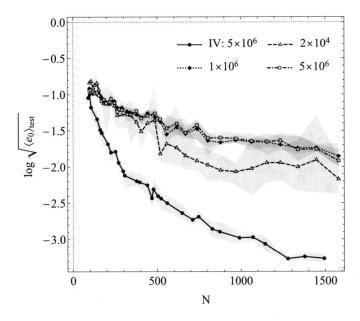

Fig. 6. Approximation of 5D function. Base 10 logarithm of root mean square error on the test set. Solid line is for the extended cost ($s = 4$), which utilizes derivatives up to the IV order.

for the classical and 4000 for all networks for the extended cost. The difference is not quite significant. It was verified that even if the extended training had the same number of epochs as the classical one, it would not bring any noticeable changes to results.

Another possibility arises from a specific property of RProp. If the landscape of a cost function has a lot of oscillations, then steps for many weights will be reduced to zero quite quickly, and the procedure will not go along those directions. If the inclusion of high order derivatives creates such "ripples", the training could effectively be constrained to a network of much smaller capacity, therefore, it can exhibit a lower overfitting. To investigate this, the following measurement was made: the amount of non zero steps δ was observed for $2 \cdot 10^4$ network during the classical and the 4th order training for 5D function approximation with $N = 449$. In both cases the procedure was run for 6000 epochs. It resulted in $\sqrt{\langle e_0 \rangle_{\text{train}}} = 1 \cdot 10^{-4}$ and $\sqrt{\langle e_0 \rangle_{\text{test}}} = 3 \cdot 10^{-2}$ for the classical and $\sqrt{\langle e_0 \rangle_{\text{train}}} = 2 \cdot 10^{-3}$, $\sqrt{\langle e_0 \rangle_{\text{test}}} = 4 \cdot 10^{-3}$ for the extended training. In both cases $\delta \sim 300$ except for small intervals of epochs close to the beginning or resetting of steps. The ratio $\delta_{\text{extended}}/\delta_{\text{classical}}$ was calculated with the following results: minimum 0.45, the lower 3-quantile 0.8, and median 0.82. The value δ is relatively small, so another measurement was made: an average probability of a weight to be changed within 10 epochs is estimated as 11% for the classical and 14% for the extended training. It does not seem possible that the overfitting

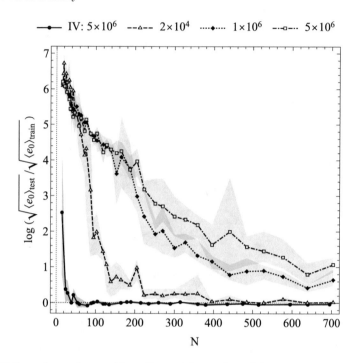

Fig. 7. 2D Boundary value problem. Base 10 logarithm of the ratio between root mean square of residuals on the test and the train grids is plotted against the equivalent number of grid points. The solid line is for the extended cost function ($s = 4$, marked IV) and the rest are for $s = 0$.

ratios of 300 and 2 are due to the δ being 20% less. The inclusion of an additional derivatives does not restrict training to a smaller subset of weights.

Another simple explanation of overfitting ratio can be obtained as follows: consider a 1-dimensional function f approximated by a network $\mathcal{N}$ trained on $n \gg 1$ points uniformly distributed on $[0, 1]$. The cost e on the test set can be estimated as a difference between network's value $\mathcal{N}$ and the target f in the middle of an interval between two neighboring training points a and b. Its length is $1/n$, and from Taylor expansion in a one can write

$$e_{a+\frac{1}{2n}} \sim [\mathcal{N} - f]_{a+\frac{1}{2n}} \simeq \left[\mathcal{N} - f + (\mathcal{N} - f)'\frac{1}{2n} \right]_a$$

After dividing this expression by $\mathcal{N} - f$ one obtains the overfitting ratio between e on train and test sets

$$\frac{e_{\text{test}}}{e_{\text{train}}} \sim 1 + \frac{(\mathcal{N} - f)'}{\mathcal{N} - f}\frac{1}{2n} \tag{6}$$

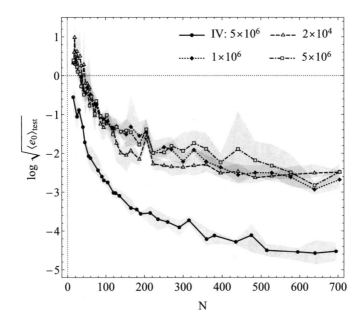

Fig. 8. 2D Boundary value problem. Base 10 logarithm of root mean square residual on the test set plotted against the equivalent number of grid points. The solid line is for the extended cost ($s = 4$), which utilizes derivatives up to the IV order.

If one chooses to split n points into two parts, distributes half of them uniformly and then uses the rest to calculate derivatives of the target in each point using a finite difference stencil, the error between points and the ratio will be written as

$$e_{a+\frac{1}{n}} \simeq \left[\mathcal{N} - f + (\mathcal{N} - f)' \frac{1}{n} \right]_a$$

$$\frac{e_{\text{test}}}{e_{\text{train}}} \sim 1 + \frac{(\mathcal{N} - f)'}{\mathcal{N} - f} \frac{1}{n} \qquad (7)$$

The term $(\mathcal{N} - f)$ is being minimized by the classical cost and its derivative $(\mathcal{N} - f)'$ is what is included in the extended cost with $s = 1$. From numerical experiments it was found that if a neural network is trained with the classical cost and shows the train set precision ε, the first derivative precision on this set is about 10ε (this factor also depends on the network size). For the extended cost, the values of precision for different trained derivatives are approximately the same ϵ, and the first non trained derivative is similarly 10 times less accurate. Therefore, one can estimate the test to train ratio (6)

$$\frac{e_{\text{test}}}{e_{\text{train}}} \sim 1 + \frac{10\varepsilon}{\varepsilon} \frac{1}{2n} = 1 + \frac{5}{n}$$

and the ratio (7)

$$\frac{e_{\text{test}}}{e_{\text{train}}} \sim 1 + \frac{\epsilon}{\epsilon} \frac{1}{n} = 1 + \frac{1}{n}.$$

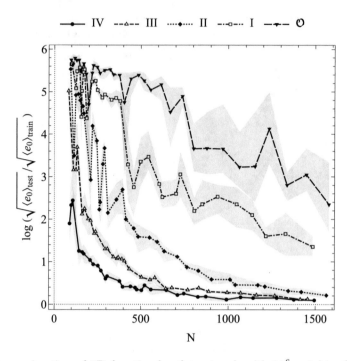

Fig. 9. Approximation of 5D function by the network with 10^6 weights. Overfitting ratio lowers as the order of included derivatives increases. Horizontal axis is the equivalent number of grid points. Curves are marked according to the maximum order of trained derivatives.

For rather small n used in this paper, the difference between logarithms of those expressions is about 0.2. The same analysis with high order Taylor expansion applied to 2D and 5D cases can more or less explain the difference between the curves for the classical and extended training shown in Fig. 9, provided the ratios $(\mathcal{N} - f)^{\{k\}}/(\mathcal{N} - f)$ are obtained from experiment. Thus, the overfitting ratio can be seen as a consequence of two reasons: the extended training being simply a higher order approximation and the existence of synergy between derivatives, which happens when inclusion of higher order terms enhances precision of the low order ones. However, this description is by no means complete. Note the solid line in Fig. 7: from the initial number of 139 grid points down to 11 the logarithm of the overfitting ratio for the network with $5 \cdot 10^6$ weights does not exceed 0.03. Meanwhile, the logarithm of the test precision $\log \sqrt{\langle e_0 \rangle_{\text{test}}}$ represented by the solid line in Fig. 8 increases from -4.5 to -2.0. It means that a network with a few millions of connections have not established them in a way that would approximate the target e_0 in 11 training points more precisely than anywhere else in Γ. A similar situation can be spotted in 2D and 5D cases.

5   Conclusion

Unexpectedly low overfitting was observed when the derivatives of target up to the 4th order were used in the cost function. The cost values on the train and test sets were very similar whether the neural network approximated an analytical expression or solved a partial differential equation. Increasing the capacity of a network from thousands up to millions of weights barely affected the results. Values of the test and train precision were less than 8% different for a network with $5 \cdot 10^6$ weights up until only 23 points left for 2D function approximated on $[-1, 1]^2$ square and 11 points left for 2D Poisson equation solved inside a unit circle. Whereas approximating an analytical expression can be regarded as highly artificial example, solving partial differential equations on a very sparse grids can have more practical applications, since the sparsity exceeds capabilities of finite difference methods. The inclusion target derivatives in the cost functions creates a seamless training process that is not prone to overfitting, does not forces weights into a smaller subspace, does not include any tunable parameters. It is also about one order of magnitude more precise than the regular training provided they both used the same amount of information about the target function. The overfitting itself as seen in Figs. 1, 4 and 7 can successfully be avoided by many techniques provided their hyperparameters are properly tuned. In the presented method however the absence of overfitting was also accompanied with a persistent increase of train set precision as seen in Figs. 3, 6 and 8 which did not occur for common regularization methods.

Generalizing this technique for complicated functions like pattern classifiers is a challenging task. Even though such mappings can be represented by a smooth function with an arbitrary precision, the necessity to know its derivatives brings a lot of uncertainty. Imagine two classifiers trained by the classical cost. Suppose they both have similar performance measured as the deviation ϵ of the output from the target. The deviation of their outputs from each other is of the same order ϵ. However, the first derivatives of their outputs with respect to inputs are much more far away (about 10ϵ) from each other and the deviation or higher derivatives grow exponentially. Since the extended training requires both: values of network and their derivatives to be close to some "ideal" target, we now have to decide which classifier has the right derivatives (or perhaps they are both wrong). This is impossible to do without some additional restrictions. Perhaps some answers can be provided by synthetic data generators [10].

References

1. Avrutskiy, V.I.: Enhancing approximation abilities of neural networks by training derivatives. arXiv preprint arXiv:1712.04473 (2017)
2. Bengio, Y.: Gradient-based optimization of hyperparameters. Neural Comput. **12**(8), 1889–1900 (2000)
3. Domhan, T., Springenberg, J.T., Hutter, F.: Speeding up automatic hyperparameter optimization of deep neural networks by extrapolation of learning curves. In: Twenty-Fourth International Joint Conference on Artificial Intelligence (2015)

4. Drucker, H., Le Cun, Y.: Improving generalization performance using double back-propagation. IEEE Trans. Neural Networks **3**(6), 991–997 (1992)
5. Flake, G.W., Pearlmutter, B.A.: Differentiating functions of the Jacobian with respect to the weights. In: Advances in Neural Information Processing Systems, pp. 435–441 (2000)
6. Hassibi, B., Stork, D.G., Wolff, G.J.: Optimal brain surgeon and general network pruning. In: IEEE International Conference on Neural Networks, pp. 293–299. IEEE (1993)
7. He, K., Zhang, X., Ren, S., Sun, J.: Delving deep into rectifiers: surpassing human-level performance on imagenet classification. In: Proceedings of the IEEE International Conference on Computer Vision, pp. 1026–1034 (2015)
8. Hinton, G.E., Srivastava, N., Krizhevsky, A., Sutskever, I., Salakhutdinov, R.R.: Improving neural networks by preventing co-adaptation of feature detectors. arXiv preprint arXiv:1207.0580 (2012)
9. Hu, H., Peng, R., Tai, Y.-W., Tang, C.-K.: Network trimming: a data-driven neuron pruning approach towards efficient deep architectures. arXiv preprint arXiv:1607.03250 (2016)
10. Jaderberg, M., Simonyan, K., Vedaldi, A., Zisserman, A.: Synthetic data and artificial neural networks for natural scene text recognition. arXiv preprint arXiv:1406.2227 (2014)
11. Krizhevsky, A., Sutskever, I., Hinton, G.E.: ImageNet classification with deep convolutional neural networks. In: Advances in Neural Information Processing Systems, pp. 1097–1105 (2012)
12. Kumar, M., Yadav, N.: Multilayer perceptrons and radial basis function neural network methods for the solution of differential equations: a survey. Comput. Math. Appl. **62**(10), 3796–3811 (2011)
13. Lagaris, I.E., Likas, A., Fotiadis, D.I.: Artificial neural network methods in quantum mechanics. Comput. Phys. Commun. **104**(1–3), 1–14 (1997)
14. Lagaris, I.E., Likas, A., Fotiadis, D.I.: Artificial neural networks for solving ordinary and partial differential equations. IEEE Trans. Neural Netw. **9**(5), 987–1000 (1998)
15. Lawrence, S., Giles, C.L.: Overfitting and neural networks: conjugate gradient and backpropagation. In: Proceedings of the IEEE-INNS-ENNS International Joint Conference on Neural Networks, IJCNN 2000, vol. 1, pp. 114–119. IEEE (2000)
16. Lawrence, S., Giles, C.L., Tsoi, A.C.: Lessons in neural network training: overfitting may be harder than expected. In: AAAI/IAAI, pp. 540–545. Citeseer (1997
17. LeCun, Y., Denker, J.S., Solla, S.A.: Optimal brain damage. In: Advances in Neural Information Processing Systems, pp. 598–605 (1990)
18. Malek, A., Shekari Beidokhti, R.: Numerical solution for high order differential equations using a hybrid neural network optimization method. Appl. Math. Comput. **183**(1), 260–271 (2006)
19. Mendoza, H., Klein, A., Feurer, M., Springenberg, J.T., Hutter, F.: Towards automatically-tuned neural networks. In: Workshop on Automatic Machine Learning, pp. 58–65 (2016)
20. Reed, R.: Pruning algorithms-a survey. IEEE Trans. Neural Networks **4**(5), 740–747 (1993)
21. Riedmiller, M., Braun, H.: A direct adaptive method for faster backpropagation learning: the RPROP algorithm. In: IEEE International Conference on Neural Networks, pp. 586–591. IEEE (1993)
22. Sarle, W.S.: Stopped training and other remedies for overfitting. Comput. Sci. Stat., 352–360 (1996)

23. Shirvany, Y., Hayati, M., Moradian, R.: Multilayer perceptron neural networks with novel unsupervised training method for numerical solution of the partial differential equations. Appl. Soft Comput. **9**(1), 20–29 (2009)
24. Simard, P., LeCun, Y., Denker, J., Victorri, B.: Transformation invariance in pattern recognition–tangent distance and tangent propagation. In: Neural Networks: Tricks of the Trade, pp. 549–550 (1998)
25. Srivastava, N., Hinton, G., Krizhevsky, A., Sutskever, I., Salakhutdinov, R.: Dropout: a simple way to prevent neural networks from overfitting. J. Mach. Learn. Res. **15**(1), 1929–1958 (2014)
26. Suganuma, M., Shirakawa, S., Nagao, T.: A genetic programming approach to designing convolutional neural network architectures. In: Proceedings of the Genetic and Evolutionary Computation Conference, pp. 497–504. ACM (2017)
27. Tetko, I.V., Livingstone, D.J., Luik, A.I.: Neural network studies. 1. comparison of overfitting and overtraining. J. Chem. Inf. Comput. Sci. **35**(5), 826–833 (1995)
28. Thomas, J.W.: Numerical Partial Differential Equations: Finite Difference Methods, vol. 22. Springer Science & Business Media, New York (2013)
29. Zaremba, W., Sutskever, I., Vinyals, O.: Recurrent neural network regularization. arXiv preprint arXiv:1409.2329 (2014)
30. Zúñiga-Aguilar, C.J., Romero-Ugalde, H.M., Gómez-Aguilar, J.F., Escobar-Jiménez, R.F., Valtierra-Rodríguez, M.: Solving fractional differential equations of variable-order involving operators with mittag-leffler kernel using artificial neural networks. Chaos Solitons Fractals **103**, 382–403 (2017)

Machine Learning for the Identification and Classification of Key Phrases from Clinical Documents in Spanish

Mireya Tovar Vidal[1]([⊠]), Emmanuel Santos Rodríguez[1],
and José A. Reyes-Ortiz[2]

[1] Faculty of Computer Science, Benemerita Universidad Autonoma de Puebla,
14 sur y Av. San Claudio, C.U., Puebla, Puebla, Mexico
mtovar@cs.buap.mx
[2] Universidad Autonoma Metropolitana, Av. San Pablo Xalpa 180,
Azcapotzalco, 02200 Mexico City, Mexico
jaro@correo.azc.uam.mx

Abstract. The key phrases play a very important role because they allow us to characterize the content of a text in a short way and even answer questions related to it. Due to the above, the extraction and classification of these words are a competent problem in different areas of knowledge such as Information Retrieval, Natural Language Processing, among others. This research presents a proposed solution for the identification and classification of key phrases through automatic learning algorithms, in electronic documents related to health topics written in Spanish. According to the experimental results, the proposed algorithm achieves 94% of correctly classified key phrases and 72% of precision for the identification phase.

Keywords: Key phrases extraction · Natural Language Processing · Machine learning

1 Introduction

Natural Language Processing (NLP) is a computational approach to analyzing text, as well as, a set of theories and technologies. Likewise, it can be defined as a variety of techniques to analyze and represent natural language texts in one or more levels of linguistic analysis. NLP has the purpose of achieving, in a human way, the processing of language for a variety of tasks or applications (since it provides both theory and implementations). Especially, any application that uses text such as Information Retrieval (IR), Information Extraction (IE), automatic translation, Question-Answering systems and dialogue systems [1]. Between these areas of application is the extraction of keywords, which can be defined as a task that automatically identifies a set of terms that best describe the subject of a document [2]. The extraction of keywords can be done manually or automatically, but the first way to do it is time-consuming and expensive. Thus, there is a need for a process that extracts these keywords automatically [3]. Keywords are widely used for searches in information retrieval systems

© Springer Nature Switzerland AG 2020
K. Arai et al. (Eds.): FTC 2019, AISC 1069, pp. 164–174, 2020.
https://doi.org/10.1007/978-3-030-32520-6_13

because they are easy to define, review, remember and share. The extracted keywords can be used to construct automatic indexes for a collection of documents or to be used for the representation of a document in classification tasks. Among the different applications of NLP or IR are the following: automatic indexing, document management, high-level semantic description, classification or clustering of text, documents or websites, construction of specific domain dictionaries, recognition of named entities and topic detection [2].

Semantic Analysis at SEPLN (TASS 2018) [11] proposed an evaluation task concerned with the classification of semantic relations and key phrases extraction in health domain. In this task, three subtasks are proposed:
Subtask A: Identification of keyphrases. The goal of this subtask is to identify all the keyphrases from eHealth documents written in Spanish.
Subtask B: Classification of keyphrases. The subtask focuses on assigning a label to each of the keywords extracted in the previous subtask.
Subtask C: Setting semantic relationships. The purpose of this subtask is to recognize all relevant semantic relationships between the entities recognized.

In this paper, an approach for subtask B is proposed. The key words were previously obtained by extraction [12] and are classified into one of two classes: concept or action. In the approach, several supervised machine learning methods are used for classification.

The document is structured as follows: Sect. 2 presents a description of some works related to the extraction and classification of key words; Sect. 3 presents a description of the classifiers used; Sect. 4 the solution methodology is shown; Sect. 5 discusses the results obtained, and Sect. 6 presents the conclusions.

2   Related Work

There are different proposals for the extraction of candidate terms and keywords in scientific documents. Therefore, below, various solutions are shown for the problem mentioned above.

Stauffer et al. [4] proposed an approach to keyword detection, based on templates, on historical manuscript documents. This approach uses different graphical representations for images of segmented words and a corresponding method. On the other hand, the performance of this system is at the level and even surpasses other methods based on templates or machine learning.

CHANEL is a system that processes sequences of words through a local parser that identifies semantically relevant noun phrases and then passes them through a forest of Keyword Classification Trees (KCTs), where each one generates different aspects of semantic representation [7].

SwiftRank is an unsupervised stochastic statistical approach for classifying key phrases and identifying main sentences within a single document for the generic abstraction extraction. This method perceives the highlighted information of a unit of text, which is related to its corresponding title and its influence depending on the position of the sentence in the text [5].

The work of Menaka and Radha [8] focused on extracting keywords using TF-IDF and WordNet. A limited number of words per document are selected and, according to these, the documents are classified using different computer learning techniques such as Naïve Bayes, decision trees and KNN (K-Nearest Neighbor).

3   Machine Learning Algorithms Description

Next, a brief description of the machine learning algorithms used is given [6]:

k-Nearest Neighbor
kNN is based on the principle that instances of a data set will generally exist near other instances that have similar properties. If the instances are labeled, then the value of the label of a non-classified instance can be determined by observing the class of its closest neighbors. kNN locates the k instances closest to the search and determines its class by identifying the label of the closest class.

Decision Trees
They classify instances in order, based on feature values. Each node in the tree represents a feature in an instance to be classified and each branch represents a value that the node can assume. The instances are classified starting at the root node and are ordered according to their characteristic values.

Multilayer Perceptron
A multilayer neural network consists of a large number of units (neurons) linked in a pattern of connections. The units in the network are usually separated into three classes: input units, which receive the information to be processed; output units, where the results are found and intermediate units known as hidden units. First, the network is trained in a paired data set to determine the input-output assignment. The weights of the connections between the neurons are fixed and the network is used to determine the classification of a new data set.

Naïve Bayes
They are very simple Bayesian networks that are composed of acyclic graphs, directed by a single parent (representing the node without observing) and multiple children (corresponding to the observed nodes) with a strong assumption of independence between the child nodes in the context of the father.

Bayesian Networks
A Bayesian network is a graphical model for probabilistic relationships between a set of variables (features). The S structure of the Bayesian network is a directed acyclic graph and the nodes in S correspond one by one with the features X. The arcs represent the causal influence between the features while the lack of possible arcs in S encodes conditional independence.

4 Approach

This section is divided into four parts. The description of the features used by the algorithm is presented. Then, the algorithms for identifying and classifying key phrases from clinical documents is show, respectively. Finally, a use case as an example is presented.

The algorithm for (identifying and) classifying key phrases from clinical documents receives as input the list of key phrases discovered in sub-task A, which are obtained from the algorithm described in Sect. 4.2 (reported in [12]).

The feature extraction of each key phrase is carried out and different machine learning algorithms are applied to classify the key phrases in two classes: concept or action. Finally, an example of its use in a sentence is presented, the extracted key phrases and the corresponding class.

It is important to highlight that both algorithms were implemented in Python by using the Spacy library [9].

4.1 Features for Key Phrases Classification

The classification of the key phrases was performed by using the term frequency (tf), inverse document frequency (idf) and term frequency - inverse document frequency (tf-idf) as features. In addition, we use a free open-source library for Natural Language Processing in Python (Spacy) [9] to assign a grammatical label for each key phrase. Based on these statistical weights and the grammatical information as attributes, the classification of the key words was made into two categories: concept or action. The machine learning algorithms used were: kNN, Naïve Bayes, Bayesian networks, multilayer perceptron and decision trees [6].

4.2 Algorithm for Key Phrase Extraction

Algorithm 1 shows the proposed solution for the identification of key phrases from clinical documents.

Algorithm 1: Key phrase extraction
Input: Plain text clinical documents.
Output: A file with the key phrases and where they start and end for each document.

```
function createFile(keyphrase_index, token_index, to-
ken_length, filename)
 open filename for appending
  write keyphrase_index, token_index, token_length
 close filename
end function

Begin
 index<-1
 for each document in documents
  open document for reading
  for each sentence in document_sentences
   for token in sentence
   if token_dependence is adjectival modifier or to-
ken_head_dependency is object
     createFile(index, token_head_index, to-
ken_head_index+length(token_head_text+" "+token_text,
"output_A_*.txt")) //(*) is the name of the subject in
the file
     index<-index+1
    end if
    if token_pos_tag is noun or token_pos_tag is proper
noun or token_dependecy is nominal subject and to-
ken_head_pos_tag is verb
     createFile(index, token_index, to-
ken_index+length(token_text),"output_A_*.txt") //(*) is
the name of the subject in the file
      index<-index+1
    end if
    if token_dependence is nominal subject or to-
ken_head_pos_tag is verb
     createFile(index, token_head_index, to-
ken_head_index+length(token_head_text), "out-
put_A_*.txt")) //(*) is the name of the subject in the
file
     index<-index+1
    end if
    end for
   end for
  end for
End
```

Algorithm results are the input for the next algorithm.

4.3   Algorithm for Feature Extraction

Algorithm 2 shows the proposed solution for the feature extraction of key phrases from clinical documents.

Algorithm 2: Feature extraction
Input: Clinical documents with key phrases, corpus.
Output: A file with the features of each key phrase.

```
function createFile(tf, idf, tfidf, token, pos_tag, file-
name)
  open filename for appending
   write tf, idf, tfidf, token, pos_tag
  close filename
end function

Begin
for each document in documents
 open document for reading
  for each sentence in document_sentences
    pos_tag<-spacy_pos_tagging(sentence) //call to spacy
library
    tf<-calculate_tf(sentence) //term frequency calcula-
tion
    idf<-calculate_idf(sentence) //inverse document fre-
quency calculation
    tfidf<-calculate_tfidf(sentence) // term frequen-
cy*inverse document frequency calculation
    createFile(tf, idf, tfidf, pos_tag, "output_B_(*)")
(*) is the name of the subject in the file
   end for
  close document
 end for
End
```

4.4   Example

The algorithm described above was applied to the sentence "Asthma affects the respiratory tract" [*El asma afecta las vías respiratorias*]. A function of the Spacy library [9] was used to show the dependency tree generated for such sentence, as shown in Fig. 1. The key phrases obtained from [12] are: asthma [*asma*], affects [*afecta*], respiratory tract [*vías respiratorias*]. At the end of Algorithm 1 the results indicate that asthma [*asma*] and respiratory tract [*vías respiratorias*] are concepts and affects [*afecta*] is an action.

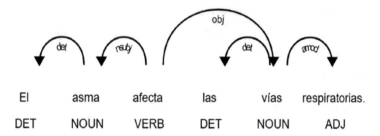

Fig. 1. Dependency tree and grammatical tagging for a Spanish sentence.

5   Results and Discussion

In this section, a description of the used data is analysed and the results obtained are detailed and explained.

5.1   Corpus Description

To carry out subtask B of task 3, eHealth Knowledge Discovery, proposals in TASS (Semantic Analysis Workshop of the SEPLN), the documents of the training data set were used. It is made up of six documents (social and family information, kidneys and urinary system, skin, hair and nails, diagnostic tests, safety issues and personal health data) [11].

5.2   Corpus Analysis

The corpus consists of six health domain documents. Table 1 shows three attributes identified in the corpus: vocabulary, tokens and lexical diversity. Tokens are sequences of characters; vocabulary are considered as different words in the document set; and lexical diversity is the average that a word is used [10].

Table 1. Vocabulary, tokens and lexical diversity of the corpus files

Document	Vocabulary	Tokens	Lexical diversity
Social and family information	418	1056	2.5263
Kidneys and urinary system	679	2780	4.0942
Skin, hair and nails	998	3642	3.6492
Personal health data	111	178	1.6036
Diagnostic tests	378	1187	3.1402
Safety issues	307	680	2.2149

5.3 Metrics

Metrics are provided for performing task 1 of TASS. They determine if the key phrases obtained by the algorithm reported in [12] are correct key phrases. The used metrics are defined as follows:

Correct: matches exactly with a corresponding text span in the gold file (file provided for the correct evaluation of the task) both in START and END values.
Partial: matches are reported when two intervals [START, END] have a non-empty intersection
Missing: matches are those that appear in the gold file but not in the *dev* file (file generated by our algorithm).
Spurious: matches are those that appear in the *dev* file but not in the gold file.

The precision, recall, and a standard F_1 measure are given in terms of the above definitions [11], calculated as follows.

$$precision = \frac{correct + \frac{1}{2}partial}{correct + spurious + partial} \tag{1}$$

$$F_1 = 2 * \frac{precision * recall}{precision + recall} \tag{2}$$

$$recall = \frac{correct + \frac{1}{2}partial}{correct + missing + partial} \tag{3}$$

5.4 Key Phrase Extraction

The algorithm reported in [12] achieved promising results, as shown in Table 2, obtaining 76% average accuracy. The standard F_1 measure, which can be interpreted as the harmonic average of precision and recall, reaches the 72% as a global value 72%.

Table 2. Results of key phrases [12]

Document	Precision	Recall	F_1
Social and family information	0.75	0.63	0.68
Kidneys and urinary system	0.79	0.74	0.77
Skin, hair and nails	0.80	0.71	0.75
Personal health data	0.76	0.71	0.73
Diagnostic tests	0.71	0.76	0.73
Safety issues	0.78	0.64	0.70
Average	0.765	0.6983	0.7267

5.5   Key Phrase Classification Results

Features for key phrases were extracted as attributes in order to perform the classification of them. Four attributes were proposed for the classification: tf, idf, tf-idf and the grammatical tag of the key phrase. This classification was done through cross-validation, in Weka. Five machine learning algorithms were used and the results obtained by them were compared. It should be noted that the training and test data for the classification are unbalanced, since there are more *concepts* than *actions*. There are 2274 concepts and 437 actions in the corpus. The results are shown in Table 3.

Table 3. Results of the machine learning algorithms applied to the key words of the corpus documents

Classifier	Correctly classified percentage	Incorrectly classified percentage
MLP	94.0981%	5.9019%
Naïve Bayes	62.9657%	37.0343%
Bayes Network	94.0612%	5.9388%
kNN	94.3563%	5.6437%
Decision tree	94.7621%	5.2379%

According to Table 3, it is observed that kNN with K = 3 and decision trees obtain the highest precision with respect to the other classifiers.

Table 4. Precision, recall and standard F_1 measure for *Concept* class

Classifier	Precision	Recall	F-Measure
MLP	0.954	0.977	0.965
Naïve Bayes	0.957	0.584	0.726
Bayes Network	0.953	0.977	0.965
kNN	0.951	0.983	0.967
Decision tree	0.954	0.985	0.969

However, for the classification of the Concept category, Naïve Bayes obtains a slightly higher precision compared to the other algorithms, but has the lowest recall value (see Table 4). This is due to it correctly classifies as Concepts most of the relevant instances that it actually found.

Table 5. Precision, recall and standard F_1 measure for *Action* class

Classifier	Precision	Recall	F-Measure
MLP	0.864	0.753	0.804
Naïve Bayes	0.286	0.865	0.430
Bayes Network	0.863	0.751	0.803
kNN	0.894	0.737	0.808
Decision tree	0.909	0.751	0.822

With respect to *Action* classification, in Table 5 is observed that the decision tree achieves the highest precision, followed by kNN and multilayer perceptron. However, Naïve Bayes has the lowest precision in Action class because it is the class with the least instances, which makes the probability biased toward the class with more instances (Concept) and therefore lowers the precision [13].

6   Conclusions

In this paper, an algorithm has been proposed to extract relevant features in order to achieve the classification of key phrases. In this aspect, a good percentage of properly classified instances was reached. Except by the Naïve Bayes classifier whose accuracy and recall are lower. It is due to the skewed probability related to the data used for the evaluation as well as the naïve nature of the classifier itself. It is observed that KNN and decision trees are competitive classifiers to perform a binary classification, compared with multilayer perceptron. This is because they find a better relevant proportion of the data in the test set, which is expressed in the results of precision and recall.

The main benefits of this paper are to present a proposal to extract relevant features of key phrases for classification, and to show a comparison on the performance of various machine learning techniques, when performing the task mentioned above. However, it is important to point out that this model is constrained to unstructured Spanish texts.

As a future work, some balancing technique could be applied to adjust the *Concepts* class since there is more quantity than *Actions*. We can also continue with subtask C that focuses on recognizing all the relevant semantic relationships. Through this, we can use these semantic relationships to build a knowledge base, such as an ontology, for an application oriented towards the medical domain. Besides, we aim to develop a deep learning model to carry out the tasks discussed in this paper.

In addition, the importance of work related to Natural Language Processing oriented to the health domain and medical text in Spanish should be highlighted since most of the works are focused on the English. Finally, this topic will become more important in the future because we have a growing flow of information and it is essential to extract significant knowledge from medical texts in Spanish.

Acknowledgment. This work is supported by the Sectoral Research Fund for Education with the CONACyT project 257357, and partially supported by the VIEP-BUAP project.

References

1. Liddy, E.D.: Natural language processing. In: Encyclopedia of Library and Information Science, 2nd edn., N.Y. (2001)
2. Beliga, S.: Keyword extraction: a review of methods and approaches (2014)
3. Siddiqi, S., Sharan, A.: Keyword and keyphrase extraction techniques: a literature review. Int. J. Comput. Appl. **109**, 18 (2015)

4. Stauffer, M., Fischer, A., Riesen, K.: Keyword spotting in historical handwritten documents based on graph matching. Pattern Recogn. **81**, 240–253 (2018)
5. Lynn, H., Lee, E., Choi, C., Kim, P.: SwiftRank: an unsupervised statistical approach of keyword and salient sentence extraction for individual documents. Procedia Comput. Sci. **113**, 472–477 (2017)
6. Kotsiantis, S.B.: Supervised machine learning: a review of classification techniques. Informatica **31**, 249–268 (2007)
7. Kuhn, R., De Mori, R.: Learning speech semantics worth keyword classification trees. In: IEEE International Conference on Acoustics, Speech, and Signal Processing, vol. 2, pp. 55–58 (1993)
8. Menaka, S., Radha, N.: Text classification using keyword extraction technique. Int. J. Adv. Res. Comput. Sci. Softw. Eng. **3**, 734 (2013)
9. Honnibal, M., Montani, I.: Spacy 2: natural language understanding with Bloom embeddings, convolutional neural networks and incremental parsing. (2017)
10. Bird, S., Klein, E., Loper, E.: Natural Language Processing with Python. O'Reilly Media Inc., Sebastopol (2009)
11. eHealth Knowledge Discovery. TASS. https://tass18-task3.github.io/website/taskA.html. Accessed 31 Jan 2019
12. Tovar Vidal, M., Santos Rodríguez, E., Contreras González, M.: Extracción de palabras clave en documentos no estructurados utilizando Spacy. Coloquio de Investigación Multidisciplinaria **6**, 1782–1789 (2018)
13. Rennie, J.D., Shih, L., Teevan, J., Karger, D.R.: Tackling the poor assumptions of Naive Bayes text classifiers. In: Proceedings of the Twentieth International Conference on Machine Learning, vol. 3, pp. 617–618 (2003)

Automatic Grasping Using Tactile Sensing and Deep Calibration

Masoud Baghbahari$^{(\boxtimes)}$ and Aman Behal

Central Florida Research Park and NanoScience Technology Center,
Department of Electrical and Computer Engineering, University of Central Florida,
Orlando, FL 32826, USA
masoud@ece.ucf.edu, abehal@ucf.edu

Abstract. Tactile perception is an essential ability of intelligent robots in interaction with their surrounding environments. This perception as an intermediate level acts between sensation and action and has to be defined properly to generate suitable action in response to sensed data. In this paper, we propose a feedback approach to address robot grasping task using force-torque tactile sensing. While visual perception is an essential part for gross reaching, constant utilization of this sensing modality can negatively affect the grasping process with overwhelming computation. In such case, human being utilizes tactile sensing to interact with objects. Inspired by, the proposed approach is presented and evaluated on a real robot to demonstrate the effectiveness of the suggested framework. Moreover, we utilize a deep learning framework called Deep Calibration in order to eliminate the effect of bias in the collected data from the robot sensors.

Keywords: Automatic grasping · Tactile sensing · Deep calibration · Assistive robot

1 Introduction and Related Work

In the world of active agents, an agent desires to adjust its behavior during interaction with its environment. Such an interaction consists of three main components: sensation, perception, and action. Depending on the task and the sensed modality, these three elements have to be implemented in coordination with each other to have the best performance. Perception is an intermediate step and is responsible for extracting useful information from the sensed data. The quality of extracted information from the sensed data is dependent on the power of designed perception algorithm to mine the data effectively. On the other hand, more powerful algorithms can consume a lot of time and energy from a light robot either in training phase or in inference phase. A promising solution to address these issues is integrated interactive perception approach. In such an approach, the main attention is in minimizing the perception computational processing by coordination of the three components of sensation, perception, and action. Such a

© Springer Nature Switzerland AG 2020
K. Arai et al. (Eds.): FTC 2019, AISC 1069, pp. 175–192, 2020.
https://doi.org/10.1007/978-3-030-32520-6_14

coordination highlights the topic of active perception and is referred as a learned policy in the context of reinforcement learning. As is well know, learning this interactive perception from gathered experiments is computationally expensive for light robots; moreover, the learned policy can not be interpreted easily.

In this paper, we aim to propose an interactive tactile perception with use of force-torque sensing for a grasping task. Grasping task can be initiated by either visual or tactile perception. Although the first one is powerful in terms of object recognition, a continuous processing of visual data during interaction is not a simple task for a light robot. In real world, a human also barely utilizes vision ability in close proximity of the object to be grasped; indeed, tactile perception ability is more useful in the vicinity of object [16,19].

Grasping is an essential and complex daily activity. Through this task, humans show an intention to affect surrounding environment in a controllable manner. Humans primarily utilize a combination of control strategy and learning from repetitive experiments to anticipate grasping in different situations [22]. In this regard, the properties of an object such as size, shape, and contact surface are important parameters during grasping task [18].

Tactile sensation is a very informative feature to recognize object properties. In [7], the authors proposed a tactile perception strategy to measure tactile features for mobile robots. Tactile sensing also is used to propose a robust controller for reliable grasping [11] and slipping avoidance [8]. Visual sensing and tactile sensing are complementary in robot grasping. A combination of both of them through deep architecture is a promising solution in [13,14]. However, processing these high-dimensional data is not an easy task and a meaningful compact representation would be needed [12,24]. A robot can learn the manipulation using tactile sensation through demonstrations [5,6,24].

A lower dimensional representation of tactile data is also more useful for object material classification [15]. With more processing approaches such as bag-of-words, identification of objects would be possible in advance [23]. Extraction of object pose via touch based perception can be used for manipulation [21]. Moreover, localization will be improved by contact information gathered by tactile sensor [9,17]. The robot can control and adjust the pose of hand with stability consideration after evaluating the tactile experiences [3,10].

Tactile feedback and interactive perception are very important components of the grasping task. An introduction of predictive force control and reactive control strategies in this domain is provided in recent years [20]. Interactive perception has been introduced as a potential field of study in recent years [4]. Such importance emerges from the fact that the perception can be facilitated by interaction with the environment.

In this paper, we pursue following objectives and contributions: (1) We show that it is feasible to propose a more human-like grasping using tactile touch sensing. (2) We describe the useful mathematical framework to extract the important tactile sensing information from robot joints. (3) We present the procedure to accurately calibrate the robot joints data in order to successfully mine the tactile sensing. (4) We propose an approach by avoiding sequential logical rules.

This paper is organized as follow. The approach on how to extract tactile sensation data is presented in Sect. 2 with grasping steps in Sect. 3. Experimental results are demonstrated in Sect. 4 while conclusions are provided in Sect. 5.

2   Tactile Sensing

To implement our proposed approach on automatic grasping based on tactile sensing, we first describe how to extract the tactile data from the robot joint sensors data.

2.1   The Six-Axis Force/Torque Tactile Data

Grasping is a physical interaction with environment. During such interaction, the exchanged data between the object and robot would be the force and torque data. The robot experiences torque data τ_f as a consequence of inserted force f as shown in Fig. 1.

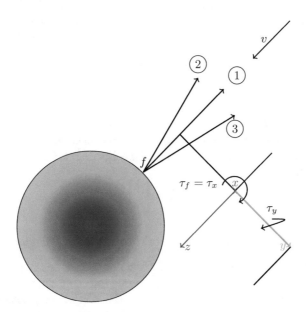

Fig. 1. Robot hand interaction with object and force/torque tactile sensing

In this figure, three directions for the sensed force are marked to be recognizable. The direction of this force (①, ② and ③) is dependent of the direction of robot hand movement toward the object v and it is in opposite direction of movement after contact with the object surface:

$$sign(f) = -sign(v) \tag{1}$$

where the sign function returns the sign of the data. The sensed force vector can be decomposed into the robot hand frame x, y and z in Fig. 1. So, the 3-axes of tactile force data are the set of $f_e = \{f_x, f_y, f_y\}$.

The remaining part of tactile sensation is the torque data. From Fig. 1, the resulting sensed torque τ_f tends to rotate the hand in the clockwise direction. This direction would be valid for all the marked force directions in the figure and since it is actually around the x coordinate:

$$\tau_x = \tau_f \tag{2}$$

This generated torque can be expressed by cross product of the lever arm vector and the sensed force vector as follows:

$$\tau_x = \mathbf{r} \times \mathbf{f} = ||r|| ||f|| sin\theta \tag{3}$$

where θ is the angle between the lever arm vector and the force vector. In three dimensional space, the interaction force f can also result in torque around y direction denoted by τ_y. The last part of torque tactile sensing set is τ_z. Any movement in x and y coordinates during contact with object would generate a torque around z axis. This torque is the consequence of friction force during movement on the object surface and the friction force is dependent on the magnitude of normal force inserted on the surface f. If we assume a constant friction coefficient μ for surface we can expand (3) as:

$$\tau_z = \mathbf{r} \times \mathbf{f} = \mu ||r|| ||f|| sin\theta \tag{4}$$

As a consequence, the torque tactile data consists of the set of $\tau_e = \{\tau_x, \tau_y, \tau_z\}$. All the six-axis force/torque tactile data set is essential for grasping task. These set of tactile data F_e are useful to guide the robot hand during interaction with the object surface:

$$F_e = \begin{bmatrix} f_e \\ \tau_e \end{bmatrix} \tag{5}$$

2.2   Force/Torque Tactile Data and Robot Joint Sensors

The external interaction with robot hand tends to act against the changes in Cartesian position and/or orientation via force and torque response. The force prevents the robot from further changes in the Cartesian position whilst the torque is against the change in orientation. Such external effects can sensed as torque in each joint of the robot. The relationship between the robot measured joint torque sensors and the interaction force torque in robot base frame (the frame attached to the first joint holding the whole robot arm) can be stated by the transpose of the Jacobian matrix J:

$$\tau_{int} = J^T F \tag{6}$$

where τ_{int} is the interaction torque data sensed by robot joint torque sensors and F is the vector of six elements encapsulating the end-effector interaction force f_b and torque τ_b expressed in the base frame:

$$F = \begin{bmatrix} f_b \\ \tau_b \end{bmatrix} \tag{7}$$

Assuming a Jacobian matrix with number of rows equal or greater than 6, the interaction force-torque F on end-effector can be retrieved by the Moore-Penrose inverse of the Jacobian matrix:

$$F = (JJ^T)^{-1}J\tau_{int} \tag{8}$$

Since the Jacobian is a joint dependent matrix, the inverse term in some specific joint positions known as singularities is not defined. In our case, we assume that our robot with 6 joints never works in such configurations of the robot arm. Although the force f_b and the torque τ_b magnitude are independent of the frame, their directions are completely dependent on the expressed frame. As shown in Fig. 2, transformation of these data from base frame (b) to end-effector frame (e) is possible using the relative Rotation Matrix R_b^e:

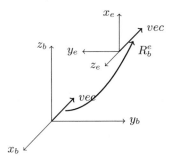

Fig. 2. A typical vector vec coordinate transformation by rotation matrix R_b^e. The magnitude of vector is same in both coordinates.

$$\begin{bmatrix} f_e \\ \tau_e \end{bmatrix} = \begin{bmatrix} R_b^e & 0 \\ 0 & R_b^e \end{bmatrix} \begin{bmatrix} f_b \\ \tau_b \end{bmatrix} \tag{9}$$

where f_e and τ_e are the force and the torque in end-effector frame obtained as force torque tactile data in (5) as shown in Fig. 1. Using both (8) and (9), it is possible to retrieve the six-axis force/toque tactile data from the joint sensed interaction torques data τ_{int}.

2.3 Deep Calibration: Robot Joint Sensors Calibration Using Deep Learning

In light of (8), we assume that the robot joint sensors data τ_s is purely related to the interaction with object. However, this data needs extra processing of calibration to accurately calculate the tactile data. In this section, we propose to utilize a deep learning technique to calibrate the joint torque sensors. These torques capture many effects, related to the robot motion and gravity as well as the interaction with the object. As a consequence, this bias is required to be removed to extract the interaction tactile data successfully.

To remove the motion-related torque data, we proposed to model the motion bias torque data. The interaction bias-free joint torque data will be obtained by subtracting the measured joint torque data τ_s from this bias model τ_{bias}:

$$\tau_{int} = \tau_s - \tau_{bias} \tag{10}$$

In this regard, we record the robot joints sensor torque data as well as the joints position q and velocity $\dot{q}$ during robot free-motion in space. Our primary experiments demonstrate that the motion-related torque data change considerably with two parameters: joint movement direction $\dfrac{\dot{q}}{|\dot{q}|}$ and the joint angle q. These two variables are fed as inputs to a three-layer fully-connected sequential multilayer perceptron deep network to train and inference the bias value from the output of the model. In fact, the model captures the significant changes in the collected bias data through fitting a regression to the bias data [1].

3 Grasping Using Tactile Sensing

In this section, we present the procedure to define the grasping steps for a robot using the six-axis tactile data acquired in the previous section.

3.1 Feedback Control Command

In this subsection, we present the procedure to define the grasping steps for a robot using the six-axis tactile data in previous section. We can use the interaction force/torque data to guide the robot hand in the vicinity of the object. These data are informative enough to define the required grasping steps. Our approach is based on feedback tactile sensation to adjust the robot hand velocity in corresponding direction during its interaction with the object (see Fig. 1). With this goal, we assume dynamics for linear velocity v and angular velocity ω of robot hand as follows [2]:

$$\dot{v} + b_v v = u_v \tag{11}$$

$$\dot{\omega} + b_\omega \omega = u_\omega \tag{12}$$

where u_v is the command to adjust the linear velocity and u_ω to adjust the angular velocity in which the direction of linear and angular velocities are expressed in the robot hand frame itself (Fig. 1). The constant coefficients b_v and b_ω are the system damping to inertia ratio and determine the time profile of linear and angular velocity in response to the adjustment command respectively.

According to Fig. 1, any change in hand location around the object surface will be effected by commanding the linear velocity. Meanwhile, any rotation around the object to point the hand fingers toward the object can be accomplished via influencing the angular velocity. As the robot fingers are interacting with the object surface, the adjustment in the commands is needed to change the hand location or the angle between the hand and the object. Our approach is based on implementing the tactile sensation to adjust the hand linear and angular velocity during its interaction with the object. The control command for linear velocity is defined as:

$$u_v = \alpha_v(f_f - f_e) \tag{13}$$

where f_e can be any element of force tactile sensation set $f_e = \{f_x, f_y, f_z\}$ and α_v is a necessary scaling constant to scale two different domains (force in several *Newton* and velocity typically in less than several mm/s). The desired value f_f is the desired final value that we consider for that direction. For instance, if the robot needs to touch the object, it has to move directly toward the object (see Fig. 1). In such case, the control command has to be designed for linear velocity in hand z direction with $\alpha_{vz} < 1$:

$$u_{vz} = \alpha_{vz}(f_{zf} - f_z) \tag{14}$$

Executing (11) with this command, the robot hand moves with a constant velocity v_{dz} given as

$$v_{dz} = \frac{\alpha_{vz} f_{zf}}{b_z} \tag{15}$$

toward the object until the force tactile sensation f_z converges to f_{zf}. Nevertheless, the velocity of movement is dependent on the final touch force f_{dz}. With a small modification, we replace the control command with following alternative:

$$u_{vz} = b_v v_{dz}(1 + \alpha_{vz} f_z) \tag{16}$$

where v_{dz} can be selected freely and the final contact force is adjusted by α_{vz} as follows:

$$f_{zf} = -\frac{1}{\alpha_{vz}}. \tag{17}$$

Similarly, to rotate the hand around each axis, the corresponding control command can be defined by a scaling coefficient

$$u_\omega = \alpha_\omega(\tau_d - \tau_e) \tag{18}$$

where $\tau_e = \{\tau_x, \tau_y, \tau_z\}$ is the feedback torque tactile sensation. As another example, if we need to align the hand by rotating the hand around its z axis, we can consider:

$$u_{\omega z} = \alpha_{\omega z}(\tau_{dz} - \tau_z) \tag{19}$$

Inserting this command into (12) after a transient time imposed by the system damping to inertia ratio in the direction $b_{\omega z}$, the rotation with a constant angular velocity ω_z equals

$$\omega_z = \frac{\alpha_{\omega z} \tau_{dz}}{b_{\omega z}} \tag{20}$$

which will continue till the corresponding tactile sensing measurement τ_z reaches the level of desired torque τ_{dz}. We also can define a desired value as a function of other sensed data. As an example, if the robot needs to rotate around the z direction as a function of force in z direction, (19) changes to:

$$u_{wz} = \alpha_{wz}(\beta_{wz} * f_z - \tau_z) \tag{21}$$

where β_{wz} is scaling factor and the rotation is effected until the desired force in z direction converges to the desired value f_{dz}. If the desired value is related to other tactile sensing data, the adjustments would be coupled to one another. This is a very important property since we can define a sequence of actions for the grasping task. We define an intuitive grasping paradigm according to Fig. 1 as follows:

- Move directly toward the object in z direction
- Stop movement when the fingers touch the object
- Keep a constant inserted force in the z direction
- Rotate around the contact point using the torque tactile data

All these defined steps can be executed by both proposed commands in (16) and (19) with some minor modifications. The three first steps are encapsulated in (16). To make sure that the inserted force is in z direction, we need to modify (16) as follow for both x and y directions:

$$u_{xz} = \alpha_{vx} f_x \tag{22}$$

$$u_{yz} = \alpha_{vy} f_y \tag{23}$$

where α_{vx} and α_{vy} are scaling coefficients in those directions. Under the implementation of these two extra commands, the robot always inserts the force in z direction. Finger elasticity is the main reason for sensing force in x and y directions. Such forces can be eliminated easily to help the robot contact with the object in just the z direction. As a result, we can plan for x and y directions to facilitated the grasping in advance settings.

To rotate around the contact point using tactile torque data, we propose the following commands:

$$u_{wx} = \alpha_{wx}(\tau_x) \tag{24}$$

$$u_{wy} = \alpha_{wy}(\tau_y) \tag{25}$$

where α_{wx} and α_{wy} are the scaling factors for the corresponding directions. The sensed torques facilitate alignment of each direction with the object surface at the end of the rotation.

In the next section, we prove the stability of above dynamic equations with designed control commands.

3.2   Stability Analysis

Since the proposed approach is based on feedback data, we need to prove the stability of the real-time dynamic system under implementation of the proposed control commands. We first investigate the stability of (11) in z direction with (16):

$$\dot{v}_z + b_z v_z = b_z v_{dz}(1 + \alpha_{vz} f_z) \tag{26}$$

Before interaction, the force sensing data is zero and the system follows the following dynamic equation to move the robot toward the object:

$$\dot{v}_z + b_z(v_z - v_{dz}) = 0 \tag{27}$$

This is a stable dynamic equation and the error converges to zero after a settling time greater than $t_s = \dfrac{4}{b_z}$:

$$(v_z - v_{dz}) \rightarrow 0 \tag{28}$$

which implies $v_z \rightarrow v_{dz}$ when the sensed force $f_z = 0$. After contact with the object, the sensed force is no longer zero. Due to flexibility of fingers, the sensed force can be modeled as a spring force with coefficient K_z. It is proportionally dependent on the level of displacement in the finger joints as well as the object deformation expressed into a unified term Δz in z direction:

$$f_z = -K_z \Delta z \tag{29}$$

We also can model the unified term Δz as the displacement from equilibrium point of the surface (see Fig. 3) and insert it into (26):

$$\dot{v}_z + b_z v_z = b_z v_{dz}(1 - \alpha_{vz} K_z \Delta z) \tag{30}$$

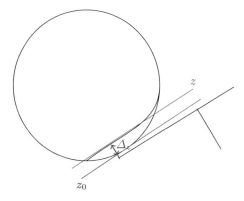

Fig. 3. Displacement model springiness term during interaction with object

$$\Delta z = z - z_0 \tag{31}$$

where z is the location of finger tip and z_0 is the equilibrium point of surface, both of them in the z direction. Rearranging (30), we have the following:

$$\dot{v}_z + b_z v_z + \alpha_{vz} K_z \Delta z = b_z v_{dz}. \tag{32}$$

Moreover, we know that the velocity is the rate of change in location

$$v_z = \dot{z} \tag{33}$$

such that we can modify (32):

$$\ddot{z} + b_z \dot{z} + b_z v_{dz} \alpha_{vz} K_z (z - z_0) = b_z v_{dz} \tag{34}$$

This is a second order stable dynamic equation. The velocity error and the acceleration error become zero after a settling time around $t_s = \dfrac{4}{b_z}$ and the sensed force converges to

$$f_z = -K_z(z_d - z_0) = -\frac{b_z v_{dz}}{b_z v_{dz} \alpha_{vz}} = -\frac{1}{\alpha_{vz}} \tag{35}$$

where z_d is the final value when the robot hand finger can not move any further. This proof can be easily extended to other directions. The results are extendable for sensed torque as well. In such a case, we utilize the mathematical relationship for sensed torque when there is an angle θ between the force vector $\mathbf{f}$ and lever arm vector $\mathbf{d}$:

$$\tau = \mathbf{d} \times \mathbf{f} = |d||f| sin(\theta). \tag{36}$$

For rotation around the hand z direction, the sensed force $\mathbf{f}$ is perpendicular to the lever arm vector $\mathbf{d}$. With a friction coefficient μ_z during rotation on the surface, the sensed torque would be:

$$\tau_z = d\mu_z f_z sin(90°) = d\mu_z f_z \tag{37}$$

After utilizing this equation in (19) and inserting the sensed torque into (12), we have

$$\dot{\omega}_z + b_{wz} \omega_z = \alpha_{wz}(-d\mu_z f_z + \beta_{wz} f_z) = \alpha_{wz} f_z(-d\mu_z + \beta_{wz}). \tag{38}$$

According to this equation, rotation around the z direction will be effected as long as the sensed force $f_z \neq 0$. The rotation by commanding the angular velocity ω_z continues on the surface until it detects the edges. In this case, the robot stops rotating around the z-axis when the following condition is satisfied

$$\beta_{wz} = d\mu_z. \tag{39}$$

If we select β bigger than $d\mu_z$, the robot only reaches the edges where μ_z is bigger than the surface friction coefficient. In this case, the sensed torque converges to:

$$\tau_z = \beta_{wz} f_{zf} \tag{40}$$

when the object is inside the robot hand and $f_z \rightarrow f_{zf}$.

4   Experimental Results

In this section, we present the experimental settings, the time profile of tactile sensation set for a typical grasping, and the successful rate for a set of different objects.

4.1   Settings

To validate the proposed approach, we use a two fingered Mico robot with 6 degrees of freedom. Except for singular configurations and some specific configurations in which the robot is unable to reach, the Jacobian matrix used in (8) results in valid tactile sensation data. The deep learning is trained with ReLU activation function for the neurons during 50 epochs and a batch size of 20 to minimize the mean square error for each joint torque bias data. The configuration of final layer is the output estimating the bias value corresponding to the current joint angle and movement direction.

The sensor data is collected through robot movement without any interaction with external environment. By this way, the gathered data is only related to the current joint angle and the direction of the movement completely reflecting the dependency of bias torque data to these inputs.

4.2   Grasping Paradigm Using Tactile Sensation

We first evaluate the deep calibration performance. In Figs. 4, 5, 6, 7, 8 and 9, the recorded torque data and the model are visualized for all six joints to illustrate the performance of proposed deep calibration in details. There are two blue lines on the vibrating noisy data and each of them is related to the direction of joint movement. In prediction phase, the torque data is extracted from the network corresponding to the joint angle and its movement direction.

Next, we apply the grasping paradigm in this paper to grasp a canned. In Table 1, we provide parameters involved in the design. Using these parameters, the commands executed in the grasping process are as follows:

$$u_{xz} = 0.0025 f_x \tag{41}$$

$$u_{yz} = 0.0025 f_y \tag{42}$$

$$u_{vz} = v_{dz}(1 + 0.4 f_z) \tag{43}$$

$$u_{\omega x} = (0.0025 * \tau_x) \tag{44}$$

$$u_{\omega y} = (0.0025 * \tau_y) \tag{45}$$

$$u_{\omega z} = (0.025 * f_z - \tau_z) \tag{46}$$

According to these commands, the tactile force sensing in approaching direction z would be

$$f_z = -\frac{1}{\alpha_{vz}} = -\frac{1}{0.4} = -2.5N \tag{47}$$

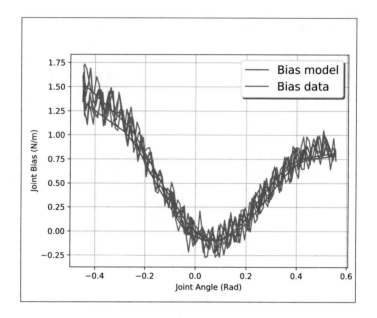

Fig. 4. Bias data and fitted model of first joint

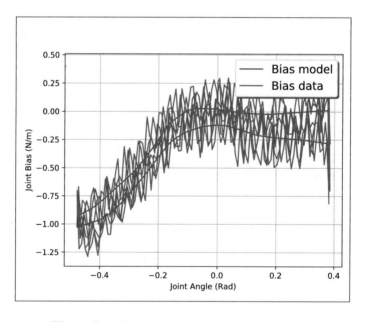

Fig. 5. Bias data and fitted model of second joint

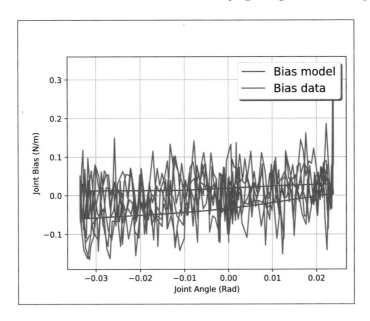

Fig. 6. Bias data and fitted model of third joint

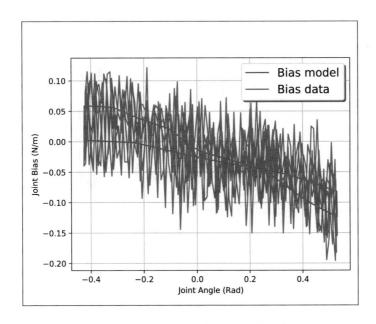

Fig. 7. Bias data and fitted model of forth joint

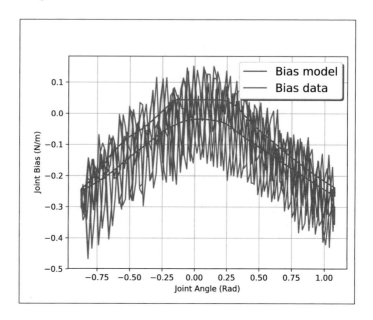

Fig. 8. Bias data and fitted model of fifth joint

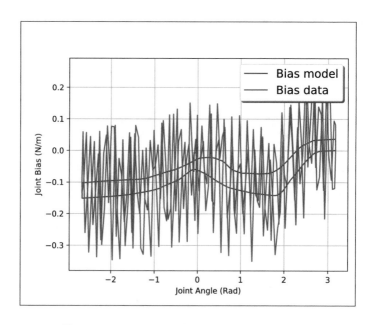

Fig. 9. Bias data and fitted model of sixth joint

Table 1. Parameters

Parameter	Value
v_{dz}	0.0055
α_{vx}	0.0025
α_{vy}	0.0025
α_{vz}	0.4
$\alpha_{\omega x}$	0.25
$\alpha_{\omega y}$	0.0025
$\alpha_{\omega z}$	1
$\beta_{\omega z}$	0.025
b_x	1
b_y	1
b_z	1
$b_{\omega x}$	1
$b_{\omega y}$	1
$b_{\omega z}$	1

which is consistent with the final value of force tactile sensing in Fig. 10. Moreover, the final values of force sensing in x and y directions are zero. As shown in the figure, the robot reaches the object surface after 6000 time samples which is equal to 6 sec since the sample time is 0.001 sec. Nevertheless, the grasping task is completed after 35 sec of touching the object surface since the robot struggles to adjust the hand with the object surface before grasping the object completely. It is hard for the robot to keep a constant contact with the object while is suffering from mechanical vibration during movement. Several jumps in Fig. 10 demonstrate this phenomenon.

During this typical interaction, the torque tactile sensing reaches zero for both x and y directions and it converges to a constant final value:

$$\tau_{zf} = \beta_{\omega z} * f_{zf} = -0.025 * 0.25 = -0.0625 \tag{48}$$

in the z direction when it detects the object edges. All these explanations can be observed in Fig. 11. As can be inferred from these figures, the sensed data is prone to mechanical vibrations during the robot movement as well as noise. We applied a threshold filter on the data to reduce this undesirable effect after removing the bias using deep calibration. This is the reason that the sensed data is zero before contact with the object. This filter has no effect on the data vibration after it passes the threshold. According to presented results, our proposed approach grasp an unknown object similar to human being. Deep calibration is useful to provide the accurate tactile sensing data. This data can be mined from robot joint sensors by a mathematical formulation and the grasping strategy implemented by feedback tactile data works smoothly without implementation any logical rules. A video of the experiment is available online https://youtu.be/y7ZBFr1IpVw.

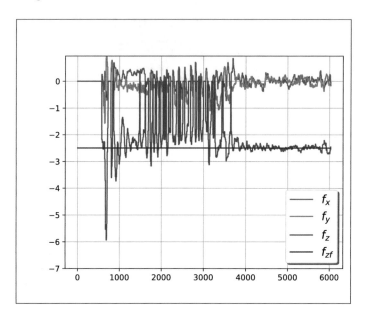

Fig. 10. Force tactile sensation profile during grasping the canned

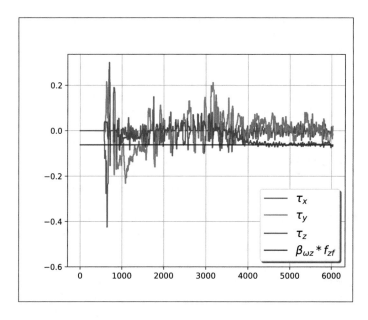

Fig. 11. Torque tactile sensation profile during grasping the canned

5   Conclusions

In this paper, we proposed a human-like automatic grasping of an object using tactile sensing without vision. The tactile sensing data is extracted from the robot joint torque sensors and calibration process is needed to rid the tactile data of the bias related to motion. To calibrate the bias data, we provided a customized deep learning structure with effective inputs to fit a model on bias data. The suggested grasping paradigm in this paper is applied to grasp an unknown shaped canned and the time profile of tactile sensing data is presented during execution of grasping with deep learning alongside the feedback tactile data.

As part of future works, our goal is to propose enhancement strategies for manipulation capability of robots using tactile sensing. In this regard, harder manipulation tasks will be addressed through a touch-based interactive perception.

References

1. Baghbahari, M., Hajiakhoond, N.: Evaluation of estimation approaches on the quality and robustness of collision warning systems. In: SoutheastCon 2018, pp. 1–7. IEEE (2018)
2. Baghbahari, M., Hajiakhoond, N., Behal, A.: Real-time policy generation and its application to robot grasping. arXiv preprint arxiv:1808.05244 (2018)
3. Bekiroglu, Y., Laaksonen, J., Jørgensen, J.A., Kyrki, V., Kragic, D.: Assessing grasp stability based on learning and haptic data. IEEE Trans. Rob. **27**(3), 616 (2011)
4. Bohg, J., Hausman, K., Sankaran, B., Brock, O., Kragic, D., Schaal, S., Sukhatme, G.S.: Interactive perception: leveraging action in perception and perception in action. IEEE Trans. Rob. **33**(6), 1273–1291 (2017)
5. Chebotar, Y., Hausman, K., Su, Z., Sukhatme, G.S., Schaal, S.: Self-supervised regrasping using spatio-temporal tactile features and reinforcement learning. In 2016 IEEE/RSJ International Conference on Intelligent Robots and Systems (IROS), pp. 1960–1966. IEEE (2016)
6. Chebotar, Y., Kroemer, O., Peters, J.: Learning robot tactile sensing for object manipulation. In: 2014 IEEE/RSJ International Conference on Intelligent Robots and Systems, pp. 3368–3375. IEEE (2014)
7. Chitta, S., Sturm, J., Piccoli, M., Burgard, W.: Tactile sensing for mobile manipulation. IEEE Trans. Rob. **27**(3), 558–568 (2011)
8. Cirillo, A., Cirillo, P., De Maria, G., Natale, C., Pirozzi, S.: Control of linear and rotational slippage based on six-axis force/tactile sensor. In: 2017 IEEE International Conference on Robotics and Automation (ICRA), pp. 1587–1594. IEEE (2017)
9. Corcoran, C., Platt, R.: A measurement model for tracking hand-object state during dexterous manipulation. In: 2010 IEEE International Conference on Robotics and Automation, pp. 4302–4308. IEEE (2010
10. Dang, H., Allen, P.K.: Stable grasping under pose uncertainty using tactile feedback. Auton. Rob. **36**(4), 309–330 (2014)

11. Gómez Eguíluz, A., Rano, I., Coleman, S.A., Martin McGinnity, T.: Reliable object handover through tactile force sensing and effort control in the shadow robot hand. In: 2017 IEEE International Conference on Robotics and Automation (ICRA), pp. 372–377. IEEE (2017)

12. Finn, C., Tan, X.Y., Duan, Y., Darrell, T., Levine, S., Abbeel, P.: Deep spatial autoencoders for visuomotor learning. In: 2016 IEEE International Conference on Robotics and Automation (ICRA), pp. 512–519. IEEE (2016)

13. Gao, Y., Hendricks, L.A., Kuchenbecker, K.J., Darrell, T.: Deep learning for tactile understanding from visual and haptic data. In: 2016 IEEE International Conference on Robotics and Automation (ICRA), pp. 536–543. IEEE (2016)

14. Guo, D., Sun, F., Liu, H., Kong, T., Fang, B., Xi, N.: A hybrid deep architecture for robotic grasp detection. In: 2017 IEEE International Conference on Robotics and Automation (ICRA), pp. 1609–1614. IEEE (2017)

15. Kroemer, O., Lampert, C.H., Peters, J.: Learning dynamic tactile sensing with robust -based training. IEEE Trans. Rob. **27**(3), 545–557 (2011)

16. Lee, M.A., Zhu, Y., Srinivasan, K., Shah, P., Savarese, S., Fei-Fei, L., Garg, A., Bohg, J.: Making sense of vision and touch: self-supervised learning of multimodal representations for contact-rich tasks. arXiv preprint arXiv:1810.10191 (2018)

17. Li, R., Platt, R., Yuan, W., ten Pas, A., Roscup, N., Srinivasan, M.A., Adelson, E.: Localization and manipulation of small parts using gelsight tactile sensing. In: 2014 IEEE/RSJ International Conference on Intelligent Robots and Systems, pp. 3988–3993. IEEE (2014)

18. Lin, Y., Sun, Y.: Robot grasp planning based on demonstrated grasp strategies. Int. J. Rob. Res. **34**(1), 26–42 (2015)

19. Murali, A., Li, Y., Gandhi, D., Gupta, A.: Learning to grasp without seeing. arXiv preprint arXiv:1805.04201 (2018)

20. Nowak, D.A., Glasauer, S., Hermsdörfer, J.: Force control in object Manipulation'a model for the study of sensorimotor control strategies. Neurosci. Biobehav. Rev. **37**(8), 1578–1586 (2013)

21. Petrovskaya, A., Khatib, O., Thrun, S., Ng, A.Y.: Touch based perception for object manipulation (2007)

22. Romano, J.M., Hsiao, K., Niemeyer, G., Chitta, S., Kuchenbecker, K.J.: Human-inspired robotic grasp control with tactile sensing. IEEE Trans. Rob. **27**(6), 1067–1079 (2011)

23. Schneider, A., Sturm, J., Stachniss, C., Reisert, M., Burkhardt, H., Burgard, W.: Object identification with tactile sensors using bag-of-features. In: IROS, vol. 9, pp. 243–248 (2009)

24. van Hoof, H., Chen, N., Karl, M., van der Smagt, P., Peters, J.: Stable reinforcement learning with autoencoders for tactile and visual data. In: 2016 IEEE/RSJ International Conference on Intelligent Robots and Systems (IROS), pp. 3928–3934. IEEE (2016)

An Emotion-Based Search Engine

Yazid Benazzouz$^{(\boxtimes)}$ and Rachid Boudour

Embedded Systems Laboratory, Computer Science Department,
Badji Mokhtar University, Annaba, Algeria
yazid_benazzouz@hotmail.com, racboudour@yahoo.fr
http://www.univ-annaba.dz/

Abstract. With the advancement in computing hardware and cloud services, many applications, imagined in the past and rejected as complicated or unfeasible, are becoming achievable. Artificial intelligence was able via new computing architecture to overcome big time consuming and insufficient data storage which permitted rapid models development and real-time applications. In this paper, we show via a challenging application how artificial intelligence can change the way we interact with our devices. Particularly, the paper attempts to develop an emotion-based search engine. In this sense, the user emotional features are used to select best Internet search results or to adapt them to user emotion. In this case, many scenarios are possible such as preventing bad influence of the search results on the user emotion. The idea presented in this solution can be adapted to other applications and brings new research challenges.

Keywords: Google · Artificial intelligence · Web engineering · Emotion inference · Search engine

1 Introduction

Nowadays, artificial intelligence emerged as a vision of the present and the road to flow for the future. This is reflected by the new objects that have invaded our environment such as robots, conversational devices, intelligent cameras, smart cars and many others. It is just an incomplete view of what intelligence is, but very promising and attractive world. Many applications have already integrated challenging abilities like vision and speech. However, emotion and smelling, particularly, they have not gained much attention in our daily life objects.

Emotion based applications are interesting because they will offer a new way to interact with human people and to understand their feeling, consequently, applications can adapt its behavior. Of course understanding human situation can be more complicate and involve more sensing environment devices and tools.

Interaction is a big challenge of the future. The real world will continue to be fused with the virtual environment (Internet of the future), starting by interconnecting daily life objects including some parts of the human body. Having

© Springer Nature Switzerland AG 2020
K. Arai et al. (Eds.): FTC 2019, AISC 1069, pp. 193–203, 2020.
https://doi.org/10.1007/978-3-030-32520-6_15

this in mind, it will be clear that emotion and smelling will be sooner integrated to make our presence in the real or virtual world completely transparent.

At this stage, an emotion-based search engine seems a promising application of artificial intelligence and a way to consider emotional data in every user interaction with the virtual world. Its also a step towards a unique digital real world where feeling are sensed and transmitted. Moreover, web applications are able to track user data to provide personalized services, to exploit user experience for new recommendation or to just to adapt their behavior and interface to users' usage. Data cover user profile, preferences and his environment which can be perceived via different connected sensors. Unfortunately, advances in sensing user emotion have been realized but they are not exploited in nowadays smart applications. Emotion detection can be a very challenging and interesting key for future application such as tracking people emotion at home to avoid dangerous stress situation or children protection, and evaluating people stress level in transportation systems, etc.

We focus our work on the most user interaction activity on the Internet. It is particularly clear that searching information is the most expected action from Internet users and Google is the most commonly used search engine. It is invoked from everywhere and through various devices.

The first need for our application is the human emotion detection. Human expresses emotion in different ways including facial expression, speech, gestures and written text. It is a key attribute for human interaction with other people and systems. In our daily life, we go through different states of feeling and it was essential to find a simple way to track this states during the mobility of the user. As the technology progresses, the internet is now commonly used on PCs, tablets, and smart phones. This way, a web cam in any of these devices can be a good start up to generate the user emotion in real time, of course with his agreement. The primary emotion levels should be enough for our application namely; Happy, Anger, Disgusted, Sadness, Fear and Surprise.

To develop our emotion-based search engine prototype, we were also interested in emotion extraction from text. It is a field of research that is closely related to Sentiment Analysis. Sentiment Analysis aims to detect positive, neutral, or negative feelings from text, whereas Emotion Analysis aims to detect and recognize types of feelings through the expression of texts, such as anger and happiness. For some types of applications, emotion extraction has become very hard to achieve because of a huge amount of data, especially textual data. It was impossible to manually analyze all the data for a specific purpose. New research directions have emerged from automatic data analysis like automatic emotion analysis to avoid such heavy work, for example, security agencies can track emails/messages/blogs etc. and detect suspicious activities; business communities can stimulate the customer's emotions to buy some products or services, etc.

This article presents an emotion-based search engine where emotion of the user gathered from his Webcam device is aligned with the emotion of search results retrieved from Google search engine. The goal is to apply some filtering mechanisms. It is a first attempt to exploit user emotion for personalized interaction between a user and him device. The idea can be generalized to many type of applications where emotion is a key.

The rest of paper is organized as follow. Section 2 is dedicated to background, where main tools of emotion detection are presented and compared. Then, a use case is detailed in Sect. 3 and used in the experimentation of the developed prototype. Next Sect. 4, focuses on the software architecture of the prototype. At this point, we move forward to the validation and tests in Sect. 5. Finally, we support our proposal with a discussion followed by a conclusion in Sect. 6.

2   Background

This section describes numerous efforts in emotion detection from text and video. Each part includes some research methods from the literature and tools that can be used to implement emotion based applications.

2.1   Text Based Emotion Detection

Text emotion detection refers to the procedure of identifying the type of emotion reflected by a text. Machine learning is a common approach to achieve this task. The process consists on a training procedure of a data-sets of labeled texts to produce a classifier. This later will be evaluated on a test corpus to determine its accuracy. In many cases, the classifier is trained again to meet a higher number of correct responses. Table 1 presents different research techniques for emotion detection from text. The accuracy can pitched up to 90% which is an acceptable value to start working on real application and get user experience. Besides, Table 2 presents more referenced tools for text based emotion detection. They are all commercial except Synesketch which is open-source and lightweight, thus, better positioned to be part of the application expected by this paper. From a technical point of view, text to emotion should be a component in the server side of the search engine platform.

2.2   Video Based Emotion Detection

Video based emotion is more mature subject and various solutions exist. The accuracy can pitch up to 99,6% which is a very good result. However, there is a technical problem with those solutions because emotion will be inferred from video stream of the web cam plugged in user device. In this sense, Javscript and WebRTC based solution will be favoured to ensure a better streaming quality. In the following, some solutions and research projects from the state of the art are presented.

Nviso: Nviso is specialized in emotion video analytics. It uses 3D facial imaging technology to monitor many different facial data points and to produce likelihoods for seven main emotions states. Nviso claims to provide a real-time imaging API. They have a reputation, awarded for smarter computing in 2013 by IBM.

Table 1. Research methods for emotion detection from text

Method	Principle	Performance	Accuracy	Reference
Deep learning (The SWAT System)	A system where words are mapped and scored according to multiple labels	Very low	53.20%	[5]
Deep learning (Recurrent Neural Networks)	Words are presented as vectors and use matrix operators for implementing deep learning techniques	Low	80,7%	[3]
Fuzzy logic	Recognize complex human emotions from textual contents using fuzzy logic rules	Considerable performance but still cannot handle the difficulty of English language	Not mentioned	[6]
Neural networks and fuzzy logic	The extraction of information is done using the fuzzy logic and classification of emotions is done using neural network	Very well	90%	[4]

Kairos. This product analyze and understand emotion, demographics and attention in most videos and images. It takes person faces and produce facial features and expressions using a proprietary facial analysis algorithms. The result enclosed the six universal emotional states, age, gender, and other useful meta data about the faces.

Imotions. Imotions is a biometrics research platform that provides software and hardware for monitoring many types of bodily cues. Imotions syncs with Emotient's facial expression technology and adds extra layers to detect confusion and frustration. The Imotions API can monitor video live feeds to extract valence or can aggregate previously recorded videos for emotions analysis. Imotion software has been used by Harvard, Procter & Gamble, Yale, the US Air Force, and was even used in a Mythbusters episode.

CrowdEmotion. CrowdEmotion offers an API that uses facial recognition to detect the time series of the six universal emotions. Their online demo analyzes facial points in real-time video, and respond with detailed visualizations. They offer an API sandbox, along with free monthly usage for live testing.

FacioMetrics. Founded at Carnegie Mellon University (CMU), FacioMetrics is a company that provides SDKs for incorporating face tracking, pose and gaze tracking, and expression analysis into apps. Their demo video outlines some creative use cases in virtual reality scenarios. The software can be tested using the Intraface iOS app.

Table 2. Tools for emotion detection from text

Tools	Website	Commercialization	Technology
CrowdFlower	https://www.crowdflower.com/solutions/how-sentiment-analysis-works	Price varies based on complexity, volume, and quality requirements	Machine learning
Lexalytics	https://www.lexalytics.com/technology/sentiment	Commercial, free limited use	Natural language processing and artificial intelligence
Google cloud prediction APIs	https://cloud.google.com/natural-language/docs/sentiment-tutorial	Commercial api see the pricing guide at Google	Machine learning capabilities and pattern matching
Brandwatch	https://www.brandwatch.com/2011/04/how-does-sentiment-analysis-work/	Commercial	Machine learning
Algorithmia	https://blog.algorithmia.com/benchmarking-sentiment-analysis-algorithms/	Commercial	Natural language processing
IBM watson tone analyzer API	https://console.bluemix.net/developer/watson/documentation	Commercial	Natural language processing
Qemotion	http://www.qemotion.com/api/documentation	Commercial	NLP, AI and Affective computing technology
AYLIEN API	http://docs.aylien.com/docs/usage	Commercial and free limited use	Machine learning, and natural language
PreCeive API	http://www.theysay.io/sentiment-analysis-api/	Commercial	Not mentioned
Synesketch	http://krcadinac.com/synesketch/	An open source library for sentence-based emotion recognition	Hybrid keyword spotting technique

Clmtrackr. Clmtrackr is a javascript library for fitting facial models to faces in videos or images. It implements a constrained local model fitted by regularized landmark mean-shift, as described in Saragih's paper [7]. Clmtrackr tracks a face and outputs the coordinate positions of the face model as an array. The library provides some generic face models that were trained on the MUCT database and some additional self-annotated images (see Tables 3 and 4).

Table 3. Research methods for emotion detection from video

Methods	Principe	Performance	Accuracy	References
Facial emotion recognition in real time	Convolutional neural network for classifying human emotions from dynamic facial expressions in real time	Good	Training accuracy of 90.7% and test accuracy of 57.1%	[2]
DeXpression	Deep Convolutional neural network	Very good	99.6% for CKP dataset and 98.63% for MM Dataset	[1]

Table 4. Tools for video based emotion detection

Nviso	http://www.nviso.ch/technology.html	Paid Api
Kairos	https://www.kairos.com	Paid Api
Imotions	https://imotions.com/blog/facial-action-coding-system/	N/A
CrowdEmotion	https://www.crowdemotion.co.uk	Free
FacioMetrics	http://www.faciometrics.com/	N/A
Clmtrackr	https://github.com/auduno/clmtrackr	MIT licence

3   Use Case

Before dealing with the software implementation of the emotion-based search engine, this section aims to motivate this application through different usage scenarios. As simple as they are, these scenarios bring many technical challenges and open the discussion about the feasibility of such applications regarding the success or not of their usage by people. This is why, the goal is not to find the perfect scenario but some of them which leads to confusion. At the end, these scenarios will serve as testing examples of the prototype.

3.1   Internet Search for Suicide

This scenario is particularly interesting because of the great number of suicide among younger people in the world and the attempts reported by news papers.

Scenario: Imagine a young girl, feeling upset about everything in the world especially about her parents and she thinks that they don't deserve her, so the best way to get her revenge is to put end to her life. So, she will Google about a manner to get suicide.

Expectation: With Google she is going to have the answer she wants but an emotion-based search engine will respond according to user emotional states. The system will evaluate her emotional states and eliminate the results with negative polarities. Moreover positive answers are favored.

Discussion: Of course, one can think that it is simple to add filtering mechanisms to the search engine using specific keywords but this will affect all the information interested by normal people. Even if the user choose to put off the camera, the most search results with negative polarities will be filtered. Because, the goal of this prototype is not to prevent the user from information but to do not let it a tool for accomplishing horrible scenarios.

3.2   Syrian War

The Syrian civil war has affected many people around the world. Many of its people left the country and immigrate in foreign countries.

Scenario: An Syrian immigrant, feeling sorry and sad for his country, wants to know about the war news.

Expectation: The information, he will get from the internet, returns fear, guilt, and desperation which tempts him to despair. The system aims to filter these results by considering his emotional states and profile. Moreover only encouraging news about the war are displayed.

Discussion: This particular scenario requires the determination of the user identity. This challenging task can be handled by existing data analysis, email information or subscription details. So, the question is why using emotional filtering? this is because the identity of a person does not reflect necessary his emotional states.

4   Software Architecture

The architecture of emotion-based search engine rely on client/server architecture. It uses Google search engine as back-end to retrieve the list of pages corresponding to a specific keywords. In contrast, it provides its own interface for making a search request. Thus, every request is routed to Google, then processed by the emotion-based server to filter the data according to gathered emotion from the user camera. To this end, the front-end interface provides a way to enable the usage of user camera for emotion detection.

4.1   Implementation Technology

The prototype is developed as a Single-page web applications. It a relatively new idea. Instead of a website requiring a network request every time where the user navigates to a different page, a single-page web application downloads the entire site (or a good chunk of it) to the client's browser. After that initial download, navigation is faster because there is little or no communication with the server. Single-page application development is facilitated by the use of popular frameworks such as Angular, Ember and Express.

4.2   Video Emotion Detection

The emotion-based search engine detects the user emotions from the web cam using Clmtrackr library, mentioned before. Then, it arranges them to six basic emotions: anger, disgust, fear, sad, surprised and happy. The library allow also user gender detection from the data which can be useful for other scenarios. The emotion data is saved into mongo-db databases in Json format like in Fig. 1. It is continuously stored into databases until the user push the stop button controlling the emotion inference process.

```
{
  "emotions" : [
    {
      "_id" : "5c334d24eeef01107b11e709",
      "emotion" : "angry",
      "value" : 0.3226875823661089
    },
    {
      "_id" : "5c334d24eeef01107b11e708",
      "emotion" : "disgusted",
      "value" : 0.2672899757175178
    },
    {
      "_id" : "5c334d24eeef01107b11e707",
      "emotion" : "fear",
      "value" : 0.003865816104344745
    },
    {
      "_id" : "5c334d24eeef01107b11e706",
      "emotion" : "sad",
      "value" : 0.037147926225450704
    },
    {
      "_id" : "5c334d24eeef01107b11e705",
      "emotion" : "surprised",
      "value" : 0.001917727769096578
    },
    {
      "_id" : "5c334d24eeef01107b11e704",
      "emotion" : "happy",
      "value" : 0.06804228659360168
    },
    {
      "_id" : "5c334d24eeef01107b11e703",
      "emotion" : "female",
      "value" : 0.6225982825012165
    },
    {
      "_id" : "5c334d24eeef01107b11e702",
      "emotion" : "male",
      "value" : 0.37740171749878354
    }
  ]
}
```

```
{
  "pages" : [
    {
      "_id" : "5c334d35eeef01107b11e734",
      "displayLink" : "Syrie : Toute l'actualité sur Le Monde.fr.",
      "title" : "https://www.lemonde.fr/syrie/",
      "snippet" : "Syrie - Découvrez gratuitement tous les articles, les vidéos et les
         infographies de \nla rubrique Syrie sur Le Monde.fr.",
      "emotions" : "[{\"disgust\":0},{\"fear\":0},{\"anger\":0},{\"sad\":0},{\"happy\":0}
         ,{\"surprise\":0}]"
    },
    {
      "_id" : "5c334d35eeef01107b11e733",
      "displayLink" : "Actualités Syrie en direct : info du jour, faits divers, actu en
         continu",
      "title" : "https://www.20minutes.fr/monde/syrie/",
      "snippet" : "Syrie : retrouvez toute l'actualité de syrienne en continu et en direct
         via nos \narticles, infographies et vidéos.",
      "emotions" : "[{\"disgust\":0},{\"fear\":0},{\"anger\":0},{\"sad\":0.032},{\"happy\":0}
         ,{\"surprise\":0}]"
    },
    {
      "_id" : "5c334d35eeef01107b11e732",
      "displayLink" : "Syrie : actualités en direct - Ouest-France",
      "title" : "https://www.ouest-france.fr/monde/syrie/",
      "snippet" : "La République arabe syrienne, au Proche-Orient, partage ses frontières
         avec la \nTurquie, le Liban, l'Irak et la Jordanie. Présidée par Bachar El-Assad, la
         Syrie a ...".
      "emotions" : "[{\"disgust\":0},{\"fear\":0},{\"anger\":0},{\"sad\":0},{\"happy\":0}
         ,{\"surprise\":0}]"
    },
    {
      "_id" : "5c334d35eeef01107b11e731",
      "displayLink" : "Syrie : l'actualité sur ce pays - Le Point",
      "title" : "https://www.lepoint.fr/tags/syrie/",
      "snippet" : "Consultez les derniers articles du Point sur le pays : Syrie et restez
         informés en \nsuivant l'actualité sur le site Le Point.fr.",
      "emotions" : "[{\"disgust\":0},{\"fear\":0},{\"anger\":0},{\"sad\":0
         .022857142857142857},{\"happy\":0},{\"surprise\":0}]"
    }
  ],
  "keywords" : "syrie"
}
```

Fig. 1. Example of the Web cam inferred emotion

Fig. 2. Example of emotion inferred from Google search results

4.3   Text Emotion Detection

To accomplish the process of emotion detection from text, two components are used: Google search API and Synesketch tool.

Google Search API. Google website describes Google custom search API to enable developers creating a search engine for their website, blog, or a collection of websites. Also, it allows to configure the engine for searching both web pages and images, as well as many other options.

Synesketch. Synesketch is a Web free open-source software for textual emotion recognition and artistic visualization, designed and developed by Uroš Krčadinac. Synesketch algorithms analyze the emotional content of text sentences in terms of emotional types (happiness, sadness, anger, fear, disgust, and surprise), weights (how intense the emotion is), and a valence (is it positive or negative). The recognition technique is grounded on a refined keyword spotting method which employs a set of heuristic rules, a WordNet-based word lexicon, and a lexicon of emoticons and common abbreviations. Synesketch was awarded by the International Digital Media and Arts Association, Canada, and the Belgrade Chamber of Commerce, Serbia.

The Overall Process. When a user types a keyword into the emotion-based search engine, the information coming from Google, thanks to the Google Custom Search API, is processed by Synesketch tool. Each page has a description attribute and only this one is used to inference the text emotion. After that the search result including the corresponding emotions are stored in the mongo-db databases in Json format like presented by Fig. 2.

Now we have two vectors the first one is the video emotion vector and the second one is text emotions vector. So, in order to filter the search results, a first approach is to use vectors based measure distance such as Euclidean distance to find the convergence or divergence of both emotion states. This is the general usage of the prototype and depending on the use case, large-scale field test was conducted as will be explained on the next section.

5   Experimental Results

The goal from this experiment is to test the functioning of the prototype and the to check to user acceptability of the system usage. This is a main question that generally developers are faced when artificial intelligence products are put into real world. In addition, a research question we aim to answer is the robustness of an intelligent systems to face simulated behaviour of users. For example, the user might be sad but express happiness. This is a general limitation for artificial intelligence systems where the user adopt a false behaviour to trigger the system to the wrong decision.

In order to evaluate the performance of the system, we have taken ten (10) persons: three females and seven males and we have limited the search responses to (25). For video emotion inference, we considered only the average of the last (10) vectors of gathered states. Each vector is calculate every 100 ms. This supposes that the emotional state of the user is stable during (1) second. Then, we asked them to use the prototype to query for some information while adopting a specific emotion state. The keywords are chosen depending in each scenario presented above. We considered the following keywords for each one:

1. Scenario of Sect. 3.1 [suicide, revenge, person suicide, person revenge]
2. Scenario of Sect. 3.2 [syrian war, syrian refugee, syrian aleb, syrian deir ez-zor]

Figure 3 represents the case where the users have an emotional state varying between sadness, anger, fear and disgust. In fact, this state necessitates usually a number of repetitions from the user to meet this requirement. The graph represents the number of response filtered for the keyword *suicide* requested by the (10) persons. The figure shows that up to 70% of pages are blocked by the prototype.

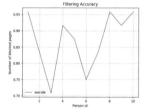

Fig. 3. Filtered information for Suicide keyword

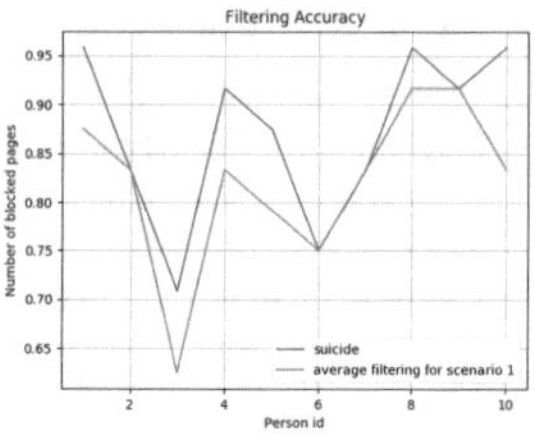

Fig. 4. Average of filtered information for scenario 1

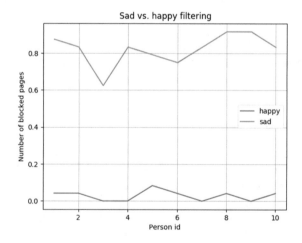

Fig. 5. Sad search results vs. Happy emotion

We evaluate also the impact of the keyword *suicide* over all the keywords considered for the scenario of Sect. 3.1. Figure 4 shows that this keyword sustained the most blocking search results which indicates the big emotional state driven by this word.

The next evaluation is presented by Fig. 5. It consists of comparing the behaviour of the prototype when the user have bad or positive emotions. The evaluation is conduct with the (10) persons for this two different emotional states: *sad* and *happy* during a search request for the keyword *suicide*. The result is as expected. There is a lower number of filtered pages when the user express a positive emotional state.

6 Discussion

Emotion play a great role in the user behaviour. It can influence easily the decisions he takes. Considering that and with the advancement technology for human body sensing devices, it becomes possible to develop innovative applications. Through this mitigated experiment, we have noticed that human emotion is very sensitive and hard to infer a stable state during a short time. If the emotion changes rashly the search results of the prototype can be out of scope.

For the emotion inference from text, We find that web pages description doesn't contains enough information for emotion detection. A recommendation from this study is to complete web pages description in search engines with this kind of information and may be more to allow artificial intelligence application to exploit this information in their tasks. Besides emotion information, a web page description could include some additional description that help to identify a class of users allowed to consult a specif web page.

7 Conclusion

In this paper, we present a prototype developed to experiment an emotion-based internet search. After some experimental tests, we confirmed that many challenges need to be addressed. First, we recommend to expand internet Web page descriptions by additional data to meet new needs of artificial intelligence applications. Such data could be, for example, smelling, taste and emotion. One can imagine, feeling the wind while he is living an immersion in the internet virtual world. We think, we are not so far from this reality. Second, we agreed that this prototype should be supported by a more complete study with end-users to bring it real improvement. Finally, this work has led the way towards other applications such as internet chat rooms and might inspire some other new applications.

References

1. Burkert, P., Trier, F., Afzal, M.Z., Dengel, A., Liwicki, M.: Dexpression: deep convolutional neural network for expression recognition. CoRR, abs/1509.05371 (2015)
2. Duncan, D.L., Shine, G., English, C.: Facial emotion recognition in real time (2016)
3. Hong, J., Fang, M.: Sentiment analysis with deeply learned distributed representations of variable length texts (2015)
4. Kanger, N., Bathla, G.: Recognizing emotion in text using neural network and fuzzy logic. Indian J. Sci. Technol. 10(12), 1–6 (2017)
5. Katz, P., Singleton, M., Wicentowski, R.: SWAT-MP: the SemEval-2007 systems for task 5 and task 14. In: Proceedings of the 4th International Workshop on Semantic Evaluations, SemEval 2007, Stroudsburg, PA, USA, pp. 308–313. Association for Computational Linguistics (2007)
6. Qamar, S., Ahmad, P.: Emotion detection from text using fuzzy logic. Int. J. Comput. Appl. 121, 29–32 (2015)
7. Saragih, J.M., Lucey, S., Cohn, J.F.: Deformable model fitting by regularized landmark mean-shift. Int. J. Comput. Vision 91(2), 200–215 (2011)

Internet Memes: A Novel Approach
to Distinguish Humans and Bots
for Authentication

Ishaani Priyadarshini$^{(\boxtimes)}$ and Chase Cotton

University of Delaware, Newark, USA
{ishaani, ccotton}@udel.edu

Abstract. More than half the web traffic is believed to be made of bots. Over the last few years, bots have grown increasingly sophisticated to carry out repetitive jobs, gain control over systems, run automated scripts and can also play games. Recently, bots have been capable of authenticating themselves as human beings. They have been known to conquer text based CAPTCHA (Completely Automated Public Turing test to tell Computers and Humans Apart), image based CAPTCHA as well as several other authentication bypassing mechanisms. Further, other techniques like neural networks and machine learning have been instrumental in imparting enough training to bots to behave as humans, which has made it very difficult to distinguish bot behavior from human for the purpose of authentication. In this article, we present the concept of the Internet Memes which may successfully tell a bot and human apart. Considering the ever dynamic nature of Internet Memes, this may be one of the strongest techniques to distinguish between a human and a bot based on conscience and interpretation.

Keywords: Bots · Authentication · Internet memes · Neural networks · Machine learning

1 Introduction

The internet harbors information related to financial personal details, sensitive personal data, medical data, etc. which hold a lot of value thus motivating adversaries to vandalize the networks in form of cyberattacks. The domain of cybersecurity demands that the organizations keep their networks secure by enabling authentication. It is the simple process of determining whether someone is who they claim to be. The idea is to ensure access controls for a system by validating the user credentials against the credentials initially stored in a database. Authentication may be single factor, that makes use of passwords or multi factor that uses a combination of credentials. A user may be authenticated by something they have (like a credit card or a mobile phone to receive one time passwords), something they know (like a password or a PIN number) and something they are (biometrics). With more than half of the web traffic being dominated by bots, it is difficult to [1] identify legitimate traffic. Sometimes these bots access a secure system by impersonating humans. With artificial intelligence making rampant progress, bots are trained using machine learning, neural networks, deep learning etc.

© Springer Nature Switzerland AG 2020
K. Arai et al. (Eds.): FTC 2019, AISC 1069, pp. 204–222, 2020.
https://doi.org/10.1007/978-3-030-32520-6_16

As more and more training data is fed to these bots, it becomes more and more difficult to tell whether a user is legitimate or not. With enough knowledge to crack passwords and even break captchas, these bots imitate as valid users thereby depreciating the concept of authentication, which is imperative to the world of cybersecurity. Moreover, malicious bots may pose as humans to access into systems incorporating sensitive data and may disclose, modify or delete it. In different sectors, people with specific roles are given specific privileges for data access. For example, in finance sectors, transactions may be authorized by the chief personnel or in the healthcare industry, access to patient records may be only restricted to doctors. A bot gaining access into these systems could lead to catastrophic results. Therefore it is necessary to distinguish a human from a bot so that authorization is something that may only be granted to humans. Several techniques like captchas, 'I am not a robot' verification tests, Turing tests [24] have been presented in the past to authenticate legitimate users (humans). However, with advancement in technology, bots have become sophisticated enough to bypass these measures along with password cracking mechanisms. One things that distinguishes a human from a robot is the ability to think, which fortunately can only be fed to the bot since bots do not have the capability to think on their own. Psychology is something that is very much confined to the human mind. Cyberpsychology is the latest addition to user authentication [25]. Over the last few years, psychology has been used as a means to identify humans from robots. Although cognitive science is still being introduced to bots to imitate humans, it may still lack judgment capabilities. IBM Watson, the famous question answering computer system is capable of answering sophisticated questions, however, despite impressive capabilities may not be actually capable of thinking [2]. Watson's Cognitive deficiencies were well highlighted by the fact that it was unable to answer questions related to jeopardy and philosophy which requires reasonable thinking. Another aspect that an Artificial Intelligence system may be fed but may not become aware of is sarcasm. However, recent studies on sarcasm detection using convolutional networks [3] suggest that in future, artificial intelligence systems may be able to respond to a sarcastic comment in sarcastic manner, thereby ruling out the fact that sarcasm may be a way of distinguishing a human from a bot.

To tell a human and a bot apart, it is required that we introduce something that a human mind understands and comprehends in much more advanced way than a bot may ever be trained to do so. If there exists a mechanism that is associated with tremendous amount of data increasing with time, such that the system (bot) may never be trained enough to bypass the authentication mechanism, of which the method forms a part, the mechanism could easily be one of the most successful and robust means of distinguishing a bot from a human. Internet Memes clearly may prove to be one of the most successful methods to distinguish humans from bots when it comes to authentication. The concept of 'Meme' dates back to 1976 defined as the unit of cultural transmission [4]. Internet Memes may be defined as phenomenon of content or concepts that spread rapidly among users in form of pictures, films, video clips, phrases, emails, etc. [5]. The content usually is in form of news, websites, catch phrases, images, jokes, underground knowledge etc. Considering memes as an authentication mechanism to tell humans and robots apart is something that has never been done before. **The novelty of the paper lies in introducing Internet Memes as a method of distinguishing bots and humans for the sole purpose of authentication**. Memes might be one of the sureshot ways to differentiate humans from robots because of the following factors:

1. Memes are specific to regions, groups, sectors and age groups. Even if a bot is trained enough with respect to data about geographical locations, groups of people (the sentiments they share) and people of certain age groups, it may still not be able to respond to a meme as accurately as a human being. This is because a specific meme can be represented with respect to multiple situations (Fig. 1). A single meme applied to different situations or environments will generate different speculations. Similarly a single meme applied to various sectors or various age groups will have different impact on the humans belonging to those sectors or age groups. While humans of different sectors and age groups may be well aware of the hidden message carried in the meme, for a system it may be nothing more than training data that is overlapping in a way, thus creating ambiguity. Therefore, for authentication purposes, specific memes belonging to a particular situation or environment must be used.
2. There is no end to news, catch phrases, jokes, etc. and combining them with various meme templates will generate millions of memes, which might be impossible to train. And even if a computer with good capabilities is trained with the huge datasets, it will become futile after a point since meme content are subject to change with continuously changing news, underground knowledge etc. Thus a bot may never be trained enough to interpret all memes.
3. Computers are only able to interpret languages that are not ambiguous. Every sentence, regardless of the context, should have one meaning. Interpreting information from a single template in different scenarios may lead to ambiguity thereby, making Internet Memes one of the best techniques to tell humans and bots apart.

Fig. 1. Specific meme template representing different situations

Figure 1 depicts how a specific meme template may be used across different situations to infer different meanings. The leftmost meme specifies a situation where a washing machine still works even though a phone went through the wash. It was probably expected that the phone would damage the machine. The meme at the centre maybe specific to someone who is dealing with the transaction and has achieved success at this point. Finally, the meme on the extreme right is about a kid who studied nothing and yet got good grades. As is evident, the same template has been used to depict various scenarios. Even though the meme template is same, all the three memes must be considered as individual data for training a bot. It is quite possible that the memes belong to people from different domains. Based on the fact that a single meme may be interpreted in several ways with or without modifications, if represented in different situations, we propose authentication of a user on the basis of understanding a meme, which may be a challenge for a robot considering different situations.

The rest of the paper is organised as follows. Section 2 highlights the related work to distinguish humans and bots followed by a summarized critical evaluation. Section 3 describes the proposed research work while Sect. 4 takes into account the evaluation methodology. Sections 5 and 6 briefly describe Conclusion and Future Work.

2 Literature Survey

In the past researchers have come up with various techniques that may contribute to tell a robot and a human apart. In this section we will delve into some of the techniques that were proposed in the past. Initially we will examine some of the general bot detection techniques. In the later part of this section, we will examine specific techniques that have been relied on for bot detection, for each of which we will perform a critical evaluation, so as to strengthen our approach.

Several researches conducted in the past aim at detecting bots from the web traffic. Duskin and Feitelson [20] conducted a research on how to distinguish between human and bot users from web search log. Based on web search behavior, they classify users to check sensitivity. Generative models of web search have also been created to explain observed behavior and to perform evaluations. Offutt et al. [21] attempted to explore the limits of human observational proofs (HOP) based bot detection systems. An evasive bot system based on human behavioral system was developed. This proposed system was capable of mimicking human behavior by achieving similarities between human users and evasive bots. Winslow [22] introduced the technique of mouse mapping for bot detection. Clicks, mouse movements, typing, and screen size are some of the features taken into consideration. Based on number of move segments generated per page, number of distinct mouse motions, average time of mouse motions, variance in speed of mouse movements, etc. humans and bots are distinguished. Walt and Eloff [23] relied on machine learning to detect fake identities. The predicted results give an F1 score of approximately 50% which is not optimal. These are some of the generalized bot detection techniques. Subsequently, we'll take a look at some techniques that can easily detect humans and bots and how they failed.

Vidya et al. [6] proposed a text based Captcha for security in web applications. The idea is to generate a challenge based framework. A random value of string code generates a random code, based on which texts are concatenated along with noise. Machine Learning techniques have been known to break text based Captchas. Zhao et al. [7] implemented deep learning neural networks by training it with sufficient handwritten digits, achieving 99.8% accuracy. The single captcha recognition algorithms taken into account were k-means clustering, support vector machines and convolutional neural networks, whereas the multi captcha recognition algorithms considered were moving window algorithm and multi-CNN algorithm.

Nejati et al. [8] suggested an image based Captcha based on Depth Perception. A merge sort algorithm has been used on 3D images to create new appearances for the objects at a multiplication factor of 200. Humans have the ability to apply rapid and reliable object recognition to solve the Captcha, while the systems may still struggle at it. Although 3D captchas were initially introduced to overcome the limitations of 2D captcha, an approach to breaking 3D captchas, known as Teabag 3D was introduced by Nguyen et al. [9].

Athanasopoulos and Antonatos [10] introduced another dimension to the field of captcha in order to make it intelligible to human beings and not computers. The newly added dimension was that of animation. The idea behind animated captchas is that they do not have a static answer, and may be exposed to laundering. However, Nguyen et al. [11] used simple techniques to extract information from animation frames of an animated Captcha which significantly reduces the animated captcha into a traditional single image captcha challenge.

The IBM Watson is a question answer computer system that answers questions posed in natural language [12]. Jeopardy style questions deal with questions not being questions but formulated as answers to which the answer has to be reverse formulated as question. The sentences may be full of clues, jokes and nuances and it is not surprising that even humans are sometimes not able to decrypt and understand. Watson works by extracting data and searching the relevant text in its database. Algorithms are used to generate responses to the questions, such that the analysis results in a score, which is weighed according to a level of confidence. The process is iterated until the response is found probable. Philosophical enquiries were made to Watson and it was inferred that Watson did not actually understand anything, although the focus was its ability to understand the language i.e. syntax and semantic language. Understanding the syntax requires a deep understanding of the grammatical structures, whereas semantics deal with meanings beyond grammar. It was deduced that Watson may be well prepared and equipped to understand the syntax of a language, but since it lacks self-consciousness to decrypt semantics, it does not understand an inquiry in the human sense. As of 2015, Watson has been involved in building codes, teaching assistantships, weather forecasting, tax preparation, advertisement etc. With enough training data, IBM Watson may be able to answer questions related to philosophy and jeopardy (MEME: overlapping, not in Watson).

Contextual double meaning words, jokes, sarcasm or contextual variations to the base language have played an important role in telling humans and robots apart in the past [13]. This is because computers are able to interpret languages that do not seem ambiguous. The way computer languages build up, according to the syntax, one sentence should generate only one meaning. Since human languages are difficult, it might be difficult for a system to interpret sarcasm. However, recently, Deep Convolutional Neural Networks have been known to detect sarcasm in text data [14]. Moreover, in the past, systems have been proposed that manifest how ontological semantics have been instrumental in imparting computers the understanding of humor [15]. Moreover Siri, Apple's Intelligent Personal Assistant has now learnt the ability of talking back with humor [16].

Google's reCaptcha was introduced to protect websites from spam and abuse by means of a risk analysis engine and adaptive challenges. The idea is to keep the automated software from engaging in abusive activities on the website, while allowing valid users to gain access. However, Sivakorn et al. [17] developed a low cost attack using deep learning for semantic annotation of images. 70.78% of the reCaptcha challenges were solved by the proposed method, requiring only 19 s per challenge. The attack model was applied to Facebook Captcha and yielded 83.5% accuracy.

Lemos and Gomes [18] suggested an enhanced image captcha. The system incorporates captcha for web security, DeepCaptcha, and FaceDCaptcha. It relies on a secure database and human ability to distinguish between real world objects. However, bots like CaptionBots and DrawingBots are capable of identifying the objects in images. Huang et al. [19] integrated CaptionBot and DrawingBot to yield an image to text generator and vice versa. It uses semi supervised learning and improves the performance for both CaptionBot and DrawingBot.

Based on the specific techniques that assist in differentiating humans from bots, we present the summarized critical evaluation of the related work is as follows (Table 1):

Table 1. Summarized critical evaluation

Author and year	Proposed methodology	Technique	Counter methodology
Vidya et al. [6]	Text based Captcha for the security in web applications	Generated a challenge based framework for text Captcha equipped with noise and strings	Zhao et al. [7], implemented deep learning neural networks by training it with sufficient handwritten digits, achieving 99.8% accuracy
Nejati et al. [8]	Image based Captcha based on Depth Perception	Merge sort algorithm used on 3D images to create new appearances for the objects at a multiplication factor of 200	Nguyen et al., presented approach to break 3D captcha using scheme Teabag 3D [9]
Athanasopoulos and Antonatos [10]	Animation Captcha	Dynamic Application, perform an action, randomly animate possible answers	Nguyen et al. [11], Simple techniques to reduce animated captcha into traditional single image captcha challenge

(continued)

Table 1. (*continued*)

Author and year	Proposed methodology	Technique	Counter methodology
Louise Beltzung [12]	IBM Watson, Question answer computer system that answers questions posed in natural language	Extracting data and searching the relevant text, using Algorithm to generate responses, results in a weighted score, according to a level of confidence. The process is iterated until the response is found probable	Could not answer philosophical enquiries, but capable of weather forecasting, teaching assistantships, tax preparation. With enough training, may be trained to answer philosophical questions
Stefan Weijers [13]	Double meaning words, jokes, sarcasm or contextual variations to the base language may tell human and robot apart	Language, Syntax	Deep CNN can extract sarcasm in text data [14] Ontological Semantics can impart humor understanding [15] Personal Assistant Siri can reply sarcastically
Google	Google reCaptcha to protect website from spam and abuse	Advanced risk analysis engine and adaptive challenges	Sivakorn et al. [17], low cost attack using deep learning to defeat Google reCaptcha with 70.78% accuracy
Lemos and Gomes [18]	Enhanced image captcha	Takes into account web security, DeepCaptcha, and FaceDCaptcha	Huang et al. [19], integrated CaptionBot and DrawingBot to improve their performance

3   Proposed Methodology

In the previous section, we observed some of the existing works that have been done in the past that may well differentiate humans and bots. We have also seen how captchas have been broken and intelligent systems have failed. With technology progressing rapidly, and with machine learning and neural networks being proficient in training the bots, it is becoming increasingly difficult to tell bots and humans apart when it comes to authentication. We rely on the novel approach of Internet memes for the process of authentication. As mentioned earlier, memes are dynamic and ever changing due to certain factors, which we briefly mentioned in the introduction section. This section is divided into two parts. In the first part, we give an overview of the classification of

Internet Memes, while in the second part, we present the schematic diagram of how we anticipate our system to work.

3.1 Classification of Internet Memes

Internet Memes are spreading rapidly among Internet users and are becoming increasingly popular day by day [5]. We identified certain groups into which the memes may be classified. Figure 2 represents the overview of classification of Internet memes.

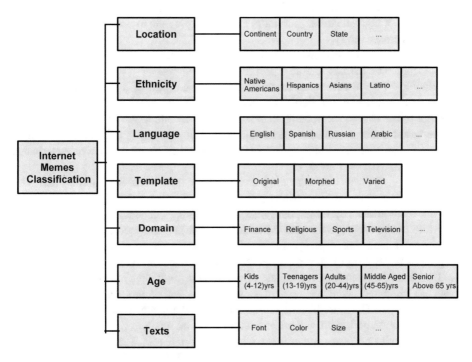

Fig. 2. Overview of classification of internet memes

As we know, due to the dynamic nature of Internet Memes, a given meme template may find place in various environments. These environments are groups (Location, Ethnicity, Language, etc.) that we have mentioned in the figure above. These groups may be further classified into sub groups which may be classified further depending on the target audience for the specific meme. Such a classification of Internet Memes makes it very specific to a given number of people. As we know, bots are trained to interpret languages that are not ambiguous. A single template across multiple locations or a single meme across various domains may carry varying information, thus creating ambiguity. A common meme template may map to several sub groups. This will make it difficult for a bot to gain a clear understanding of the meme, hence during the process of authentication, a human may pass the authentication test, while a bot may not be able to do so. Based on the figure above, we will explore the groups and subgroups in detail

1. Location: Since memes incorporate information about news, websites, catch phrases, underground knowledge, etc. which may be specific to a particular location, location is a significant classification. Memes may vary across Continents, Countries, States, etc. There are images, movies, video clips and documentaries that are specific to various regions. Internet Memes generated using the reference of these is popular among people based in the given location. The sub groups are as follows:

   1.1. Continents: People belonging to a particular continent are usually aware of the News, Sports, Climate pertaining to the given continent, memes highlighting such topics might be easily perceived by people in a given continent.

   1.2. Countries: Memes that form part of folklore, television shows, news, sports etc. of a given country are more easily interpretable to people belonging to the specific country.

   1.3. States: States may have videos, languages, culture, etc. of their own, so there might be memes that are specific to given states.

2. Ethnicity: People belonging to an ethnic group are part of a social group tied by national or cultural tradition. They may also be identified by racial, religious and other traits. Due to the specific beliefs shared in an ethnic group, for some memes, the target audience might be a section of people belonging to a particular ethnic group. Some of the ethnic groups listed in the given figure are native Americans, Hispanics, Asians, Latino, etc.

3. Language: With memes spreading across all continents via the Internet, and with memes being specific to various regions, a single meme template may be represented in various languages, only intelligible and comprehensible by those who are aware of the language the meme bears. Some phrases and quotes are confined to a particular language, thereby making the memes language specific. The few languages listed in the figure are English, Spanish, Russian, Arabic, etc.

4. Template: A given meme template may have the following forms:

   4.1. Original: The meme content stays as it is and is circulated across various groups. The idea behind the meme stays the same, however based on the situation and environment, inferences may be made.

   4.2. Morphed: The meme content may slightly changed according to any given environment, and the inferences are relatively different for each environment.

   4.3. Varied: The meme content is significantly modified and the inference is remarkably different across various environments.

5. Domain: Understanding a meme is greatly affected by the thought process of a human being. People belonging to different domains may have varying thinking capabilities. There can be multiple domains, however in the given figure we have mentioned only few domains. These domains may also have their own sub domains. Increasing the specificity makes it further difficult for the bots to interpret the memes. Few domains listed are as follows:

   5.1. Finance: People belonging to this specific sub group may have better idea about topics related to stock markets, investment banking, transactions etc. So a meme describing these particular topics would be easier for this sub group to understand.

5.2. Religious: People belonging to this specific sub group may be well versed with the concepts of mythology, spirituality, theology, each further a sub group in its own, so memes relating to these topics will be better understood by these group of people.

5.3. Sports: People belonging to this specific sub group may be aware of team names, sports personalities, current sports news etc. thereby making it effortless for them to understand memes corresponding to this section.

5.4. Television: People who are a part of the television industry or audience who keeps their knowledge up to date with what happens in the television industry may understand memes specific to it.

6. Age: People with different age groups have different capabilities of interpreting memes. We have identified five age groups, which are as follows:

   6.1. Kids (4–12) yrs: Simple picture memes or memes related to cartoons may be something that this group can easily interpret.

   6.2. Teenagers (13–19) yrs: Memes related to various television shows, music, education etc. are something most teenagers would be able to comprehend.

   6.3. Adults (20–44) yrs: Memes related to news articles, television, websites, general knowledge, job specific, sports etc. might be few areas of meme that would interest adults.

   6.4. Middle Aged (45–65) yrs: Memes related to what they are aware of happening around them.

   6.5. Above 65 yrs: Memes related to what they are aware of happening around them.

7. Texts: The text content of Internet Memes may vary according to the groups they belong to. Even though a system may be trained to understand memes, the moment there is a slight change in the template, be it the color, font, or size, it will be unintelligible to the bot and will be treated as an altogether new training data. Some of the sub groups are as follows

   7.1. Font: Changing font of the text of a given meme, even though the content stays same may make it difficult for a system to interpret the meme even if the same meme has been fed into the system for training.

   7.2. Color: Similarly, changing the color of the text of a given meme, even though the content is same means an altogether new meme for the system.

   7.3. Size: The size of texts may be varied to make sure that the bot is unable to interpret the meme correctly as humans.

It is evident from the above classification that a given Internet Meme may be produced in various types of groups and sub groups either in original form or modified forms. The template and sometimes the text content being the same across several environments (groups and platforms) might pose a challenge for the bot to comprehend the meme, whereas humans in a specific group may still infer the underlying information. Our approach is authenticating a legitimate user based on the user's interpretation of a meme. Therefore, for authenticating a user the meme should be specific to the environment where authentication is taking place. For example: The authentication system in a law firm should have memes whose content the lawyers must be familiar with. Asking a lawyer to interpret a meme related to healthcare might pose a challenge to the lawyer even though the lawyer is a legitimate user. The detailed idea of the authentication mechanism is presented in the next section.

3.2   Proposed System and Schematic Diagram

The process of authentication primarily involves an authentication service wherein the user is supposed to enter credentials like username and password. The underlying mechanism is matching of username and password (sometime hashed) to a database in order to authenticate the user. Authentication based on Internet Meme follows the same mechanism along with a few modifications. The idea is a inspired from multi-factor authentication, where the user has to validate its identity multiple times. For our authentication system, the user initially asks for the username and password, answering which correctly, the user is redirected to another page that displays a meme. As discussed in the previous section, the meme is specific to the environment, so that the user has some knowledge about it. The second page displays a meme and provides a list of options, from which the user must choose the correct option. The correct option is matched against the database and the user is granted access.

We assume that the system is secure and that only one meme template is used per authentication. This would ensure that the user (human or bot) has never come across the meme before, so is not aware of the correct answer. The human user would be able to choose the answer based on conscience and cognizance while the bot will be able to choose the answer based on how it is trained. The same template may be used in the same environment only if it undergoes some modifications. The Internet Meme Authentication process may be expressed as a series of the following steps:

Step 1: User is asked to enter credentials (username and password) in Page 1

Step 2: User enters the credentials

Step 3: System checks the credentials entered against the credentials stored in database in plaintext or hashes.

Step 4: If the credentials match, user is directed to Page 2

Step 5: The Interface displays an Internet Meme with multiple choice options. Based on the meme, the user interprets the underlying message to choose one of the options.

Step 6: The Internet Meme database consists of Internet Memes (denoted by an Identifier in the database), the corresponding options and the correct option of the form Meme (Identification Number, (List of options), Correct Option). Once the user enters the option, it is checked against the correct option in the database.

Step 7: If the option is correct, the user is granted access.

Step 8: The specific Internet meme used, the options and the correct option are deleted from the database.

The schematic diagram for the same has been presented below (Fig. 3):

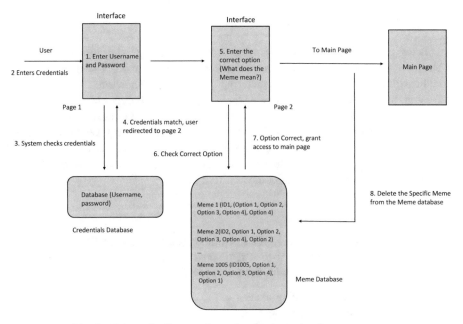

Fig. 3. Schematic diagram for authentication using internet memes

For the system to successfully implement Internet memes as authentication mechanism to distinguish humans and bots, we have made two assumptions which are mandatory.

1. We are assuming that the system is secure and that it may not be vandalized by an adversary to infer the correct options beforehand.
2. Once a Meme has been used for the process of authentication irrespective of whether the answer chosen is correct or not, it is automatically deleted. This makes sure that one meme is used per authentication and that the user is not aware of the correct option beforehand. Although, the database may contain the same template with varying texts, fonts, color etc. This is because using the same template even with the slightest changes may not be that challenging for a human to understand, but for a bot it will be demanding.
3. Since this is a particular authentication system, we may also assume that the authentication service is for a particular group or subgroup. This would make sure that the users have at least some knowledge related to the memes in question.

4 Proposed Evaluation Technique

In the previous section, we took into account the classification of Internet memes as well as the Schematic Diagram depicting the working of the overall system. In this section, we will propose an evaluation technique for the same.

Owing to the number of memes generated with every passing second, it may be difficult to present proper evaluation of the proposed method, but our goal is to introduce the approach and the underlying mechanism. Therefore, we present the evaluation technique and justify why Authentication using Internet Memes for human and bot distinguishing might be a successful approach. We justify the proposed approach by means of two case studies. Each of the case studies takes into account a specific meme template that has undergone some modifications, as to how different options have been generated and how a human being would be much more comfortable in choosing the correct option as compared to a bot.

4.1 Case Study 1

Consider the following image (Fig. 4). For our proposed authentication system, once the credentials are matched, the user is redirected to a page that shows an image and asks the meaning of the meme with the following four options

Option 1: The child is running away from Coffee
Option 2: The child is scared of coffee
Option 3: The child is holding coffee
Option 4: The child wants coffee

Fig. 4. Meme 1

Observation

Based on the given meme and the options supporting it, we can make the following:

1. As human beings we are aware of the necessity of coffee in our lives therefore, we choose Option 4. However, a bot trained to recognize images and expressions may not be able to comprehend the same. A bot that seeks to interpret a meme has definitely undergone some amount of training. Even though a bot may be trained with several memes, it may still be insufficient for it to comprehend the underlying meaning of a meme. A human may perceive the image in question as the need to get coffee. But what a bot might capture from the picture is a child who is running, looks anxious and is holding something in hand. Therefore it may be challenging for a bot to choose the correct option out of the existing options.
2. Further since it has undergone some training before, same captcha applied to various groups and sub groups will definitely lead to ambiguity.
3. If the same template is again used by the particular meme database for the process of authentication, with few altered fonts, texts and colors, even though keeping the underlying message the same, it may still deceive the bot because its altogether new data for the bot. Therefore we insist on using one meme per authentication.
4. The same meme in a region where coffee is banned may altogether change the underlying idea. Thus classification has a great impact on the meme interpretation.

The same meme may be represented with modified content and the interpretation is completely different.

4.2 Case Study 2

In the previous subsection we considered a situation wherein an Internet meme has been used for authentication. We mentioned in the Schematic Diagram section that a given template may be used in the same database only if it has been modified to some extent, so as to make it even more challenging for a bot to interpret the correct option. In this section we present the modified version of the meme considered in the previous section.

Consider a situation when the same meme template refers to different situations. The image in Fig. 5 may list out a set of options as follows:

Option 1: The child is scared of the announcement
Option 2: The child is excited about free refreshments
Option 3: The child wants to go somewhere else
Option 4: The child is running away from the church

Fig. 5. Meme 2

The same template modified with new text content may ask a user to choose the correct option from a set of options that look like (Fig. 6):

Fig. 6. Meme 3

Option 1: The child is excited that the phone is at one percent
Option 2: The child is scared of the phone
Option 3: The child is rushing to get the charger
Option 4: The child sees something interesting

Observation

From Fig. 5, it may be interpreted that the child is excited about the refreshments being served after church and is running to get the refreshments, whereas, Fig. 6 may be thought of as the child in an anxious state because the phone battery is almost dead and the phone might switch off soon unless it is plugged to a charger. Again, these two have the same template but different text content because of the varying situations. Each of them is accompanied by a set of four different multiple choice options out of which one will be correct. A human may easily interpret the underlying message even though the template is applied to different situations or environments, but it will be nearly impossible for a bot to guess the underlying message since the images overlap. There will be some kind of ambiguity from the training perspective of the bot. Internet Memes being dynamic pose great challenge to bots to imitate as human beings. **Thus Internet Memes, being highly extensive may be one of the most successful methods that could distinguish humans and robots for the purpose of authentication.**

Further, since no amount of training data would ensure that the bot completely comprehends the meme in various situations for different groups and subgroups, it would be impossible for a bot to gain authentication, should Internet Memes be used as an Authentication Mechanism. **This may be by far the most robust authentication system to tell a human and a bot apart, thereby making it almost impossible to break due to the unending number of memes and modifications that may be applied.** Memes change with the times, Contents (news, websites), with respect to Groups (Locations, Languages, etc.). Therefore a given meme may be represented in hundreds of thousands of ways, each meme being an altogether new training data, thereby making it one of the most powerful authentication methods.

Based on the case studies and their respective observations, we may infer that the suggested approach is one of the most novel and feasible methods of distinguishing a human from a robot. Internet Meme Authentication may be one of those methods that would be almost impossible for an adversary to break, due to the following reasons:

(1) Memes are based on day to day activities (news, website information, underlying knowledge, advertisements, etc.) which means memes might never come to an end. They belong to various domains which could result in generation of hundreds of thousands of memes in a given day.

(2) Due to memes being generated on such a scale, being applied to so many groups, and having the potential of being modified, there will never be enough training data for a bot to completely understand meme due to its ever dynamic nature.

(3) Since memes are specific to groups, the same template or content may be applied to various groups, or modified accordingly. A bot that is being trained to interpret memes that have common content applied to different platforms may face ambiguity and therefore may not be able to take decisions like humans.

(4) Slight change to a particular meme may lead the meme to profess an altogether different message. This slight change is nothing but new training data for a bot that has to be trained, because it will be completely different to the memes that have already been fed to it.

(5) Training a bot to break an authentication system like the one proposed may be futile in sense that the authentication system eradicates the memes used for authentication. So every time a bot is fed meme data, it is completely new data.

5 Conclusion

With more than half of the internet governed by bots, that are capable of impersonating as human beings and carrying out malicious activities, it is necessary to develop a mechanism that would distinguish a bot from a human. This would serve in eradicating the unnecessary bot traffic as well as the malicious activities bots are capable of performing by gaining access into systems as humans. Conscience and ability to think play a key role in differentiating a human and a bot. One of the ways to test the thinking capability is the act of interpreting memes. In this article, we introduced Internet Memes as the novel technique of authenticating users, such that a given meme is understood by humans but not robots. Memes are specific to specific group of people based on which we performed an overall classification of memes. Based on region, ethnicity, language etc., various memes are introduced on a daily basis. Moreover the same meme template may be applied to several groups thereby generating a new meaning. Since Internet Memes are ever changing and so dynamic, using them for authentication purpose limits the training of data. Even though bots are trained to interpret memes, a single template may undergo modifications in multiple ways, thereby making the training futile. The system we have proposed uses only one meme per authentication, thereby eradicating the correct answer, so that the bot may not be trained further. The proposed evaluation technique justifies as to why the proposed approach could be one of the most successful and robust approaches for distinguishing humans and bots. We may conclude that the groups into which the memes have been classified have a remarkable effect on the understanding of meme. More the bot trains on a specific template, more the chances of ambiguity. Using one meme per authentication further strengthens the authentication mechanism.

6 Future Work

In this paper, we have proposed the concept of introducing Internet Memes as an Authentication Mechanism to distinguish Humans and Robots. Extracting data from images may not be that simple at this stage due to the lack of sophisticated tools. Currently there are no public datasets that incorporate memes specific to groups. We would like to implement the Internet Meme Authentication using real data for which we would be collecting memes specific to groups. This paper aims at giving the overall idea of the proposed method and tries to justify its feasibility by analyzing case studies.

Initially we may train a bot with memes specific to a group and test if it is capable of recognizing the underlying message accompanying slight changes. As of now, the authentication mechanism proposed using the schematic diagram is restricted to a single group or subgroup. Since memes are extensively large in number, the authentication mechanisms for several groups and sub groups may be performed using a distributed network. Since the system is robust, we can definitely increase its scalability.

References

1. Gorwa, R., Guilbeault, D.: Unpacking the social media bot: a typology to guide research and policy. Policy Internet (2018)
2. Martinez, M.: Future Bright: A Transforming Vision of Human Intelligence. Psychology. Oxford University Press, Oxford (2013)
3. Sivaprakasam, L., Jayaprakash, A.: A study on sarcasm detection algorithms. Int. J. Eng. Technol. Sci. Res. (2017)
4. Diaz, C.: Defining and characterizing the concept of Internet Meme. CES Psicología **6**, 82–104 (2013)
5. Bauckhage, C.: Insights into internet memes. In: Conference: Proceedings of the Fifth International Conference on Weblogs and Social Media (2011)
6. Vidya, N., Naika, S.: Simple text based CAPTCHA for the security in web applications. Int. J. Comput. Sci. Mob. Comput. (2015)
7. Zhao, N., Liu, Y., Jiang, Y.: CAPTCHA Breaking with Deep Learning. Stanford University (2017)
8. Nejati, H., Cheung, N., Sosa, R., Koh, D.: DeepCAPTCHA: an image CAPTCHA based on depth perception. In: Conference: Proceedings of the 5th ACM Multimedia Systems Conference (2014)
9. Nguyen, V., Chow, Y., Susilo, W.: Breaking a 3D-based CAPTCHA scheme. In: ICISC 2011 Proceedings of the 14th International Conference on Information Security and Cryptology, pp. 391–405 (2011)
10. Athanasopoulos, E., Antonatos, S.: Enhanced CAPTCHAs: using animation to tell humans and computers apart. In: Communications and Multimedia Security, pp. 97–108. Springer, Heidelberg (2006)
11. Nguyen, V., Chow, Y., Susilo, W.: Breaking an animated CAPTCHA scheme. In: Proceedings of the 10th International Conference on Applied Cryptography and Network Security (2012)
12. Beltzung, L.: Watson Jeopardy! A Thinking Machine. Vienna University of Technology (2013)
13. Weijers, S.: Exploring Human-Robot Social Relations. University of Twente (2013)
14. Mehndiratta, P., Sachdeva, S., Soni, D.: Detection of sarcasm in text data using deep convolutional neural networks. Scalable Comput. **18**(3), 219–228 (2017)
15. Hempelmann, C., Raskin, V., Triezenberg, K.: Computer, tell me a joke … but please make it funny: computational humor with ontological semantics. In: Proceedings of the Nineteenth International Florida Artificial Intelligence Research Society Conference (2006)
16. Fowler, G.: Are Smartphones Becoming Smart Alecks? Wall Street J. (2011). https://moodle.cs.huji.ac.il/cs12/file.php/67842/SiriJokes.pdf
17. Sivakorn, S., Polakis, J., Keromytis, A.D.: I'm not a human: Breaking the Google reCAPTCHA (2016)

18. Lemos, N., Gomes, J.: Enhanced image captcha. Int. J. Eng. Res. Technol. 2014
19. Huang, Q., Zhang, P., Wu, D., Zhang, L.: Turbo learning for captionbot and drawingbot. In: Proceedings of NeurIPS 2018 (2018)
20. Duskin, O., Feitelson, D.: Distinguishing humans from bots in web search logs. In: WSCD 2009 Proceedings of the 2009 Workshop on Web Search Click Data, pp. 15–19 (2009)
21. Offutt, J., Jin, J., Zheng, N., Mao, F., Koehl, A., Wang, H.: Evasive bots masquerading as human beings on the web. In: 2013 43rd Annual IEEE/IFIP International Conference on Dependable Systems and Networks (DSN) (2013)
22. Winslow, E.: Bot Detection via Mouse Mapping. Stanford University (2009)
23. Walt, E., Eloff, J.: Using Machine Learning to Detect Fake Identities: Bots vs Humans. IEEE Access (2018)
24. Turing, A.: Computing Machinery and Intelligence. Computing Machinery and Intelligence (1950)
25. Priyadarshini, I., Cotton, C., Wang, H.: Some cyberpsychological techniques to distinguish human and robot authentication. In: Advances in Intelligent Systems and Computing. Springer, Cham (2019)

Estimating the Ground Temperature Around Energy Piles Using Artificial Neural Networks

Mohamad Kharseh[1(✉)], Mohamed El koujok[2],
and Holger Wallbaum[1]

[1] Architecture and Civil Engineering, Chalmers University of Technology,
Cothenburg, Sweden
mohamad.kharseh@chalmers.se
[2] CanmetENERGY-Natural Resources Canada, Varennes, Canada

Abstract. Ground source heat pump (GSHP) systems are using vertical ground heat exchangers, known as Borehole Heat Exchangers (BHEs), as a heat source or sink. The performance of the GSHP system strongly relies on the ground temperature surrounding the BHEs. This temperature depends on many parameters and varies during the operating time. Therefore, the determination of the ground temperature is crucial to define the design and the proper size of the BHEs so that the performance of the GSHP system can be kept at the desired level. The current study aims to formulate a complex structure of artificial neural network (ANN) model in a mathematical equation that expresses the change in the ground temperature around BHEs due to heat injection in the long run. To fulfill this aim, a numerical model of BHEs was created using the ANSYS (Analysis System) software to generate data. The generated data was then used to train the ANN model, which was built for this study. The simulation results show that the ANN model estimates the ground temperature (T_g) in the target GSHP system with higher accuracy.

Keywords: Artificial neural network · Ground heat exchanger · Ground source heat pump

1 Introduction

Heating and air-conditioning (A/C) systems account for about 33% of the world's total energy consumption [1–3]. There is a global trend to improve the heating and cooling systems, toward reducing carbon emission. According to the literature, ground source heat pump system (GSHP) seems to be an excellent option to reduce the energy consumption of heating and cooling systems, which leads to a substantial reduction in greenhouse gas emission (GHG) [3–5]. Several studies have been conducted to investigate the feasibility of GSHP systems in residential and commercial buildings [6–8]. Moreover, the GSHP system is considered as one of the fastest growing applications of renewable energy, with the installed capacity yearly growth of 12.3%, as reported in [8]. However, for a given condition, the benefits of GSHP systems in saving energy strongly depends on the temperature of the ground surrounding the ground heat exchanger (GHE). A lower ground temperature means a better performance of the

© Crown 2020
K. Arai et al. (Eds.): FTC 2019, AISC 1069, pp. 223–229, 2020.
https://doi.org/10.1007/978-3-030-32520-6_17

GHSP in the summertime and vice-versa in wintertime. Therefore, a proper design of the GHE means the determination of the spacing between the boreholes in which the supplied temperature from GHE is kept at the desired value. Indeed, the ground temperature varies during system operation due to heat extraction and injection in the winter and summer time, respectively. Therefore, predicting the changes in the ground temperature around the GHE is crucial for the designers and operators as well. The objective of the current work is to apply artificial neural network (ANN) to derive a mathematical equation that assesses the ground temperature around a borehole. For this aim, a fluent-based model has been developed to generate data that is used to train the ANN model and extract the corresponding mathematical equation.

2   Application of ANN for Modeling Ground Heat Exchanger

ANN is considered as a universal approximator for many unknown nonlinear functions (whether simple or complex) [9, 10]. The structure of ANN is not an easy task to determine, that is to say, there is no general rule to define the number of hidden layers and neurons. This depends on the application at hand. However, one hidden layer with several neurons seems to be sufficient and has been used in many different applications for its predictive modeling capabilities [11]. ANN was used to predict the fluid temperature at the outlet of buried pipes for ground source heat pumps. Linear and tangent sigmoid functions were used as the transfer function in output and hidden layers, respectively. Levemberg Marquardt algorithm was used with different numbers of hidden neurons. The best combination was achieved by using 18 neurons in the hidden layer with $R^2 = 0.7714$ (see [12]). The ANN was also used in [13] to simulate a large rectangular cross-sectional area for Earth-to-Air heat exchangers (ETAHE) to preheat/precool the ambient air. Based on simulation data obtained from the computational fluid dynamics (CFD), Zhang and Haghighat utilized ANN to predict ten values of Nusselt number in order to design an optimal ETAHE. In that work, three layers were used, each with 40 neurons. The comparison between the results obtained from the ANN and the results obtained from CFD showed that the ANN model is more successful.

ANN was used as well in another study to predict the performance of a ground-coupled heat pump system [14]. The input parameters were the ground temperature, the inlet and outlet air temperature of condense. The output parameter was the coefficient of performance (COP) of the systems. In this study, the training of the ANN model was done by using 25 data patterns from a total number of 38 data patterns. In that work, three different algorithms were tested to conclude that the best algorithm was the Levenberg-Marquardt (LM) algorithm. The tangent sigmoid function was used for both output and hidden layers with seven neurons resulting in an accuracy of $R^2 = 0.999$. The ANN method was used to compare the performance of vertical and horizontal ground source heat pump system for greenhouse heating in [15]. Three known algorithms were tested. The best algorithm was found to be LM with three neurons in the hidden layer with tangent sigmoid as a transfer function in both layers. The suggested model resulted in $R^2 = 0.9998$ and the minimum root mean square (RMS) error.

3   Methodology

In this work, a fluent based model was developed to simulate the heat exchange between the GHE and the surrounding ground. The model was used to calculate the temperature change of the soil due to heat-injecting into the ground. Figure 1 is a two-dimensional projection of the borehole filed. The simulations were carried out by changing several working parameters. The considered working parameters are the velocity of groundwater "u" (m/s), ground diffusivity "a" (m^2/s), ground porosity "p" (%), heat injection rate per unit length of the borehole "q" (w/m), distance from the boreholes "r" (m), and time elapsed since the heat injection started "t" (h). In addition, the upper and lower limits of each working parameter are considered (see Table 1).

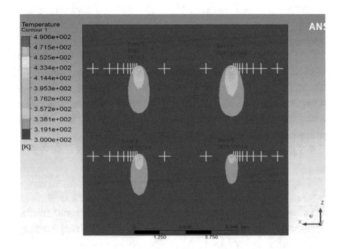

Fig. 1. Temperature distribution obtained from the fluent model.

Table 1. Upper and lower limits of considered working parameters.

Parameter	U (m/s)	A (m^2/s)	P (%)	Q (w/m)	R (m)	t (h)	T$_g$ (K)
Maximum	50E−07	16.8E−07	0.4	30	2.25	100	519.9
Minimum	1E−07	9.6E−07	0.1	15	0.1	0	300.0

The simulation was run for 100 h, and the temperature changes were determined at different distances from the GHE every five minutes. The generated data, i.e., the various working parameters and the corresponding ground temperature, were used to train the ANN. The ANN has been built to create a formula that expresses the relationship between the working parameters and the corresponding ground temperature. It is worth to mention that, in this work, the LM algorithm is used to tune the ANN weights. The LM is a combination of two minimization methods; the gradient descent method and the Gauss-Newton method [16, 17]. It is known by the robustness and rapid convergence to capture the nonlinearity of a complex process [18]. LM algorithm was used with different numbers of neurons in the hidden layer by the trial and error process. Also, the various transfer functions in the hidden layer and the output layer were examined toward obtaining the highest prediction accuracy.

4   Results and Discussions

The Matlab software was used to build the ANN model and to extract a simplified equation. A data set with 900,000 points was used to train the network, and another data set with 78,954 points was used for model testing and validation. To increase the accuracy of the model, different transfer functions in both the hidden layer and the output layer with different neurons were tested. The simulations show that the ten neurons with tangent sigmoid function in the hidden layer and the output layer lead to very high accuracy.

Figure 2 shows the regression curves for training, validation, and testing which is R = 0.99985 for all combined curves. Also, Fig. 2 shows the comparison between the ground temperature values calculated using the fluent base model with the results obtained by the ANN model. As can be seen, the developed ANN model can predict the ground temperature with 99% accuracy. In order to write the equation of ANN, firstly, we determined the scaled value for each parameter. For example, the scaled value of the groundwater movement speed (u) is given as:

$$u = \frac{2 \cdot (U - min(U))}{(max(U) - min(U))} - 1 \tag{1}$$

where U is the value of the groundwater movement, max(U) and min(U) is the maximum and the minimum value of the groundwater movement, respectively, see Table 1.

Secondly, the hidden layer inputs of ANN model can be represented by the following equations:

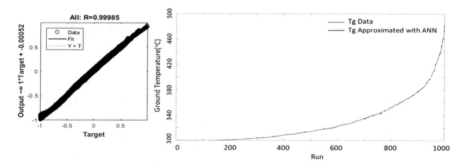

Fig. 2. The regression curves using ten hidden neurons along with the ground temperature calculated using the fluent model as compared with the results obtained by the ANN model.

$$X_1 = b_{1.1} + (w_{1_{1.1}} * u) + (w_{1_{1.2}} * a) + (w_{1_{1.3}} * p) + (w_{1_{1.4}} * q) + (w_{1_{1.5}} * r) + (w_{1_{1.6}} * \tau)$$
$$X_2 = b_{1.2} + (w_{1_{2.1}} * u) + (w_{1_{2.2}} * a) + (w_{1_{2.3}} * p) + (w_{1_{2.4}} * q) + (w_{1_{2.5}} * r) + (w_{1_{2.6}} * \tau)$$
$$X_3 = b_{1.3} + (w_{1_{3.1}} * u) + (w_{1_{3.2}} * a) + (w_{1_{3.3}} * p) + (w_{1_{3.4}} * q) + (w_{1_{3.5}} * r) + (w_{1_{3.6}} * \tau)$$
$$X_4 = b_{1.4} + (w_{1_{4.1}} * u) + (w_{1_{4.2}} * a) + (w_{1_{4.3}} * p) + (w_{1_{4.4}} * q) + (w_{1_{4.5}} * r) + (w_{1_{4.6}} * \tau)$$
$$X_5 = b_{1.5} + (w_{1_{5.1}} * u) + (w_{1_{5.2}} * a) + (w_{1_{5.3}} * p) + (w_{1_{5.4}} * q) + (w_{1_{5.5}} * r) + (w_{1_{5.6}} * \tau)$$
$$X_6 = b_{1.6} + (w_{1_{6.1}} * u) + (w_{1_{6.2}} * a) + (w_{1_{6.3}} * p) + (w_{1_{6.4}} * q) + (w_{1_{6.5}} * r) + (w_{1_{6.6}} * \tau)$$
$$X_7 = b_{1.7} + (w_{1_{7.1}} * u) + (w_{1_{7.2}} * a) + (w_{1_{7.3}} * p) + (w_{1_{7.4}} * q) + (w_{1_{7.5}} * r) + (w_{1_{7.6}} * \tau)$$
$$X_8 = b_{1.8} + (w_{1_{8.1}} * u) + (w_{1_{8.2}} * a) + (w_{1_{8.3}} * p) + (w_{1_{8.4}} * q) + (w_{1_{8.5}} * r) + (w_{1_{8.6}} * \tau)$$
$$X_9 = b_{1.9} + (w_{1_{9.1}} * u) + (w_{1_{9.2}} * a) + (w_{1_{9.3}} * p) + (w_{1_{9.4}} * q) + (w_{1_{9.5}} * r) + (w_{1_{9.6}} * \tau)$$
$$X_{10} = b_{1.10} + (w_{1_{10.1}} * u) + (w_{1_{10.2}} * a) + (w_{1_{10.3}} * p) + (w_{1_{10.4}} * q) + (w_{1_{10.5}} * r) + (w_{1_{10.6}} * \tau)$$

where u, a, p, q, r, and τ are the scaled inputs of the ANN; w_l is the weight factors; b_l is the bias factors which are responsible for linking the input parameters to the input hidden layers. The values of weight and bias factors are listed in Table 2.

Thirdly, the hidden layer outputs are given as:

Table 2. Values of weight and bias factor of 10 hidden neurons model of the current study.

b1	W1						b2	W2
1.141	0.044	−0.080	0.063	−0.009	−0.011	−0.017	6.78	13.988
0.557	−0.169	−1.146	−0.051	−0.104	0.200	−0.677		1.653
2.350	4.662	−3.022	−0.054	−0.045	0.070	0.133		1.064
−2.147	−0.073	−1.917	0.099	0.030	−0.041	0.082		45.856
−0.250	−1.074	1.551	0.114	0.125	−0.158	−0.317		−0.739
2.295	0.071	2.202	−0.078	−0.028	0.050	−0.087		19.661
5.866	−0.208	1.636	−0.162	0.334	0.753	0.520		−10.889
20.270	24.199	−5.973	−0.019	−0.042	0.029	−0.036		31.066
13.918	−0.431	9.713	−0.634	−0.193	0.403	0.383		−41.281
2.012	**0.077**	**1.686**	**−0.129**	**−0.029**	**0.033**	**−0.086**		**24.102**

$$Y_1 = tansig(X_1), \quad Y_2 = tansig(X_2), \quad Y_3 = tansig(X_3), \quad Y_4 = tansig(X_4), \quad Y_5 = tansig(X_5)$$
$$Y_6 = tansig(X_6), \quad Y_7 = tansig(X_7), \quad Y_8 = tansig(X_8), \quad Y_9 = tansig(X_9), \quad Y_{10} = tansig(X_{10})$$

where tansig stands for the hyperbolic tangent sigmoid transfer function, which is as [12]:

$$tansig(n) = \frac{2}{1 + e^{-2 \cdot n}} - 1$$

Fourthly, the output layer input is

$$X_{2-1} = b_{2.1} + (w_{2_{1.1}} * Y_1) + (w_{2_{1.2}} * Y_2) + (w_{2_{1.3}} * Y_3) + (w_{2_{1.4}} * Y_4) + (w_{2_{1.5}} * Y_5)$$
$$+ (w_{2_{1.6}} * Y_6) + (w_{2_{1.7}} * Y_7) + (w_{2_{1.8}} * Y_8) + (w_{2_{1.9}} * Y_9) + (w_{2_{1.10}} * Y_{10})$$

where w_2 are the weight factors; b_2 is the bias factor. Again, the w_2 and b_2 are responsible for linking the hidden layers with the output parameter. Thus, the scaled value of the ground temperature (θ_g) can be calculated with the following equation:

$$\theta_g = tansig(X_{2-1})$$

Finally, the ground temperature (T_g) becomes:

$$T_g = min(T_g) + \frac{(1 + \theta_g) * (max(T_g) - min(T_g))}{2}$$

where $max(T_g)$ and $min(T_g)$ is in respective the maximum and the minimum value of the ground temperature obtained by fluent-base model, see Table 1.

5   Conclusion

The ground temperature plays a crucial role in determining the thermal performance of a ground source heat pump system. In a borehole filed, the ground temperature is affected by ground properties, the spacing distance between the boreholes, and the heat injection rate. This work aims at applying an ANN model to derive a mathematical equation that expresses the ground temperature around a borehole heat exchanger as a function of the working parameters. The considered parameters are the heat injection rate, the thermal diffusivity and the porosity of the ground, the groundwater speed, and the distance from the borehole. For this goal, the ANN model of two layers and ten neurons was built in Matlab. The transfer function in the hidden and output layers was chosen to be a tangent sigmoid function. Data that have been generated previously in the fluent model were used to train and to test the ANN model. The comparison between the measured data (given by fluent model and simulated data generated by the suggested ANN model shows that there is an excellent match between the calculations and the simulation. In other words, the proposed ANN model seems to be trustful to help the designers in determining the ground temperature of borehole filed for different working parameters. For the continuing research, more operating parameters should be considered in the simulation (e.g., borehole diameter or ground heat exchanger design). Also, a broader range of working parameters covers more general conditions.

References

1. Wong, S.L., Wan, K.K.W., Li, D.H.W., et al.: Impact of climate change on residential building envelope cooling loads in subtropical climates. Energy Build. **42**, 2098–2103 (2010)
2. IEA. Renewables for Heating And Cooling. International Energy Agency, Paris (2007). http://www.iea.org/textbase/nppdf/free/2007/Renewable_Heating_Cooling_Final_WEB.pdf
3. Seyboth, K., Beurskens, L., Langniss, O., et al.: Recognising the potential for renewable energy heating and cooling. Energy Policy **36**, 2460–2463 (2008)
4. Al Jaber, S.A.,, Amin, A.Z., Clini, C., et al.: Renewables 2011 Global Status Report. Renewable Energy Policy Network for the 21st Century, Paris (2011). http://www.ren21.net
5. Michopoulos, A., Papakostas, K.T., Kyriakis, N.: Potential of autonomous ground-coupled heat pump system installations in Greece. Appl. Energy **88**, 2122–2129 (2011)
6. Esen, M., Yuksel, T.: Experimental evaluation of using various renewable energy sources for heating a greenhouse. Energy Build. **65**, 340–351 (2013)
7. Balbay, A., Esen, M.: Experimental investigation of using ground source heat pump system for snow melting on pavements and bridge decks. Sci. Res. Essays **5**, 3955–3966 (2010)
8. Lund, J.W., Freeston, D.H., Boyd, T.L.: Direct utilization of geothermal energy 2010 worldwide review. In: World Geothermal Congress, Bali, Indonesia, pp. 1–23 (2010)
9. Bishop, C.M.: Neural Networks for Pattern Recognition. Oxford University Press, Oxford (1995)
10. Beale, M.H., Hagan, M.T., Demuth, H.B.: Neural Network Toolbox^{TM} User's Guide (2017)
11. Hornik, K., Stinchcombe, M., White, H.: Multilayer feedforward networks are universal approximators. Neural Netw. **2**, 359–366 (1989)
12. Demir, H., Atayilmaz, O.S., Agra, O., et al.: Application of artificial neural networks to predict heat transfer from buried pipe for ground source heat pump applications. In: ASME 2013 Heat Transfer Summer Conf. Collocated with the ASME 2013 7th International Conference on Energy Sustainability and the ASME 2013 11th International Conference on Fuel Cell Science, Engineering and Technology, HT 2013. Epub ahead of print 2013. https://doi.org/10.1115/ht2013-17729
13. Zhang, J., Haghighat, F.: Development of artificial neural network based heat convection algorithm for thermal simulation of large rectangular cross-sectional area earth-to-air heat exchangers. Energy Build. **42**, 435–440 (2010)
14. Esen, H., Inalli, M., Sengur, A., et al.: Performance prediction of a ground-coupled heat pump system using artificial neural networks. Expert Syst. Appl. **35**, 1940–1948 (2008)
15. Benli, H.: Performance prediction between horizontal and vertical source heat pump systems for greenhouse heating with the use of artificial neural networks. Heat Mass Transf und Stoffuebertragung **52**, 1707–1724 (2016)
16. Tingleff, O., Madsen, K., Nielsen, H.B.: Methods for non-linear least squares problems. Lect Note Comput Sci 02611 Optim Data Fitting (2004)
17. Lourakis, M.I.A.: A brief description of the Levenberg-Marquardt algorithm implemented by levmar. Found. Res. Technol. **4**, 1–6 (2005)
18. Kermani, B.G., Schiffman, S.S., Nagle, H.T.: Performance of the Levenberg-Marquardt neural network training method in electronic nose applications. Sens. Actuators B Chem. **110**, 13–22 (2005)

Emerging Technology of Man's Life-Long Partnership with Artificial Intelligence

Nicolay Vasilyev[1]($\boxtimes$), Vladimir Gromyko[2], and Stanislav Anosov[3]

[1] Fundamental Sciences, Bauman Moscow State Technical University,
2-d Bauman str., 5, b. 1, 105005 Moscow, Russia
nik8519@yandex.ru
[2] Computational Mathematics and Cybernetics, Lomonosov Moscow State
University, Leninskye Gory, 1-52, 119991 Moscow, Russia
gromyko.vladimir@gmail.com
[3] Public Company Vozrozhdenie Bank, Luchnikov per., 7/4, b. 1,
101000 Moscow, Russia
sanosov@cs.msu.su

Abstract. Computer networks and electronic era dissemination marked end of Gutenberg's epoch big data of knowledge being recorded in Internet (I_{Net}) files. Trans-disciplinary activity impels man to learn and cognate world satiated by wholesome system meanings. Serious problem is revealed: how to strengthen significantly the thinking code of a man for natural life in system-informational culture (SIC). Investigation showed that rational consciousness auto-molding occurs based on fundamental concepts descript in language of categories. Besides, SIC subject incarnation can be achieved only in life-long partnership with artificial intelligence (I_A). Corresponding technology emerges which consists of I_A adaptation to self-reflecting subject. Their reciprocal universal tutoring results in deep-learned I_A and rational man neurophenomenology. Artificial neuro-object will assist man to identify and understand universalities using system axiomatic method on personal cogno-ontological knowledge base grounds.

Keywords: Trans-disciplinary activity · Deep-learned artificial intelligence · Consciousness auto-building · Language of categories · Meaning · System axiomatic method · Cogno-ontological knowledge base · Neuro-object

1 Introduction

Socio-cultural evolution and technical possibilities of SIC inspired deep-learned artificial intelligence (DLI_A) model as universal meanings co-processor adaptive to man [1–3]. Our approach leans on philogenetic trend in anthropogenesis and differs significantly from wide spread ultra-artificial intelligence bionic brain [4]. With the help of sets theory deterministic tools, its modeling restricts the abilities of bionic brain. For the reason, it acts in pre-definite situation controlled by algorithms imitating human behavior. It helps man to solve problems of decisions making, pattern recognition, control, et cetera [5–8]. On the contrary, DLI_A is open system with dynamic organization having interpretive scheme of functioning. It can assist man in intellectual labor itself on the base of system axiomatic method (AM_S) usage. Corresponding emerging

© Springer Nature Switzerland AG 2020
K. Arai et al. (Eds.): FTC 2019, AISC 1069, pp. 230–238, 2020.
https://doi.org/10.1007/978-3-030-32520-6_18

technology in SIC consists of I_A adaptation to self-reflecting mind – natural intelligence (I_N). Discovered laws of gender evolution can be implanted in DLI_A in order that it was capable to assist subject to anticipate and pursue the perspective of intellectual development [1–3]. Life-long partnership of I_N with I_A supposes continuous inter-connection in mutual universal tutoring. Due to it, man will investigate system world issuing from fundamental concepts coordinated with his abilities. Subject's rational auto-development will allow him entering hermeneutics circle of SIC [3, 9]. True general directives and outlines propounded by DLI_A will rule man in meanings iden-tification, study, and comprehension. Only DLI_A is able to support continuously rational consciousness auto-molding for cognition of sophisticated trance-disciplinary culture. Contributed technology is to untwist man's intellectual processes with the help of dynamic cogno-ontological knowledge base (CogOnt) equipped with philogenetic universal essences expressed in language of categories (L_C) [10, 11]. In the paper its substantiation and explanatory examples are given. CogOnt realization is based on theories category description. Working place for I_N and I_A partnership organization will require huge power of super computer calculations because I_A must supply I_N adap-tation to universalities of multi-disciplinary educational space (libraries - M_DL_E).

2   Double Helix of Consciousness

Meanings are the utmost mathematical abstractions manifesting themselves in gener-alizations. Their identification and understanding launch ontological mechanisms of consciousness which auto-build its rational part. De docta ignorantia and knowledge without premises (axiomatic approach) principles are enriched in SIC by linguistic tools discovered in mathematics. There is language of categories (L_C) – strict ontological

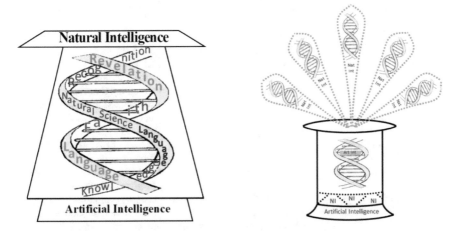

Fig. 1. Consciousness double helix: duality of natural L_N (*faith, revelation*) and natural sciences languages (*knowledge recognition* – identification). On the left - crossing over processes (*double lines*). On the right – universalities auto-folding in L_C (*fractal structure*) as knowledge personal objectization with gender bifurcation root.

means of universalities description. They are ideals expressing functional unity of systems. On their basis, DLI_A is able to assist man to search, identify and understand the most general concepts [12–14].

They are geometrical presentations, system integrity meaning, formalization, calculability, duality, category, limit constructions, interrelations: discrete – continuous, finite – infinite, static – dynamic, division – factorization; and means of coordinatization: number, algebra, function, equivalence, order and relation [15–25]. Man gets to know about his intellectual restrictions by means of auto-reflection. Bringing up adequate aims and right natural laws of thinking form consciousness. Gender enriches, constantly natural languages by scientific meanings [13], see Fig. 1. Thinking code is supported by languages and significantly strengthened by L_C [3, 13]. That is why DLI_A emerging technology leans on model of consciousness double helix (CDH) presented in Fig. 1. Over-crossing of natural, natural sciences, and super natural mathematical languages reconstruct human mind. It happens especially if man pursues aims of auto-development. Inter-discipline activity impels him to do it. SIC subject is occupied by models comparison. CDH auto-folding happens by meanings factorization (compression) with the help of L_C. The operation is especially conspicuous in mathematics. It widely uses Intercourse I_N with DLI_A because ontogenesis repeats anthropogenesis. See Table 1. Man's auto-molding happens by meanings inheritance. Different forms of knowledge apperception answer to the next levels of axiomatic method development.

Table 1. Themes of universal tutoring: CDH stimulation with the help of CogOnt.

Obviousness of abstractions AM_I: categorical theories (function)	Semantic naturalness AM_M: open theories (morphism)	Compulsory identification AM_S: theories in the form of categories (functor)
Space: Eucleidus Descarte Hilbert Weyl Kolmogorov **Algebra:** group, ring, module; structure – Skorniakov Shafarevich **Analysis:** non–standard – Hilbert; standard – Newton Leibniz **Calculation:** machine – Turing; language – McCarty; insoluble - innumerable – Post Church **Means of expression:** sentence – Boole; set – Cantor; predicate – Frege Hilbert; **Factorization in CogOnt:** propeudevtic courses in L_N	**Number**: Hamilton Artin Pontriagin **Space**: non – Archimedes– Hilbert; non – Eucleidus – Lobachevski, Riemann **Universal algebra:** Van der Warden; Kourosh **Analysis**: discrete – Knuth Scott; elemental – Tarski; non-standard – Leibniz Robinson; functional – Fourier Schwarz; probabilities – Kolmogorov; measurable – Lebesgue **Computation**: effective; multi-processor **Means of expression**: multi valued logics; **Factorization in CogOnt:** philogenetic courses of general algebra	**SIC:** I_{NET}, M_D-theca, Applic. **Informatics**: S_I, OOP; ATD; I_{S_i}; D_B; K_B; T_C, T_{I_i}; I_A; M_{DLE}; CogOnt; networks; SC; I-devices **Algebra**: categories; free constructions **Category theories of:** logic; algorithms; linear algebra; analysis; casual values; measure **Means of expression:** ZFC – Zermelo Fraenkel; L_C – McLane, TOPOI – Lauvere **Methods of:** closure – Cantor; evaluation – Carry; transfinite induction – Zorn meta method– Gödel **Factorization in CogOnt:** universalities in the form of L_C commutative diagrams

Partnership of I_N with DLI_A can use intellectual interface (T_{I_i}) in L_C and integrated computer systems I_{S_i}. Cloud technology (T_C) can support semantic processing of data bases (D_B) for knowledge bases (K_B) building. DLI_A realization can lean on object-oriented programming (OOP), abstract types of data (ATD) usage in interpretive scheme of meanings co-processing. Gender always created I_A on constructive grounds. Computer systems (S_I) are built inductively but needs conceptual grasping.

In its functioning, personal adaptive form of CogOnt can use verified tutoring approaches. It must apply true consequences from sciences origins [16, 17] expressed universally with the help of functional language [19, 20]. Only courses having philo-genetic significance must be applied in CogOnt, for instance [15–25]. Knowledge presented in axiomatic form allows different models comparison supporting unity of real tools and ideal presentations, see Table 1.

3 Axiomatic Method

Axiomatic method is modern means of knowledge presentation in the form of verifiable rational minutes, see Table 1. In SIC, AM was put forward by inter-discipline activity. Man must be literate enough to anticipate the outlook of his investigation and perceive at once related fundamental laws. Axiomatic approach requires new quality of thinking. Unity of inductive and within reach means and ontological receptive essences is provided on genomic level helping forward man's complexity, see Fig. 1, Table 1. On the basis, DLI_A acts applying useful linguistic tools, geometrical interpretations, and algebraic operations. Man can lean on maps universality. The approach is crowned by morphisms in mathematics. *Mathesis universalis* of L_C is adequate axiomatic means of modeling in SIC [10, 11].

Example 1. *Thinking in terms of functions* forces man to reason strictly and clearly. Maps description points out their domains and co-domains and defines their general properties. For instance, fundamentality of linear function $f : U \to V$ is consequence and generalization of values direct proportionality. In outline and it is the most important, f is morphism in category of vector spaces over a field K. The presentation corresponds to CDH auto-folding causing some questions. Is this property will be conserved after field K extension? Is axiom $(\forall u, u')f(u + u') = f(u) + f(u')$ sufficient for f linearity? Is linear function are necessary continuous in case of infinite dimensional space over total field? Traditional teaching pays usually main attention to values calculation that cannot result in mind development.

Example 2. DLI_A *and inverse way of tutoring.* Trigonometrical functions were discovered while segments and angles were measured. The maps introduce equivalence relations on real line R:

$$R_f = \{(x_1, x_2) \in R^2 : f(x_1) = f(x_2)\}.$$

In *universum* of all categories the notion can be expressed universally in the form of kernel relation R_f. It is defined by means of next commutative diagram (Fig. 2) [10]:

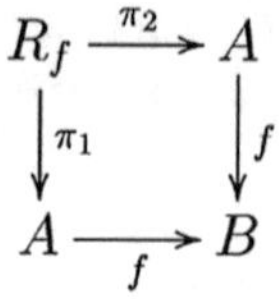

$$R_f \xrightarrow{\pi_2} A$$
$$\downarrow{\pi_1} \qquad \downarrow{f}$$
$$A \xrightarrow{f} B$$

$$R_f = A \underset{B}{\times} A = \{(x,y) : x,y \in A\};$$
$$(x,y)\pi_1 = x, (x,y)\pi_2 = y;$$
$$\pi_1 f = \pi_2 f.$$

Fig. 2. Commutative diagram of kernel relation.

It is particular case $(g = f)$ of inverse image of maps f, g. Due to relation R_f, natural map $f_{R_f} : A \rightarrow A/R_f$ and factor-object A/R_f emerge. The latter corresponds to initial object A division in equivalence classes. The factorization (Table 1) can be expressed otherwise by means of projections π_1, π_2 co-equalizer f_{R_f}[10, 11] (see Fig. 3):

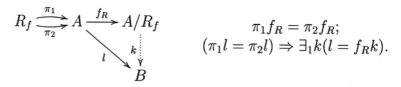

$$\pi_1 f_R = \pi_2 f_R;$$
$$(\pi_1 l = \pi_2 l) \Rightarrow \exists_1 k(l = f_R k).$$

Fig. 3. Commutative diagram of co-equalizer.

Displayed generalization perspective answering to system AM (AM_S) becomes stable being complemented by real tools of initial AM (AM_I). There is intermediate ideal on the level of modern AM (AM_M) which is simplier than category one, see Table 1. In mathematics, particular always represents general. Therefore, factorization analysis for function $f = \sin$ helps to understand these universal constructions. Elements of factor-set $R/R_f \tilde{=} S_1$ are points of unit circle $[a] \in S_1$ answering to equivalent numbers $a' R_f a$. They are classes

$$[a] = \{a' : (\exists k \in Z)a' = \varphi_a + 2\pi k\}, f_{R_f}(a) = [a] \tilde{=} \varphi_a, \varphi_a \in S_1.$$

Mathematical glottogenesis widens little by little its field of action, see Table 1.

Example 3. Equivalence relation on a set A is introduced:

(i)   on elements basis – level of AM_I:

$$\rho \subset A \times A; \ a,b,c \in A : \{a\rho a, \ (a\rho b) \Rightarrow (b\rho a), \ (a\rho b) \wedge (b\rho c) \Rightarrow (a\rho c)\}.$$

(ii) in relational data bases algebra – level of AM_M:

$$\tau \subseteq \rho, \ \rho^{-1} \subseteq \rho, \ \rho^2 \subseteq \rho \Leftrightarrow \tau \subseteq \rho, \ \rho^{-1} = \rho, \ \rho^2 = \rho. \tag{1}$$

Algebraic systems language $\langle \mathfrak{P}(A \times A) \mid \cup, \cap, \subseteq, ^-, \emptyset, \widehat{\ }, \circ, ^{-1}, \tau \rangle$ is used in (1). Basic operations of completing, transitive closure, production, inverse, and unit are applied in (1) to arbitrary relations.

There is inductive self-obviousness of equivalence definition (i). Besides, nature of set elements is not important. AM_M level (ii) enriches approach (i) by presentation about equivalence inner closure respectively operations (1). Algebraic view requires higher level of thinking than (i) and suggests future investigation. Are axioms (ii) independent? Can other relations $(\rho_1 \odot \rho_2, \overline{\rho}, \rho^{-1}), \odot \in \{\cup, \cap, \circ\}$ be continued up to equivalence? Thinking in L_C on AM_S level applies visually clear universal constructions of inverse image and co-equalizer, see Example 2.

Factorization is used for new objects building, see Table 1. For instance, china theorem on residues for mutually prime elements [18, 23] can be expressed universally in category of rings. Moreover,

Example 4. Concept of mutually prime elements p, q in a ring W can be defined by next commutative diagram (Fig. 4):

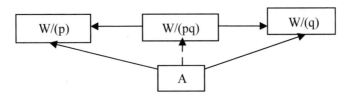

Fig. 4. Universal construction of objects product is used here [10].

4 Artificial Intelligence Technology

DLI_A is technical system capable to meanings generation and co-processing. Semantic level of natural and artificial intelligences partnership is maintained by *CogOnt* built in L_C [1]. It contains system of universal constructions expressed in the form of commutative diagrams, see Examples 2 and 4. By means of factorization, data-base (D_B) is transformed in system of super natural mathematical knowledge. It is done by functor $CogOnt : D_B \to SS_{K_{S_N}}$. All levels of AM are applied to achieve the goal, see Examples 2–4, Table 1. Aim of DLI_A functioning is to turn man's auto-reflection on rational consciousness auto-development by means of CDH auto-folding process, see Fig. 1. It directs person's energies in universal trend of learning in frames of his inter-discipline activity. It coordinates man's intellectual processes with his abilities, see Figs. 5 and 6.

- Person's auto-reflection controls results of interaction with I_A. Cognition success is intellectual break to meanings understanding. It manifests itself if studied knowledge becomes self-obvious. DLI_A is occupied by true interpretation of arising contexts (Ct) from interaction with man, see Figs. 5 and 6. The latter are conserved in artificial neuro-structure in semantic form. Due to it, DLI_A can display universal mathematical constructions (UMA) on examples selected from different categories already studied by man.

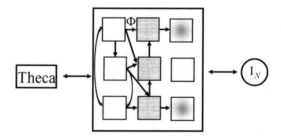

Fig. 5. Neuro-object: versatile architecture of cognatoms network.

- Main subject's objective is to learn to identify UMA and think in L_C [14]. For the purpose, internal *CogOnt* processes are to be specialized by neuro-structure acting adaptively to man. On the base of abstract types of data (ATD) DLI_A can untwist person's thinking code, see Fig. 6. Cognatoms states and their emergent neuro-connections $Cog_a \to Cog_{a'}$ coordinate neuro-object functioning, see Figs. 5 and 6. They are modified by functors $\Phi_a : Cat_i^a \to Ct_a$ of specialization establishing neuro-object inner connections. Their transformations are used for continues man's tuning in universalities.

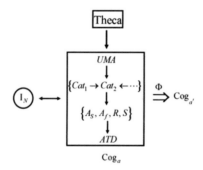

Fig. 6. Cognatom: universalities (*UMA*), (quasi-)categories net ($\{Cat_i\}$), algebraic system (A_s), free algebra (A_f), relations (R), sets (S), tools - ATD.

5   Conclusions

Emerging technology of man's partnership with DLI_A is contributed in order to grow up person capable for successful trance-disciplinary activity. Universalities problem can be resolved personally with the help of DLI_A. It will supply subject with third world objectization. Man's auto-molding is directed on his rational sophistication. With the help of DLI_A person's self-reflection allows promoting adaptive universal tutoring for cognogenesis support. Rational consciousness is to be equipped with functional language of categories thus strengthening and enriching man's thinking code for SIC meanings fostering. De *docta ignorantia* principle took form of axiomatic method as means of knowledge identification, presentation, and study. It is used for cogno-ontological bases building. Auto- and self-reflection removes divergence between ideal ontological constructions of thinking and real inductive means of knowledge implementation in computer systems. Partnership with DLI_A allows this synchronization connecting true fundamental presentations with real work. Their tuning auto-molds subject's rational consciousness and results in his natural existence in sophisticated system world. In future, the unity of natural and artificial intelligences will become possible and SIC will be informatics for mathetics.

References

1. Gromyko, V.I., Kazaryan, V.P., Vasilyev, N.S., Simakin, A.G., Anosov, S.S.: Artificial intelligence as tutoring partner for human intellect. J. Adv. Intel. Syst. Comput. **658**, 238–247 (2018)
2. Gromyko, V.I., Vasilyev, N.S.: Mathematical modeling of deep-learned artificial intelligence and axiomatic for system-informational culture. Int. J. Robot. Autom. **4**(4), 245–246 (2018)
3. Vasilyev, N.S., Gromyko, V.I., Anosov, S.S.: On inverse problem of artificial intelligence in system-informational culture. J. Adv. Intel. Syst. Comput. Hum. Syst. Eng. Des. **876**, 627–633 (2019)
4. Shaikh, M.S.: Defining ultra artificial intelligence (UAI) implementation using bionic (biological-like-electronics) brain engineering insight. MOJ Appl. Bio. Biomech. **2**(2), 127–128 (2018)
5. Deviatkov, V.V., Lychkov, I.I.: Recognition of dynamical situations on the basis of fuzzy finite state machines. In: International Conferences on Computer Graphics, Visualization, Computer Vision and Image Processing and Big Data Analytics, Data Mining and Computational Intelligence, pp. 103–109 (2017)
6. Fedotova, A.V., Davydenko, I.T., Pförtner, A.: Design intelligent lifecycle management systems based on applying of semantic technologies. J. Adv. Intell. Syst. Comput. **450**, 251–260 (2016)
7. Volodin, S.Y., Mikhaylov, B.B., Yuschenko, A.S.: Autonomous robot control in partially undetermined world via fuzzy logic. J. Mech. Mach. Sci. **22**, 197–203 (2014)
8. Svyatkina, M.N., Tarassov, V.B., Dolgiy, A.I.: Logical-algebraic methods in constructing cognitive sensors for railway infrastructure intelligent monitoring system. Adv. Intell. Syst. Comput. **450**, 191–206 (2016)
9. Hadamer, G.: Actuality of Beautiful. Art, Moscow (1991)
10. McLane, S.: Categories for Working Mathematician, Phys. Math. Ed., p. 352. Moscow (2004)
11. Goldblatt, R.: The Categorical Analysis of Logic. North-Holland Publishing Company, Amsterdam, New York, and Oxford (1979)
12. Husserl, A.: From idea to pure phenomenology and phenomenological philosophy. In: General Introduction in Pure Phenomenology, Book 1, Acad Project, Moscow (2009)
13. Pinker, S.: Thinking substance. Language as window in human nature. Librokom, Moscow (2013)
14. Kassirer, E.: Philosophy of symbolical forms. Language. Univ. book, Moscow, St.-Pet. 1 (2000)
15. Courant, R., Robbins, G.: What is Mathematics? Moscow Center for Continuous Mathematical Education, Moscow (2017)
16. Euclid, "Elements", GosTechIzd., Moscow – Leningrad (1949–1951)
17. Hilbert, D.: Grounds of geometry, Tech.-Teor. Lit., Moscow, Leningrad (1948)

18. Kirillov, A.: What is the Number? Nauka, Moscow (1993)
19. Artin, E.: Geometric Algebra. Nauka, Moscow (1969)
20. Bachman, F.: Geometry Construction on the Base of Symmetry Notion. Nauka, Moscow (1969)
21. Maltsev, A.: Algebraic Systems. Nauka, Moscow (1970)
22. Maltsev, A.I.: Algorithms and Recursive Functions. Nauka, Moscow (1986)
23. Shafarevich, I.R.: Main Notions of Algebra. Reg. and chaos dynam., Izhevck (2001)
24. Kourosh, A.: Lecture Notes on General Algebra. Phys.-mat., Moscow (1962)
25. Engeler, E.: Metamathematik der Elementarmathematik. Mir, Moscow (1987)

Making Food with the Mind: Integrating Brain-Computer Interface and 3D Food Fabrication

Nutchanon Ninyawee[1], Tawan Thintawornkul[1], Pat Pataranutaporn[1,2(✉)],
Bank Ngamarunchot[1], Sirawaj Sean Itthipuripat[1], Theerawit Wilaiprasitporn[3],
Kotchakan Promnara[1], Potiwat Ngamkajornwiwat[1],
and Werasak Surareungchai[1]

[1] Futuristic Research Cluster (FREAK Lab), Bangkok 10150, Thailand
ninyawee@freaklab.org
[2] Massachusetts Institute of Technology (MIT), Cambridge, MA 02139, USA
patpat@media.mit.edu
[3] Vidyasirimedhi Institute of Science and Technology (VISTEC), Rayong, Thailand

Abstract. We presented "Mind-Controlled 3D Printer" that translates
brain signals from the user into 3D printed food. This system integrated
an EEG recording device that measures neural activities in real-time with
a machine learning algorithm that classify emotional valence and arousal
levels, which determine the shape and size of the food fabricated by the
food 3D printer. This research introduced the opportunity for combining
brain-computer interface (BCI), affective computing, and additive manufacturing technology, which will ultimately enable the automation of
mind to matter materialization. We demonstrated three use cases and
envisioned the future research on BCI and food fabrication.

Keywords: Brain-Computer Interface · 3D fabrication · Food
printing · Edible material · Cognitive Food · Affective computing

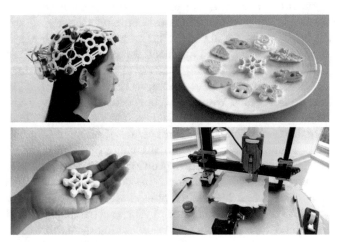

© Springer Nature Switzerland AG 2020
K. Arai et al. (Eds.): FTC 2019, AISC 1069, pp. 239–247, 2020.
https://doi.org/10.1007/978-3-030-32520-6_19

1   Introduction

1.1   Food and Digital Interfaces

Identifying novel opportunities for computer to weave into human daily activities such as preparing and consuming food is one of the emerging focuses of HCI (Human-Computer Interaction) research community. The ubiquitous nature of computing allows researchers to develop technology that enhances human's experiences using food as the medium of digital computation; adding new functionality and meanings to them. For example, Maynes-Aminzade has introduced the concept of "Edible User Interface (EUI)", a multi-sensory design paradigm that utilizes edible materials to create novel interfaces such as "TasteScreen", an LCD screen that allows the user to taste different flavors by licking liquid residue released on the screen [9]. Murer et al. designed "LOLLio", an edible interface made of lollipop for playing games [12]. Narumi et al. designed "Meta Cookie", an augmented reality system that changes the perceived taste of food by overlaying visual and olfactory information onto a real food [14]. Finally, Burneleit et al. presented "Impatient Toaster", a food machine that could motivate the user to "eat more often and in regular intervals" by showing nervous and hungry movements if the user has not used it in a while. This project investigated how the life-like behaviour could increasing user's sympathy for everyday object [1].

1.2   Food Fabrication

Beyond food as digital interfaces, researchers have explored the process for food fabrication using digital technology. Zoran and Coelho introduced the concept of "digital gastronomy", which introduced the idea on integrating digital fabrication technologies into kitchen, allowing the human to design and manipulate food with precision [31]. Lee et al. demonstrated the concept of digital gastronomy using additive manufacturing technology such as 3D printer to create custom tools that assist the process of cooking dumplings in various designed shapes [8]. Recently, the advancement of 3D printing technology allows the fabrication of edible materials to be possible. The most common technique is extruding the material layer by layer into a 3D shape [21]. Companies such as, Foodini uses the prepared food pastes as the ingredient for printing, while PancakeBot^{TM} extrudes pancake dough on a heating plate to create edible drawings [22].

Since then, HCI researchers have embraced the intersection of food and HCI as an emerging design space. Wang et al. purposed the futuristic dining experience, where the foods can change their shapes. They developed "Transformative Appetite", a digitally fabricated 2D films made of edible materials that could transform into customized 3D foods when absorbing water [27]. Wei et al. studied the user's perception on the use of food printing as a way to send messages. The results show a strong acceptance of food messaging as an alternative message channel and a new way of gifting [28]. Khot et al. developed "EdiPulse", a system that print chocolate treats based on user physical activities [5]. The shape of the 3D printed chocolates represent the heart rate aiming to support the user's physical activity. We were inspired by the spectrum of research that demonstrates how computer could play a role in mediating and augmenting human experiences and interactions with food. As research in HCI have expanded beyond designing and evaluating work-oriented applications towards dealing with leisure-oriented applications [3], research around affective computing that recognizes human affects becomes increasingly important since emotion plays an essential role not only in human creativity and intelligence but also in decision-making [19]. We wonder if the future of food technology can be affective, where the fabrication of food is controlled directly from our mind and be responsive to our emotion.

2   Our Design

We present "Mind-Controlled 3D Printer" that translates EEG signals from the brain into 3D printed food (see Fig. 1). The system used a machine learning algorithm trained on the EEG data set to classify human's emotional valence and its arousal level from the user to determine the shape and size of the printed food. We implemented three use cases: "interpretative food", "responsive food", and "talk back food" to demonstrate the applications of the mind-controlled 3D printer. This research makes the following contributions to the HCI community: (1) We introduced the opportunity of combining brain-computer interface (BCI), affective computing, and additive manufacturing technology, which will ultimately enable the automation of mind to matter materialization. (2) We demonstrated three applications of the mind-controlled 3D printer, which could guide the future research on BCI and food fabrication. Finally, (3) we conceptualize "Cognitive Food", our vision on the future of the mind, machine and food. The mind-controlled 3D printer consists of three major parts: (1) the brain-computer interface device, (2) the emotion recognition system, and (3) the 3D fabrication platform.

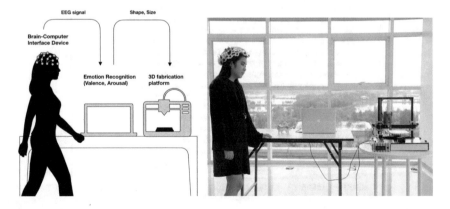

Fig. 1. The diagram and the overall system of the Minded Controlled 3D Printer.

2.1   BCI and Emotion Recognition

Table 1. DEAP data set representation for each participant

Array name	Array shape	Array contents
Data	40 × 40 × 8064	Video/trial × channel × data
Labels	40 × 2	Video/trial × label (valence, arousal)

Electroencephalography (EEG), a non-invasive method for monitoring brain electrophysiological signals was deployed to access the user's emotional state. To read EEG signals from human users, we used an open source brain-computer interface device OpenBCI with 16 channels of electrode inputs. The data of the user is live stream through OpenBCI software to the emotion recognition system based on the machine learning classifier and data pipeline presented by Tripathi et al. [24]. We used the preexisting EEG data set recorded from 32 human adults while they were watching stimulative videos that evoked different emotional states (DEAP data set, see Table 1) [6] to train the machine learning algorithm. To make our model suitable for real-time classification, we improved upon the work of Tripathi et al. by first, deploying Gradient Boosting Decision Tree (GBDT) [4] classifier on two variables: valence and arousal. We created two GBDT models on the full readings (10 batches) and summed the important feature scores for each model (see Fig. 2).

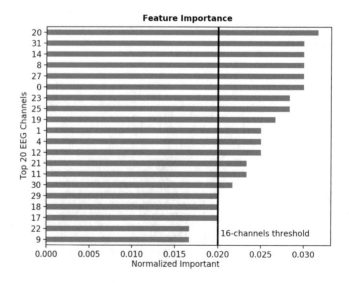

Fig. 2. The feature importance score for the emotion recognition model

We found that the first 24 s of the readings can provide the same accuracy as the full readings while shortening the input period. Therefore, we optimized the EEG signal reading frame from the 60 s to 24 s (from 10 batches to 4 batches), removed all non-EEG channels (33–40), and finally removed participant id and experimental id from DEAP data set. We optimized the number of input channels from 32 inputs to 16 inputs by ranking the weight of each channel on the accuracy of the prediction. Our results showed that the GBDT and GBDT optimized for real-time classification could perform better compared to the original model (see Table 2).

The results from the EEG classifier on valence and arousal are used for determining emotion. We use Thayer's emotion plane, a commonly adopted approach for defining the class of emotion in terms of arousal (exciting/calming) and valence (positive/negative) [23]. Combining the two inputs, the system could classify the user into four emotional states: angry (high arousal, negative valence), happy (high arousal, positive valence), bored (low arousal, negative valence), and peaceful (low arousal, positive valence). Finally, these four quadrants represent the output that the emotion recognition system used to communicate with the 3D fabrication process.

Table 2. The accuracy scores from different machine learning models

Model	Valence accuracy	Arousal accuracy
Tripathi et al. DNN model	58.44%	55.70%
Tripathi et al. CNN model	66.79%	57.58%
Our GBDT	79.75%	66.71%
Our real time GBDT	59.87%	61.10%

2.2   3D Fabrication Platform and Food Creation

We developed the 3D food printer by adding a semi-liquid extruder, a mechanical device that pushes the syringe containing edible materials to form the three-dimensional shape. We explored various edible materials for printing, but finally decided to use royal icing as it is highly customizable (in terms of shape, color, and density), easy to prepare, and wildly adopted as part of desserts such as cookie, cake, etc. Prior to connecting the 3D printer with the BCI device, the condition for printing was optimized in term of the ingredient, printing speed, extrusion rate, layer-height, and nozzle diameter to accommodate the physical properties of royal icing. The food printer was operated on the Modified Marlin open-source firmware and connected with the emotion recognition system through Serial Terminal using Printrun Pronsole via ASCII code. When the user put on the OpenBCI and the device started picking up the signals, the emotion recognition system would send the initial code to initialize the printer. When the process of the emotion classification is completed, the algorithm would send the second code with the value of the emotion for the printer to start printing food accordingly.

3   Use Cases

We proposed three potential use cases for the mind-controlled 3D printer by implementing three processes of interpreting EEG signal into the food (see Fig. 3).

Fig. 3. (Left) The 3D food printing process using edible material (Right) Different shapes of food created based on user's emotional states

First, "responsive food" uses a degree of arousal and valence to determine the portion quantity of the printed food. This process serves as the way to personalize the intake of individuals based on their cognitive states. For example, researchers had identified the effects of stress on food choices and intakes [17,18,29] showing that "stress could induce secretion of glucocorticoids, which increases motivation for food, and insulin that promote food intake and obesity" [2]. Therefore, using technology to detect and forecast the level of stress [25] in order to fabricate the right quantity of food to counteract the condition could be impactful for the user's health and well-being.

Second, "interpretative food" uses the four emotional states to determine the shape of the 3D printed food in order to reflect the emotional state of the user. For example, the user could program the 3D food printer to print a flower-shape breakfast if the device sense the positive emotion in the morning. Researchers had identified how different shapes and forms could embody certain emotions through connecting the visual design space with the affective space [10,11]. Applying this principle to the edible materials, the printed food could be use as the embodiment of emotion for daily reflection and self-realization.

Finally, "talk back" food responded to the user by generating texts to counter the user's emotion. For instance, if the user is having a sad emotion, the machine could print cheering edible messages such as "Good day ahead!" for the user. Building on top of prior research in HCI that show a strong acceptance of food as an alternative medium for messaging [28] and communication [5], this approach aims to use food as the medium to not only reflect but also transform human's emotion for the well-being of the user.

4   Conclusion and Our Vision for the Future

As Brain-Computer Interface become more ubiquitous, commercialized [13,30], and intergrated with our everyday life. The applications of BCI and affective computing would enhance our interaction with smart environments [7]. In the long run, the mind-controlled food printer allowed us to envision the future of "Cognitive Food", where food technology that can be affective and controlled by the user's mind. Food as the edible medium of computation could be used to counteract emotionally challenging situations and enhance everyday interaction. We also envision the use of 3D printer for the fabrication of various tastes [26] that have different psychological effects on human perception to also play a critical role in the future of food HCI [15,16].

The cognitive food technology presented in this paper showed a range of possibilities from using food as a mental reflection to a tool aiming for changing human's emotional state. This project demonstrated how technology could seamlessly weave itself into human daily life and play a role during our dining experience, engaging with emerging fields of neuroscience and assistive augmentation [20]. Therefore, we believe that this work would be significant and inspiring for designing the future of food technology, 3D fabrication technology, and brain-computer interface research.

References

1. Burneleit, E., Hemmert, F., Wettach, R.: Living interfaces: the impatient toaster. In: Proceedings of the 3rd International Conference on Tangible and Embedded Interaction, pp. 21–22. ACM (2009)
2. Dallman, M.F.: Stress-induced obesity and the emotional nervous system. Trends Endocrinol. Metab. **21**(3), 159–165 (2010)
3. Höök, K.: Affective computing (2012)
4. Ke, G., Meng, Q., Finley, T., Wang, T., Chen, W., Ma, W., Ye, Q., Liu, T.Y.: LightGBM: a highly efficient gradient boosting decision tree. In: Advances in Neural Information Processing Systems, pp. 3146–3154 (2017)
5. Khot, R.A., Pennings, R., Mueller, F.: EdiPulse: supporting physical activity with chocolate printed messages. In: Proceedings of the 33rd Annual ACM Conference Extended Abstracts on Human Factors in Computing Systems, pp. 1391–1396. ACM (2015)
6. Koelstra, S., Muhl, C., Soleymani, M., Lee, J.S., Yazdani, A., Ebrahimi, T., Pun, T., Nijholt, A., Patras, I.: DEAP: a database for emotion analysis; using physiological signals. IEEE Trans. Affect. Comput. **3**(1), 18–31 (2012)
7. Kosmyna, N., Tarpin-Bernard, F., Bonnefond, N., Rivet, B.: Feasibility of BCI control in a realistic smart home environment. Front. Hum. Neurosci. **10**, 416 (2016)
8. Lee, B., Hong, J., Surh, J., Saakes, D.: Ori-mandu: Korean dumpling into whatever shape you want. In: Proceedings of the 2017 Conference on Designing Interactive Systems, pp. 929–941. ACM (2017)
9. Maynes-Aminzade, D.: Edible bits: seamless interfaces between people, data and food. In: Conference on Human Factors in Computing Systems (CHI 2005)-Extended Abstracts, pp. 2207–2210. Citeseer (2005)

10. Melcer, E., Isbister, K.: Motion, emotion, and form: exploring affective dimensions of shape. In: Proceedings of the 2016 CHI Conference Extended Abstracts on Human Factors in Computing Systems, pp. 1430–1437. ACM (2016)
11. Mothersill, P., Bove, Jr., V.M.: The emotivemodeler: an emotive form design cad tool. In: Proceedings of the 33rd Annual ACM Conference Extended Abstracts on Human Factors in Computing Systems, CHI EA 2015, pp. 339–342. ACM, New York (2015). https://doi.org/10.1145/2702613.2725433, http://doi.acm.org/10.1145/2702613.2725433
12. Murer, M., Aslan, I., Tscheligi, M.: LOLLio: exploring taste as playful modality. In: Proceedings of the 7th International Conference on Tangible, Embedded and Embodied Interaction, pp. 299–302. ACM (2013)
13. Musk, E., et al.: An integrated brain-machine interface platform with thousands of channels. BioRxiv p. 703801 (2019)
14. Narumi, T., Kajinami, T., Tanikawa, T., Hirose, M.: Meta cookie. In: ACM SIGGRAPH 2010 Emerging Technologies, SIGGRAPH 2010, pp. 18:1–18:1. ACM, New York (2010). https://doi.org/10.1145/1836821.1836839, http://doi.acm.org/10.1145/1836821.1836839
15. Obrist, M.: Don't just look – smell, taste, and feel the interaction. In: Proceedings of the 26th ACM International Conference on Multimedia, MM 2018, pp. 182. ACM, New York (2018). https://doi.org/10.1145/3240508.3267343, http://doi.acm.org/10.1145/3240508.3267343
16. Obrist, M., Marti, P., Velasco, C., Tu, Y.T., Narumi, T., Møller, N.L.H.: The future of computing and food: extended abstract. In: Proceedings of the 2018 International Conference on Advanced Visual Interfaces, AVI 2018, pp. 5:1–5:3. ACM, New York (2018). https://doi.org/10.1145/3206505.3206605, http://doi.acm.org/10.1145/3206505.3206605
17. Oliver, G., Wardle, J.: Perceived effects of stress on food choice. Physiol. Behav. **66**(3), 511–515 (1999)
18. Oliver, G., Wardle, J., Gibson, E.L.: Stress and food choice: a laboratory study. Psychosom. Med. **62**(6), 853–865 (2000)
19. Picard, R.W.: Affective Computing. MIT Press, Cambridge (2000)
20. Shilkrot, R., Huber, J., Boldu, R., Maes, P., Nanayakkara, S.: Assistive Augmentation. Springer, Singapore (2018)
21. Sun, J., Zhou, W., Yan, L., Huang, D., Lin, L.Y.: Extrusion-based food printing for digitalized food design and nutrition control. J. Food Eng. **220**, 1–11 (2018)
22. Tan, C., Toh, W.Y., Wong, G., Lin, L.: Extrusion-based 3D food printing–materials and machines (2018)
23. Thayer, R.E.: The Biopsychology of Mood and Arousal. Oxford University Press, New York (1989)
24. Tripathi, S., Acharya, S., Sharma, R.D., Mittal, S., Bhattacharya, S.: Using deep and convolutional neural networks for accurate emotion classification on deap dataset. In: Twenty-Ninth IAAI Conference (2017)
25. Umematsu, T., Sano, A., Taylor, S., Picard, R.: Improving students' daily lifestress forecasting using LSTM neural networks. In: IEEE-EMBS Biomedical and Health Informatics 2019 (2019)
26. Vi, C.T., Ablart, D., Arthur, D., Obrist, M.: Gustatory interface: the challenges of 'how' to stimulate the sense of taste. In: Proceedings of the 2Nd ACM SIGCHI International Workshop on Multisensory Approaches to Human-Food Interaction, MHFI 2017, pp. 29–33. , ACM, New York (2017). https://doi.org/10.1145/3141788.3141794, http://doi.acm.org/10.1145/3141788.3141794

27. Wang, W., Yao, L., Zhang, T., Cheng, C.Y., Levine, D., Ishii, H.: Transformative appetite: shape-changing food transforms from 2D to 3D by water interaction through cooking. In: Proceedings of the 2017 CHI Conference on Human Factors in Computing Systems, pp. 6123–6132. ACM (2017)
28. Wei, J., Ma, X., Zhao, S.: Food messaging: using edible medium for social messaging. In: Proceedings of the SIGCHI Conference on Human Factors in Computing Systems, pp. 2873–2882. ACM (2014)
29. Zellner, D.A., Loaiza, S., Gonzalez, Z., Pita, J., Morales, J., Pecora, D., Wolf, A.: Food selection changes under stress. Physiol. Behav. **87**(4), 789–793 (2006)
30. Zhang, B., Wang, J., Fuhlbrigge, T.: A review of the commercial brain-computer interface technology from perspective of industrial robotics. In: 2010 IEEE International Conference on Automation and Logistics, pp. 379–384. IEEE (2010)
31. Zoran, A., Coelho, M.: Cornucopia: the concept of digital gastronomy. Leonardo **44**(5), 425–431 (2011)

Neuroscience of Creativity in Human Computer Interaction

Ali Algarni[(✉)]

University of North Carolina at Charlotte, Charlotte, NC 28223, USA
aalgarnl@uncc.edu

Abstract. Human computer interaction (HCI) has a rapid growth in designing creative systems and AI agents. The neuroscience of creativity uses different imaging systems, such as electroencephalogram (EEG), to visualize and interpret the brain activations with respect to creativity. This paper presents a survey on the contribution of the neuroscience of creativity in HCI research. It covers two HCI areas, which are computational creativity and brain computer interfaces (BCI). Computational creativity includes two categories: creativity support tools (CST) and co-creative agents. The discussion section compares several studies in term creativity patterns, creative tasks, and neuroscience efforts. There are gaps between neuroscience of creativity and HCI that need more contributions. Future works include evaluating or developing CSTs, improving creative tasks used in computational creativity, or proposing neuroscience of creativity theories in designing creative systems.

Keywords: Human computer interaction · Computational creativity · Brain computer interface · Creativity · Cognitive science · Neuroscience

1   Introduction

Human computer interaction (HCI) is a computer science field that develops and designs computer interfaces. It aims to create an interactive environment between humans and software agents. It combines different disciplines, such as computer programming, artificial intelligence (AI) and cognitive science, to build collaborative and interactive tools. This paper focuses deeply on HCI, creativity, and neuroscience. The survey also includes brain computer interfaces (BCI) and computational creativity systems.

Creative cognition is the study of creativity from a cognitive science perspective. It investigates the cognitive processes and mechanisms that occur in creativity such as defocused attention and high sensitivity. In addition, creativity can also be validated using some cognitive theories that are associated with generating novel ideas. The neuroscience of creativity examines creativity from a neuroscience perspective. It studies the functions and structures of the nervous system and brain with respect to creativity. Brain activations are changes in the brain states when there are a set of stimuli received. The activations are visualized using several imaging systems. These neuroscience tools are different based on the technologies embedded, data acquisition and analysis algorithms, brain areas to be captured, and the study needs. Brain activations

© Springer Nature Switzerland AG 2020
K. Arai et al. (Eds.): FTC 2019, AISC 1069, pp. 248–262, 2020.
https://doi.org/10.1007/978-3-030-32520-6_20

can be linked to creative cognition theories for creativity analysis. In summary, creativity analysis needs both cognitive mechanisms and neuroscience activations results.

Computational creativity is a multidisciplinary endeavor to study and model human creativity using a computer. Davis reported that computational creativity has three domains: creativity support tools (CST), generative computational creativity systems, and computer colleagues [2]. CSTs develop software and user interfaces that empower users to be more productive and creative. Generative computational creativity is used to produce creative artifacts. Computer colleagues, or co-creative agents, are artificial intelligence (AI) agents that collaborate with a user as a partner. They encourage creativity of the user and generate creative work as well. Computational creativity can achieve enhancements in design and development using neuroscience. Brain computer Interfaces (BCI) have been used in creativity. BCI was investigated in this survey because it uses neuroscience as an essential part in data acquisition. This paper collected several works in creativity support tools (CST) [2, 9, 10, 19, 20], computer colleagues/co-creative agents [2, 24] and brain computer interface [28–30].

In the rest of this survey, Sect. 2 talks about creativity in detail. Section 3 discusses brain networks and lobes, neuroscience of creativity theories and brain imaging technologies. Section 4 covers HCI and neuroscience of creativity. Section 5 discusses computational creativity. Section 6 includes the discussion and future works, and Sect. 7 is the conclusion.

2   Creativity

Creativity has become a crucial topic in several disciplines such as cognitive science [5–7], neuroscience [6, 9, 13, 16, 17] and human-computer interaction (HCI) [12, 15, 19]. Creativity is defined as the production of novel and useful ideas [16, 18]. Another work has added "surprising" to the definition of creativity, because the authors considered the difference of cognitive processes that underpin creativity [14]. Dietrich also explained creativity as "the ability to break conventional or obvious patterns of thinking, adopt new and/or higher order rules, and think conceptually and abstractly is at the heart of any theory of creativity" [16]. Srinivasan noted that creativity maintains some features such as fluency, flexibility, originality, and elaboration [4]. Fluency describes the amount of ideas to be generated by a person. Flexibility means finding different solutions to solve the problem. Originality means producing novel and new ideas that were not used before, and elaboration is the ability to add more details or expand novel ideas.

Creativity has several modes and stages. Dietrich proposed four creativity modes [16]. The modes were categorized using two factors: memory structure (either cognitive or emotional) and information processing modes (deliberate or spontaneous). Deliberate situations use prefrontal cortex (PFC) because they need more executive functions to produce insights. Deliberate-cognitive structure activates the prefrontal cortex to retrieve relevant information from the temporal, occipital and parietal lobes (TOP). This structure needs more experience and knowledge to produce creativity. Deliberate-emotional structure also activates the prefrontal cortex, but information comes from the emotional structure. Emotion information can be retrieved either from the limbic system (basic emotions) or from cingulate cortex and the ventromedial prefrontal cortex (complex social emotions). Spontaneous–cognitive structures come from the temporal, parietal and occipital (TOP) lobes, but it does not use prefrontal cortex. He claimed that modes do not represent distinct creativity types, and creativity could be a result from several modes. Creativity can take numerous patterns, and that depends on the type of the task as well the personality and experience. Creativity occurs through phases from starting to think about a problem until finding novel ideas. Creativity comes through four stages: preparation, incubation, illumination and verification [27]. Some works suggested two stages: the idea generation and the idea evaluation. These stages work in an iterative way.

In addition, creativity has been examined using a cognitive science perspective. It is important to understand the cognitive mechanisms that are involved in creativity. Defocused attention has been used to identify the occurrence of creative in many studies. It is defined as a broad attention to overcome mental fixation and come up with new ideas [3]. Kaufman et al. linked between defocused attention and associative thinking, when they implied to memory elements (information that is stored in memory) activated simultaneously [25]. It happens when the brain does not filter the irrelevant information of current stimulus, and this causes more ideation. High sensitivity is also used to describe creative people. It is defined as "sensitivity to subliminal impressions; stimuli that are perceived but of which one is not conscious of having perceived" [3]. It means that creatives are more sensitive to stimuli than average individuals [3, 27]. For example, creative people experience drawing tasks in different way, and they are stimulated differently by colors or shapes. High sensitivity shares the same mechanism of defocused attention, which is the lack of filtering irrelevant stimuli. Low cortical arousal is another mechanism that describes creativity. It was reported that low cortical arousal is associated with creative ideas [3, 20].

When examining creativity, we must pay a close attention to the tasks and study structure. Divergent thinking has been used to evaluate creativity [1, 4, 6, 13, 22]. It is defined as an open-ended and ill-defined problem that motivates people to generate many solutions [4, 13]. Abraham and Windmann defined divergent thinking as the process of generating many ideas for situations that do not have right or wrong solutions [6]. Alternative uses (AU) is a test used to evaluate divergent thinking problems. It uses fluency, flexibility, originality and elaboration to score ideas [4]. Artwork has been widely used to investigate creativity. Artistic tasks range among visual art [5, 11, 20, 23], music performance [5, 29] and poetry [5]. There are different types of visual art tasks such as doodling, coloring and free drawing. Visual art tasks have been used to evaluate creativity of well-trained and expert people, and it confirmed the role of knowledge and experience to generative creativity [23]. Convergent thinking has been used in creativity studies. It is opposite of divergent thinking, which asks participants to generate a correct and specific answer to a problem. Convergent thinking includes two common tasks: remote associates test (RAT) [1, 22] and insight problems [1, 26] RAT tasks measure the ability to find relationships between things, such as word association task [22]. Insight problems encourage participants to change their perspective and think in a novel way [26].

3   Neuroscience of Creativity

Neuroscience uses brain imaging tools to capture brain activations. These systems vary based on technologies embedded, data acquisition used, brain areas to be captured, and the study needs. This survey covers three types of brain imaging tools used in creativity. They are electroencephalograms (EEG), functional magnetic resonance imaging (fMRI) and functional near-infrared spectroscopy (fNIRS). Before reviewing these tools, I will talk briefly about brain lobes, brain networks and theories of the neuroscience of creativity.

3.1   Brain Networks and Lobes

Several neuroscience of creativity works have focused on the brain networks during creative tasks. The brain has two networks: default mode network (DMN), and executive network. Beaty et al. reported that the default mode network includes the medial prefrontal cortex (mPFC), the posterior cingulate cortex (PCC), the precuneus, and the bilateral inferior parietal lobes [5, 7]. Ellamil et al. added the temporoparietal junction (TPJ) as a part of the default mode network [11]. The default mode network plays a role in the idea generation [7], but it has also been implicated in the idea evaluation as well [5, 11]. It is responsible for internally-directed attention and spontaneous cognition [7]. Several works found that the creative users had a greater connection between the default mode network and the other networks such as the left inferior frontal gyrus (IFG) [7] and the executive network [5, 11]. The executive network has been examined in creativity studies. It is responsible of the executive functions such as judgment and decision making. Creativity works have found that the executive network plays a role in the idea evaluation [5, 11].

The brain has four lobes: frontal, temporal, occipital and parietal. These lobes are not explicitly used to describe creative cognition. However, they are used to describe the structure of the brain networks. Temporal and parietal lobes, in addition to the small areas from both the frontal and occipital lobes, make the default mode network [7]. The frontal lobe is the largest lobe in human brain. The frontal lobe and prefrontal cortex (PFC) are key players in creativity. The PFC is the cerebral cortex that covers the front part of the frontal lobe. It is considered the center of the executive network and analytical/convergent thinking. The PFC has two components: ventromedial (VMPFC) and dorsolateral (DLPFC) [16].

3.2   Creativity Theories

This section focuses on the neuroscience of creativity theories. They are used to find association between creativity and the brain activation patterns. These theories include blind variation and selective retention, contextual focus and associative mode, representational change theory, interhemispheric and cognitive disinhibition theories.

Blind Variation and Selective Retention (BVSR). BVSR describes the cognitive processes involved in a creative work. Jung et al. reported that BVSR describes the development of creative thoughts [18]. It is jumping from struggling in fixation state (blind state) to solving the problem in novel way (selective retention). Beaty et al. reported the relationship between BVSR and brain networks [7]. The default mode network has *blind variation* state that associates to the idea generation and divergent thinking. The executive network is responsible for *selective retention*, which associates with the creative evaluation and convergent thinking [7].

Contextual Focus and Associative Mode. Gabora focused her study on the contextual focus, which is the shift between the associative thoughts (defocusing attention) and convergent thoughts (focusing attention) [3]. In her work, she claimed that creativity uses both associative/divergent and analytical/convergent thinking [3]. The associative mode is more about ideation (generation of different ideas), while the analytic mode synthesizes thoughts created by the associative mode. The associative mode can be measured using the *associative richness* or the *flat associative hierarchies,* which assess the deepness and steepness of the associative mode [27].

Representational Change Theory: Goel has discussed the importance of representational change to find the solution of insight problems [26]. Insight problems encourage the user to change the perspective and view the problem in a novel way. Representational change has two approaches: constraint relaxation and chunk decomposition [26]. Constraint relaxation helps the user to solve unfamiliar problem and break the fixation state. Goel showed the equilateral triangles problem example, which aims to build four equilateral triangles using six sticks of equal length [26]. By relaxing space dimensionality and changing it from 2D to 3D, the solution presented itself. Chunk decomposition is another representational change technique [26]. Wu et al. defined chunks as "collection of elements having strong associations with one another, but weak associations with elements within other chunks" [31]. A chunk actually contains information that is coded in long term memory from previous experiences. Chunks are decomposed to their main features or elements (D-process), and then regrouped (R-process) in a different meaningful manner.

Interhemispheric (Bi-laterality) and Hemispheric Asymmetry (Laterality): There is an argument about the hemispheric activation in the creativity. There are some works that confirmed hemispheric asymmetry, while the other works noted the bilateral hypothesis. The hemispheric asymmetry hypothesizes that creativity results from one hemisphere (laterality) either the right or left hemisphere, while the Interhemispheric assures the activation of both hemispheres. Kaufman et al. found some works that claimed the hemispheric asymmetry hypothesis [25]. Yoruk and Runco claimed that divergent thinking presents more activations in the left hemisphere because of *the semantic memory traces* activation [13]. Semantic memory traces are located in the left side of the brain, and they are activated in divergent thinking situations [13]. Geol found hemispheric dissociation in the activations of the prefrontal cortex (PFC) [26]. He reported the left PFC is dominant over the right, and each hemisphere tries to inhibit the other one. He collected split-brain patients' studies, when participants, who have a lesioned hemisphere, used the other hemisphere to solve the problem [26]. In contrast, several works emphasized both of the right and left hemispheres in creativity. Gibson et al. claimed that the interhemispheric activation is essential to generate novel ideas [22]. They conducted a divergent thinking study, and the experiment's sample included both well trained musicians and non-musicians. The musicians used both brain hemispheres, while non-musicians only used the left hemisphere. Srinivasan claimed that creativity needs large scale activations in the brain, which include regions from both hemispheres [4].

Cognitive Disinhibition: Cognitive disinhibition simply means not filtering irrelevant information/stimulus. Kaufman et al. summarized some works that investigated creativity and cognitive disinhibition [25]. They found that the low cortical arousal is accompanied with high cognitive disinhibition/low inhibition. When the cortical arousal is low, this allows different mechanisms from different regions to occur at the same time [25]. Cognitive disinhibition is related to different creativity mechanisms such as low cortical activation, associative thinking, high sensitivity and defocused attention [25].

3.3 Electroencephalograms (EEG)

EEG is an electrophysiological imaging system that records the brain electrical activities. EEG has a set of electrical conductors called *electrodes*, which are installed over different positions of the skull to collect the signals. EEG-creativity investigation extracts brain rhythms from the electrical signals. Brain rhythms are frequencies that are produced by different brain states, which include delta (0.5–4 Hz), theta (4–8 Hz), alpha (8–13 Hz), beta (>13–30 Hz) and gamma (more than 30 Hz) [4].

Carroll and Latulipe investigated the temporal creativity of participants, (in-the-moment creativity or 'ITMC') using a triangulation method [20]. They used *Emotiv EPOC* headset, which is a mobile EEG system that is easy to handle and use. The experiment used sketching tasks, because there are plenty of tools that can be used to run a sketching experiment, and the task did not require an artistic background. The experiment took about 30 min for each participant to complete the task. Alpha and beta waves were extracted because they are important to investigate creativity and the cortical arousal [20]. Self-report questionnaires, such as creative behavior inventory (CBI), were used to group users to either creative or less creative groups. CBI is a questionnaire that asks participants about the frequency of doing creative activities in the adult life. In this study, participants with CBI scores more than 50% were considered as creative. Creative moments were assessed by a video application that asked the participants to watch their drawing and find the creative moments. It was found that creative participants presented higher alpha wave and lower cortical arousal in the creative moments [20].

Srinivasan has reviewed several EEG efforts in creativity [4]. He explained the scientific way to investigate creativity using EEG. The article briefly noted the importance of choosing a task to detect creativity. EEG data preprocessing steps includes artifact removal/filtering and features extraction [4]. Artifact is the noisy data that is not related to cerebral activity such as respiration and heart data. EEG has different data analysis methods that include spectral analysis, coherence analysis, non-linear signal analysis, principal component analysis (PCA) and independent component analysis (ICA). However, creativity can be investigated using the spectral analysis and coherence analysis [4]. The spectral analysis extracts EEG rhythms to identify the cortical arousal level in the brain [4]. It decomposes EEG signal of each electrode to different waves using different algorithms such as Discrete Fourier Transform (DFT), autocorrelation function, power spectral density (PSD), and autoregressive model (AR) [4]. The coherence analysis is used to find several brain regions that have the spectral analysis at the same moments. If the coherence analysis value is higher, it means that more brain regions are activated through the task [4]. The coherence analysis is related to the associative mode, because it presents number of associated brain areas in the creative moments.

Dietrich and Kanso have investigated the cognitive processes that underlie novel thoughts [1]. They evaluated creativity in three tasks: divergent thinking, creative artistic work, insightful problems. Evaluating tasks used the brain hemispheres activation and the cortices roles in creativity [1]. Insightful problems were more consistent than both divergent thinking and artistic work [1]. There was no consensus about the brain hemispheres activations in divergent thinking, because neuroscience did not find agreement among the results [1]. They found the old studies support the right hemisphere, few works supported the left hemisphere, and the recent works claimed the need of both hemispheres. Artistic tasks showed a similar result to divergent thinking. These studies could not confirm the significance of the PFC in creative thoughts [1]. Some works claimed the presence of PFC in novel ideas, while others reported the *spontaneous creativity* that did not require the PFC activation. The paper concluded that creativity does not associate with particular brain region and mechanism. There are several regions from both hemispheres that were implicated in creative work [1].

Fink et al. have used EEG to examine creativity [8]. They reviewed creativity-EEG studies to find the potential brain correlates of creativity. They conducted an experiment that used both divergent and convergent thinking tasks. The tasks were insight task (IS), utopian situation (US), alternative uses test (AU) and the word ends task (WE). In IS, the participants faced unusual situations that need explanation, and the task required different solutions based on different circumstances. Utopian situation task (US) put the participants in utopian situations, which are imaginary situations, to create many original ideas. Alternative uses test (AU) measures divergent thinking by generating many unusual uses of conventional objects. Word ends task (WE) provided the users with a suffix and asked them to find words that are related to that suffix. The experiment focused on EEG alpha power changes. It was found that IS, US, and AU tasks showed significant increases in alpha power when compared to WE [8]. To justify the variance of results, authors assumed these tasks used different cognitive functions [8]. WE is more intelligence-related, while IS, US and AU are more associative thinking mode [8]. This explanation may imply that creativity takes different types, and each type is underpinned by distinct mechanisms.

Yoruk and Runco have studied the neuroscience efforts in divergent thinking [13]. The authors collected different works about EEG and divergent thinking, and they examined creativity analysis using EEG such as brain waves, coherence and event related activity [13]. Event related activity describes the brain activation changes from the resting state. It takes two types: Event-related Synchronization (ERS) and Event-related Desynchronization (ERD) [13]. The ERS means the increase in neural activations, compared to the rest activation level, when a participant focuses on a task, while ERD is the opposite [13]. Divergent thinking presented both of the ERS and ERD, but the ERS was more dominant [13]. The authors found a relationship between the occurrence of ERS, the existence of alpha wave and the occurrence of the low cortical arousal. In addition, ERS was stronger the right hemisphere, while ERD was greater in both of the left hemisphere and brain central regions [13]. This actually supports both hemispheric asymmetry and cognitive inhibition theories that confirmed the dominance of one hemisphere (usually right hemisphere) in some creativity situations [13].

3.4 Functional Magnetic Resonance Imaging (fMRI)

fMRI is a brain imaging system that measures the cerebral activity by detecting changes associated with blood flow using blood-oxygen-level dependent (BOLD) technique. It covers most of the brain regions and lobes. fMRI has the edge over both EEG and fNIRS in term of visualization. It shows the activations from different angles and views. fMRI, however, has a limited number of studies in creativity because it is expensive, complex and does not have portable versions like EEG and fNIRS.

Beaty et al. have investigated creativity and the default mode network in divergent thinking tasks [7]. The divergent thinking had two parts: alternate uses tasks and instances tasks. The alternate uses tasks asked the participants to generate several uses of three common objects: a can, a knife, and a hairdryer [7]. The instances tasks asked the participants to generate creative solutions to the following problems: what can make noise? what can be elastic? and what can be used for speedy travel [7]? The creativity scores were rated by three raters, and raters scores were then averaged to make the final score for each answer [7]. The authors found that the left inferior frontal gyrus (IFG) and the entire default mode network were strongly connected in the high creative group [7]. They also found that the right IFG had a greater connectivity with the bilateral inferior parietal cortex and the left dorsolateral prefrontal cortex in the high creative group [7]. They emphasized that creativity is related to the strength of connectivity between the inferior prefrontal cortex and the default mode network as well the connectivity between the left and right hemispheres [7].

Ellamil et al. have conducted fMRI study to investigate creative modes and explore the neural correlates of each mode [11]. They hypothesized that creativity needs both generative and evaluative modes. Each mode has a unique neural activation and both modes are not activated at the same time [11]. The experiment had two conditions (generate and evaluate) and two baseline periods (trace-g, trace-e). *Generate* asked the users to draw what they read in the short description and *evaluate* refined the results of the generate condition and added more details [11]. Blocks *trace-g* and *trace-e* were used to separate the two conditions and to distinguish the activation of each condition. *Generate* condition showed activations in medial temporal lobe (MTL) and other areas such as parahippocampus, bilateral inferior parietal lobule (IPL) and bilateral premotor area (PMA) [11]. On the other hand, the *evaluate* task reported coupling between the executive and default mode networks [11].

Beaty et al. investigated the default mode and executive networks in creativity [5]. The study reviewed the brain networks connectivity in two tasks: domain-general creative problem solving (divergent thinking) and domain-specific artistic performance [5]. In divergent thinking problems, they found coupling among regions from the both networks [5]. Domain-specific artistic problems showed different results [5]. In the poetry composition study, the default mode and executive networks were not connected in the spontaneous ideas' generation. When evaluating generated ideas, there was a connectivity between the networks [5]. In musical improvisation, constraining performance showed more activations in the executive network only, but emotional playing music showed a coupling of both networks [5]. Visual art reported activations of the default mode network in the idea's generation and both networks in the idea's evaluation [5].

3.5   Functional Near-Infrared Spectroscopy (fNIRS)

fNIRS is a multi-wavelength optical spectroscopy technique introduced as a non-invasive brain activity monitoring modality [21]. It uses hemodynamic responses of the blood to identify the PFC activations. fNIRS uses a sensor pad that is installed over the subject's forehead to collect data. Analyzing creativity using hemodynamic responses is still complex topic, because there are limited works in fNIRS activations and creativity theories. fNIRS has advantages over fMRI, because it is lightweight, cheaper, and easier.

Kaimal et al. have used fNIRS to assess reward perception in creative drawing tasks [23]. The assessment of reward perception focused on the activation of the medial prefrontal cortex (mPFC). The study hypothesized that reward activation will be greater in the artists. Study structure included expression conditions and control conditions. The expression conditions were coloring, doodling and free drawing, while the control conditions were four resting periods with eyes closed between the artistic tasks [23]. The authors used *five-item questionnaire* to assess users' creativity. The questionnaire asked users about their abilities to "have new ideas; have good ideas; have a good imagination; have novel ideas; and solve problems" [23]. They used two-way repeated measures ANOVA to show the difference between both expression and control conditions. Results presented significant activations of three conditions regardless the skills level [23]. Doodling showed the highest mPFC activation over the other conditions, but the differences among tasks were not significant [23]. When comparing between the artists and non-artists, the artists had more activations in the doodling and less activations in the coloring tasks [23]. Free drawing did not show any difference between the two groups [23].

Gibson et al. used the near-infrared spectroscopy (NIRS) to examine creativity of the expert users (trained musicians) using both of the convergent and divergent thinking [22]. The study hypothesized that the musicians have an enhanced divergent thinking and more bilateral PFC activation. The sample included 20 students, and all of them were females. Nine students were from the school of music with experience greater than 8 years, and the other 11 students were non-musicians with no experience after the high school [22]. The study ran two experiments. First experiment collected data about the personality and intelligence tests in respect to creativity. The questionnaires included Wechsler abbreviated scale of intelligence (WASI), Gough personality scale and schizotypal personality questionnaire (SPQ). The second experiment asked the participants to solve two creative tasks, which are remote association Test (RAT) and divergent thinking test (DTT) [22]. The paper reported significant results of musicians when compared to non-musicians. Musicians came up with a greater number of the alternative uses. The average scores of the alternative uses was 36.2 for the musicians and 22.3 for the averaged individuals, with significant *p-value* equals to 0.02 [22]. The study also found more bilateral PFC on the musicians, while the non-musicians showed a dominance activation of the left hemisphere [22].

4 Neuroscience of Creativity and Human Computer Interaction (HCI)

Creative cognition and neuroscience have been used in human computer interaction (HCI) research. Brain Computer Interface (BCI) is a computer program that uses human brain states, either as a control (direct input) or as a supplementary input, to interact with the interface. BCI has two types: active BCI and passive BCI. The active BCI uses the brain signals to interact with the system directly instead of using hands or hardware devices. It is mainly designed for disabled people, who cannot use hands to interact. On the other hand, passive BCI uses the brain signals as additional information, but the user can interact normally with the interface.

Todd et al. conducted an initial investigation of creativity using an active BCI [28]. The brain signals were received as inputs by signal processing modules and then matched with corresponding commands created by the users [28]. They used BCI-EEG system to investigate users' perceptions. The experiment used three tasks: one brain drawing task (TASK 1) and two brain painting tasks (TASK 2 and 3). In drawing task, participants were asked to draw squares. In painting tasks, participants were asked to draw a circle and two stars in TASK 2 and do free drawing in TASK 3 [28]. Creativity was evaluated using questionnaires about the participants performance of all tasks. It was noted that users feel more creative in painting over drawing [28]. It was also reported that TASK 3 (unstructured painting task) was more rewarding [28]. However, the study did not report EEG activations or creativity mechanisms such as the low cortical arousal and the alpha rhythm.

The implicit BCI, or the passive BCI, has been used to investigate creativity. Yuksel et al. have used a passive BCI to develop a novel musical interface [29]. The passive BCI collected brain signals as additional information about the brain states (like high and low workload), while the users used a piano connected to the BCI. The users played music harmonies while fNIRS collected their mental workload. The study proposed BRAAHMS, which is passive real-time BCI, to help users achieve more creative musical improvisation. BRAAHMS used the brain signals to add or remove musical harmonies [29]. The study ran two experiments. Experiment 1 was conducted to check the feasibility of the study whether the brain data correspond with high and low difficulty levels. Participants were asked to play easy and hard musical pieces, and each piece were about 30 s. Experiment 2 used BRAAHMS to evaluate musical improvisation of users. The study recruited 20 participants to perform four musical conditions [29]. *BCI1* condition adds musical harmonies when brain signals correspond with low cognitive workload and removes musical harmonies when brain signals correspond with high cognitive workload. *BCI2* condition adds musical harmonies are added when brain signals correspond with high cognitive workload and removes musical harmonies when brain signals correspond with low cognitive workload. *Constant* condition means musical harmonies are always present. *Non-adaptive* condition means no musical harmonies at all [29]. There users completed a survey about their performances in all conditions. The results showed more creativity in both *BCI1* and *BCI2* over the other conditions. Some participants reported that they were creative, because they did novel and new performance when they interacted with BRAAHMS [29]. The authors found users with less experience were more creative when using BRAAHMS, while experts were hindered [29].

5   Computational Creativity

Computational creativity designs computer programs to emulate, study, stimulate and enhance human creativity. It covers two areas: creativity support tools (CST) and co-creative agents. Since there is incomplete neuroscience contribution in computational creativity, this section has a limited number of works that may imply to the future works.

Carroll and Latulipe conducted a neurological study to evaluate CSTs using in-the-moment creativity (ITMC) [20]. ITMC has three approaches: self-report, external raters and EEG. The study used both sketching software for drawing and video recording application to evaluate users' performances. The study did not allow free improvisation because participants were prepared before running the experiment. EEG data included both alpha and beta waves. The participants whose creative behavior inventory (CBI) scores over 50% were classified as creative. The creative participants exhibited higher alpha wave power comparing to the averaged individuals (p-value <0.05), which means they had the lower cortical arousal and more creativity. The article did not report details about the brain activations that were detected by EEG.

CST has also been used in cognitive science to understand novice creativity. Davis et al. proposed new theories, which are derived from cognitive theories, to improve CSTs and make novice people more creative [19]. They have explored some related works about evaluating the CST tools. They found that most CST tools are designed and used by professional or creative people, while novices have some difficulties to use such tools. They derived three creativity theories from cognitive ones; they proposed embodiment creativity (from embodiment cognition), situated creativity (from situated cognition), and distributed creativity (from distributed cognition) [19]. They evaluated different CSTs in computer programming, music and storytelling [19]. However, the study did not report the neuroscience implication.

Another topic of computational creativity is *co-creative systems or agents*. A co-creative agent is an AI-based program that interacts with the users to achieve creativity. There are few co-creative agents that are only available for research goals such as the *Apprentice web application* and *PlayPartner* [2]. However, there are no works that have used the neuroscience of creativity to evaluate co-creative agents.

6 Discussion and Future Works

By looking to the results of several works, it is hard to identify the center of creativity in the human brains. There are several brain regions and networks that are activated in creative tasks. Several studies found different hemispheric activations, some works reported either the absence or presence of the PFC, and some of them reported a spread activation over most of brain regions. In addition, it is not reported whether all parts of these regions contributed in creativity or not. For example, the prefrontal cortex plays a role in generating novel ideas, but it is not clear whether all areas in the prefrontal cortex were implicated or not.

The brain networks have been used to investigate creativity in fMRI. The default mode network is responsible for spontaneous and self-generated thought such as mind-wandering [3, 5]. The default mode network is responsible for the idea's generation and evaluation [5, 11], while the executive network is only responsible for the ideas' evaluation. Both networks are required to produce creative thoughts, and this is emphasized by some theories such as blind variation and selective retention (BVSR) [7, 18]. EEG uses the brain rhythms to evaluate creativity. These rhythms correspond to different brain states. For example, alpha wave represents a relaxing state that shows a low cortical arousal and defocused attention. Beta rhythm represents more attention and focus. Gabora reported that creativity occurs when beta is decreased, and alpha is increased [3]. In addition, the lower and higher alpha rhythms are also considered when investigating creativity [1, 9].

The study design may affect the producing of novel thoughts. Some studies used structured drawing tasks such as drawing particular objects [28], while others used free drawing style [20, 28]. Some studies were a goal-directed, such as investigating creativity for the purpose of reward perception [23]. In that study, authors examined a part of the PFC (optode 7 in fNIRS band) and ignored the remaining episodes because the optode 7 has a relationship with reward perception. Some studies hindered the users' improvisation by directing them to come prepared with ideas before running the study [11] However, some studies claimed that the improvisation is an essential to generate a creative work [17, 29].

Knowledge and skills may play a role to generate creative works. Gibson et al. explained that musicians and artists are more creative, because they have more skills that refine their ideas. The knowledge may have a significant impact on neural circuits that are associated with creativity [22]. Dietrich reported that the experience and knowledge can produce creative ideas [16]. He explained that creativity can be a deliberate-cognitive structure, which depends on the experiences and knowledge of the users. Some creative studies reported more creative outcomes achieved by musicians [5, 22] and artist [23].

In terms of computational creativity, this literature review has discovered a gap between neuroscience investigations of creativity and computational creativity systems. Most computational creativity works did not show a neuroscience contribution in detecting creative processes. This review aims to combine computational creativity with neuroscience by first understanding creativity, neuroscience theories and tools, and then evaluating computational creativity using the neuroscience literature. Future work may focus on investigating cognitive neuroscience theories that underlie creativity with respect to the computational creativity. It may recruit neuroscience of creativity in designing creative tools that encourage creativity such as using both convergent and divergent thinking as one task or using problems that trigger associative thinking mode.

7 Concluding Remarks

This survey reviews numerous works that investigated the neuroscience of creativity in HCI. Creativity does not have specific regions or mechanisms. Creativity has different modes, and it is evaluated using different tasks. Idea generation showed more activations in TOP and the default mode network, while the idea evaluation presented activations in both of default mode and executive networks. creativity can be detected using several cognition mechanisms such as the defocused attention, low cortical arousal and interhemispheric activations. EEG has the edge over other neuroscience tools, because we can use several creativity theories and mechanisms to evaluate neurological results. Neuroscience of creativity theories can add more contributions in designing and evaluating HCI systems. However, the computational creativity and HCI still need more efforts from the neuroscience of creativity.

References

1. Dietrich, A., Kanso, R.: A review of EEG, ERP, and neuroimaging studies of creativity and insight. Psychol. Bull. **136**(5), 822–848 (2010)
2. Davis, N.: An enactive approach to facilitate interactive machine learning for co-creative agents. In: Proceedings of the 2015 ACM SIGCHI C&C 2015, pp. 345–346 (2015)
3. Gabora, L.: Revenge of the 'Neurds': characterizing creative thought in terms of the structure and dynamics of memory. Creat. Res. J. **22**(1), 1–13 (2010)
4. Srinivasan, N.: Cognitive neuroscience of creativity: EEG based approaches. Methods **42**, 109–116 (2007)
5. Beaty, R.E., Benedek, M., Silvia, P.J., Schacter, D.L.: Creative cognition and brain network dynamics. Trends Cogn. Sci. **20**(2), 87–95 (2017)
6. Abraham, A., Windmann, S.: Creative cognition: the diverse operations and the prospect of applying a cognitive neuroscience perspective. Methods **42**(1), 38–48 (2007)
7. Beaty, R.E., Benedek, M., Wilkins, R.W., Jauk, E., Fink, A., Silvia, P.J., Hodges, D.A., Koschutnig, K., Neubauer, A.C.: Creativity and the default network: a functional connectivity analysis of the creative brain at rest. Neuropsychologia **64**, 92–98 (2014)
8. Fink, A., Benedek, M., Grabner, R.H., Staudt, B., Neubauer, A.C.: Creativity meets neuroscience: experimental tasks for the neuroscientific study of creative thinking. Methods **42**(1), 68–76 (2007)
9. Shneiderman, B.: Creativity support tools. Commun. ACM **45**(10), 116–120 (2002)
10. Shneiderman, B.: Creativity support tools: accelerating discovery and innovation. Commun. ACM **50**(12), 20–32 (2007)

11. Ellamil, M., Dobson, C., Beeman, M., Christoff, K.: Evaluative and generative modes of thought during the creative process. Neuroimage **59**(2), 1783–1794 (2012)
12. Fox, K., Girn, M., Parro, C., Christoff, K.: Functional neuroimaging of psychedelic experience: an overview of psychological and neural effects and their relevance to research on creativity, daydreaming, and dreaming. In: The Cambridge Handbook of the Neuroscience of Creativity, pp. 92–113 (2018)
13. Yoruk, S., Runco, M.A.: The neuroscience of divergent thinking. Activitas Nervosa Super. **56**(1–2), 1–16 (2014)
14. Simonton, D.K.: Quantifying creativity: can measures span the spectrum? Dialogues Clin. Neurosci. **14**(1), 100–104 (2012)
15. Farooq, U., Carroll, J.M., Ganoe, C.H.: Supporting creativity in distributed scientific communities. In: Proceedings of the 2005 International ACM SIGGROUP Conference on Supporting Group Work, pp. 217–226 (2005)
16. Dietrich, A.: The cognitive neuroscience of creativity. Psychon. Bull. Rev. **11**(6), 1011–1026 (2004)
17. Sawyer, K.: The cognitive neuroscience of creativity: a critical review. Creat. Res. J. **23**(2), 137–154 (2011)
18. Jung, R.E., Mead, B.S., Carrasco, J., Flores, R.A.: The structure of creative cognition in the human brain. Front. Hum. Neurosci. **7**, 330 (2013)
19. Davis, N., Winnemöller, H., Dontcheva, M., Do, E.Y.L.: Toward a cognitive theory of creativity support. In: Proceedings of the 9th ACM Conference on Creativity & Cognition, pp. 13–22 (2013)
20. Carroll, E.A., Latulipe, C.: Triangulating the personal creative experience: self-report, external judgments, and physiology. In: Proceedings of Graphics Interface, pp. 53–60 (2012)
21. Ayaz, H., Izzetoglu, M., Bunce, S., Heiman-Patterson, T., Onaral, B.: Detecting cognitive activity related hemodynamic signal for brain computer interface using functional near infrared spectroscopy. In: 3rd International IEEE/EMBS Conference on Neural Engineering, pp. 342–345 (2007)
22. Gibson, C., Folley, B., Park, S.: Enhanced divergent thinking and creativity in musicians: a behavioral and near-infrared spectroscopy study. Brain Cogn. **69**(1), 162–169 (2009)
23. Kaimal, G., Ayaz, H., Herres, J., Dieterich-Hartwell, R., Makwana, B., Kaiser, D.H., Nasser, J.A.: Functional near-infrared spectroscopy assessment of reward perception based on visual self-expression: coloring, doodling, and free drawing. Arts Psychother. **55**, 85–92 (2017)
24. Davis, N., Hsiao, C.P., Singh, K.Y., Lin, B., Magerko, B.: Creative sense-making, quantifying interaction dynamics in co-creation. In: Proceedings of the 2017 ACM SIGCHI Conference on Creativity and Cognition - C&C 2017, pp. 356–366 (2017)
25. Kaufman, A.B., Kornilov, S.A., Bristol, A.S., Tan, M., Grigorenko, E.L.: The neurobiological foundation of creative cognition. In: The Cambridge Handbook of Creativity, pp. 216–232 (2010)
26. Goel, V.: Creative brains: designing in the real world. Front. Hum. Neurosci. **8**, 241 (2014)
27. Gabora, L.: Cognitive mechanisms underlying the creative process. In: Proceedings of the 4th Conference on Creativity & Cognition (2002)
28. Todd, D.A., McCullagh, P.J., Mulvenna, M.D., Lightbody, G.: Investigating the use of brain-computer interaction to facilitate creativity. In: Proceedings of the 3rd Augmented Human International Conference, p. 19 (2012)
29. Yuksel, B.F., Afergan, D., Peck, E.M., Griffin, G., Harrison, L., Chen, N.W., Chang, R., Jacob, R.J.: BRAAHMS: a novel adaptive musical interface based on users' cognitive state. In: Proceedings of the International Conference on New Interfaces for Musical Expression, pp. 136–139 (2015)

30. Andujar, M., Crawford, C.S., Nijholt, A., Jackson, F., Gilbert, J.E.: Artistic brain-computer interfaces: the expression and stimulation of the user's affective state. Brain-Comput. Interfaces 2(2–3), 60–69 (2015)
31. Wu, X., He, M., Zhou, Y., Xiao, J., Luo, J.: Decomposing a chunk into its elements and reorganizing them as a new chunk: the two different sub-processes underlying insightful chunk decomposition. Front. Psychol. 8 (2017)

Optimal Mapping Function for Predictions of the Subjective Quality Evaluation Using Artificial Intelligence

Lukas Sevcik[1]($\boxtimes$), Miroslav Uhrina[2], Juraj Bienik[2], and Miroslav Voznak[1]

[1] IT4Innovations, VSB - Technical University of Ostrava, 17. listopadu 15/2172, 708 33 Ostrava - Poruba, Czech Republic
lukas.sevcik@vsb.cz
[2] University of Zilina, Univerzitna 8215/1, 010 07 Zilina, Slovakia

Abstract. With the growth of QoE interest, IPTV providers need a method to control QoE. The paper describes the correlation between the results of objective and subjective methods in video quality assessment. The authors proposed the optimal mapping function for predictions of the subjective quality evaluation based on the objective evaluation to determine the perception of the video quality by the human brain. Our model using artificial intelligence, it is based on a neural network which can simulate and predicts the subjective quality of the scene. It also can predict subjective or objective video quality for video sequences defined by spatial, temporal information, which is the critical and key variable of a given scene, and by the qualitative parameters of the scene. The results from the model are verified by comparing predicted video quality using the proposed classifier with the required value. The two most common statistical parameters related to express performance are Pearson's correlation coefficient and Root Mean Square Error.

Keywords: Neural network · Objective quality evaluation · QoE · Spatial information · Subjective quality evaluation · Temporal information · Video quality assessment

1 Introduction

Services as the store, display and transmit of the video content are part of our day. Thus digital video needs to improve the quality by the service providers. It includes a quality evaluation of the video sequences, which can be divided into two types, objective and subjective. The first one uses mathematical estimation quality evaluation. Controlling the quality of video sequences requires a perceptual assessment of the video by subjective estimation. It is time-consuming, but service providers get opinion directly from the end customers. Computer's computing can't replace the perception of the human brain. Evaluation directly from the customers is more reliable. The service providers can adapt qualitative setting and bitrate of the video sequences for the requirements of the people. Subjective evaluation can help to set these parameters for the optimal quality of the video suitable for providers and also for the perception of the end

© Springer Nature Switzerland AG 2020
K. Arai et al. (Eds.): FTC 2019, AISC 1069, pp. 263–275, 2020.
https://doi.org/10.1007/978-3-030-32520-6_21

customers. Nowadays, researchers discuss QoS and QoE. Many providers take into account the opinion of the customers. As each technology develops every day of life, also video services need to improve for the actual customer needs. Services as the video of the demand, online streaming and gaming require sufficient quality assurance and customer's satisfaction. It is the reason why QoE is a little bit over the QoS. More information about QoE is defined in [1, 2].

Video sequences are based on a variety of dynamics. For evaluation and prediction of the video quality is essential to have a big database of the video scenes. Our paper presents twelve video scenes which were selected depending on the content, and which were selected based on spatial (SI) and temporal information (TI). Subjective and also objective metrics evaluate the quality of the video sequences. Correlation between these two types of approach for quality assessment is described. The main aim of this paper is to propose new classifier for prediction of the subjective quality based on the objective expression.

2   State of the Art

Many papers deal with exploring video quality. Authors examined methods of evaluation of the audio and video based on the perception in the paper [3]. There were introduced and analysed the main existing methods, and the experimental results concerning subjective assessment were presented. A no-reference quality assessment framework was presented in [4]. SI and TI were used for evaluation of the quality. Author in [5] introduced an assessment of the perceived video quality. A method to derive information contents of the video sequences from the SI and TI is proposed. Prediction of the quality for the no-reference metric is approved by the subjective data in the paper [6]. The presented model uses SI and TI for better performance of MOS (Mean Opinion Score) correlation. Mapping of the subjective metric to the objective by nonlinear regression and inverse interpolation is presented in [7].

Authors of this paper have already published several papers on this topic. In [8] is evaluated the impact of the configuration of the Quality of Service for LTE networks. In [9], we examined the effect of data out of order and packet loss to quality of the triple-play. Also, the influence of the delays to the quality of these services was studied in [10]. Network traffic simulation was described in [11–15], where we introduced a model for prediction of subjective evaluation affected by packet loss.

3   Methodology

Evaluation of the video quality and classification of the video sequences are described in this section.

3.1   Subjective Evaluation of Quality

The procedures and conditions for subjective tests are described in recommendations [16, 17]. These define the requirements for the source signals (reference sequences), quality assessors, initial instructions submitted before testing, procedure for video rating, results documentation process, analysis of results and factors that may negatively impact the subjective test. The rating group must consist of at least 15 evaluators. All evaluators need to receive the same guidelines, which must be clear. This eliminates a potential source of bias in the subject. The guidelines should describe the rating cycle, quality scale, how to record ratings, system behaviour and playback. It should include a warning that users not evaluate the quality of the scene according to the feelings and opinions expressed by the video or the people who appear in it. The instructions are followed by a video training section that demonstrates the task and familiarises the respondent with the range of quality and type of degradation in the experiment. Videos included in the training section are not used for regular testing.

Subjective testing was conducted in an adapted room. It complied with ITU-T Recommendations P.910 and ITU-T BT.500-13 so that evaluators had the correct lighting conditions to Keep the monitor at a certain distance. The rating was conducted on 24" Dell P2415Q Ultra HD monitors that meet recommendations [16, 17].

ACR was selected as a subjective method, because the evaluation is based on a visible video without being able to compare with a reference video. It is the same as real-time TV viewing. ACR is a subjective method based on a respondent's assessment of quality. The five-degree MOS scale defines quality. The video can only be rated once with one specific value. The benefit of this subjective method is the speed of execution. It did not take as much time as with DSIS or DSCQS.

3.2 Objective Evaluation of Quality

The MSU Video Quality Measurement Tool [25] was used to evaluate the quality of individual videos. This tool compares the reference and test sequences based on the following parameters: the name of the reference video, resolution, name of the comparison sequence, its resolution, and the metrics for evaluation.

3.3 Spatial a Temporal Information

This information is used for classification of the video scenes according to the detail and motion.

Spatial Information
Spatial information is based on the Sobel filter (the Sobel operator) that use the edge of the frames for taken information about frequency which determines SI. The standard deviation over pixels is calculated. It is repeated on all frames of the sequence. Maximum value from all SI in the time series represents the entire sequence, and it is expressed as (Fig. 1):

$$SI = max_{time}\{std_{space}[Sobel(f_n)]\} \tag{1}$$

Fig. 1. Spatial information

Temporal Information

This information determines differences in the movement of the frames or consecutive times, which defines the function *(n)*:

$$M_n(i,j) = F_n(i,j) - F_{n-1}(i,j) \tag{2}$$

where $F_n(i,j)$ is the pixel of the image F at the *i-th* row, *j-th* column, and *n-th* frame in time. The standard deviation is calculated from the function above, and the maximum value represents the sequence (Fig. 2):

$$TI = max_{time}\left\{std_{space}[M_n(i,j)]\right\} \tag{3}$$

Fig. 2. Temporal information

4 Measurement

For the simulation of the neural network, we need to prepare a testing video database. Individual video scenes are encoded to different compression format, bitrate and qualitative settings. The video scenes are characterized by SI, TI and their quality.

4.1 Video Database

In our measurement, different types of video sequences, high and slow-motion were used. It covers twelve content types depending on the detail of the image, texture, intensity of the light, object movement. We used scenes in an uncompressed format (*.yuv) created by Media Lab of the Shanghai Jiao Tang University [18] and by Laboratory of Pervasive Computing from the Tampere University of Technology (Ultra Video Group sequences; UVG) [19]. Basic parameters of these sequences are:

- Resolution: Ultra HD
- Chroma subsampling: 4:4:4
- Bit depth: 10 bits (per channel)
- Framerate: 30 fps (frames per second)
- Duration: 10 s (300 images). Ten-second length is sufficient for quality estimation; longer sequences do not change the rating.

These video sequences were encoded to the required format using FFmpeg tool [20–23]. The encoding parameters are shown in Table 1.

Table 1. Encoding parameters

Parameter	Description
Resolution	Full HD, Ultra HD
Type of codec	H.264 (AVC), H.265 (HEVC)
Bitrate [Mb/s]	1, 3, 5, 8, 10, 12, 15, 20
Chroma subsampling	4:2:0
Bit depth	10
Framerate	30 fps
Duration	10 s

4.2 Classification of the Video Sequences

Spatial and temporal information was used for identification of the video sequences. This information classifies the video content of each scene in our database. We used the Mitsu tool [24] for the evaluation of this information. SI and TI for H.265 codec and both resolution with two types of bitrate (three and twenty Mb/s) are displayed in the graphs 3.1.–3.4. As we can see, these parameters are different for each type of the encoded video; it changes by resolution, bitrate and compression standard. It means that it is possible to use this for the classification of the scenes. Our database contains a wide range of video sequences from the perspective of the TI and SI; thus, the results can be used in practice (Fig. 3).

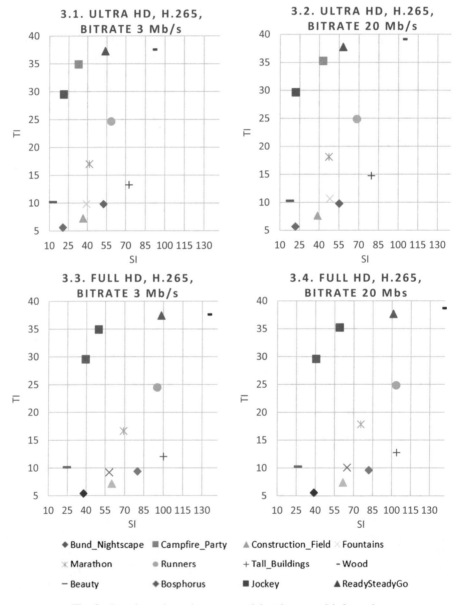

Fig. 3. Interdependence between spatial and temporal information

5   The Statistical Methods of Evaluation

Any data analysis is divided into specific steps: "data clearing", correct statistical evaluation and interpretation of the results. It is necessary to detect irrelevant data and determine those that do not correspond to the confidence interval compared to other samples. These data were not included in the final evaluation. The most commonly used method to evaluate the relevance of data is the Pearson correlation coefficient. It assumes that data have a normal distribution. If the scale is small, Pearson's correlation coefficient can be substituted with Spearman's correlation coefficient. For a short range, it is hard to assume that the division is close to normal (especially when most of the answers are close to a particular scale boundary).

Thirty-five respondents participated in the evaluation; data from five respondents were not accepted because they deviated from the other assessments. The final database contains data from thirty evaluators, including fifteen men aged 21–44 and fifteen women aged 19–34. The average age of the evaluators is 25 years (Table 2).

Table 2.   Number of evaluators

	Number of men	Number of women
Full HD + Ultra HD	15	15
Age	21–44	19–34
Average age	25	

6   Classifier Based on Artificial Intelligence

We proposed new classifier for prediction of quality expressed by MOS based on objective metric SSIM, the qualitative parameters and scene characterization by SI and TI. Our model using artificial neural network created in Matlab for a prediction. It capable of performing complex calculations similar to routine operations in the human brain. In principle, neural networks can calculate any countable function. Neural networks are mostly used to classify and functionally approximate or map functions using a lot of training data, where systems with clearly defined rules.

The new model also can predict objective or subjective evaluation based on bitstream, SI and TI. Back-propagation algorithm is implemented in our model. It can be implemented in many modes. We tested three of them using gradient method ('traingd' - Batch Gradient Descent; 'traingdm' - Batch Gradient Descent with Momentum and 'traingdx' - Gradient Descent with Variable Learning Rate).

A pattern is presented to the neural network, using feedback as it is for biological neurons. Based on the current setting, the actual result is determined, compared to the desired result, and then the error is calculated. After correction, weights and biases are adjusted, resulting in a decrease of the error value. Learning continues until the minimum error is reached, and then we can say that the network is adapted. This process needs to define training data, learning our network object and simulation of the neural network with test data.

7   Subjective Quality Prediction

This section describes the prediction of the subjective quality evaluation based on the defined objective metric and scene characterization. Determining the right topology of the learning were defined in this part. The right set of the number of hidden layers and iteration leads to the trained neural network. Five primary topologies arising from the inputs were defined. Five inputs characterize the video, which leads to the one output (prediction of the quality) as a result of the proposed model. Hence two to four hidden layers 5-1, 5-1-1, 5-3-1, 5-3-2-1 and 5-5-3-1 were defined.

Each simulation was taken ten times with randomly generated of weights and biases. Database of the video sequences is divided between training and testing set at a ratio of 70% for training collection.

Twenty-five percent of the validation set is generated from the training collection.

After the neural network is trained, simulation is taken another ten times with generating of the training and validation set each time.

Table 3 shows the simulation results of the primary layers for Ultra HD. It is optimal from the perspective of the average Pearson's Correlation Coefficient of the testing collection and the mean values of iterations and time reached for optimal result.

For the selection of the best topology is the most significant correlation with small error, subsequently the time and amount of periods for optimal result and the number of neurons.

The two-layer neural network reached the best results from the primary topologies of the neural network. These were achieved from both compression format, as shown in Table 3. Regarding it, two-layer topology was used for the next simulation and looked for more optimal outcomes. The simulation was performed with a gradual increase in the number of neurons. The best results from them are displayed in Table 3 (part of each compression format separately) and compared with primary topologies.

Table 3. The best topologies of the subjective prediction

	Topology	R^2_{TRAIN} [-]	Time [s]	Num. of epochs [-]
H.264 + H.265	5-1	0.97	64.962	681
	51-25	**0.996**	**12.446**	**302**
	5-1-1	0.974	99.480	762
	5-3-1	0.977	108.160	965
	5-3-2-1	0.974	91.926	820
	5-5-3-1	0.98	106.551	924
H.264	5-1	0.934	93.305	853
	31-15	**0.993**	281.610	273
	5-1-1	0.953	91.826	963
	5-3-1	0.934	135.354	1101
	5-3-2-1	0.957	152.567	1118
	5-5-3-1	0.953	94.658	888
H.265	5-1	0.971	96.904	808
	39-19	**0.992**	30.651	278
	5-1-1	0.969	91.443	836
	5-3-1	0.975	116.228	1037
	5-3-2-1	0.984	103.205	827
	5-5-3-1	0.989	118.936	918

The best topologies were selected for next data simulation; thus the reliability of the results is achieved. Table 4 shows results in the form of MSE and correlation expressed by Pearson correlation coefficient for the training, validation and test set. The results are expressed for both resolutions and compression standards together. In the case of Ultra HD, it is also separately for each codecs.

Table 4. Simulation of the MOS prediction for the best topologies

Resolution	Codec	Topology	MSE	R^2_{TRAIN} [-]	R^2_{VAL} [-]	R^2_{TEST} [-]
Ultra HD	**H.264 + H.265**	**51-25**	0.002	0.996	0.996	0.997
	H.264	**31-15**	0.004	0.994	0.987	0.999
	H.265	**39-19**	0.004	0.993	0.998	0.997
Full HD + Ultra HD	**H.264 + H.265**	**47-23**	0.006	0.986	0.991	0.993

The best topologies were achieved by setting of parameters to the neural network described in Table 5.

Table 5. Neural network setting

Parameter	Description
Algorithm	traingdx
Data scale interval	<−1; 1>
Activation function	tansig

Algorithm traingdx needs significantly less number of iteration for training the network and the associated shorter period to achieve it.

8 Model Verification

Simulated result need to be confirmed by statistical evaluation. Three steps for verification of the prediction subjective quality based on the objective metric, spatial information, temporal information and qualitative information were performed for simulation of Ultra HD for both compression standard using 95% confidence interval.

1. Comparison of the descriptive statistics between reference and test set (Table 6)

Table 6. Descriptive statistics of the reference and test values

	Reference set	Test set
Average value [-]	3.481	3.501
Minimum [-]	1.333	1.382
Maximum [-]	4.900	4.920
Median [-]	3.700	3.850
Confidence interval [-]	<3.067; 3.896>	<3.091; 3.910>
Pearson's correlation coef. [-]	0.999	
Spearman's correlation coef. [-]	0.996	

2. The Root Mean Square Error (RMSE) and the correlation diagram - Comparison of the reference and test data by the correlation diagram is very similar. It is confirmed by the RMSE when it represents 1.32% of the predictive error. RMSE is a measure of how much of the deviation is between original and streamed data set using a regression curve (Fig. 4).

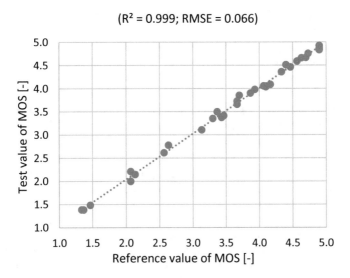

Fig. 4. Correlation diagram

3. Probability density function (PDF) defines the probability that a continuous random variable falls in the place between some borders equal to the area under the probability curve between these borders (Fig. 5).

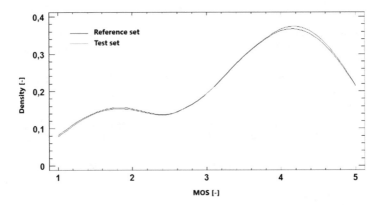

Fig. 5. Probability density function

9   Conclusion

The paper dealt with the prediction of subjective quality. After the introduction and description of the papers that relate to the issue, the methodology is described. Evaluation of the quality by both objective and subjective metrics are defined. Spatial and temporal information is used for scene description.

Section 4 explains creating of video database, and evaluate results from classification of the video scenes. The statistical methods of evaluation are defined in Sect. 5, followed by the proposed model based on the classifier of the video in Sect. 6. The main aim of the paper is the prediction of subjective quality describes in Sect. 7, followed by the validation in the next section.

The created database contains twelve types of video sequences depending on content. Full HD and Ultra HD resolutions with the compression standards H.264 and H.265 (HEVC) were tested. A wide range of the bitrate was used for encoding. Video samples were evaluated by SSIM and ACR metrics. The correlation between these metrics was described. Based on it, the proposed model can predict the subjective quality expressed by the MOS. Artificial intelligence was used as a tool for developing the system model. Verification of the model confirms the accuracy of the prediction in high score.

Acknowledgment. This work was supported by The Ministry of Education, Youth and Sports from the Large Infrastructures for Research, Experimental Development and Innovations project "IT4Innovations National Supercomputing Center – LM2015070" and partially received a financial support from grant No. SGS SP2019/41 conducted at VSB Technical University of Ostrava, Czech Republic.

References

1. Le Callet, P., Möller, S., Perkis, A.: Qualinet white paper on definitions of quality of experience. In: European Network on Quality of Experience in Multimedia Systems and Services (COST), version 1.2, pp. 1–17 (2013)
2. Recommendation ITU-T P.10/G.100 Amendment 2: New definitions for inclusion in Recommendation ITU-T P.10/G.100. Vocabulary for performance and quality of service (2008)
3. You, J., Reiter, U., Hannuksela, M.M., Gabbouj, M., Perkis, A.: Perceptual-based quality assessment for audio–visual services: a survey. Signal Process. Image Commun. **25**(7), 482–501 (2010)
4. Yang, F., Wan, S., Xie, Q., Wu, H.R.: No-reference quality assessment for networked video via primary analysis of bit stream. IEEE Trans. Circ. Syst. Video Technol. **20**(11), 1544–1554 (2010)
5. de la Cruz Ramos, P., Vidal, F.G., Leal, R.P.: Perceived video quality estimation from spatial and temporal information contents and network performance parameters in IPTV. In: Proceedings of 2010 Fifth International Conference on Digital Telecommunications, pp. 128–131 (2010)
6. Romaniak, P., Janowski, L., Leszczuk, M., Papir, Z.: Perceptual quality assessment for H.264/AVC compression. In: Proceedings of 2012 IEEE Consumer Communications and Networking Conference (CCNC), pp. 597–602 (2012)
7. Moldovan, A.-N., Ghergulescu, I., Muntean, C.H.: VQAMap: a novel mechanism for mapping objective video quality metrics to subjective MOS scale. IEEE Trans. Broadcast. **62**(3), 610–627 (2016)
8. Sevcik, L., Tomala, K., Frnda, J., Voznak, M.: QoS of triple play services in LTE networks. Intell. Data Anal. Appl. **2**, 25–33 (2014)

9. Sevcik, L., Frnda, J., Voznak, M.: Degrading effect analysis, packet loss and out of order data on various tips and video resolution. In: Proceedings of Networking and Electronic Commerce Research Conference (NAEC 2008), pp. 130–138 (2008)
10. Frnda, J., Sevcik, L., Uhrina, M., Voznak, M.: Network degradation effects on different codec types and characteristics of video streaming. Adv. Electr. Electron. Eng. 12(4), 377–383 (2014)
11. Sevcik, L., Voznak, M., Frnda, J.: QoE prediction model for multimedia services in IP network applying queuing policy. In: Proceedings of International Symposium on Performance Evaluation of Computer and Telecommunication Systems (SPECTS 2014), pp. 593–598 (2014)
12. Frnda, J., Voznak, M., Sevcik, L.: Network performance QoS prediction. In: Intelligent Data analysis and its Applications, vol. 1, pp. 165–174 (2014)
13. Frnda, J., Voznak, M., Sevcik, L., Fazio, P.: Prediction model of triple play services for QoS assessment in IP based networks. J. Netw. 10(4), 232–239 (2015)
14. Frnda, J., Voznak, M., Rozhon, J., Mehic, M.: Prediction model of QoS for triple play services. In: 21st Telecommunications Forum Telfor, TELFOR 2013 - Proceedings of Papers, art. no. 6716334, pp. 733–736 (2013)
15. Sevcik, L., Behan, L., Frnda, J., Uhrina, M., Bienik, J., Voznak, M.: Prediction of subjective video quality based on objective assessment. In: 26th Telecommunications Forum (TELFOR) (2018)
16. Recommendation ITU-T P.910. Subjective video quality assessment methods for multimedia applications (2008)
17. Recommendation ITU-R BT.500–13. Methodology for the subjective assessment of the quality of television pictures (2012)
18. SJTU Media Lab: Database of the test sequences (2013). http://medialab.sjtu.edu.cn/web4k/index.html
19. Ultra Video Group: Database of the test sequences. http://ultravideo.cs.tut.fi/#testsequences
20. FFmpeg. A complete, cross-platform solution to record, convert and stream audio and video. https://www.ffmpeg.org/
21. Fröhlich, P., Egger, S., Schatz, R., Mühlegger, M., Masuch, K., Gardlo, B.: QoE in 10 seconds: are short video clip lengths sufficient for quality of experience assessment? In: Proceedings of Fourth International Workshop on Quality of Multimedia Experience, pp. 242–247 (2012)
22. Romaniak, P., Janowski, L., Leszczuk, M., Papir, Z.: Perceptual quality assessment for H.264/AVC compression. In: IEEE Consumer Communications and Networking Conference (CCNC), pp. 597–602 (2012)
23. Bech, S., Zacharov, N.: Perceptual Audio Evaluation_Theory, Method and Application (2006)
24. MitsuTool, Video Quality Indicators. http://vq.kt.agh.edu.pl/metrics.html
25. MSU Video Quality Measurement Tool. Program for objective video quality assessment. http://www.compression.ru/video/quality_measure/video_measurement_tool.html

Negative Log Likelihood Ratio Loss for Deep Neural Network Classification

Hengshuai Yao[1(✉)], Dong-lai Zhu[2], Bei Jiang[3], and Peng Yu[3]

[1] Huawei Hi-Silicon, Edmonton, AB, Canada
hengshuai.yao@huawei.com
[2] Huawei Noah's Ark Lab, Edmonton, AB, Canada
donglai.zhu@huawei.com
[3] Department of Mathematical and Statistical Sciences,
University of Alberta, Edmonton, AB, Canada
{beil,pyu2}@ualberta.ca

Abstract. In deep neural network, the cross-entropy loss function is commonly used for classification. Minimizing cross-entropy is equivalent to maximizing likelihood under assumptions of uniform feature and class distributions. It belongs to generative training criteria which does not directly discriminate correct class from competing classes. We propose a discriminative loss function with negative log likelihood ratio between correct and competing classes. It significantly outperforms the cross-entropy loss on the CIFAR-10 image classification task.

Keywords: Loss function · Cross entropy · Likelihood ratio · Deep neural network

1 Introduction

Deep neural network (DNN) has achieved remarkable success in classification tasks such as image classification [1]. The network output can mimic the posterior probabilities of target classes for the input observation when the non-linear activation function in the output layer is defined as a soft-max function [2]. The learning objective is to minimize the difference between the predicted distribution and the true data-generating distribution. In information theory, the cross entropy between two probability distributions over a common event set of events measures the average number of bits needed to identify an event if coding follows a learned probability distribution rather than the true but unknow distribution [3]. Therefore, cross entropy is a reasonable loss function for the DNN-based classification.

However, in practice the true data-generating probability distribution is unknown and replaced by the empirical probability distribution over a training set where each sample is drawn independently and identically distributed (i.i.d.) from the data space [4]. Under assumptions of uniform distributions of feature and label spaces, minimizing cross-entropy is equivalent to maximum likelihood, i.e., the learning problem aims to maximize likelihood of correct class for each of training samples [2].

© Springer Nature Switzerland AG 2020
K. Arai et al. (Eds.): FTC 2019, AISC 1069, pp. 276–282, 2020.
https://doi.org/10.1007/978-3-030-32520-6_22

Maximum likelihood is a generative training criterion by which the model learns the likelihood of correct class for the observation. The model makes predictions by using Bayes rules to calculate posterior probabilities of target classes for the observation and then select the most likely class. In contrary, discriminative training criteria discriminates the posterior probability of correct class against the competing classes. The model makes predictions with posterior probabilities directly. Therefore, discriminative training criteria are usually preferred to generative training criteria [4].

In this paper we propose a discriminative loss function based on the negative log likelihood ratio (NLLR) between the correct and competing classes for an input feature. It aims to discriminate the posterior probability of correct class against the competing classes. In the DNN learning, the stochastic gradient descent (SGD) algorithm is utilized to minimize the NLLR loss [5]. We evaluate the recognition result on two image classification tasks (MNIST and CIFAR-10) [6, 7] and compare with the cross-entropy (CE) loss. On the MNIST task, NLLR obtained comparable results as CE because the probability distributions of correct and competing classes are well separated in this relatively easy task. On the CIFAR-10 task, NLLR outperforms CE significantly implying that the discriminative criterion is superior to the generative criterion in classification.

2 Loss Function for Classification

A classification task is to categorize an observation to one of the target classes. Suppose x is the feature vector representing the observation, and y is the class that the observation belongs to. The goal is to find a function $y = f(x)$ that can predict y from x. However, in practical classification tasks there exists the misclassification risk due to missing information, noise interference or probabilistic process. Then the learning problem is to minimize expected risk [4], defined as

$$R(f) = \mathbf{E}_{X,Y \sim p(x,y)} L(f(x), y) \tag{1}$$

where $L(f(x), y)$ is the loss function measuring the cost of misclassification, and $p(x, y)$ is the probability density function over the feature space X and the class space Y. Because $p(x, y)$ is true but unknown data-generating probability distribution, we draw independently and identically distributed (i.i.d.) samples from the data space and generate a training set

$$S = \{(x_1, y_1), \cdots, (x_m, y_m)\} \tag{2}$$

which consists of m training samples. Then the expected risk can be approximated by the empirical risk

$$R_S(f) = \frac{1}{m} \sum_{i=1}^{m} L(f(x_i), y_i) \tag{3}$$

The loss function $L(f(x), y)$ can be flexibly defined to suit the model parameter optimization. For example, in deep neural networks (DNN), the cross entropy is ubiquitous because it measures the difference between the empirical distribution and the predicted distribution and can be minimized using the stochastic gradient descent (SGD) methods [5]. Suppose there are C target classes in the classification task. The DNN takes the feature vector x as input, and outputs C nodes each representing the score for the corresponding class. When the non-linear activation function in the output layer is defined as a soft-max function, the C outputs can mimic the posterior probabilities of classes $\{\hat{p}(y_c|x); c = 1, \cdots, C\}$ [8]. The cross-entropy loss function is defined as

$$L(f(x), y) = E_{p(y|x)}[-\log \hat{p}(y|x)] = -\sum_{c=1}^{C} p(y_c|x) \log \hat{p}(y_c|x) \tag{4}$$

where $p(y_c|x)$ is the empirical distribution of the training set, and $\hat{p}(y_c|x)$ is the predicted distribution from the DNN model. In classification, each training sample is commonly labeled as the correct class it belongs to, i.e.,

$$p(y_c|x) = \begin{cases} 1 & \text{if } x \in y_c \\ 0 & \text{otherwise} \end{cases} \tag{5}$$

Then Eq. (4) can be simplified as

$$L(f(x), y) = -\log \hat{p}(y_c|x) \tag{6}$$

Since the feature distribution $p(x)$ and the class distribution $p(y)$ are irrelevant to the model parameters and assumed to be uniform distributions, according to the Bayesian inference, Eq. (6) can be converted to

$$L(f(x), y) = -\log \hat{p}(x|y_c) \tag{7}$$

Equation (7) shows that the loss function is negative log likelihood of sample x. Therefore, minimizing the empirical risk Eq. (3) is equivalent to maximizing the likelihood. Maximum likelihood is a generative training criterion in which the likelihood score of each training sample is measured. In contrary, the discriminative training criteria attempt to discriminate correct class score from competing class scores for training samples and was preferred as a better direct solution for classification problems [4].

In binary classification task where there are two target classes, a model can be designed with one single output for which the empirical probability $p(y_c|x)$ equals 1 for one class and 0 for the other class. Then Eq. (4) can be simplified to a binary cross-entropy loss function [9] as

$$L(f(x), y) = -p(y_c|x) \log \hat{p}(y_c|x) - [1 - p(y_c|x)] \log[1 - \hat{p}(y_c|x)] \tag{8}$$

In condition of Eq. (5), the binary cross-entropy loss function can be extended to the multi-class classification task as

$$L(f(\boldsymbol{x}), y) = -\log \hat{p}(y_c|\boldsymbol{x}) - \sum_{k=1, k \neq c}^{C} \log[1 - \hat{p}(y_k|\boldsymbol{x})] \qquad (9)$$

It can be rewritten as

$$L(f(\boldsymbol{x}), y) = -\log \left[\hat{p}(y_c|\boldsymbol{x}) \prod_{k=1, k \neq c}^{C} \hat{p}(\overline{y_k}|\boldsymbol{x})\right] \qquad (10)$$

where $\hat{p}(\overline{y_k}|\boldsymbol{x}) = 1 - \hat{p}(y_k|\boldsymbol{x})$ represents the probability of $\boldsymbol{x}$ not classified as y_k. Comparing Eq. (10) and the cross-entropy loss function Eq. (6), it shows that each $\hat{p}(y_c|\boldsymbol{x})$ is weighted by a factor $\prod_{k=1, k \neq c}^{C} \hat{p}(\overline{y_k}|\boldsymbol{x})$ which is product of probabilities of not belonging to each of competing classes. Although Eq. (10) consists of probabilities of competing classes, it does not discriminate the correct class probability from the competing ones. We will show that the extended binary cross-entropy loss function cannot improve the performance on multi-class classification tasks.

3 Negative Log Likelihood Ratio Loss Function

In this paper we propose an inverse likelihood ratio loss function as a discriminative training criterion. As discussed in the previous section, in practical classification design, minimizing the cross-entropy loss function is equivalent to maximizing likelihood, which is a generative training criterion meaning that only the probability of the correct class is measured for each training sample. It cannot learn to optimize the discrimination between the correct class probabilities and the competing ones. To overcome the shortage of the generative training criterion, we propose to measure the ratio between the predicted correct-class probability and the competing ones in the loss function, defined as

$$L(f(x), y) = -\log \frac{\hat{p}(y_c|x)}{\sum_{k=1, k \neq c}^{C} \hat{p}(y_k|x)} \qquad (11)$$

The denominator on the right side is sum of competing-class probabilities representing the probability of not belonging to the correct class, noted as $\hat{p}(\overline{y_c}|\boldsymbol{x})$. Assuming the feature distribution $p(\boldsymbol{x})$ and the class distribution $p(y)$ are uniform distributions, Eq. (11) can be written as

$$L(f(x), y) = -[\log \hat{p}(x|y_c) - \log \hat{p}(x|\overline{y_c})] \qquad (12)$$

It shows that the loss function is negative log likelihood ratio between correct and competing classes [7]. It directly learns to discriminate correct class from the competing classes for each training sample and achieved significant improvement in our experiments.

4 Experiments

We did experiments on two popular image classification tasks. One is the MNIST task to classify each of 28 × 28 handwritten digit images into one of 10 digits [6]. The MNIST database has a training set of 60,000 samples, and a test set of 10,000 samples. The second task is on the CIFAR-10 dataset which consists of 60000 32 × 32 color images in 10 classes, with 6000 images per class [7]. The images are divided into 50000 training images and 10000 test images. For both tasks we train the deep convolutional neural networks (CNN) with the same topologies as the Keras examples [10]. Specifically, for MNIST the CNN topology is defined as two convolutional layers followed by two fully-connected layers with intermediate max-pooling and dropout layers; for CIFAR-10 the CNN topology is defined as four convolutional layers followed by two fully-connected layers with max-pooling and dropout layers. In both networks, the activation function of the output layer adopts the soft-max function to predict the probabilities of target classes on the output nodes.

We compare training and test losses and accuracies over the training epochs for the three loss functions: cross-entropy (CE), binary cross-entropy (BCE), and negative log likelihood ratio (NLLR). For CE and BCE, we train the models with 100 training epochs because MNIST training converges within 100 epochs and CIFAR-10 training encounters overfitting issue within 100 epochs (i.e., accuracy on the test set has started to degrade). For NLLR we train the models with 500 epochs to investigate the generalization performance of the NLLR loss function.

Figure 1 shows the training and test losses and accuracies on the MNIST task. The loss values of NLLR are much lower than the loss values of CE and BCE because NLLR calculates the probability ratio between the correct class and competing classes. MNIST is a relatively easy task where the predicted correct and competing probabilities are mostly close to one and zero respectively, resulting the low loss values. For CE and BCE, the training loss and accuracy curves converge quickly within 100 epochs; the test loss and accuracy converge as well. For NLLR, the training loss and accuracy achieve similar convergence and keep stable in 500 epochs, indicating the robustness of the NLLR loss function. The test loss and accuracy are also convergent and stable within 500 epochs, indicating that the overfitting issue is trivial. Compared the CE and BCE, NLLR obtained comparable results. Because the correct probability is much bigger than the competing probabilities for most of samples, using NLLR to discriminate the correct probability from the competing probability becomes unimportant for these samples. For more challenging classification tasks where the correct and competing probabilities could be easily confusing, the NLLR should show different performance from the CE or BCE losses.

CIFAR-10 is a more challenging task than the MNIST. Figure 2 shows the losses and accuracies on the CIFAR-10 task. For CE and BCE, the training loss and accuracy curves converge within 100 epochs; the test loss and accuracy degrade after certain epochs indicating that the training suffers overfitting problems. For NLLR, the training loss and accuracy keep improving over the 500 epochs, indicating that the NLLR loss

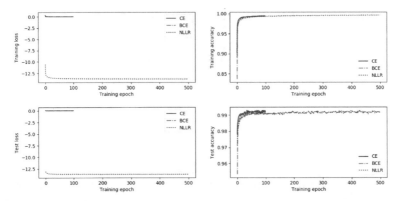

Fig. 1. Training and test loss and accuracy of three loss functions (CE, BCE and NLLR) over training epochs on the MNIST task. NLLR is trained with 500 epochs while CE and BCE trained with 100 epochs.

function well fits the SGD optimization; the test loss and accuracy also keep improving over the 500 epochs, indicating that the training does not encounter the overfitting problem. In comparison with CE and BCE, NLLR achieved significant accuracy improvement.

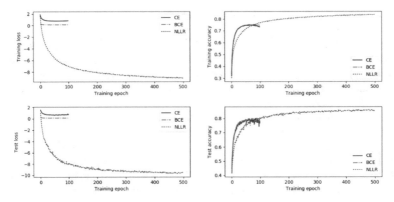

Fig. 2. Training and test loss and accuracy of three loss functions (CE, BCE and NLLR) over training epochs on the CIFAR-10 task. NLLR is trained with 500 epochs while CE and BCE trained with 100 epochs.

Table 1 summarizes the test accuracies of the three loss functions (CE, BCE and NLLR) on the MNIST and CIFAR-10 tasks. Compared to CE and BCE, NLLR obtained comparable accuracy on the MNIST task and achieved significant improvement on the CIFAR-10 task. It proves that in challenging tasks where correct probabilities are not largely superior to the competing probabilities, the NLLR loss function forms a discriminative criterion yielding better model parameter optimization.

Table 1. Test accuracies of three loss functions (CE, BCE and NLLR) on MNIST and CIFAR-10 tasks.

	CE	BCE	NLLR
MNIST	99.28%	99.27%	99.28%
CIFAR-10	79.92%	80.01%	86.01%

5   Conclusion

In this paper we first discussed that minimizing cross-entropy is equivalent to maximum likelihood given assumptions of uniform feature and class distributions. It is a generative training criterion where only correct class is measured for each feature sample. We proposed a discriminative loss function NLLR which is a negative log likelihood ratio between correct and competing classes. It aims to maximize the correct class probability and minimize the competing class probabilities simultaneously. Compared to cross-entropy and binary cross-entropy, NLLR obtained comparable results on the MNIST task and significant improvement on the more challenging CIFAR-10 task.

References

1. He, K., Zhang, X., Ren, S., Sun, J.: Deep residual learning for image recognition. In: CVPR (2016)
2. Goodfellow, I., Bengio, Y., Courville, A.: Deep Learning. MIT Press, Cambridge (2016)
3. Boer, D., et al.: A tutorial on the cross-entropy method. Ann. Oper. Res. **134**, 19–67 (2005)
4. Vapnik, V.: The Nature of Statistical Learning Theory. Springer, Heidelberg (2000)
5. Rumelhart, E., Hinton, G., Williams, J.: Learning representations by back-propagating errors. Nature **323**, 533–536 (1986)
6. LeCun, Y., Bottou, L., Bengio, Y., Haffner, P.: Gradient-based learning applied to document recognition. Proc. IEEE **86**(11), 2278–2324 (1998)
7. Casella, G., Berger, R.L.: Statistical Inference, Duxbury Thomson Learning (2002)
8. Krizhevsky, A.: Learning Multiple Layers of Features from Tiny Images (2009)
9. Murphy, K.: Machine Learning: A Probabilistic Perspective. MIT Press, Cambridge (2012)
10. Keras examples. https://github.com/keras-team/keras/tree/master/examples

When Inclusion Means Smart City: Urban Planning Against Poverty

Jean-Claude Bolay[⌧]

Center for Cooperation and Development (CODEV),
Swiss Federal Institute of Technology (EPFL), 1015 Lausanne, Switzerland
jean-claude.bolay@epfl.ch

Abstract. A majority of human beings live in cities, half of them in cities of less than 500.000 inhabitants, and around 1 billion of dwellers live today in slums (meaning 25% of urban residents in the global South). Having these numbers in mind, the main question for urban planners and decision makers is: why our forecasts of a sustainable urban development have not been reached, at least not for all people, what may we change in order to create and enhance "smart cities" offering access to infrastructures, services and environment of quality to all, without any exception? Based on a conceptual approach of urban planning and on case studies in Africa and Latin America, we shall focus on 3 dimensions of a renewed urban planning: base the planning on a diagnosis of human and material reality specific to each city – involve the planning in a critical perspective of financial and human available resources – implement innovative technologies in a participatory approach including inhabitants and all urban stakeholders.

Keywords: Urban Planning · Smart cities · Poverty

1 Poverty, Urban Development and Planning

A reflection on urban development and poverty has to be considered as a priority for cities in Southern countries; with a special focus on small and medium-sized cities, knowing that these settlements gather 50% of the world's urban population [1, 2]. These cities are those with the highest rates of population growth; and their authorities do not manage enough financial and human resources to address the material, economic and technical questions linked with the urban integration of these new dwellers. The indirect impact of this inability to handle this demographic trend is translated in marginality, precariousness and instability for a huge amount of poor inhabitants [3]. The agency of United Nations dedicated to habitat and cites teaches us that more than 30% of the world urban population live in slums in 2010 [4].

The medium-sized cities play a role of intermediation between rural population and the urban network, at national and even sometimes at international levels. These cities are under particularly strong pressure without having the means to face it, in terms of financial funds, local budgets and external endowments, as in terms of competent staff. Moreover, these "ordinary cities", to use the terminology of Robinson, are generally not under the political attention of the central government, and their public budget are

© Springer Nature Switzerland AG 2020
K. Arai et al. (Eds.): FTC 2019, AISC 1069, pp. 283–299, 2020.
https://doi.org/10.1007/978-3-030-32520-6_23

relatively low to tackle all the urgent questions to resolve and investing for the future [5, 6]. In addition, just as important, the public investments, however small, do not prioritize poverty reduction. They are more in relation with opportunities and under pressure of powerful stakeholders.

As result of this dichotomy between needs, requests and offers, urban planning, as implemented in a majority of intermediate cities of Southern countries, only partially address the problems facing urban populations. It is partial on the spatial level, covering only certain territories of the city, generally abandoning poor and peripheral neighborhoods. It is biased on the socio-economic level, focusing primarily on the areas invested in by the privileged social actors, as business districts and high-level residential areas. Precarious living conditions, at different levels, are reflected by a continuous expansion of slums and by the increasing number of people living there. The slums reflect both a territorial exclusion, through the lack of equipped spaces accessible to the poor, and the tensions arising from the occupation by the poor of land that sometimes becomes very attractive for the market. Which does not prevent the inhabitants of the slums to greatly contribute to the economic and social dynamics of the city: job creation, income generation, community organization, and social and political participation.

Taking the example of South Africa, and in reference with many other developing countries, Watson [7] considers that unplanned urban growth causes a concentration of poverty, but also spatial inequalities in cities. And, unfortunately, urban planning, when applied by local authorities, is unable to apprehend with anticipation and to solve a multitude of intertwined problems. There is a discrepancy between, on the one hand the local needs, of each family, of each community, of each neighborhood, and on the other hand, the production by the specialists of plans, programs and projects, with unclear aims and a poor coordination. For this author, planning is primarily viewed in a purely technical posture, with little concern for local urban history. The stakeholders involved in the process do not share a societal vision taking into account all of the urban community, in its various components, especially the poor who, as is often the case in cities in South, represent the majority of citizens. The result is often that investments that are often poorly targeted and do not address the crucial questions that the majority of urban dwellers are facing. The question arises in terms of equipment choices to focus on, but also, and most often, in terms of conditions of accessibility, as costs are not adapted to the financial conditions of the most underprivileged segments of the population. According to Yftachel [8], the alternative would be to question five key dimensions of urban development: land-use and its allocation criteria; the policies put in place to fight against segregation; decision-making procedures; the consideration of the socio-economic conditions of urban dwellers as a whole and particularly groups in the disadvantaged population; as well as the financial impact of urban transformations in order to tackle the risk of gentrification of rehabilitated neighborhoods.

The first variable to consider in appropriate planning in these contexts is the alleviation of urban poverty. Tannerfeldt and Ljung [9] emphasize that urban poverty is multidimensional. Financial and economic criteria are not the only ways to assess urban poverty. In addition, indicators based on health, education, environmental quality, violence and insecurity, represent a multitude of risks that these poor families must confront. Tacoli, McGranahan and Sattethwaite [10] designate urban sectors where the poor are systematically ostraziced: insecure land tenure, poor quality housing and a lack of public provision for infrastructure.

2   Smart Cities: Which Concept for Which Reality?

The question of smart cities is a topic that has gained momentum in recent years for two reasons. The first is that technology has advanced, allowing us to approach in an innovative way the urban problems (e.g. mobility and transportation, maintenance of infrastructures, development of public spaces, housing, etc.) [11]. The second is related to the rapid extension of urban population, particularly in South countries. The cities are becoming increasingly vulnerable to environmental risks and climate change. This is valid for all cities, regardless of the location or size. The desire for smart sustainable development merits questioning when it comes to South countries, given the many challenges and lack of financial and human resources these cities face.

The United Nations has established a new list of priorities in view of sustainable development for the planet. Of the 17 objectives targeted for 2030,[1] one focuses specifically on sustainable cities and communities[2] with the goal of making them inclusive, safe, resilient, sustainable places where people can live decently and be part of the city's production dynamic, creating shared prosperity and social stability without harming the environment. The ten priority areas that should frame urban action for the coming years are housing, basic services and slum upgrading, transportation, urban planning and management, protection of cultural and natural heritage, mitigation and reduction of risks associated with disasters, environmental protection, democratic access to green spaces for everyone, links between rural and urban areas, and technical/financial assistance for the use of local materials in construction.

These intentions are commendable and relevant with regard to the many unresolved urban issue. However, they more closely resemble a vast, loosely-designed catch-all. Despite their generality - which allows for innumerable interpretations by urban actors - it is surprising that topics as universal as insecurity (real or as perceived by inhabitants), drinking water supply, treatment of wastewater and solid waste, informality and the urban economy in general are not mentioned, though they are among the urban MDGs in 2000 [12]. Though merely in foresight, there was nonetheless a notable step backwards between the 2015 SDGs and the 2000 MDGs. No further quantified information is provided, making assessing progress in achieving these objectives more difficult. Given this fixed framework for sustainability at the global level, in what way is this other urban referential that is the concept of the "smart city" useful?

The concept of "smart city" was first debated in the 1990s. At that time, the focus was on the significance of new ICT in setting up new urban infrastructures. However, the concept was criticized for being too focused on technological aspects and too limited in this definition. According to Albino et al. [13], several authors broadened the topic by including both the managerial role of the city (which is also a component of

---

[1]   http://www.undp.org/content/undp/en/home/sustainable-development-goals.

[2]   https://www.un.org/sustainabledevelopment/cities/.

the "smart city") and a warmer strategy based on a more user-friendly approach than terms like 'smart city' or 'digital city' denote. Of the different definitions taken up by the author in his article, the one of Marsal-Llacuna et al. [14] is comprehensive and includes the idea of a "modernity" comprised of cutting-edge technologies, intersectorial collaboration and user-friendliness. The key idea here is less technology for its own sake or for better designed and controlled infrastructure, and more connectivity and inter-connectivity between individuals and between individuals and new communication technology objects. This is what emerges from a definition by another group of authors, Bakici et al., who specifically emphasize these social, institutional and technological exchanges: "Smart city as a high-tech intensive and advanced city that connects people, information and city elements using new technologies in order to create a sustainable, greener city, competitive and innovative commerce, and an increased life quality" [15, p. 139].

While the technological contribution is indispensable, its strength and relevancy depend on "people" and their creativity and desire to "play with technology" in order to participate in the local life, generate new knowledge and thus help improve living conditions. Social capital is therefore a fundamental dimension of projects. These principles are interesting and explain the attractiveness of "smart city" projects. However, we must not be too naive given that behind the enthusiasm for many new forms of communication lies the interests of private companies, which know and predict the profitability of investments in urban technologies in cities.

Although many authors who have worked on the "smart city" concept approach the field according to a specific grid, Giffinger and Gudrun have broken this down into six components to describe the different aspects of this type of city. They are smart economy, smart mobility, smart environment, smart people, smart living and smart governance, the unifying goal being to improve the quality of life. For each component, a battery of indicators was set up to calculate the level of "smartness" and, consequently, the points to focus on for each city studied. Albino et al. express how these elements combine with one another in a synthetic way: "A city has networked infrastructure that enables political and social efficiency and cultural development; an emphasis on business-led urban development and creative activities for the promotion of urban growth; social inclusion of various urban residents and social capital in urban development; the natural environment as a strategic component for the future" [16, p. 13].

The smart city concept is not completely out of line with SDGs. To follow Ahvenniemi et al. [17] parallel between sustainable cities and smart cities, the latter concept includes the idea of environmental sustainability by reducing greenhouse gas emissions in urban areas with the use of appropriate technologies. Such 143 projects were implemented in 2012, including 47 in Europe and 30 in the United States. As stated at the time, sustainable development aims to balance the environmental, social and economic dimensions of development. However, in analyzing its global or local implementation, it is clear that environmental preservation remains central.

Applied to the city, sustainable development adds concern with regard to a balanced distribution of human activities across the territory. The case studies show that sustainable urban development focuses more on environmental and social dimensions (with entry points such as the management of waste, water or the built environment for the ecological dimension and on health, safety and well-being for the social dimension), somewhat abandoning certain economic aspects of sustainable development. Urban analysis based on the "smart city" concept combines both ICT-related technological dimensions and the social dimension, with inroads related to education, culture, science and innovation, leaving little room for environmental issues. This can be done by taking into account the strong economic implication of new technologies in urban management and the involvement of private companies in this area. Monfaredzadeh and Berardi further emphasize that "Amid profound economic, social and technological changes, cities around the world are facing the challenge of reconciling the needs of immediate competitiveness with long-term sustainable development" [18, p. 1064]. This apparent dichotomy between competitiveness and sustainable development is problematic, as these elements appear on the agendas of many cities indiscriminately and without any further explanation.

Comparing sustainable cities and smart cities, the authors agree that urban problems must be tackled using a multidimensional approach. They also bring draw a parallel between the two concepts to redefine urban studies in a more coherent way based on the strengths and weaknesses inherent in both approaches.

3 Urban Planning and Smart Cities in Developing Countries: A Fruitful Approach?

The work we have done over the years has largely been inspired by sustainable urban development with a focus on questions of inclusion, social equity, combating poverty, environmental preservation and economic redistribution. The goal here is to foster urban regeneration to counter precarious housing and infrastructure and the socioeconomic exclusion that the slums so perfectly symbolize. These elements should guide us in renewing urban planning in a way that responds to the challenges of each specific urban context and aims at integrating populations into their urban, built and natural environments. Three recent case studies from Burkina Faso, Brazil and Argentina show that these intentions are completely innovative compared to what is really happening in those countries. In the best case scenario, planning is essentially spatial and material and has little interest in social and economic dimensions; in the worst case scenario, there is simply no medium or long-term projection, and urban planning authorities simply work opportunistically with the available funding based on political priorities, with no sustainable vision [19–21].

3.1 Urban Practices: The Cases of Koudougou, Montes Claros and Nueve de Julio

At first glance, these three cities have little in common. Nueve de Julio, in Argentina, is a medium-sized city in an agricultural region, with just over 50,000 inhabitants. Koudougou, with 100,000 inhabitants, is a provincial capital in Burkina Faso. Montes Claros, an industrial and academic city in the state of Minas Gerais in Brazil, has 400,000 inhabitants. Radically different in their geographies and histories, these three intermediate cities are similar in that they play a crucial role in their regional environments and face major urban planning challenges. All three are reliant on the globalization of trade and exogenous factors for their development: Nueve de Julio is a marketing center for Argentinian agricultural export products; Montes Claros is home to many Brazilian industries and/or branches of foreign companies that target domestic and foreign markets; and Koudougou occasionally benefits from support from international cooperation, development agencies and NGOs.

In all three cities, more or less successful attempts at urban planning have been made in recent years with results that cast doubt on the intentions, approaches and methods used (Figs. 1, 2 and 3).

Fig. 1. A poor downtown area of Nueve de Julio (photo by CODEV: T. Wilkinson, 2017).

Like many medium-sized cities in the Global South, Nueve de Julio could epitomize the negation of urban planning. Indeed, up until the creation of a scientific and technical cooperation organized by the Center for Development and Cooperation of the Swiss Federal Institute of Technology (CODEV/EPFL[3]), the various city governments were always reluctant to develop planning tools and the human resources to manage them.

Over the past 25 years, the city's inhabited territory (i.e. new human settlements) has expanded in an uncontrolled manner. Agricultural lands have made way for a suburb of private properties with no connection to public electricity and water networks. The extension of these networks and other urban services subsequently generates very high additional costs for the whole collectivity. The centric area denominated "Ciudad Nueva" is now inhabited by more than 20% of the urban population, which is mostly poor and lives in self-built homes that are poorly integrated into the urban agglomeration. A third element in Nueve de Julio is an electricity cooperative that has gradually taken over many of the urban services, as a substitute for the municipality. The current political majority is in line with the liberal perspective

---

[3] https://cooperation.epfl.ch/sustainablehabitatandcities.

Fig. 2. Suburb of Koudougou (photo by JC Bolay, 2014).

Fig. 3. Implementation of a new settlement funded by the public program "Minha Casa, Minha Vida" in Montes Claros (photo by JC Bolay, 2015).

championed by the national government and, above all, is seeking to invest in promising, short-term projects that give its work visibility. Urban management is treated as an emergency (for example, to combat flooding due to the defective drainage system) and without the skills required for longer-term planning. The main reason for these shortcomings is a lack of understanding of the benefits of long-term foresight and a real lack of political will. This results in an inability to rationally control the city's spatial expansion and to foster the socio-economic and urban inclusion of disadvantaged populations.

Koudougou has experimented with numerous sectoral and general development plans along the last 15 years, supported by foreign cooperation agencies, with the aim to sustain the city's public activities in a well-thought-out, long-term framework. Due to the numerous technical issues that must be resolved, everything immediately becomes a priority, without any particular criteria to justify the choices made. A special attention must be paid to the conditions of production of urban plans in Koudougou, as for the other main agglomerations in Burkina Faso. These urban plans are generally designed by the national government and "imposed" on the municipalities. In addition, the municipal administration is not endowed with the competent human resources to participate in the supervision and monitoring of local plans. Moreover, the municipal budget makes it impossible for the local authorities to follow the guidelines of the plans and implement rationally the projects. This projection only serves to attract external funders and doesn't guide a coherent urban development.

The Brazilian city of Montes Claros is a typical example of an intermediate city. Although it cannot be considered a poor city, the urban population is highly segregated. 30% of its poor. And the wealthy - who represent 20% of the population - control 66% of the local wealth. This discrimination of the social fabric is also reflected in the more than 140 unregulated neighborhoods and 50,000 people living in substandard housing conditions. In addition to these socio-economic and territorial disparities, two other major issues prevail. The first is the depletion of natural resources due to advanced deforestation and the resulting periodic flooding of central neighborhoods. This environmental deterioration results from the implementation of new peripherical settlements (social and luxury ones) without preliminary studies of ecological impact. According to the legal obligations specific to Brazil, the local administration was in 2015 in the process of developing a new master plan that aims to create a sustainable city (urban land rights, housing, infrastructure, services, transport and employment for all). In Montes Claros, the municipal administration in charge of the exercise brought together a panel of professional and academic experts to define the guidelines for the plan. And the public participation to this important definition of the future of the city took the form of four presentations in several neighborhoods. An on-site survey found that few people were actually aware of the ongoing process. The current process raises many questions. To begin, it was impossible to perceive in the documents produced at this time a global and long-term visions of the future of Montes Claros. Furthermore, the exercise does not meet the ambitions of an urban planning conscious of future challenges since the plan focuses only on the central areas of the city without including its periphery; and this while the main issues of planning and organization of the territory and its inhabitants are located on the fringes of the city.

These three cases prove that planning cities and city regions is no easy task, especially in South countries. Though we cannot generalize, certain elements found in the case studies can help in deciphering key dimensions of urban planning for Southern cities: only sincere political will, both locally and nationally can ensure the effectiveness of these processes. This political will must be then translated into financial and technical means that will allow the public authorities and their private partners to develop a rigorous, long-term framework. Explicit framework conditions must provide for the training and recruitment of professionals who are specialized in the field and committed to monitoring these planning actions in the long term. Finally, only true participation by the people and the urban stakeholders in determining sectoral and spatial priorities can enable planning to meet citizens' approval and ultimately gain their support.

3.2 Smart and Sustainable, What Criteria for Planning the Development of South Cities?

South cities are diverse; their characteristics cannot be generalized unambiguously and categorically. That said, a number of points allow us to look beyond their distinctive characteristics in features to establish a framework of similarities and differences.

The first feature common to South cities is strong demographic growth, particularly in Africa and Asia (which, according to the United Nations' statistics [22] will see 90% of the urban growth worldwide) and to a lesser extent Latin America.

The second specificity of South cities is that this demographic growth has led to spatial fragmentation and socio-economic segregation in the form of increasing mass poverty and the expansion of slums, which today are home to nearly a billion poor people [23, 24]. This figure is expected to increase significantly to almost two billion in the next 30 years [25].

Environmental degradation is, unfortunately, also the consequence of poorly controlled urban growth, in which the overexploitation of natural resources (water, air and land) deteriorates the population's living conditions, notably affecting the health and well-being of the most destitute [26, 27].

South cities are also marked by the *informalization* of a large proportion of their economies, leaving workers - who, as entrepreneurs, are often without job security or on-site safety standards - outside legal and regulatory frameworks. This unstable labor sector can account for 50% to 80% of active workers depending on the city [28].

In Southern countries, the intermediate cities, with less than 500.000 inhabitants, play a crucial role in the interface between rural and urban development. These regional spots have, today, the highest population grow rates. They are also those that most lack human and financial resources for coping with these issues through critical infrastructure and services to offer inhabitants a dignified life [29].

Given these indicators, determining what the priorities are in terms of sustainable urban development is essential for establishing guidelines that can serve as a basis for public policy.

The starting point is the fact that urban growth continues in the Global South - particularly Africa and Asia. This is reflected in an increase in population and territorial expansion, both of which necessitate the provision of services, equipment and infrastructure. In most cases, the supply does not meet the demand, leading to increased urban poverty. Three concomitant processes fuel the gradual impoverishment of South cities. The first is unemployment and underemployment, which combine with economic liberalism and strengthen informal urban work practices. The second are environmental issues, which are not addressed prior to urban development operations and result in the degradation of natural resources. Finally, access to quality housing is limited by the number of units available and urban dwellers' financial means, leading to an increase in slums and other forms of instability.

Urban planning must therefore serve as a means to organize the territory and human activities in a rational, coherent way in order to sustainably include all citizens in key sectors that will (1) offer them decent living conditions (i.e., housing and infrastructure), (2) promote an economy that provides jobs and stability for working people, and (3) ensure respectful use of the natural resources.

Making the link between urban planning for sustainable development and cities' "smartness" means reconsidering the indicators through the lens of technological innovation and citizen participation.

Hence we could combine the issue of sustainable urban development with criteria for identifying smart cities in order to evaluate hypothetical compatibilities (Table 1).

From this first comparative analysis emerges evidence that gateways exist between the concepts of sustainable cities and smart cities. However, in the South city context - especially in small and medium-sized cities in developing countries - it is above all the disparities among dwellers and territories that characterize urban dynamics.

When it comes to the smart city, we must first distinguish two of its key features: The first is technological innovation, with a focus on new information and communication technologies (NICT).[4] The second is key areas for organizing the city, such as environmental technologies, mobility management and traffic, for example. As Lea argues, "a smart city is not just a city that leverages new technologies, but is a complex ecosystem made up of many stakeholders including citizens, city authorities, local companies and industry and community groups. It should be emphasized that the geographical boundaries of what is called a smart city may be wider than the city itself, gathering multiple governance bodies and municipalities to define services at the metropolitan or regional scale" [30, p. 3] (Fig. 4).

As the author says, the city is a complex ecosystem that, in order to be more efficient, involves multiple and interconnected technologies. These technologies are supposed to greatly improve the urban environment and the quality of life for those who live and work there. Numerous examples are cited on the Internet, including: less polluting collective and individual mobility, better controlled use of energy sources (heating, transportation, etc.), sensors to track air quality,[5] waste collection, leakage

---

[4] Mahashreveta Choudhary, Six technologies crucial for smart cities, 18.04.2018. (https://www.geospatialworld.net/blogs/six-technologies-crucial-for-smart-cities/).
Teena Maddox, Smart cities: 6 essential technologies. (https://www.techrepublic.com/article/smart-cities-6-essential-technologies/).

[5] Oshin. What is a Smart City and How Will They Help the Environment? 06.11.2017. (https://greenerideal.com/guides/smart-city-will-help-environment/).

Table 1. Comparison between sustainable urban development and smart cities

Sustainable urban development	Smart city
Demographic growth	*Smart mobility, smart economy, smart people* In terms of mobility, urban-rural relations largely depend on transportation and ICT infrastructure outside of urban centers that facilitate commuting versus emigration. All of this results in local and regional accessibility Economically speaking, the overwhelming issue of informality is intrinsically linked to the globalization of the market, which fosters opportunistic and costly solutions in the long term, particularly as far as the environment is concerned (without environmental protection or environmental costs included in product prices). Given complementarity of economic liberalism and job insecurity, it is equally clear that the informal sector is a gold mine for the development of innovative spirits, entrepreneurship and productivity. In the vast majority of South cities, the economy flows from large cities to smaller ones, concentrating in shopping centers and industrial outskirts at the expense of rural areas As for the people, it has been clearly shown that in the pull factors that lead to urban immigration, it is the more creative and flexible individuals who choose this option for the more obvious reason of employability, but also to improve their children's level of education. Education, health and work are the three sectors to work on Urban growth is a heavy trend in the world's evolution. It is by better understanding its causes that cities will become more successful in controlling its consequences. This inevitably implies a multi-scale analysis and not merely a local/intramural focus
Territorial fragmentation and slums	*Smart living, smart governance, smart environment* Due to its precarious nature, lack of infrastructure and the poverty of its inhabitants, slums may seem like the antithesis of smart living. However, they can also be seen as a bottom-up alternative for low-income urban dwellers who put their energy and know-how to use to build neighborhoods that help integrate them into the urban system and increase their potential for individual and collective development. What is certain is that the slum reflects a mode of urban governance that is anything but smart, given that it divides, segregates and marginalizes more than it includes and unites To be effective and legitimate, urban public policies in the Global South must take into account people's needs and demands. This requires context-specific urban planning and multi-stakeholder management that enables a democratic participation in the projects' choice as well as monitoring their implementation A combination of bottom-up approaches (as they prevail in slums) and top-down approaches (which are often initiated by urban authorities) would allow for better integration of the territory. The environmental issues cities face (water, soil and air) can be duly addressed, with a positive impact on social inclusion and quality of life for inhabitants
Environmental degradation	*Smart environment, smart living and smart governance* Pollution of natural resources in urban areas is of particular concern in most Southern cities, with a dual impact in urban areas themselves (with consequences in terms of individuals' health) and in outlying areas (through contaminated water and air) that extend to the urban fringes, hinterlands and more distant rural areas. Natural resource protection must be efficiently managed by the government and adopted both city dwellers and public and private businesses. A legislative framework and financial means are imperative, as is stakeholders' political will to preserve the environment.

(continued)

Table 1. (*continued*)

Sustainable urban development	Smart city
	This is one of the most fundamental challenges to take into account in smart governance, as it is one of the pillars of sustainable development. Without its resolution, smart living is simply not feasible Two complementary elements must also be considered. Firstly, considering environmental issues also means considering climate change, which, depending on the city's geographical location and spatial configuration, is more or less urgent (rising water levels, drought, clean water supply, hurricanes, etc.). Secondly, the smart environment concerns not only the natural environment but also the built environment, which means buildings and infrastructures whose maintenance and embellishment contribute to overall urban well-being and thus to smart living
Economic informality	*Smart economy, smart environment, smart living, smart governance* A smart economy is an economy that is both innovative and productive. However, from a sustainability perspective, it must also provide quality jobs and an environmentally friendly economy in terms of its production and waste recycling. It is also a constituent element of sustainable urban development in balance with social and environmental dimensions In South countries, the urban economy must face the informality of entire sectors of production. In certain cities, as much as 80% of the workforce functions informally, working freelance with no safety standards or job security. Without tackling this issue head on, there can be no smart urban economy, just as there can be no smart living without solutions that combat insecurity (employment in this case) and poverty (in terms of income) Due to its social and environmental impact, the smart economy transcends individual sectors. In the Global South, it can only truly take hold and spread within a framework of consultations between partners that aims to implement smart governance of the city and of human activities
Small and medium-sized cities	*Smart living, smart governance, smart people* As we have already mentioned, one of the characteristics of the ongoing growth of intermediate cities. These cities are now those that most suffer from a lack of urban management, both in terms of human skills and financial resources. Conversely, studies on small and medium-sized cities show that quality of life, conviviality and sense of belonging were stronger in these cities and thus formed a substrate of smart living Some indicators of smart people might, at first glance, not seem to apply to this type of urban agglomeration (e.g. cosmopolitanism, open-mindedness, level of qualifications, social and ethnic diversity). Others would be more appropriate to translate this urban localism and sense of heritage belonging (notably, participation in public life) One of the major challenges to the universal application of the smart city concept is the ability to include all types of cities in all kinds of contexts, both North and South

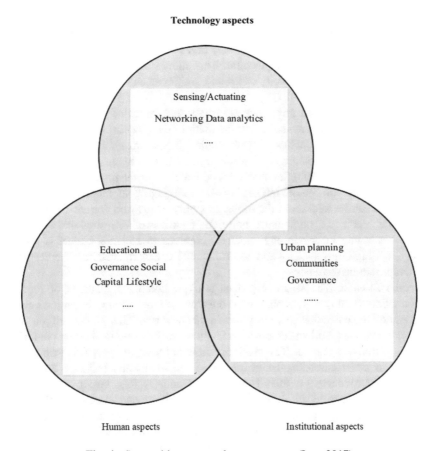

Technology aspects

Sensing/Actuating

Networking Data analytics

....

Education and

Governance Social

Capital Lifestyle

.....

Urban planning

Communities

Governance

......

Human aspects   Institutional aspects

Fig. 4. Smart cities as complex ecosystems (Lea, 2017).

detection, automating water adduction,[6] smart street lights, microgrids and monitoring environmental conditions.[7] Numerous other projects and ideas are also cited.

What also emerges from this first comparative analysis is that the smart city is built based on "implicits" that, more than likely will propose models and approaches that are not in line with the reality of development in the majority of cities marked by poverty and social exclusion.

It is therefore necessary to alert promoters of smart cities to be particularly attentive to the framework conditions in which such initiatives can be drivers of a sustainable urban future by drawing parallels between SDGs and the contributions of the new technologies.

---

[6] Deloitte. Environment. (http://smartcity.deloitte.com/domains/environment/).

[7] Matthes Speer. The benefits to the environment of smart city living. (http://www.isustainableearth. com/sustainable-living/benefits-of-smart-city-living) 28.02.2018.

Though a fundamental factor in unequal, segregative and exclusive development both territorially and socio-economically, the smart city totally ignores the demographic growth many South cities face. It is urgent to take this reality and the risks it poses to smart technologies into account. This should help in redefining what smart people are (which the vast majority of urban immigrants and other commuters, in fact, are) based on reality.

Smart living and smart environments imply that issues of urban precariousness, poverty and territorial fragmentation are the focus of the process and are in line with the SDG's desire to promote inclusive, livable cities. Integrating slums and creating equal access to quality infrastructures and services are inherent to smart environments and require appropriate technologies that are available to all citizens. Yet, no one talks about this material and human integration when talking about smart cities.

The same can be said about the urban economy. Under the current circumstances, the informal sector is a particularly notorious example of man exploiting man. The "modern" economy in developing countries involved in globalized trade is possible, but only at a high cost socially and environmentally. Smart, sustainable cities should also incorporate these issues.

A final thought. The smart city does not question the modalities of its future transformation. Again, in an implicit way, the areas of intervention, the projects cited as examples and the cities that serve as pilots in this new trend in urban development are, for the most part, large urban centers located almost exclusively in Western countries or highly-industrialized regions. The challenge now is trying to apply this technological modernity to poor cities, small and medium-sized towns and ordinary South cities (Robinson, 2006) based on their respective advantages (e.g. easier control over a spatially and demographically-limited territory, sociability and a sense of belonging to a place and a history, un urban-rural relationship that makes these cities a major interface for urban and regional dynamics, modest investments with a strong leverage effect in the three key areas of sustainable development (i.e. environmental, social and economic)).

4   Conclusion: The Smart City and Urban Planning in the Global South: Bridging the Gaps

The smart city is a popular concept that resonates across the urban world; it is one with which we want to identify, linking us via our smartphones, laptops and Bluetooth connections to our immediate environment and the vast, virtual global village to which we all belong.

In this bustling, generous emulation that aims to putting new technologies at the service of quality of life, well-being and better management of existing resources, implicit prejudices exist that could challenge the universality of the concept. More specifically, it could challenge its adaptation to such specific contexts as South cities, which are often poor in that they lack resources, human skills and financial means.

Urban planning in the Global South is widely regarded as a failure when we sporadically apply a model of urban organization - created decades ago in North countries - to South cities as a "guidance tool." These technical and managerial tools obviously lack perspective and comprehension of social and territorial issues, and must be integrated into multi-stakeholder urban governance based on the needs identified and the legitimization of social demands. All of this must be done with a view to sustainable urban development aimed at social inclusion, the fight against poverty and the protection of a highly degraded natural and built environment.

Innovation in urban planning can serve as a complementary component of the smart city provided that it considers South cities in their diversity and the issues they face.

This implies that the links between innovative technologies and friendlier and greener cities have to do with the priorities identified in terms of urban management. The latter can be achieved through the social and economic integration of citizens and by reorganizing the territory using innovative, robust, inexpensive and appropriate technologies whose upkeep is easy. All of this technological and society's innovation will then take on new meaning: sustainable and inclusive smart cities.

We are in the infancy of the smart city concept and its application to real cities. Hence, we do not propose transferring these technologies to the Global South without paying close attention to how they are used. While they appear to be highly profitable for many high-tech companies around the world, it is crucial that this profitability extend to Africa, Latin America and Asia, and the world's poorest countries and cities first and foremost. Turnkey export of these technologies to favor international cooperation must help in building local skills, supporting businesses, creating jobs and strengthening the basic sectors of community life. This requires smart governance that associates all urban actors, who then must choose the common good over special interests.

These priorities remain to be established on the world agenda. Failure to do so may further widen the gap between rich and poor countries, between the privileged and the underprivileged, and between urban precariousness and elite overconsumption. So why not approach this major challenge for the future head on by putting the issues on the table rather than feigning globalized urban conviviality and attempting to mask the inequalities of access to urban resources? The right to the city Lefebvre demanded in 1968 is still valid today, even if technology has evolved and offers us new opportunities to better share (or divide) the urban territory. It's up to us to choose, and work towards this goal.

References

1. Bolay, J.-C., Kern, A. L.: Intermediary cities. In: Wiley-Blackwell Encyclopedia of Urban and Regional Studies. Wiley-Blackwell, Hoboten (2019). https://doi.org/10.1002/9781118568446.eurs0163
2. United Nations, Department of Economic and Social Affairs, Population Division: World Urbanization Prospects: The 2014 Revision, Highlights. United Nations, New York (2014). https://esa.un.org/unpd/wup/publications/files/wup2014-highlights.Pdf
3. Bolay, J.-C., Chenal, J., Pedrazzini, Y.: Learning from the Slums: The Habitat of the Urban Poor in the Making of Emerging Cities. Springer, Paris (2016). https://doi.org/10.1007/978-3-319-31794-6

4. UN-Habitat: State of the World's Cities 2010/2011. Bridging the Urban Divide. UN- Habitat & Earthscan, London: (2010)
5. Robinson, J.: Ordinary Cities: Between Modernity and Development. Routledge, London (2006)
6. Parnell, S., Robinson, J.: (Re)theorising cities from the global south: looking beyond neoliberalism. Urban Geograph. 33(4), 593–617 (2012). https://doi.org/10.2747/0272-3638. 33.4.593
7. Watson, V.: Seeing from the south: refocusing urban planning on the globe's central issues. Urban Stud. 46, 2259–2275 (2009). https://doi.org/10.1177/0042098009342598
8. Yiftachel, O.: Ethnocracy. Land and Identity Politics in Israel/Palestine. University of Pennsylvania Press, Philadelphia (2006)
9. Tannerfeldt, G., Ljung, P.: More Urban Less Poor. An Introduction to Urban Development and Management. Sida & Earthscan, London (2006)
10. Tacoli, C., McGranahan, G., Satterthwaite, D.: Urbanisation, rural-urban migration and urban poverty. IIED, London (2015)
11. Bolay, J.-C., Kern, A.L.: Technology and cities: what type of development is appropriate for cities of the south? J. Urban Technol. 18(3), 25–43 (2011). https://doi.org/10.1080/10630732.2011.615563
12. Bolay, J.-C.: What sustainable development for the cities of the South? Urban issues for a third millennium. Int. J. Urban Sustain. Dev. 4(1), 76–93 (2012). http://www.tandfonline.com/doi/abs/10.1080/19463138.2011.626170
13. Albino, V., Berardi, U., Dangelico, R.M.: Smart cities: definitions, dimensions, performance, and initiatives. J. Urban Technol. 22(1), 3–21 (2015). https://doi.org/10.1080/10630732.2014.942092
14. Marsal-Llacuna, M.-L., Colomer-Llinàs, J., Meléndez-Frigola, J.: Lessons in urban monitoring taken from sustainable and livable cities to better address the Smart Cities initiative. Technol. Forecast. Soc. Change Part B 90, 611–622 (2015). https://doi.org/10.1016/j.techfore.2014.01.012
15. Bakıcı, T., Almirall, E., Wareham, J.: A smart city initiative: the case of barcelona. J. Knowl. Econ. 4(2), 135–148 (2013). https://doi.org/10.1007/s13132-012-0084-9
16. Giffinger, R., Haindlmaier, G.: Smart cities ranking: an effective instrument for the positioning of cities? ACE: Architecture, City and Environment = Arquitectura, Ciudad y Entorno, year IV, Num. 12, pp. 7–25 (2010). http://www-cpsv.upc.es/ace/Articles_n10/Articles_pdf/ACE_12_SA_10.pdf
17. Ahvenniemi, H., Huovila, A., Pinto-Seppä, I., Airaksinen, M.: What are the differences between sustainable and smart cities? Cities 60, 234–245 (2017). https://doi.org/10.1016/j.cities.2016.09.009
18. Monfaredzadeh, T., Berardi, U.: How can cities lead the way towards a sustainable, competitive and smart future? WIT Trans. Ecol. Environ. 191, 1063–1074 (2014). https://doi.org/10.2495/SC140902
19. Bolay, J.-C.: Planning the intermediate city, or how to do better with little: the case of the city of Nueve de Julio. Argentina. Curr. Urban Stud. 6(3), 366–400 (2018). https://doi.org/10.4236/cus.2018.63020
20. Bolay, J.-C.: Prosperity and social inequalities: montes claros, how to plan an intermediary city in Brazil. Curr. Urban Stud. 4, 175–194 (2016). https://doi.org/10.4236/cus.2016.42013
21. Bolay, J.-C.: Urban planning in Africa: which alternative for poor cities? the case of Koudougou in Burkina Faso. Curr. Urban Stud. 3, 413–431 (2015). https://doi.org/10.4236/cus.2015.34033

22. United Nations, Department of Economic and Social Affairs, Population Division: World Urbanization Prospects: The 2018 Revision, Key Facts. United Nations, New York (2018). https://population.un.org/wup/Publications/Files/WUP2018-KeyFacts.pdf
23. UN-Habitat: Slum Almanac 2015–2016. Tracking Improvement in the Lives of Slum Dwellers. UN-Habitat, Nairobi (2016). https://unhabitat.org/slum-almanac-2015-2016/
24. World Bank. Global Monitoring Report: Rural-:Urban Dynamics and the Millennium Development Goals. World Bank, Washington, DC (2013). https://doi.org/10.1596/978-0-8213-9806-7
25. Bolay, J.-C.: Slums and urban development: questions on society and globalisation. Eur. J. Dev. Res. **18**(2), 284–298 (2006). https://doi.org/10.1080/09578810600709492
26. Hardoy, J.E., Mitlin, D., Satterthwaite, D.: Environmental Problems in an Urbanizing World. Finding Solutions for Cities in Africa, Asia and Latin America. Earthscan, London (2010)
27. Satterthwaite, D.: The links between poverty and the environment in urban areas of Africa, Asia, and Latin America. Ann. Am. Acad. Polit. Soc. Sci. **590**, 73–92 (2003). https://www.jstor.org/stable/3658546
28. Cohen, M.: Urban economic challenges and the new urban agenda. UN-Habitat, Nairobi (2016)
29. United Nations Human Settlements Program (UN-Habitat): Urbanization and Development: Emerging Futures. World Cities Report 2016. UN-Habitat, Nairobi (2016)
30. Lea, R..: Smart Cities: An Overview of the Technology Trends Driving Smart Cities. IEEE, Piscataway, USA (2017). https://www.researchgate.net/publication/326099991_Smart_Cities_An_Overview_of_the_Technology_Trends_Driving_Smart_Cities

AntiOD: A Smart City Initiative to Fight the Opioid Crisis

Claudia B. Rebola and Sebastian Ramirez Loaiza$^{(\boxtimes)}$

University of Cincinnati, Cincinnati, OH 45221, USA
rebolacb@ucmail.uc.edu, ramirej3@mail.uc.edu

Abstract. In the United States the opioid epidemic is on the rise with over 70.000 deaths by overdose in the last 20 years. Naloxone, also known as Narcan, is an effective antidote that can be administered to people who are suffering an overdose. Currently, the public relies on Quick Response Teams, which are overly saturated with requests. The country and counties are facing a drug-abuse problem of epidemic proportions, where it is imperative for disciplines like design to bring about new methods to save lives. The purpose of this paper is to describe AntiOD, a smart community-access naloxone program that provides tools, training, and awareness to empower bystanders to rescue victims of overdose. The project proposing a two-part solution: (1) empowering the public to take action through communication, training, and outreach; and (2) providing the public with access to the means. The goal is to help alleviate quick responses teams and empower the community to become first responders.

Keywords: Smart cities · Opioid crisis · Access design · Social design

1 Introduction

From 1999 to 2017, over 700.000 in the United States died due to a drug overdose [1]. Moreover, in 2017 the number of deaths by overdose involving either prescript opioids or illegal opioids like heroin was six times higher than in 1999 [1]. Furthermore, the rate of overdose deaths increased significantly by 9.6% from 2016 to 2017 [1]. In 2016, the five states with the highest rates of death due to drug overdose were West Virginia, Ohio, New Hampshire, Pennsylvania, and Kentucky. Ohio, as the second highest state, rate 46.3 deaths per 100,000 inhabitants [2]. Moreover, Ohio has three of the top ten deadliest cities in the United States: #1 Dayton, #6 Cincinnati and #10 Toledo [3]. Specifically, in Cincinnati, the Emergency Medical Services of Cincinnati presented in 2017 and an outstanding number of 2582 responses just in the category of a heroin overdose [4].

Naloxone, also known as Narcan, is an effective antidote that can be administered to people who are suffering an overdose [5]. The drug reverses the overdose effects, therefore, resuscitating a victim from an overdose. Access to Narcan is problematic from different aspects. First, Narcan may not be available to the public either by prescription or standing prescription orders. Even though if a prescription is not required for its purchase, it is still a highly regulated medicament which hinders its access. Because of the fact that it doesn't require prescriptions, significant figures, like

© Springer Nature Switzerland AG 2020
K. Arai et al. (Eds.): FTC 2019, AISC 1069, pp. 300–305, 2020.
https://doi.org/10.1007/978-3-030-32520-6_24

the Surgeon General of the United States encourage any community members to have this overdose antidote at reach [6]. Second, the public might not have awareness about the drug or education on how to administer the drug. Lastly, victims of overdose cannot self-administer the drug themselves due to unconsciousness. Even if an individual has personal use Naloxone, he or she cannot self-administer if rendered unconscious by the overdose. Altogether with the lack of resources, the public relies on Quick Response Teams, which are overly saturated with requests. The country and counties are facing a drug-abuse problem of epidemic proportions, where it is imperative for disciplines like design to bring about new methods to save lives.

The purpose of this paper is to describe AntiOD, a smart community-access naloxone program that provides tools, training, and awareness to empower bystanders to rescue victims of overdose [6]. The project proposing a two-part solution: (1) empowering the public to take action through communication, training, and outreach; and (2) providing the public with access to the means. The goal is to help alleviate quick response teams and empower the community to become first responders.

2 Background

There are a number of programs and initiatives that have been developed to help with the overdose crisis. Project Dawn, in Cincinnati, is a community-based overdose education and naloxone distribution program, in which members of the community must receive training on recognizing the signs and symptoms of overdose, distinguishing between different types of overdose, performing rescue breathing, calling emergency medical services, and administering intranasal Naloxone [7]. The critical element for Narcan distribution is the required training prior getting access to the drug.

A number of lists and articles have been published online for the public to learn about naloxone access. For example, the State of Ohio Board of Pharmacies lists Ohio licensed pharmacies which dispense naloxone pursuant to OAC 4729-5-39 [8]. Similarly, local newspaper help spread information about Narcan access [9]. Naloxbox was a collaborative project held in Providence, RI that sought to empower common citizens and give them access to Naloxone in order to save lives [10]. The main goal was to impact on the social dynamics of the community and propose technology-based solutions that would address the problem and help it understand it.

Ohio's comprehensive approach to fighting drug abuse and overdose deaths includes different actions [11]. These include items such as stepping up law enforcement drug interdiction efforts; increasing the penalties for trafficking fentanyl; improving access to addiction treatment; expanding prescribers' use of Ohio's opioid prescribing guidelines and the prescription drug monitoring program, the Ohio Automated Rx Reporting System (OARRS); and expanding and recognizing the number of schools that have implemented the Start Talking! youth drug prevention initiative, or a similar initiative [12]. But the two most important items where the public and communities have an opportunity to help with the crisis are: (1) continuing to work with communities to enhance local efforts through the Health Resources Toolkit for Addressing Opioid Abuse; and (2) increasing the awareness and availability of naloxone to reverse opiate overdoses and save lives.

3   AntiOD

AntiOD wants to address the opioid crisis through design aiming to 3 main goals. The first aim is to raise awareness on the issue, communicating the magnitude of the problem and its impact on the nation and the city. The second aim is to educate and train bystanders to take actions in an emergency situation where the community has delegated responsibility on quick response teams. But more importantly, the subsequent aim is to destigmatize addiction, emphasizing that it is a disease that knows no distinction and that victims are humans (see Fig. 1). Lastly, the third aim is to grant better access (public/semipublic) to the drug Naloxone-Narcan.

Learning from NaloxBox [10], AntiOD was developed as version 2.0 specifically designed for the city of Cincinnati. The dispensing unit, described below (see Fig. 2), is ultimately a smart platform as a sustainable solution that gathers data for public health analysis, and contributes to preventive methods and builds on smart cities initiatives for social problems of any kind. In addition to designing the dispensing unit, the project engaged in developing promotional materials to engage the public. The campaign emphasizes a humanized approach, portraying stories as well as educational materials for recognizing overdose victims.

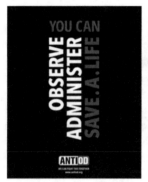

Fig. 1. AntiOD campaign

Besides the campaign, the design is a smart dispenser that contains single package units. Each package contains 2 doses of Narcan/Naloxone in nasal spray option, Narcan information sheet, gloves, and a CPR mask. The package itself is printed with the instructions on how to administer the medication. This pack is set on a smart dispenser that includes a pocket for a card-sized booklet with expanded education/training information. Lastly, a 7-inches tablet that showcases visually and animated the steps for administering Narcan as well as providing CPR. Overall, the design provides multiple venues for accessing information in administering Narcan in the moment of emergency or need.

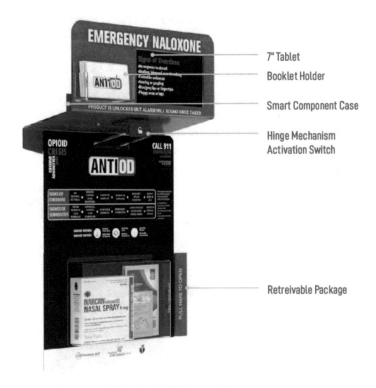

7" Tablet

Booklet Holder

Smart Component Case

Hinge Mechanism
Activation Switch

Retreivable Package

Fig. 2. AntiOD smart dispenser

3.1 Smart City Features

The smart cabinet features a custom electronic design component that operates like a phone linked to a lock mechanism that keeps the packages secured. In the event of an emergency, any person could retrieve the medicine just texting their name and zip code to the number of the smart cabinet. The technology gathers the information required to comply with the Board of Pharmacy requirements for Narcan distribution.

In addition, partnering with local smart city initiates, this project joined efforts with Cincinnati Bell, to strengthen the system for meeting the aforementioned aims. Via text-messaging and geo-location, AntiOD uses text message technology to aid in the emergency situation.

The first stage is to provide text messaging support to bystanders who are seeking information on locations (see Fig. 3). The project proposes to engage the city in placing AntiOD numbers in lighting post for easy access to contact numbers. Whereas the person texted after seeing an urban AntiOD sign or after taking the Naloxone package out of the dispenser, the system will send automatic SMS in response based on key-words of location, antidote administration, and overdose recognition.

The Second Stage is the safety lock and data gathering. The medicine package is secured in the smart dispenser. In order to unlock it, a bystander can send a text message with their name and zip code to the unique phone number in each dispenser.

This data will be monitored and displayed where overdoses are occurring. In addition, a retrieval of a single-use package will alert the AntiOD team for restocking purposes and/or Narcan expiration dates.

Lastly, the third stage is bridging Quick Response Teams to the bystander. After the product is unlocked and retrieved, an automatic text message is sent to the Quick Response Teams to bridge the gap between the person assisting the victim and the specialized help. The inputted phone number to unlock the system will be sent as a text to the emergency service, where the bystander will receive a call for sending the response team (Fig. 4).

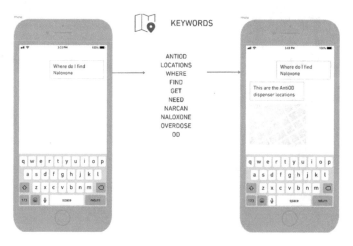

Fig. 3. Example of location SMS

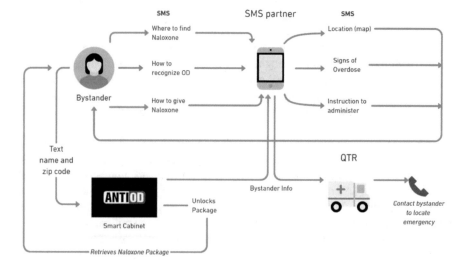

Fig. 4. AntiOD SMS system

4   Conclusion

AntiOD is a community-based program to educate and empower the public to become first responders. The project, as a smart city program, focuses on designing access. Designing access includes designing accessible information by developing bilingual training materials, pocket sheets, video trainings, and text messaging support. It is about visible awareness, encouragement, empowerment, and de-stigmatization through campaigns, education strategies, and visual materials. And it is about providing semipublic access to naloxone. Technology-based solutions open the possibilities to map complex situations supported on data, but just as important, they provide the chance to predict their course (peaks, ripple effects, etc.) to take actions towards friendlier, healthier cities.

References

1. Center for Disease Control and Prevention, Understanding the Epidemic. https://www.cdc.gov/drugoverdose/epidemic/index.html. Accessed 14 Jun 2019
2. Center for Disease Control and Prevention, Drug Overdose Deaths. https://www.cdc.gov/drugoverdose/data/statedeaths.html. Accessed 14 Jun 2019
3. Muray, V.: Livestrong, The 10 Deadliest U.S. Cities for Drug Overdoses. https://www.livestrong.com/slideshow/13399806-the-10-deadliest-cities-for-drug-overdoses/. Accessed 14 Jun 2019
4. City of Cincinnati, Heroin Overdose Responses. https://insights.cincinnati-oh.gov/stories/s/dm3s-ep3u. Accessed 14 Jun 2019
5. Adapt Pharma, What is Narcan® (Naloxone) Nasal Spray. https://www.narcan.com. Accessed 14 Jun 2019
6. U.S. Department of Health and Human Services, U.S. Department of Health & Human Services. Surgeon General's Advisory on Naloxone and Opioid Overdose. https://www.hhs.gov/surgeongeneral/priorities/opioids-and-addiction/naloxone-advisory/index.html. Accessed 14 Jun 2019
7. Ohio Department of Health, Project DAWN (Deaths Avoided With Naloxone). https://odh.ohio.gov/wps/portal/gov/odh/know-our-programs/violence-injury-prevention-program/projectdawn/. Accessed 14 Jun 2019
8. State of Ohio Board of Pharmacy, Ohio Pharmacies Dispensing Naloxone Without a Prescription. https://pharmacy.ohio.gov/Licensing/NaloxonePharmacy.aspx. Accessed 14 Jun 2019
9. DeMio, T.: The enquirer, Surgeon general says get naloxone. That's easy. Here's what to do, 05 Apr 2009
10. Capraro, G.C., Rebola, C.B.: The NaloxBox program in Rhode Island: a model for community-access Naloxone. Am. J. Public Health 108(12), 1649–1651 (2018)
11. Ohio Department of Health, New Strategies to Fight Opiate and Fentanyl 2016-17
12. State of Ohio, Start Talking! Building a Drug-Free Future. https://starttalking.ohio.gov. Accessed 14 Jun 2019

Common Data Format in Visual Assembling Guidance Using Augmented Reality

Dawid Pacholczyk$^{(\boxtimes)}$ and Mariusz Trzaska

Polish-Japanese Academy of IT, Warsaaw, Poland
{dpacholczyk, mtrzaska}@pjwstk.edu.pl

Abstract. Augmented Reality (AR) gains increased attention over the past few years. It becomes recognized as a valuable tool for a casual user and for a business/industry customer. One of the biggest problems of AR is lack of a common data protocol that would allow creating a data scheme independently of technology that will be used by the end user. This factor slows down the evolution of augmented reality as an everyday technology. ARAssembler is a project that is using our data format for assembling instructions used in augmented reality aided software. Instructions are shared between multiple client applications. The scheme can be used on any device regardless of the technology that it is using. The only task that needs to be done by the client application is to use a common data model and prepare actions/animations based on the scheme. The scheme can be used in any way, to present an object picker, or to create a visual version of paper instruction. Assembly tasks and AR technology are currently a very important topic in this area of research, that is why we decided to focus our prototype on the possibility of creating a step-by-step guidebook and reconstructing it on a client mobile application. Both ends of the project (recorder, client) are based on a mobile device as those are currently the most available augmented reality platform.

Keywords: Augmented reality · Data-information layer · Protocol · Generic platform · Assembly manual · Guidebook

1 Introduction

Assembly is the process in which a set of activities are performed to join two or more elements to get an artifact [1]. An assembly manual is used to guide the tasks that are required to achieve the goal. The assembly process is well known and understood in the industry (factories, manufactures) and also for a casual person (e.g. furniture). The properly formatted guidebook should deliver all necessary information in a clear and understandable way. Augmented reality is a perfect tool that can support this process. Thanks to the possibility of working with 3D objects, looking at the element from a different angle or using animations to simulate proper movement, it already presents a significant advantage over a standard, paper manual. The main problem is the lack of methods to create such interactive manual using an easy, fast and low-cost method that could be reused on different platforms related to AR [6]. Hence, we decided that this will be the best area to test our thesis. We believe that we can create a common data

© Springer Nature Switzerland AG 2020
K. Arai et al. (Eds.): FTC 2019, AISC 1069, pp. 306–317, 2020.
https://doi.org/10.1007/978-3-030-32520-6_25

layer that will help in creating a multiplatform software based on augmented reality. Thanks to our Augmented Reality Data Layer (ARDL) we will be able to lower the time and cost of creating such software.

The goal of our research is to create a data format and strategy used for sharing information between the server and a client layer. Such format will help the creators to prepare a common scheme that they will be able to use independently of the technology utilized as a core of their application. A common source of data for multiple client applications is something that we can call a standard approach in the software engineering but we can observe that in terms of the AR data we are missing a similar solution. This is due to the large dispersion of the AR technology. We can find a lot of different methods, devices, platforms, approaches to the augmented reality [2]. In our opinion, this is one of the key factors affecting the rate of adaptation of augmented reality in everyday activities. In our research, we want to not only prove that it is possible to create such an approach, but to prepare a prototype and publish it for everybody.

In this paper, we present our approach that can be the solution for strong fragmentation of this area of technology. It is currently on a late stage of development. We want to describe, "work in progress", our concept, our approach, and future development. We will explain why we took such a route, and what we want to achieve at the end of the road. In addition, we want to prove that as long as we contain and operate with data that are independent of technology, then they can be used to recreate prefabricated step-by-step instruction.

ARAssembler is a prototype project focused on creating simple step-by-step guidebooks based on one or more 3D objects set in a specific sequence. We selected this area because it is a natural branch of augmented reality development [3, 4].

Our protocol along with a fast-growing segment of AR devices can have a significant and positive impact on the adoption rate of this technology. In addition, the phenomenon of reusing the same data (in this case step-by-step instruction) are well known and very desirable in the IT world. The biggest advantage of this approach is lowering the time and cost that is needed for developing the software. Based on different reports [4, 5] the cost of creating a fully immersive AR application can go up to $250,000. This is a very important entry threshold. If a company plans to deliver a similar experience on different platforms than we must increase this cost. That is why our goal is to find a way that can – at least in some part – reduce the amount of work, time, and in the effect, the cost of developing augmented reality software. During our research in the area of assembly software, cooperation aided AR and AR object recognition, we observed, that most of those solutions are implemented in a different way even when authors want to receive similar effect. We think that it is an ineffective and wrong approach. With our ARDL we want to create a tool that will be able to standardize the data layer of each application. This will significantly decrease the time that is needed for future development and will allow making better experiments as we will be able to collate different prototypes/projects with the same data format.

The rest of the paper is structured as follows. In Sect. 2 we discuss the background of the. We present how researchers and market are facing data communication and assemble tasks using AR. In Sect. 3 we are presenting the concept of our prototypes and the ARDL as a whole project. We want to describe the key values that are standing behind our work, our approach and planned strategy. In Sect. 4 we are presenting the current stage of our research. As this paper is describing the "work in progress", we want to be sure that we deliver the understanding of the concept and current stage of our prototype. Section 5 is focused on future development. You will find there a description of next steps related to our prototype and plans.

2 Background

Over the past few years, we can observe increased interest in augmented reality both in commercial, business [23], and industrial usage [1]. Involvement of top IT companies with their devices and software like ARCore [7], ARKit [8], HoloLens [9] or Magic Leap One [10]. These kind of projects are opening new paths and presenting new possibilities for AR.

On the other hand, Virtual reality (VR) is a well-known technology. We understand its capabilities in areas like training new employees at the workplace, and in assembly process [11]. An important advantage of VR technology is the fact that it can reconstruct an environment that would be normally very hard to get. However, we need to remember that virtual reality completely cuts us off from the key environmental incentives like:

- weather,
- limitations of the workplace,
- time pressure,
- senses,
- the texture of the item,
- the sense of space.

Those are the most important elements in terms of assembly process at the workplace.

Another negative aspect of using VR is related to the scene itself. Currently, we struggle with the limitation of its reality, and lack of complex haptic response from the virtual elements to the user. All those elements combined are limiting our cognitive abilities and our perception [12] and those are major factors in terms of the learning process.

Augmented reality has an undoubted advantage in this area. As a technology, it is ideal for helping in an assembly process because AR is based on extending user perspective without cutting him of the real world. It allows for simultaneous observation of the real and virtual object improving human cognition [13].

Augmented reality is seen as a technology that can improve processes related to areas like training [15], education [14], assembly [16] and more. For the greatest profit of using AR in those areas, we need to ensure that the time and cost that is needed for developing software – regardless of the selected platform – are as low as possible. We believe that creating a common data format and our scheme creation application will help in achieving this goal and will be a big contribution on the track to popularization augmented reality.

On the other side, we have the process of creation augmented reality aided instructions and manuals. During our research, we found multiple projects related to a similar theme. There are two most common approaches. The first one is related to a very specific and narrow area of interest. It is used when there is no need for usage of a wide range of technologies. An example can be found in [20] were thanks to limiting the area of interest, authors are able to use OpenCV [21] and create a possibility of tracking the current progress of the assembly process. This interesting feature brings a very big limitation which is the fact that we need to ship the scheme, assets, visual models, and the whole rest along with the software. Similar, and even more advanced approach can be found in [22]. In that paper we can see that authors are also dealing will predefined structure, they are also able to monitor the current progress, but thanks to keeping all information about the prefabricated structure in scale 1:1 they are also able to check the distances between each element to analyze to the correctness of the process.

The second approach related to creating an AR aided instruction is based on preparing a 3D, animated model or a whole animation. In terms of the client application, we receive a fairly similar effect to ours which is a sequence of operations that need to be done from first to the last step. At the current stage, we agree that such an approach where we can use for example Unity can give more complex animations. But we need to remember that it is not lowering the cost of development and it is very strongly related on the platform that we want to use. Such a solution does not meet the goals we set to ourselves and which are described in the further part of this paper.

3   Concept

In this part, we want to answer a question: why we are building such data information scheme? To do that, we need to understand why augmented reality is not a common thing. We already have the technology that is well tested in maintenance [17], industry [18], education [14], and more. We have devices that are available for a common user like smartphones with ARCore/ARKit onboard. Unfortunately, the one thing that we are missing is the standard of communication between different devices and technologies. Every time when we are creating a software based on AR we need to start everything from scratch (besides assets like 3D models). The goal of our project is to change that. Our data scheme is something that we can name "AR data layer" that will be common for every application on every platform. It is fully technology independent. It does not matter if the author is using JAVA to create an ARCore application or is he/she using C# to create a mixed reality software for HoloLens.

Our standard is based on a few principles:

• Available regardless of technology – the whole scheme is saved in JSON [19]. Thanks to that it is light, fast, and well-known for every software engineer.

• Independent from AR frameworks – in the scheme, we do not save anything that is related to any specific language, technology, framework, etc. We save only generic data that can be reused in a different situation or with most suitable tools.

- Asset distribution – data saved in the scheme are related to specific assets. Those assets can (but there is no "must") be used by the developer. The decision of using them is on his/her side.
- Scheme creator – the AR Data Layer will be distributed with a special mobile application for recording the scheme. In this case "recording" means handling the process of positioning AR object and gathering data related to their relative movement.

Those 4 points above are the key principles that are the fundament of AR Data Layer. Every decision must be made in the line with them. We will try to explain the roots of each of them.

3.1 Available Regardless of Technology

During our research and development of different prototypes, we've noticed that every device, technology, the framework is forcing their own solutions and approaches. This is generating a problematic situation in which every time we had to start development from the beginning (of course except situation when we stick to the same platform). There is no easy, and smooth way to move from one to another technology. Every key concept has its own implementation that is significantly different one from another. During our interviews with multiple developers, this was marked as a number one cause of slow development of AR software, and slow absorption of new technologies both by the users and industry in terms of widely available software. It is caused by the fact that every time the developer needs to commit to the specific approach, a set of tools and strategies of development, from the level of the framework through handling assets to the level of information exchange.

Based on that, we understood that if we want to make augmented reality more available and by that, more usable for a common user, we need to create a layer that will be available in every approach independently of the technology that is used by the core. It must be well known and understood by developers and researchers to reduce the time that is needed for its adoption. It must be possible to handle it in every common technology that is used for augmented reality software, and – what is even more important – it won't be outdated in near future. AR Data Layer is an idea that is created not for "here and now", but for "here, now and there in the future". If we want to make augmented reality more available, our approach needs to be resistant to fast changes that are very common in the IT world especially in the area of languages and technologies.

After gathering everything together we have decided that our data layer will be based on JSON as it is the technology with an established position. It is well known for every software engineer regardless of the technology that he/she is using. It has an implementation in every common language. It is easy to use and propagate over the web.

3.2   AR Technology Independence

As we have mentioned in the point above, currently we`re in a situation in which every company/producer has its own unique approach of resolving the same problems. That is why our scheme must be 100% free of any tool/framework/ technology specific data. Regardless of what technology we use to make our service and/or recording application working, we must remember that none of its specific elements are part of the scheme created and stored by the AR Data Layer. Because of that, we use a number of transformations that allow us to convert technology/tool specific data to generic information based on mathematical and text data based on the purpose.

The key goal of this fundament is to assure that every developer that will reach for our idea today or next year will be able to use it. He/she will not have to focus on the tools that we have used and he/she will not be forced to change the technology that he is using for his/her software. Furthermore, it will improve the development of the same software on multiple platforms. Thanks to our concept, the AR information's will be handled in the same way on each platform. We believe that this will have a significant impact on the time that is needed to develop augmented/mixed reality software and will reduce the time that is needed for porting an application from one to another platform. If we assume that

$$COST = TIME * AMOUNT\ OF\ WORK$$

Just by reducing one of those two elements (time, amount of work) we can significantly reduce the cost. We believe that our data layer will be able to reduce both of those factors and have an even bigger impact on the development process. This will lead us to the situation where creating AR software will be easier and faster. We believe that thanks to that, there will be more AR software and this will directly translate into its faster utilization and increased availability.

3.3   Asset Distribution

It's hard to talk about AR software and not to mention assets (e.g. 3D objects/models). If there will be no correlation between data that are stored, and assets that are used by the software, then we can forget about creating a common, and generic layer for each platform. Because of that, we decided to base our idea on a System as a Service (SaaS) architecture. It will be built around a web creator (web application) in which user will describe basic metadata of his/her data scheme and he/she will be able to load necessary 3D assets. Those objects will be available in the recording application and in the client software.

On the recording side, they will be placed by the creator (user). He/she will able to move them and they will become a part of data scheme. Each asset will be related to his/her specific step(s), movement and placement in the space.

On the client side, they will be used to recreate the recorded scheme, for example as an instruction for assembling a device or a wardrobe in someone's room. The possibilities are endless.

3.4   Scheme Creator

We already mentioned the application for creating the scheme. Scheme creator will be built from two parts. The first one is a web application for creating a necessary description of the scheme (metadata along with assets). Technically it is possible to build it as a standalone web-application but it would force us to make a very strong commitment in a specific technology and we want to avoid that. Secondly, after gathering feedback from potential users, we agreed that working with 3D objects and loading them for the need of the scheme, will be much easier from a PC level.

The second part is a recording application. Our approach is much different from those described in the previous part of this paper. We decided to use the biggest advantage that augmented reality can give, which is the connection between the real and virtual world. Instead of creating a software that is separate from our world (even though we are building something that will be located in it), we decided to build an AR creator mobile application.

Our proof of concept (POC) is based on an assembly task. It is much more natural when we are able to create a step by step instructions actually looking on it and placing elements in real time rather than describing the movement as a UML style block or as a Unity animation and checking each change. We do not need a "test iteration" after changing each param to verify is the result as we expected. Our POC and the concept of AR Data Layer allow us to work with the 3D elements as we would work with real objects. Then the recording application is saving the whole operation in a unified data scheme that can be shared with client applications in real time or can be part of a bigger software with a preloaded scheme. There is no limitation in how the developer will use that information, and we believe that this is the true value and contribution for the AR world.

3.5   Prototype Concept

We decided to base our prototype client application on a process of object assembly. We believe that this is the best scenario to present the capabilities of AR Data Layer because of a few reasons:

1. Assembly process is a well-known augmented reality research area. Many researchers – including us – believe that AR can significantly improve the experience of instructing people on how to assembly specific objects.
2. We believe that AR Data Layer can truly contribute to this area and provide a new quality for preparation step by step virtual instructions. We want to show the biggest value of augmented reality, which is working with 3D objects in real time instead of working with text data and/or in separate software like Unity.
3. Assembly task is able to present every aspect of our concept. Those are:
   a. Possibilities to work in real-time with 3D objects and create step-by-step instructions using augmented reality. Thanks to that the person that is creating the instruction will have exactly the same perspective as the user that will be using this instruction. We believe that will significantly reduce the time needed for such operation.

b. Possibility to share the scheme with client applications regardless the technology that is used.

c. Possibility to connect each step with proper asset and recreate the scheme.

The prototype will be based on a mobile platform to make it affordable, and available for as wide as possible range of users. It will help us in gathering valuable feedback during the next iteration of this project.

To present the whole concept we will build a system based on three different parts. The first one is the Scheme Creator. A web application that will allow the user to create Scheme metadata. Those data will describe specific scheme and deliver 3D objects. The second part is the Scheme recorder. A mobile application used for recording data for a specific scheme. Using it, the user will be able to download needed assets and all scheme metadata and build a step-by-step instruction around them. The instruction will be built from steps, on each step user can place a selected object, and if needed, select for it his/her start and end position (it will allow creating an animated movement of the object if needed). Scheme recorder is saving the instruction based on our data format that will be shared with the client applications. The third part is the Scheme Player. In this part there are no technological limitations, every developer can select the platform that he/she needs in current time. He/she only needs to understand our data format and convert them according to selected technology.

4 Current State

In this part, we will describe where we are with our "work in progress".

As we are building a proof of concept, we decided to simplify the whole architecture as it is not the core of our thesis. For the current moment we resigned from the Scheme creator, and we are working with predefined scheme metadata. The end result is exactly the same, but we reduced the time and cost of delivering the POC. We are currently focused on the Scheme recorder and the client application.

The second important decision that we've made is the selection of a platform for the client application. Again, for reducing time and cost, we have decided to base it on the Android OS. Even though we are using the same system for recorder and client we are strongly separating elements related with the framework from our data format so at the end we will be able to prove our thesis.

For the current moment, we are focused on proving that such scheme recorder and the proper client application can truly be developed. The main problems that we need to overcome are:

- Create a generic format of saving data that can be reused in the client application.
- Overcome different reference systems: both the recording app and client app are working with its own reference system so the data cannot be reused without proper transformation.

At the moment of writing this paper, we finished the Scheme recorder, preview can be found on Fig. 2. We are able to make a step-by-step instruction (steps of creating such scheme is presented on Fig. 1). In each step user can place an object, set its start

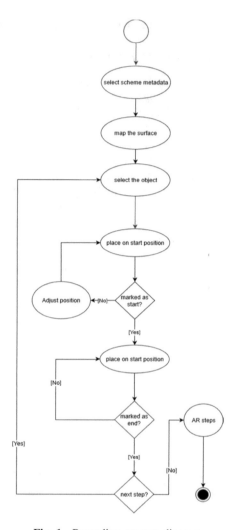

Fig. 1. Recording process diagram

and end position (they can be exactly the same), and thanks to that build a flat structure from multiple components (we still have some problems with proper element positioning on different height levels). Everything is based on augmented reality; this is a huge adv`antage because the user can see the exact same structure as the client will see. This can significantly reduce the time needed for creating an assembly instruction.

We can convert the recorded structure to our data format. Data format structure contains:

- Basic information about the scheme for which we are building the instruction (Scheme metadata)

- Information about each step:
  - Order number
  - Object key: needed to map step on a proper object in the client app
  - Starting vector: x, y, z coordinates based on the recording app reference system
  - End vector: x, y, z coordinates based on the recording app reference system.

This structure is read and transformed into a data model in the client part of the application. Currently, we are able to reproduce over 80% of the examples that were recorded. Unfortunately, we are still working on some issues related to the transformation of the vectors from one to another reference system and creating a proper animation from it. About 20% of the examples have some imperfections during the whole process and due to that fact, they are disqualified by us. However, this 80% makes us sure that we will achieve our goal in the coming months.

Fig. 2. Example of Scheme Recorder usage

5 Future Development

Current and future development will be focused on delivering a fully working prototype. We want to make it available for a wide audience. We believe that only clashing the prototype with real case scenarios and real user problems can prove the validity of our thesis.

In next month, we will be shipping Scheme Creator, Scheme Recorder, and full documentation with examples or the client application. Scheme Creator will be available as a publicly accessible webpage. Scheme Recorder will be available in Google Play for all ARCore supporting devices. All of those elements will be free to use, the only "cost" will be related to creating an account and declaring future participation in our surveys. We will be gathering data related to real usage and problems related to our concept. We will analyze those data and describe conclusions in the next paper.

Based on the survey's results, we will decide on the future of the prototype. For sure we won't stop our research as we believe that it is the area that can have a significant impact on the augmented reality as a technology and as a part of the IT world.

References

1. Hou, L., Wang, X., Bernold, L., Love, P.E.D.: Using animated augmented reality to cognitively guide assembly. J. Comput. Civil Eng. **27**(5), 439–451 (2013)
2. Peddie, J.: Types of augmented reality. In: Augmented Reality. pp. 29–46. Springer, Cham (2017)
3. Radkowski, R., Kanunganti, S.: Augmented reality system calibration for assembly support with the microsoft HoloLens. In: Volume 3: Manufacturing Equipment and Systems, College Station, Texas, USA, p. V003T02A021 (2018)
4. De Amicis, R., Ceruti, A., Francia, D., Frizziero, L., Simões, B.: Augmented reality for virtual user manual. Int. J. Interact. Des. Manuf. (IJIDeM) **12**(2), 689–697 (2018)
5. Eisenberg, A.: How Much Does It Cost to Develop an AR App? appreal-vr.com, 19 October 2018
6. Tecsynt Solutions, How Much Does Augmented Reality App Development Cost in 2018? 30 January 2018. medium.com
7. Google, ARCore by Google, 02-sie-2018. https://developers.google.com/ar/discover/
8. Apple, ARKit. https://developer.apple.com/arkit/
9. Microsoft, HoloLens. https://www.microsoft.com/en-us/hololens
10. Magic Leap, Magic Leap One. https://www.magicleap.com/magic-leap-one
11. Ritchie, J.M., Robinson, G., Day, P.N., Dewar, R.G., Sung, R.C.W., Simmons, J.E.L.: Cable harness design, assembly and installation planning using immersive virtual reality. Virtual Reality **11**(4), 261–273 (2007)
12. Wang, X., Dunston, P.S.: Compatibility issues in Augmented Reality systems for AEC: an experimental prototype study. Autom. Constr. **15**(3), 314–326 (2006)
13. Salonen, T., et al.: Demonstration of assembly work using augmented reality. In: Proceedings of the 6th ACM International Conference on Image and Video Retrieval - CIVR 2007, Amsterdam, The Netherlands, pp. 120–123 (2007)
14. Akçayır, M., Akçayır, G.: Advantages and challenges associated with augmented reality for education: a systematic review of the literature. Educ. Res. Rev. **20**, 1–11 (2017)
15. Gavish, N., et al.: Evaluating virtual reality and augmented reality training for industrial maintenance and assembly tasks. Interact. Learn. Environ. **23**(6), 778–798 (2015)
16. Blattgerste, J., Strenge, B., Renner, P., Pfeiffer, T., Essig, K.: Comparing conventional and augmented reality instructions for manual assembly tasks. In: Proceedings of the 10th International Conference on PErvasive Technologies Related to Assistive Environments - PETRA 2017, Island of Rhodes, Greece, pp. 75–82 (2017)
17. Henderson, S., Feiner, S.: Exploring the benefits of augmented reality documentation for maintenance and repair. IEEE Trans. Vis. Comput. Graph. **17**(10), 1355–1368 (2011)
18. Caudell, T.P., Mizell, D.W.: Augmented reality: an application of heads-up display technology to manual manufacturing processes. In: Proceedings of the Twenty-Fifth Hawaii International Conference on System Sciences, Kauai, HI, USA, vol. 2, pp. 659–669 (1992)

19. Marrs, T.: JSON at Work: Practical Data Integration for the Web, 1st edn. O'Reilly Media, Sebastopol (2017)
20. Okamoto, J., Nishihara, A.: Assembly Assisted by Augmented Reality (A3R). In: Bi, Y., Kapoor, S., Bhatia, R. (eds.) Intelligent Systems and Applications, vol. 650, pp. 281–300. Springer International Publishing, Cham (2016)
21. Wikipedia, OpenCV, 29 August 2018. https://en.wikipedia.org/wiki/OpenCV
22. Cuperschmid, A.R.M., Grachet, M.G., Fabricio, M.M.: Augmented reality as a tutorial tool for construction tasks (2016)
23. Olshannikova, E., Ometov, A., Koucheryavy, Y., Olsson, T.: Visualizing Big Data with augmented and virtual reality: challenges and research agenda. J. Big Data 2(1), 22 (2015)

Determining a Framework for the Generation and Evaluation of Ambient Intelligent Agent System Designs

Milica Pavlovic[1,2], Sotirios Kotsopoulos[1(✉)], Yihyun Lim[1],
Scott Penman[1], Sara Colombo[1], and Federico Casalegno[1]

[1] Massachusetts Institute of Technology, Design Lab,
Cambridge, MA 02139, USA
{milicap,skots}@mit.edu
[2] Politecnico di Milano, Interaction and Experience Design Research Lab,
20158 Milan, Italy

Abstract. The design and realization of Ambient Intelligence (AmI) systems using Artificially Intelligent (AI) agents is a rising field of research. However, the absence of clearly defined working criteria, supporting the generation and evaluation of AmI agent system designs, is a conspicuous obstacle to their advancement. The contribution of this paper is that we determine and test a framework for the generation and evaluation of AmI system designs, based on user experience and business criteria. Specifically, the process of designing a personal lighting AI agent, in collaboration with a leading lighting design company, is used as a case study to determine and test a framework for the generation and evaluation of AmI system designs based on feasibility and acceptability. First, we use storytelling videos to describe and communicate the user values and design scenarios to the stakeholders. Second, we generate design proposals for a lighting AmI agent based on five distinct systemic factors, namely: (a) the context of interaction; (b) the required system data; (c) the required sensing input; (d) the required user input; and (e) the desired system output. Finally third we determine an evaluation framework that is based on three distinct levels of in-built system intelligence, from lower to higher. The three levels reflect the feasibility and acceptability of the system. Feasibility is what a specific company is capable of producing, and in what timeframe. Acceptability is the potential of familiarity and trust that the users can feel while interacting with the AI agent.

Keywords: Ambient Intelligence · User Experience · Design vision ·
Generation framework · Evaluation framework

1 Introduction

Ambient Intelligence (AmI) refers to environments that are sensitive and responsive to people. They integrate a variety of devices operating in concert to support human activity in an unobtrusive and intelligent way, using intelligence that is hidden in the network connecting them. AmI experiences can be provided by autonomous Artificial

© Springer Nature Switzerland AG 2020
K. Arai et al. (Eds.): FTC 2019, AISC 1069, pp. 318–333, 2020.
https://doi.org/10.1007/978-3-030-32520-6_26

Intelligence (AI) agents or not, in response to perceived needs, or user input. Autonomous AmI agents know how and when to provide a functionality.

The original AmI vision [1], builds upon the concepts of pervasive computing, ubiquitous computing, profiling, context awareness, and human-centric computer interaction. It is characterized by networked devices that are: *embedded*, integrated into the same environment; *context aware*, able to recognize the users and their situational context; *personalized*, tailored to the user needs; *adaptive*, able to change states in response to users' needs; and *anticipatory*, able to anticipate user desires without explicit user input. As AmI devices grow smaller, more connected, and more integrated into the environment, the technology will disappear and only the user interface will remain perceivable by the users. A typical context of AmI experimentation is home, but applications may also be extended to work in public spaces, with technologies such as smart streetlights, and hospital environments. The first generation of intelligent agents includes personal software assistants with a certain degree of autonomy.

Today AmI is a futuristic vision that promises to transform the role of technology in everyday life and to change the way people live, work, relax and use their leisure time. The AmI vision differs from earlier technology visions due to its explicit human-centered goals. Unlike other visions of technology, which can be deterministic, shaped by what a specific technology can do, the AmI vision is open-ended. To realize their full potential AmI systems with AI agents need to be sensitive to the needs and the micro-contexts of their potential users. For this purpose, sensing and communication components must be always open and receptive to user input and other contextual variables. Ultimately, the acceptance of AmI systems depends on demographic and personal preferences regarding privacy, security, trust, individualism, diversity, mobility and lifestyle that affect the structure of communities and the way people live and work.

AmI design is influenced by user-centric methods where the user is placed at the center of the design activity and asked to give feedback through evaluations and tests to improve the design, or even co-create the design with a group of designers, or users. The challenge of designing and implementing an AmI system is the lack of models enabling the analysis of the system requirements while designing the system, and of verification and testing methods when the system is implemented. Designing an AmI system involving AI agents requires a different approach from traditional system design. While in traditional systems performance and interaction are determined in advance and remain fixed, in AmI agent systems interactions are contextual and open ended, triggered by the unrestricted activity of the users within the environment. This becomes possible through the integration of ICT components in the background. Furthermore, since AmI aims to build experiences that are entirely new, there is no proper framework for situating and evaluating them within existing user contexts, as there is no framework for aligning them to the existing production strategies of companies.

The absence of clearly defined working criteria, supporting the generation and evaluation of AmI agent systems, is a conspicuous obstacle to their advancement. In this paper, we determine and test a framework for the generation and evaluation of AmI system designs, based on user experience and business criteria. We adopt the Research-Through-Design (RTD) method [2] based on which the necessary insights for advancing a theoretical framework are extracted through design practice. More specifically, the process of designing a personal lighting AI agent in collaboration with a leading lighting design company is used as a case study to determine and test a framework for the generation of AmI system design proposals, and for the evaluation of their implementation roadmap, based on parameters of feasibility and acceptability.

The project, called *Connected Lighting for a Caring City*, includes research, design, and evaluation. First, primary and secondary research on the significance of the notion of "caring" within different user contexts is used to extrapolate user values for the *caring city* concept. Tech research is used to identify relevant emerging technology trends and possibilities. Second, a personal lighting AI agent is envisioned based on the user values and the technologies. Alternative design proposals are described and communicated in video storytelling format. The design alternatives are based on five systemic factors, namely: (a) the context of interaction; (b) the required system data; (c) the required sensing input; (d) the required user input; and (e) the desired system output. Finally third, a structured framework is used for the evaluation of implementation roadmaps for each alternative design vision. Visions are evaluated by assessing three parameters: the complexity of the enabling technologies, the availability of these technologies in the partner company, and the prospect of advantageous business partnerships.

This research contributes a novel framework for the generation and evaluation of AmI systems involving AI agents, based on three distinct levels of in-built system intelligence. The first level corresponds to one-step notifications and simple system outputs, requiring simple data input (i.e. reading a scheduled event in the user's calendar). The second level corresponds to hardware requirements for the sensing and processing routines of the system. The third level corresponds to emotional intelligence [3], which is the higher level that an AI agent can reach, where the system acts as a personal assistant. The three levels of in-built intelligence in combination reflect the feasibility and acceptability of the system. Feasibility is what a specific company is capable of producing and in what timeframe. Acceptability is the potential of familiarity and trust that the users can feel while interacting with the AI agent.

After a brief exposition of recent papers on existing AmI design approaches, and related UX methodologies, the case study *Connected Lighting for a Caring City* is exposed in detail. The case study is used to determine a framework for the generation and evaluation of AmI systems design. In the generation section we present the iteration steps of the case study and the extraction of user values based on UX principles, which lead to the formation of the design proposal. In the evaluation section, we present the evaluation framework and the validation steps. In the following section of the results, we analyze how the framework was used by the design team and by the partner company. The presentation ends with a brief discussion on the potential of this generation and evaluation framework, and a brief reference to future research.

2   Background and Related Work

Ambient Intelligence (AmI) is defined as a specific class of ICT applications enabling physical environments to become sensitive, adaptive, and responsive to human activities [4]. Beyond the integration of ICT devices into the physical environment, the AmI paradigm promotes the creation of new, enhanced user experiences [5]. Cook et al. [6] explain that AmI systems are sensitive, responsive, adaptive, transparent, ubiquitous, and intelligent. Building on the ideas of ubiquitous computing by Marc Weiser [7] who envisioned a digital world in which ICT components form a distributed network, AmI systems aim to supply an enhanced physical environment that strengthens the prospect of well-being, improves productivity and creativity, and augments the enjoyment of leisure time. At the same time, AmI systems introduce new levels of complexity and new challenges. Interactions cease to be human-to-machine. They are ubiquitously distributed within the living environment, where new challenges emerge regarding front-end communication and avatar interaction [8]. AmI systems can also involve AI agents [9, 10] and perform as autonomous systems [11]. These AI agent-based systems are recommendation systems that interpret the user's state and habits and initiate proper responses [12].

Because AmI systems must be sensitive, adaptive, and responsive to people, they must be aware of their preferences, intentions, and needs [13]. Furthermore, AmI agent systems need to be unobtrusive and easy to live with [14]. Their interfaces must be context aware, natural, and acceptable from an ethical point of view [15]. Streitz [16] argues in favor of a transition from Human–Computer Interaction to Human–Environment Interaction, which leads to responsive environments [17]. Koskinen [18] points out that recent design examples dematerialize design. Material configuration becomes a secondary issue, and social aspects become the main focus of the design process. Recent examples in using methods of civic engagement in the design process of smart cities, support this view [19, 20]. Design is focusing to configuring complex sociotechnical networks involving services and human experience [20, 21].

In the field of Human-Computer interaction (HCI), User Experience (UX) has emerged as a new paradigm for the generation and evaluation of designs. UX shifts the focus from utility and task-based performance, to user experience and interaction value in different contexts [22]. In evaluating AmI systems via UX, the pragmatic and enjoyment aspects become equally important [23, 24]. MacDonald and Atwood [22] argue that UX evaluation methods is an approach that leads to real-world design. Usability is not enough [25]. Furthermore, because user experience is mutable, affecting both the perception and the behavior of the user, it should be evaluated within a specific timeframe and context [26].

If we narrow the focus of HCI research to AmI systems, we can point existing methods for evaluating UX in AmI system design [27]. Human activity recognition and biofeedback analysis are common practices [28]. For example, applications of AmI systems focusing on Ambient Assisted Living (AAL) involve monitoring and supporting elderly and disabled people in their homes [29, 30]. In this context, Bono-Nuez et al. [31] propose a model for evaluating the quality of life of these individuals, based on activity monitoring, while Ntoa et al. [32] present a UX evaluation framework measuring how AmI systems anticipate and satisfy the user needs. This method employs video recording and measuring during experiments with users. O'Grady et al. [9] propose six quantifiable dimensions to measure software quality from the end-user's point of view, namely: efficiency, affect, helpfulness, control, learnability and global usability. These methods are then used in evaluating AmI systems during prototyping.

Different methods apply in the evaluation of early design concepts. The assessment is based on the potential of a system to get broadly adopted, rather than usability criteria. Video-fiction prototyping is used for this purpose [33]. Gaggioli [34] proposes an Experience Sampling Method to evaluate UX in AmI systems, focusing on how user attention selects specific information from the environment. Gaggioli defines an "optimal experience" as a flow of psychological processes producing various states of consciousness to the user.

Additional research focuses on the user journey and the relationship between system operations and user expectations to assess user confidence [35]. In this case, the design vision and the user's point of view are often not identical. Forest et al. [36] base the assessment of AmI concepts on this paradox. Cabitza et al. [37] propose Event-Condition-Action rules as a novel conceptual framework for designing complex socio-technical systems, and support the users to propose trigger-action rules. Besides the potential of system adoption, UX assessment relies on the anticipation of social impact [38] and ethical norms [39]. Wiegerling [40] poses the question of ethics in AmI, observing that autonomous systems incapacitate the users if they are not in proper control. As the spatial context within which the interaction between humans and system is no longer static, single-user and location-independent, but a dynamic multi-user, situated environment, there is a need for reconsidering the implications and proposing new design methods for intelligent systems [41]. And because data is a core resource in designing AmI systems [42] the acquisition of data demands broader social consensus [43].

In the featured case study *Connected Lighting for a Caring City* we envision a personal, intelligent lighting agent for a city that "cares" for its residents. The case study is the result of collaboration with a world leading company in lighting, aiming to develop new design strategies to guide the company practices [44, 45]. In the case study we follow the Value Sensitive Design (VSD) theoretical approach [46] based on which the value of the design process is determined by the priorities of a person or a group. We also adopt the Research through Design (RtD) method [2] where the design practice is used as a generator of knowledge. Main contribution of this research is that it determines and tests a framework for the generation and evaluation of AmI systems involving AI agents, based on three distinct levels of in-built system intelligence. The three levels reflect the feasibility and acceptability of the system. Hence, the evaluation is equally based on user experience and business value. The user journeys are mapped with use case scenarios [47] and a roadmap for incremental implementation of each design proposal is provided [48]. Although evaluation on the basis of specific parameters is a common method, determining the implementation roadmap based on the specific three levels of in-built system intelligence is a new idea. To our knowledge there are no existing assessment methods of AmI agents that are based on parallel account of intelligence levels and user interaction modalities.

3 Designing an AmI Agent System

The *Connected Lighting for a Caring City* was a yearlong project, developed in collaboration with the company Signify. The objective was to design an artificial lighting AI agent that would provide its services in a "caring" manner. The agent involves a complex AmI system and interactions that take place in diverse urban contexts. This design vision was developed with the potential of implementing it in the next decade. With this timeframe in mind, the research followed six phases: (1) Secondary research, (2) Primary research, (3) Identification of user values, (4) Definition of the design concept, (5) Storytelling of the design concept, and (6) Evaluation of the design concept and roadmap construction.

Secondary research targeted three topics: the impact and effects of light throughout historical periods in various cultural contexts, the current related technological trends, and the anticipated future behavioral and lifestyle trends of urban living. Primary research focused on collecting user insights related to the notion "caring". We interviewed select individuals across generations to record what "caring" means for them, and how a city can become caring. Six key user values were identified: invited, accepted, acknowledged, accompanied, assisted, and protected, which shaped our design process in the subsequent phase. In the design generation phase, we created personas of potential users and we envisioned alternative concepts for a lighting AmI system involving an AI personal assistant. The lighting personal assistant caters the needs of the city dwellers in various indoor and outdoor contexts. We then produced storytelling prototypes to demonstrate certain scenarios of use. Finally, we developed explicit tools to perform the evaluation of the design alternatives, and to build the corresponding implementation roadmaps based on the availability of relevant technologies in the partner company.

3.1 Generation Framework

Current advancements in the field of AmI systems include sensing and actuating networks, lighting systems supporting data gathering and transmission, and light embedded or enabled materials. Further research includes user-system inputs (such as sensing modalities), and interaction modalities, enabling natural user interaction (such as touch, gesture and sound). We performed secondary research to understand the state of the art in the fields of interaction, lighting, and the related technologies. Then, the focus of our primary research was to identify what makes city dwellers feel cared for, across generations.

In secondary research we examined the emerging behavioral and social trends of overusing tech devices and the isolation phenomena of urban living. We identified a need for human-scale service systems that could support the future shaping of "caring cities". Main characteristic of these systems would be to make urban inhabitants feel they are cared for. User values were extrapolated by observing the feelings of people in various activities and urban contexts. We observed the change that the adoption of digital technologies and interactions causes across generations, in order to envision inclusive design alternatives for an AmI lighting agent.

In primary research we conducted face-to-face semi-structured interviews to identify what makes city dwellers feel cared for, across generations. The selection of participants was based on diversity in age, gender, and cultural background, as well as familiarity with urban living. Their occupations ranged from students, part-time or full-time employees, to retired elderly. The data obtained by the interviews were divided in two thematic areas: (1) what are the characteristics of various generations of urban dwellers, and (2) what makes them feel cared for.

Based on the first thematic area of interviews we outlined four personas, corresponding to four age groups of targeted city inhabitant types. The personas incorporate common characteristics among the interviewees. They belong to different age spans and reflect their thoughts, daily activities, considerations, and needs.

Based on the second thematic area of interviews we identified six key user values setting guidelines towards alternative design visions for a caring city. The designs evolved around Phil, an intelligent lighting agent that enables citizens to feel: Invited, Accepted, Acknowledged, Accompanied, Assisted, and Protected. Phil senses the user activities, identifies, learns, and generates information through AI algorithms enabling the actuation of the proper light sources for each occasion (see Fig. 1).

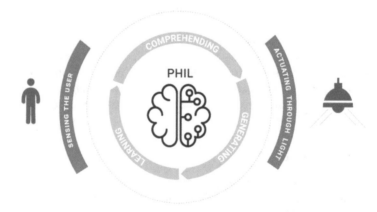

Fig. 1. Overview of the AmI agent system.

To sense the user, Phil relies on sensors that are embedded in materials, for real-time activity detection. Phil can also access cloud stored user profiles, preferences, and APIs. Connected with the light sources Phil assists the user in outdoor and indoor activities, at home, work, leisure, and transportation. Phil is capable to identify the real-time context in which the user acts by accessing their profile, the location and time of the activity, and the social dynamics. Phil learns people's daily routines, their needs, preferences, interests, and what they consider optimal ambiance settings. Then, Phil caters the user needs with illumination, ambiance lighting, social lighting triggers, way finding, and event notification lighting.

We designed an interaction language for communicating with Phil, based on gestural inputs and visual outputs. The interactions would become possible through smart materials [49] and electroactive fabrics [50], enabling capacitive touch sensitive surfaces. These materials could be embedded in the furniture, the wall surfaces, and the clothing, permitting seamless interactions.

We produced video storytelling prototypes based on use cases depicting the system performance in daily activities by determining the following general parameters:

- the potential users through personas (i.e. their personal goals, needs, and issues across generations);
- the user values (i.e. the six identified values for feeling cared-for);
- the contexts of use (i.e. their spatial-temporal contexts);
- the data input and props;
- the interaction modalities.

3.2   Evaluation Framework

At the end of the design process, we conducted a workshop with the partner company to evaluate the use cases of the system and to explore potential business strategies for its implementation (see Fig. 2).

Fig. 2. Workshop material and layout of the working area.

The use case scenarios were presented in video form, and were analyzed and evaluated based on specific moments of interaction. This analysis has focused on points such as: The spatial context within which the scene was enacted; the required data for the functionality of the system; and the required user inputs, or the inputs from sensors.

Specific moments of interaction, called *scenes*, were documented with the aid of *analysis cards* (see Fig. 3). To evaluate a scene with the partner company, we evaluated the perceived *feasibility* of the interaction, and the user *acceptability* this interaction could entail. Feasibility was determined by what the company was capable of producing, and in what timeframe. Acceptability was determined by the potential for familiarity and trust that the users could feel while interacting with a particular AI agent design. Hence, a two-axis diagram was constructed with x axis representing acceptability, and y axis representing feasibility scores. Analysis cards were numbered, and numbers were placed on the two-axis diagram (matrix) indicating the overall score of the feasibility/acceptability evaluation (see Fig. 4).

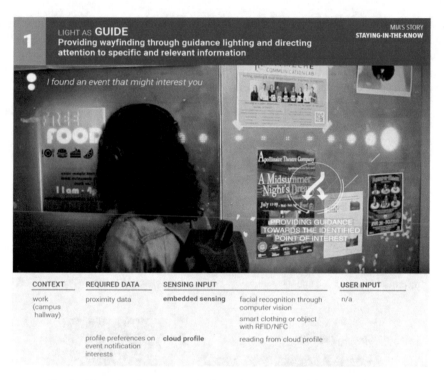

Fig. 3. Example of a scene analysis card that was used in the workshop.

Fig. 4. Scene evaluation matrix with card placements, as it was used in the workshop.

Based on the evaluation results of each AmI agent design we outlined the associated development and implementation steps in a strategic *roadmap*, aiming at the gradual introduction of the AmI agent in order to reduce the risk of disruption. The evaluation results emphasized the importance of enforcing trust to ensure the acceptability of the system by the users. Trust includes both the preservation of privacy and familiarity. A system that relates to familiar, existing services or products is more likely to be trusted. And furthermore, the implementation of each design concept was envisioned to progress based on the pace of technological progress. The AmI agent system was designed based on existing solutions and services, with the vision to incorporate more sophisticated technologies when they become available. The use scenes were grouped, evaluated and placed on a timeline, based on three stages corresponding to the three levels of system intelligence, namely: (a) one-step notification, (b) context aware interaction, and (c) personalized assistant (see Fig. 5).

ONE-STEP NOTIFICATION   CONTEXT AWARE INTERACTION   PERSONALIZED ASSISTANT

Fig. 5. Concept implementation roadmap stages, as emerged from the evaluation of scenes.

The first stage of the roadmap is *one-step notification*. It uses off-the-shelf technology including luminaires linked to an API, which can trigger a notification as required. The scenes in this group rely on a simple, reactive system, where lighting is the primary medium of notification. One-step notification infrastructure can also apply to settings in which in addition to API, simple gestural inputs can serve as triggers.

The second stage of the roadmap is *context aware interaction*. Additional hardware, containing appropriate sensors, needs to be embedded in the outdoor and indoor environments for this type of interaction. Along these lines, we examined the designing of "clip-on" items to be added to the existing infrastructure. This would enable the context awareness of the system (space, time, movement, activity) based on sensory input. In comparison to the first stage, this second stage incorporates additional elements of remote interaction and awareness, which synthesize a living context for each user.

Finally, the third stage of the roadmap is *personalized assistant*. At this stage the system obtains a complex set of features and the scenes are tailored to unique user experiences. The AI agent becomes a personal assistant that can track the physiological states, lifestyle, habits, needs, and desires of the user, and act accordingly. At this third stage, the system incorporates machine learning and is capable of emotion tracking and empathy, thus reaching the highest level of intelligence.

4   Validation of Tools

The design process included concept generation and evaluation. In both we followed techniques that supported the design practice. In the generation phase we followed the Design Thinking [51] framework. In the evaluation phase we used a framework that was developed for the needs of the AmI agent evaluation. It was based on a three-stage analysis, corresponding to three levels of in-built system intelligence. Specific tools were employed to support this analysis, namely: (1) use cases videos, (2) scene-analysis cards, (3) scene-evaluation diagram (matrix), (4) scene-implementation roadmap. The videos and analysis cards were made by our design team. The evaluation and implementation roadmap was produced in collaboration with the partner company during the workshop session. These tools supported the design process and provided a platform of communication between the design team and the company. They also served the alignment of user values and business values.

To validate our evaluation framework and tools, we set up a survey to collect feedback from the partner company. The survey was delivered online. First, it validates the adequacy of video storytelling as a tool of *description* of designs and the underlying user values. Then, it validates the adequacy of the scene-analysis cards as a tool of *design analysis*. And finally, it validates the adequacy of the evaluation matrix and the implementation roadmap as *business value demonstrations* of the designs.

4.1 Survey Results

Four participants replied to the survey: Head of Research (24 years in the company); a Principal Scientist (18 years in the company); an Industrial Designer (17 years in the company); and the Head of Design (3 years in the company).

All respondents agreed that *video storytelling* offers sufficient means to communicate the design vision and specifically the user values, the concepts, the enabling technologies, and the interaction modalities. One participant claimed that "video storytelling is efficient to convey a vision within the organization". However, other respondents argued that the videos do not adequately represent the design as a whole, because each time they capture only a fragment of the story. All respondents agreed that video storytelling is sufficient to convey the user values. They also appreciated that the videos were "well anchored in research" allowing the company team to "see the impact of the system on people".

We also collected feedback related to the adequacy of the parameters in the *scene-analysis cards*. The cards made the vision concrete by analyzing the involved technologies, the criticalities and the feasibility of the design scenarios. We considered parameters such as, the context of interaction, the required data, the inputs from sensors and from the user, and the system outputs. All respondents agreed that the scene-analysis cards were useful in making the vision concrete and understandable. However, only two respondents agreed that the parameters were useful in determining the feasibility and the criticalities of the design vision.

All the respondents agreed that the 2×2 matrix of the *scene-evaluation diagram* was useful to assess the feasibility of the design concepts, based on the current technological capabilities of the company. The respondents observed that the strengths of the scene-evaluation diagram are the simplicity in communication, the intuitiveness, and the simplification of the decision-making process. However, some respondents have pointed two shortfalls: the openness of the matrix to personal interpretation, and the fact that the matrix could "oversimplify the decisions without mapping the full set of implications". These respondents claimed that introducing additional matrices to the already proposed one, could strengthen the validity of the tool. A participant suggested that it is important to acknowledge "the people's role in the company" when capturing their evaluations on the 2 × 2 matrix.

Three of the respondents claimed that the scene-evaluation diagram was useful in determining the *scene-implementation roadmap* in association to the strategic framework of the company. The roadmap was based on the following parameters: the complexity of the enabling technology; the availability of the technology within the company; and the potential for productive partnerships. Three of the respondents have found these parameters useful. However, not all the respondents agreed that these parameters are sufficient. One suggestion was to add more parameters, such as: "internal strategic direction, annual operating plans, external highest demand, and piloting opportunities".

5 Conclusion

AmI systems involving AI agents promise to radically transform the role of technology in everyday life. The AmI vision has explicit human-centered goals. To realize these goals AmI systems need to be sensitive to the needs and the micro-contexts of their potential users. Hence, AmI system design is influenced by user-centric methods where the user is placed at the center of the design activity. The challenge of designing an AmI system involving AI agents is the lack of proper models that enable the analysis of the system requirements while designing the system, and of verification and testing methods, when the system is prototyped. This absence of clearly defined working criteria, supporting the generation and evaluation of AmI agent system designs, is an obstacle to their advancement.

In this paper, we determined and tested a framework for the generation and evaluation of AmI system designs, based on user experience and business criteria. We adopted the Research-Through-Design method based on which the necessary insights for advancing a theoretical framework are extracted through design practice. The process of designing a personal lighting AI agent in collaboration with a leading lighting design company was used as case study to determine and test a framework for the generation and evaluation of AmI system designs based on feasibility and acceptability criteria.

First, primary and secondary research on the significance of a "caring city" within different user contexts was used to extrapolate user values, and tech research identified related, emerging technology trends. Second, a personal lighting AI agent was proposed based on the user values. Alternative designs were described and communicated in video storytelling format. The design alternatives were based on five systemic factors, namely: (a) the context of interaction; (b) the required system data; (c) the required sensing input; (d) the required user input; and (e) the desired system output. Finally third, a structured framework was used for the evaluation of each alternative design proposal. The evaluation was based on the complexity of the enabling technologies, the availability of these technologies in the partner company, and the prospect of advantageous business partnerships.

Main contribution of this research is the new proposed framework for the generation and evaluation of AmI systems involving AI agents, based on three distinct levels of in-built system intelligence. Although evaluation on the basis of parameters is a common method, defining the implementation roadmap based on specific levels of in-built intelligence is a new method. The first level of intelligence corresponds to one-step notifications and simple system outputs, requiring simple data input. The second level of intelligence corresponds to the hardware requirements for the sensing and processing routines of the system. The third level of intelligence corresponds to the capacity for emotional interaction, which is the higher state that an AI agent can reach. The three levels of in-built intelligence in combination reflect the feasibility and acceptability score of the system. Feasibility is what a specific company is capable of producing, and in what timeframe. Acceptability is the potential of familiarity and trust that the users could develop while interacting with the AI agent. To our knowledge there are no other existing assessment methods for AmI agent systems that account in parallel for intelligence and user interaction modalities.

The tools that we used in the evaluation framework – the video storytelling, the scene-analysis cards, the scene-evaluation diagram, and the scene-implementation roadmap – were validated with a survey that was distributed within the partner company, and overall, they received positive ratings.

The presented process of generation and evaluation of AmI agent system designs emphasizes that the transition from general human-centered considerations towards specific user centered results is never a simple or straightforward process. It reflects the diverse nature of the creative and analytical considerations of design. In the generation phase, the design team considered the human values related to the daily routines of people, in different contexts. In the evaluation phase the use cases depicted city dwellers as users. Collaborating with the specific partner company directed this analysis to focus on certain favorable user values and company requirements.

The potential of the presented evaluation framework lies in the necessity to understand the role of different intelligence levels of AmI systems. We claim that the different levels of in-built intelligence can be a significant factor and design-aid in generating and evaluating AmI systems. Furthermore, we argue that understanding the role of different intelligence levels is significant both for envisioning new user interactions, and from a business point of view. Hence, this research contributes new means of evaluation and communication, which can be equally useful to designers and to companies. In future stage, we plan to validate the presented evaluation framework with additional companies in more case studies involving AmI agent system design.

Acknowledgments. This research is a result of collaboration between the MIT Design Lab and Signify. Doctoral research of the author Milica Pavlovic has been funded by TIM S.p.A., Services Innovation Department, Joint Open Lab Digital Life, Milan, Italy.

References

1. Zelkha, E., Epstein, B.: From devices to ambient intelligence. In: The Transformation of Consumer Electronics, Presentation at the Digital Living Room Conference, Philips (1998)
2. Zimmerman, J., Forlizzi, J.: Research through design in HCI. In: Ways of Knowing in HCI, pp. 167–189. Springer, New York (2014)

3. Picard, R.W.: Affective computing: challenges. Int. J. Hum. Comput. Stud. **59**(1–2), 55–64 (2003)
4. Mukherjee, S., Aarts, E., Doyle, T.: Special issue on ambient intelligence. Inf. Syst. Front. **11**(1), 1–5 (2009)
5. Aarts, E., Encarnaçao, J. (eds.): True Visions: Tales on the Realization of Ambient Intelligence. Springer, Heidelberg (2006)
6. Cook, D.J., Augusto, J.C., Jakkula, V.R.: Ambient intelligence: technologies, applications, and opportunities. Pervasive Mob. Comput. **5**(4), 277–298 (2009)
7. Weiser, M.: The computer for the twenty-first century, pp. 94–100. Scientific American, September Issue (1991)
8. Hanke, S., Tsiourti, C., Sili, M., Christodoulou, E.: Embodied ambient intelligent systems. In: Recent Advances in Ambient Assisted Living - Bridging Assistive Technologies e-Health and Personalized Health Care. IOS Press, Tepper Drive Clifton (2015)
9. O'Grady, M.J., O'Hare, G.M., Poslad, S.: Smart environment interaction: a user assessment of embedded agents. J. Ambient Intell. Smart Environ. **5**(3), 331–346 (2013)
10. Burr, C., Cristianini, N., Ladyman, J.: An analysis of the interaction between intelligent software agents and human users. Mind. Mach. **28**(4), 735–774 (2018)
11. Gams, M., Gu, I.Y.H., Härmä, A., Muñoz, A., Tam, V.: Artificial intelligence and ambient intelligence. J. Ambient Intell. Smart Environ. **11**(1), 71–86 (2019)
12. Rasch, K.: An unsupervised recommender system for smart homes. J. Ambient Intell. Smart Environ. **6**(1), 21–37 (2014)
13. Plötz, T., Kleine-Cosack, C. Fink, G.A.: Towards human centered ambient intelligence. In: European Conference on Ambient Intelligence, pp. 26–43. Springer, Heidelberg (2008)
14. Airaghi, A., Schuurmans, M.: ISTAG scenarios for ambient intelligence in 2010. European Commission Community Research (2001)
15. Brey, P.: Freedom and privacy in ambient intelligence. Ethics Inf. Technol. **7**(3), 157–166 (2005)
16. Streitz, N.A.: From human–computer interaction to human–environment interaction: ambient intelligence and the disappearing computer. In: Universal Access in Ambient Intelligence Environments, pp. 3–13. Springer, Heidelberg (2007)
17. Alves Lino, J., Salem, B., Rauterberg, M.: Responsive environments: user experiences for ambient intelligence. J. Ambient Intell. Smart Environ. **2**(4), 347–367 (2010)
18. Koskinen, I.: Agonistic, convivial, and conceptual aesthetics in new social design. Des. Issues **32**(3), 18–29 (2016)
19. Forlano, L.: Decentering the human in the design of collaborative cities. Des. Issues **32**(3), 42–54 (2016)
20. Hill, D.: The city is my homescreen. In: Proceedings of the 2018 ACM International Conference on Interactive Surfaces and Spaces, p. 1. ACM (2018)
21. Lou, Y.: The idea of environmental design revisited. Des. Issues **35**(1), 23–35 (2019)
22. MacDonald, C.M., Atwood, M.E.: Changing perspectives on evaluation in HCI: past, present, and future. In: CHI Extended Abstracts, Paris, France (2013)
23. Hassenzahl, M.: The thing and I: understanding the relationship between user and product. In: Funology: From Usability to Enjoyment, pp. 31–42. Kluwer Publishers, Norwell (2003)
24. Forlizzi, J., Battarbee, K.: Understanding experience in interactive systems. In: Conference on Designing Interactive Systems. Cambridge (2004)
25. Arhippainen, L., Hickey, S., Pakanen, M., Karhu, A.: User experiences of service applications on two similar 3D UIs with different 3D space contexts. MindTrek, Helsinki (2013)
26. Karapanos, E., Zimmerman, J., Forlizzi, J., Martens, J.: User experience over time: an initial framework. CHI, Boston (2009)

27. Pavlovic, M., Colombo, S., Lim, Y., Casalegno, F.: Designing for ambient UX: case study of a dynamic lighting system for a work space. In: Proceedings of the 2018 ACM International Conference on Interactive Surfaces and Spaces, pp. 351–356. ACM (2018)

28. Treur, J.: On human aspects in ambient intelligence. In: European Conference on Ambient Intelligence, pp. 262–267. Springer, Heidelberg (2007)

29. Salvi, D., Montalva Colomer, J.B., Arredondo, M.T., Prazak-Aram, B., Mayer, C.: A framework for evaluating ambient assisted living technologies and the experience of the universAAL project. J. Ambient Intell. Smart Environ. 7(3), 329–352 (2015)

30. Veronese, F., Masciadri, A., Trofimova, A.A., Matteucci, M., Salice, F.: Realistic human behaviour simulation for quantitative ambient intelligence studies. Technol. Disabil. 28(4), 159–177 (2016)

31. Bono-Nuez, A., Blasco, R., Casas, R., Martín-del-Brío, B.: Ambient intelligence for quality of life assessment. J. Ambient Intell. Smart Environ. 6(1), 57–70 (2014)

32. Ntoa, S., Margetis, G., Antona, M., Stephanidis, C.: UXAmI observer: an automated user experience evaluation tool for ambient intelligence environments. In: Proceedings of SAI Intelligent Systems Conference, pp. 1350–1370. Springer, Cham (2018)

33. Kymäläinen, T., Kaasinen, E., Hakulinen, J., Heimonen, T., Mannonen, P., Aikala, M., Lehtikunnas, L.: A creative prototype illustrating the ambient user experience of an intelligent future factory. J. Ambient Intell. Smart Environ. 9(1), 41–57 (2017)

34. Gaggioli, A.: Optimal experience in ambient intelligence. Ambient Intell. 3543(5), 35–43 (2005)

35. Corno, F., Guercio, E., De Russis, L., Gargiulo, E.: Designing for user confidence in intelligent environments. J. Reliable Intell. Environ. 1(1), 11–21 (2015)

36. Forest, F., Mallein, P., Arhippainen, L.: Paradoxical user acceptance of ambient intelligent systems: sociology of user experience approach. In Proceedings of International Conference on Making Sense of Converging Media, p. 211. ACM (2013)

37. Cabitza, F., Fogli, D., Lanzilotti, R., Piccinno, A.: End-user development in ambient intelligence: a user study. In: Proceedings of the 11th Biannual Conference on Italian SIGCHI Chapter, pp. 146–153. ACM (2015)

38. Little, L., Briggs, P.: Designing ambient intelligent scenarios to promote discussion of human values. In: Interact: Workshop on Ambient Intelligence, September 2005, Rome (2005)

39. Colombo, S.: Morals, ethics, and the new design conscience. In: Rampino, L. (eds.) Evolving Perspectives in Product Design: From Mass Production to Social Awareness. Franco-Angeli (2018)

40. Wiegerling, K.: The question of ethics in ambient intelligence. In: Ubiquitous Computing in the Workplace, pp. 37–44. Springer, Cham (2015)

41. Streitz, N., Charitos, D., Kaptein, M., Böhlen, M.: Grand challenges for ambient intelligence and implications for design contexts and smart societies. J. Ambient Intell. Smart Environ. 11(1), 87–107 (2019)

42. Arslan, P., Casalegno, F., Giusti, L., Ileri, O., Kurt, O.F., Ergüt, S.: Big Data as a source for Designing Services. Web (2017)

43. Pavlovic, M., Botto, F., Pillan, M., Criminisi, C., Valla, M.: Social consensus: contribution to design methods for AI agents that employ personal data. In: International Conference on Intelligent Human Systems Integration, pp. 877–883. Springer, Cham (2019)

44. Parmar, A.J.: Design attitude. Des. Issues 32(3), 116–118 (2016)

45. Sheppard, B., Sarrazin, H., Kouyoumjian, G., Dore, F.: The business value of design. McKinsey design. McKinsey Quarterly. https://www.mckinsey.com/business-functions/mckinsey-design/our-insights/thebusiness-value-of-design. Accessed from 11 Jan 2019

46. Friedman, B.: Value sensitive design. In: Bainbridge, W.S. (ed.) Encyclopedia of human-computer interaction, pp. 769–774. Berkshire Publishing Group, Great Barrington (2004)
47. Kalbach, J.: Mapping Experiences: A Complete Guide to Creating Value Through Journeys, Blueprints, and Diagrams. O'Reilly Media, Inc., Sebastopol (2016)
48. Norman, D.A., Verganti, R.: Incremental and radical innovation: design research vs. technology and meaning change. Des. Issues **30**(1), 78–96 (2014)
49. Barrett, G., Omote, R.: Projected-capacitive touch technology. Inf. Display **26**(3), 16–21 (2010)
50. Syduzzaman, M., Patwary, S.U., Farhana, K., Ahmed, S.: Smart textiles and nanotechnology: a general overview. J. Text. Sci. Eng. **5**, 1000181 (2015)
51. Dorst, K.: The core of 'design thinking' and its application. Des. Studies **32**(6), 521–532 (2011)

Virtual Reality Rendered Video Precognition with Deep Learning for Crowd Management

Howard Meadows[✉] and George Frangou

Massive Analytic Limited, IDEALondon, 69 Wilson Street, London, UK
howard.meadows@massiveanalytic.com
http://www.massiveanalytic.com

Abstract. We describe an AI driven model based on the Nethra Video Analytic platform, optimised for overhead detection and designed specifically for CCTV and which can detect and categorize people from any angle. The model runs in real time in crowded and noisy environments and can be in- stalled on devices as in edge analytics or applied directly to existing video feeds. By mapping an entire space, we link together individual camera feeds and data points to calculate the total number of people to assist with capacity planning and to pin point bottlenecks in people flows. Solving the problem of aggregating multiple 360-degree video camera feeds into a single combined rendering, we further describe a novel use of interactive Virtual Reality. This model renders St Pancras International Station in VR and can track people movement in real time. Real people are rep- resented by avatars in real-time in the model. Users are able to change their viewpoint to look at any angle. The movement of the avatars exactly mirrors what can be seen in the cameras.

Keywords: Video · Virtual Reality · Mixed Reality · Deep learning · Crowd management · CCTV · City · Railway station · Airport · Shopping · Precinct · Metro · webGL · Massive Analytic · Nethra · Precognition · Path analysis · People detection · People counting · GPU

1 Introduction

An often-cited study in the monitoring industry concludes that "After 12 min of continuous video monitoring, a person will often miss up to 45% of screen activity. After 22 min, up to 95% is overlooked." In today's digital world where we capture millions of hours of footage every day, the demand on operators is too high to rely on manual viewing of video feeds. Nethra which means 'eye' in Sanskrit [1], is a video analytics platform created by Massive Analytic that automatically detects a multitude of different events in real time from video feeds, drastically reducing the demand on manpower and delivering quicker and better outcomes.

© Springer Nature Switzerland AG 2020
K. Arai et al. (Eds.): FTC 2019, AISC 1069, pp. 334–345, 2020.
https://doi.org/10.1007/978-3-030-32520-6_27

Nethra is a 'precognitive' video platform that uses sophisticated deep-learning algorithms to analyse patterns and see possible outcomes in real-time. Nethra transcends the limitations of the human eye and what a human being can process at one time. The philosophy behind Nethra as applied to machine vision is that its functioning reflects the two fundamental characteristics of human thinking. The creative, holistic, and theory-building intelligence (often identified with the right hemisphere of the brain) is mirrored by the rule induction capabilities using possibility theory described as coarse tuning. The strictly logical, deductive processes (often attributed to the left hemisphere of the brain) have their analogy in the machine learning process described as fine tuning. This unique feature of combining the two types of reasoning, the creative with the deductive, is described as artificial precognition, and is protected by ten international patents [2]. Artificial Precognition (AP) provides a significant level of inferential thinking and communication. This goes far beyond what has until now been possible by going some way to toward mimicking the way humans use a combination of stored memories and fused data set input to interpret events as they occur and anticipate (cognize) likely scenarios. Therefore, AP correlates directly with a fundamental of human intelligence and the way visual information is processed and interpreted.

Firstly, we will discuss how Nethra tracks and counts crowds of individuals in a busy subway station from multiple video feeds. Existing solutions focus on facial recognition or pose estimation in order to track people, however many CCTV installations are in overhead positions where the face and the body are hardly ever fully visible. Furthermore, in crowded scenarios the top of people's heads may be the only thing visible as people move closer together. These alternate solutions are therefore incapable of accurately tracking people across multiple frames in crowded scenarios from an overhead viewpoint. Even when a model or program has been trained to recognise movement alone, this is not sufficient, as in very crowded scenarios, crowds start to move as one, and individuals' movements are no longer decipherable from that of the crowd [3]. We will then illustrate the platform's support in a variety of use cases such as crowd management, people movement analytics and security.

Secondly, we take this concept to the next level and built a Virtual Reality (VR) model of St Pancras international train station in London, allowing a 360-degree view of people movement and crowd density – enabling live management and event detection of busy public spaces.

2 Overhead CCTV Head Model

2.1 Description and Use Case from a Large City Subway

Used deep learning tools including as tensorflow and pytorch a model was developed for detecting human heads with a high degree of accuracy. Custom low-level software code was written to track heads in real time across video frames. Deep learning for video requires that image frames are tagged (labelled and boxed) with the objects you are trying to get the deep learning neural network to learn

[4]. This therefore required the preparation of thousands of images over several datasets. Figure 1 shows a streaming video augmented with what we call our head model and path results from a large city Subway. The red boxes represent the head model results, while the other coloured boxes represent the tracking of these across frames. The tracking result e.g. a person's path across video frames, is represented by the coloured lines following the boxes. From the tracking result we are also able to tell whether it is a man, woman or child [7]. The blue lines are what we call lines of interest and are added to each video/camera setup. These may be on particular gangways, or entrances, etc. and can be at any angle. The lines of interest are the focal point for the video analysis, drawn on these lines of interest are counters showing the number of people crossing the line, based on the head tracking. You can see numbers for the number of people crossing the line as either Up/Down or Left/Right total figures. All statistics, including head positions and unique identifiers, can be streamed in real time as raw data and will work on a live video feed from an IP camera.

Fig. 1. Showing the basic features of the head tracking model, with person tracking and counter lines and statistics

Each video can have a slightly different scenario in terms of crowd entries and exits, and demonstrates the head model's ability to cope with various issues. Some of these issues include people being reflected in walls and railings or as shadows on the floor Fig. 2. Multiple exit/entry points can be covered in a single video, for example, people moving to the right either from a stairwell or an entrance on the top. It can also cope well with sudden changes from static to dynamic movement, as with people queuing and then boarding a train [10].

As well as tracking movement these images also demonstrate Nethra's coat colour matching capability. Each box contains a colour swatch matching the coat or jacket colour of the particular person. This capability has many applications, one example is in security and policing where it can help identify a suspect or perhaps alert station staff to large groups of sports fans. Nethra can also detect whether people are carrying bags and with a little more development this could be expanded to recognising particular brands from their shopping bags on a busy high street for marketing or advertising use cases [7].

Where this becomes really exciting is when we combine feeds to get a clearer understanding of real-time people movement in a space, whether that's a train station, airport or even a busy street.

Fig. 2. Showing the head model performance in a crowded video scene

2.2   Technical Challenges

There are many issues when training a deep learning model to recognise particular objects. Shadow and reflections in the training videos from shiny floors, mirrors on walls and reflective railings, for example, all create artefacts the model needs to learn to ignore [5]. There is also the problems of scale, if a model has been over trained on one set of videos at one scale it may fail to recognise the same things at a different scale, in practical terms this means that training videos have to be tagged at all scales you are interested in if you want any hope of success. We include a Head model C++ code sample at Fig. 3.

```
void init_common()
{
    trace_color = rand_color();
    num_detections = 0;
    track_valid = false;
    ticks = 0;
    colors.reserve(500);
}


void update(const Rect2f &bbox, int frame_number)
{
    last_frame_seen = frame_number;
    bounding_box = bbox;
    update_frame_number = frame_number;
    num_detections++;
}


bool tick(size_t frame_number)
{
    // it might have updated this frame or not
    // can check by checking frameNumber and updateFrameNumber
    // if it has been updated then the crit. to recognize the same person has passed
    // or the person is new
    ticks++;
    bool detected = (frame_number == update_frame_number);

    kalman.tick(detected, bounding_box);
    bounding_box = kalman.predRect;

    Point2f center;
    center.x = kalman.predRect.x + kalman.predRect.width/2;
    center.y = kalman.predRect.y + kalman.predRect.height/2;
    trace.push_back(center);
    bbox_trace.push_back(kalman.predRect);

    if(ticks%10 == 0 && !colors.empty() && (frame_number - color_updated_framenumber) < 3)
    {
        dominant_colors = find_dominat_colors(colors, 3);
    }
    return detected;
}
```

Fig. 3. C++ code sample people tracking across frames

The head detection and tracking also needs to be able to run in real time, so the model and associated algorithms may work but just not be fast enough to deploy onto real CCTV camera feeds as the results will always lag. Ideally the solution should be capable of being installed on an actual IP camera with an embedded microcontroller and GPU. Other issues include the general deep learning problems of overall accuracy and over fitting the model.

2.3 Overcoming the Technical Challenges

The head model was created using primarily tensorflow and over-fitting was prevented by using some of the inbuilt tools. Tracking software was all written in C++ with extensive memory leak checking during the development process to enable fast real time operation.

In order to demonstrate results videos were augmented with head model and path results as well as raw jsonlines text output. In the augmented video red boxes represent the head model results, other coloured boxes represent the tracking of these across frames. The tracking results, a person's path across video frames, are represented by coloured lines following the boxes.

Multiple video scenarios in terms of crowd entries and exits demonstrated the head model's ability to cope with issues of people reflected in walls and railings as well as floor shadow. An advanced application of Kalman filters before the video frame is shown to the head model removed most of the shadow artefacts [8, 11]. Multiple exit/entry points can be covered in a single video, for example people moving to the right either from a stairwell or an entrance on the top. The application can also cope well with sudden changes from static to dynamic movement, for example, with people queuing and then boarding a train.

3   Multiple 360-Degree CCTV 3D Mixed Reality Rendering in St Pancras International

3.1   Combine Multiple CCTV and Other Feeds Rendered in a VR (Virtual Reality) Scape

In this section we describe how to combine multiple CCTV and other feeds for a single environment, and render it all in a VR (Virtual Reality) scape that can also be extended to AR (Augmented Reality) depending on the end use case. The environment could be a shopping precinct, railway station or airport.

Real world problems, for example, threat detection, lifesaving and labour time spent watching CCTV could all be automated away once a bespoke mapping and alerting system had been setup for the space in question The potential of mixed input feeds combined with the mapping of a whole space has not been sufficiently utilized previously, in particular to demonstrate the power of a converged video and mapping data has never been fully realized [9]. Taking standard video feeds from 360-degree CCTV cameras placed in gangways, we modelled the St Pancras International train station in London (Fig. 4). Similar to the city subway, we have the same head tracking and coat colour recognition as before. In addition, however, we have combined data from multiple camera feeds with a webGL model to create a VR view of the station that shows people movement in real time. Real people are represented by avatars in real-time in this model. This model is interactive and the users can change their viewpoint to look at any angle. The movement of the avatars matches what can be seen in the camera.

The VR model contains head model data imported from CCTV videos combined with a model of the entire space. People in the areas covered by CCTV are colour coded according to their jacket colour which has come from the head model video data. Via the browser-based interface, a user can easily switch views correlated to camera positions as well as move around the entire area as required.

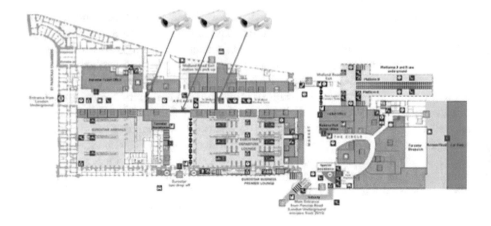

Fig. 4. CCTV camera placement in gangways at St Pancras International

So that data is in a usable state for export and blending we augment the CCTV video to compensate for spherical distortion before it is fed into the person tracking, this gives even more accurate position numbers that are perfect for feeding into the webGL model as a geospatial transform. Whilst this process effectively distorts the bodies of the people of we are tracking the associated deep learning models have proven are robust enough to withstand this distortion without a drop in overall accuracy. The process schematic used for rendering 360-degree CCTV 3D Mixed Reality is shown at Fig. 5.

Figures 6 and 7 are VR renderings from the webGL model.

In both Figs. 6 and 7, on the left side of the display is the crowd density meter which uses a separate crowd density algorithm to arrive at overall crowd size by taking as input a wide-angle camera. The output is correlated with the figures given by the head model and can be used to give an accurate number of people in shot at any given time. This allows us to give total numbers outside of the normal CCTV areas which are focused on specific gangways.

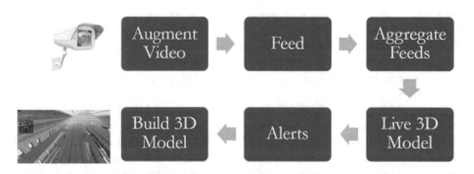

Fig. 5. Multiple 360-degree CCTV 3D Mixed Reality rendering in St Pancras process schematic

Fig. 6. The full model of St Pancras with the tracked areas on the left

Fig. 7. Model position correlated with one of the CCTV cameras

As well as illustrating the movements of individuals, the CCTV feeds to the VR model contain predictive information about potential events that can be used to create alerts in the display. A lot of the additional information relies on specific algorithms per use case, for example, a clustering algorithm works out when peoples' combined movements start to behave as a group which allows the display to highlight identifiable 'clusters' of people (the large green boxes in Fig. 8 below).

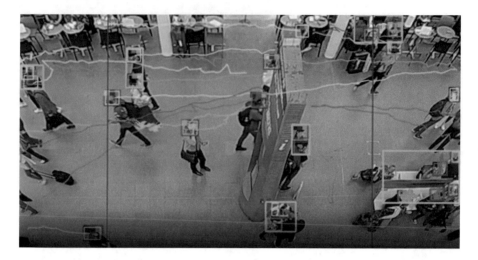

Fig. 8. Tracking video with correction for spherical distortion (note the floor tiles are perfect squares) as well as group tagging (the large green square tags)

The data feeds from the CCTV cameras were pre-aggregated by a software utility to minimise the size of file preloaded into a browser. Instead of every event the utility broke people positions' down to an average for every second, still plenty of granularity for a VR display.

By modelling an entire space, we are able to stitch together unique identifiers across CCTV feeds, really get a feel for overall movement, crowd control pain points, and custom alerts. In a live situation the model would be the first point of call with specific footage and alerting available to what would be an AR environment.

Detecting groups of people is already available in the feed. This package focuses on counting individuals but can also include some of MAL Nethra's other statistics that can be made available such as height and pose estimation (so bodies/skeletons plus heads plus person tracking), and can linkup to our pre-existing Nethra action detection for fight and other alerts as a single package [6].

3.2 The Challenges of Building a Browser Based VR Model

To map the area, we needed to build a model in webGL, a browser based technology, see Fig. 9 for code sample. For that to be viable in terms of rendering speed, it can't have too many vectors in its' construction. More than one technique was tried and tested, including the use of 3D cameras, these contained many visual artefacts (it was impossible to remove people and objects we didn't want in the frame whilst taking the footage) and by the time these were cleaned up the resulting point clouds models still had too many vectors to be viable for web-based demo.

The data feeds from each camera (in text based jsonlines format) output for people positions and attributes included all the events for every person tracked, to display and process all the results from the feeds seemed to require too much processing power to be viable for a web-based demo. An aggregation process was applied so data points for every second was applied instead of every frame. Our usage of 3D cameras to create the model could have eventually been overcome if we had more control over the space we were modelling (like we would have had for a specific customer deployment). The model we decided to build from scratch based on significant resources being already online for the target environment. The area of St. Pancras International station is already clearly mapped in Google streetview for example. St Pancras also makes an authentically pleasing backdrop compared to somewhere like Euston Station in London for example which has next to no pre-existing resources that can be found online. By building the model from scratch the problem of too many vectors was overcome.

```
function animate() {

    requestAnimationFrame( animate );

    var r = clock.getElapsedTime();

    roundedTime = Math.round(r);

    if (roundedTime < maxTime) {

        if (tempTime !== roundedTime) {

            pplCountTemp = pplCount[Math.floor(roundedTime/12)];
            updateParams();
            movePeople(roundedTime);
            tempTime = roundedTime;
        }
    }

    render();
    stats.update();
    TWEEN.update();
}

function render() {
    renderer.render( scene, camera );
}
```

Fig. 9. Sample webGL code animating the person avatars based on clock time

4   Conclusion

We have described an AI driven model based on the Nethra Video Analytic platform, optimised for overhead detection and designed specifically for CCTV and which can detect and categorise people from any angle. The model runs in real time in crowded and noisy environments and can be installed on devices as in edge analytics or applied directly to existing video feeds. By mapping an entire space, we link together individual camera feeds and data points to calculate the total number of people to assist with capacity planning and to pin point bottlenecks in people flows. In the augmented video red boxes represent the head model results, other coloured boxes represent the tracking of these across frames. We show how we solved of the technical challenges when training a deep learning model to recognise particular objects.

Solving the problem of aggregating multiple 360-degree video camera feeds into a single combined rendering, we further describe a novel use of interactive Virtual Reality, modelling the St Pancras International train station in London to track people movement in real time. The VR model contains head model data imported from CCTV videos combined with a model of the entire space. Real people are represented by avatars in real-time in the model. The movement of the avatars exactly mirrors what can be seen in the cameras.

References

1. http://spokensanskrit.org/index.php?mode=3&script=hk&tran_input=nethra&direct=se&anz=100
2. Frangou, G.J.: Apparatus for Controlling a Land Vehicle which Is Self-Driving or Partially Self-Driving. U.S. Patent No. 9,645,576 Chinese Patent 105189237, Japanese Patent 2016520464, Israel Patent 241688, European Patent Application 2976240, Korean Patent Application 20150138257, PCT Patent Application 2014147361 (2013)
3. Cao, Z., Simon, T., Wei, S.-E., Sheikh, Y.: Realtime Multi-Person 2D Pose Estimation using Part Affinity Fields. The Robotics Institute, Carnegie Mellon University. arXiv:1611.08050v2 [cs.CV]. Accessed 14 Apr 2017
4. Karpathy, A., Toderici, G., Shetty, S., Leung, T., Sukthankar, R., Fei-Fei, L.: Large-scale Video Classification with Convolutional Neural Networks (2014)
5. Marcenaro, L.: Access to Data Sets, ERNCIP Thematic Group on Video Surveillance for Security of Critical Infrastructure. European Commission, JRC Science Hub, University of Genoa, Italy (2016)
6. Arroyo, R., Javier Yebes, J., Bergasa, L.M., Daza, I.G., Almazan, J.: Expert video surveillance system for real-time detection of suspicious behaviors in shopping malls, expert systems with applications (2015)
7. Bleser, G., Stricker, D.: Computer Vision: Object and People Tracking, Proceedings of Seminar and Project, Winter semester 2013/14, University of Kaiserslautern and DFKI GmbH
8. Romera, E., Bergasa, L.M., Arroyo, R.: A real-time multi-scale vehicle detection and tracking approach for smartphones. In: 2015 IEEE 18th International Conference on Intelligent Transportation Systems (2015)

9. Romera, E., Alvarez, J.M.A., Bergasa, L.M., Arroyo, R.: Efficient ConvNet for real-time semantic segmentation. In: 2017 IEEE Intelligent Vehicles Symposium IV (2017)

10. Maddalena, L., Petrosino, A.: Background subtraction for moving object detection in RGBD data: a survey. J. Imaging **4**, 71 (2018)

11. Laaraiedh, M.: Implementation of Kalman Filter with Python Language. The Python Papers (2009)

Food, Energy and Water (FEW) Nexus Modeling Framework

Yemeserach Mekonnen, Arif Sarwat$^{(\boxtimes)}$, and Shekhar Bhansali

Florida International University, Miami, FL, USA
{ymeko001,asarwat,Sbhansa}@fiu.edu

Abstract. As the global population soars from today's 7.3 billion to an estimated 10 billion by 2050 of which 400 million to the US, the demand for Food, Energy and Water (FEW) are expected to more than double. Such an increase in population and consequently, in the demand for FEW resources will undoubtedly be a global challenge. Food, energy and water for smart sustainable cities involve a multi-scale challenge problem. The three dynamic interacting infrastructures require a mathematical framework for analyzing such a large complex system. Technology innovation at the nexus and quantifying the nexus are two critical solutions in this research area. This paper focuses on quantifying and modeling the nexus by proposing a Leontief input-output model. It further uses network analysis to investigate the networks of FEW interdependencies from the input-output model.

Keywords: FEW systems · Modeling · Nexus · Energy · Input-output · Graph theory · Network analysis

1 Introduction

Currently, almost 70% of the global fresh water is being used of agriculture and along being used to transport and produce energy in different forms [1]. The demand for water is set to increase to 55% by 2050 [2]. Similarly, 30% of total global energy consumption is spent on producing, transporting and distributing food as well as in the application of pumping, extracting, treating and transporting water [3,4]. Global energy consumption is projected to increase to 80% by 2050 [5,6]. As the demand for food soars to 60% by 2050, food security along with water and energy supply poses key issues in the availability, accessibility, and utilization of these resources. The interlinkage and interdependency of the water, energy and food systems known as the nexus is a key concern globally for all involved stakeholders in pursuing and meeting their sustainable development strategies. Increased population growth, economic development, and urbanization are the driving factors in the demand for food, energy and water resources

The work is an outcome of the research supported by the U.S. National Science Foundation under the grant number 1745829 and CAREER-1553494.

© Springer Nature Switzerland AG 2020
K. Arai et al. (Eds.): FTC 2019, AISC 1069, pp. 346–364, 2020.
https://doi.org/10.1007/978-3-030-32520-6_28

more than ever [7] as illustrated in Fig. 1. Another challenge has been urban population expansion where the solution to solve food demand has to be more innovative with the use of technology such as building modular and vertical farming accounting for land scarcity. The conventional way of thinking about these intertwined problems focuses on the "peace-meal approach" where decisions are made in one of the nexus areas of water, energy, and food without making an allowance for the consequences on the other areas [5–7]. The nexus approach provides decision makers with better information through optimization of synergies and trade-offs. The three main objectives of the nexus approach are addressing resource scarcity and security as opposed to environmental impacts, developing synergy and collaboration between stakeholders directly influencing the nexus, and development of modeling tools to support integrated decision making [7,8]. In this paper, the third objective will be further explored. Quantitative assessment of specific interventions, with the aim of analyzing how they perform from a nexus perspective. The paper explores the FEW nexus using the Leontief input-output model to quantify the FEW interdependencies. It further presents a graph theory framework to analyze a FEW interdependence matrices as a network.

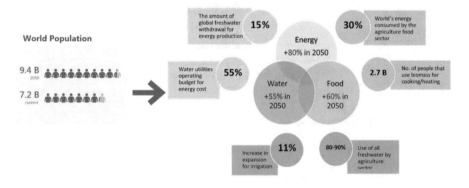

Fig. 1. The future of FEW resources [2].

The paper is organized as follows: Sect. 2 presents the nexus by delving into a detailed coupled dependency of each system. Section 3 describes the FEW nexus challenges. Section 4 discusses the modeling challenges and presents the Leontief IO model as applied to the FEW nexus.

2   The Nexus

In addition to increased population, rapid urbanization and industrialization, further complicating these resources challenges are that they are interlinked. Water is needed to produce energy; energy is essential in sourcing, treating and distributing water; and both the use of water and energy are required to produce food as shown in Fig. 2. The nexus approach encourages addressing these resources' scarcity jointly, as decisions taken with regard to one resource are likely to influence the other two. Recent works have shown that the FEW security can be improved through an integrated management approach that will bring all involved sector which is called nexus [7]. A nexus approach can also support the transition into a circular economy where renewable sources utilization and byproduct/ waste recycling are encouraged. Although it is a long way from achieving food, water and energy security for all, the solution can be facilitated through an approach that integrates management and governance across all sectors and scales.

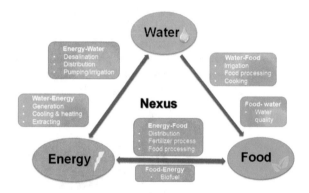

Fig. 2. The FEW nexus

2.1   Energy-Food Nexus

Agricultural production consumes energy directly in the form of fuels for land preparation, crop and pasture management, and transportation or electricity supply. Indirectly, the use of fertilizers and pesticides which are energy intensive inputs. A huge chunk of the energy is spent on the food supply chain which includes the processing and distribution of the food products [1, 2]. Energy link to food by sector varies among developed and developing countries stated in Fig. 3. To alleviate the food security problem, some countries are heavily invested in the use of bioenergy. In the face of climate change and rising energy security, the demand for more viable renewable energy use for food is at a critical point. Over 20% of total Green House Gas (GHG) emission comes from the food sector [9].

2.2   Water-Energy Nexus

The use of water for energy currently accounts to 8% of global water withdrawals, in industrialized countries, this number reaches to 45% in developing countries (Europe) [10]. Water is used for extraction, mining, processing, refining and residual disposal of fossil fuels and growing biofuels for energy generation. Water is used in renewable and fossil fuel energy sources. Bio-fuels and

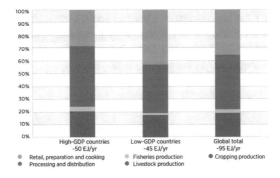

Fig. 3. Direct and indirect energy use link to food sector [1]

hydropower are the two renewable energy sources that require a large number of water [9]. However, Biofuels are more water intensive than fossil fuels using 10,000–100,000 L/GJ of energy [11]. Comparing this number to the fossil fuel production in oil and gas, fossil fuels only use 0.01% of biofuel's water consumption [12]. Overall energy sources that require the use of water are Biofuel, Hydropower, non-conventional fossil fuels such as fracking [13]. The use of water in the electricity market for hydropower production is evident as water is used as a primary source for generating power as shown in Fig. 4. Hydropower currently provides 16% of global electricity generation accounting to 86% of the global renewable energy. This number is very far below the feasible potential like in Africa where only it taps 5% of its potential [14,15]. Another renewable energy sources that uses water is a Photo-voltaic thermal (PVT) unit that still widely used in parts of the world [16,17].

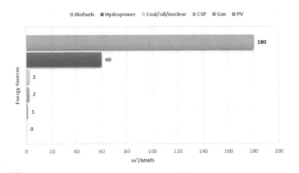

Fig. 4. Water use by various energy source in electricity production [9].

Energy is required for lifting, moving, distributing and treating water. The majority of energy which is about 40% total energy is used for pumping groundwater. Energy is also utilized in the desalination process which is expected to grow by 500% in 2030 especially in Asia [18]. In addition, energy is used directly and indirectly in irrigation practices in large scale farming practices. Generally, various systems of irrigation are currently in practice such as rain-fed agriculture, sprinkler, and drip irrigation. Drip irrigation practices are more energy intensive since the water must be pressurized [19].

2.3 Food-Water Nexus

The demand for a FEW resources is estimated to grow by 30–50% over the next two decades due to economic and population growth [20]. Food is the largest consumptive of water use. Agricultural production is projected to increase by 60% in 2050 causing an increased water consumption for irrigation to 11% [21]. This increase is especially noticed in an area where water is already scarce not fulfilling the demand. Growth in agricultural production has to meet the demand for feeding 9 billion population in 2050 by increasing crop yields and expanding arable land areas. This is where innovative methods are adapted to provide sustainable solutions to increase crop yield productivity. With the advent of IoT and sensor technologies, it has become possible to monitor crop health, yield, environmental parameters which further can be used in future prediction [22–24]. In addition, to meet such demand the use of fertilizer and pesticides has a direct effect on water quality through pollution. Water contamination from a discharge of pollutants originating from pesticides and fertilizers as a result of poor agricultural practices is one of the risks associated with the food-water nexus. Furthermore, one of the big issues facing the food system more than the shortage is the accessibility of existing food reaching to consumers [1]. In addition, the demand for agricultural goods is directly dependent on consumption patterns, market variability, policies, and the economy. Almost 40% of food produced is wasted in transportation from distribution to consumers [10]. This, directly and indirectly, means the waste of water and energy that was embedded in the production, transportation, and processing of this food.

3 FEW Challenges

3.1 Towards a Circular Economy

There are many inputs and outputs of the agriculture, water and food systems that should be designed and managed to minimize inputs and inequality, and maximize outputs. By closing the loop and adopting a more circular economy, the system will be immune to wastage and negative environmental impacts. Targeting the life cycles of a FEW resources, a solution to close the loop in the urban and rural communities can be implemented. Managing the FEW systems from an integrated perspective will be key to the successful loop closure where waste products from one resource can be used as an input in other resources. For example, food waste can be used as a biofuel as an energy source to power irrigation or satisfy other energy demands. However, this requires diverse stakeholders working together across various sectors in terms of governance and policy. Currently, with the advent of technology and its impact on shaping the future of the smart city policy, major global alliances are pushing the idea from "linear to circular" economy [25]. The idea of a circular economy that closes the loop across the production of FEW systems is presented in Fig. 5. Integrative management of wastes in a FEW systems is presented in [26] by averting one system's by-products to satisfy the need of another. It further explains how different FEW systems can be optimized to design a closed-loop production system.

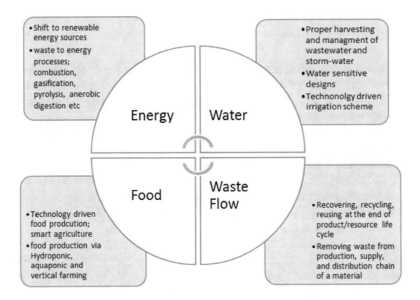

Fig. 5. Circular economy linkage [3].

3.2   Spatio-Temporal Variation in FEW Systems

There is a divide between time and space among the supply and demand of FEW resources. The use of water and energy is embedded in the production and delivery of food which makes it hard to quantify. In addition to this, the spatial disparity in food is observed in changing demands between the urban and rural areas. Harvested food is now transported for longer duration and making the supply chain process complex as to preserve and make the food fresh especially on green produces, fruits and seafood. Although food waste can be eliminated as a result of such complex supply chain processes and it solves the spatio-temporal disconnect between food produced and food waste, it intensifies the demand for energy [27]. In [28], energy use in the agricultural sector is linked to increased drought year by 17.5%, in part as higher energy is needed to maintain a chilled transportation within the supply chain of food. The spatio-temporal disconnect in water and energy sectors vary in frequency and complexity. Water demand and supply divides are in seasonal and geographic location. For energy, the demand for electricity is at its peak in the afternoon where in most cases currently it is met by fossil fuels. The push in implementing renewable energy sources like solar and wind presents a challenge as these resources can be intermittent and may not supply the peak time demand [29]. Alleviating the spatio-temporal disparity in FEW systems require innovative solution that will be able to address the integrative optimization of these resources.

4   Modeling Review

The FEW nexus modeling involves and requires the multi-scale level challenge that comes with various dimensionality and time scale variance within the systems. This presents unpredicted consequences and complexity to decision making. The three dynamic interacting infrastructures require a mathematical framework for analyzing such a large complex system. Part of the challenges for the modeling is directly related to the data acquisition gaps such as visualization of these data models and gaps. Another case is most of the data collected are over different space and time scales with missing values. There are different types of models depending on the sectors, stakeholders and area of the expertise, models can be data-driven [30], process-driven [31] or cost-driven [32].

The increasing acceptance of the interconnectedness between the FEW systems has led to the demand for a tool or methodology to model these systems as a nexus. One of the biggest challenges in the FEW nexus modeling is the lack of interlinkages data, access to data within private sector stakeholders (e.g. energy data), the inconsistency of collected data with the requirements of the nexus tool and nonexistence of interdisciplinary exchanges [23]. Due to the early stage research in FEW nexus, most of the analytical tools and the overall research focuses on a macro level. However, the purpose of this paper is to direct the FEW nexus modeling approaches from a bottom to top-down scale through the understanding of the FEW interactions on a small scale smart farm test-bed.

Data accessibility and acquisition is a huge challenge in nexus modeling. Often times, there exists a lack of data on the footprint among the FEW resources making hard to quantify and model the nexus [30]. Understanding this gap, the first segment of this research is focused on designing a smart farm prototype where all the three FEW elements are involved [22]. The test-bed is optimally designed to maximize crop yield, minimize energy consumption and extreme environmental effects through real-time sensor data. From this physical model, the data collected will be structured into three different FEW components through a robust data acquisition infrastructure. Data obtained from here will be used in the model that will be introduced in Sect. 4.1.

4.1   Leontief Input-Output Model

Input-Output (IO) analysis was first proposed by Nobel prize economist Wassily Leontief. The Leontief input-output analysis has been used to explore the direct and indirect intersectoral linkages systems specifically for economic and cost-driven models [33]. The model is great for the economy that has a number of interrelated industries whose output is dependent on the output on other industries. It has been used in supply chain linkage, energy input-output, environmental pollution impact, and many other interdependent applications. The algorithm contains three major components:

(1) An intersectoral use matrix from sector i to sector j usually denoted as Z.
(2) A final demand vector matrix f of sector i.
(3) Total produced or utilized vector matrix x noted as output.

Given there are n sectors, the way sector i distributes to the other sectors can be expressed:

$$x_i = Z_{i1} + ... + Z_{ij} + Z_{in} + f_i = \sum_{j=1}^{n} Z_{ij} + f_i \tag{1}$$

where x_i is the total of sector i output. For multi-sectoral instances the output economic quantity expressed below for $i = 1, 2,... n^{th}$ sectors:

$$x_1 = Z_{11} + ... + Z_{1j} + Z_{1n} + f_1$$

$$\vdots$$

$$x_i = Z_{i1} + ... + Z_{ij} + Z_{in} + f_i$$

$$\vdots$$

$$x_n = Z_{n1} + ... + Z_{nj} + Z_{nn} + f_n \tag{2}$$

The above system of equation can be expressed in matrix form as follows:

$$\underbrace{\begin{bmatrix} x_1 \\ \vdots \\ x_n \end{bmatrix}}_{\text{output}} = \underbrace{\begin{bmatrix} Z_{11} & ... & Z_{1n} \\ \vdots & \ddots & \vdots \\ Z_{n1} & ... & Z_{nn} \end{bmatrix}}_{\text{component linkage}} + \underbrace{\begin{bmatrix} f_1 \\ \vdots \\ f_n \end{bmatrix}}_{\text{demand}} \tag{3}$$

The direct input coefficient or technical coefficient is defined as:

$$a_{ij} = \frac{Z_{ij}}{x_j} \tag{4}$$

a_{ij} is a measure of the fixed relationship between sectors output and its input. Rearranging the above main output function of Eq. (2) with a_{ij}:

$$x_1 = a_{11}x_1 + ... + a_{1j}x_j + a_{1n}x_n + f_1$$

$$\vdots$$

$$x_i = a_{i1}x_1 + ... + a_{ij}x_j + a_{in}x_n + f_i \qquad \Rightarrow$$

$$\vdots$$

$$x_n = a_{n1}x_1 + ... + a_{nj}x_j + a_{nn}x_n + f_n$$

$$\begin{aligned} x - Ax &= f \\ (I - A)x &= f \qquad (5) \\ x = (I - A)^{-1}f &= Lf \end{aligned}$$

Where $L = (I - A)^{-1}$ is known as the Leontief inverse matrix indicating the total requirements. The Leontief inverse can further be used to predict future sector output based on final demand changes. If final demand for one sector increases, the final output (x) and intersectoral use (Z) can be calculated:

$$x^{new} = Lf^{new} \qquad (6)$$

$$Z^{new} = A\hat{x}^{new} \qquad (7)$$

4.2 FEW IO Mathematical Framework

The FEW nexus can be represented with the IO model. An equivalent of Z vector matrix will explore the interdependence of one resource in the other FEW resources i.e. F–F, F–E, F–W, E–F, E–E, etc. The A matrix will indicate the different sources' intensity of use in others. Finally, the L matrix also known as the Leontief inverse will be used to track and find how an increase in demand will change final output and intercomponent usage. In this paper will use the input-output model to investigate the interlinkage of FEW nexus. Using the foundation of the input-output model, the FEW nexus will be framed as follows:

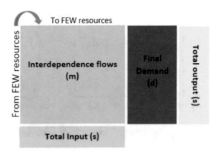

Fig. 6. FEW IO framework

Furthermore the above framework in Fig. 6 can be explicitly described in Table 1 below. The interdependence flow from one source to the other is denoted

Table 1. FEW interlinkage based on IO model

	FEW Intersectoral Inputs			Final	Total
	Food (F)	Energy (E)	Water (W)	Demand (d)	Output (y)
Food (F)	m_{ij}^{ff}	m_{ij}^{fe}	m_{ij}^{fw}	d_i^f	y_i^f
Energy (E)	m_{ij}^{ef}	m_{ij}^{ee}	m_{ij}^{ew}	d_i^e	y_i^e
Water (W)	m_{ij}^{wf}	m_{ij}^{we}	m_{ij}^{ww}	d_i^w	y_i^w

by the m matrices with $i \times j$ dimensions. d_i^f, d_i^e, d_i^w are the final demand vector each sector respectively. The total output matrix for each resources are denoted by y_f^i, y_e^i, y_w^i. Mathematically the FEW component linkage matrix which is equivalent to the inetersectoral inputs mentioned in Table 1 can be expressed as follow in Eq. (7). Here the V_i and the U_i are simply row and column labeling indicating the different FEW sources involved.

$$
\begin{array}{c}
\begin{array}{cccc} U_1 & U_2 & U_{j-1} & U_j \end{array} \\
\begin{array}{c} V_1 \\ V_2 \\ \\ V_{i-1} \\ V_i \end{array}
\left[
\begin{array}{cccc}
m_{11} & m_{12} & \dots & m_{1,j-1} & m_{1j} \\
m_{21} & m_{22} & \dots & m_{2,j-1} & m_{2j} \\
\vdots & \vdots & \ddots & \vdots & \vdots \\
m_{i-1,1} & m_{i-1,2} & \dots & m_{i-1,j-1} & m_{i-1,j} \\
m_{i1} & m_{i2} & \dots & m_{i,j-1} & m_{i,j}
\end{array}
\right]
\end{array}
\tag{8}
$$

$$\underbrace{\qquad\qquad}_{\text{FEW component linkage}}$$

Where, both V_i and U_j are $\in \{F_1, ..., F_p, E_1, ..., E_q, W_1, ..., W_r\}$, which are different food, energy, and water sources. p, q, and r are the number of FEW sources, respectively. $\{V_1, V_2, ..., V_{i-1}, V_i\}$ are the different FEW sources being used in the same $\{U_1, U_2, ..., U_{j-1}, U_j\}$ sources creating m_{ij} interdependence flow. For i^{th} and j^{th} sources, m_{11} is the use of V_1 source in U_1; m_{21} is the use of V_2 source in U_1, m_{ij} is the use of ith source in j^{th} source, etc.

Applying the Leonteif equation Eq. (3) to Table 1 framework, the general FEW system IO can be expressed as follows:

$$
\underbrace{
\begin{bmatrix}
m_{11} & m_{12} & \dots & m_{1j} \\
m_{21} & m_{22} & \dots & m_{2j} \\
\vdots & \vdots & \ddots & \vdots \\
m_{i1} & m_{m2} & \dots & m_{ij}
\end{bmatrix}
}_{\text{Interdependence flow matrix}}
+
\underbrace{
\begin{bmatrix} d_f \\ d_e \\ d_w \end{bmatrix}
}_{\text{demand}}
=
\underbrace{
\begin{bmatrix} y_f \\ y_e \\ y_w \end{bmatrix}
}_{\text{output}}
\tag{9}
$$

However, for a given computation it is possible to have many different sources of food, energy or water. Given p number of food sources, q number of energy sources, and r number of water sources, Eq. (9) expression can be mathematically formulated below as the total output of each FEW sources. Therefore, the total food source output given p number of food sources based on Leontief IO Eq. (2)

formulation is:

$$\sum_{j=1}^{p} m_{ij}^{ff} + \sum_{j=1}^{q} m_{ij}^{fe} + \sum_{j=1}^{r} m_{ij}^{fw} + \sum_{i=1}^{p} d_i^f = \sum_{i=1}^{p} y_i^f, \text{ for } i = 1, 2, ..., p \qquad (10)$$

Similarly, given q number of energy sources, the total energy source output is:

$$\sum_{j=1}^{p} m_{ij}^{ef} + \sum_{j=1}^{q} m_{ij}^{ee} + \sum_{j=1}^{r} m_{ij}^{ew} + \sum_{i=1}^{q} d_i^e = \sum_{i=1}^{q} y_i^e, \text{ for } i = 1, 2, ..., q \qquad (11)$$

Finally, the water source output given r number of water sources is:

$$\sum_{j=1}^{p} m_{ij}^{wf} + \sum_{j=1}^{q} m_{ij}^{we} + \sum_{j=1}^{r} m_{ij}^{ww} + \sum_{i=1}^{r} d_i^w = \sum_{i=1}^{r} y_i^w, \text{ for } i = 1, 2, ..., r \qquad (12)$$

The above three equations can be rewritten with the technical coefficient as indicated in Eq. (4). The technical coefficient measures the direct linkage of each source revealing the dependence of one source on another. It is the interdependence flow divided by the total output of the receiving sector. Ideally, within the FEW input-output flow there will be nine FEW technical coefficient expressed as follows:

food for food, $\gamma_{ij}^{ff} = \frac{m_{ij}^{ff}}{y_j{}^f}$; food for energy, $\gamma_{ij}^{fe} = \frac{m_{ij}^{fe}}{y_j{}^e}$; food for water, $\gamma_{ij}^{fw} = \frac{m_{ij}^{fw}}{y_j{}^w}$; energy for food, $\beta_{ij}^{ef} = \frac{m_{ij}^{ef}}{y_j{}^f}$; energy for energy, $\beta_{ij}^{ee} = \frac{m_{ij}^{ee}}{y_j{}^e}$; energy for water, $\beta_{ij}^{ew} = \frac{m_{ij}^{ew}}{y_j{}^w}$; water for food, $\alpha_{ij}^{wf} = \frac{m_{ij}^{wf}}{y_j{}^f}$; water for energy, $\alpha_{ij}^{we} = \frac{m_{ij}^{we}}{y_j{}^e}$; water for water, $\alpha_{ij}^{ww} = \frac{m_{ij}^{ww}}{y_j{}^w}$.

Replacing the interdependence flow variables with the above FEW technical coefficient:

$$\sum_{j=1}^{p} \gamma_{ij}^{ff} y_j^f + \sum_{j=1}^{q} \gamma_{ij}^{fe} y_j^e + \sum_{j=1}^{r} \gamma_{ij}^{fw} y_j^w + d_i^f = y_i^f, \text{ for } i = 1, 2, ..., p \qquad (13)$$

$$\sum_{j=1}^{p} \beta_{ij}^{ef} y_j^f + \sum_{j=1}^{q} \beta_{ij}^{ee} y_j^e + \sum_{j=1}^{r} \beta_{ij}^{ew} y_j^w + d_i^e = y_i^e, \text{ for } i = 1, 2, ..., q \qquad (14)$$

$$\sum_{j=1}^{p} \alpha_{ij}^{wf} y_j^f + \sum_{j=1}^{q} \alpha_{ij}^{we} y_j^e + \sum_{j=1}^{r} \alpha_{ij}^{ww} y_j^w + d_i^w = y_i^w, \text{ for } i = 1, 2, ..., r \qquad (15)$$

where γ, β, and α are the FEW nexus technical coefficient associated with food, energy and water, respectively. Furthermore, the FEW systems can be expressed as follows:

$$\underbrace{\begin{bmatrix} \gamma^{ff} & \gamma^{fe} & \gamma^{fw} \\ \beta^{ef} & \beta^{ee} & \beta^{ew} \\ \alpha^{wf} & \alpha^{we} & \alpha^{ww} \end{bmatrix}}_{\text{FEW coefficient matrix}} \begin{bmatrix} y^f \\ y^e \\ y^w \end{bmatrix} + \begin{bmatrix} d^f \\ d^e \\ d^w \end{bmatrix} = \begin{bmatrix} y^f \\ y^e \\ y^w \end{bmatrix} \qquad (16)$$

Let A be the $n \times n$ denote the FEW technical coefficient matrix, Eq. (16) can be arranged in a matrix notation,

$$AY + d = Y \tag{17}$$
$$Y = Ld$$
$$\text{where, } L = (I - A)^{-1} \tag{18}$$

If the demand of the FEW resources (d) are prespecified and assuming the per unit amount of the food, energy and water outflow used in sector j which is the A matrix remains unchanged, new total FEW output can be calculated:

$$Y' = Ld' \tag{19}$$

where Y' and d' are the new total output, and changes in final demand, respectively.

Furthermore, the changes in interdependence flow matrix Z' due to changes in final demand (d') can be traced and calculated per Eq. (7),

$$Z' = AY' \tag{20}$$

The Leontief inverse matrix summarizes the network effects generated when final output changes. A single coefficient of matrix entails the direct and indirect effects created in resource i to supply a single unit of final demand for resource j. The challenge of such a model in a FEW resources is the absence of flow data for all components in such application. In addition the input-output system is an ideal test-bed for network science. The future work will further extend the FEW the technical coefficient in understanding the FEW nexus from a network perspective using graph theory [34].

5 Data

In this example a hypothetical FEW interdependence flow matrix is provided in Table 2.

(a) Food inflow includes two sources (f_1) vegetables and (f_2) is poultry all measured in $ton/year$.
(b) Energy sources include electricity from solar (e_1), and diesel (e_2) measured in $toe/year$.
(c) Water inflow include sources from rain harvested (w_1), and dehumidification (w_2) measured in $m^3/year$.

Using the interdependence flow values (m_{ij}), the technical coefficient values (A_{ij}) can be computed. The total output (y_1) for f_1 is the sum of all its outflow to all j^{th} resource and the final demand (d_1).

The technical coefficient matrix that represent an ith resource footprint in the production or use of jth resource is presented in Table 3.

Table 2. A hypothetical FEW input-output table

FEW resources supply (Outlfow)	FEW resources consumption inflow						Final demand (d)	Total output (Y)
	Food (ton)		Energy (toe)		Water (m^3)			
	f_1	f_2	e_1	e_2	w_1	w_2		
f_1	150	500	50	25	75	200	45	1000
f_2	200	100	400	200	100	150	50	1150
e_1	300	500	50	60	40	120	200	1070
e_2	75	100	60	200	250	140	500	825
w_1	50	25	25	150	100	145	50	495
w_2	90	150	420	500	200	140	120	1500

Food for Food: The total output of all food resources is m^{ff} 2150 ton/year. The total footprint of food outflow into other food sources γ^{ff} is 2.19.

Food for Energy: The total footprint food use in energy is γ^{fe} 0.69 ton/toe. This means 0.69 ton of food have been used per unit toe of energy.

Food for Water: The total direct effect of food in water sources is γ^{fw} 0.59 ton/m^3.

Energy for Food: The total footprint of energy in food use is 2.31 toe/ton. Solar energy use in irrigated vegetable has the highest footprint with β^{ef} 1.43 toe per unit ton of vegetables.

Energy for Energy: The total use of energy to energy is β^{ee} 0.42 toe.

Energy for Water: Energy for water footprint is β^{ew} 0.76 toe/m^3. This scenario is evident in diesel hybrid microgrids, PV thermal plants and hydro power plants.

Water for Food: Food sources use of water (α^{wf}) is 0.82 m^3/ton.

Water for Energy: Water footprint in the energy use is at α^{we} 1.204 m^3/toe.

Water for Water: The use of water for water intensity is α^{ww} 0.796. The total water outflow in the production of all the other water resources m^{ww} is 585 m^3.

In the FEW input-output framework, computing the total FEW requirement in each sources which will be referred in this article as FEW footprints or intensity is similar to computing the total cost requirement or Leontief inverse of the traditional input-output model. The different food, energy, and water sources intensity are combined together and summarized in Table 3. These values are the direct FEW requirement matrix (A_{ij}). Table 4 presents the water and energy source intensity in food production. The intensity values indicate the requirement of resources in per unit linkage in the production of vegetables and poultry. Furthermore, the intensity interconnection among the FEW sources can further be visualized as a network of three systems as presented in Fig. 7.

Table 3. FEW intensity coefficients

Nexus	FEW intensity (γ, β, α)
F to F	0.84 ton/ton
F to E	0.524 ton/toe
F to W	0.537 ton/m^3
E to F	0.859 toe/ton
E to E	0.283 toe/toe
E to W	0.693 toe/m^3
W to F	0.28 m^3/ton
W to E	0.841 m^3/toe
W to W	0.726 m^3/m^3

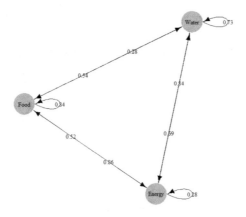

Fig. 7. Graphical representation

The FEW Leontief inverse matrix is computed in R programming based on the mathematical framework given in Sect. 4.2. These values shown below in Table 5 are also known as multipliers that captures in each of its element all of the infinite series of round by round direct and indirect effects that new final demands have on the outputs of each FEW resources. It will thereby enable to forecast for new total output (Y') and the new interdependence flow (m'_{ij}) as

Table 4. Water and energy footprints in food production

Footprints	Vegetables (f_1)	Poultry (f_2)
β_{1j}^{wf} (m³/ton)	0.048	0.021
β_{2j}^{wf} (m³/ton)	0.086	0.125
α_{1j}^{ef} (toe/ton)	0.287	0.417
α_{2j}^{ef} (toe/ton)	0.072	0.083

Table 5. The FEW Leontief inverse matrix values

	f_1	f_2	e_1	e_2	w_1	w_2
f_1	2.12	1.65	0.99	0.98	1.72	0.8
f_2	1.26	2.49	1.24	1.15	1.9	0.82
e_1	1.35	1.79	2.05	1.05	1.77	0.8
e_2	0.75	0.95	0.68	1.9	1.74	0.6
w_1	0.44	0.53	0.39	0.55	1.94	0.39
w_2	1.35	1.75	1.44	1.64	2.56	1.98

noted in Eqs. (19) and (20). The column and row sums of this matrix are a measure of the total backward and forward linkage of the FEW resources.

Assuming there will be a demand increase of 60% in poultry (f_2), 80% in solar energy (e_1), and 55% in dehumidification (w_1) as a result of increase in population and other stress drivers. The new final total output can be computed as stated per Eq. (19). The percent change in the total output of all the FEW sources are presented in Fig. 8 with highest percent increase induced in solar energy (e_1) at 34%, poultry (f_2) at 27% and dehumidification (w_2) at 25%. Similarly, the percent change in the FEW interactions flow is presented in Fig. 9. These values are the delta between the original and new flow. The highest percent increase inflow is to solar energy (e_1).

6   Future Work

Complex networks can be used to characterize a series of different systems where multiple sources interact with each other through their relationships. This can be represented as nodes for resources and their relationship as edges. The FEW IO system describes a network, a mathematical object defined by a set of nodes or vertices and a set of edges connecting them. The FEW input-output networks as shown in Fig. 10 can be formulated where nodes are specific FEW resources and the edges as the interdependence flows (technical coefficient) between resources. The IO table which will be equivalent to is an adjacency matrix whose entry (i, j) represent the flow from node i to node j. The future work will use a FEW

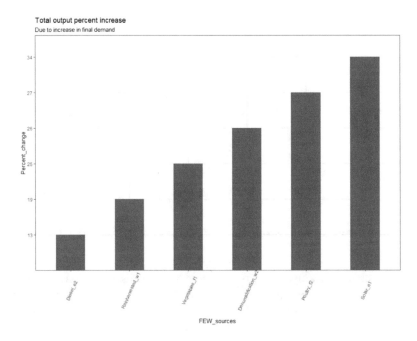

Fig. 8. Percent increase in the total FEW output as result of change in demand.

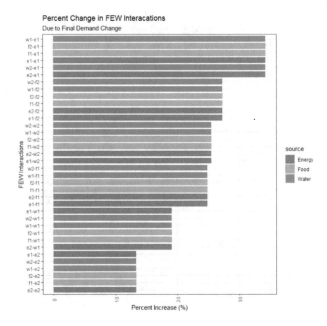

Fig. 9. A change in FEW flow interaction as a result of change in final demand.

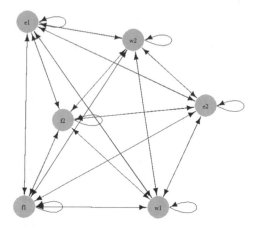

Fig. 10. A directed FEW interaction as a network.

nexus coefficients obtained from the IO model to represent the edges with magnitude and direction from one node to another. Each node will be specific FEW sources with size proportional to its total degree. The edges will be directed and weighted representing the use of one resource in the production of the other. The complex FEW networks will be analyzed on assortativity, clustering coefficient, and degree and strength distribution. Assortativity measures the propensities of nodes to attach among each other. The clustering coefficient measures the degree to which each node (FEW resources) in a graph tend to cluster together. The degree distribution and strength show the direct and total linkages of the FEW resources. All these properties will be vital understanding the complex FEW network systems.

7   Conclusion

The FEW nexus problems involve and require the multi-scale challenge problem that comes with various dimensionality and time scale variance within the systems which presents unpredicted consequences and complexity to decision making. The three dynamic interacting infrastructures require a mathematical framework for analyzing such a large complex system. FEW IO model can account for demand as a result of stressors. It allows for the computation of intersectoral usage of various FEW components. Technical coefficient allows for the direct and indirect effect of resource on each other with the ability to trace back. This paper presented a framework for a FEW nexus based on Leontief input-output model and how it further will be developed to a networked model as interdependent systems. Quantifying the FEW nexus is more important than ever as the demand for FEW resources will be exacerbated in the quest to achieve a sustainable smart city.

References

1. FAO: Energy Smart Food for People and Climate. Food and Agriculture Organization of the United Nations (2011)
2. IRENA: Renewable Energy in the Water, Energy and Food Nexus. International Renewable Energy Agency, January 2015
3. Lehmann, S.: Implementing the urban nexus approach for improved resource-efficiency of developing cities in southeast-asia. City Culture Soc. **13**, 46–56 (2018)
4. Hussey, K., Pittock, J., et al.: The energy-water nexus: managing the links between energy and water for a sustainable future (2012)
5. FAO: The water-energy-food nexus - a new approach in support of food security and sustainable agriculture. Food and Agriculture Organization of the United Nations (2014)
6. FAO: Walking the Nexus Talk: Assessing the Water-Energy-Food Nexus in the Context of the Sustainable Energy for all Initiative. Food and Agriculture Organization of the United Nations (2014)
7. Hoff, H.: Understanding the nexus. Background paper for the Bonn2011 conference (2011)
8. Weitz, N., Nilsson, M., Davis, M.: A nexus approach to the post-2015 agenda: formulating integrated water, energy, and food sdgs. SAIS Rev. Int. Aff. **34**(2), 37–50 (2014)
9. IEA: Water energy nexus: excerpt from the world energy outlook 2016 (2016)
10. SOER: State of the environment, cross-sectoral assessment of the agriculture, energy, forestry and transport sectors (2010)
11. Gerbens-Leenes, W., Hoekstra, A.Y., van der Meer, T.H.: The water footprint of bioenergy. Proc. Natl. Acad. Sci. **106**(25), 10219–10223 (2009)
12. World Economic Forum Water Initiative, et al.: Water Security: The Water-Food-energy-Climate Nexus. Island Press (2012)
13. Gaffigan, M.: Energy-Water Nexus: A Better and Coordinated Understanding of Water Resources Could Help Mitigate the Impacts of Potential Oil Shale Development. DIANE Publishing Company (2011)
14. Mekonnen, Y., Sarwat, A.I.: Renewable energy supported microgrid in rural electrification of Sub-Saharan Africa. In: 2017 IEEE PES PowerAfrica, pp. 595–599. IEEE (2017)
15. Jafari, H., Mahmoudi, M., Fatehi, A., Naderi, M.H., Kaya, E.: Improved power sharing with a back-to-back converter and state-feedback control in a utility-connected microgrid. In: 2018 IEEE Texas Power and Energy Conference (TPEC), pp. 1–6. IEEE (2018)
16. Olowu, T.O., Sundararajan, A., Moghaddami, M., Sarwat, A.: Fleet aggregation of photovoltaic systems: a survey and case study. In: 2019 IEEE Power Energy Society Innovative Smart Grid Technologies Conference (ISGT) (2019)
17. Jafari, H., Mahmodi, M., Rastegar, H.: Frequency control of micro-grid in autonomous mode using model predictive control. In: Iranian Conference on Smart Grids, pp. 1–5. IEEE (2012)
18. IEA: World Energy Outlook 2009 (2009)
19. Cooley, H., Christian-Smith, J., Gleick, P.H.: More With Less: Agricultural Water Conservation and Efficiency in California. Pacific Institute, Oakland, California, September, 30:2011 (2008). Accessed May
20. Kaddoura, S., El Khatib, S.: Review of water-energy-food nexus tools to improve the nexus modelling approach for integrated policy making. Environ. Sci. Policy **77**, 114–121 (2017)

21. FAO: Global agriculture towards 2050. High level expert forum - how to feed the world in 2050. Food and Agriculture Organization of the United Nations (2009)
22. Mekonnen, Y., Burton, L., Sarwat, A., Bhansali, S.: IoT sensor network approach for smart farming: an application in food, energy and water system. In: 2018 IEEE Global Humanitarian Technology Conference (GHTC), pp. 1–5. IEEE (2018)
23. Burton, L., Mekonnen, Y., Sarwat, A.I., Bhansali, S., Jayachandran, K.: Exploring wireless sensor network technology in sustainable okra garden: a comparative analysis of okra grown in different fertilizer treatments. CoRR, abs/1808.07381 (2018)
24. Onibonoje, M.O., Olowu, T.O.: Real-time remote monitoring and automated control of granary environmental factors using wireless sensor network. In: 2017 IEEE International Conference on Power, Control, Signals and Instrumentation Engineering (ICPCSI), pp. 113–118. IEEE (2017)
25. Michelini, G., Moraes, R.N., Cunha, R.N., Costa, J.M.H., Ometto, A.R.: From linear to circular economy: PSS conducting the transition. Procedia CIRP **64**, 2–6 (2017)
26. Davis, S.C., Kauneckis, D., Kruse, N.A., Miller, K.E., Zimmer, M., Dabelko, G.D.: Closing the loop: integrative systems management of waste in food, energy, and water systems. J. Environ. Stud. Sci. **6**(1), 11–24 (2016)
27. James, S.J., James, C.: The food cold-chain and climate change. Food Res. Int. **43**(7), 1944–1956 (2010)
28. Karan, E., Asadi, S., Mohtar, R., Baawain, M.: Towards the optimization of sustainable food-energy-water systems: a stochastic approach. J. Cleaner Prod. **171**, 662–674 (2018)
29. Olowu, T., Sundararajan, A., Moghaddami, M., Sarwat, A.: Future challenges and mitigation methods for high photovoltaic penetration: a survey. Energies **11**(7), 1782 (2018)
30. Eftelioglu, E., Jiang, Z., Ali, R., Shekhar, S.: Spatial computing perspective on food energy and water nexus. J. Environ. Stud. Sci. **6**(1), 62–76 (2016)
31. Garcia, D.J., You, F.: The water-energy-food nexus and process systems engineering: a new focus. Comput. Chem. Eng. **91**, 49–67 (2016)
32. Martinez, P., Blanco, M., Castro-Campos, B.: The water-energy-food nexus: a fuzzy-cognitive mapping approach to support nexus-compliant policies in Andalusia (Spain). Water **10**(5), 664 (2018)
33. Miller, R.E., Blair, P.D.: Input-Output Analysis: Foundations and Extensions. Cambridge University Press (2009)
34. Bondy, J.A., Murty, U.S.R., et al.: Graph Theory with Applications, vol. 290. Citeseer (1976)

End-to-End Drive By-Wire PID Lateral Control of an Autonomous Vehicle

Akash Baskaran[1($\boxtimes$)], Alireza Talebpour[2($\boxtimes$)], and Shankar Bhattacharyya[3($\boxtimes$)]

[1] Texas A&M University, College Station, TX 77840, USA
akashbaskaran@tamu.edu
[2] Zachary Department of Civil Engineering, Texas A&M University,
College Station, USA
atalebpour@tamu.edu
[3] Electrical and Computer Engineering, Texas A&M University,
College Station, USA
spb@tamu.edu

Abstract. The number of autonomous vehicles with advanced driver assistance systems have been increasing multi-fold. These technologies have reduced the work of the driver and have increased the safety of roads. Though a lot work has been done on development of autonomous vehicles, not much attention has been given to the millions of existing cars without these features. In this paper, we propose a method to implement level 2 autonomy in vehicles without Advanced Driver assistance systems. In this work, steering control of vehicles using voltage spoofing (can be extended to throttle and braking modules), development of PID controllers for the modules, and implementation of end-to-end driving to enable autonomous applications have been discussed. By searching for the stabilizing set to find the controller parameters K_p, K_i, and K_d, the system response has been improved and by implementing transfer learning, training data required has been reduced, and thus end-to-end driving with comparable results have been obtained.

Keywords: End-to-end drive · PID controller · Transfer learning

1 Introduction

Though the idea of self driving cars can be traced back to the early 1990's, the advent of the DARPA grand challenge in 2004 brought it to prominence. In the past decade, with the resurgence of Neural Networks, the field of Autonomous vehicles has grown by leaps and bounds. Cars with advanced driver assistance are invading the roads with about 10 million self driving cars expected to hit the roads by 2020. This number is just going to get higher in the next decade. But a very big question which arises is, what about the fate of existing cars that do not have advanced driver assistance features? In this work, we propose a solution to retrofit cars which have come into the market in the past decade with drive by

© Springer Nature Switzerland AG 2020
K. Arai et al. (Eds.): FTC 2019, AISC 1069, pp. 365–376, 2020.
https://doi.org/10.1007/978-3-030-32520-6_29

wire features and discuss a method to enable lateral autonomy using end to end driving.

Traditional vehicle control and planning problems involve a lot of sub-tasks such as lane detection [1,2], obstacle detection [9], control logic [8] to name a few. These sub-tasks rely mainly on feature detection and extraction from image data. The huge problem with such an approach is that the features detected are manually defined and may not be the optimal from a machines perspective. Errors in measurement accumulate over time and may lead to huge margin of errors. Whereas in an end-to-end approach [4], the neural network tries to learn the important features and aspects needed from the data-set solely to create a self-optimized algorithm. The lack of manual interference and optimized feature detection leads to better performance with this strategy. Driving data is an important aspect which is needed for training any neural network and to enable end-to-end driving, the amount of data needed is enormous. Bojarski et al. [4] in their work use data which has been collected over a period of a few months to get a specified accuracy. Due to the absence of such vast amount of data, transfer learning [11] is used and it can be seen in later sections that with just a few hours of training data, reasonable results are obtained.

The rest of this work is divided into six sections. The first section describes the problem statement. With the problem statement explained, the next section deals in detail with the system setup and on retrofitting a vehicle to enable by wire drive functions. The third section of this work describes the data pre-processing which is done on the camera and Control Area Network (CAN) message feed. The fourth section of this paper explains the methodology of the Convolutional Neural Network (CNN) and describes the network architecture involved. The fifth section of the work deals with the experiments conducted and the results obtained. The final section includes some concluding remarks and future advancements and advantages of this work.

2   Problem Statement

The main problem which had to be tackled was to create a system to enable autonomy on cars which lack Advanced Driver Assistance Systems (ADAS) and to achieve lateral control of the vehicle at constant speed. More specifically, the task is to analyze CAN messages from the vehicle (more information about CAN messages can be found in the next section) to deduce the setting for the steering angle of the car, and also to process the camera feed from the on-board cameras of the car, when a human drives and emulate the same using convolutional neural networks to drive autonomously. The problem statement can be further subdivided into tasks which involve implementing a drive by wire system, reading CAN messages, reading image values from the camera using Robot Operating System (ROS), modelling and developing a neural network to predict the output under different scenarios.

3   Drive by Wire

Drive by wire in the automotive industry refers to the use of electrical signals and electro-mechanical systems to implement functionalities which were traditionally fulfilled by mechanical linkages. In the last decade, all vehicles which have come out to the market use by-wire systems for steering, throttle and braking functions. In this work, we particularly lay emphasis on the steering module and it can be noted that the implementation remains the same for the throttle as well as the braking modules. Though the modules are implemented by wire, traditionally the auto manufacturers do not support user control of these modules. This is where the system described below comes into the picture.

3.1   CAN Messages

A Controller Area Network (CAN) bus is a robust vehicle bus standard designed to allow micro controllers and devices to communicate with each other in applications without a host computer. It is a protocol designed originally at BOSCH in 1985 for multiplex electrical wiring. But over the years, automotive manufactures have adopted CAN message based communication to reduce bulky wiring systems and to improve the responses of the system. ISO 11898 was established in 1993 to adopt CAN as the international standard for communication in automobiles.

To read from the CAN Bus of the vehicles, a device called as CAN sniffer is used. A CAN sniffer is just as it sounds; a tool that monitors, or sniffs out the data flowing over a control area network in real time. In our case a device called as KVASER Leaf light V2 is used to read the CAN messages of the car. The KVASER Leaf Light V2 is connected to the car using a On-Board Diagnostic (OBD) connector and is connected to the computer using a DB9 connector. The messages from the CAN bus which correspond to the steering angle are read and are published to a topic in ROS from which the steering angle can be computed.

3.2   Vehicle

The whole system has been developed on a 2014 KIA Soul. This car uses a CAN Bus based communication system. Being an older model, the car is not equipped with features to write CAN messages into the ECU of the vehicle. Most of the cars which have come out in the last decade have similar configurations. These vehicles use analog signals to send torque sensor readings (which correspond to the current steering angle) to the Engine Control Unit (ECU). By splicing the wires which correspond to the torque sensor measurements and spoofing in voltages in the range of $0-5$ V (subject to change based on the manufacturer) using an Arduino DUE, the steering can be moved around based on the input supplied (Fig. 1).

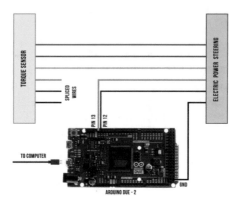

Fig. 1. Wiring diagram for steering module control system setup.

3.3 Controller

To get the steering to move to any required angle and hold its position, it is important to have a controller in place. The controller should have fast settling time and really low overshoot. In the past various theories and methods have been investigated for solving the control problem such as the Proportional Integral Derivative (PID) controller [3], fuzzy controller [7], model reference adaptive controller [12], fractional order control method to name a few. Due to the simplicity, the ease of implementation and the robustness of the PID controller, it has been used here. Several common approaches have been adopted to simulate the model and again for simplicity we use the bicycle model [5]. Based on the equations of motion obtained from the bicycle model, the state space system [13] can be described as:

$$\dot{\mathbf{X}} = \mathbf{AX} + \mathbf{BU} \tag{1}$$

where:

$$\mathbf{A} = \begin{bmatrix} a_{11} & a_{12} \\ a_{21} & a_{22} \end{bmatrix}, \mathbf{B} = \begin{bmatrix} b_1 \\ b_2 \end{bmatrix}, \mathbf{X} = \begin{bmatrix} v \\ r \end{bmatrix}, \mathbf{U} = \delta_f(t)$$

and,

$$a_{11} = -\frac{C_f + C_r}{u_c m}, a_{12} = -u_c - \frac{aC_f - bC_r}{u_c m}$$

$$a_{21} = -\frac{aC_f - bC_r}{u_c I}, a_{22} = -\frac{a^2 C_f - b^2 C_r}{u_c I}$$

$$b_1 = \frac{C_f}{m}, b_2 = \frac{aC_f}{I}$$

where a and b represent the distance of the front and rear axles from the center of gravity and u be the velocity of the vehicle. Usually the longitudinal component of velocity is larger than the lateral component of velocity. Therefore we can represent u as:

$$u = u_c + \Delta u \tag{2}$$

We consider the vehicle to moving at a uniform velocity and with a small disturbance. Hence Δu can be ignored. Since the vehicle has been driven at a speed of around 10 miles per hour, u is approximated to 4.47 m/s. For the given vehicle, the approximate pertinent vehicle parameters [6] are as follows:

Table 1. Pertinent vehicle parameters

Parameter	Value
Vehicle mass, m(kg)	1524
Vehicle Yaw moment inertia, $I(kg.m^2)$	2535
Wheelbase (m)	2.57
Longitudinal position of front wheel from center of gravity, a(m)	1.30
Altitudinal center of gravity, h_g(m)	0.55
Cornering stiffness of front tyre, $C_f(N.ran^{-1})$	80,000
Cornering stiffness of rear tyre, $C_r(N.ran^{-1})$	96,000

By substituting the values from Table 1, we have:

$$\begin{bmatrix} \dot{v} \\ \dot{r} \end{bmatrix} = \begin{bmatrix} -25.835 & -2.121 \\ 1.411 & 1.306 \end{bmatrix} \begin{bmatrix} v \\ r \end{bmatrix} + \begin{bmatrix} 52.493 \\ 41.025 \end{bmatrix} \delta_f(t) \tag{3}$$

In the equation above, the two state variables are yaw rate and lateral velocity. The lateral path error E, can be defined as a function of the lateral, longitudinal velocity and the heading θ [13].

$$\dot{E} = v + u_c\theta \tag{4}$$

$$\dot{\theta} = r \tag{5}$$

The augmented state space model can be specified as:

$$\begin{bmatrix} \dot{v} \\ \dot{r} \\ \dot{\theta} \\ \dot{E} \end{bmatrix} = \begin{bmatrix} -25.835 & -2.121 & 0 & 0 \\ 1.411 & 1.306 & 0 & 0 \\ 0 & 1 & 0 & 0 \\ 1 & 0 & 4.47 & 0 \end{bmatrix} \begin{bmatrix} v \\ r \\ \theta \\ E \end{bmatrix} + \begin{bmatrix} 52.493 \\ 41.025 \\ 0 \\ 0 \end{bmatrix} \delta_f(t) \tag{6}$$

The state space equation can be described as

$$\mathbf{Y=CX+DU} \tag{7}$$

and C matrix is described as:

$$\mathbf{C} = \begin{bmatrix} 0 & 0 & 0 & 1 \end{bmatrix}, \mathbf{D} = 0 \tag{8}$$

The open loop transfer function can be thus calculated by:

$$G(s) = \mathbf{C}(s\mathbf{I} - \mathbf{A})^{-1}\mathbf{B} + \mathbf{D} \tag{9}$$

$$= \frac{52.5s^2 + 27.8s + 5068.7}{s^4 + 24.529s^3 - 30.747s^2} \tag{10}$$

The transfer function is thus used to calculate the characteristic equation [8]which is used to search for the stabilising set for K_p, K_i and K_d [3].

3.4   System Configuration

NVIDIA Drive PX2, a scalable open autonomous vehicle platform serves as the brain of the autonomous vehicle. Robotics Operating System (ROS) is a middleware used for heterogeneous computer clusters. ROS is predominantly used for communication between the various functionalities/platforms on the computer. It is used to enable time synchronized communication between the CAN Bus, the camera and the drive PX2 (see Figs. 2 and 3).

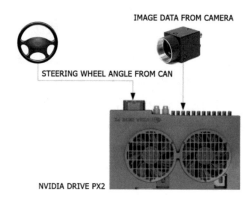

Fig. 2. High-level view of the data collection system

4   Data Collection and Pre-processing

The input from the camera comes in at around speeds from 7–30 frames per second (FPS) and that from the CAN bus comes in at around 11 FPS. The frame rates are matched based on the timestamp and time synchronized camera and steering angle data are saved as .bag files. These .bag files can be re-played later to extract the images and the steering angles. The training data was collected by driving on the tracks present in the university. 90 min of driving data was collected, which yielded about 37,800 images. During the process of data collection, the vehicle was driven at a speed of about 10 MPH in the circuit.

Since the vehicle is moving at a really slow speed, sampling at a speed of 30 FPS would result in redundant images. Hence the images are sampled from the video at a reduced frame rate of 7 FPS. Moreover to reduce the computation involved and hence the increase in efficiency, the images are cropped. It can be noticed from the image obtained from the camera that, the hood of the car occupies a small part of the whole image. Hence the image is cropped by about

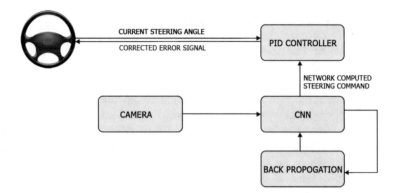

Fig. 3. Overall system setup

15% from the bottom. In the same way it is cropped at around 35% from the top to remove the sky which occupies a major portion of the image. Moreover the data obtained was flipped horizontally to obtain more training data for the network. It has to be noted that, this is an optional step to reduce the computational effort required (see Figs. 4 and 5).

Fig. 4. Uncropped camera feed

The obtained data was also augmented with random brightness changes, random linear and angular shifts to make sure the trained model is robust. The obtained images are then resized to a size of (200,66) and is feeded to the neural network.

Fig. 5. Cropped image feeded to the Neural network

5   Convolutional Neural Network

Convolutional neural networks are primarily made of multiple convolutional layers, fully connected layers and pooling layers which are stacked in different ways to give us the desired output. In a typical case, the input layer is followed by multiple convolutional layers, followed by fully connected layers, which are then connected to the output layer. The input which is fed into the network is a raw image acquired from the font facing camera. The required features are learnt by the network during training and these help to achieve end-to-end learning. The trouble with CNN's is the abundance of parameters which can be trained. Due to this, a huge dataset of images with steering angle is needed to predict the desired output. To reduce the size of the dataset needed and the computational time, transfer learning is used.

5.1   Transfer Learning

As mentioned previously, while it is better to train a model from scratch, due to the amount of data needed and time it takes to train the model, it can be cumbersome.

By using a pre-trained image classification model, we reduce the time and data by using the pre-trained network as a mere feature extractor and use these features to train a model. For our current problem, a pre-trained VGG16 model which was trained on imagenet dataset to classify 1000 classes has been used. These kinds of pre-trained models are efficient for most of the computer vision tasks and achieve acceptable accuracy.

5.2   Network Architecture

The transfer learning model was derived from the VGG16 model [10] which has a depth of around 16–19 layers. All the convolutional and max pooling layers are used for transfer learning. On top of these layers, a flattening layer and two dense layers are added to obtain results pertaining to our application. Followed by the VGG16 model, the model suggested by [4] is appended to complete the task suggested (see Figs. 6 and 7).

The neural network is now retrained with the collected data to produce a new convolutional neural network. The inputs are first normalized using a Keras Lambda layer to normalize and recenter the data. To introduce non linearity to the network, exponential linear units are used as the activation function.

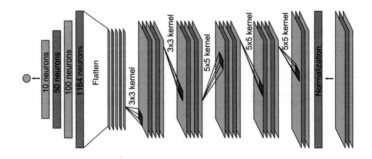

Fig. 6. Network architecture used by Bojarski et al. [4]

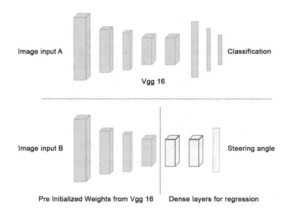

Fig. 7. Transfer learning overview

6 Experiments

6.1 Training Strategy

The model was trained and validated on data collected at multiple times to ensure the model is not over fitting. The data being feeded into the network was continuously shuffled for each epoch and also for each batch. The drop rate becomes an important parameter in such training methods. A moderate drop rate of 0.2 has been used. ADAM optimizer was used to aid the learning process. It used moving averages of the parameters to allow a larger effective step size when required. As is the standard practice, 80% of the data was used to train the neural network and the rest was used to validate the results of the neural network (see Fig. 8).

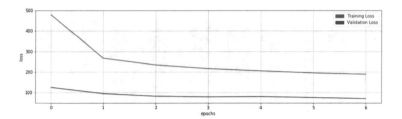

Fig. 8. Training, validation loss vs time

6.2   Testing

Data for testing was collected in tracks unseen by the vehicle. These images were sent to the model and the predicted results noted. In the works by Bojarski et al. at NVIDIA [4], an error is said to be a situation where manual driver inter-ference is needed to correct the vehicle to stay on track. Accuracy is calculated as the ratio of the number interventions multiplied by time taken to get back on track(assumed to be 6 s) to the total driving time. But this metric is not emper-ical and cant be used in cases were just the neural network performance is to be analyzed. We define a new metric as a bench mark for prediction accuracy. Since the vehicle is moving at a speed of 10 MPH, a comparable alternative would be to consider a deviation of up to $\pm 15°$ (equivalent to vehicle deviating by 2 feet for a unit time step) as a working solution and anything beyond that as an error. Based on this, an accuracy of about 78.19% has been achieved.

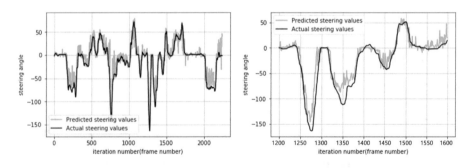

Fig. 9. Plot of predicted steering angle vs actual steering angle

Moreover the testing data has been collected with a human driver behind the wheel, and hence just looking at numerical data can be a major source of error. A better conclusion can be achieved by plotting the predicted steering angle vs. the actual steering angle by a human driver. Figure 9 depicts the graph of predicted steering angle vs the actual steering angle. It can be observed closely from the figure that both the actual and the predicted values are nearly the same and at a few places the neural network optimizes the turns. The predicted error has a mean of 9.6215° and a standard deviation of 11.1808°. Figure 10 gives a brief insight about the predictions given by the network and the actual steering angle recorded.

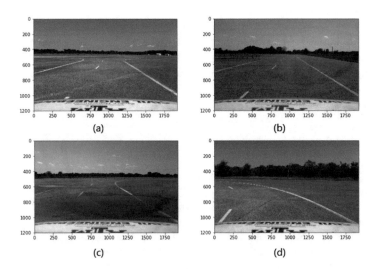

Fig. 10. Prediction results (a) Predicted value: 14.114615°, Actual value: 14.600015° (b) Predicted value: 0.225915°, Actual value: 1.015° (c) Predicted value: −19.993115°, Actual value: −25.515° (d) Predicted value: −55.663215°, Actual value: −67.90015°

7   Conclusion

In this work, we present a method for implementing level 2 autonomy on vehicles which lack Advanced driver assistance features. Voltage spoofing techniques and a PID controller is used to implement a drive by wire system. The proposed Neural network used transfer learning to reduce the amount of data needed to train the end-to-end drive system. While the other works in this field [4] have used data collected over a period of months, our work used data collected in 1.5 h. With a higher training data set, the accuracy of the network can be increase a lot further and it can be noted that even in such cases, the amount of training data needed will be much lesser than where transfer learning is not used. Though the implementation of throttle and braking control modules have not been discussed in this work, the methodology remains the same. With these implemented, advance driver assistance features will be accessible to every driver. Bringing in level 2 autonomy for all vehicles not only contributes to reducing the fatalities on the roads but also reduces the stress involved with driving.

Due to the huge contrasts in traffic discipline in countries such as the United states and most Asian countries, driver assistance systems which are effective in countries with high traffic discipline might prove to be ineffective and often hazardous in such countries. Systems developed on lane based control and speed based control can prove to be ineffective. End-to-end driving on the other hand, if trained on data sets collected on such conditions can help the vehicle learn new rules to drive instead of trying to stick with existent ones. More work can be done to improve the robustness of the system by training the model based on data obtained from various weather and road conditions. Moreover the system can be trained to adapt to various speeds if the throttle measurements are also given as an input to the neural network. In scenarios where an autonomous vehicle switches from autonomous mode to manual mode or vice verse, instead of it being a sudden impulsive switch, once the neural network is trained along with throttle values, It can be a gradual phase by phase switch where in the throttle control first comes to the driver followed by the steering.

References

1. Aly, M.: Real time detection of lane markers in urban streets. In: 2008 IEEE Intelligent Vehicles Symposium, pp. 7–12. IEEE (2008)
2. Bertozzi, M., Broggi, A.: Gold: a parallel real-time stereo vision system for generic obstacle and lane detection. IEEE Trans. Image Process. **7**(1), 62–81 (1998)
3. Bhattacharyya, S.P., Keel, L.H., Datta, A.: Linear Control Theory: Structure, Robustness, and Optimization. CRC Press, Boca Raton (2009)
4. Bojarski, M., Del Testa, D., Dworakowski, D., Firner, B., Flepp, B., Goyal, P., Jackel, L.D., Monfort, M., Muller, U., Zhang, J., et al.: End to end learning for self-driving cars. arXiv preprint arXiv:1604.07316 (2016)
5. Campion, G., Bastin, G., D'Andrea-Novel, B.: Structural properties and classification of kinematic and dynamic models of wheeled mobile robots. In: 1993 Proceedings IEEE International Conference on Robotics and Automation, pp. 462–469. IEEE (1993)
6. Garrott, W.R., Monk, M.W., Chrstos, J.P.: Vehicle inertial parameters-measured values and approximations. Technical report, SAE Technical Paper (1988)
7. Hessburg, T., Tomizuka, M.: Fuzzy logic control for lateral vehicle guidance. IEEE Control Syst. Mag. **14**(4), 55–63 (1994)
8. Rajamani, R.: Vehicle Dynamics and Control. Springer, Heidelberg (2011)
9. Redmon, J., Divvala, S., Girshick, R., Farhadi, A.: You only look once: unified, real-time object detection. In: Proceedings of the IEEE Conference on Computer Vision and Pattern Recognition, pp. 779–788 (2016)
10. Simonyan, K., Zisserman, A.: Very deep convolutional networks for large-scale image recognition. CoRR, abs/1409.1556 (2014)
11. Torrey, L., Shavlik, J.: Transfer learning. In: Handbook of Research on Machine Learning Applications and Trends: Algorithms, Methods, and Techniques, pp. 242–264. IGI Global (2010)
12. Wang, W., Nonami, K., Ohira, Y.: Model reference sliding mode control of small helicopter XRB based on vision. Int. J. Adv. Robotic Syst. **5**(3), 26 (2008)
13. Zhao, P., Chen, J., Song, Y., Tao, X., Xu, T., Mei, T.: Design of a control system for an autonomous vehicle based on adaptive-pid. Int. J. Adv. Robotic Syst. **9**(2), 44 (2012)

Cryptography in Quantum Computing

Pam Choy, Dustin Cates$^{(\boxtimes)}$, Florent Chehwan, Cindy Rodriguez,
Avery Leider, and Charles C. Tappert

Seidenberg School of Computer Science and Information Systems, Pace University,
Pleasantville, NY 10570, USA
{pc96111n,dc04381n,fc87943n,cr84102n,aleider,ctappert}@pace.edu

Abstract. Quantum cryptography is an area of intense interest, as quantum computers contain the potential to break many classical encryption algorithms. With so much on the line, it is imperative to find a new quantum encryption method before quantum technology catches up with current cryptography. This study examines one of the very few experiments on encryption that has already been conducted and analyzes the results of the tests run on the IBM Cloud Server. It attempts to recreate the sample experiment and make comprehensive adjustments for a real-world environment. This study also looks at the possible application of quantum public key encryption and the theoretical importance of quantum key distribution.

Keywords: Quantum computers · Quantum cryptography · Quantum encryption · Encryption algorithm · Quantum public key encryption · Quantum key distribution

1 Introduction

The world of quantum computers is based on quantum theory, which largely deals with the behavior of particles at the atomic and subatomic levels. At these infinitesimal scales, the laws of physics, as scientists have grown accustomed, almost seem not to exist, or at least are drastically different.

In the case of quantum computers, the atomic particles are actual information. In a traditional computer, information is organized into series of transistors that, depending on the patterns of which they are on or off, form bits that can then be translated into binary. Each transistor can only represent a single state at a time: a 0 or a 1. In quantum computers however, a quantum bit (also known as a qubit) can be a 0, a 1, both, or somewhere in between. It is even capable of representing multiple states at the same time [18]. This state is known as superposition. The superposition of qubits is the fundamental factor that makes quantum computers vastly more powerful in comparison to a classical computer. However, the measurement of these states can oftentimes be difficult due to this

---

C.C. Tappert—Thanks to the IBM Faculty Award that made this research possible.

© Springer Nature Switzerland AG 2020
K. Arai et al. (Eds.): FTC 2019, AISC 1069, pp. 377–393, 2020.
https://doi.org/10.1007/978-3-030-32520-6_30

superposition quality [13]. In quantum computers, the state of a qubit is stored in an atomic or subatomic particle using a variety of experimental methods.

Quantum mechanics, and therefore quantum computers, are anything but simple. Despite the complexity, the age of quantum computers is coming. With it will come a phenomenon popularly called the quantum break [5], which is a theoretical time when quantum computing will evolve to a point where it will render many common encryption algorithms obsolete [12]. Think about it. Many current encryption algorithms are based upon multiplying two very large prime numbers. This process, however, is reversible. These types of algorithms are not unbreakable but more accurately, not feasibly breakable because it would take current computers an unreasonable amount of time (sometimes decades or centuries) to break an algorithm or find a factorization. Because quantum computers are theoretically capable of computing at exponentially higher speeds, it is only a matter of time before today's strongest algorithms can be solved within minutes, therefore rendering these types of encryption obsolete.

While the quantum break has not yet occurred, theorists suggest that recent advances in quantum computing means that the quantum break is a highly likely scenario in coming years. With more and more of individuals' private data, e-commerce, and banking being done strictly through computers, the impending quantum break creates an urgent need for the knowledge and implementation of quantum encryption methods.

The goal of this research is to test a quantum algorithm cipher, using a simple example, on a real quantum computer, and then to identify what the obstacles are to understanding it. There are many jobs available, and unfilled, for quantum computing technologists who are capable of this task [1]. There are also very few courses of instruction offered, from PhD level to Master's level to Baccalaureate Level because of the complexity of the subject.

The next section of this paper describes project requirements. A research study then continues with the next section on literature review related to this focus.

2   Literature Review

As more and more everyday transactions containing sensitive data migrate to a digital format, ensuring the safety of information has become more important than ever. As of now, complex algorithms are being used to secure information stored on computers and servers all over the world. Many, however, believe the age of quantum computing will bring this to an end.

As explained earlier, the emergence of the quantum computer will soon make most current methods of algorithmic encryption obsolete and insecure. One of the biggest problems security specialists are facing is the issue of key distribution [20]. Heavy investment is being made by major governments worldwide, such as the United States, European Union, and China. The United States Department of Defense has announced an $899 million research budget claiming the majority of that will be dedicated to quantum computing [15]. The European Union has announced a ten-year, billion-Euro plan to fund research that will turn quantum techniques into commercial products [2]. While official budgets are kept confidential, China has announced their plans for a billion dollar quantum research facility to be finished in 2020. They also claim to have launched the first satellite capable of quantum communications [16]. All three of these are showing clear goals to make significant advancements in the field of quantum mechanics and with these advancements, develop a new method to protect data from the enhanced abilities of quantum computers.

To date, while many theories exist, researchers have been unable to produce any concrete findings. Currently, it appears that the answer may lie in quantum public-key encryption (QPKE). In fact, significant strides have been made in a new method of encryption, known as quantum public-key cryptosystem. As detailed in "A Practical Quantum Public-key Encryption Model", a quantum public-key cryptosystem has the potential to be very efficient on a quantum machine, if implemented in conjunction with the ease of current asymmetric key distribution methods [17]. As "The Race Towards Quantum Security" explains, quantum encryption through quantum key distribution (QKD) becomes a necessary step to ensure the security of information. In the purposed model, a QKD scheme would be implemented through the storage of information in quantum data carriers, known as qubits. This information would then be transmitted between two qubits via a quantum channel [9].

The term, "Quantum channel" refers to the media for quantum information to be transferred during quantum entanglement. Quantum entanglement is when two qubits reflect the same state at the same time despite being separated by, theoretically, infinite distance. This would mean that even after the qubits are separated, if one qubit is changed, the second qubit will instantaneously reflect the same changes. Correlations between the two separated qubits cannot be explained without quantum mechanics. The simplest way to visualize quantum entanglement is through the use of Bell states.

As shown above, Bell States can be represented by four different states: Phi+, Phi−, Psi+, and Psi− (Fig. 1a–d). Out of these, the most studied state is Phi+. The expression Phi+, basically means that a qubit held by Alice (subscript "A") can be either "0" or "1". If Alice were to measure the qubit, the result would be random, meaning either possibility ("0" or "1") would have the probability of $1/2$. Similarly, if Bob (subscript "B") were also to measure his qubit at that moment, he would always get the same outcome as Alice. Therefore, when Alice and Bob communicated, they would find that, although their individual outcomes seemed random, they would be correlated with one another. This is partially why entanglement is believed by many to be the key to using quantum computing for encryption [3].

Theoretically, a Bell State expression could be used to securely create a key and send it from Alice to Bob using QKD protocol (Fig. 2). Alice and Bob would start by sharing number pairs of entangled qubits. Alice would then send the qubits to Bob in different directions (vertical, horizontal, or diagonal "left" or "right"). Alice and Bob would then randomly pick and make local measurements

(a)

$$|\Phi^+\rangle = \frac{1}{\sqrt{2}}\left(|0\rangle_A \otimes |0\rangle_B + |1\rangle_A \otimes |1\rangle_B\right)$$

(b)

$$|\phi-\rangle = \frac{1}{\sqrt{2}}\left(|0\rangle_A \otimes |0\rangle_B - |1\rangle_A \otimes |1\rangle_B\right)$$

(c)

$$|\psi^+\rangle = \frac{1}{\sqrt{2}}\left(|0\rangle_A \otimes |0\rangle_B + |1\rangle_A \otimes |0\rangle_B\right)$$

(d)

$$|\psi-\rangle = \frac{1}{\sqrt{2}}\left(|0\rangle_A \otimes |1\rangle_B - |1\rangle_A \otimes |0\rangle_B\right)$$

Fig. 1. a: Phi+ Bell State b: Phi− Bell State c: Psi+ Bell State d: Psi− Bell State

of the qubits to ensure maximally entangled state (correlation). If they are not in correlation, then the photons will be discarded. In essence, the photons in the correct state would be translated into bits (for example, horizontal = "1" and vertical = "0") for Bob and thus produce a set of quantum keys for decryption. Even with the random outcomes, there is always correlation [3].

Additionally, quantum secure direct communication, also known as QSDC, is another proposed method of encrypting messages between Alice and Bob. QSDC is different from the QKD protocol mentioned above because, while both methods would generate a private key for the two parties to use in communication, QSDC does so without the need for key generation in advance [19]. With a QSDC scheme, a controlled-not (CNOT) gate is the primary method of encoding and decoding the messages passed between Bob and Alice. The key, in this case, is a "sequence of two-photon pure entangled states ... [which are] reused with an eavesdropping check" so that an eavesdropper, such as Eve in Fig. 2, would be detected [19]. Each of these transmitted photons could theoretically carry one bit of the message being communicated.

It is also believed that quantum key encryption could be used to bolster the already quantum proof symmetric cipher schemes such as Advanced Encryption Standard (AES) [11]. However, a problem arises when considering key distribution for asymmetric cipher schemes which are currently expected to be broken by future quantum computers. To compensate for this problem, Wang and She propose that private keys be held by actual individuals and correspond to a single public-key that is created by a trusted third-party using the quantum computers already proven ability to generate truly random numbers [7,17]. Through this process, the information would be secure and the plaintext would be unrecoverable from the cipher text. To break the cryptosystem, an individual would need

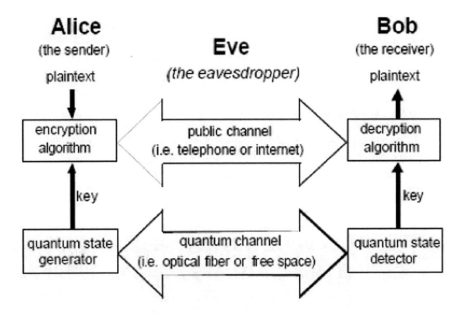

Fig. 2. Quantum key distribution

to recover the private key from the public key, potentially giving them access to all messages of the intended recipient [17].

This intended attack however, would be theoretically impossible due to a law of quantum mechanics known as Heisenberg's uncertainty principle. This law states that the position and velocity of an object, an electron for example, can not be measured precisely at the same time [4]. In the context of QKD and QSDC, this means that a third party would be unable to time the signal exactly in order to obtain the secured information needed to break the encryption.

Heisenberg's uncertainty principle is not the only law of physics that is being relied upon to guarantee security. Lindsay believes that the "no-cloning theorem of quantum informatics" will also play a role in securing or detecting potential attacks [11]. The no-cloning theorem is a widely accepted rule of quantum mechanics and essentially states that an arbitrary quantum state cannot be cloned [14]. This coincides with the even more widely accepted theory that one can change a quantum state simply by looking at it. As Lindsay applies it, this would mean that a quantum encrypted message could potentially be intercepted but the copying or reading of this message would actually change the state of the information, greatly increasing the likelihood that an eavesdropper would be detected.

While the quantum computer's ability to generate random numbers would make the creation and distribution of cryptographic keys more secure, most current QKD models would suffer from a low rate of key generation on current machines, which would greatly limit the possibly of widespread distribution [10]. This limit is caused by the operation of current quantum computers which encode a single qubit to a single photon. As Lai *et al.* explain, there is currently a solution to this problem involving changes to the QKD protocol itself but it is far from cost effective. Despite multiple theories floating around the computer science and physics communities, a cost effective solution to QKD has yet to emerge.

3   Project Requirements

1. QC Set-up Instructions/QISKit
2. Registration and API key from IBM's Quantum Experience
3. Operation Systems: Windows and Mac OS
4. Software Development Kit: QISkit and The Composer

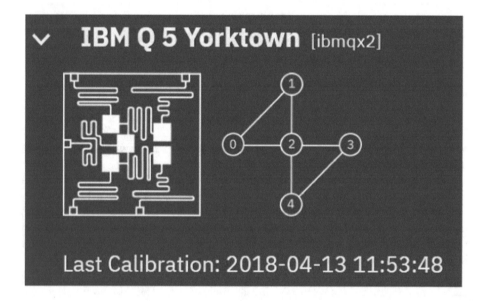

Fig. 3. IBM Qx5

In order to conduct this project, it was necessary to have access to quantum computers. The experiment made use of the public access recently granted by IBM to a 5-qubit superconducting device (Fig. 3) via their "Quantum Experience" cloud service. This service made use of two software development kits in the initial phase of the project: the IBM composer and QISkit. The Composer is IBM's graphical user interface for programming a quantum processor while QISkit is an open-source framework for quantum computing with Python. The end experiment most heavily utilized the composer but QISkit was also important to the experience (see Fig. 4 to activate QISkit Error Message).

Once successfully installed, a user must register with IBM's Quantum Experience website in order to start using the IBM Q Experience and QISkit to

```
Requirement already satisfied: cffi>=1.7 in /anaconda3/lib/python3.6/site-packages (from cryptography>=1.3
->requests_ntlm->IBMQuantumExperience>=1.9.6->qiskit) (1.11.5)
Requirement already satisfied: pycparser in /anaconda3/lib/python3.6/site-packages (from cffi>=1.7->crypto
graphy>=1.3->requests_ntlm->IBMQuantumExperience>=1.9.6->qiskit) (2.18)
Building wheels for collected packages: IBMQuantumExperience
  Running setup.py bdist_wheel for IBMQuantumExperience ... done
  Stored in directory: /Users/pamchoy/Library/Caches/pip/wheels/8c/25/a5/3cdd1e8bde0b66de4c02a3124fb183d52
c49765798a68466b5
Successfully built IBMQuantumExperience
distributed 1.21.8 requires msgpack, which is not installed.
Installing collected packages: ntlm-auth, requests-ntlm, IBMQuantumExperience, qiskit
Successfully installed IBMQuantumExperience-2.0.3 ntlm-auth-1.2.0 qiskit-0.5.7 requests-ntlm-1.1.0
You are using pip version 10.0.1, however version 18.0 is available.
You should consider upgrading via the 'pip install --upgrade pip' command.
Pams-Air:~ pamchoy$ pip install --upgrade pip
Collecting pip
  Downloading https://files.pythonhosted.org/packages/5f/25/e52d3f31441505a5f3af41213346e5b6c221c9e086a166
f3703d2ddaf940/pip-18.0-py2.py3-none-any.whl (1.3MB)
    100% |████████████████████████████████| 1.3MB 1.2MB/s
distributed 1.21.8 requires msgpack, which is not installed.
Installing collected packages: pip
  Found existing installation: pip 10.0.1
    Uninstalling pip-10.0.1:
      Successfully uninstalled pip-10.0.1
Successfully installed pip-18.0
[P           $ conda create -y -n QISKitenv python=3 pip sicpy jupyter
Solving environment: failed

PackagesNotFoundError: The following packages are not available from current channels:

  - sicpy

Current channels:

  - https://repo.anaconda.com/pkgs/main/osx-64
  - https://repo.anaconda.com/pkgs/main/noarch
  - https://repo.anaconda.com/pkgs/free/osx-64
  - https://repo.anaconda.com/pkgs/free/noarch
  - https://repo.anaconda.com/pkgs/r/osx-64
  - https://repo.anaconda.com/pkgs/r/noarch
  - https://repo.anaconda.com/pkgs/pro/osx-64
  - https://repo.anaconda.com/pkgs/pro/noarch

To search for alternate channels that may provide the conda package you're
looking for, navigate to

    https://anaconda.org

and use the search bar at the top of the page.
```

Fig. 4. Activate QISkit error message

run programs on the remote back-end provided by IBM. A user can register by creating a log-in using an email address or other social media account such as Linkedin. The subscription is free for all users. After registering to the "Quantum Experience", IBM provides an API token that will be saved in a file called qconfig, which is passed to the program. This sets up a connection to the back-end server. It will then be possible to select a back-end server from those available on the IBM Q Experience, run the program, and get the results of the program.

A Quantum Computing source repository is available on Github to explore tutorials, run programs, and contribute to the development of quantum computing by addressing important issues.

4 Methodology

Resources to learn quantum computing and simple experiments are available in the IBM Quantum Experience, the Jupyter Notebooks, and the JupyterLab.

These are primordial when learning to understand the structure of the program behind the more approachable graphical representation in the IBM Composer. The three steps of the development of a quantum program are the same as a classic program with the following functions: build, compile, and run. The first step, build, consists of creating a quantum circuit composed of quantum registers and adding gates to it in order to use and manipulate the registers. These gates will perform directly on qubits. The different back-end servers available can then be chosen during the compile step if, for example, it is desirable for the code to be executed on a quantum simulator, on the local machine, or on an IBM quantum chip. After running the code in the final step, and receiving a result, the user can select a variety of options that can affect execution of the code.

In preparation for the experiment, the Composer was tested to better understand how to program the score which can be run via the simulator or on the real hardware of QISkit. One thing that made the Composer easier to learn, was its graphical representation of QASM, a simple text-format describing quantum circuits (Fig. 5). The Composer provides tabs to drag and drop gates, barriers, and measurements onto the score or to apply them to the entire QASM code. It also gives options to either run the score against the actual hardware or in simulation mode.

Quantum Circuit

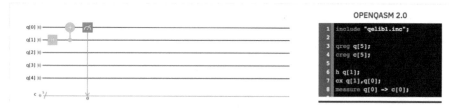

Fig. 5. Composer showing QASM code

The effect that the previously mentioned gates can have on qubits is sometimes described using a Bloch sphere. The diagram below (Fig. 6) depicts a simple Bloch Sphere where the 0 Vector ($|0>$) on the North Pole means low energy and 1 Vector ($|1>$) on the South Pole means active energy. The gates cause the qubits to go in different directions. In our experiment, the qubit was placed into superposition using a Hadamard gate, which transformed the qubit from the standard basis elements (North-South) to a new basis on the equator of the sphere (East-West) in which the probabilities of observing a "0" or a "1" would be simply 50%. Then, a Controlled-NOT gate was placed on q0 to generate entanglement. Lastly, a measurement was applied. Due to the fact that quantum computers process, but do not measure, classical bit state (0 or 1), a measurement must then be applied when running the simulator (Fig. 7).

It must also be noted that, because of the randomness of the quantum process, each instance of the experiment can have different results. The IBM Quan-

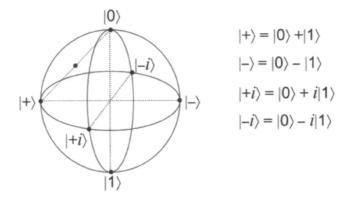

$$|+\rangle = |0\rangle + |1\rangle$$
$$|-\rangle = |0\rangle - |1\rangle$$
$$|+i\rangle = |0\rangle + i|1\rangle$$
$$|-i\rangle = |0\rangle - i|1\rangle$$

Fig. 6. Bloch sphere

Quantum Circuit

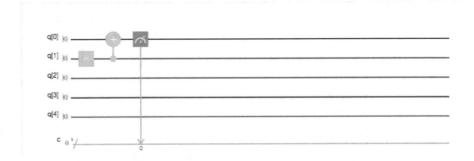

Fig. 7. Superposition entanglement and measurement

tum Experience allows users to specify the number of runs (shots) for an experiment. In this particular experiment, 100 shots were specified. Once completed, the results were displayed as a histogram (Fig. 8).

This simple simulation was useful in learning to understand how to run a quantum code and prepare for more complicated encryption experiments with the IBM Q Experience Composer.

4.1 Encryption Algorithm Experiment

While quantum encryption is still in a developing state, several studies have already been conducted. Many of them were attempts to lay a foundation for a stable encryption process. Researchers are attempting to develop an encryption model that works on a quantum machine. At this project's onset, several different criteria were examined in order to decide which experiment would be the best for analysis and testing. The biggest concerns were namely the algorithm and

Fig. 8. Quantum state result

the process itself. Many different encryption methods were found throughout the literature. Nonetheless, similarities were found in terms of process and actions taken on the input Table 1. It was found that, in order to perform quantum encryption, it is necessary to use Quantum gates performing unitary operations, such as polarization and rotation of the qubits. Upon comparison, homomorphic encryption seemed to stand out as the most widely explored type of encryption and the most widely experimented on across the different studies. The main reason for this is that it is a reversible operation which, as a result, makes it possible to verify that the right operation was performed on the input during the encryption simply by looking at the output. Since it is still an emerging field, that is constantly evolving, the date of the publication in relation to current technology was an important factor in choosing the study that would be experimented on. The last and most important aspect that needed to be considered was the capability of testing the algorithm in real-life using the different tools available for this project.

A quantum algorithm developed by researchers from the University of Shanzhou and the University of Science and Technology of China, presented the first encoding algorithm performed on the IBM Quantum Computer. This addressed the concerns of the capability to test the experiment. Similar to most of the other encryption experiments available, this study performs a homomorphic encryption, which is a type of encryption that allows computation on encrypted data [8]. It serves as a scope to show possible future uses and purposes of quantum computers. This encryption generates an encrypted result that, once decrypted, matches the result of the operations performed on the input.

Starting from a quantum algorithm created to solve linear equations proposed by Harrow *et al.* in 2009, the encryption model encodes the problem to test homomorphic encryption [6]. In order to take into account the issue of security, this study presents a different approach by compiling an encrypted version of the input in the first step of the algorithm instead of encoding the quantum circuit on the servers directly. In this case, the server can only deal with encrypted data. After examining the model presented in their study, it became possible to attempt to encode the circuit on IBM's cloud Quantum Computers thanks to the methodology. Among the challenges we faced in our experiment, the most notable was to translate the annotations for the gates and unitary operations presented in the model of the researchers into the corresponding gates and operations available in the IBM Composer following the guidelines of the examined study (Table 1).

Table 1. Annotation comparisons

Zhengzhou University Annotation	Transformation (If applicable)	v	Pace University Annotation
R-Gate R+ Gate (Phase Estimation)	R-Gate-CNOT Gate-R+Gate	v	Hadamard gate-CNOT-hadamard Gate
Rotation Ry(θ)	CNOT- Ry($-\theta/2$) - CNOT Ry($\theta/2$)	v	CNOT - combination of 7 H gates and 7 T gates - CNOT - Combination of 7 T+ gates and 7 H gates
R-Gate R+ Gate (Inverse Phase Estimation)	R-Gate-CNOT Gate-R+Gate	v	Hadamard gate-CNOT-hadamard Gate
Measurement	Measurement	v	Measurement gate

To understand the different phases of the algorithm, it is important to fist understand a few concepts, such as the fact that quantum computers only use operations that are their own inverses. As a result, if you apply the operation twice, you will get the same value as the input value. In other words, an operation performed on a qubit can be performed on this qubit the opposite way. In order to write non-reversible functions in a reversible way, quantum computing uses two qubits. One qubit, the input qubit, stays unchanged and the value of the function is written to the output qubit, also called the target qubit. This process makes it possible to rewire operations (Fig. 9).

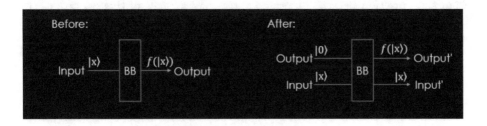

Fig. 9. Reversible function for two circuits

The gates perform operations on qubits. The gates are capable of slightly changing the qubits without collapsing them. In order to understand this concept, one could imagine that the qubits move around the unit circle of a Bloch sphere depending on the operations performed on them. They collapse once they are measured. For example, X gates flip the qubits around the unit circle. An X gate would change the qubit state from 1 to 0. On the other hand, an H gate, also called Hadamard gate, would change the state of 1 or 0 to a flip coin state, which is a state of superposition.

The operations performed by the other gates included in the computation will be examined later in this study. The algorithm studied here uses three qubits. The first qubit is the input qubit. The state of this qubit should not change. The second qubit is a spare qubit. It is needed due to the way quantum circuits work. For instance, we are using the second qubit because CNOT gates operate on pairs of qubits, they cannot operate on a single qubit. This is also represented in the IBM Quantum Composer. The third qubit named in Huang's study, the ancilla operator, is the output qubit. This qubit is the one measured in the new experiment because it was this qubit that got affected by the operations used by the gates. The algorithm would only be successful if the value of the ancilla operator was equal to 1 [8].

The equation in this study can be encoded as the following: $A|x>=|b>$ (1) This equation was considered given the input being $|x>$, the matrix A and the output being vector $|b>$. In order to adapt to quantum requirements, the study considered that $||x|| = ||b|| = 1$.

Huang's algorithm is composed of three different steps. The first step is called the Phase Estimation. It is described in the study as the encryption process. It takes place through the entanglement of the input qubit and the spare qubit. Entanglement occurs by putting the input qubit through the Hadamard gate, which puts it in superposition, and then using a CNOT gate. The input qubit becomes the control qubit and the spare qubit becomes the target qubit. As a result of this process, these two qubits will be in superposition. They will be coordinated, and cannot be separated. Because of the superposition state, their state will be neither 1 or 0. They will have a 50% chance to collapse to 0 and a 50% chance to collapse to 1. Nonetheless, another Hadamard Gate is added to the input qubit. This other Hadamard gate will change the state of the qubit from superposition back to its original state. As a result of this operation, the two qubits are in the original state of the input qubit. The CNOT gate, that is then operated on the spare qubit, sets the spare qubit as the control qubit and the output qubit as the target qubit. The state of the spare qubit, which is flipped to the output qubit. At this point of the experiment, the output qubit is in the same state as the input qubit. This process takes us back to the fact that quantum computing operations need to be reversible and that the output qubit state should be the same as the input qubit state before operating gates on the output qubit.

The second phase of the experiment is named in Huang's study as the rotation phase. It will operate a succession of Hadamard gates that will change the qubit into a superposition state. Phase gates, called T gates, in the IBM Composer that will rotate the output qubit around the Z axis and T+ gates which are the inverse of the phase gates. The first combination of Hadamard gates and Phase gates will put the qubit into superposition around the different axis of the Bloch sphere before rotating the qubit around the Z axis. The second combination is made of Hadamard gates and T+ gates. This process can be explained as doing the opposite rotation previously done. The goal of this algorithm is showing that the qubit can travel around the Bloch sphere with a high probability for getting an output similar to the input.

The last part of the algorithm is called the inverse phase estimation. This process is described in the model study as the disentanglement of the input qubit and spare qubit, or the decryption process.

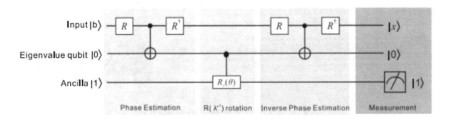

Fig. 10. Huang's experiment

At first Huang's study was strictly followed (see Fig. 10), by setting the input qubit as Q(0), the spare qubit as Q(1) and the output qubit as Q(2) (Fig. 11). After running the first tests, results were found to be the opposite of those from Huang's study (Fig. 12). Further work on the experiment concluded that, in order to perform this algorithm on the IBM Composer, the qubits needed to be flipped to operate the CNOT gates. As a result, Q(0) becomes the output qubit, Q(1) becomes the spare qubit and Q(2) becomes the input qubit. The model in Fig. 13 shows the final algorithm version executed in our experiment.

Fig. 11. First encryption algorithm IBM Q

4.2   Results

Before revealing the results, it is important to understand how the measurements work. The probability calculated is the probability that a qubit collapses to 0 or 1. The measurement gate will fire through 0 and 1 and calculate this probability.

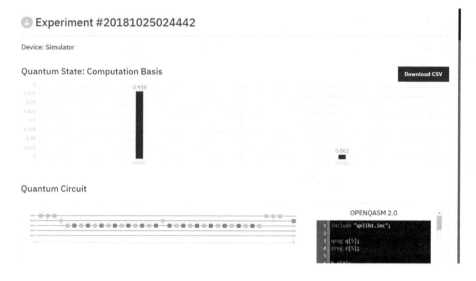

Fig. 12. First algorithm results

Fig. 13. Final encryption algorithm IBM Q

The results of this study, after running the experiment on the real quantum computer and in the simulators, were different from the results of the original study. Some factors and implications needed to be considered in order to draw conclusions from this experiment. For instance, the previous study did not include any details regarding the number of shots executed. When changing the number of shots executed in the new tests the results change. These changes were not significant enough to point to a determining factor (Fig. 12). As shown in Table 2 our results oscillated between 0.400 and 0.500 for the qubit to collapse to 1. It represents a 40 to 50% chance for the experiment to get the expected result. These results are far from the results shown in Huang's study. According to their study, the output states were between 0.920 and 0.992. It means that they never fell under a 92% chance to get the expected output. The results of our experiment show probabilities that are closer to a superposition state (Fig. 14). Furthermore, when measuring the other qubits alone to check if the conditions of the experiment were met, the spare qubit is in the expected state of 1 as it is simply used to conduct the CNOT operations and the input qubit is in a state close to our output qubit due to the last Hadamard gate used in the algorithm, thus proving that the disentanglement process worked.

Table 2. Results comparison

Results of Huang's Study	Results of Pace University
Between 0.400 and 0.500 for expected result	Between 0.920 and 0.992 for expected result

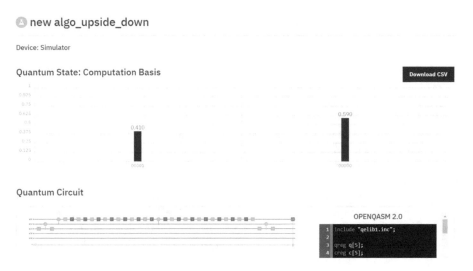

Fig. 14. Final algorithm results

It is difficult to say that what was discovered is a strong encryption model that can be tested on the IBM Quantum Composer following the provided methodology. It may yet be too early to compute this algorithm using the IBM composer and some recommendations could be made. Indeed, it might be better to test this algorithm using Qsharp (Q#), the Microsoft platform for performing Quantum computing simulations. Using this tool would allow one to test inputs while utilizing the same gates as Huang's experiment on the qubits. One limitation of the Microsoft tool is that it only runs simulations and it does not run on a real quantum computer. Also, Huang's experiment uses an eigenvalue in the matrix transformation. This eigenvalue should be passed through the spare qubit in order to be included in the rotation process. This may be possible by using a physical gate called the U2 gate in the IBM Composer. Huang's experiment does not refer to this gate, so it was excluded from this new experiment. One should understand that the goal behind using so many gates is to cover a larger area of the Bloch sphere. Nonetheless, it should also be considered to reduce the number of gates used in the process. By testing with fewer H and T gates the results may differ as the number of operations on the qubits decreases.

5   Conclusion

Quantum cryptography, just like quantum computing, is still very much in the experimental stages with only a handful of actual working machines in existence, and even these have some extreme requirements for operation. Scientists all over the world are busy trying to accomplish any kind of quantum algorithm, let alone one as complex as encryption. Experiments across the literature, agree on the need of operations such as vertical and horizontal polarization in order to achieve what could be a suitable quantum encryption process. However, the application of these experiments have proven to be theoretical, and very difficult to reproduce, rather than scalable proof-of-principle demonstration. Viability may still be a long way off. While the power of quantum computing cannot be denied when it comes to cryptography, the uncertainties around quantum computing, as well as the lack of available tools, remain constant issues in the field, and quantum encryption is no exception. In fact, many tools have to be created solely to complete certain experiments. According to Lindsay, error correction in quantum computing is not as easy as in traditional computing, because "random errors in the physical media tend not to be correlated, but correlation, is the whole point of quantum entanglement" [11]. When trying to solve simple algorithms, stability in qubits is easy to maintain, however, as the algorithm gets more complex, stability becomes a major issue. Another challenge pointed out by Lindsay, is that some quantum computers actually struggle to "maintain coherence long enough to complete useful calculations" [11]. Any small disturbances (for example, a small beam of light) can change the direction of the photon even slightly and cause a build up of errors.

While many believe quantum computing will eventually evolve to a point where it will render many common encryption algorithms obsolete, there is still a question as to when. Some theorize that it could be achieved within the next few years, while others believe it will not happen within our generation's lifetime [5]. Regardless if this were to happen or not, it is likely that the entire global internet ecosystem would need to be restructured, as well as human thinking about computing, before the technology could truly be applied [3].

Acknowledgment. The authors would like to thank Prof Charles Tappert, PhD and Prof Avery Leider.

References

1. Quantum computing report: Where qubits entangle with commerce. https://quantumcomputingreport.com/players/universities/. Accessed 29 Nov 2018
2. Cartlidge, E.: Europe's€ 1 billion quantum flagship announces grants (2018)
3. Chen, S.: Quantum cryptography explained. https://youtu.be/UiJiXNEm-Go
4. Furuta, A.: One thing is certain: Heisenberg's uncertainty principle is not dead. Scientific American (2012)
5. Grimes, R.: The quantum break is coming will you be ready? https://youtu.be/jB47_xoeB4o

6. Harrow, A.W., Hassidim, A., Lloyd, S.: Quantum algorithm for linear systems of equations. Phys. Rev. Lett. **103**(15), 150502 (2009)
7. Herrero-Collantes, M., Garcia-Escartin, J.C.: Quantum random number generators. Rev. Mod. Phys. **89**(1), 015004 (2017)
8. Huang, H.-L., Zhao, Y.-W., Li, T., Li, F.-G., Du, Y.-T., Fu, X.-Q., Zhang, S., Wang, X., Bao, W.-S.: Homomorphic encryption experiments on IBM's cloud quantum computing platform. Front. Phys. **12**(1), 120305 (2017)
9. Ismail, Y., Petruccione, F.: The race towards quantum security. In: 2018 IST-Africa Week Conference (IST-Africa), pp. Page–1. IEEE (2018)
10. Lai, H., Luo, M.-X., Pieprzyk, J., Zhang, J., Pan, L., Li, S., Orgun, M.A.: Fast and simple high-capacity quantum cryptography with error detection. Sci. Rep. **7**, 46302 (2017)
11. Lindsay, J.: Why quantum computing will not destabilize international security: The political logic of cryptology (2018)
12. Lo, H.-K., Lütkenhaus, N.: Quantum cryptography: from theory to practice, arXiv preprint quant-ph/0702202 (2007)
13. Moller, M., Vuik, C.: On the impact of quantum computing technology on future developments in high-performance scientific computing. Ethics Inf. Technol. **19**(4), 253–269 (2017)
14. Patel, D., Patro, S., Vanarasa, C., Chakrabarty, I., Pati, A.K.: Impossibility of cloning of quantum coherence, arXiv preprint arXiv:1806.05706 (2018)
15. Prisco, J.: The quantum computing race the US can't afford to lose. https://thenextweb.com/contributors/2018/09/01/quantum-race-united-states-must-compete/. Accessed 4 Dec 2018
16. Segal, A., Nilekani, N., Dixon, H., Flournoy, M., Sulmeyer, M., Mayer-Schiinberger, V., Ramge, T.: When China rules the web. Foreign Aff. **97**(5), 10–18 (2018)
17. Wang, Y., She, K.: A practical quantum public-key encryption model. In: 2017 3rd International Conference on Information Management (ICIM), pp. 367–372. IEEE (2017)
18. Woodford, C.: Quantum computing: A simple introduction (2012). https://www.explainthatstuff.com/quantum-computing.html
19. Li, X.-H., Li, C.-Y., Deng, F.-G., Zhou, P., Liang, Y.-J., Zhou, H.-Y.: Quantum secure direct communication with quantum encryption based on pure entangled states. Chin. Phys. **16**(8), 2149–2153 (2007)
20. Yuan, Y., Wang, F.-Y.: Blockchain: the state of the art and future trends. Acta Automatica Sin. **42**(4), 481–494 (2016)

A Stealth Migration Approach to Moving Target Defense in Cloud Computing

Saikat Das$^{(\boxtimes)}$, Ahmed M. Mahfouz, and Sajjan Shiva

Department of Computer Science, The University of Memphis,
Memphis, TN 38152, USA
{sdasl,amahfouz,sshiva}@memphis.edu

Abstract. A stealth migration protocol is proposed in this paper that obfuscates the virtual machine (VM) migration from intruders and enhances the security of the MTD process. Starting by encrypting the VM data and generating a secret key that is split along with the encrypted data into small chunks. Then the fragments are transmitted through intermediate VMs on the way to the destination VM. As a result, the chances of an intruder detecting the VM migration is reduced. The migration traffic is maintained close to normal traffic by adjusting the chunk size, thereby avoiding the attention of the intruder. Finally, the normal and migration traffic patterns are analyzed with the proposed protocol.

Keywords: Stealth migration · Cloud computing · Moving Target Defense · Secure MTD · Live migration

1 Introduction

Traditional cyber networks tend to be static, giving an attacker plenty of time to study them. During this time, the attacker can generate footprints, determine vulnerabilities, and decide when, how, and where to attack the network. Besides, once the attacker gains a privilege, he can maintain it for an extended period without being discovered by the system administrators. Various detection-based security approaches have emerged, but they struggle to detect attacks with accuracy and precision [1]. A new promising security technique, Moving Target Defense (MTD), has been adopted as a solution to overcome the challenge of network stat`ic nature. The idea came from the tactic of the battlefield where occasionally fighters change their positions and resources so that their enemies get confused and have difficulty attacking them. Likewise, in computer systems, moving the target resources could increase the complexity and uncertainty for attackers. In cloud computing, the virtualization techniques solely rely on distributing virtual machines (VMs) across different host machines around the world. Live Migration of VM instances from one physical device to another can be treated as MTD since the targeted host offloads its resident VM instances to a different host [12]. However, the process of VM live migration itself is not secure enough and is vulnerable to various active and passive attacks [3]. But, if the live migration is performed in a trusted and secure environment [8], it could be considered as a guaranteed MTD strategy against several types of attacks. In this paper, a stealth migration protocol is proposed that obfuscates the VM migration from the intruder and enhances the security of MTD.

© Springer Nature Switzerland AG 2020
K. Arai et al. (Eds.): FTC 2019, AISC 1069, pp. 394–410, 2020.
https://doi.org/10.1007/978-3-030-32520-6_31

The rest of the paper is organized as follows: In Sect. 2, an overview of MTD along with Live Migration is provided. Section 3 summarizes the related work. In Sect. 4, the threat model is discussed. The Stealth Migration Protocol is proposed in Sect. 5 to secure the Live Migration process and evaluate it by analyzing the normal vs. migration traffic patterns in Sect. 6. Finally, Sect. 7 concludes the paper and provides direction for future research.

2 Moving Target Defense

2.1 What is MTD?

MTD is the concept of controlling change across multiple system dimensions to increase uncertainty and apparent complexity for attackers, reduce their window of opportunity, and increase the costs of their probing and attack efforts [2]. MTD mainly aims to rebalance cyber-security by integrating dynamism, randomization, and diversification techniques in computer networks to hide the properties that an attacker can exploit to compromise the system. By changing the status of the system to be more dynamic and less deterministic, MTD boosts the system immunity to different attacks and increases the workload for the adversary.

MTD utilizes a wide range of systems security techniques and strategies [10, 11] that include Software Diversification, Runtime Diversification, Communication Diversification, and Dynamic Platform Techniques [12]. All these techniques intend to mutate the target system, making it unfamiliar to the attacker and force him to learn about the target repeatedly and newly, thus decreasing the probability of discovery and making the attacks costlier or unachievable.

2.2 Why is MTD Insecure?

MTD normally doesn't happen across fully secure networks. It is probable that the MTD traffic movement paths span across multiple networks and significant geographic distances [4] that allow attackers to identify the traceroute and discover the network footprint. Moreover, a compromised cloud system employing MTD can facilitate untrusted access to the moving VMs [9]. The ability to view or modify data associated with MTD or influence the movement services on the source and destination hosts raises several important security questions [5].

2.3 Live Migration

Live Migration of VM in cloud computing is the process of moving the VM from one physical host to another without affecting the service availability to users. It requires the transfer of the complete state of a VM from a host to another that comprises all the resources the VM uses in the source device. The resources include volatile storage, permanent storage, the internal state of the virtual CPUs, and connected devices (e.g., LAN Cards) [12]. Since the network-attached storage provides permanent room in the data center, it is not required to move the permanent storage during the VM migration process. The internal states of the virtual CPUs are usually only a few kilobytes of data and does not take considerable amount of time to be transferred. More extended periods are required to move the volatile memory contents which affect the performance of the live migration process and hence more attention is given to improve the transfer of volatile memory from the source to the destination [12].

3   Related Work

Several solutions have been proposed and implemented to secure the MTD approach over different live migration protocols. Some of these solutions are listed below.

Isolating Migration Network: This approach separates the virtual LAN consisting of the source and the destination hosts from the migration traffic of other networks, thus reducing the risk of exposure of migration to the whole network.

Network Security Engine Hypervisor (NSE-H): Xianqin et al. [13] proposed a secure VM migration framework that is based on hypervisors included with NSE. The framework provides an extension to the hypervisor by allowing the functionality of firewall and IDS/IPS to secure the migration from external attacks. It can also check the network for any intrusion and generate an alarm in case of any intrusion detection. The drawback of this framework is that the migration data may not remain unmodified during the transmission process because the data is not hashed or encrypted.

Secure VM-vTPM Migration Protocol: Berger et al. [14] classify the requirements and propose a new design of a virtual Trusted Platform Module (vTPM). The module consists of various steps starting from authentication, attestation, and data transfer stage. It also checks the integrity of the source and the destination. Only after verifying the integrity, the source VM starts the transfer to the destination VM. The file sent by the source VM is encrypted at vTPM and transferred to the destination VM. After completion of the transfer, the data at the vTPM is deleted. In the improved version, the source VM and target VM first authenticate each other to establish the trusted channel and then verify the integrity. Both the source and the destination negotiate keys with each other using DH key exchange algorithm. After the channel is established, the transfer begins. A specific mechanism for detecting and reporting suspicious activities is missing in this research.

SSH Tunnel: SSH tunnel is established between the proxies for secure movement. The proxy server at the source and the destination cloud communicate with each other and hide the details of both source and destination VMs [6]. In this research, an attacker can still examine the payloads in the flow by applying algorithms that are based on statistical characteristics to perform traffic analysis.

IPSec Tunnel: IPSec tunnel protects the data flow at server-to-server levels or from the edge router to edge router [15]. If MTD were done through IPSec tunnel, IP packets would be encrypted making it challenging to sniff data and trace. However, this approach slows down the migration process resulting in increased live migration downtime.

The above approaches have significant impact on securing the migration process. However, their overhead and differing attacking intention of intruders do not converge to a comprehensive solution. A protocol is proposed in this paper whose primary goal is to secure the migration process. Besides, the aim of this paper is to minimize the overhead and migration downtime.

4 Threat Model

In the threat model, an attacker is considered who is using Internet route hijacking to perform a man-in-the-middle attack on routes over the network to capture appropriate data from the traffic. Such an attack is emerging and represents a real security threat [16, 17]. The pre-condition to perform this attack is that the attacker can recognize the exact movement of the VM on the cloud network. Even with applying security measures like using encryption, secure tunnels or onion routing, an attacker can still use traffic characteristics to perform traffic analysis and detect the VM movement [7]. The attacker can view critical information such as the traffic speed, size, duration, and the involved endpoint hosts [18, 19] and launch his attack on the movement flow. Once the attacker detects the VM movement and the destination address, he can continue to attack the VM. This attack type can be reduced by utilizing secret sharing and moving VMs using multiple chunks.

5 Proposed Solution

Here, the new stealth migration technique is proposed that secures the live migration process at both the application and the live migration levels. First, a brief introduction on secret sharing encryption technology is given that has been used in the protocol, followed by the description of the protocol design.

5.1 Secret Sharing

Secret sharing is the method of dividing a secret among a group of participants where each participant allocates one secret share. The secret can only be reconstructed only when a certain number of shares are combined. Individual shares are of no use by themselves.

Shamir's secret sharing [10] is a popular way of splitting the secret. It divides the secret S (for example, the combination to a secure lockable box) into n pieces of data: $S_1, S_2, ... S_n$ in such a way that the knowledge of any k-1 or fewer S_i pieces leaves S completely undetermined, in the sense that the possible values for S seem as likely as with knowledge of 0 pieces.

(i) The knowledge of any k or more secret pieces (S_i) makes the secret (S) computable. That is, the construction of complete secret S can be done from any combination of k pieces of data [10].

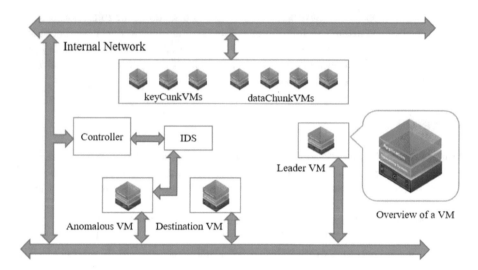

Fig. 1. A top-level view of stealth migration protocol

(ii)   The knowledge of any k-1 or fewer Si pieces leaves S completely undetermined, in the sense that the possible values for S seem as likely as with knowledge of 0 pieces. Said another way, the secret S cannot be reconstructed with fewer than k pieces [10].

This scheme is called (k, n) threshold scheme. When $[k = n]$, then every piece of the original secret S is required to reconstruct the secret.

Secret sharing encoding scheme is used in this stealth migration approach to split and hide the key information over the network. The key will be distributed by k/n pieces where k is the minimum number of shares to construct the secret.

5.2   Stealth Migration Protocol

Our stealth migration protocol is now discussed along with different algorithms used to migrate the VM using our stealth migration protocol.

A top-level view of the stealth migration protocol is shown in Fig. 1. The blue lines indicate the internal network by which all the five components communicate with each other. A virtual machine is also shown on the popup in detail. The stealth migration protocol is designed with five major components:

(i)   Controller: Controller of the virtual machine manager (VMM) that controls each component.
(ii) $VM_{anomalous}$: The virtual machine that is identified as affected/anomalous.
(iii) VM_{leader}: Controller marks a VM as leader who leads all intermediate VMs.
(iv) $VM_{available}$: The virtual machines that accept the data chunks and send to VM_{leader}.
(v) $VM_{destination}$: The virtual machine where the affected VM's data will be transferred finally.

In earlier research [20, 21], collaborative runtime monitors as IDS were created. They were attached to each running VM and continuously monitor VMs activity. As soon as the VM acts abnormally, IDS raises an alarm and sends the necessary VM information to the controller for further analysis. Figure 2 shows the workflow of the stealth migration protocol.

Each of the components and their tasks are briefly discussed next.

The Controller Component: The controller is the heart of the VMM, which has the full control over any component of this stealth migration protocol. Typically, it receives notifications (signals) from VMs, intrusion detection systems (IDS) and starts VM migration and manages all components to finish the migration. After successful completion of a migration, it maintains the connectivity of all components on the network. The controller works according to the following algorithm:

Algorithm (1): Maintain the connectivity among VMs and control over all VMs to operate the VM migration.

Input: Notifications (From IDS, VMs)

Output: Controlling signal, assigning task

```
while(true):
    notification = push_notification_from_IDS()
    if notification = anomalous
        normalTraffic = normal_traffic_size()
        VMsize = check_size(VManomalous)
        VMstate = state_of_VM(VManolamous)
        chunkSize = calculate_chunk_size()
        availableVMs = check_available_VMs()
        VMs = choose random 9VM from availableVMs
        VMdestination = VMs[random(1-9)]
        dataChunkVMs, keyChunkVMs = split_VMs()
        VMleader = random(from dataChunkVMs)
        send_instruction_to_VMdestination()
        secretKey = cerate_secret_key(256 bit)
        send_instruction_to_VManomalous()
        send_instruction_to_VMleader()
        for each VM in VMs
send_instruction_for_all_VMs()
        if notification (found from VM)
make_available(VM)
        if notification (found from VMdestination)
make_VManomalous_honeypot()
disconnect_VManomalous_from_network()
connect_VMdestination_to_network()
notify_IDS()
end while
```

The utility and usage of some of the methods used in the above algorithm are discussed below.

normal_traffic_size(): By using this method, the controller gets the normal traffic size from the IDS. The IDS provides the normal traffic size (packet transfer per second) to the controller by analyzing its normal network traffic activity.

check_size(VM$_{anomalous}$): Uses the IDS to determine the actual size of the affected virtual machine including OS, applications, dirty pages, etc.

state_of_VM(VM$_{anolamous}$): IDS provides the current state of the anomalous VM to the controller. State can be 'STARTING', 'RUNNING', 'STOPPED', 'SUS-PENDED', etc.

calculate_chunk_size(VMsize, 5n, normalTraffic): Calculates the chunk size by using the normal Traffic and VM$_{size}$. Since five intermediate VMs are used to migrate the data, the method determines and adjusts the chunk size by multiples of five to maintain the migration traffic similar to the normal traffic. For an example, the normal traffic is 6 MB/s and the VMsize is 500 MB. If the 100 MB data are transferred over the network at a single instance to five different VMs, an intruder can easily detect it as a migration traffic. So, if the chunk size can be maintained at approximately 6 MB, it can easily be transferred to intermediate VMs that eventually retains the normal traffic pattern. The purpose of maintaining the chunk size is to obfuscate the migration from intruder.

check_available_VMs(): Returns all VMs that are currently available.

split_VMs(): Here, nine VMs are used to operate our whole migration process. This method splits two types of VM lists from the available VM list except for the one that has already been marked as the destination VM where the affected VM's data will be moved eventually. The two lists are dataChunkVMs (where data segment will be transferred) and keyChunkVMs (where key segment will be transferred).

send_instruction_to_VM$_{destination}$(): Sends an instruction to destination VM address. The instruction instructs the VM to open a port and listen to the data coming from the VM$_{leader}$ address.

send_instruction_to_VM$_{anomalous}$(VMs address, VM$_{destination}$ address, chunkSize, secretKey): This method sends chunkSize, secretKey, VM$_{destination}$ address, data-ChunkVM lists (where to send data chunk), keyChunkVM lists (where to send the key chunk) as parameters to VM$_{anomalous}$ address. It also sends other instructions that will be discussed later in VM$_{anomalous}$ component section.

send_instruction_to_VM$_{leader}$(VM$_{anomalous}$ address, dataChunkVMs, keyChunkVMs address): Sends the VM$_{anomalous}$ address, dataChunkVMs list, keyChunkVMs list as parameter to VM$_{leader}$ address.

send_instruction_for_all_VMs(VM$_{anomalous}$ address, VM$_{leader}$ address): Sends the VM$_{anomalous}$ address, VM$_{leader}$ address as a parameter to all VMs. The instruction commands to transmit the data that has already been received from the anomalous VM to the leader VM address.

make_available(VM): This method marks the VM as available for further use.

disconnect_VM$_{anomalous}$_from_network(): Disconnects the anomalous VM from the network.

make_VM$_{anomalous}$_honeypot(): Marks the anomalous VM as a honeypot to play further games with the intruder.

connect_VM$_{destination}$_to_network(): Connects the destination VM to the real network.

Notify_IDS(): Finally, the controller notifies the IDS (from where it raised an anomalous activity alarm). The notification signal consists of monitoring the same VM or monitoring a new assigned VM.

The VM$_{anomalous}$ Component: In this component, an anomalous VM receives the instruction from the controller and performs the tasks that have been assigned to it by the controller. The tasks are conducted in accordance with the following algorithm:

Algorithm (2): Data wrap up, encryption, split data and key, send.

Input: VM$_{destination}$, VM8Addresses[], chunkSize, secretKey.

Output: Send data chunks to 8 VMs, notify controller.

```
while(true)
        if instruction (found from the controller)
        wrapData = wrap_data(VMData, VMdestination address)
        encryptedData = encrypt_data(wrapData, secretKey)
        dataChunks[dataSegment]=split_data_chunks()
        dataChunks[keySegment]=split_key_chunks()
        for VMAddress in VM8Addresses
      send_to_each_VMAddress(dataChunks[i])
  notify_controller(sending successful)
end while
```

The goal of this component is to send the data chunks and key chunks to the intermediate VMs in a normal traffic fashion. To do so, the VM wraps all of its VM data and merges it with the destination VMaddress. VM wraps data into the host machine by executing secure instructions provided by the controller. The merged data is then encrypted by using the secret key that is provided by the controller. After the encryption, the encrypted file is split according to the chunk size that is given by the controller. Finally, data chunks are sent one by one to the intermediate VMs listed on dataChunk VM list. The secret key is then split by *2/3* using the secret sharing scheme where 2 is the minimum share to construct the key and 3 is the total number of shares. Some of the methods used in the VM$_{anomalous}$ component are discussed below:

wrap_data(VMData, VM$_{destination}$ address): The affected VM uses this method to wrap up its whole content and adds the destination VM address with it to make a new dataset.

encrypt_data(wrapData, secretKey): Using this method, affected VM encrypts the whole dataset with the secret key given by controller.

split_data_chunks(chunkSize, encryptedData): After the encryption from the encrypt_data() method, affected VM splits the encrypted data into small chunks using the chunkSize. Since five VMs are used to migrate the affected VM intermediately, this method splits the whole file in such a way that it can be equally distributed to five VMs without affecting the normal traffic flow.

split_key_chunks(secretSharing scheme 2/3): This method splits the secret key into 2/3 size, where 2 is the minimum share to construct the complete secret and 3 is total share size.

send_to_each_VMAdress(dataChunks[i]): Affected VM sends each data chunk and key chunk to the specified VM addresses using this method. The method has the list of datachunkVMs, keychunkVMs and transfers the data chunks and key chunks to different VM lists.

notify_controller(sending successfully): Finally, using this method, affected VM sends the successful task completion signal to the controller.

The VMdatachunks and VMkeychunks Component: VMdataChunks and VMkeyChunks components are the intermediate state of this stealth migration protocol. The affected VM transfers the chunks (dataChunks, keyChunks) to the VMs that are listed to the dataChunk or keyChunkVM list as an intermediate storage. The basic difference between those two components is marking the chunked segment either by data or key. All VMs, receive chunks (dataChunk or keyChunk), mark them as data segment or key segment and send them to the VM_{leader} address. The following algorithm does the tasks:

Algorithm (3): Receive datachunk, mark it either as data segment or key segment and send it to the VM_{leader} address.

Input: $_{VManomalous}$ adress, dataChunks, VM_{leader} address

Output: Send data chunks to VM_{leader} address.

```
while(true)

        if instruction (found from the controller)
            open_port()
            listen(from VM_anomalous address)
        if recv()
            // mark segment for VMdatachunks component
            mark_as_data_segment(dataChunks)
            // mark segment for VMkeychunks component
            mark_as_key_segment(dataChunks)
            send_to_VM_leader()
            wipe(whole memory)
            make_available(make its available)
                notify_controller(availability)

end while
```

The methods used in the above algorithm are described below:

mark_as_data_segment(dataChunks): In this method, the VMs mark their data chunks as a data segment which is necessary to do for the further computation on VM_{leader} section. It helps the leader VM to identify the data chunks and the key chunks to construct the data and the key.

mark_as_key_segment(dataChunks): Here, the VMs mark their data chunks as a key segment, then transfer the data to the leader VM and after successfully transmitting the data, the VM wipes itself.

send_to_VM_{leader}(dataChunks, VM_{leader} address): By using this method, VMs send their received data chunks to the leader VM.

wipe(full memory): The VMs wipe their whole memory along with the data chunks in preparation for the future migration operation.

make_available(makeavailable)&notify_controller(availability): Each VM marks itself available and notifies the controller about its availability by using these two methods.

The VM_{leader} Component: In this component, leader VM receives all data and key chunks and constructs the secret key from the key chunks by using secret sharing encoding scheme. After successfully retrieving the secret key, it merges all data chunks and decrypts the complete data. After decoding the data, this component reveals the destination address for the data. Those tasks utilize the following algorithm:

Algorithm (4): Construct secret key, reveal destination VM address after decrypting the data and send data to the destination VM.

Input: dataChunks[], VMAddresses[]

Output: send data to $VM_{destination}$ address

```
while(true)
        if instruction (found from the controller)
                open_ports (for 7 VM address)
                listen_from_ports(VMAddresses[])
                if rccv()
        dataSegments[],keySegments[]=separate_data_or_ke
        y_chunks(dataChunks)
        secretKey = construct_key(keySegment[])
        data = construct_data(dataSegments[])
        VMdestinationAddress,VmData = decrypt_data()
        send_data(VMData, VMdestinationAddress)
        wipe(whole memory)
        make_available(make its available)
                notify_controller(availability)
end while
```

The details of the methods used in the above algorithm are discussed below:

dataSegments[ ], keySegments[ ] = separate_data_or_key_chunks(dataChunks): In this method, the leader VM separates the data segments and key segments from chunks that is received from the open port. This method differentiates the data chunks and key chunks as the previous component. Every chunk was marked either data or key when it was sent from intermediate VMs.

construct_key(keySegment[ ]): Secret sharing encoding scheme is used in this method to construct the key from the key segments. Since 2/3 encoding scheme is used, only 2 or more shares can reveal the key. Less than two shares never reveal the key according to the secret sharing encoding scheme. Shamir's secret sharing [10] scheme is used to construct the key.

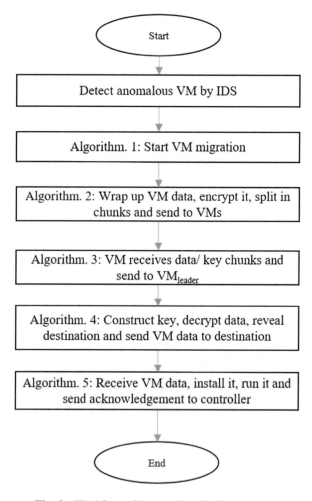

Fig. 2. Workflow of the stealth migration protocol

construct_data(dataSegments[ ]): Here, leader VM merges the data from the data segments.

decrypt_data(data, secretKey): In this method, complete data is decrypted through a decryption process. From the data, VM leader separates the destination VM address and the VM data content to be transferred.

send_data(VMData, VM$_{destination}$Address): This method is used to send the data to destination VM address.

Methods *wipe()*, make *available()*, and *notify_controller()* have been described earlier.

The VM$_{destination}$ Component: In this component, the destination VM receives the data from leader VM, installs it, and then makes it runnable before notifying the controller. The following algorithm performs the tasks:

Algorithm (5): Receives data, install it, notify controller after successful running.

Input: VMData, VM$_{leader}$ Address

Output: Notify controller

```
while(true)
       if instruction (found from the controller)
           open_port()
           listen (from VM_leader Address)
           if recv()
       installation(VMData)
       checking_feasibility()
       test_all_use_cases()
       if running () = successful
           send_ack(to controller)
end while
```

Methods used in the above algorithm are described below:

installation(VMData): In this method, destination VM installs the VM data for further processing.

checking_feasibility() and test_all_use_cases(): After installing the VMData, destination VM does the feasibility checking and checks all the test cases to make the VM workable. If VM works perfectly, it sends an acknowledgment to the controller with its current VM state.

6 Experimental Results

A cloud environment with OpenStack has been setup for this experiment. The testbed has one controller node, and four compute nodes. All the nodes are Dell Optiplex 960 machine with 16 GB RAM, 500 GB Hard Disk, and 3 GHz processor. Each of the nodes has two gigabit-Ethernet cards. One of the Ethernet network interfaces is connected to all other nodes via a switch. The other Ethernet interface is connected to the internet with public IP address. All the nodes are running on the Ubuntu 12.4 LTS server. The VM live migration is implemented with OpenStack for Moving Target defense as directed by the guidelines in [1]. Network-attached Storage was implemented among the three nodes via the local area network. In this current OpenStack implementation, the software does not support automated live migration. Only the user with admin privilege can issue the command for live migration either from the command prompt or from the horizon dashboard webpage. So, to automate the live migration, a python script is written which runs in controller node and any time it gets a notification from the IDS that is attached to the possible anomalous VM, it automatically instructs the affected VM to live migrate through the stealth migration protocol.

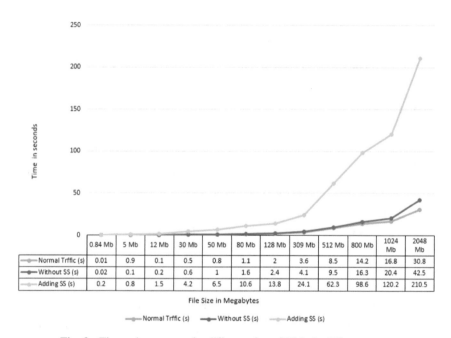

	0.84 Mb	5 Mb	12 Mb	30 Mb	50 Mb	80 Mb	128 Mb	309 Mb	512 Mb	800 Mb	1024 Mb	2048 Mb
Normal Trffic (s)	0.01	0.9	0.1	0.5	0.8	1.1	2	3.6	8.5	14.2	16.8	30.8
Without SS (s)	0.02	0.1	0.2	0.6	1	1.6	2.4	4.1	9.5	16.3	20.4	42.5
Adding SS (s)	0.2	0.8	1.5	4.2	6.5	10.6	13.8	24.1	62.3	98.6	120.2	210.5

File Size in Megabytes

Normal Trffic (s)   Without SS (s)   Adding SS (s)

Fig. 3. Time taken to transfer different size of VMs in different ways

In the experimentation, VMs of different sizes ranging from 840 KB to 2048 MB were migrated and measured the time taken to transfer the VM over the network. The time taken for file transfer was measured in three different ways: normal traffic, traditional migration traffic, and migration traffic using our stealth migration protocol. In typical traffic scenario, a similar size file (not a VM) is transferred through the application and recorded the time duration to transfer it completely. In traditional migration traffic scenario, a VM is migrated through the traditional OpenStack live migration process and recorded the time duration to move a VM from one physical address to another physical address. Finally, similar size VMs were migrated from one location to another through our stealth migration protocol. Figure 3 shows the time taken to transfer different size VMs for three different scenarios. Figure 3 shows that as long as the VM size stays between 840 KB to 512 MB, there are no major significant time changes. When the size of the VM increases from 512 MB, a spike is shown on the graph, which means that the cost of time needs to be compromised for the sake of the system security.

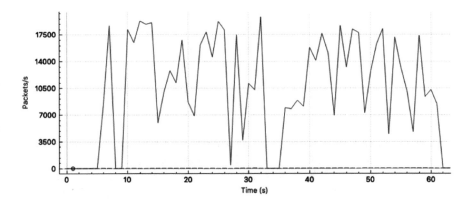

Fig. 4. Network analysis of a normal traffic

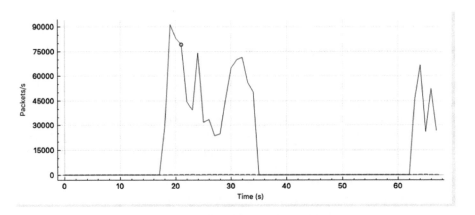

Fig. 5. Network analysis of a migration traffic

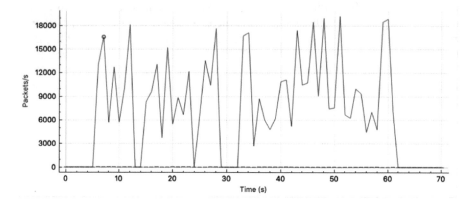

Fig. 6. Network analysis of migration traffic using Secret Sharing scheme and stealth migration protocol

Figures 4, 5 and 6 show respectively the normal traffic, traditional migration traffic, and migration traffic pattern that used stealth migration while it was transferring 900 MB of data or a VM. It is obvious that in Fig. 5, an attacker can easily distinguish migration traffic while he was analyzing the VM traffic pattern. The proposed protocol converts the distinguishable migration traffic pattern into a typical traffic pattern, which eventually obfuscates it from the attackers. Figures 4 and 6 are pretty much similar in terms of packets transferred per second which indicates that the attackers would not be able to distinguish the migration traffic from the normal traffic when the migration goes through the stealth migration protocol.

7   Conclusion and Future Work

The security concerns of VM live migration and deduced state of the art to secure MTD are discussed here. In this research, a stealth migration protocol is designed to minimize the risk and reduce the chance of being identified by an intruder to enhance the security of MTD. The protocol was implemented on the development testbed using "OpenStack" and the migration traffic was made indistinguishable from the normal traffic to help hide the migration information from the intruder. In experimentation, different sizes VMs are used for live migration purpose, and it can be concluded that if the size of VM is increased beyond 500 MB, the cost of time becomes prominent. Since the primary goal of this proposed protocol is to migrate the VMs securely while obfuscating it from the intruder, the cost of security turns into migration downtime. Investigation of those issues to minimize the overhead regardless of VM size and plans to adjust the migration traffic chunk size at a variable rate that could be more accurate to hide the migration traffic from the intruder are continuing. Further, machine-learning techniques to detect the migration traffic anomalies and to reduce the false positive alarms during the migration process are also being investigated.

References

1. Yackoski, J., et al.: Mission-oriented moving target defense based on cryptographically strong network dynamics. In: Proceedings of the Eighth Annual Cyber Security and Information Intelligence Research Workshop. ACM (2013)
2. https://www.dhs.gov/science-and-technology/csd-mtd
3. Ahmad, R.W., Gani, A., Hamid, S.H.A., Shiraz, M., Xia, F., Madani, S.A.: Virtual machine migration in cloud data centers: a review, taxonomy, and open research issues. J. Supercomput. **71**(7), 2473–2515 (2015)
4. Oberheide, J., Cooke, E., Jahanian, F.: Empirical exploitation of live virtual machine migration. In: Proceedings of BlackHat DC Convention (2008)
5. Kozuch, M., Satyanarayanan, M.: Internet suspend/resume. In: Proceedings of Fourth IEEE Workshop on Mobile Computing Systems and Applications. IEEE (2002)
6. Duncan, A., et al.: Cloud computing: Insider attacks on virtual machines during migration. In: 2013 12th IEEE International Conference on Trust, Security and Privacy in Computing and Communications (TrustCom). IEEE (2013)
7. Achleitner, S., et al.: Stealth migration: hiding virtual machines on the network. In: INFOCOM 2017-IEEE Conference on Computer Communications (2017)
8. Suetake, M., Kizu, H., Kourai, K.: Split migration of large memory virtual machines. In: Proceedings of the 7th ACM SIGOPS Asia-Pacific Workshop on Systems. ACM (2016)
9. Deshpande, U., et al.: Fast server deprovisioning through scatter-gather live migration of virtual machines. In: 2014 IEEE 7th International Conference on Cloud Computing (CLOUD). IEEE (2014)
10. https://en.wikipedia.org/wiki/Shamir%27s_Secret_Sharing
11. Clerk Maxwell, J.: A Treatise on Electricity and Magnetism, 3rd edn, vol. 2, pp. 68–73. Clarendon, Oxford (1892)
12. Polash, F., Shiva, S.: Automated live migration in openstack: a moving target defense solution. J. Comput. Sci. Appl. Inform. Technol. **2**(3), 1–5 (2017). https://doi.org/10.15226/2474-9257/2/3/00119
13. Xianqin, C., Xiaopeng, G., Han, W., Sumei, W., Xiang, L.: Application-transparent live migration for virtual machine on network security enhanced hypervisor. China Commun. **8**, 32–42 (2011). Research paper
14. Berger, S., Caceres, R., Goldman, K.A., Perez, R., Sailer, R., Doorn, L.: Virtualizing the trusted platform module. In: USENIX Security, pp. 305–320 (2006)
15. Tamrakar, A.: Security in live migration of virtual machine with automated load balancing. Int. J. Eng. Res. Technol. (IJERT) **3**(12) (2014)
16. Bgp hijacking. https://www.blackhat.com/docs/us-15/materials/us-15-Gavrichenkov-Breaking-HTTPS-With-BGP-Hijacking-wp.pdf
17. Hierarchy token bucket theory. http://research.dyn.com/2013/11/mitm-internet-hijacking/
18. Fu, X., et al.: On effectiveness of link padding for statistical traffic analysis attacks. In: Proceedings of 23rd International Conference on Distributed Computing Systems. IEEE (2003)
19. Houmansadr, A., Brubaker, C., Shmatikov, V.: The parrot is dead: observing unobservable network communications. In: 2013 IEEE Symposium on Security and Privacy (SP). IEEE (2013)

20. Dharam, R., Shiva, S.G.: Runtime monitors for tautology based SQL injection attacks. In: 2012 International Conference on Cyber Security, Cyber Warfare and Digital Forensic (CyberSec). IEEE (2012)
21. Shiva, S., Das, S.: CoRuM: collaborative runtime monitor framework for application security. In: 2018 IEEE/ACM International Conference on Utility and Cloud Computing Companion (UCC Companion). IEEE (2018)

The First Quantum Co-processor Hybrid for Processing Quantum Point Cloud Multimodal Sensor Data

George J. Frangou[1]([⊠]), Stephane Chretien[2], and Ivan Rungger[2]

[1] Massive Analytic Ltd, IDEALondon, 69 Wilson Street,
London EC2A 2BB, UK
george.frangou@massiveanalytic.com
[2] National Physical Laboratory, Hampton Road, Teddington TW1 0LW, UK

Abstract. The large-scale multimodal sensor fusion of the internet of things (IoT) data can be transformed into an N-dimensional classical point cloud. For example, the transformation may be the fusion of three imaging modalities of different natures such as LiDAR (light imaging, detection, and ranging), a set of RGB images, and a set of thermal images. However, it is not easy to process a point cloud because it can have millions or even hundreds of millions of points. Classical computers therefore often crash when operating a point cloud of multimodal sensor data. Quantum Point Clouds (QPC) address the problem of uncertainty in multi-modal sensor data, such that precognitive/predictive models can be derived with outcomes of greater certainty than classical information processing methods. This paper presents early experiments of the first application of a quantum co-processor hybrid for processing quantum point cloud multimodal sensor data from an autonomous racing car. Applied to the more complex case of cave mapping, it then describes the first hybrid classical-quantum co-processor, comprising a graphical processing unit, differential pulse code modulator and a quantum computer. The graphical processing unit comprises a multiple input/output data interface, transformation means for transforming a fused depth bitmap of the multi-modal sensor data into a point cloud representation with world coordinates, control logic that manages the multiple input/output data interface, and the differential pulse code modulator. The quantum co-processor comprises an assembly of quantum computing chips.

Keywords: Quantum computing · Quantum point clouds · Multimodal sensor · Data fusion · Data uncertainty measures · Autonomous systems

1 Introduction

1.1 The Limitations of Classical Computing with Multimodal Sensor Data

The development of quantum computing hardware is proceeding at a fast pace and current quantum computers exist, with the number of quantum-bits (qubits) per computer steadily increasing. Developments in quantum sensors and devices are also showing promise but are not yet commercially available.

© Springer Nature Switzerland AG 2020
K. Arai et al. (Eds.): FTC 2019, AISC 1069, pp. 411–426, 2020.
https://doi.org/10.1007/978-3-030-32520-6_32

Realizing quantum computing capability demands that hardware efforts are being augmented by the development of quantum software to obtain optimized quantum algorithms able to solve application problems of interest. Transforming multimodal data, such as sensor data or generating big data, into a point cloud allows its analysis with classical algorithms, and we aim to extend this to quantum computing architectures. Applications can include precision medicine, surveillance, cybersecurity, new types of sensors and better, more secure communication systems.

1.2   What Are the Advantages of a Quantum Computer over a Classical Computer?

A binary digit, characterized as 0 and 1, is used to represent information in classical computers. A binary digit can represent up to one bit of Shannon information, where a bit is the basic unit of information and where the word bit is synonymous with a binary digit.

In classical computer technologies, a processed bit is implemented by one of two levels of low DC voltage. Whilst switching from one of these two levels to the other, a so-called forbidden zone must be passed as fast as possible. This is because electrical voltage cannot change from one level to another instantaneously.

There are two possible outcomes for the measurement of a qubit—usually taken to have the value "0" and "1", like a bit or binary digit. However, whereas the state of a bit can only be either 0 or 1, the general state of a qubit according to quantum mechanics can be a coherent superposition of both. Moreover, whereas a measurement of a classical bit would not disturb its state, a measurement of a qubit would destroy its coherence and irrevocably disturb the superposition state. It is possible to fully encode one bit in one qubit. However, a qubit can hold more information, for example up to two bits using superdense coding [8].

For a system of n components, a complete description of its state in classical physics requires only n bits, whereas in quantum physics it requires $2n - 1$ complex number.

1.3   New AI and Robotics Capabilities from QPC Processing

Quantum computers can be used to operate point clouds based on their more powerful processing capabilities than classical computers. Processing of a quantum point cloud with quantum computers enables new artificial intelligence and robotics products to be developed. For example [9–14]:

(a) Improving situational awareness, including ensuring safety during transportation in hazardous conditions such as darkness, fog or dust.
(b) Increasing productivity in deployment, improvement or maintenance of buildings or national infrastructure.
(c) Improving identification and understanding of states and features that is impossible to see by conventional means, particularly in medical, environmental and security applications.
(d) Supporting secure peer-to-peer transfer of data and information, such as across smart cities and environments.

1.4   Classical-Quantum Computer Hybrid

We propose a hybrid approach, where outputs from classical sensor technologies are modified to enable early applications of current state-of-the-art quantum computers of the order of tens of qubits.

The large-scale multimodal sensor fusion of the internet of things (IoT) data can be transformed into an N-dimensional classical point cloud. For example, the transformation may be the fusion of three imaging modalities of different natures such as video, LiDAR (light imaging, detection, and ranging), a set of RGB images, and a set of thermal images. However, it is not easy to process a point cloud because it can have millions or even hundreds of millions of points. Classical computers therefore often crash when operating a point cloud of multimodal sensor data.

Emerging quantum computing technology can help users to solve the multimodal sensor point cloud processing problem more efficiently. Quantum computation is therefore expected to become an important and effective tool to overcome the high real-time computational requirements. In order to operate point clouds in quantum computers, there are two problems to be solved, and these are quantum point cloud representation and quantum point cloud processing [1]. Quantum representations are well described in the literature [2, 5]. However, there is a distinct paucity of methods to express a three-dimensional image using quantum representation. Furthermore, to provide a quantum computing based application for fused multimodal sensor data the representation and processing need to be further generalized to N-dimensional quantum point clouds.

1.5   Classical Point Cloud Generation Experimental Setup

We are developing quantum representation and processing algorithms for point cloud data that can be run on a noisy intermediate-scale quantum (NISQ) computer. We, therefore, reduce our point cloud data volumes to enable processing with available quantum computers to prove the feasibility of our approach. We will then add information on how we expect the methods to scale as future quantum computers become larger in terms of their number of qubits. These research findings will be presented in a future paper. The first stage of the project was therefore devoted to extensive discussions and exchange of knowledge to determine which types of systems can potentially be suitable as test systems for a NISQ quantum computer. In such a near term device the quantum processing unit (QPU) will typically have 50–200 qubits, which will not be fault-tolerant but will instead be affected by a finite amount of noise during the processing of data. We evaluated the current state of the art for demonstrations of concepts on quantum computers for point clouds [1] and quantum image processing [2], and typically the prototype demonstration of the first quantum co-processor hybrid for processing quantum point cloud multimodal sensor data is done for around 10 data points or pixels using video data from an autonomous racing car developed by the UK Institute of Mechanical Engineers. Instead of using just an arbitrary set of 10 data points, we aimed to find a system which can compress a realistic classical point cloud to have only around 10 final points to then be passed to the quantum computer.

We chose a dataset that was simple enough to test the method and sufficiently realistic to demonstrate real-world applicability. The chosen system is a point cloud dataset from an 'Environment Perception for an Autonomous Formula Student Car' project, which is a collaborative project between Massive Analytic Limited (MAL) and University College London (UCL) for the simulation of autonomous racing cars. As the car drives around a track bounded by a set of cones at the edge of the road, video footage is recorded. Buildings were also added around the track to give the simulation more features (Fig. 1(a)). The video footage is first converted into a point cloud, and then the position of the cones delimiting the road is extracted. This corresponds to a large data compression, and for each position of the car one can pass the 6–10 cones closest to the car as compressed CPC to the quantum computer.

Two compressed point-clouds were extracted from the video footage, the first denoting the positions of the cones, and the second the position of the car as it moves around the track. A modified ORB-SLAM algorithm was used to turn the video into point clouds, with world-coordinates used to identify the position of the car in each frame.

Fig. 1a. Image from simulated environment showing cones.

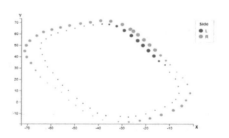

Fig. 1b. CPC representation: position of cones with z co-ordinate represented by size of point (larger is higher).

Applying MAL's AI, artificial precognition (AP) to the point clouds to remove outliers and predict where missing cones would be, MAL used two clustering algorithms (stage 1 - coarse tuning): Density-Based Spatial Clustering of Applications with Noise (DBSCAN) and K-Means, to ensure a regular density of cones around the whole track and to classify the cones as either left or right. To predict or preconize where missing cones would be and ensure a regular density of cones around the whole track, two ellipsoid regression models (stage 2 - fine tuning) were fitted for each of the left set of cones and the right set. The final compressed CPC representation of the cones is shown in Fig. 1(b). This 3-dimensional CPC was then used by the National Physical Laboratory (NPL) for quantum representation.

2 Theoretical Considerations of Classical and Quantum Point Clouds

2.1 What Is a Point Cloud?

A point cloud is a set of data points in a three-dimensional coordinate system to represent the external surface of an object [3]. Point clouds can be acquired from hardware sensors such as for example stereo cameras, 3D scanners, LiDAR, CCTV, or time-of-flight cameras [4]. Alternatively, the point clouds can be generated synthetically from a computer program [5]. As the scanned initial data, point clouds have been widely used in the protection of cultural relics [6], topographic surveys, body scanning and industrial designing. Point cloud examples are shown in Fig. 2 below.

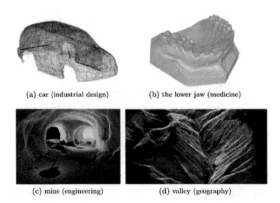

(a) car (industrial design)   (b) the lower jaw (medicine)

(c) mine (engineering)   (d) valley (geography)

Fig. 2. Point Cloud examples Point clouds. **a–d** cited from [3–5] respectively

2.2   Mathematical Representation of a Classical Point Cloud

Figure 3 shows a simple point cloud example and its three views, in which a cube represents a point. The point cloud has six points. If attribute A is the point's color, including R, G, and B, then the point cloud can be represented as:

$$P = \{$$
$$
\begin{aligned}
p_0 &= \langle\ 1,\ 1,\ 3,\ 255,\ 0,\ 0\ \rangle, \\
p_1 &= \langle\ 1,\ 2,\ 2,\ 255,\ 0,\ 255\ \rangle, \\
p_2 &= \langle\ 2,\ 1,\ 2,\ 255,\ 255,\ 0\ \rangle, \\
p_3 &= \langle\ 1,\ 0,\ 2,\ 0,\ 255,\ 0\ \rangle, \\
p_4 &= \langle\ 0,\ 1,\ 2,\ 0,\ 0,\ 255\ \rangle, \\
p_5 &= \langle\ 1,\ 1,\ 1,\ 0,\ 255,\ 255\ \rangle
\end{aligned}
$$
$$\}$$

$$(1)$$

Where the first three digits of the matrix are real space coordinates and the other four are color or other properties of the point.

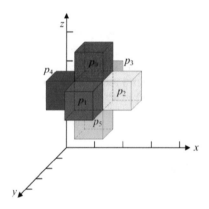

Fig. 3. Simple point-cloud example

We can transform fused large-scale multimodal IoT sensor into a classical point cloud (CPC). For example, the transformation may be the fusion of three imaging modalities of different natures such as LiDAR, a set of RGB images, and a set of thermal images. To combine imagery from different data sources a variety of SLAM (Simultaneous localization and mapping) techniques for sensor fusion can be used including scale-invariant feature transforms (SIFT), occlusion detection, manifold fusion, Extended Kalman filters, and visual odometry [7].

2.3   Representing a Multimodal Sensor Network as a Quantum Point Cloud

First, we prepare the classical point cloud for quantum processing by first compressing it. The purpose is to reduce requirements for the number of quantum gates and the total network complexity. Using the point cloud in Fig. 3 as an example and selecting p_0 as a data point, the point cloud can then be expressed as:

$$
\begin{aligned}
P = {} & \{p_0, p_1, p_2, p_3, p_4, p_5\} \\
= {} & \{ \\
& 1,\ 1,\ 3,\ 255,\ 0,\ 0,\ 1 \\
& 1,\ 2,\ 2,\ 0,\ 0,\ 1,\ 0 \\
& 2,\ 1,\ 2,\ 0,\ 1,\ 0,\ 0 \\
& 1,\ 0,\ 2,\ 2,\ 1,\ 0,\ 0 \\
& 0,\ 1,\ 2,\ 2,\ 0,\ 1,\ 0 \\
& 1,\ 1,\ 1,\ 2,\ 1,\ 1,\ 0 \\
& \}
\end{aligned}
\tag{2}
$$

This is a modified representation post data reduction and compression using differential pulse code modulation (see Sect. 3.1, Differential Pulse Code Modulator). Secondly, based on the compression above, a generalized QPC can be represented as:

$$
\begin{aligned}
|P\rangle &= \frac{1}{2^{\frac{m_i}{2}}} \sum_{i=0}^{N-1} |i\rangle \otimes |X_i\ Y_i\ Z_i\rangle \otimes |A_i\rangle \otimes |f_i\rangle \\
|i\rangle &= |i_{m_{i-1}}\ i_{m_{i-2}} \ldots i_0\rangle, i_j \in \{0,1\} \\
|X_i\ Y_i\ Z_i\rangle &= \left|x_i^{m_x-1} \ldots x_i^0\right\rangle \left|y_i^{m_y-1} \ldots y_i^0\right\rangle \left|z_i^{m_z-1} \ldots z_i^0\right\rangle, x_i^j, y_i^j, z_i^j \in \{0,1\} \\
|A_i\rangle &= \left|a_i^{m_a-1} a_i^{m_a-2} \ldots a_i^0\right\rangle, a_i^j \in \{0,1\} \\
f_i &\in \{0,1\}
\end{aligned}
\tag{3}
$$

where, i is the counting number, X_i, Y_i, Z_i, are the 3D coordinates, A_i, is the point's attribute, f_i, is the flag and

$$
m_i = \begin{cases} \lceil \log_2 N \rceil, & N > 1 \\ 1, & N = 1 \end{cases}
$$

We have represented a fused multimodal sensor data as an N-dimensional quantum point cloud (QPC). By generalizing this quantum representation from three to N-dimensions, we have derived, a numerical solution suitable for QPC processing. If a point has no attribute, the QPC can be simplified to

$$
|P\rangle = \frac{1}{2^{\frac{m_i}{2}}} \sum_{i=0}^{N-1} |i\rangle \otimes |X_i\ Y_i\ Z_i\rangle
\tag{4}
$$

2.4   Example of a QPC

Equation (5) gives the quantum representation of the example shown in Fig. 3.

$$
|P\rangle = \frac{1}{2^{\frac{3}{2}}} (|000\rangle \otimes |010111\rangle \otimes |11111111\ 00000000\ 00000000\rangle \otimes |1\rangle
$$

$$
\begin{aligned}
&+ |001\rangle \otimes |011010\rangle \otimes |00000000\ 00000000\ 00000001\rangle \otimes |0\rangle \\
&+ |010\rangle \otimes |100110\rangle \otimes |00000000\ 00000001\ 00000000\rangle \otimes |0\rangle \\
&+ |011\rangle \otimes |010010\rangle \otimes |00000010\ 00000001\ 00000000\rangle \otimes |0\rangle \\
&+ |100\rangle \otimes |000110\rangle \otimes |00000010\ 00000000\ 00000001\rangle \otimes |0\rangle \\
&+ |101\rangle \otimes |010101\rangle \otimes |00000010\ 00000001\ 00000001\rangle \otimes |0\rangle)
\end{aligned}
$$

$$
\text{Distinguisher} \mid \text{Position} \mid \qquad\qquad \text{Color} \qquad\qquad \mid \text{Difference flag}
$$

$$(5)$$

The main resource in quantum preparation is the number of quantum gates instead of the number of qubits, because the number of qubits used in a QPC cloud is far less than the number of bits used in a CPC. If we can decrease the number of quantum gates in the preparation, it will be even more efficient.

Therefore, the main task of QPC compression is to reduce the number of gates used during QPC preparation. Furthermore, the number of quantum gates can be used to indicate the network time complexity because in quantum network, each quantum gate is an operation which needs a certain amount of time to do it.

In Fig. 4, we can see that the main part of the QPC preparation circuit is m_i-CNOT gates. The total is 34 qubits (just about the size of current quantum computers).

This is not efficient in terms of requirements for the number of qubits. Amplitude for each term in the sum is equal: we should use the amplitude to store information.

Does that make processing more difficult or even impossible? This will be evaluated in the second part of the project. Position information is not encoded efficiently as this requires 2^n qubits, while ideally, we only need n qubits. A large number of qubits are used to store color information: we suggest that we put that information in the amplitude.

2.5   Quantum Point Clouds (QPC) and Uncertainty in Multi-modal Sensor Data

We have hypothesized that Quantum point clouds (QPC) may address the problem of uncertainty in multi-modal sensor data better than CPCs, such that precognitive or predictive models can be derived with outcomes of greater certainty than classical information processing methods. Furthermore, we can show that quantum computers can help users to solve the multimodal sensor point cloud processing problem more efficiently than classical computers. The experimental results of this research will be published in a subsequent paper.

In order to operate point clouds in quantum computers, as stated above there are two problems to be solved. These are quantum point cloud representation and quantum point cloud processing. This is similar to quantum image processing. Research in the

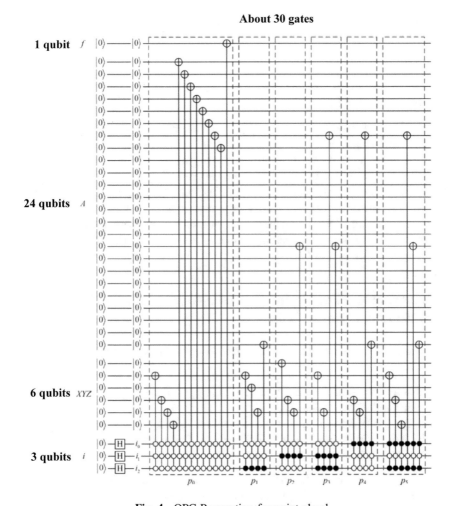

Fig. 4. QPC Preparation for point cloud

field of quantum image processing started with proposals on quantum image representations. The quantum images are two-dimensional arrays of qubits, called a qubit lattice or a quantum state. While quantum representations of two-dimensional images are well described in the literature, there is a paucity of methods to express a three-dimensional image using quantum representation.

There is a need to overcome the high real-time computational requirements of classical digital image processing. Also, there is a need to process large-scale multi-modal sensor fusion of IoT data as a QPC encoded as qubits, i.e. as quantum infor"-mation. The field includes quantum computation, and cryptographic and communication protocols like quantum key distribution and dense coding.

Quantum computation is becoming an important and effective tool to overcome the high real-time computational requirements of classical digital image processing [20], of which multimodal sensor fusion is just one example.

2.6   Experimental System for Processing Quantum Point Clouds of Multimodal Sensor Data

It is envisaged that processing of a QPC with quantum computers will enable new artificial intelligence and robotics products to be developed. Some of these are described in the next sections. Importantly, we present a detailed description of an experimental system we have built for processing QPCs derived from multimodal sensor data. The system comprises a graphical processing unit and a quantum co-processor. As described above, recent experiments have included using the hybrid system representing a QPC of multimodal sensor data from an autonomous racing car. We now describe a more complex application of the quantum co-processor hybrid for processing QPC multimodal sensor data applied to cave mapping.

3   Experimental Set-Up for Processing Quantum Point Cloud Multimodal Sensor Data

3.1   System for Processing a QPC Derived from Multimodal Sensor Data

With reference to Fig. 5, we propose an experimental system for processing a QPC derived from multimodal sensor data [15]. The system supports quantum computing chips with qubits greater than 20.

The system comprises three main subsystems a Graphical Processing Unit, Differential Pulse Code Modulator and Quantum Co-processor. The system is now described in detail.

Graphical Processing Unit (GPU). The GPU has a multiple input/output data interface with a peripheral component interconnect express bus. The data interface allows input of multiple multimodal sensor data feeds, and transforms those inputs into a predefined fused depth bitmap with world coordinates (x, y, z). The fused depth bitmap of the multimodal sensor data is transformed into a CPC. The output interface uses quantum encoding and includes control logic and circuitry that manages the multiple input/output data interface between the graphical processing unit and the quantum co-processor. We used a GPU in the NVIDIA Tesla range. The input interface of the multiple input/output data allows multiple multimodal sensor data inputs from multi-sensors.

By way of an example, the multi-sensors applied to cave mapping can include LiDAR, digital imaging and surveying. The output interface allows the processing of a CPC using quantum computing, which is more efficient that classical methods.

Differential Pulse Code Modulator (DPCM). The DPCM improves the GPU by compressing the CPC in a lossy manner. The purpose is to reduce the number of quantum gates required during QPC preparation and to reduce the total network complexity.

The digital pulse code modulator operates with a compression factor of R [1]; the greater is R then the better the compression effect is *mi*-CNOT gates before compression, and in which the digital postcode modulator operates with a GA which is the number of *mi*-CNOT gates after compression.

Quantum Co-processor (QC). The QC has an assembly of computing chips with:

(a) QPC preparation which uses quantum mechanical processes to process the QPC information which is first stored as a quantum state;
(b) QPC preparation circuits m_i -CNOT gates (m_i control bit);
(c) control logic that manages the input/output interfaces between the QC and GPU; and
(d) a quantum register which stores the quantum data.

4 Example: Quantum Co-processor Hybrid for Processing Quantum Point Cloud Multimodal Sensor Data Applied to Cave Mapping

4.1 QPC Preparation: CPC Compression

We now describe by way of example and with reference to Fig. 5 how the quantum co-processor hybrid processes point cloud multi-modal sensor data is applied to cave mapping.

The system in Fig. 5 supports the representation of a 3D hybrid classical point cloud (CPC). As already stated in the previous sections, it takes into account the realization that, in order to operate point clouds in quantum computers, there are two problems to be solved: quantum point cloud representation and quantum point cloud processing. This is similar to quantum image processing (QIP).

In classical image processing, however, three-dimensional (3D) image representation model is very different from a two-dimensional (2D) model.

In order to provide a higher order quantum point cloud (QPC), the CPC must first be improved instead of being transposed into quantum computers directly. Compression is important in reducing the quantum resources used to prepare a QPC. The apparatus shown in the drawing uses Differential Pulse Code Modulation [13]. We use Differential pulse code modulation (DPCM) because the difference between the actual point-cloud value and its predicted value is quantized and then encoded forming a digital value.

QPC compression is the procedure that reduces the quantum resources used to prepare quantum point clouds. The main resource in quantum preparation is the number of quantum gates not the number of qubits; the number of qubits in a QPC cloud is far fewer than the number of bits used in a classical point cloud (CPC). Furthermore, the number of quantum gates can be used to indicate the network time complexity because in a quantum network, each quantum gate is an operation which requires a specific processing time. This is referred to as network complexity. Therefore, the main task of QPC compression is to reduce the number of gates used during QPC preparation. Basic quantum gates, NOT, CNOT and Toffoli are preferably used.

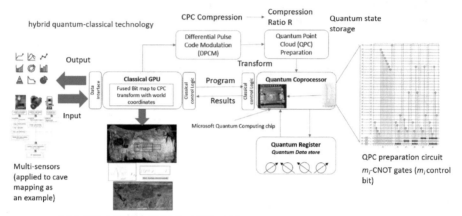

Detailed 3D hybrid Classical Point Cloud (CPC) model

Fig. 5. Classical-Quantum co-processing applied to multimodal sensor cave mapping

4.2 Hybrid Quantum-Classical Technology

The system shown in Fig. 5 presents a hybrid quantum-classical technology (HQCT) system for processing a quantum point cloud (QPC) of multimodal sensor data. The basis of the system is a classical Graphical Processing Unit (GPU) such as an NVIDIA Tesla range, working cooperatively with a Quantum co-processor such as Microsoft, IBM or Google quantum computers supporting qubits >20.

By exploiting the inherent advantages of using quantum mechanics to predict and infer outcomes, the purpose of the HQCT is to output a numerical solution to the QPC, to enable extremely large and noisy multimodal sensor data to be analyzed with greater precision, efficiency and certainty than is possible than with the numerical analysis of a classical point cloud (CPC).

Simply as an example, the HQCT is shown applied to the analysis of cave mapping. We have selected this application because of the large number of different sensor types involved. There are myriad other applications where the analysis of fused multimodal data is required. Another notable example is autonomous vehicles.

In the cave mapping example, a Multiple Input Multiple Output (MIMO) Data Interface to the classical GPU allows the input of multiple multimodal sensor data LiDAR, Digital Imaging and Surveying inputs) which it converts into a predefined fused bitmap format. The fused depth bitmap of the multi-modal sensor data is then transformed into a detailed 3D hybrid classical point cloud representation with world coordinates.

Mathematically, the hybrid classical point cloud representation with world coordinates can presented as an N-point matrix (if a cube represents a point then N = 6).

The function of the differential pulse code modulator (DPCM) is to improve the CPC by compressing in a lossy manner (with a compression factor R). Its main purpose is to reduce the number of quantum gates during quantum point cloud preparation, and to reduce total network complexity. The greater R is, the better the compression effect. R is the number of m_i-CNOT gates before compression, and GA is the number of m_i-CNOT gates after compression.

4.3   QPC Preparation and the Quantum Co-processor

There are classical control logic devices that manage the input/output interfaces between the classical GPU and the quantum co-processor (Program - Query and Results – Return).

The quantum co-processor shown uses an assembly of quantum computing chips such for example Microsoft, IBM or Google supporting qubits >20. For illustration only, Fig. 5 shows a photographic image of a single Microsoft quantum computing chip. In practice there will be a number of these chips connected together.

The next stage of the process of transforming the CPC to a QPC is called QPC preparation. The QPC preparation which uses quantum mechanical processes to process the QPC information which is first stored as a quantum state. An image of a QPC preparation circuit with mi-CNOT gates (mi control bit) within the quantum computing chip is shown. The quantum register which is analogous to a classical computing register stores the quantum data.

The QPC preparation provides for a quantum matrix representation of the CPC matrix, to which a numerical solution has greater precision, efficiency and certainty than with the numerical analysis of the classical point cloud (CPC).

4.4   QPC Numerical Solution

The numerical solution to the QPC is effectively a multi-dimensional number matrix that can be analyzed. This matrix is presented as an agent response ('Program' in the diagram) from the classical control logic device on the classical GPU to the classical control logic device on the quantum co-processor. The output on the GPU data interface enables the multi-dimensional number matrix from the quantum co-processor to be visualized as a set of charts. Charts from the CPC analysis can be compared with charts from the QPC to quantify the improvement in accuracy and analytical outcome certainty gained by representing the fused multimodal sensor data as a QPC.

5   Summary and Conclusions

We are developing quantum representation and processing algorithms for point cloud data that can be run on a noisy intermediate-scale quantum (NISQ) computer. We evaluated the current state of the art for demonstrations of concepts on quantum computers for point clouds [1] and quantum image processing [2], and typically the prototype demonstration of the first quantum co-processor hybrid for processing quantum point cloud multimodal sensor data is done for around 10 data points or pixels using video data from an autonomous racing car developed by the UK Institute of Mechanical Engineers. We generated a compressed CPC representation of cones on a racing track as shown in Fig. 1(b).

We have adapted our considered system sizes to available quantum computers to prove the feasibility of our approach. We will then add results from further research on how we expect the methods to scale as future quantum computers become larger in terms of their number of qubits.

We describe the first quantum co-processor hybrid for processing point cloud multimodal sensor data. The system shown in Fig. 5 presents a hybrid quantum-classical technology (HQCT) system for processing a quantum point cloud (QPC). The basis of the system is a classical Graphical Processing Unit (GPU) such as an NVIDIA Tesla range, working cooperatively with a Quantum co-processor such as Microsoft, IBM or Google quantum computers supporting qubits >20.

By exploiting the inherent advantages of using quantum mechanics to predict and infer outcomes, the purpose of the HQCT is to output a numerical solution to the QPC, to enable extremely large and noisy multimodal sensor data to be analyzed with greater precision, efficiency and certainty than is possible than with the numerical analysis of a classical point cloud (CPC).

Simply as an example, the HQCT is shown applied to the analysis of cave mapping. There are myriad other applications where the analysis of fused multimodal data is required. Examples are in agriculture, medicine, automotive, and aerospace. In the cave mapping example, a MIMO Data Interface to the classical GPU allows the input of multiple multimodal sensor data LiDAR, Digital Imaging and Surveying inputs) which it converts into a predefined fused bitmap format. The fused depth bitmap of the multimodal sensor data is then transformed into a detailed 3D hybrid classical point cloud representation with world coordinates. Mathematically, the hybrid classical point cloud representation with world coordinates can presented as an N-point matrix (if a cube represents a point then $N = 6$).

QPC compression is the procedure that reduces the quantum resources used to prepare quantum point clouds. The main resource in quantum preparation is the number of quantum gates not the number of qubits; the number of qubits in a QPC cloud is far fewer than the number of bits used in a classical point cloud (CPC). Furthermore, the number of quantum gates can be used to indicate the network time complexity because in a quantum network, each quantum gate is an operation which requires a specific processing time. This is referred to as network complexity. Therefore, the main task of QPC compression is to reduce the number of gates used during QPC preparation. Basic quantum gates, NOT, CNOT and Toffoli are preferably used.

The function of the differential pulse code modulator (DPCM) is to improve the CPC by compressing in a lossy manner (with a compression factor R). Its main purpose is to reduce the number of quantum gates during quantum point cloud preparation, and to reduce total network complexity. The greater R is, the better the compression effect. R is the number of mi-CNOT gates before compression, and GA is the number of *mi*-CNOT gates after compression.

The next stage of the process of transforming the CPC to a QPC is called QPC preparation. The QPC preparation which uses quantum mechanical processes to process the QPC information which is first stored as a quantum state. An image of a QPC preparation circuit with *mi*-CNOT gates (mi control bit) within the quantum computing chip is shown. The quantum register which is analogous to a classical computing register stores the quantum data.

The QPC preparation provides for a quantum matrix representation of the CPC matrix, to which a numerical solution has greater precision, efficiency and certainty than with the numerical analysis of the CPC.

The numerical solution to the QPC is effectively a multi-dimensional number matrix that can be analyzed. This matrix is presented as an agent response ('Program' in the diagram) from the classical control logic device on the classical GPU to the classical control logic device on the quantum co-processor. The output on the GPU data interface enables the multi-dimensional number matrix from the quantum co-processor to be visualized as a set of charts. Charts from the CPC analysis can be compared with charts from the QPC to quantify the improvement in accuracy and analytical outcome certainty gained by representing the fused multimodal sensor data as a QPC.

6   Further Work

MAL was awarded a UK government research grant under the UK government funded Analysis for Innovators programme (A4I) to collaborate with NPL. A CPC of multimodal sensor data is the starting point, analyzed with MAL's artificial precognition ML and adaptive control algorithms including artificial precognition (AP) using adaptive (model) cognized control (APACC) [18]. The research project involves the metrological comparison between a generalized N-dimensional classical and quantum point cloud to measure data quality such as point density, point numbers, shadow removal, noise, precision, uncertainty and errors. Quantum enabled technologies have been shown to significantly impact metrology [19]. This project is a metrological evaluation involving the comparison of various measures between a CPC and a QPC, including how long does it take to prepare and process the quantum point cloud and the accuracy and uncertainty of the results. Therefore, a novel means of modelling uncertainty of sensor and IoT networks, robotic and autonomous systems can be developed.

Future planned applications of our classical-quantum processing approach include driverless cars and autonomous drones.

References

1. Jiang, N., Hu, H., Dang, Y., Zhang, W.: Quantum point cloud and its compression. Int. J. Theor. Phys. **56**, 3147–3163 (2017)
2. Luo, Z., Zheng, W., Li, J., Zhao, M., Peng, X., Suter, D.: Quantum image processing and its application to edge detection: theory and experiment. Phys. Rev. X **7**, 031041 (2017)
3. https://www.sciencedirect.com/topics/engineering/point-cloud
4. (2015). http://pointclouds.org/about/

5. Venegas-Andraca, S.E., Bose, S.: Storing, processing and retrieving an image using quantum mechanics. In: Proceedings of the SPIE Conference on Quantum Information and Computation, pp. 137–147 (2003)

6. Cao, M., Wang, P., Wu, L., Lu, Q., Lu, Z., Lu, Q.: The research on the online publishing platform of point clouds of chinese cultural heritage based on LIDAR technology: a case study of chen clan academy in Guangzhou, Guangdong Province. In: IOP Conference Series Materials Science and Engineering, vol. 452, p. 032019, December 2018

7. Fuentes-Pacheco, J., Ruiz-Ascencio, J., Rendón-Mancha, J.M.: Artif. Intell. Rev. **43**, 55 (2015). https://doi.org/10.1007/s10462-012-9365-8F

8. Satyajit, S., Srinivasan, K., Behera, B.K., et al.: Quantum Inf. Process. **17**, 212 (2018). https://doi.org/10.1007/s11128-018-1976-9

9. Gustavson, T.L., Bouyer, P., Kasevich, M.A.: Precision rotation measurements with an atom interferometer gyroscope. Phys. Rev. Lett. **78**, 2046–2049 (1997)

10. Chou, C.W., Hume, D.B., Rosenband, T., Wineland, D.J.: Optical clocks and relativity. Science **329**, 1630–1633 (2010)

11. Shah, V., Knappe, S., Schwindt, P.D.D., Kitching, J.: Subpicotesla atomic magnetometry with a microfabricated vapour cell. Nat. Photon. **1**, 649–652 (2007)

12. Aasi, J., et al.: Enhanced sensitivity of the LIGO gravitational wave detector by using squeezed states of light. Nat. Photon. **7**, 613–619 (2013)

13. Cutler, C.C.: Differential quantization of communication signals. U.S. patent 2,605,361 (filed 1950, issued 1952)

14. Wasilewski, W., et al.: Quantum noise limited and entanglement-assisted magnetometry. Phys. Rev. Lett. **104**, 133601 (2010)

15. Frangou, G.J.: Great Britain Provisional Patent Application 1816049: Processing Quantum Point Cloud Multimodal Sensor Data (2018)

16. Wang, J.: QRDA: quantum representation of digital audio. Int. J. Theor. Phys. **55**(3), 1622–1641 (2016)

17. Vlatko, V., Adriano, B., Artur, E.: Quantum networks for elementary arithmetic operations. Phys. Rev. A **54**(1), 147–153 (1996)

18. Frangou, G.J.: U.S. Patent No. 9,645,576 Chinese Patent 1,051,892,37, Japanese Patent 2,016,520,464, Israel Patent 2,416,88, European Patent Application 2,976,240, Korean Patent Application 2,015,013,8257, PCT Patent Application 2,014,147,361: Apparatus for Controlling a Land Vehicle which is Self-Driving or Partially Self- Driving (2013)

19. Leibfried, D., et al.: Toward Heisenberg-limited spectroscopy with multiparticle entangled states. Science **304**, 1476–1478 (2004)

20. Zhang, Y., Lu, K., Gao, Y.H., Wang, M.: NEQR: a novel enhanced quantum representation of digital images. Quantum Inf. Process. **12**(12), 2833–2860 (2013)

Low Power High Performance Computing on Arm System-on-Chip in Astrophysics

Giuliano Taffoni, Sara Bertocco$^{(\boxtimes)}$, Igor Coretti, David Goz, Antonio Ragagnin, and Luca Tornatore

National Institute of Astrophysics, Astronomical Observatory of Trieste, via G.B. Tiepolo 11, Trieste, Italy
{giuliano.taffoni,sara.bertocco}@inaf.it

Abstract. In this paper, we quantitatively evaluate the impact of computation on the energy consumption on Arm MPSoC platforms, exploiting both CPUs and embedded GPUs. Performance and energy measures are made on a direct N-body code, a real scientific application from the astrophysical domain. The time-to-solutions, energy-to-solutions and energy delay product using different software configurations are compared with those obtained on a general purpose x86 desktop and PCIe GPGPU. With this work, we investigate the possibility of using commodity single boards based on Arm MPSoC as an HPC computational resource for real Astrophysical production runs. Our results show to which extent those boards can be used and which modification are necessary to a production code to profit of them. A crucial finding of this work is the effect of the emulated double precision on the GPU performances that allow to use embedded and gaming GPUs as excellent HPC resources.

Keywords: Arm · GPU · MPSoC · HPC · Energy-to-solution · Energy Delay Product

1 Introduction

In the last decade, energy efficiency has become a main concern in the High Performance Computing (HPC) sector. These systems are built using power hungry high performance systems, and their high energy consumption poses major hedges for achieving exascale computation. Energy efficiency is both a fundamental requirement of large scale platforms and one of the main challenges for future processors, interconnect and storage design [17]. In fact, the eligibility of a exascale computing system must not only pass through performance assessment of its hardware but also of its energy usage.

Commodity single board computers are an interesting case of heterogeneous systems to be utilized for energy efficiency studies. These are low cost single circuit board computers that embed CPUs, GPUs, memory, storage, general purpose I/O ports for external devices and expansions (e.g. SD card connects,

© Springer Nature Switzerland AG 2020
K. Arai et al. (Eds.): FTC 2019, AISC 1069, pp. 427–446, 2020.
https://doi.org/10.1007/978-3-030-32520-6_33

USB, PCIe, HDMA, etc.). The HPC community is already studying the use of those low-powered System-on-Chip (SoC) architectures in large-scale HPC systems, trying to reach production-ready solutions. Additionally, various companies are also studying one-board computers equipped with different hardware solutions and based on Multi-processing System-on-Chip (MPSoC).

This work was done in the framework of ExaNeSt and EuroExa European funded project aiming at the design and development of a prototype of an exascale HPC facility based on low power Arm SoC and FPGAs as accelerators [13,14].

Although the performance of these machines has been profiled in the context of benchmarking tools [19], in this work we study the performance on a real code, an N-Body solver for astrophysical simulations. Our goal is to investigate the trade-off between time-to-solution and energy-to-solution when using real full production runs, and the problems that a developer can face when approaching this kind of platforms. In doing this study, we analyze the computing capabilities and the relative power efficiency between the CPU (single-core, dual-core, multi-core) and the GPU on a MPSoC produced by Rockchip's Firefly-RK3399. We further compare these results with a "standard architecture" based on an Intel server with a GPGPU.

To our knowledge, this paper provides the first comprehensive evaluation of a real astrophysics application on single board computers and in particular on the MPSoC Rockchip Firefly RK3399.

The paper is organized as follows. In Sect. 2, we introduce previous results in the literature that shaped this work. In Sect. 3, we describe the code and we discuss strategies adopted in order to port and optimize a state-of-the-art N-body code on heterogeneous platforms. In Sect. 4, we present the single board MPSoC computers that we have identified for our tests and we discuss our choice to use the Firefly-RK3399 board. In Sect. 5, we discuss the methodology we used to make the performance and energy tests, including some considerations on our choices of architectures for HPC and the role of double precision arithmetic. In Sect. 6, we discuss the performance measurements for all the platforms. Section 7 is dedicated to the power consumption analysis including a description of the experimental setup. Energy measurements results are discussed in Sect. 8. Last sections are dedicated to the conclusions.

2   Related Works

SoC devices are experiencing a growing interest because of their versatility, low-power consumption and their low cost. This is showed, for instance, by the success of one of the first single board computers: the Raspberry Pi [31]. Today there exists a large number of alternatives to Raspberry Pi, and various companies are investing on boards that are equipped with different hardware solutions and are based on MPSoC architectures (in Sect. 4, we will present some of them).

MPSoC integrated circuits are composed of asymmetric multi-core systems combined with graphic-processing units (GPUs) aiming to optimize the energy-to-performance ratio. MPSoC are mainly deployed for the mobile market, although they have been recently utilized in sectors where traditional resources would not be appropriate or in situations where a standard computing platform would not be suitable: for instance, educational purposes [8,30], HPC or Cloud [1,9,25], expendable computers [32], sensors networks [15,20] and Fog computing [5].

MPSoC devices are particularly interesting as they implement heterogeneous architectures where multi-core CPUs and GPUs are coupled with a unified memory system (UMS) where expensive copy operations that exchanges data between the host and the device are not required.

In the last years, GPUs became widely used in scientific programming in order to accelerate computational demanding applications with extensive data-level parallelism because they offer high floating-point throughput and memory bandwidth. In the past, they have had limited device memory, until recently, when their on-board capacity has grown up to several GBs. Though, in general, the capacity of a GPU memory is significantly lower than its host memory. For this reason programmers are obliged to work with two memory spaces and move data from one to the other memory space with an impact in performances and energy consumption. This is crucial for applications striving to solve larger and larger problems. There exists already some solutions to solve this limitation (e.g. POWER8 with NVLink CPUs with four NVIDIA[1]), however the UMS of MPSoC boards may represent an interesting low power and low price solution.

Some authors already analyze MPSoC performance and Energy consumption using standard benchmarks (e.g. HPL, HPCG, DGEMM) [24]. Instead of benchmarking our target platforms using standard suites tuned to measure peak performances. Instead, we are probing the platforms by using real scientific applications and real production runs. In fact, it is well known that, in some case, codes are able to use only a few percentage of the peak Floating point operations/second (FLOPs) [16], in particular memory bound codes as numerical cosmological simulations or data reduction analysis programs.

3   *N*-body Astrophysical Codes

In astrophysics, the *N*-body problem is the problem of predicting the individual motions in a group of celestial objects interacting with each other gravitationally. This applies mainly to the study of the dynamic of star clusters and globular clusters [27].

The numerical solution of the direct *N*-body problem is still considered a challenge despite the significant advances in both hardware technologies and software development. The main drawback related to the direct *N*-body problem relies on the fact that the algorithm requires $O(N^2)$ computational cost. There are some *N*-body codes designed for real scientific production in astrophysics

---

[1] https://www.ibm.com/blogs/systems/ibm-nvidia-present-nvlink-server-youve-waiting/.

using CPUs or GPUs [4, 7, 12, 18, 21, 26]. None of the above has been ported or optimized for embedded GPUs.

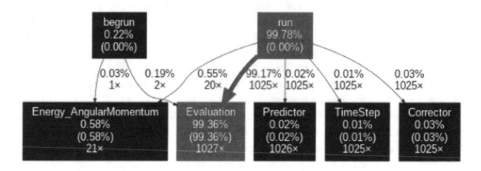

Fig. 1. Call graph of HY-NBODY profiled using *gprof* tool. The figure shows nodes and edges above the threshold 0.01.

3.1   The HY-NBODY Code

Our HY-NBODY code [11] is a modified version of a GPU N-body algorithm [7, 26], based on high order Hermite integration schema [22] using a block time-stepping.

In our implementation, the GPU is fed by the host CPU with the gravity equation of data in the form of coordinates, velocities and masses of particles, and it handles calculating the forces for the data points. Differently from other N-Body codes, we design the algorithm to fully exploit the compute capabilities of heterogeneous architectures. The Hermite schema is implemented and optimized using OpenCL kernels, allowing to test the code on any OpenCL-compliant device (e.g. CPUs and GPUs). We use a fine grained parallelization approach: the host code is parallelized with hybrid MPI+OpenMP programming, while the device code is parallelized with OpenCL. The user is allowed to choose at compile time if the application uses MPI or OpenMP, or both, or neither. The Hermite integration is performed on the selected OpenCL-compliant device(s) and all kernels of the application have been vectorized improving memory bandwidth and reducing the number of loads/stores.

3.2   Profiling HY-NBODY

Code profiling reveals that 99% of the time is spent on the *Evaluation* stage of the 6th order Hermite integration schema (serial application when I/O is disabled), as shown in Fig. 1.

The floating point operations (FLOPs) of the HY-NBODY code have been evaluated through the Performance Application Programming Interface (PAPI) tool [28], allowing us to estimate the arithmetic intensity (ratio of FLOPs to the memory traffic) of the *Evaluation* kernel as $I \simeq 1.5 \cdot 10^7/N$ [FLOPs/byte], with N the number of particles.

Following the *Roofline Model*[2], each kernel is going to be either memory-bound or compute-bound on a specific architecture, since performance is upper bounded by both the peak flop rate, and the product of streaming bandwidth and the flop to byte ratio. The peak performance of a platform can be usually derived from architectural manuals, while the peak bandwidth, which references to peak DRAM bandwidth to be specific, is instead obtained via benchmarking. However, both code profiling and arithmetic intensity estimate suggest that the *Evaluation* kernel is compute-bound on every architecture with $N \sim 10^4 - 10^5$, which is the typical number of particles assigned to a device during a production run.

4   A Single Board Computer for HPC: Firefly-RK3399

To fully exploit the MPSoC heterogeneous boards for scientific calculations, it is necessary to use hardware solutions that offers at least: (i) double-precision floating point arithmetic, (ii) options for high performance I/O and memory interface, (iii) full support for a parallel programming model as CUDA [23], OpenCL [3] or OpenACC [10].

The latest Arm MPSoC boards satisfy these requirements as they support 64-bit floating-point arithmetic precision operations and OpenCL 1.2 specifications.

Those boards are based on the so called Arm big.LITTLE architecture. It features two sets of cores: a low performance energy-efficient cluster (the LITTLE one), and a power-hungry high-performance cluster (the big one). Big and LITTLE cores support the same instruction set, so they can run the same binaries and therefore are easily combined within the same system. Even if extremely promising from the energy efficiency point of view, this kind of heterogeneous boards are extremely complex to exploit for scientific applications. They require to design codes that are able to (i) optimize the scheduling of big.LITTLE cluster, (ii) the use of GPUs, and (iii) the memory access.

Nowadays, there is a wide variety of single board computers available on the market each with its own characteristics. We analyzed some of them and identified the most promising board for our analysis. On the basis of a set of requirements, our candidate should accomplish:

- at least 4 GB of RAM: our software is quite demanding in terms of RAM in particular as we are willing to test some scientific full production runs;
- OpenCL capable GPU device as our code has been re-design in OpenCL to exploit heterogeneous platforms;
- commodity hardware with Linux OS support;

---

[2] Roofline is a visually intuitive performance model used to bound the performance of various numerical methods and operations running on multicore, manycore, or accelerator processor architectures.

- ease of expansion. A single board computer that can be expanded with PCI or USB devices (e.g. external disks or network cards);
- on-board gigabit Ethernet. This will allow future expansion towards a cluster of boards.

Table 1. A comparison of the Single Board platforms based on the MPSoC that we identified for our tests.

Board	SoC	Arch	RAM	GPU	NET
Raspberry Pi 3 B+	Broadcom BCM2837B0, Cortex-A53 @ 4 × 1.4 GHz	64-bit	1 GB	VideoCore IV	1 GB over USB
Odroid XU4	Exynos 5422 Cortex-A15 and Cortex-A7 4 × 2.1 GHz & 4 × 1.5 GHz	32-bit	2 GB	Mali-T628 MP6	1 GB
Banana Pi M64	Allwinner A64 Cortex-A53 @ 4 × 1.4 GHz	64-bit	2 GB	Mali-400 MP2	1 GB
Pine A64	Allwinner r18 Cortex-A53 @ 4 × 1.3 GHz	64-bit	2 GB	Mali-400 MP2	1 GB
Asus Tinker Board	Rockchip RK3288 Cortex-A17 @ 4 × 1.8 GHz	32-bit	2 GB	Mali-T760 MP4 OpenCL 1.1	1 GB
Firefly RK3399	Rockchip RK33399 Cortex-A53 and Cortex-A72 4 × 1.4 GHz & 2 × 2.0 GHz	64-bit	2/4 GB	Mali-T864 MP4 OpenCL 1.2	1 GB

All the boards listed in Table 1 are based on Arm SoC but only a few of them implement a big.LITTLE architecture and only one has enough memory to satisfy our requirements: the Firefly-RK3399 board.

The Firefly-RK3399 single board computer has 6 core 64-bit Arm big.LITTLE SoC architecture. The board contains a cluster of four Cortex-A53 cores with 32 kB L1 cache and 512 kB L2 cache, and a cluster of two Cortex-A72 high-performance cores with 32 kB L1 cache and 1M L2 cache. Each cluster operates at independent frequencies, ranging from 200 MHz up to 1.4 GHz for the LITTLE and up to 1.8 GHz for the big. The SoC contains 2 or 4 GB DDR3 - 1333 MHz RAM. The L2 caches are connected to the main memory via the 64-bit Cache Coherent Interconnect (CCI) 500 that provides full cache coherency between big.LITTLE processor clusters and provides I/O coherency for the Mali-T864 GPU. The peculiarity of this board is that Mali-T864 is a OpenCL-compliant Quad-Core Arm Mali GPU.

This board is an Open Source platform with excellent expansion capabilities, it is equipped with 4 USB2.0,1 USB3.0, 1 USB3.0 Type-C, a MicroSD (TF) Card Slot and an HDMI video connector. The network card is a Realtek RTL8211E 10/100/1000 RJ-45 interface and it also has a PCIe Next Generation Form Factor M.2 connector.

There are different RAM and storage sizes available, and we decide to test the 4 GB of RAM and 16 GB of High-Speed eMMC configuration.

The Board is installed with Ubuntu 16.04 LTS Linux distribution and OpenCL 1.2.

Table 2. The main characteristics of the board used in the test and of the Desktop.

Platform	Firefly-RK3399 board	Desktop
	Rockchip RK3399	ASUS P8B75-M LX
CPU	Arm A72x2 + A53x4 64-bit	Intel i7-3770x4 64-bit
GPU	Arm Mali-T864	NVIDIA GeForce-GTX-1080
RAM	4 GB DDR3	16 GB DDR3
OS	Ubuntu 16.04 LTS	Ubuntu 18.04 LTS
Compiler	gcc version 7.3.0	gcc version 7.3.0
OpenCL	OpenCL 1.2	OpenCL 1.2

The host hardware we used to develop and validate the application and to compare with the MPSoC device, is a workstation with one Intel Core i7-3770x4 running at 3.40 GHz and one Nvidia GeForce-GTX-1080 graphics card in the PCI Express (16x) bus. The workstation runs a Linux Ubuntu SMP kernel version 4.15.0-20 generic and graphics card driver NVIDIA 390.48. In Table 2 we describe the main characteristic of the MPSoC platform used in our tests.

5   Considerations and Methodology

The big.LITTLE architecture has been conceived for mobile devices where applications are developed and optimized in order to stick to low energy consumption for routinely workloads and to benefit to more performant cores on demand. At odds, a scientific code is designed to work on homogeneous core sets and to steadily extract maximum performance from them. Our application is in fact conceived to run on homogeneous platforms and typically operate by dividing the workload on even units. Executing these equal work units on an asymmetric system is expected to degrade the overall performance due to load imbalance.

In our case, we limit the load imbalance binding the OpenMP threads to a defined CPU cluster, setting explicit core affinity. We use Linux `taskset` command to choose the affinity. This way, we avoid performance degradation due to the thread migration from a cluster to the other and we measure the performances and energy of each one of the two CPU clusters homogeneously. Anyway our code is designed to maximize the performance, so we cannot benefit of the cluster migration features for energy saving.

While ARM cores does not have frequency throttling ability, in our experiments we freeze the frequencies at a given level using the `performance` scaling governor, in order to prevent the dynamical scaling of the cores frequencies during runtime. Furthermore, to stabilize the results and to average possible system fluctuation, each run of the code is repeated 10 times and the results presented later in the paper are always the averaged values. Errors are reported explicitly when greater than 1%.

On the MPSoC architectures GPUs, the global and local OpenCL address spaces are mapped to main host memory. This means that explicit data copies from global to local memory and associated barrier synchronizations are not necessary. Thus, using local memories as a cache can waste both performance and power. For this reason, specific Arm-optimized version of all kernels of HY-NBODY has been implemented in which the local memory is not used.

5.1 Floating Point Arithmetic Considerations

The Hermite 6th order integration schema requires double precision (DP) arithmetic in the evaluation of inter-particles distance and acceleration in order to minimize the round-off error. Full IEEE-compliant DP-arithmetic is efficient in market available CPUs, but it is still extremely resource-eager and performance-poor in other accelerators like gaming or embedded GPUs. The theoretical best case for DP performance is 1:2 SP, simply because it involves computing with double the number of bits as FP32. However, this ratio can me much lower for many devices[3].

As an alternative, the extended-precision (EX) (or emulated double precision) numeric type [29] can represent a trade-off in porting HY-NBODY on devices not specifically designed for scientific calculations. An EX-number provides approximately 48 bits of mantissa at single-precision exponent ranges. HY-NBODY can be compiled using DP, EX or single precision (SP) arithmetic (user-defined at compile time).

The energy E and the angular momentum L of the N-body system during the simulation are constant quantities and we use them to evaluate the effect of the arithmetic on the accumulation of the round-off error. The stability of the computation is verified with an error lower than 10^{-8} for DP and EX, while SP calculations do not conserve neither E nor L, and therefore we do not consider them suitable for real scientific full production runs. Since in this work we are interested in benchmarking real scientific cases, we do not include any SP result.

6 Computational Performances: CPUs and GPUs

First we measure and investigate the CPUs speedup, i.e. the ratio of the serial execution time to the parallel execution time utilizing multiple cores by means of OpenMP threads. We run our code varying the number of particles in the N-body integration for different OpenMP threads, pinning the processes first to the 4 cores of the Arm Cortex-A53 and then to the 2 cores of the Cortex-A72.

Figure 2 shows the speedup for both Arm Cortex-A53x4 and Cortex-A72x2 CPUs varying the number of OpenMP threads as a function of the number of particles. As expected, the best performance is achieved when each core handles one thread. Time-to-solution saturates when the number of OpenMP threads exceeds the available cores. In the case of Arm Cortex-A53x4, we observe super-linear scalability that may be due to an optimized use of the caches.

---

[3] https://www.geeks3d.com/20140305/amd-radeon-and-nvidia-geforce-fp32-fp64-gflops-table-computing/.

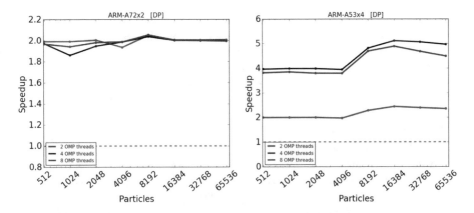

Fig. 2. Host speedup for DP-arithmetic as a function of the number of particles. We vary the number of OpenMP threads. Left panels for Arm Cortex-A72x2 and right panels for Arm Cortex-A53x4.

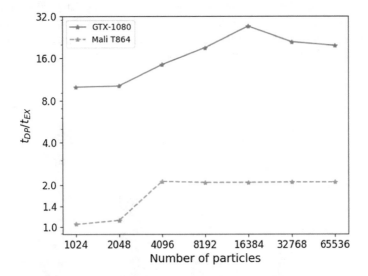

Fig. 3. The ratio between the execution time of EX arithmetic and DP arithmetic as a function of the number of particles for both Mali-T864 and Nvidia GeForce-GTX-1080 GPUs.

As for the measure of the Mali GPU performance, we first note that, as already mentioned, we never use the local memory to avoid the cost of memory copy. Following the Arm OpenCL Developer Guide [2], we test the effects of vectorizing the code increasing the size of work-group. We run HY-NBODY varying the work-group size from 4 up to 256 and we measure the execution time increasing the number of particles from 1024 up to 65536.

Kernel execution times on the GPU have been obtained by means of OpenCL's built-in profiling functionality, which allows the host to collect runtime information. Despite Arm recommends for best performance using a work-group size that is between 4 and 64 inclusive, our measurement shows that the execution time is not driven by any specific work-group size (the execution time is constant within 3%). The work-group size is not even affecting the energy consumption so, from now on, we fix the work-group size at 64.

Finally, we measure the DP and EX performance of both Mali-T864 and GeForce-GTX-1080 and compare it with the SP performance executing a set of runs varying the number of particles from 1024 to 65536. The DP/FP ratio for the GTX-1080 is ~1/32 as also discussed by Geeks3D blog, while for the Mali GPU it is ~1/10 (see Table 3 for more details).

On the other hand the effect on performance of the EX arithmetic is extremely important. In Fig. 3, we present the result of a set of simulations increasing the number of particles. The effect of the EX becomes evident from 2048 particles and it stabilizes from 4096 particles on. The performance improvement is ~2 for the Mali GPU and ~20 for GTX-1080. For the GTX-1080, tests has been done using a work-group size of 64, however, differently from Mali GPU, the size of the work-group significantly affects the execution time; with EX arithmetic the best performances are obtained for work-group size of 64 and 128.

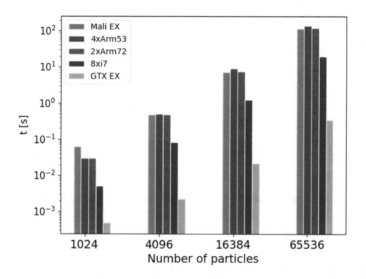

Fig. 4. The execution time in seconds for different devices as a function of the number of particles.

In Fig. 4, we compare the time-to-solution in seconds for different devices. For each device we plot the configuration that optimizes the performances in terms number of cores, work-group size, DP for CPUs and EX for GPUs. A summary of the configuration and results is presented in Table 3. We note that, from a pure performance point of view the Nvidia GPU is the most powerful device tested while the MPSoC performance is two order of magnitude lower. That being a somewhat expected result, the key question to be investigated in the next sections is whether or not such gap is compensated by the lower energy consumption of MPSoC.

7 Power Consumption Measurements

In this section we discuss our work to estimate and compare the instantaneous power, the total energy consumption, the execution time and the energetic cost for a simulation using the HY-NBODY code for the various devices listed in Table 3.

We also estimate the energy impact of our code in terms of Energy Delay Product (EDP). The EDP proposed by Cameron [6], is a "fused" metric to evaluate trade-off between time-to-solution and energy-to-solution. The EDP is defined as:

$$EDP = E \times T^w \tag{1}$$

where E is the total energy consumed during the run, T is the time-to-solution and w is a parameter to weight performance versus power. Common value of this parameter are $w = 1, 2, 3$ ($w = 3$ was suggested by Cameron), the larger is w the greater the value of performance.

As discussed in Sect. 3.2, the *Evaluation* kernel is the most computational demanding part of our code, it is strongly compute-bound on every architecture so it is excellent to make energy tests. Relying on these profiling results, we measure the energy consumption during the execution of the *Evaluation* kernel, in an infinite while loop.

In order to minimize the inaccuracies in estimating the current consumed by the CPU and the GPU while running the kernel, we apply two different methodologies, one for the Firefly KR-3399 and one for the Intel desktop with Nvidia GPU.

The Firefly RK-3399 board has been powered by a DC power supply (a Keysight E3634A) to avoid the power draw by the AC-to-DC transformer, which makes the readings more noisy and spread out. After booting up the platform, we measure its stable current while the system is in idle. This gives us the $I_{baseline}$ consumption by the system.

$I_{impl,baseline}^{device}$ is the current consumed by the system running a given code implementation using a particular device (CPU or GPU).

I_{impl}^{device} is the current that we are interested in:

$$I_{impl}^{device} = I_{impl,baseline}^{device} - I_{baseline} \tag{2}$$

$I_{impl,baseline}^{device}$ and $I_{baseline}$ are the mean values over a range of three minutes.

The energy consumed by a given implementation of the kernel (energy-to-solution) is

$$E_{impl}^{device} = V \times I_{impl}^{device} \times T_{impl}^{device} \tag{3}$$

where V and T_{impl}^{device} are the voltage and the kernel running time (time-to-solution averaged over ten runs), respectively (voltage is constant, namely $V = 12$ V).

On the Intel desktop we set the frequency governor to performance level and the electric power draw is measured by means of a power meter (Yokogawa WT310E).

After booting up the platform, we measure the watts hours consumed in idle during a period of three minutes, giving us the $W_{baseline}$ of the system. $W_{impl,baseline}^{device}$ is the electric power drawn by the system running a given code implementation using a particular device (CPU or GPU) over a period of three minutes (ΔT_3). The power drawn by the dedicated GPU (Nvidia GeForce-GTX-1080) is also monitored by a current probe (Fluke i30ss).

The watts hours (energy-to-solution) that we are interesting in are:

$$W_{impl}^{device} = (W_{impl,baseline}^{device} - W_{baseline}) \times T_{impl}^{device}/\Delta T_3 \tag{4}$$

where T_{impl}^{device} is the kernel running time (time-to-solution averaged over ten runs).

7.1 Experimental Setup

Fig. 5. Experimental setup at the Astronomical Observatory of Trieste - electronic laboratory.

To measure the current consumption of the devices under test, two simple setups were used, depending on the power supply type of the device (see Fig. 5).

– Devices powered by Direct current:

- Benchtop Laboratory Power Supply, Keysight model E3634A;
- Benchtop Multimeter, Hewlett Packard model 34401A;
- AC/DC Current clamp, Fluke model i30s;
- Digital Storage Oscilloscope, Keysight model MSOX3024T.

The benchtop laboratory power supply was set at the nominal supply voltage for the system and the multimeter was connected in series to measure the current flow. The output used in our test is the mean value of 450 measurements taken at each run with a sample rate of 2,5 Hz (1 sample every 400 ms). The oscilloscope was used to measure the dynamic behaviour of the current consumption taken by means of the current clamp. These devices were used just to monitor that the measurements are taken under almost constant load.

- Devices powered by Alternate current (mains supply):
  - Digital Power Meter, Yokogawa model WT310E;
  - AC/DC Current clamp, Fluke model i30s;
  - Digital Storage Oscilloscope, Keysight model MSOX3024T.

In this case, the systems were powered by their own power supply and the measurements were taken at the 230 V mains input. The Power meter integrates the total power used during the chosen time period. Also in this case, the oscilloscope and the current clamp were used to monitor the dynamic behaviour of the current consumption, but this was possible only with a limited subset of the tests, since only the auxiliary power supply input of the GPU could be intercepted. The discrete GPU is also supplied by the PCIe connector and thus the measurements taken with the current clamp are not reliable.

Table 3. The main characteristics of the configuration used in the test including the arithmetic's capacity of the accelerators. The last two columns are the device energy measured in [Watt/h]. The $E_{baseline}$ is the idle energy on 3 min of evaluation and the E_{impl}^{device} is the energy for 3 min of HY-NBODY kernel continuous execution.

Device	Core workgroup	Arithmetic	DP/FP	EX/FP	$E_{baseline}$	E_{impl}^{device}
					[Watt/h]	
i7-3770	4	DP	1		3.95	6.92
Arm A53	4	DP	1		0.15	0.23
Arm A72	2	DP	1		0.15	0.34
Arm Mali-T864	64	EX	1/10	1/5[a]	0.15	0.29
GeForce-GTX-1080	64	EX	1/32	5/8	3.95	8.40

[a] Arm Optimized.

8   Energy Results

Energy-to-solutions and time-to-solutions are obtained running the *Evaluation* kernel using both DP and EX arithmetic using 65536 particles. That number of particles has been chosen in order to keep busy the device for a reasonable amount of time and therefore to make robust measurements.

For each device we estimate both the $\mathbf{E_{baseline}}$ and the $\mathbf{E_{impl}^{device}}$, i.e. the energy for 3 min of HY-NBODY kernel continuous execution. Our energy consumption measurements are reported in Table 3. Not surprisingly, the most energy consuming devices are the i7 and the Nvidia GPU that absorb more than twenty times the Firefly-RK3399.

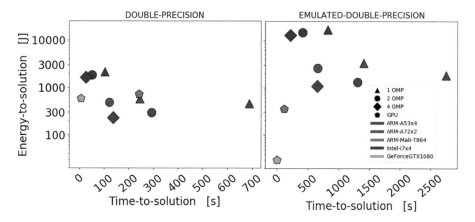

Fig. 6. Energy-to-solution (in Joule) as a function of time-to-solution (in second) for DP (left panel) and EX (right panel) arithmetic. Blue symbols for Arm-A53x4 CPU, red symbols for Arm-A72x2 CPU, green symbol for Arm Mali-T864, violet symbols for Intel-i7x4 CPU and orange symbol for Nvidia GeForce-GTX-1080. Triangle up for 1 OMP thread (serial calculation), circle for 2 OMP threads, diamond for 4 OMP threads, pentagon for GPUs with work-group size of 64.

Time-to-solution and energy-to-solution results are plotted in Fig. 6 for both DP and EX arithmetic.

The most effective device, both in terms of time-to-solution and energy-to-solution, is the dedicated Nvidia GeForce-GTX-1080 GPU. Regarding CPUs, the time-to-solution scales linearly with the number of cores exploited, and saturates when the number of OpenMP threads exceeds the available cores, as expected. Multi-core implementation is always the most effective solution, both in terms of time-to-solution and energy-to-solution. It is worth noting that dual-core Arm-Cortex A72, running at 1.80 GHz, is 4 times more power-efficient than the single-core Intel-i7, running at 3.40 GHz. Moreover, CPUs do not benefit at all of the EX arithmetic. Indeed, their performances degrade when using EX (note that right and left panels in Fig. 6 have different x-scales): for this reason on Fig. 7

we plot the "best results" in terms of computational performances, namely DP arithmetic for CPUs and EX arithmetic for GPUs.

When using EX arithmetic, the performances are improving dramatically for the GPUs.

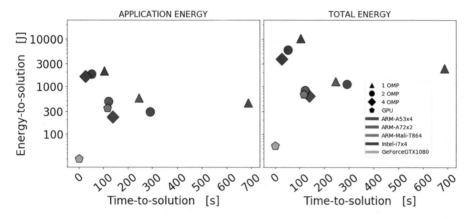

Fig. 7. Energy-to-solution (in Joule) as a function of time-to-solution (in second) for best configuration i.e. DP for CPUs and EX for GPUs. On the left panel we plot the energy-to-solution for the running kernel excluding the energy baseline, on the right panel we plot the total electric power drawn by the system running the kernel. Symbols and lines are the same as described in Fig. 6.

As we are also interested in the total energy impact (including boards base line), this is the real energy consumption of the platforms when making computations. In Fig. 7 we present the best cases plot for the total energy (right panel). The effect of the higher baseline is affecting mainly the computations on i7 and GPGPU, while the CPUs and GPUs on the Firefly have roughly the same power consumption and energy behaviour.

Finally we compare our devices in terms of EDP. In Fig. 8, we plot the EDP for 3 values of the w parameter comparing the different devices. For the CPUs we vary the number of cores involved in the calculation, while for GPUs we fixed the of work-group size (no significant energy difference has been noticed modifying the work-group size).

As expected, for CPUs the EDP factor is a function of the cores used, the most is the occupancy of the CPU the best is the energy impact. When performances are highly valued, the Intel processors and the GTX-1080 GPU are the devices with best trade-off between time-to-solution and energy-to-solution (in particular the GTX-1080 GPU), even if their instantaneous power is higher than the Firefly-RK3399 board.

In Fig. 9 we measure the EDP when using the EX arithmetic. In this case both the Mali GPU and the GTX-1080 are favoured in terms of energy and computing time as also shown in Fig. 7.

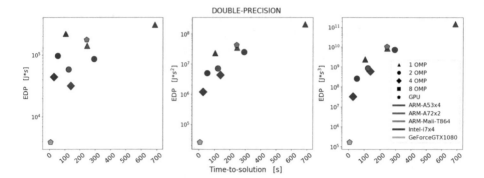

Fig. 8. EDP as a function of time-to-solution (in second) for double-precision arithmetic varying the weight of the execution time ($w = 1, 2, 3$). Symbols and lines are the same as described in Fig. 6.

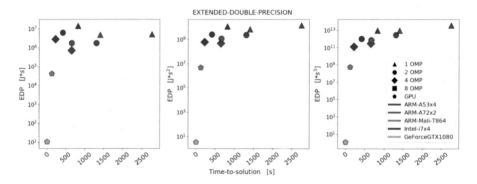

Fig. 9. EDP as a function of time-to-solution (in second) for emulated double-precision arithmetic varying the weight of the execution time ($w = 1, 2, 3$). Bottom-left is better. Symbols and lines are the same as described in Fig. 6.

As final remark, we have also measured the efficiency of the AC power supply of the Firefly-RK3399 board provided by the vendor. The AC power supply efficiency is $\simeq 85\%$ in idle, while is $\simeq 91\%$ at full workload.

9   Conclusion and Future Developments

The energy footprint of scientific applications will become one of the main concerns in the HPC sector. SoC technology is specifically designed to optimize the energy-to-performance ratio. In this work, we begin to explore the impact of the software design of a scientific application on its energy-to-solution and time-to-solution footprints exploiting low-cost SoC-based platforms. We compare the relative power efficiency between the CPU (single-core, dual-core, multi-core) and the GPU on SoC using a real scientific application.

The code we used is conceived to minimize the MPI communications, so that we could evaluate the impact of the computation, on both GPUs and CPUs, on the power consumption. Given the negligible role of networking and the absence of I/O, our results are sufficiently robust for the discussion and the conclusions that we present. Hence, they can be considered representative of arithmetic-intensive scientific codes with a small need of MPI communication.

Furthermore, the effectiveness of EX arithmetic to exploit commercial gaming-class GPUs in scientific calculations is also a significant outcome of this work that may have a general interest.

Therefore, we identify an application able to exploit both the CPUs and GPUs, and then we study the electric power consumption at specific instant during a run.

We use the Firefly-RK3399 Arm MPSoC based board that meets the requirements identified to successfully run a scientific HPC code in particular: 4 GB of RAM and excellent expansion capacity. The results obtained with this board are compared with the ones from a general purpose x86 desktop and PCIe GPGPUs, the Nvidia GeForce-GTX-1080. We deliberately use a consumer grade desktop and a commodity GPGPU to compare "similar" platforms (single boards computers are not designed to compare with high-end HPC servers).

We analyze how to optimize our code to benefit of heterogeneous platforms with a big.LITTLE architecture and integrated GPU. This is complicated mainly by two problems: the efficient scheduling of the algorithms on big.LITTLE architecture and the use of full double precision arithmetic in order to provide real scientific results. The first affects the way a developer can run a HPC code on the MPSoC, the latter involves the efficient use of consumer grade GPUs.

Our results show that even if the energy consumption of the single board computers is orders of magnitude lower that the one form a desktop (and then also lower than a server), when evaluating the energy-to-solution of an HPC designed core, the best device seems to be the Nvidia GeForce-GTX-1080 at least for this specific code. However, also the low power Mali embedded GPU performs much better even of the i7, as also demonstrated by the EDP analysis.

Even if not comparable with the GPGPU, the overall single board computer (CPUs and GPUs) is extremely promising both in terms of performances and energy consumption, in particular if applications will be able to use at the same time the two CPU clusters and the associated GPU. This requires to re-engineer the applications completely.

Furthermore, a crucial finding of this work is the effect of the emulated double precision on the GPU performances. Our tests using DP arithmetic have demonstrated the poor results of those devices. This is not an unexpected behaviour, since consumer grade GPUs (and also MPSoC Mali GPUs) are not built for high performance DP. This is because they are targeted towards games and game developers. So vendors commonly do not cram DP compute cores in their GPUs. However the introduction of EX arithmetic highly improves the performance filling the gap with the SP capability and opening the path for a success full and cheap use of those devices also on HPC area.

In conclusion, we have shown that SoC technology is emerging as a promising alternative to "traditional" technologies for HPC, which are more focused on peak-performance than on power-efficiency. That being especially true when a significant effort was spent in re-engineering the computaitonal parts and the communication schemes, since it is expected that to achieve small enough times-to-solution a large number of low-power-consumption devices will be needed.

How a more complex communication pattern may affect the time-to-solution and energy-to-solution is not a focus of this preliminary work, although that is obviously a major point in HPC. We leave this to a forthcoming work in which we use different, more complex codes that implements a large variety of different algorithms. For instance, a tree-code - suited for a different class of problems than a direct N-Body - would challenge quite differently both the cache hierarchy and the interconnect. Also, on one hand, the implementation of tree-related routines for a GPU are far more efficient than a simple vector-based series of instructions, and on the other hand the execution pattern of more complex codes that include more physical modules, will change heavily the role balance between CPU and GPU.

This analysis has been done using an Astrophysical code, however our results are valid also for any computing intensive code that requires double precision arithmetic: in practice the majority of codes in science. Furthermore, the use of EX arithmetic to exploit commercial GPUs is an important result that opens to the possibility to use gaming GPUs also for HPC.

Our future plan is to assess the energy footprint of other aspects of this application, such as network and I/O and compare clusters of MPSoCs with HPC resources, where multi-node MPI communication becomes an important aspect of a simulation.

Acknowledgments. This work was carried out within the ExaNeSt (FET-HPC) project (Grant no. 671553), the ASTERICS project (Grant no. 653477) and EuroExa (FET-HPC) project (Grant no. 754337) funded by the European Union's Horizon 2020 research and innovation programme.

References

1. Ammendola, R., Biagioni, A., Cretaro, P., Frezza, O., Cicero, F.L., et al.: The next generation of Exascale-class systems: the ExaNeSt project. In: Euromicro Conference on Digital System Design (DSD), Vienna, pp. 510–515 (2017). http://dx.doi.org/10.1109/DSD.2017.20
2. Arm Mali GPU OpenCL Developer Guide, Version 3 (2016). http://infocenter.arm.com/help/topic/com.arm.doc.100614_0300_00_en/arm_mali_gpu_opencl_developer_guide_100614_0300_00_en.pdf
3. Gaster, B., Howes, L.W., Kaeli, D.R., Mistry, P., Schaa, D.: Heterogeneous Computing with OpenCL - Revised OpenCL 1.2 Edition. Morgan Kaufmann (2013)
4. Berczik, P., Nitadori, K., Zhong, S., Spurzem, R., Hamada, T., Wang, X., Berentzen, I., Veles, A., Ge, W.: High performance massively parallel direct N-body simulations on large GPU clusters. In: International conference on High Performance Computing, Kyiv, Ukraine, 8–10 October 2011, pp. 8–18 (2011)

5. Bonomi, F., Milito, R., Zhu, J., Addepalli, S.: Fog computing and its role in the internet of things. In: Proceedings of the First Edition of the MCC Workshop on Mobile Cloud Computing - MCC -12, p. 13. ACM Press, New York (2012). http://dx.doi.org/10.1145/2342509.2342513

6. Cameron, K.W., Ge, R., Feng, X., Varner, D., Jones, C.: High-performance, power-aware distributed computing framework. In: Proceedings of the International Conference on High Performance Computing, Networking, Storage, and Analysis (SC). ACM/IEEE (2004)

7. Capuzzo-Dolcetta, R., Spera, M.: A performance comparison of different graphics processing units running direct N-body simulations. Comput. Phys. Commun. **184**, 2528–2539 (2013)

8. Doucet, K., Zhang, J.: Learning cluster computing by creating a Raspberry Pi cluster. In: Proceedings of the SouthEast Conference, ACM SE 2017, pp. 191–194 (2017). http://dx.doi.org/10.1145/3077286.3077324

9. Durand, Y., Carpenter, P.M., Adami, S., Bilas, A., Dutoit, D., et al.: EUROSERVER: energy efficient node for European micro-servers. In: 17th Euromicro Conference on Digital System Design, Verona, pp. 206–213 (2014). https://doi.org/10.1109/DSD.2014.15

10. Farber, R.: Parallel Programming with OpenACC, 1st edn. Morgan Kaufmann Publishers Inc., San Francisco (2016)

11. Goz, D., Tornatore, L., Bertocco, S., Taffoni, G.: Direct N-body code designed for heterogeneous platforms. In: INAF-OATs Technical Report, vol. 223, July 2018. http://dx.doi.org/10.20371/INAF/PUB/2018_00002

12. Harfst, S., Gualandris, A., Merritt, D., Spurzem, R., Portegies, Z.S., Berczik, P.: Performance analysis of direct N-body algorithms on special-purpose supercomputers. New Astron. **12**, 357–377 (2007)

13. Katevenis, M., Chrysos, N., Marazakis, M., Mavroidis, I., Chaix, F., Kallimanis, N., et al.: The ExaNeSt project: interconnects, storage, and packaging for exascale systems. In: 2016 Euromicro Conference on Digital System Design (DSD), Limassol, pp. 60–67 (2016)

14. Katevenis, M., Ammendola, R., Biagioni, A., Cretaro, P., Frezza, O., Lo, C.F., et al.: Next generation of Exascale-class systems: ExaNeSt project and the status of its interconnect and storage development. Microprocess. Microsyst. **61**, 58–71 (2018)

15. Keller, M., Beutel, J., Thiele, L.: Demo abstract: mountainview precision image sensing on high-alpine locations. In: Pesch, D., Das, S. (Eds.) Adjunct Proceedings of the 6th European Workshop on Sensor Networks, EWSN, Cork, pp. 15–16 (2009)

16. Kobayashi, H.: Feasibility study of a future HPC system for memory-intensive applications: final report. In: Resch, M., Bez, W., Focht, E., Kobayashi, H., Patel, N. (eds.) Sustained Simulation Performance 2014. Springer, Cham (2014)

17. Kogge, P., Bergman, K., Borkar, S., Campbell, D., Carson, W., Dally, W., Denneau, M., Franzon, P., Harrod, W., Hill, K., et al.: Exascale computing study: technology challenges in achieving exascale systems. Technical report, University of NotreDame, CSE Department (2008)

18. Konstantinidis, S., Kokkotas, K.: MYRIAD: a new N-body code for simulations of star clusters. Astron. Astrophys. **522**, A70 (2010)

19. Mantovani, F., Calore, E.: Performance and power analysis of HPC workloads on heterogeneous multi-node clusters. J. Low Power Electron. Appl. **8**(2) (2018). http://www.mdpi.com/2079-9268/8/2/13

20. Martinez, K., Basford, P.J., DeJager, D., Hart, J.K.: Using a heterogeneous sensor network to monitor glacial movement. In: 10th European Conference on Wireless Sensor Networks, Ghent, Belgium (2013)
21. Nitadori, K., Aarseth, S.J.: Accelerating NBODY6 with graphics processing units. MNRAS **424**, 545–552 (2012)
22. Nitadori, K., Makino, J.: Sixth- and eighth-order Hermite integrator for N-body simulations. New Astron. **13**, 498–507 (2008)
23. Nickolls, J., Buck, I., Garland, M., Skadron, K.: Scalable parallel programming with CUDA. Queue **6**(2), 40–53 (2008). https://doi.org/10.1145/1365490.1365500
24. Ou, Z., Pang, B., Deng, Y., Nurminen, J., Yla-Jaaski, A., Hui, P.: Energy- and cost-efficiency analysis of ARM-based clusters. In: 12th IEEE/ACM International Symposium on Cluster, Cloud and Grid Computing, CCGrid 2012, pp. 115–123 (2012)
25. Rajovic, N., Rico, A., Puzovic, N., Adeniyi-Jones, C., Ramirez, A.: Tibidabo: making the case for an ARM-based HPC system. Future Gener. Comput. Syst. **36** 322–334 (2014). http://dx.doi.org/10.1016/J.FUTURE.2013.07.013
26. Spera, M.: Using Graphics Processing Units to solve the classical N-body problem in physics and astrophysics. ArXiv e-prints 1411.5234 (2014)
27. Spera, M., Capuzzo-Dolcetta, R.: Rapid mass segregation in small stellar clusters. Astrophys. Space Sci. **362**(12), 12 (2017). article id 233
28. Terpstra, D., Jagode, H., You, H., Dongarra, J.: Collecting performance datawith papi-c. In: Muller, M.S., Resch, M.M., Schulz, A., Nagel, W.E. (eds.) Tools for High Performance Computing 2009, pp. 157–173. Springer, Heidelberg (2009)
29. Thall, A.: Extended-precision floating-point numbers for GPU computation, p. 52 (2006). https://doi.org/10.1145/1179622.1179682
30. Turton, P., Turton, T.F.: Pibrain'a cost-effective supercomputer for educational use. In: 5th Brunei International Conference on Engineering and Technology, BICET 2014, pp. 1–4 (2014)
31. Upton, E., Halfacree, G.: Raspberry Pi User Guide, 4th ed. Wiley (2016)
32. Yoneki, E.: Demo: RasPiNET: decentralised communication and sensing platform with satellite connectivity. In: Proceedings of the 9th ACM MobiCom Workshop on Challenged Networks - CHANTS -14. ACM Press, New York, pp. 81–84 (2014). http://dx.doi.org/10.1145/2645672.2645691

Quantum Computer Search Algorithms: Can We Outperform the Classical Search Algorithms?

Avery Leider$^{(\boxtimes)}$, Sadida Siddiqui, Daniel A. Sabol, and Charles C. Tappert

Seidenberg School of Computer Science and Information Systems, Pace University,
Pleasantville, NY 10570, USA
{aleider,ss05380p,dsabol,ctappert}@pace.edu

Abstract. Quantum Computers are not limited to just two states. Qubits, the basic unit of quantum computing have the power to exist in more than one state at a time. While the classical computers only perform operations by manipulation of classical bits having two values 0 and 1, quantum bits can represent data in multiple states. This property of inheriting multiple states at a time is called superposition which gives quantum computers tremendous power over classical computers. With this power, the algorithms designed on quantum computers to solve search queries can yield result significantly faster than the classical algorithms. There are four types of problems that exist: Polynomial (P), Non-Deterministic Polynomial (NP), Non-Deterministic Polynomial Complete (NP-complete) and Non-Deterministic Polynomial hard (NP-hard). P problems can be solved in the polynomial amount of time like searching a database for an item. However, when the size of the search space grows, it becomes difficult to compute solutions even for P problems. Quantum algorithms like Grover's algorithm has reduced the time complexity of some of the classical algorithm problems from N to $\sqrt{N}$. Variants of Grover's algorithm like Quantum Partial Search propose changes that yield not exact but closer results in time even lesser than Grover's algorithm. NP problems are the problems whose solution if known can be verified in polynomial amount time. Factorization of prime numbers which is considered to be an NP problem took an exponential amount of time when solved using the classical computer while the Shor's quantum computing algorithm computes it in polynomial time. Factorization is also a class of bounded-error quantum polynomial time (BQP) problems which are decision problems solved by quantum computers in polynomial time. There are problems to which if a solution is found can solve every problem Of NP class, these are NP-complete problems. The power of Qubits could be exploited in the future to come up with solutions for NP-complete problems in the future.

Keywords: Qubit · Superposition · Search algorithms · Grover's algorithms · Quantum computing · Shor's algorithm

© Springer Nature Switzerland AG 2020
K. Arai et al. (Eds.): FTC 2019, AISC 1069, pp. 447–459, 2020.
https://doi.org/10.1007/978-3-030-32520-6_34

1   Introduction

A search algorithm is an algorithm that retrieves the desired information or desired path (the path from source to destination) from a particular search space (database or map) in which instances can be categorical (nominal) or continuous (interval level). The efficiency of an algorithm is decided by its time complexity and space complexity. There are three types of time complexity: best time complexity denoted by Ω, average case denoted by Θ, worst case denoted by O. The worst case scenario is always considered first while choosing the best option from a set of algorithms. This is because the time complexity in worst case scenario tests the algorithm when the pattern of arrangement of the instances is most unfavorable or an example when the sought after instance is the last instance of the database list as opposed to best case scenario where the sought after instance is the first instance of the database.

1.1   Classical Algorithms

Imagine the following situation: there are N items in a database (say A1, A2...AN). One of the items is marked. The challenge is to find out which item is marked with the minimum number of computations. In the worst case scenario, if AN is the item which is marked then the classical computer will have to perform N operations to arrive at solution using the simple classical linear search algorithm and hence, in the worst case scenario, the time complexity of linear search is denoted by $O(N)$ where N is the total number of instances in search space. It can become very difficult to arrive at the solution if the total number of instances N approaches infinity. A database can also be searched using binary search algorithm in which instances are of interval level and are already sorted before searching and it is comparatively faster than the linear search as at every stage it divides the search space into half and disposes the other half and this repeats until the search space reduces to 1. This gives us the worst case time complexity to be $O(\log N)$ which is significantly better than $O(N)$.

There are other classical algorithms like Breadth First Search (BFS) and Depth First Search tree (DFS) which are used to find the path to a specific destination node from the source. This is done mainly by searching and visiting different nodes one after the other until the desired node is visited and then the path to the desired node is traced back. The time complexity for search using DFS and BFS is $O(V+E)$ where V represents the total number of nodes and E represents the total number of edges. This classical search concept has been used to solve the maze problem.

1.2   Searching with Quantum Computing

Qubit. A Quantum bit or Qubit can be considered as the basic unit of quantum computing similar to its classical computing counterparts. However, unlike classical bits that possess either one of the two fixed states 0 or 1, Qubits can exist in a coherent superposition of several states between 0 and 1 at a time.

The major difference is that while measuring the value of bits, the classical bit does not disturb its state whereas the Qubit loses its coherence and its superposition state gets irrevocably disturbed. The values 0 and 1 denote spins of Qubits which represent up spin and spin down respectively. It is believed that the state of Qubit is both 0 and 1 because the state is always expressed as the superposition of spin up and spin down or it is expressed as a linear combination of 0 and 1. Based on two different probabilities, the Qubit state will have a share of both spins as shown in Fig. 1. Hence, a Qubit is able to express both 0 and 1 at the same time, unlike a classical bit which is either 0 or 1 at a given time.

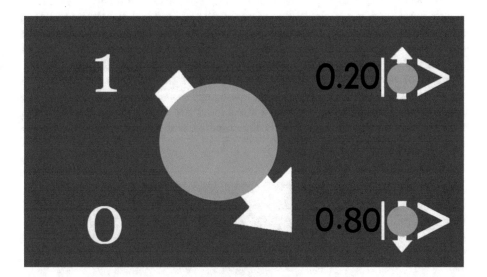

Fig. 1. State of quantum particle [3]

Principles of Quantum Computing. Quantum computing search algorithms have surpassed classical algorithm speed to a great extent using a distinct property of Qubit called superposition. Superposition is the fundamental principle of quantum mechanics. It means that quantum states can be added together to yield a new valid quantum state which means every quantum state can be expressed as a linear combination of other distinct quantum states. This property of superposition of states exhibited by qubits allows them to exist in multiple states at a time using which the algorithm is able to perform multiple operations at the same time and this has reduced the search time complexity to $O(\sqrt{N})$ which is even lower than $O(\log N)$ in the classical computing where N is the total number of instances present in the search space.

Maze Search with Quantum Computing. Quantum computers can be used to solve search problems, such as finding the correct path in a maze, faster than classical computers can [10]. However, contrary to popular thought, Qubits do not do this by trying every possible path at the same time. This would be a much more powerful device which would make the time complexity as low as $O(i)$. Qubits put different paths in superposition and then compute with that superposition. This gives quantum computers a considerable advantage over classical computers [6–8].

1.3 Superposition and the Double Slit Experiment

The behavior of Qubits exploring superposition of paths is difficult to visualize. However, it is because of the theory of quantum mechanics that light or a photon exhibits two kinds of nature: particle nature and wave nature. The double slit experiment conducted on electrons helped the scientist demonstrate this nature and explain the superposition principle. In the basic version of double slit experiment, light rays from a laser beam are allowed to pass through a plate having two parallel slits and the light coming through these slits is observed on the screen placed behind the plate. If we considered the light to be of wave nature, it would form a pattern of alternate light and dark fringes on the screen as shown in the upper half region of Fig. 2.

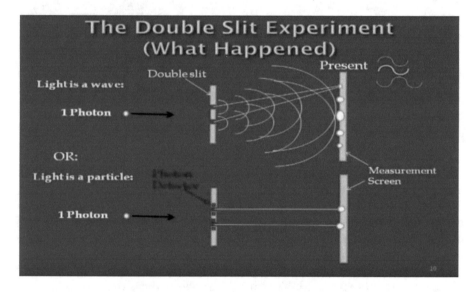

Fig. 2. Dual nature of quantum particle [3]

However, if we considered it to be of particle nature, the stream of particle would be expected to travel through one of slits, sometimes through the upper slit and sometimes through the lower slit forming two bright fringes on the screen where particles could be expected to hit as shown in the lower half region of Fig. 2. However, in reality, the pattern formed on the screen was the one which would have been expected from the wave nature of the light. To have a clear understanding of how actually the particle nature of light could generate a pattern like the one generated by wave nature, the scientist placed a detector behind one of the slits to observe how the particles were passing through the slits. They noticed that after placing the detector that measured the particles passing through the slit, the pattern obtained on screen was of particle nature with just two fringes as shown in Fig. 3.

This caused confusion and a slight modification was made to the experiment where the detector was turned off as they believed it caused interference in the pattern. After the detector was turned off, the pattern obtained on the screen was of wave-like with alternate dark and light fringes throughout the screen as shown in Fig. 4.

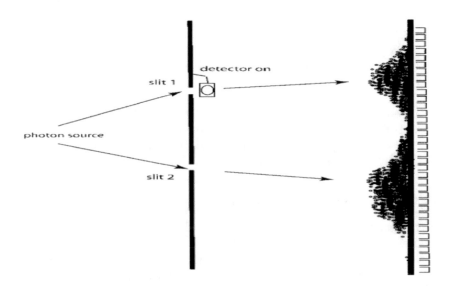

Fig. 3. Measurement with the detector turned on [9]

Hence, it was concluded the detector that measured particles led them to lose their coherence and their behavior on screen was like one shown in Fig. 3. However, they seemed to act like waves when the detector was turned off. From the pattern observed in Fig. 4, the scientists were led to believe that the particles hit the surface of the screen in a probabilistic distributed fashion shown in the upper region of Fig. 5 similar to the pattern formed by wave nature of the light where certain spots were brighter than others indicating that more particles took the path relevant to those spots to hit the screen and in between regions on screen was darker.

It is this probabilistic nature of particle taking different paths, which is exploited by Qubits while searching for a specific node in a graph or a tree.

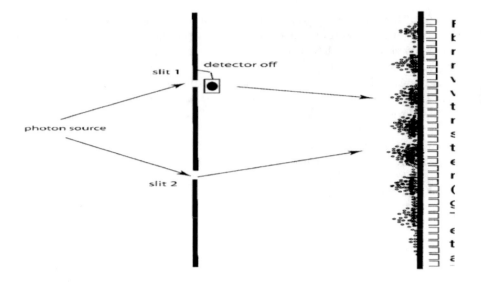

Fig. 4. Measurement with the detector turned off [9]

As opposed to classical bits used by the classical search algorithm that would explore only one path at a time, a Qubit explores several paths at a time depending on the probability distribution which leads it to arrive at a conclusion faster than the other classical algorithms.

1.4   Quantum Search Algorithms

Grover Algorithm. Grover's algorithm is a kind of search algorithm that is able to find an item in time complexity $O(\sqrt{N})$ where N is the total items to be searched. Conventionally the Linear algorithm used for this function would be able to accomplish this task in time complexity $O(N)$. For instance, the number of evaluations performed by the linear algorithm to search a particular item from the list of 10000 items would be 10000 computations however the Grover's algorithm will fetch the same result only in 100 computations [1]. For simplification, if we have n bits to represent our items, then the combinations, we can form are 2^n, so if we are dealing with 2 qubits, our total no of items can be $2^2(=4)$ namely apple, banana, orange, mango represented as 00, 01, 10, 11 respectively and let's assume that the item we are looking for is banana (10). Essentially, Grover's algorithm implements a function which returns true for the desired item (mango-"10") and false for any other input. The purpose of the algorithm is to find the one input that returns true. The algorithm performs the following steps:

Step-Using Hadamard Gates. Qubits are put through Hadamard gates to calculate amplitude for every possible input (or state or item) using superposition. It is represented in Fig. 6. In the beginning, the amplitude will be the same

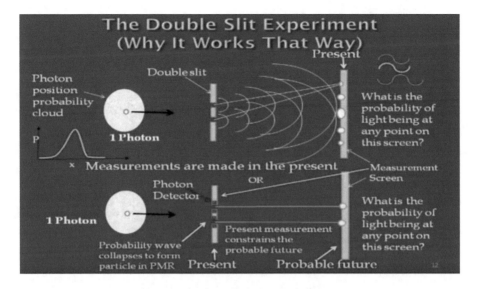

Fig. 5. Probabilistic distribution of hitting a spot on a screen [3]

for every input (item) in the search list. The probability is linearly related to amplitude. Also, the probability of every item will be the same and so for a list of total 4 items, each item will have a 25% probability Fig. 7.

$$|s\rangle = \frac{1}{\sqrt{N}} \sum_{x=0}^{N-1} |x\rangle$$

Fig. 6. Calculate amplitude of each input [5]

Step-Using Unitary Operator. The Unitary Operator U_f shown in Fig. 8 flips the amplitude of an input based on the oracle function f(x) which returns 1 when the input is the searched input (or item) and 0 if it is otherwise as shown in Fig. 9.

Step-Using Diffusion Operator. The Diffusion Operator U_f shown in Fig. 10 amplifies the amplitude of the each input reflects the input across the x axis.

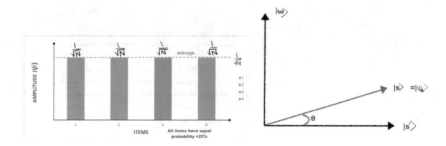

Fig. 7. Amplitude of each input [5]

$$U_f|x\rangle = (-1)^{f(x)}|x\rangle$$

Fig. 8. Unitary function U_f [5]

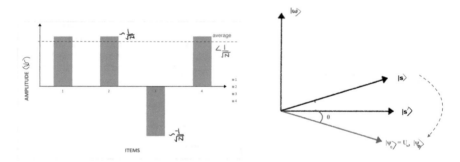

Fig. 9. Flipping of amplitude [5]

The equation of this step is summarized in Fig. 11. This operation amplifies the amplitude value of inputs however it significantly amplifies the amplitude of the desired input. This also changes the probability of every input shown in Fig. 11. This step which is shown in Fig. 12 is repeated several times until the amplitude of the desired input compared to other is such that the probability of that item reaches nearly 100%. This repetition number comes equal to $\sqrt{N}$. And the inputs are finally measured through Qubits to yield an input with the highest probability or amplitude.

Quantum Partial Search. It is a variant of Grover's algorithm. This algorithm does not find the exact location of the desired item. it basically divides the search space into chunks or blocks and yields the address of the block which might contain the desired item. For example, if one has a list of students and

$$U_s = 2|s\rangle\langle s| - 1$$

Fig. 10. Diffusion operator U_f [5]

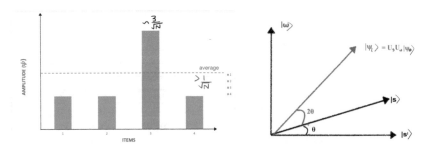

Fig. 11. Amplification [5]

$$|\psi_t\rangle = (U_s U_f)^t |\psi_0\rangle$$

Fig. 12. Summarized equation U_f [5]

one is interested to know about a particular student's exam grade, he can use the partial search algorithm to find out which particular group (e.g. A-91-100, B-81-90, C-71-90) the student belongs to instead of searching for his exact scores. If one has K blocks and N items, the size of each black would be b = N/K. If one block is selected at random and the rest of the blocks are searched using the classical algorithm for the desired item and if the desired item is not found in those blocks then it is assumed that item is present in the block not searched. This reduces the average number of computations from N/2 to (N-b)/2. The partial algorithm makes use of this concept. The Grover's algorithm yields the answer in $\sqrt{N}$ steps whereas the partial algorithm would yield result faster by a numerical factor that depends on the number of blocks K.

Shor's Algorithm. Shor's algorithm is used to for integer factorization. Decomposition of integer number into prime factors is believed to be a hard process. This belief was challenged by scientist Peter Shor when he presented Shor's algorithm for factoring in polynomial search time. It runs partially on quantum computer with time complexity $O((logn)^2(loglogn)(logloglogn))$. The pre-processing and post-processing is done on classical computers. It makes use of reduction of factorization problem to order-finding problem. It uses Quantum Fourier transformation to yield result in the polynomial time limit. The algorithm is very sufficient in case of public key cryptography which can be cracked using a sufficiently large quantum computer. RSA encryption uses a product of two prime numbers which is very large to factorize with classical computers. The time only grows as the number becomes larger and there is no classical algorithm that can accomplish this task in $O((logN)k)$ for any k.

The rest of this paper will discuss Project Requirements, Literature Review, Conclusion and Future Work.

2 Project Requirements

For this particular project, there has been considerable discussion in the literature of the possibility of building quantum computing databases. This has moved from a basic discussion about the concept of such machines through studies of the mathematical properties of logic gates that might be adequate to build them, to discussions of practical algorithms that might be run on them.

Despite doubts that have been expressed about the physical practicability of quantum computers due to the problem of decoherence, there seems to be a good reason to hope that these are soluble in light of the development of quantum error correcting code. Increasing numbers of practical suggestions for the technological implementation of quantum computers have been advanced ranging from the use of cracking large passwords to many other applications.

Although conventional computers using semi-conductors rely upon quantum effects in their underlying technology, their design principles are classical. They have a definite state vector and they evolve deterministically between states in this space. Thus the state of a classical computer with an n bit store is defined by a position in an n-dimensional binary co-ordinate state space.

In contrast, the state of a quantum computer with a store made up of n quantum two-state observables, or qubits is given by a point in dimensional space. Each dimension of this space corresponds to one of the possible values that n classical bits can assume. These possible bit patterns constitute basis vectors for space, and, associated with each such basis vector there is a complex valued amplitude. At any instant, the quantum computer is in a linear superposition of all of its possible bit patterns. It is this ability to exist in multiple states at once that is exploited by algorithms such as Shor's method of prime factorization [12]. If we abstract from the difficult technical problem of long term coherent storage of qubit vectors, this ability of the store to exist in multiple simultaneous states may be relevant to database compression.

3 Literature Review

Quantum computation has the advantage of speed [4] over its classical counterpart which makes the quantum computation more favorable. Although building a full-edged quantum computer is still far from reality, some of the research works such as [12] Shors algorithm and Grover algorithm have attracted much attention in the theoretical side. In the experimental side, some success with a small number of quantum bits has already been achieved. Peter Shor showed that it is possible for a quantum algorithm to compute factorization in polynomial-time. L. K. Grover, on the other hand, showed [2] that it is possible to search for a single target item in an unsorted database, i.e., the elements of the database are not arranged in any specific order, in a time which is quadratically faster than what a classical computer needs to complete the same task. Here time is measured in terms of the number of queries to the oracle one needs to complete a task. Grover algorithm needs $O(N)$ queries to the oracle. Although Grover algorithm cannot perform a task exponentially faster than classical computer still it is quite popular because of its wide range of applications such as a subroutine of some large algorithms in computer science.

It can be shown that the quantum algorithm of Grover is the fastest algorithm, i.e., optimal [11] search in an unsorted database. Instead of looking for the target element in the whole database at once it is sometimes natural to divide the database into several blocks and then look for the particular block which contains the target element. This is called quantum partial search algorithm, first studied by Grover and Radhakrishnan [10], which can be optimized and further generalized to the hierarchical quantum partial search algorithm. The purpose of this article is to review the basic concepts of quantum search algorithms. In our daily life, we encounter databases which contain many elements. The database may be arranged in a particular order, i.e. sorted or may not have any order at all, i.e. unsorted. For example, consider the telephone directory which has a large number of contact details of individuals. This example is particularly interesting because it serves both as a sorted and an unsorted database. When we look for the names, which are arranged in lexicographical order, then the telephone directory is an example of a sorted database. However when we look for a telephone number then the telephone directory becomes an example of an unsorted [10] database. The job of a quantum search algorithm is to find a specific element, usually called the target item or the solution from the vast number of elements in a database. Typically classical computer takes time proportional to the size of the database.

Quantum search algorithms, which are based on the principle of quantum mechanics, promise to significantly reduce the computation time for the same database search.

4   Conclusion

The tremendous computing power offered by quantum computing helped immensely to solve problems in a very small time compared to the time taken by classical computers. In the beginning, machines were designed based on the laws of quantum mechanics and they were utilized to simulate quantum mechanical systems but later with the realization that they exhibited heavy power, they began to be utilized and exploited for various other applications in cryptography, language theory, mathematics and information processing. The quantum computers were utilized to perform heavy computation with their inherent power to execute complex programs in polynomial time compared to the classical computers that consumed an exponential amount of time to process data. Algorithms that were executed on classical computers are constantly being improved to be executed on quantum computers that will yield faster results. There are extensive researches performed to design programs and revise concepts based on quantum computers that will replace the existing programs run on classical computers. The quantum algorithm like Shor's algorithm has been enabled to find prime factors of product of two very large numbers in a relatively very short time. Also, Grover's Algorithm based on the superposition of states on qubits performs search and yields the result comparatively in a very short time. In a very short time, it is expected that we might be able to find answers to a lot of questions that were difficult with the limitation of classical computers.

5 Future Work

The most relevant research is about maintaining the stability of the qubits At this point, there are many unknowns in the field of quantum computation, which makes future prediction difficult. The limitations of quantum computation have not yet been demonstrated. However, it has been shown that quantum computers are demonstrably more efficient than classical computers for some problems, so it is reasonable to assume that more such problems will continue to be discovered. Likewise, it is difficult to predict which of the current hardware paradigms if any will yield a scalable design. A significant problem is that maintaining coherence and manipulating Qubits are two necessary components of all designs which are fundamentally at odds with each other. The least promising of the current technologies is nuclear magnetic resonance, so it is unlikely that this path will yield practical quantum computers. Atom trap designs are effective at maintaining coherence and have succeeded in entangling relatively large numbers of simultaneous Qubits. Conversely, quantum dot designs have the advantage of extremely fast cycle times (measured in picoseconds), while superconductors have the advantage that they can be fabricated using well-understood methods (but they struggle with rapid decoherence). It could be argued that atom trap computers have achieved the most promising results to date. Because quantum computers are difficult to scale and there are many common applications for which classical computers are just as effective, it is possible that quantum computers might never achieve widespread consumer use. Additionally, designing quantum algorithms requires not only knowledge of computer programming, but a thorough understanding of quantum physics. Average programmers, even many very good programmers are unlikely to have the necessary training, so software engineering for quantum computers will be restricted to programmers who are also accomplished, physicists. As the technology matures, quantum computers may become useful as expensive specialty computation machines used mostly by research laboratories or large corporations, similar to mainframes. On the other hand, the research is still immature, so it is still possible that a breakthrough enabling practical consumer use could be discovered.

References

1. Aaronson, S.: Quantum computing and hidden variables. Phys. Rev. A **71**(3), 032325 (2005)
2. Brassard, G., Hoyer, P., Mosca, M., Tapp, A.: Quantum amplitude amplification and estimation. Contemp. Math. **305**, 53–74 (2002)
3. Campbell, T.: Double Slit Experiment (2010). https://youtu.be/LW6Mq352f0E
4. Cleve, R., Ekert, A., Henderson, L., Macchiavello, C., Mosca, M.: On quantum algorithms (1998)
5. Gritter, T.: Grover's algorithm (2017). https://youtu.be/hK6BBluTGhU
6. Grover, L.K.: A fast quantum mechanical algorithm for estimating the median, arXiv preprint quant-ph/9607024 (1996)
7. Helson, J.: QTM1x The quantum Internet and quantum computers: how will they change the world? (2018). https://youtu.be/8t1CNC6dqPQ
8. Grover, L.K.: A fast quantum mechanical algorithm for database search, arXiv preprint quant-ph/9605043 (1996)
9. Lederman, L., Hill, C.T.: Quantum Physics for Poets. Prometheus Books (2011)
10. Grover, L.K.: Quantum mechanics helps in searching for a needle in a haystack. Phys. Rev. Lett. **79**(2), 325 (1997)
11. Nielsen, M.A., Chuang, I.: Quantum computation and quantum information (2002)
12. Shor, P.W.: Polynomial-time algorithms for prime factorization and discrete logarithms on a quantum computer. SIAM Rev. **41**(2), 303–332 (1999)

Methodically Unified Procedures for Outlier Detection, Clustering and Classification

Piotr Kulczycki[1,2(✉)]

[1] Systems Research Institute, Centre of Information Technology for Data Analysis
Methods, Polish Academy of Sciences, Kraków, Poland
kulczycki@ibspan.waw.pl,kulczycki@agh.edu.pl
[2] Faculty of Physics and Applied Computer Science, Division for Information
Technology and Systems Research, AGH University of Science and Technology,
Kraków, Poland

Abstract. In the practice of data analysis some problems for many-sided researches are caused by the methodological variety of specific algorithms, often leading to laborious interpretations and time-consuming studies. This paper presents the concept of methodically unified procedures, based on kernel estimators, for three fundamental tasks: outlier detection, clustering, and classification. Their clear interpretation facilitates the applications and potential individual modifications. The investigated procedures are distribution-free, enabling analysis and exploration of data with any distributions, also when elements are grouped in several separated parts. The results obtained depend not only on the values of particular attributes, but above all on the complex relationships between them.

Keywords: Outlier detection · Clustering · Classification ·
Distribution free methods · Kernel estimators · Numerical algorithm

1 Introduction

The suddenness of the computer technology expansion in recent decades has caused excessive fragmentation of a methodology applied in data analysis problems [2,14]. In this situation universal concept, enabling various researches with the aid of uniform apparatus, become valuable, e.g. based on decision trees [26] or k-Nearest Neighbor concept [4]. The subject of this paper is the presentation of a coherent concept of using the methodology of kernel estimators [13,28,30] for the three main tasks of data analysis: detection of atypical elements (outliers) [1], clustering [24], and classification [7]. The application of a uniform apparatus for all three fundamental problems facilitates comprehension of the material and, in consequence, the creation of individualized modifications in the latter phase of designing a personal application. The use of nonparametric kernel estimators

© Springer Nature Switzerland AG 2020
K. Arai et al. (Eds.): FTC 2019, AISC 1069, pp. 460–474, 2020.
https://doi.org/10.1007/978-3-030-32520-6_35

frees the results from data distribution – this concerns not only the shape of their grouping, but also the possibility of their partition in separate incoherent parts. The results obtained take into account not only the values of specific attributes but the complex relationships between them. The methodology investigated in this paper is practically parameter-free, i.e. the calculation of parameter values is not required from the user, although it is possible to optionally modify them in order to achieve the specific desired properties.

In successive parts of this paper are described the kernel estimators methodology (Sect. 2) and further its use to the tasks of atypical elements detection (Sect. 3), clustering (Sect. 4), and classification (Sect. 5). The method worked out here was applied to the creation of a mobile phone operator's marketing support strategy, which is presented in the framework of the summary (Sect. 6). This paper constitutes a condensed conference version of the material [15] synthesizing investigations by the research group of the author from previous articles [16–23], with indispensable supplements, updates, and corrections.

2 Kernel Estimators Methodology

Consider the m-elements set of n-dimensional vectors with continuous attributes:

$$x_1, x_2, \ldots, x_m \in \mathbb{R}^n. \tag{1}$$

The density of its distribution can be characterized by the kernel estimator $\hat{f} : \mathbb{R}^n \to [0, \infty)$ defined in the following manner:

$$\hat{f}(x) = \frac{1}{m} \sum_{i=1}^{m} K(x, x_i, h), \tag{2}$$

denoting

$$x = \begin{bmatrix} x_1 \\ x_2 \\ \vdots \\ x_n \end{bmatrix}, \quad x_i = \begin{bmatrix} x_{i,1} \\ x_{i,2} \\ \vdots \\ x_{i,n} \end{bmatrix} \quad \text{for} \quad i = 1, 2, \ldots, m \quad \text{and} \quad h = \begin{bmatrix} h_1 \\ h_2 \\ \vdots \\ h_n \end{bmatrix}, \tag{3}$$

where the positive constants h_j for $j = 1, 2, \ldots, n$ are called smoothing parameters for particular coordinates, while the kernel $K : \mathbb{R}^n \to [0, \infty)$ in contemporary applications is commonly used in the product form

$$K(x, x_i, h) = \prod_{j=1}^{n} \frac{1}{h_j} K\left(\frac{x_j - x_{i,j}}{h_j}\right), \tag{4}$$

i.e. as the product of one-dimensional kernels $K : \mathbb{R} \to [0, \infty)$ being a measurable function, fulfilling the condition $\int_{\mathbb{R}^n} K(y) \, dy = 1$, symmetric with respect to zero and having a weak global maximum at this point. The choice of the kernel form and the calculation of the smoothing parameter value is made most often with the criterion of the mean integrated square error.

In the following, the normal kernel

$$K(x) = \frac{1}{\sqrt{2\pi}} \exp\left(-\frac{x^2}{2}\right) \tag{5}$$

is generally held as basic. For special purposes the uniform

$$K(x) = \begin{cases} \frac{1}{2} & \text{for} \quad x \in [-1,1] \\ 0 & \text{for} \quad x \notin [-1,1] \end{cases} \tag{6}$$

will be also used later in Sect. 3.

From the practical point of view, the establishment of the smoothing parameter is of crucial importance. When its value is too small, too many local extremes of the estimator $\hat{f}$ appear, which is contrary to the actual properties of real populations. On the other hand, when it is too big, this results in overflattening this estimator, hiding specific properties of the distribution under investigation. In the procedures proposed in the following, the approximate method with linear calculation complexity will be used, consisting in the direct application to particular coordinates of the formula

$$h = \left(\frac{8\sqrt{\pi}}{3} \frac{W(K)}{U(K)^2} \frac{1}{m}\right)^{1/5} \hat{\sigma}, \tag{7}$$

where

$$\hat{\sigma} = \sqrt{\frac{1}{m-1} \sum_{i=1}^{m} (x_i)^2 - \frac{1}{m(m-1)} \left(\sum_{i=1}^{m} x_i\right)^2}. \tag{8}$$

In specific cases one can alternatively propose the effective *plug-in* method with quadratic calculation complexity [13, Section 3.1.5], [30, Section 3.6.1]. The constant appearing in the above procedures $W(K) = \int_{-\infty}^{\infty} K(y)^2 \, dy$ and $U(K) = \int_{-\infty}^{\infty} y^2 K(y) \, dy$ equal $1/2\sqrt{\pi}$, $1/2$ and 1, $1/3$, respectively[1] for normal (5) and uniform (6) kernels.

Particular applications may also use specific concepts optionally fitting the model to a considered reality and requirements. A so-called modification of the smoothing parameter [13, Section 3.1.6], [28, Section 5.3.1] is presented below and applied later in Sect. 4. This procedure relies on introducing the positive modifying parameters on particular kernels

$$s_i = \left(\frac{\hat{f}_*(x_i)}{\bar{s}}\right)^{-c} \quad \text{for } i = 1, 2, \ldots, m, \tag{9}$$

where $c \in [0, \infty)$, $\hat{f}_*$ denotes the kernel estimator without modification (2), $\bar{s}$ is the geometrical mean of the numbers $\hat{f}_*(x_1), \hat{f}_*(x_2), \ldots, \hat{f}_*(x_m)$, and finally, defining the kernel (4) in the more general form:

---

[1]   In the book [30] with the following notation changes: $W(K) \overset{into}{\to} R(K)$ and $U(K) \overset{into}{\to} \mu_2(K)$.

$$K\left(\boldsymbol{x}, \boldsymbol{x}_i, h\right) = \prod_{j=1}^{n} \frac{1}{s_i h_j} K_j \left(\frac{x_j - x_{i,j}}{s_i h_j}\right). \tag{10}$$

Thanks to the above procedure, the areas in which the kernel estimator assumes small values, are additionally flattened, and the regions connected with large values – peaked, which allows individual properties of a distribution to be better revealed. The parameter c stands for the intensity of the modification procedure; a large value increases, a small value decreases it. Note also that $c = 0$ implies $s_i \equiv 1$ and in consequence lack of modification – kernel estimator (10) reduces to the basic form (4). According to the criterion of the integrated mean square error, the value $c = 0.5$ can be suggested as standard. The above algorithm is of quadratic calculation complexity; therefore, for $m > 1,000$, and especially for $m > 10,000$, this should be used with care.

Detailed information on the kernel estimators methodology can be found in the classic books [13, 28, 30].

3   Detection of Atypical Elements

The basic idea of the investigated procedure stems from the significance test proposed in the work [23]. Suppose dataset (1) containing elements representative for the considered population. Based on the material presented in Sect. 2, the kernel estimator (2) with (4) can be calculated; normal kernel (5) is proposed here. Then, consider also the set of its values for elements of dataset (1), therefore

$$\hat{f}_{-1}\left(x_1\right), \hat{f}_{-2}\left(x_2\right), \ldots, \hat{f}_{-m}\left(x_m\right), \tag{11}$$

where $\hat{f}_{-i}$ means the kernel estimator $\hat{f}$ calculated excluding the i-th element of the dataset. Next, define the number $r \in (0,1)$ establishing the sensitivity of the procedure for identifying atypical elements. This number will determine the assumed proportion of atypical elements in relation to the total population; therefore, the ratio of the number of atypical elements to the sum of atypical and typical elements. In practice

$$r = 0.01, 0.05, 0.1 \tag{12}$$

is the most often used. Next, for set (11) one can calculate the positional estimator for the quantile of the order r [12, 25] given by the formula

$$\hat{q}_r = \begin{cases} z_1 & \text{for} \quad mr < 0.5 \\ (0.5 + i - mr) z_i + (0.5 - i + mr) z_{i+1} & \text{for} \quad mr \geq 0.5 \end{cases}, \tag{13}$$

where $i = [mr + 0.5]$, while $[d]$ denotes an integral part of the number $d \in \mathbb{R}$, and z_i is the i-th value in size of set (11) after being sorted; thus

$$\{z_1, z_2, \ldots, z_m\} = \left\{\hat{f}_{-1}\left(x_1\right), \hat{f}_{-2}\left(x_2\right), \ldots, \hat{f}_{-m}\left(x_m\right)\right\} \tag{14}$$

with $z_1 \leq z_2 \leq \ldots \leq z_m$. Generally, there are no special recommendations concerning the choice of sorting algorithm [6] used for specifying set (14). However, let us interpret definition (13), taking into account condition (12). So, it is enough to sort only the $i+1$ smallest values in set (14), therefore, about 1–10% of its size. One can apply a simple algorithm that subsequently finds the $i+1$ smallest elements of set (14).

Finally, if for a given tested element $\tilde{x} \in \mathbb{R}^n$, the condition $\hat{f}(\tilde{x}) \leq \hat{q}_r$ is fulfilled, then this element should be considered atypical; for the opposite $\hat{f}(\tilde{x}) > \hat{q}_r$ it is typical.

The properties of the above algorithm can be corrected for datasets (1) of small size – i.e. when $m < 10,000$, especially when $m < 1,000$ – through generating additional elements with a distribution identical to that characterizing dataset (1). The Neumann's elimination concept [8] is suggested here. This allows the generation of a sequence of random numbers of distribution with support bounded to the interval $[a, b]$, while $a < b$, characterized by the density f of values limited by the positive number c, i.e.

$$f(x) \leq c \quad \text{for } x \in [a, b].\tag{15}$$

In the multidimensional case, the interval $[a, b]$ generalizes to the n-dimensional cuboid $[a_1, b_1] \times [a_2, b_2] \times \ldots \times [a_n, b_n]$, while $a_j < b_j$ for $j = 1, 2, \ldots, n$.

First the one-dimensional case is considered. Let us generate two pseudorandom numbers u and v of distribution uniform on the intervals $[a, b]$ and $[0, c]$, respectively. Next, one should check that

$$v \leq f(u).\tag{16}$$

If the above condition is fulfilled, then the value u ought to be assumed as the desired element from a set with distribution characterized by the density f, that is

$$x = u.\tag{17}$$

In the opposite case, the numbers u and v need to be removed and steps (16) and (17) repeated, until the desired number of pseudorandom numbers x with density f is obtained.

In the presented procedure, the density f is characterized by the kernel estimator $\hat{f}$ using dependences (2) and (4). The uniform kernel will be employed, allowing easy calculation of the support boundaries a and b, as well as the parameter c appearing in condition (15). Namely:

$$a = \min_{i=1,2,\ldots,m} x_i - h, \quad b = \max_{i=1,2,\ldots,m} x_i + h\tag{18}$$

and

$$c = \max_{i=1,2,\ldots,m} \left\{ \hat{f}(x_i - h), \hat{f}(x_i + h) \right\}.\tag{19}$$

The last formula results from the fact that the maximum for a kernel estimator with the uniform kernel must occur on the edge of one of the kernels. It is also worth noting that calculations of parameters (18) and (19) do not require much effort.

In the multidimensional case, Neumann's elimination algorithm is similar to the previously discussed one-dimensional version. The edges of the n-dimensional cuboid $[a_1, b_1] \times [a_2, b_2] \times \ldots \times [a_n, b_n]$ are calculated from formulas comparable to dependences (19) separately for particular coordinates. The kernel estimator maximum is thus located in one of the corners of the kernels, therefore

$$c = \max_{i=1,2,\ldots,m} \left\{ \hat{f} \left(\begin{bmatrix} x_{i,1} \pm h_1 \\ x_{i,2} \pm h_2 \\ \vdots \\ x_{i,n} \pm h_n \end{bmatrix} \right) \right\} \qquad \text{following all combinations } \pm. \qquad (20)$$

Note that the number of these combinations is finite. Using the presented algorithm, n particular coordinates of the vector u and the subsequent number v are generated, after which, condition (16) is checked.

The paper [21] describes the method designed to represent the results in fuzzy and intuitionistic fuzzy forms and also the procedure for transposing an unsupervised version of the atypical elements detection to a convenient supervised form with equalized patterns – it allows the application of numerous and diverse classification apparatus. Example applications in biomedicine are described in the publication [22].

4   Clustering

As previously considered, suppose dataset (1) containing elements under research. Let's divide it into groups of elements (clusters) as similar as possible within these groups and dissimilar between them. The most intuitive and natural concept for this purpose is the assumption that specific clusters are related to modes (local maxima) of distribution density; thus, the 'valleys' become the borders of the resulting clusters [10]. The algorithm described in this section can be summarized as follows:

1. Parameters' values can be numerically calculated based on optimizing criteria.
2. The algorithm does not require stringent assumptions concerning the number of clusters, which enables the obtained number to be better fitted to a real data structure.
3. The parameter which is responsible for the approximate cluster number is indicated; also displayed will be how possible modifications to its value – e.g. obtained in point 1 (above) – have an influence on increases or decreases in the number of clusters but still without any necessity to determine their precise number.
4. Furthermore, the next parameter is indicated influencing the proportion of the number of clusters in dense versus sparse regions of data elements; its

value can also be obtained by the optimization criteria, with the option for possible modifications with the aim of simultaneously increasing the cluster quantity in dense regions and reducing from sparse areas of data, or *vice-versa*.

5. The suitable relationship between the parameters mentioned in points 3 and 4 enables reducing or eliminating clusters in sparse regions, virtually without affecting the cluster quantity in dense areas of data.

The characteristics from point 4, and consequently point 5, are particularly important to highlight as being practically absent in other clustering methods. In applications, one should underline the consequences of points 1 and 2, as well as possibly point 3.

Thus, having dataset (1), one can apply the material presented in Sect. 2 to create the kernel estimator $\hat{f}$. After assuming that subsequent clusters are mapped to the local maxima of the above function, the elements of set (1) can be shifted in the direction of gradient $\nabla \hat{f}$ with the suitable step. It can be carried out iteratively using the classic gradient concept [11], given as

$$x_i^0 = x_i \quad \text{for } i = 1, 2, \ldots, m \tag{21}$$

$$x_i^{k+1} = x_i^k + \boldsymbol{b} \circ \frac{\nabla \hat{f}\left(x_i^k\right)}{\hat{f}\left(x_i^k\right)} \quad \text{for } i = 1, 2, \ldots, m \text{ and } k = 0, 1, \ldots, k^*, \tag{22}$$

whereas $k^* \in \mathbb{N} \setminus \{0\}$ denotes the number of steps, while based on optimizing criterion, the following can be proposed [10]:

$$\boldsymbol{b} = \frac{1}{n+2} \begin{bmatrix} h_1^2 \\ h_2^2 \\ \vdots \\ h_n^2 \end{bmatrix}, \tag{23}$$

where the operator $\circ : \mathbb{R}^n \times \mathbb{R}^n \to \mathbb{R}^n$ is understood in the sense of products of particular coordinates of vectors from $\mathbb{R}^n$, e.g. for the case $n = 2$ one has $\begin{bmatrix} y_1 \\ y_2 \end{bmatrix} \circ \begin{bmatrix} z_1 \\ z_2 \end{bmatrix} = \begin{bmatrix} y_1 z_1 \\ y_2 z_2 \end{bmatrix}$. In the following, the kernel estimator (2) with modification of the smoothing parameter (10) will be applied (if one uses basic version (4) without modification, then $s_i \equiv 1$ should be assumed in the formulas below) using the normal kernel (5). In this case

$$\boldsymbol{b} \circ \frac{\nabla \hat{f}\left(x_i^k\right)}{\hat{f}\left(x_i^k\right)} = \frac{1}{(n+2) s_i^2} \begin{bmatrix} x_{i,1} - x_{i,1}^k \\ x_{i,2} - x_{i,2}^k \\ \vdots \\ x_{i,n} - x_{i,n}^k \end{bmatrix} \tag{24}$$

with the natural notation

$$\boldsymbol{x}_i^k = \begin{bmatrix} x_{i,1}^k \\ x_{i,2}^k \\ \vdots \\ x_{i,n}^k \end{bmatrix}. \tag{25}$$

It is assumed that algorithm (21) and (22) needs to be stopped, in the event when the following inequality is fulfilled after the subsequent k-th step

$$|D_k - D_{k-1}| = aD_0, \tag{26}$$

in which $a > 0$ and

$$D_0 = \sum_{i=1}^{m-1} \sum_{j=i+1}^{m} d(x_i, x_j), \quad D_{k-1} = \sum_{i=1}^{m-1} \sum_{j=i+1}^{m} d(x_i^{k-1}, x_j^{k-1}),$$

$$D_k = \sum_{i=1}^{m-1} \sum_{j=i+1}^{m} d(x_i^k, x_j^k), \tag{27}$$

where d denotes Euclidean metric in $\mathbb{R}^n$. The value $a = 10^{-6}$ can be suggested. Finally, when condition (26) is fulfilled after the k-th step, then $k^* = k$ and this is considered to be the last step. Additionally, to avoid of unexpected situations, the constraint $500 \leq k^* \leq 5000$ should be also assumed.

A procedure now needs to be applied for the creation of clusters and for particular elements to be assigned to them. The set comprising elements of dataset (1) submitted to k^* steps of algorithm (21) and (22) is considered to achieve this objective:

$$x_1^{k^*}, x_2^{k^*}, \ldots, x_m^{k^*}. \tag{28}$$

Now, define the set of their mutual distances

$$\left\{ d\left(x_i^{k^*}, x_j^{k^*}\right) \right\}_{\substack{i = 1, 2, \ldots, m-1 \\ j = i+1, i+2, \ldots, m}} . \tag{29}$$

Its size is $m_d = m(m-1)/2$. Based on the above set one can calculate the auxiliary kernel estimator $\hat{f}_d$ of mutual distances. Normal kernel (5) is also suggested here. Finding – with appropriate accuracy – the 'first' (in the sense of the lowest argument value) local minimum of the function $\hat{f}_d$ in the interval $(0, D)$, where D means the greatest element of set (29), is the next task. This can be implemented beginning with 0 and increasing the argument until the value where the increment changes from negative into positive[2]. Such a value – referred to as x_d hereinafter – will be treated as the smallest distance between clusters.

The final step is the creation of the clusters. This is achieved in the following way:

---

[2] In the event that such a value does not exist, the presence of one cluster should be recognized and the procedure completed. The same applies to the irrational but formally possible situation $m = 1$, when set (30) is empty.

1. Take an element of set (28) and firstly produce a one-element cluster including it.
2. Discover an element of set (28) different to elements in the cluster, and lying closer than x_d to at least one of these other elements; in the event when such an element exists, one should add this to the cluster and repeat point 2 rummaging set (28) from the beginning.
3. Add the attained cluster to a 'cluster list' and remove its elements from set (28); if this set reduced in such a way remains not empty, go to point 1, otherwise, finish the algorithm.

Note that the above procedure did not necessitate any initial, often arbitrary, assumption regarding the number of clusters – it mainly depends on the internal data structure. The parameters are calculated effectively based on optimizing criteria, however in practical applications, it can be beneficial to suitably modify the values of kernel estimator parameters – thus influencing, in such a manner, the cluster number and additionally, the proportion of their occurrence between dense and sparse regions.

Namely, as mentioned in Sect. 2, too great a value of the smoothing parameter h results in the over-smoothing of the kernel estimator; while if it is too small, this causes the appearance of too many local extremes. Therefore, the result of increasing – with respect to that calculated by the criterion of the mean integrated square error – this parameter value is the occurrence of fewer clusters; conversely, decreasing to this value yields more clusters. In both cases, one can emphasize that despite influencing the number of clusters, their number still solely depends only on the data's internal structure, without arbitrary assumptions in this matter.

The intensity of the smoothing parameter modification – as mentioned in Sect. 2 – is defined by the parameter c, where $c = 0.5$ constitutes the standard value. Its increase sharpens the kernel estimator in the dense regions of the dataset and also smooths it in the sparse areas; consequently, if this parameter value rises, then the number of clusters in dense areas increases and simultaneously decreases in sparse regions. These effects are reversed in the event of this parameter value diminishing.

However, maintaining the cluster number in dense data regions and at the same time lowering or even eliminating clusters in sparse areas (as they often contain atypical elements, which are incorrect or unimportant for further research) is frequently desired in practice. Combining the aforementioned considerations, it is appropriate to propose increases to both the change of the intensity of the smoothing parameter modification c and simultaneously, the smoothing parameter h value calculated on the basis of the optimization criterion, to the value h^* given as

$$h^* = \left(\frac{7}{4}\right)^{c-0.5} . \tag{30}$$

The combined effect of both of these factors implies a two-fold smoothing of the estimator $\hat{f}$ in the areas in which the dataset is sparse. At the same time, the above factors approximately cancel each other out in dense regions; hence, they have negligible influence on the discovery of clusters in such areas.

More information with visual aids is presented in the article [16]. Some applications in bioinformatics, management, and fuzzy control were described in the paper [17].

5 Classification

The concept presented below will be based on the Bayes approach [7]. The classifiers gained in this way work quite well in complex real-world situations and are eagerly used by practitioners, chiefly because of their robustness and low requirements for specific patterns. Assume J sets containing n-dimensional vectors with continuous attributes

$$x_1', x_2', \ldots, x_{m_1}' \tag{31}$$

$$x_1'', x_2'', \ldots, x_{m_2}'' \tag{32}$$
$$\vdots$$

$$x_1''^{\ldots\prime}, x_2''^{\ldots\prime}, \ldots, x_{m_J}''^{\ldots\prime} \tag{33}$$

being patterns for assumed classes. The sizes $m_1, m_2, \ldots, m_J$ need to be proportional to the 'contribution' of specific classes within the investigated dataset. The classified element $\tilde{x} \in \mathbb{R}^n$ must be mapped to one of the above classes.

Denote as $\hat{f}_1, \hat{f}_2, \ldots, \hat{f}_J$ kernel estimators successively calculated on the basis of sets (31)–(33) treated as sets (1) each time; normal kernel (5) with possibly modification of smoothing parameter (10) is proposed here. According to the Bayes concept, the classified element $\tilde{x}$ needs then to be mapped to that class in which the value

$$m_1 \hat{f}_1 (\tilde{x}), m_2 \hat{f}_2 (\tilde{x}), \ldots, m_J \hat{f}_J (\tilde{x}) \tag{34}$$

is the largest. By introducing the positive coefficients $z_1, z_2, \ldots, z_J$, the above can be generalized to

$$z_1 m_1 \hat{f}_1 (\tilde{x}), z_2 m_2 \hat{f}_2 (\tilde{x}), \ldots, z_J m_J \hat{f}_J (\tilde{x}). \tag{35}$$

Giving the values $z_1 = z_2 = \ldots = z_J = 1$, formula (35) reduces to dependency (34). By applying a suitable increase to the value z_i, one introduces an inversely proportional decrease to the probability of mistakenly assigning the i-th class elements to an incorrect class; however, the danger of slightly increasing the overall quantity of misclassifications then theoretically appears. Bearing this in mind, it is possible to increase the values of even a few coefficients z_i.

Due to the supervised character of the classification problem, it is possible to improve in fitting the parameter values of the classification model to the specific structure of the dataset under investigation. Let us propose the introduction of $n + 1$ multiplicative correcting coefficients, relating to the values of the intensity of modification procedure c and the smoothing parameters for particular coordinates $h_1, h_2, \ldots, h_n$, with respect to those obtained applying the procedures based on integrated square error criterion, presented in Sect. 2. These coefficients are denoted as $b_0 \geq 0$ and $b_1, b_2, \ldots, b_n > 0$, respectively. Note that $b_0 = b_1 = \ldots = b_n = 1$ implies a lack of correction. Using, for a comprehensive search, a grid with a rather sizable discretization value, one may find the most beneficial points within the context of correct classification. The last stage is a classic optimization algorithm in the $(n + 1)$-dimensional space, with starting conditions being the points found above, while the performance index is

$$J\left(b_0, b_1, \ldots, b_n\right) = \#\left\{\text{incorrect classifications}\right\}, \qquad (36)$$

where $\#$ means the number of elements within a set. The classic leave-one-out method can be applied for the calculation of the above functional value for any fixed argument. Due to this value being an integer, the Hook-Jeeves algorithm [11] was used to find a minimum.

As a result of the performed research, the assumption can be made that for every coordinate, the grid should usually have nodes at the points 0.5, 0.75, ... , 1.5. The functional (36) values are calculated for these nodes; the attained results are then sorted and the five best become starting conditions for the Hook-Jeeves procedure with the initial step value 0.2. Following completion of each of the above five executions, the values of functional (36) for the obtained end points are calculated, and that which has the smallest value is the sought vector of the parameters $b_0, b_1, \ldots, b_n$. It is worthy of note that in the procedure described above this is not necessary; however, doing so would enhance the classification quality and, moreover, would allow one to apply the convenient formula (8) with (9), to calculate smoothing parameter values.

The broader description of the above procedure, also including classification of interval information, can be found in the article [19]. The application of the sensitivity method borrowed from artificial neural networks, which allows the elimination of those pattern elements which have insignificant or even a negative influence on the correctness of results, is likewise presented there. These concepts also constitute the basis for the creation of an adaptation structure, adjusting a classifier to nonstationary data (concept drift), which is presented in the paper [20]. The task to express the result in fuzzy and intuitionistic fuzzy forms, optimal from the point of view of classification results, is currently being investigated.

6   Final Comments and Example of Application

This paper presents methodologically uniform material for three fundamental procedures of data analysis: outliers' detection, clustering, and classification. It should be clearly underlined that thanks to this uniformity and ease of interpretation of universal mathematical apparatus used here, the presented concept becomes convenient for individual adaptations towards specific conditionings of the issues under investigation. In particular, tuning of parameters values to specific requirements is particularly easy and effective. Due to the use of the nonparametric kernel estimators method, the worked out procedures are independent of the distributions of the dataset being studied. They can be multimodal and even their supports may consist of several different parts. Thanks to the properties of kernel estimators, not only do the values of the particular attributes influence the results but in the multidimensional case – above all – the relationships between them. The procedures are ready-to-use; all formulas are given in the above text, especially allowing particular parameters to be calculated (excluding the easy-to-interpret participation of atypical element r), and if necessary, clear interpretations of the implications of their potential changes are described. The correct functioning of the procedures was positively verified in numerical and practical ways. The restricted length of this text precludes a detailed presentation of the obtained results. For outliers detection see the papers [21, 22], for clustering – the publications [16, 17], and for classification the articles [19, 20] are recommended. Many additional aspects can be found there.

The current research of the author's research group concentrate on the aspects resulting from the large size of contemporary datasets, including stream type [27], and also high dimensionality [29]. Attributes of the continuous type are supplemented in these cases by categorical, nominal and ordinal [3]. The conditional approach [5] of the investigated phenomena is also examined in detail.

The material presented in this paper is universal in nature and can be employed in many various tasks of modern science, and practical applications including engineering, econometrics and management, sociology, as well as natural sciences. The applied example of using a methodologically uniform concept to create a mobile phone operator's marketing support strategy is presented below.

Due to the present highly dynamic growth within the market of mobile phone networks, there is a natural necessity for such companies to permanently focus their strategies on the aim of satisfying their clients' different requirements, while simultaneously maximizing the operator's income. To avoid the lack of coherence with regard to serving particular clients, thus resulting in their possible defection to competitors, it is necessary to ensure that services remain uniform for comparable clients. The results of investigations presented below concerns business clients and were conducted for one Polish mobile phone operator.

There exists an enormous range of characteristic quantities that can be applied in practice to describe specific subscribers. These are primarily the mean monthly incomes per SIM card, subscription length, number of active SIM cards, and possibly other quantities, appropriately adapted to the current market specifics. Therefore, each m-element within a database characterizes a successive business client, and comprises n their attributes easily available to an operator. For the above example we have

$$x_i = \begin{bmatrix} x_{i,1} \\ x_{i,2} \\ x_{i,3} \\ \vdots \\ x_{i,n} \end{bmatrix} \quad \text{for } i = 1, 2, \ldots, m, \tag{37}$$

where $x_{i,1}$ denotes the average monthly income per SIM card of the i-th client, $x_{i,2}$ – its length of subscription, $x_{i,3}$ – the number of active SIM cards, and other characteristics.

Firstly, atypical elements within the dataset were removed, according to the procedure presented in Sect. 3, with $r = 0.1$. The regularity of the data structure was thus significantly enhanced; it is worthy of note that this was achieved through the cancellation of only elements which had negligible importance on the results of the investigated procedure.

Secondly, the dataset was submitted to clustering by the procedure described in Sect. 4. The consequence was the partitioning of the dataset which consisted of particular clients, into separate groups each comprised of similar members. The results, achieved for standard values of the modification intensity and smoothing parameters obtained from the integrated mean square criterion, showed too great a number of small-sized clusters lying in low density areas of data – mostly containing irrelevant, unusual clients – and an excessively large main cluster containing more than half of all the elements. Taking into account the properties of the procedure, the value of the parameter c increased from 0.5 to 1. As a consequence, the desired effects – the significant lowering of the number of peripheral clusters as well as the splitting of the main cluster – were thus simultaneously attained. The number of clusters was then satisfactory, and possible changes to the smoothing parameters value became redundant. At this point, the dataset comprising 1639 elements was partitioned into 26 clusters with sizes 488, 413, 247, 128, 54, 41, 34, 34, 33, 28, 26, 21, 20, 14, 13, 12, 10, two containing 4-element, three of 3-elements, two of 2-elements, and two of 1-element. Note that four groups can be clearly distinguished – the first includes two large clusters of 488 and 413 elements, the following contains two medium clusters with 247- and 128-elements, the next nine are small and have 20 to 54, and finally there are 13 each with fewer than 20 elements. It is now appropriate to eliminate the last of these clusters; however, those including key or prestigious clients (14-, 13-, 12- and 10-elements) were excluded from removal. Finally, for further analysis, 17 clusters remained.

Then, in the case of each of the clusters found in this manner, an optimal scheme – with regard to anticipated operator profit – was defined for the treatment of subscribers belonging to this group. Elements of fuzzy preference theory [9] were applied due to the usually imprecise nature of expert evaluation of such problems. Details of this operation [18] lie beyond the scope of the investigations presented in this paper.

It is notable that none of the above calculations must be performed during client negotiations; however, they should merely be updated (every 1–6 months in practice).

The client with whom negotiations are conducted can be characterized through the use of – in accordance with formula (37) – an n-dimensional vector the specific coordinates of which represent the respective features of this client. Such data can be obtained from the operator data-base archive – if the client has previously been the subscriber – or alternatively, from historic invoices issued by a rival network in the case of attempting to poach a client. Mapping the client to an appropriate subscriber group during negotiations – according to clusters defined earlier – was performed by applying the Bayes classification presented in Sect. 5. Because the marketing strategies regarding specific clusters have been previously established, the above action completes the procedure of supporting the marketing strategy with regard to business clients, which was the objective of the project presented above.

The comments summarizing the above application example are symptomatic and can be treated as a recapitulation of all the material presented in this paper. Above all, the use of the methodologically uniform apparatus of kernel estimators makes the analysis of the concept and creation of a computer application significantly easy and efficient.

Acknowledgments. I would like to express my gratitude to my close associates – former Ph.D.-students – Małgorzata Charytanowicz, D.Sc., Karina Daniel, Ph.D., Piotr A. Kowalski, D.Sc., Damian Kruszewski, Ph.D., Szymon Łukasik, Ph.D., with whom the research summarized in this paper was conducted.

References

1. Aggarwal, C.C.: Outlier Analysis. Springer, New York (2013)
2. Aggarwal, C.C.: Data Mining. Springer, Cham (2015)
3. Agresti, A.: Categorical Data Analysis. Wiley, Hoboken (2002)
4. Biau, G., Devroye, L.: Lectures on the Nearest Neighbor Method. Springer, Cham (2015)
5. Billingsley, P.: Probability and Measure. Wiley, New York (1995)
6. Canaan, C., Garai, M.S., Daya, M.: Popular sorting algorithms. World Appl. Program. **1**, 62–71 (2011)
7. Duda, R.O., Hart, P.E., Storck, D.G.: Pattern Classification. Wiley, New York (2001)
8. Gentle, J.E.: Random Number Generation and Monte Carlo Methods. Springer, New York (2003)
9. Fodor, J., Roubens, M.: Fuzzy Preference Modelling and Multicriteria Decision Support. Kluwer, Dordrecht (1994)
10. Fukunaga, K., Hostetler, L.D.: The estimation of the gradient of a density function, with applications in pattern recognition. IEEE Trans. Inf. Theory **21**, 32–40 (1975)
11. Kelley, C.T.: Iterative Methods for Optimization. SIAM, Philadelphia (1999)
12. Kulczycki, P.: Wykrywanie uszkodzeń w systemach zautomatyzowanych metodami statystycznymi. Alfa, Warsaw (1998)
13. Kulczycki, P.: Estymatory jągonadrowe w analizie systemowej. WNT, Warsaw (2005)
14. Kulczycki, P., Kacprzyk, J., Kóczy, L.T., Mesiar, R., Wisniewski, R. (eds.): Information Technology, Systems Research, and Computational Physics. Springer, Cham (2020)
15. Kulczycki, P.: Kernel estimators for data analysis. In: Ram, M., Davim, J.P. (eds.) Advanced Mathematical Techniques in Engineering Sciences, pp. 177–202. CRC/Taylor & Francis, Boca Raton (2018)

16. Kulczycki, P., Charytanowicz, M.: A complete gradient clustering algorithm formed with kernel estimators. Int. J. Appl. Math. Comput. Sci. **20**, 123–134 (2010)
17. Kulczycki, P., Charytanowicz, M., Kowalski, P.A., Łukasik, S.: The complete gradient clustering algorithm: properties in practical applications. J. Appl. Stat. **39**, 1211–1224 (2012)
18. Kulczycki, P., Daniel, K.: Metoda wspomagania strategii marketingowej operatora telefonii komórkowej, Przeglągonad Statystyczny, vol. 56, no. 2, pp. 116–134 (2009). Errata: vol. 56, no. 3–4, s. 3 (2009)
19. Kulczycki, P., Kowalski, P.A.: Bayes classification of imprecise information of interval type. Control Cybern. **40**, 101–123 (2011)
20. Kulczycki, P., Kowalski, P.A.: Bayes classification for nonstationary patterns. Int. J. Comput. Methods **12**(2), 19 (2015). ID 1550008
21. Kulczycki, P., Kruszewski, D.: Identification of atypical elements by transforming task to supervised form with fuzzy and intuitionistic fuzzy evaluations. Appl. Soft Comput. **60**, 623–633 (2017)
22. Kulczycki, P., Kruszewski, D.: Detection of rare elements in investigation of medical problems. In: Nguyen, N.T., Gaol, G.L., Hong, T.-P., Trawiński, B. (eds.) Intelligent Information and Database Systems, pp. 257–268. Springer, Cham (2019)
23. Kulczycki, P., Prochot, C.: Identyfikacja stanów nietypowych za pomocą estymatorów jądrowych. In: Bubnicki, Z., Hryniewicz, O., Kulikowski, R. (eds.) Metody i techniki analizy informacji i wspomagania decyzji, pp. 57–62. EXIT, Warsaw (2002)
24. Mirkin, B.: Clustering for Data Mining. Taylor & Francis, Boca Raton (2005)
25. Parrish, R.: Comparison of quantile estimators in normal sampling. Biometrics **46**, 247–257 (1990)
26. Rokach, L., Maimon, O.: Data Mining with Decision Trees. World Scientific, New Jersey (2015)
27. Silva, J., Faria, E., Barros, R., Hruschka, E., de Carvalho, A., Gama, J.: Data stream clustering: a survey. ACM Comput. Surv. **46**, 13 (2013)
28. Silverman, B.W.: Density Estimation for Statistics and Data Analysis. Chapman and Hall, London (1986)
29. Sorzano, C.O.S., Vargas, J., Pascual-Montano, A.: A survey of dimensionality reduction techniques, arXiv, signature 1403.2877v1 (2014)
30. Wand, M., Jones, M.: Kernel Smoothing. Chapman and Hall, London (1995)

A Trust Model for Cloud: Results from a Survey

Stephen Kirkman[1(✉)] and Richard Newman[2]

[1] College of Computer and Information Sciences, Regis University,
Denver, CO 80221, USA
skirkman@regis.edu
[2] Computer and Information Science and Engineering, University of Florida,
Gainesville, FL 32611, USA
nemo@ufl.edu

Abstract. This research proposes a cloud consumer trust model that focuses on decision making and the value of our data. The value consumers place on data plays a critical role in how consumers trust. The research was validated with a questionnaire and statistical power analysis. Few cloud trust models appear to be validated against real world data, and none, that we found, accomplish an *a-priori* statistical power analysis. Results indicated consumers were interested in data location, third party access and most surprisingly unbiased/expert organization recommendations were more important than family and friends recommendations.

Keywords: Cloud trust · Statistical power analysis · Hypothesis testing · Modeling

1 Introduction

Trust theory in the social and cognitive sciences has been studied for decades in many works. Since the birth of the Internet, the study of trust has found new directions for research. Trust is hard to qualify - even with face to face interactions. Now consumers have to trust *without* the face to face interactions. Approaches to achieving trust over the Internet (also cloud computing) include reputation systems, trust management systems, hardware-based integrity approaches, location-based approaches, policy-based approaches, social network trust, or a combination of these.

This research models consumer trust in the context of cloud computing. The value consumers place on data plays a critical role in how consumers trust. This model considers expectations, data value, and includes a decision framework. In the trust research space, there is a lack of models that specialize in cloud computing. The models that do exist for calculating trust have very little 'real world' validation. How accurate are these existing models? In addition, few surveys in this research space accomplished an *a-priori* statistical power analysis.

Contributions of this research include: a trust model for the cloud, formalized accumulated trust based on past experiences, *a-priori* statistical power analysis, feedback on the cloud model from a survey, and hypothesis testing.

© Springer Nature Switzerland AG 2020
K. Arai et al. (Eds.): FTC 2019, AISC 1069, pp. 475–496, 2020.
https://doi.org/10.1007/978-3-030-32520-6_36

Section 2 covers existing formal trust models. Section 3 introduces our trust model. Section 4 discusses the existing validation in academic literature and this research validation. We tested several hypotheses in Sect. 5 for acceptance or rejection criteria. Sections 6 and 7 is the discussion, limitations, future work, and conclusion.

2   Background on Trust Models

Qualitative models are described in terms of descriptions and ideas. Quantitative models (including probabilistic models) are based on more formal methods. This research is a combination of both types of models.

2.1   Quantitative Trust Models

Marsh [1] defined three types of trust in chapter 3 of his thesis: basic, general, and situational. Basic trust is the trust inherent to one person. It is a measure of how likely one is to trust anything or anyone and derived from past life experiences (i.e., some people are more trusting than others). General trust is the trust that one individual has in another without regard to a specific situation. Situational trust adds context or a specific task. This trust is the most applicable to the work we are doing.

In 2010, *Trust Theory: A Socio-Cognitive and Computational Model* [2] attempted to quantify belief-based aspects of trust and defined a quantitative model. In chapter 3, the analysis of trust is based on the assertion that degree of trust is a function of two types of belief. The evaluated beliefs of the *attributes* (i.e., the attributes under evaluation) of the trustee upon which positive expectation is based, and the subjective certainty (*meta-belief*) of the evaluated belief. "Degree of trust of X in Y about activity τ" is represented as $DoT_{XY\tau}$. The explanation of the resulting equation is lengthy. To summarize, the trustee Y is evaluated (reference belief item 1) based on a certain task by several different reasons or sources (this is open to interpretation and flexible). Each evaluation has its own subjective certainty applied (reference belief item 2). These certainties act as weights, but they are intended to function as "how sure the source is". These evaluated beliefs (abilities) are then paired with their certainties. The final *Degree of Trust* is defined in the following equation which uses the *Degree of Credibility* from multiple dimensions. We noted inconsistencies in this work. The variables τ and α both appear to represent the task and are used in the same equation which leads to confusion. In addition, the variable p appears to represent a goal, but it has no apparent contribution. We include the equation, however, due to space constraints, we do not provide all the variable definitions:

$$DoT_{XY\tau} = C_{opp}DoC_X[Opp_Y(\alpha, p)]$$
$$\times\ C_{ab}DoC_X[Ability_Y(\alpha)] \tag{1}$$
$$\times\ C_{will}DoC_X[WillDo_Y(\alpha, p)]$$

Where Degree of Credibility DoC represents X's beliefs about Y's opportunity/ability/action to perform a task α to realize some goal p.

2.2   Probabilistic Models

Probabilistic models are also a popular approach to modeling trust. These models are built to assist in the prediction of future principals' behavior. The hallmark of Bayesian techniques, named in honor of Rev. Thomas Bayes, is to predict a variable of interest based on the causal effects of other events. The foundation of the approach is captured in the phrase: "updating probabilities in the light of new evidence" [3].

Several works in the literature have examined Bayesian techniques for trust modeling [4–8]. According to [8], Bayesian theory is intuitive because it is a sequential process; as new evidence comes to light, the evidence is processed and added to the global evidence. It allows the trustor to update her information dynamically, which happens in real life. In 2016 Sapienza [4] used classical Bayesian logic applied to evidence and opinions from several sources of information to evaluate current trust.

Thirunarayan [9] reviewed the most common trust management models including recommender systems and trust propagation frameworks. In trust propagation frameworks, a trust chain is formed from recommender to recommender. These networks are most commonly modeled as directed acyclic graphs (DAGs). In DAGs, each node is a recommender and each directed link is the trust value. Trust is accumulated probabilistically from node to node and provide a final recommended trust value.

We believe that models which propagates trust values from entity to entity does not reflect how people make decisions. In some of these DAG models, each entity's *interpretation* of trust is being propagated along the links. However, that value cannot be properly estimated unless each node in the DAG is examined individually. This is usually not possible in real life as each node is a "black box" generating its own trust value based on the entity's interpretation. Nevertheless, most DAGs that model trust propagation networks evaluate trust in this manner.

Probabilistic models are very useful for trust, however, propagating trust (as many DAG models do) is not very realistic. If it were possible to interview the entity directly there would be no "black box."

3   A Trust Model for Cloud

We model consumer trust in the context of cloud computing. The value we place on data plays a critical role in how we trust.

3.1   Five Degrees of Recommendation

This research examines the decision making process and how consumers use different sources of recommendation. These sources are placed into groups which are referred to as degrees. These degrees are part of the decision process consumers use when trusting the cloud. We assert that it is more realistic than much of the existing research and we validate this model via survey.

Figure 1 shows how consumers seek recommendations and reflects the nature of recommendation gathering in real life. These recommendation degrees are in the order people might use them (based primarily on familiarity). If the recommendation is in the first degree, the decision to trust is based on first person experience. Consumers then take recommendations from family and relatives followed by friends and colleagues. Lastly, consumers seek the advice of expert organizations and large groups of people. The larger the size of the group ratings that agree about a product or service, the stronger the recommendation and lesser chance of collusion. We call this last source the 5th degree. A special case of the 5th degree is a group of one, such as an unfamiliar third party or stranger. Under some circumstances we are willing to trust anyone if the potential benefit outweighs the risk. For example, auto accident survivors on a remote road who cannot get a response from a professional responder would accept help from anyone because the alternative is loss of life.

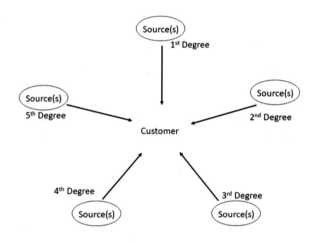

Fig. 1. Degrees of recommendation

- 1st degree: direct experience (i.e., this refers to personal experience).
- 2nd degree: family and relatives
- 3rd degree: friends/colleagues/acquaintances
- 4th degree: un-biased or expert organizations (e.g., Cloud Security Alliance - Cloud, Consumer Reports - Retail Products, US News - College Ratings)
- 5th degree: groups of unfamiliar people (e.g., ratings from Amazon, Tripadvisor)

These degrees do not work in isolation, a consumer can get recommendations from all degrees. They often work together with the person seeking the recommendation. For all degrees but the first degree, the strength of the recommendation is examined using its source's attributes. These recommendation attributes are contained in the variable ρ "rho".

$$\rho = F((\sigma * NUMSOURCES), (\epsilon * EXPERTISE), (\eta * FAMILIARITY)) \in [0.0..1.0]$$

Where σ, ϵ, η are weights assigned by the trusting individual and they must satisfy the following equation:

$$\sigma + \epsilon + \eta = 1 \tag{2}$$

Table 1 represents the possible values of each attribute. For cloud trust, the recommendation ρ leads to a decision to trust and feeds into the cloud trust model. The categories can be applied to all degrees (except direct experience). The expertise is most relevant for organizations. The number of sources is most relevant for group ratings. Finally, the familiarity is most relevant for friends or family members. However, each category can be applied to any source.

Table 1. ρ Attributes

# Sources	Value	Perceived Expertise	Value	Familiarity	Value
1	0	No Expertise	0	No Familiarity	0
2 .. 10	.	Low	.	Low	.
11 .. 100	.	Medium	.	Medium	.
101 .. 1000	.	High	.	High	.
> 10000	≤ 1	Very High	≤ 1	Very High	≤ 1

3.2 Cloud Spiral of Trust

Table 2 shows the variables used in this formal model. Trust has a cyclical nature in multiple dimensions. In Fig. 2, each loop represents a higher level trust (i.e., data value). As you move to a higher spiral of trust, the consumer is willing to entrust data of higher value to a third party.

Table 2. Definitions for cloud trust functions

Meaning	Representation
Recommendation value [0.0..1.0]	ρ
Trust threshold required for the value of consumers data	θ
Evidence: attestations, security plans, external sources	EV
Consumers' expectations, expressed policies	$EXPT$
Consumers' experience good or bad	$EXPR \in \{\gamma, \beta\}$
Trust level	τ
Evidence leading to loss of trust in a cloud (beta for bad)	β
Evidence to increase trust in a cloud (gamma for good)	γ

θ represents an important threshold decision point based on value a consumer places in her data. It is different per person and represents a boundary for each data set.

We assume the consumer has the competence to judge its sources of recommendation. Once a consumer has elected to use a provider based on recommendations ρ and her expectations $EXPT$, they begin to have direct experience with the provider and the cycle of experience, expectations, and evidence (i.e., attestations) begins. A low datum value will bring the consumer's trust threshold θ down because there is less personal risk. For example, using Instagram, Twitter, or Facebook, consumers might not mind if their data are available to the world.

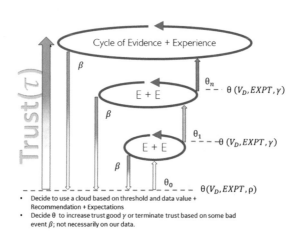

Fig. 2. Cloud spiral of trust

Each cycle feeds into trust and leads to yet another trust evaluation and decision. This decision threshold θ is based on the value of consumer's data, expectations $EXPT$, and experience $EXPR$ (which can be good γ or bad β). Trust at any level is a function of a decision, experiences (good or bad) and expectations.

$$\tau_x(Y, \alpha, t) = F(\tau_x(Y, \alpha, t - 1), EXPR(t - 1)) \qquad (3)$$

Where trust is τ that X has in Y over an action α at time t. $EXPR$ represents experience (personal and recommendations). The refining of function F is left to future work.

θ represents a trust threshold decision point based on the value V_D that a consumer places in her data. In order to take an action α, trust has to meet or exceed a threshold. The relationship of this threshold to trust is represented by Eq. 4.

$$\tau_x(Y, \alpha, t) \geq \theta(V_D, EXPT, EXPR(t)) \qquad (4)$$

In addition, while the consumer is evaluating trust, others are also evaluating their evidence (which can become part of a consumer's evidence, sometimes via out-of-band channels) as shown by "reports" in Fig. 3.

How can a trust model be evaluated for its effectiveness? The next section discusses our research in validation of our trust model.

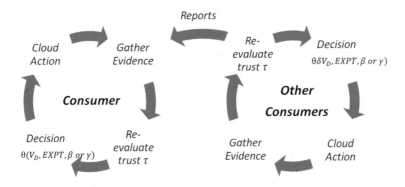

Fig. 3. Trust cycle

4 Trust Model Validation

There are existing surveys (i.e., questionnaires) on cloud trust, why propose another? Many of the existing models are not practical in light of recommended trust and few are validated [10]. Furthermore, many of the questions we asked have never been asked to the best of our knowledge. This research focuses on consumer data in the cloud. What is the location of our data? Do consumers want to grant blanket access to data when it is uploaded to the cloud? Do consumers trust one cloud over another?

Existing academic literature on trust and cloud usage was reviewed. The focus was on existing surveys that performed statistical power analysis. Our overall approach was to: (1) look for existing trust surveys, (2) determine the effect variable, sample size, and power that was used, (3) develop and submit a questionnaire to validate our model, and 4) administer the questionnaire using university processes.

4.1 Industry Surveys and Academic Survey Research

There are several annual industry sponsored surveys that target enterprise cloud usage (e.g., Forbes, Gartner, Rightscale). In 2018, Rightscale targeted primarily technical professionals at the enterprise and small business level and found that most at 97% use the cloud.[1] Also in 2018, Forbes reports that 80% of enterprises are both running apps on or experimenting with Amazon Web Services (AWS).[2]

In the academic community, Alsmadi [11] accomplished a *post-hoc* power analysis. The research asserted that there is an inconsistency between the claims of industry reports on cloud usage versus academia's reports. Alsmadi used questions from previous surveys. The chosen effect variable (Cohen's f^2) is consistent with the multiple regression testing. It is not clear that this effect variable was consistent with the research goals. The power achieved by [11] was significantly greater than 95%.

Burda [12] also accomplished a *post-hoc* power analysis to explain consumer archiving patterns in the cloud. The research goal was to prove that trust was a factor that helps uncertainty and reduces the perception of risk. They also sought to understand the effects of trust and risk in the context of cloud archiving and important drivers of trust as feedback for the cloud to improve. They had several hypotheses that addressed different aspects of trust, risk, and perception in the context of archiving data in the cloud. The effect variable and value they used to calculate statistical power was again $f^2 = .15$ for 'medium' effect (no justification was provided for using the medium effect). They reported a study power of .99, given a sample size of 229.

Horvath [13] did not conduct a power analysis. The survey question types consisted of T/F questions, Likert scale (e.g., strongly agree, neutral, disagree), and some open questions. They had 236 responses. They concluded that consumers trust cloud computing only if the risk is low and it adds convenience.

Hsu [14] conducted a *post-hoc* power analysis for their 2010 telephone survey. The goal was to examine cloud adoption in Taiwan from an industry perspective and test their research model. Their survey targeted top Taiwan companies and had approximately 200 responses. Their results appeared to show that cloud adoption was still in its initial stage (not surprising since it was 2010). They accomplished a *post-hoc* power analysis with a reported statistical power of .999. However, they did not use any null hypotheses, the effect variables were not consistent with the authors' definitions, and the definition of power analysis was inconsistent with the references.

4.2   Statistical Power Results

Since the statistical power analysis conducted in this research was an opinion poll, a test for certain proportion was used. For example, how many consumers trust the cloud? The effect index g is most appropriate for this.

Cohen's effect index g is a test for a proportion which assumes that the nominal proportion that exists in the population is a 50/50 split. These hypotheses are typically related to public opinions and the null hypothesis could take the

---

[1] https://www.rightscale.com/blog/cloud-industry-insights/cloud-computing-trends-2018-state-cloud-survey#96-percent.

[2] https://www.forbes.com/sites/louiscolumbus/2018/09/23/roundup-of-cloud-computing-forecasts-and-market-estimates-2018/#7073eb37507b.

form of a fraction of the population with certain beliefs or a defined character-istic which is $1/2$ (i.e., H_0: $P = .50$, where P is the sample proportion). For example, a majority support for a political issue [15].

Cohen [15] defines this effect index using the variable g. The g index relates to a departure from $P = .50$ in either direction. For example, a small effect is $g = .05$. At this effect definition, a division in the population would be a 55/45 split. But a small departure in one instance might be considered large in real life (e.g., a 55/45 split is considered a landslide in political elections). Therefore context has to be taken into account.

Given Power at 0.80, $\alpha = .05$, $g = .25$, Cohen's Table 5.4.1 [15] for test of 50% proportion suggests a sample size of 30 is required.

4.3   Survey Distribution

Despite the low sample size required based on the power analysis, we desired to gather a sample size of at least 150 based on the other surveys on trust. Industry surveys mostly sought tech professionals, management, and CEOs. Many of these surveys were overly positive on both cloud usage and trust leading to 'bullish' results.

An anonymous survey was conducted at the University of Florida with per-mission from the UF Institutional Review Board and permission to distribute at select departments, groups, and colleges. Since the survey distribution and the responders were anonymous, the response rate was based on the approxi-mate size of the departments and the number of undergraduates in the College of Engineering. The 35 responses represent an estimated 2% response rate. The respondents were 90% in the 18–34 age range. Almost 70% considered themselves either savvy or pro technical competence.

4.4   Selected Result Charts

This section presents selected charts and discussion. Percentages from the data were rounded to the nearest full percentage for pie chart readability. In some cases, this resulted in totals not equal to 100% and caused an empty slice. This was done to make the pie chart more readable and to avoid misrepresenting the data.

The first question in Fig. 4 was mainly informational to get the attitudes of the target population with regard to trusting the cloud. Results showed that a majority of those surveyed do trust the cloud.

We were very interested in the top cloud concerns. The process of determining initial cloud concerns came from the authors' experiences and informal polling. There was variation among the remaining options. The authors had believed that location of users' data might come in a clear second place, but it did not and this will guide future work. Table 3 shows cloud concerns ranking preferences. One represents highest concern and five represents the least concern to consumers.

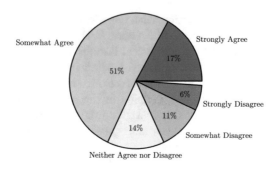

Fig. 4. Q1. I trust the cloud

Table 3. Q3. Rank these cloud concerns (1 being most important)

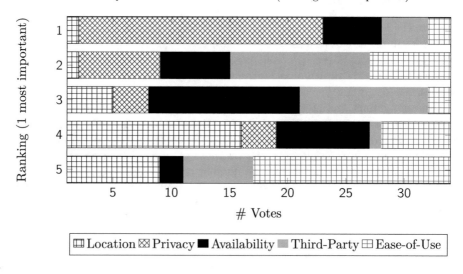

Each horizontal bar in Fig. 3 shows the number of votes each cloud concern received. In other words, not everyone ranked the cloud concerns the same. However, the overwhelming majority believed privacy was a primary concern.

How much do consumers care about third parties accessing their data without their knowledge? We were not very surprised that this kind of access was a concern to overwhelming majority of those surveyed shown in Fig. 5.

One of the major themes of this research was data location in the cloud. The next question was designed to determine the level of detail consumers want. Figure 6 shows that even though data location was not a top concern for cloud users, 80% of those surveyed want to know where their data are and 25% want to know at which data center their data resides.

One of the linchpins to this research is controlling consumers' data in the cloud via cloud policies. The cloud should seek more fine grained permission to

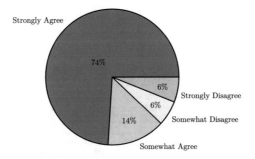

Fig. 5. Q8. I care if my cloud data are accessed by a third-party without my knowledge

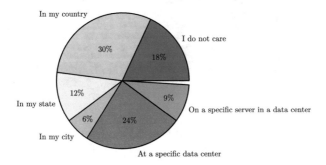

Fig. 6. Q11. At what detail would you like to know where your data resides

access consumers data. Figure 7 shows that the majority of those surveyed do care if their data are moved without their knowledge.

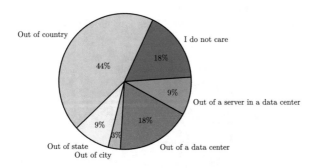

Fig. 7. Q12. I care if my cloud data is moved ___ without my knowledge

Another goal of this research is to determine how consumers decide to trust the cloud. We assert there is a specific order of recommendations people take before choosing a cloud. Table 4 shows a clear order of recommendation for the top two. While these results are not confirmed by our hypothesis, this forms the basis for further research.

Table 4. Q13. Rank these potential sources of knowledge (for a recommendation) in the order that you would use it

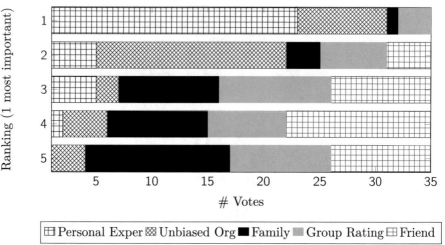

Many trust models include recommendation chains. We believe that recommendations cannot be accumulated in ways that existing trust models assert. Consumers typically do not seek recommendations from a friend of a friend without knowing the third party directly (at which point they themselves become a *friend*). Figure 8 shows that the overwhelming majority of those surveyed do not take recommendations from someone they do not know. Family member's recommendation was surprisingly low; this might signal an issue of perceived competence.

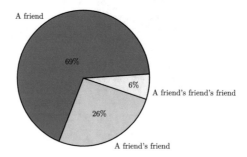

Fig. 8. Q14. This question concerns referred trust. When choosing a cloud, I would use a recommendation from ___

In our other research, we have created a cloud policy blockchain decentralized application (DApp) [16]. This DApp would require small fees to use. Will consumers pay for this kind of service? Fig. 9 shows that there is definitely some interest in paying for policies if all clouds followed it.

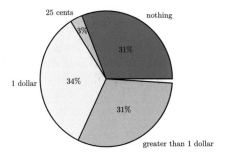

Fig. 9. Q17. I would pay __ as a one-time charge to use a personalized cloud policy, if all clouds followed it

5   Hypotheses for Cloud Trust Model

We propose to target specific areas of cloud consumer trust consistent with our proposed model. The goal is to validate elements identified as key in determining consumers' actions with regards to using the cloud. How do consumers decide to trust, are they concerned with third-parties, or data location? Most of the null hypotheses are estimates gained through the literature review, and the authors' experience. No evidence for or against any of these hypotheses was found in the literature review.

5.1   Hypothesis Test Plan

To the best of our knowledge, many of the questions we posed have never been asked. The general approach for setting up a test is to:

1. State the null and alternate hypothesis (or hypotheses), which are mutually exclusive (i.e., if one is true, the other must be false). In many of the cases, we found no equivalent statistics for the questions and took educated guesses based on industry surveys.
2. Use the 5–80 convention for power at 0.80 and $\alpha = .05$.
3. Choose the test statistic. We will use the one-sample z-test for a normal distribution. To assume a normal distribution, the following equations must be satisfied: $n\pi \geq 10$ and $n(1 - \pi) \geq 10$ where π is the proportion in the null hypothesis and n is the number of samples. Using the value 10 is a conservative rule of thumb, not a hard requirement. According to [17], if $\pi = .5$, the z-test can be used for n down to 20. Likewise for a π closer to 0 or 1, n should be much larger; at least 200.

4. Collect survey data.
5. Perform computations:
   (a) Obtain standard deviation
   (b) Compute test statistic. The test statistic is z, z can be negative. P is proportion in sample from my test and p is the proportion in the null hypothesis.

$$z = (p - P)/\sigma \qquad (5)$$

   (c) Compute P-value: The P-value is the probability of observing a sample statistic more extreme than the ones observed in the data.
6. Compare p value to alpha using rule: If $P <= \alpha$, then reject H_0. If $P > \alpha$, then fail to reject H_0.

5.2 Hypothesis 1

Table 5 contains the raw results when asking about third-party access. Approximately 89% were concerned about third-party access.

- π = the true proportion of cloud consumers that are interested in cloud third party associations.
- Null: $\pi < .25$. We estimate less than 25% of cloud consumers are interested in cloud third party associations.
- Alternate: $\pi \geq .25$. We believe 25% or more of cloud consumers are interested in cloud third party associations.

Table 5. Q8. I care if my cloud data are accessed by a third-party without my knowledge

Opinion	Average %
Strongly agree	74.29%
Somewhat agree	14.29%
Neither agree nor disagree	0.0%
Somewhat disagree	5.71%
Strongly disagree	5.71%

1. Pre-conditions to use *z-score*. The first one is not satisfied, however these the rules are conservative rules of thumb, we continue to use the normal distribution

$$n(\pi) \geq 10; 35(.25) \geq 10; 8.75 \not\geq 10$$
$$n(1 - \pi) \geq 10; 35(.75) \geq 10; 26.25 \geq 10 \qquad (6)$$

2. Obtain standard deviation σ of sampling distribution.

$$\sigma = \sqrt{(P * (1 - P))/n} = \sqrt{.24 * (1 - .24)/35} = \sqrt{.0052} = .072 \quad (7)$$

3. Compute Test Statistic:

$p = 89\%$ surveyed number of people who care about 3rd party access

$P = 24\%$ estimated less than 24% care about 3rd party access

$$z = (p - P)/\sigma = (.89 - .24)/.072 = .65/.072 = 9.03 \quad (8)$$

4. Compute P-value: A right tailed test is used since greater than 25% is used.

$$P(z > 9.03) = P(z < -9.03) = 0000 \text{ based off of the standard normal table} \quad (9)$$

5. Compare P to α using rule:

If $P \leq \alpha$, reject H_0. If $P > \alpha$, fail to reject H_0

Since $P = 0 < .05(\alpha)$, we reject the null hypothesis. $\quad (10)$

The observed sample supports the hypothesis.

5.3 Hypothesis 2

Table 6 contains the raw results when asking about referred trust (i.e., a direct friend tells me her friend knows of a great product). Any second-hand recommendation is referred trust.

- π = the true proportion of cloud consumers that use referred trust chains.
- Null: More than 50% of people use referred trust chains based on several algorithms for trust in literature we surveyed.
- Alternate: We believe less than 50% use referred trust in actuality extending beyond the person they know.

Table 6. Q14. This question concerns referred trust. When choosing a cloud, I would use a recommendation from ___

Recommend source	Average %
A friend	68.57%
A friend's friend (here and below, you do not know the individual)	25.71%
A friend's friend's friend	5.71%
A friend's friend's friend's friend	0.0%
A friend's friends' friend's friend's friend	0.0%

1. Pre-conditions to use *z-score.*

$$n(\pi) \geq 10; 35(.5) \geq 10; 17.5 \geq 10$$
$$n(1-\pi) \geq 10; 35(.5) \geq 10; 17.5 \geq 10 \qquad (11)$$

2. Obtain standard deviation σ of sampling distribution.

$$\sigma = \sqrt{(P*(1-P)/n} = \sqrt{.50*(1-.50)/35} = \sqrt{.0072} = .084 \qquad (12)$$

3. Compute Test Statistic:

$p = 31\%$ surveyed number of people who use trust chains
$P = 50\%$ research uses trust chains frequently to represent trust
$$z = (p-P)/\sigma = (.31-.50)/.084 = -.19/.084 = -2.26 \qquad (13)$$

4. Compute P-value: A left tailed test is used since less than 50% is used.

$$P(z < -2.26) = .0096 \text{ based off of the standard normal table} \qquad (14)$$

5. Compare P to α using rule:

If $P \leq \alpha$, reject H_0. If $P > \alpha$, fail to reject H_0
$$\text{Since } P = .0096 < .05(\alpha), \text{ we reject the null hypothesis.} \qquad (15)$$

The observed sample certainly supports the hypothesis, test statistic z along with the P value showed that the survey data is enough to reject the null hypothesis. The researchers assert that referred trust is a poor way to calculate trust.

5.4   Hypothesis 3

Table 7 contains the raw results when asking about trusting different data to different clouds.

- π = the true proportion of cloud consumers that have different trust thresholds.
- Null: Based on common cloud surveys (e.g., Gartner), many people and companies are putting more and more data in the cloud. Therefore, the null hypothesis is less than 25% of cloud consumers have different trust thresholds based on value of their data. This is an estimate since there is little specific discussion in research on use of the cloud depending on the sensitivity of consumers data.
- Alternate: We believe 25% or more of consumers have different trust thresholds based on their individual experiences and value of data being placed into the cloud.

Table 7. Q15. I trust more sensitive data to some clouds, but not others

Opinion	Average %
Strongly agree	20.00%
Somewhat agree	40.00%
Neither agree nor disagree	22.86%
Somewhat disagree	11.43%
Strongly disagree	5.71%

1. Pre-conditions to use *z-score*. The first one is not satisfied, however these the rules are conservative rules of thumb, we continue to use the normal distribution.

$$n(\pi) \geq 10; 35(.25) \geq 10; 8.75 \ngeq 10$$
$$n(1 - \pi) \geq 10; 35(.75) \geq 10; 26.25 \geq 10 \quad (16)$$

2. Obtain standard deviation σ of sampling distribution.

$$\sigma = \sqrt{(P * (1 - P)/n} = \sqrt{.25 * (1 - .25)/35} = \sqrt{.0054} = .073 \quad (17)$$

3. Compute Test Statistic:

$$p = 60\% \text{ surveyed people who trust different clouds with different data}$$
$$P = 25\% \text{ those who trust the cloud regardless of data sensitivity}$$
$$z = (p - P)/\sigma = (.60 - .25)/.073 = .35/.073 = 4.79 \quad (18)$$

4. Compute P-value: A right tailed test is used since greater than 25% is used.

$$P(z > 4.79) = P(z < -4.79) = 0 \text{ based off of the standard normal table}$$
$$(19)$$

5. Compare P to α using rule:

$$\text{If } P \leq \alpha, \text{reject } H_0. \text{ If } P > \alpha, \text{fail to reject } H_0$$
$$\text{Since } P = 0 < .05(\alpha), \text{we cannot reject the null hypothesis.} \quad (20)$$

The observed sample certainly supports the hypothesis.

5.5 Hypothesis 4

Table 8 contains the raw results when asking about respondents about their familiarity with cloud service level agreements.

- π = the true proportion of cloud consumers that use SLAs to gauge cloud trustworthiness.

- Null: Much of the research we surveyed uses the SLA as a basis or major contributor to consumer trust. The SLA is outdated, but there is little research on its decline in importance. Therefore, conventional wisdom suggests 25% or more of cloud consumers believe Service Level Agreements are very useful to gauge cloud trustworthiness.
- Alternate: We believe less than 25% of cloud consumers believe Service Level Agreements are used to gauge cloud trustworthiness anymore.

Table 8. Q16. I have read __ service level agreements

SLAs I have read	Average %
0	48.57%
1	11.43%
2	0.0%
More than 2	17.14%
What's a service level agreement	22.86%

1. Pre-conditions to use *z-score*. The first one is not satisfied, however these the rules are conservative rules of thumb, we continue to use the normal distribution.

$$n(\pi) \geq 10; 35(.25) \geq 10; 8.75 \not\geq 10$$
$$n(1 - \pi) \geq 10; 35(.75) \geq 10; 26.25 \geq 10 \quad (21)$$

2. Obtain standard deviation σ of sampling distribution.

$$\sigma = \sqrt{(P * (1 - P)/n} = \sqrt{.25 * (1 - .25)/35} = \sqrt{.0054} = .073 \quad (22)$$

3. Compute Test Statistic:

$p = 72\%$ surveyed number of people who have never read an SLA

$P = 25\%$ much research use the SLA as the basis of trustworthiness still

$$z = (p - P)/\sigma = (.72 - .25)/.073 = .47/.073 = 6.43 \quad (23)$$

4. Compute P-value: We have a left tailed test since we are saying less than 25%.

$$P(z < 6.43) = 1 \text{ based off of the standard normal table } P = 1 \quad (24)$$

5. Compare P to α using rule:

If $P \leq \alpha$, reject H_0, if $P > \alpha$, fail to reject H_0

$$\text{Since } P = 1 > .05(\alpha), \text{ we cannot reject the null hypothesis.} \quad (25)$$

The observed sample certainly supports the hypothesis, but the test statistic z along with the P value showed us that it is not enough to reject the null hypothesis.

6   Discussion and Limitations

From the data, just about everyone surveyed ordered from Amazon and used Facebook. The most surprising result was the location of data was not as big of a concern to cloud users as the authors expected. However, the third party access along with confidentiality of cloud data ranked high. The survey also revealed that long trust chains, as suspected, are not used in practice. Due to space constraints, not all survey questions are presented here. The remaining tables (Tables 9, 10, 11, 12, 13, 14, 15 and 16) list selected results and raw data from our survey.

Table 9. Q1. I trust the cloud

Opinion	Average %
Strongly agree	17.14%
Somewhat agree	51.43%
Neither agree nor disagree	14.29%
Somewhat disagree	11.43%
Strongly disagree	5.71%

Table 10. Q3. Rank these cloud concerns (1 being most important)

Concerns	Average rank
Location of my data	3.82
Privacy and confidentiality of my data	1.65
My data availability	2.88
Third party access to my data (cloud partners)	2.79
Ease of use of the cloud	3.85

Table 11. Q4. Do you order from Amazon?

Y/N	Average %
Y	100%
N	0%

There is broad trust in the cloud at some level. However, there is differentiated trust in cloud providers based on a consumer's data value. Consumers would be willing to pay to have a policy if clouds followed it. The location of a consumer's data is important, but not as important as privacy and availability.

Table 12. Q6. Do you use Facebook?

Y/N	%
Y	88.57%
N	11.43%

Table 13. Q7. If you use Facebook, how often do you use it?

Frequency	Average %
Daily	50.00%
Weekly	34.38%
Monthly	3.13%
Rarely	9.38%
Never	3.13%

Table 14. Q11. At what detail would you like to know where your data resides?

Locale	Average %
I do not care	18.18%
In my country	30.30%
In my state	12.12%
In my city	6.06%
At a specific data center	24.24%
On a specific server in a data center	9.09%

Table 15. Q12. I care if my cloud data is moved ___ without my knowledge (choose one)

Locale	Average %
I do not care	17.65%
Out of country	44.12%
Out of state	8.82%
Out of county	0.0%
Out of city	2.94%
Out of a data center	17.65%
Out of a server in a data center	8.82%

Table 16. Q13. Rank these potential sources of knowledge (for a recommendation) in the order that you would use it

Recommendation source	Aver rank of source
Personal experience	1.60
Unbiased organization	2.40
Family member recommendation	3.86
Large group rating	3.37
Friend's recommendation	3.77

Surveys are accomplished because it is impossible to question everyone. Even national census and elections are subject to error. This survey was limited to predominantly undergraduates at University of Florida. There are some segments of the population that might not ever use the cloud. We want to widen our scope to a wider population demographic.

7   Future Work and Conclusion

This research presented a cloud trust model using a decision based framework that was validated using real world data obtained via questionnaire. The survey results, size, and the hypothesis testing are clear indicators of further research avenues. There were interesting suggestions for new policies to explore in future work.

Knowledge and control over data movement is important to a vast majority of those surveyed. The most surprising result was that familiarity with the recommender is not most significant factor when decided to use a cloud. Furthermore perceived expertise may play a larger role than the researchers believed. Lastly, we determined that trust transitivity drops off very quickly (i.e., most people do not trust a friend's friend if they do not know them personally).

Based on the survey results, size, and the hypothesis testing, we can make educated guesses about a good direction to head. Our target number of responses was roughly 200. Due to the limited distribution allowed by the survey, and a response rate of approximately 2%, we would like to try again at a later date. We received interesting suggestions for new policies that we plan to explore.

Consumers' expectations play a significant role in framing their experiences. These expectations can be expressed via policies that are consumer-oriented and stored in a high integrity fashion; we use blockchains in a larger research project to store and evaluate these policies.

References

1. Marsh, S.P.: Formalising trust as a computational concept (1994)
2. Castelfranchi, C., Falcone, R.: Trust Theory: A Socio-Cognitive and Computational Model, vol. 18. Wiley, Hoboken (2010)

3. Pearl, J.: Bayesian networks (2011)
4. Sapienza, A., Falcone, R.: A Bayesian computational model for trust on information sources. In: WOA, pp. 50–55 (2016)
5. Wang, Y., Vassileva, J.: Bayesian network-based trust model. In: IEEE/WIC International Conference on WebIntelligence, 2003, WI 2003, Proceedings, pp. 372–378. IEEE (2003)
6. Wang, Y., Cahill, V., Gray, E., Harris, C., Liao, L.: Bayesian network based trust management. In: International Conference on Autonomic and Trusted Computing, pp. 246–257. Springer (2006)
7. Nielsen, M., Krukow, K., Sassone, V.: A Bayesian model for event-based trust. Electron. Notes Theor. Comput. Sci. **172**, 499–521 (2007)
8. Melaye, D., Demazeau, Y.: Bayesian dynamic trust model. In: International Central and Eastern European Conference on Multi-agent Systems, pp. 480–489. Springer (2005)
9. Thirunarayan, K., Anantharam, P., Henson, C., Sheth, A.: Comparative trust management with applications: Bayesian approaches emphasis. Future Gener. Comput. Syst. **31**, 182–199 (2014)
10. Kirkman, S., Newman, R.: A data movement policy framework for improving trust in the cloud using smart contracts and blockchains. In: 2018 IEEE International Conference on Cloud Engineering (IC2E), pp. 270–273. IEEE (2018)
11. Alsmadi, D., Prybutok, V.: Sharing and storage behavior via cloud computing: security and privacy in research and practice. Comput. Hum. Behav. **85**, 218–226 (2018)
12. Burda, D., Teuteberg, F.: The role of trust and risk perceptions in cloud archiving: results from an empirical study. J. High Technol. Manage. Res. **25**(2), 172–187 (2014)
13. Horvath, A.S., Agrawal, R.: Trust in cloud computing. In: SoutheastCon 2015, pp. 1–8. IEEE (2015)
14. Hsu, P.-F., Ray, S., Li-Hsieh, Y.-Y.: Examining cloud computing adoption intention, pricing mechanism, and deployment model. Int. J. Inf. Manage. **34**(4), 474–488 (2014)
15. Cohen, J.: Statistical Power Analysis for the Behavioral Sciences, 2nd edn. Lawrence Erlbaum Associates, Mahwah (1988)
16. Kirkman, S., Newman, R.: Intercloud: a data movement policy DApp for managing trust in the cloud. In: 5th Annual Conference on Computational Science and Computational Intelligence (CSCI 2018), CSCI (2018)
17. Peck, R., Olsen, C., DeVore, J.L.: Introduction to Statistics and Data Analysis. Cengage Learning (2000)

Quantum Recommendation System for Image Feature Matching and Pattern Recognition

Desislav Andreev[✉]

Technical University of Sofia, St. Kliment Ohridski Blvd. 8, 1000 Sofia, Bulgaria
desislav.andreev@gmail.com

Abstract. The clustering and the classification techniques for pattern recognition are applied in variety of ways, but the problem of clustering binary vectors has not been thorough analyzed. This paper provides a novel approach towards the problem of pattern recognition through the ORB image descriptors. An advanced clustering method is provided in order to deal with the binary image feature descriptors, thus providing the opportunity of adding new classes of recognizable objects later-on: the *k-majority algorithm* over *ORB descriptors* is applied, where the *Jaccard-Needham dissimilarity* measure is used as a distance measure step of the algorithm. It is established, that the following methodology is well suited for a quantum interpretation of the system. A detailed analysis of such transformation is conducted and the *Grover's algorithm* is proposed for providing the opportunity to search a specific feature in the available clusters, while reducing the number of iterations of the k-majority routine. In addition to the presentation of the system, described above, this paper provides also the main steps in constructing a similar recommendation system. To that the transformation from a classical to a quantum representation algorithm is described in detail. Such approach can be applied later-on in other applications. Both, the computational complexity and the verification correctness are also indicated below.

Keywords: Quantum k-means · k-majority · ORB · Grover · Jaccard · Pattern recognition · Recommendation system · Feature matching

1 Introduction

Nowadays the information seems the most esteemed and the most expensive resource in the world. Many applications use binary image feature descriptors matching or basing their functionality on visual object detection and recognition. In addition to that, there's a high demand in improving the retrieval performance, while reducing the computational complexity. Even if one doesn't look only into the feature matching in images, but in other applications for object recognition or similarity measurement, there's still a need for improvement in both hardware and software. A good example for this is the *autonomous driving,* which is still a difficult challenge, because of the computational complexity in every stage of the algorithms' evaluation [4]. Here, the computer vision part is used not only for object recognition, but also for 3D reconstruction (stereo-vision), motion tracking, robot navigation. Another commercial

© Springer Nature Switzerland AG 2020
K. Arai et al. (Eds.): FTC 2019, AISC 1069, pp. 497–511, 2020.
https://doi.org/10.1007/978-3-030-32520-6_37

application for image matching is *Google Images* [1], which is providing an opportunity to the clients to search for similar pictures, containing objects, that are similar or looking almost the same. This happens, when an image is provided as an input to the system. Of course, there're many prerequisites in order to achieve meaningful results with the given search machine: the object on the input picture is well displayed, the object is occasionally searched-for and there's much data assigned to it, e.g. there're many similar images in the database. What happens usually is, that the input image alone is really not enough, so the clients apply recommendation to the search machine: adding text description to the search and better results should be achieved.

That application is mostly directed towards the vast public, but when one gets to specific cases, like *plagiarism detection,* there'd still be a dependency on a good image matching algorithm [2]. There's not only the requirement for accuracy, but also for speed and performance of the algorithm itself. Again, the expectation would be a fast search through a large database, which is not always an easy task. As it was with the previous example, there's the possibility for an object in the image to be misplaced. Also, the image could be rotated, colors may be changed or parts of the visual data may be cropped. Therefore, a high-quality image matching algorithm is required in order to be able to detect such situations.

Based on the observations above, it becomes clear, that there's the need for a system, which produces fine image feature descriptors, clusters or classifies them effectively in order to extract features with a certain accuracy. The output result may be amplified if an interface for this is provided: for example, a person provides his own requirement: to search for a specific element, or an outer system gives this one additional data. In order to speed this up, a methodological transformation from the classical machine learning approach to a *quantum* machine learning representation of the system is designed. Even if the quantum computer is still a challenge as a hardware implementation, there are already many frameworks and computer languages, providing enough instruments to support at least the theoretical analysis for the given topic, as in [5]. One must emphasize on the fact, that the provided results are with a certain (high) probability, but as an output from a quantum computer.

The methodology for transformation between a classical and a quantum representation is already discussed in [3] and [5], where the focus is mainly on *k-means,* but the intuition would go the same way, when the *k-majority* is applied. The reason for changing the algorithm is, that in this case a *k-means-like* algorithm is expected, because the actual *mean* of the values doesn't make sense in the area of image feature descriptors' classification, because the system is dealing with a set of binary features. The term *average vector* cannot be defined. The technique of *k-medoids* can be applied, but in [6] it is stated, that it will worsen the evaluation time further. Therefore, the following method is applied, which takes a number of *k* and allows its own elements to execute the *majority voting* in order to choose the centroid of the given cluster, as done in [6].

Now a distance measure function is required, which usually is chosen as the *Euclidean distance.* Currently, a need to apply it over binary vectors of features is in effect, so the sense of a real distance is missing. The effective distance between the bits can be measured through the *Hamming distance* or similar. There is a conducted research about the best suiting distance metric in [7]: a few binary image feature descriptors like BRIEF, ORB and BRISK are analyzed with different distance measuring techniques. Their results showed a better performance if *Hamming* distance was replaced with *Jaccard-Needham* for ORB descriptors.

The recommendation system, using Grover's algorithms, is presented in [8], where the authors provide a good explanation of the way to build a system, similar to theirs, although their research is directed in the unstructured databases (for example, movie database). Still, their idea for using Grover's algorithm for faster search seems to work efficiently, and also - this algorithm is based on *quantum entanglement*, which its already discussed as a successful approach, when it goes to overall evaluation complexity, in [3]. The research in [9] also adds to the benefits of the quantum entanglement. In addition, the system, presented in [8], gives a detailed explanation of the *recommendation* part of it, which is also considered interesting to the topic of this paper. Finally, there're schematics of the given quantum system, so if a hardware implementation is existent, this can be applied successfully. A hardware realization for Grover's algorithm can be found in [10]. The authors claim, that their approach is a generic one and can be implemented in any multilevel quantum system.

In summary one can state that the quantum machine learning technique is a unification between Quantum Physics (QP) and Machine Learning (ML) algorithms. It uses the ideas of the classical ML algorithms to analyze quantum systems [3]. After applying a quantum machine learning clustering technique the resulted clusters are provided further as-are, or are fed as an input to the recommendation step. The steps of the algorithm, described in this paper, can be presented as follows:

(1) Extract ORB descriptors from binary images.
(2) Clusterize the results from (1) with a quantum k-majority algorithm, where the measurement step is replaced with Jaccard-Needham dissimilarity method.
(3) Use clusters to match an input image and recommend it as a result.
(4) Use Grover's algorithm for additional speed-up.

The present paper is organized as follows: in the next Sect. 2 the problem of fast image feature descriptors matching is defined. It also describes the input to the system and preparation for clustering. The main requirements for the design of the clustering step and its transformation to a quantum representation are proposed in Sect. 3. In Sect. 4 the functionality of the newly suggested quantum recommendation technique for separating for enhancing the end results, is described in details. This technique is discussed in Sect. 5 as an illustrative example of evaluation complexity and memory usage. The paper finishes with conclusion remarks about advantages and disadvantages of the new quantum machine learning clustering algorithm and the system's performance overall. The article ends with the cited bibliography.

2 Problem Statement

As the design and the implementation of an image matching and recommender (recommendation) system is provided further, there's the need to state, that the application of it is usually in search machines, content-based filtering, rating systems and etc. So it is expected for it to work over large and unstructured datasets. The trivial approach is to apply k-means (or kNN) or Factorization Matrix. Harder task will be to implement Association Rules and Deep Learning. It is obvious, that the first two approaches are expensive for large datasets: the k-means itself has $O(ndk)$ computational complexity, where n is the number of samples, d is the number of dimensions of the feature vector, and k - the number of clusters. But still, such an approach would be easy to implement and describe, and the quantum machine learning provides exactly the instruments to improve this. As explained in the previous section: Why would there be the need to switch to the k-majority method, there comes the requirement to prepare this technique for a quantum computation. In addition, in this paper's suggested system the feature vectors of a binary image dataset must be extracted through ORB. This data is expected to be available a priori. The further-described algorithm can be summarized as in Fig. 1 below.

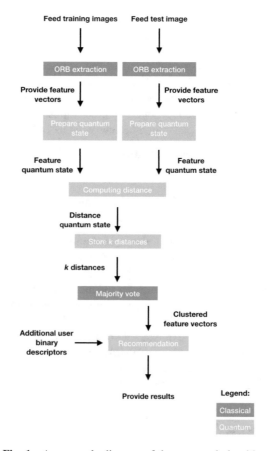

Fig. 1. An example diagram of the proposed algorithm.

3   Methodology of Interpreting a Classical to a Quantum Machine Learning Algorithm

In other words, it's not required from the current system to transform the ORB feature descriptor vectors runtime, because they will be provided as raw data from the outside, e.g. be an input to the system – there's no guarantee this is a quantum computer. Future research on the matter can tackle this problem, but since the ORB execution is not with high computational complexity in comparison with the clustering technique, it is left as in its classical representation. Another important remark is that the ORB is rotation invariant and resistant to noise [13]. Further, the quick cluster of the feature vectors is needed, in order to store a given set of patterns and later - match images to them. Afterwards, the recommendation (feature amplification) step will trigger and the final result will be displayed to the user. Its intention is to lower the number of iterations of the clustering technique, while helping it to *guess* the correct centroids faster, thus reducing the number of iterations until convergence.

In short, the quantum-enhanced versions of classical machine learning algorithms include least-squares fitting, support vector machines, principal component analysis and deep learning [3]. In addition, the adiabatic quantum machine learning seems to work for some classes of optimization problems and the stochastic models such as Bayesian decision theory or HMM find an elegant translation into the language of open quantum systems [3]. Still, the main challenge is not only to transform these algorithms from classical to quantum world, but also to represent the classical input data in a correct way and pass it to a given quantum device or method. The reason being is that the quantum information theory must provide an understanding of how fundamental laws of nature impact the ability of physical agents to learn. In order to explain the quantum representation of the algorithms there must be first an explanation of the main entities and concepts in a quantum system:

1. **Quantum bit** (or q-bit) is the quantum representation of the classical term *bit*. The difference is that the q-bit exists in a superposition of states - a given electron could be at the same time in two different orbits of the same atom. Using the Dirac's notation, this is denoted as: $|\psi\rangle = \alpha|0\rangle + \beta|1\rangle$ with $\alpha, \beta \in C$ and $|0\rangle, |1\rangle$ in the two-dimensional Hilbert space - H^2. α and β are the amplitudes of classical states $|0\rangle$ and $|1\rangle$, maintaining the property of probability conservation given by: $|\alpha|^2 + |\beta|^2 = 1$.

2. **Measured state** $||\psi\rangle$ - This means that either $|0\rangle$ or $|1\rangle$ is observed with probability $|\alpha|^2$ or $|\beta|^2$ respectively. Another representation of the state vector is: $|0\rangle = \begin{bmatrix} 1 \\ 0 \end{bmatrix}$ and $|1\rangle = \begin{bmatrix} 0 \\ 1 \end{bmatrix}$. The state measurement is irreversible, because the system *collapses* to one of the proposed states, thus losing the previous value of α and β. All other operations in the quantum mechanics are reversible and are represented with the so-called *gates*.

3. **Quantum gate** - The main unit in the quantum logical circuit. Sometimes these gates have both classical and quantum representation as the Toffoli gate, for

example. Each gate using a k-number of q-bits requires $2^k \times 2^k$ unitary matrix and the number of the input and output q-bits must be the same.

The gate's function for a specific quantum states is evaluated through a multiplication of the state vector and the gate's matrix:

for one q-bit:

$$V_0|0\rangle + V_1|1\rangle \rightarrow [V_0; V_1]$$

for two q-bits:
$$V_{00}|00\rangle + V_{01}|01\rangle + V_{10}|10\rangle + V_{11}|11\rangle \rightarrow [V_{00}; V_{01}; V_{10}; V_{11}],$$
where the $|\alpha\beta\rangle$ is the basis of the vector and the first q-bit is in $|\alpha\rangle$ state and the second – in $|\beta\rangle$.

4. **Quantum random access memory** (or QRAM) is using n q-bits to address any quantum superposition of N memory cells. For a memory call $\mathcal{O}logN$ switches need be thrown instead of the N used in conventional (classical or quantum) RAM designs. This yields a more robust QRAM algorithm, as it in general requires entanglement among exponentially less gates, and leads to an exponential decrease in the power needed for addressing.

5. **Quantum entanglement** is one of the phenomena of the quantum physics and an entity in the mathematics of the quantum mechanics, which cannot be ignored. Its sole purpose is to represent a system of particles, which have to be taken as a whole and not each particle as a separate one. This characteristic of the quantum computation world is usually used to enhance the memory's capacity [11].

6. **Quantum register** is similar to the classical register, e.g. it is constructed by multiple qubits and can be expressed with the following example for n qubits:

$$|\psi\rangle = |\psi_0\rangle \otimes |\psi_1\rangle \otimes |\psi_2\rangle \otimes \ldots \otimes |\psi_{n-1}\rangle \tag{1}$$

It has an exponential capacity, e.g. the amount of classical bits, contained in a quantum register, is 2^n, where n is the number of qubits. This works in the opposite way, too: in order to simulate a quantum computation, the classical machine must have an exponential amount of resources. In order words 16 qubits will require 512 KB of memory. Furthermore, the quantum registers can be represented by quantum logic circuits as mentioned above.

7. **Hadamard gate** *introduces* the superposition of the quantum states and it is represented as:

$$H = \frac{1}{\sqrt{2}} \begin{pmatrix} 1 & 1 \\ 1 & -1 \end{pmatrix} \tag{2}$$

The use of Hadamard gate results in the amplitude values being equal to the absolute value $\frac{1}{\sqrt{2^n}}$ for all states, taken by the specified quantum register. It is mostly used for initialization and preparation of states.

8. **Swap test** [14] is based on the *swap* gate [5], which is used for swapping two qubits, e.g. $|\alpha\rangle|\beta\rangle \rightarrow |\beta\rangle|\alpha\rangle$. The goal of the swap test is two measure the proximity of two quantum states: $\langle\alpha|\beta\rangle$, which is actually their *Cosine* similarity. The quantum logical circuit is shown in Fig. 2:

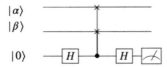

Fig. 2. Quantum logic scheme representation of a swap test.

It consists of two Hadamard gates, a swap gate and an ancillary bit - the actual measurement of probability. The probability of measuring $|0\rangle$ and $|1\rangle$ in the ancilla bit is respectively:

$$P(|0\rangle) = \left(\frac{1 + |\langle\alpha|\beta\rangle|^2}{2}\right) \tag{3.1}$$

$$P(|1\rangle) = \left(\frac{1 - |\langle\alpha|\beta\rangle|^2}{2}\right) \tag{3.2}$$

Revisiting what's already observed in [3], one can briefly mention the following: The *k-means* algorithm is alternating constantly between two main steps - assign step and update step:

Assign each observation x_p to the cluster whose mean has the least squared *Euclidean* distance (nearest mean) and the observation is assigned to exactly one $S^{(t)}$:

$$S_i^{(t)} = \left\{ x_p : \left\|x_p - m_i^{(t)}\right\|^2 \leq \left\|x_p - m_j^{(t)}\right\|^2 \forall j, 1 \leq j \leq k \right\} \tag{4}$$

Since this is the classical approach, some different distance measurements, like *Mahalanobis* or *Closest-string,* can be used instead of the Euclidean - it depends on the application [3]. The same goes for the quantum representation of the assignment step, although the idea here is to measure the similarity between two qubits (and their states): The *Hamming* distance may be applied as in [8] or [12], or the *Jaccard-Needham* as discussed above. Simply said, the Hamming distance deals with the number of bits between two registers of the same size. The Jaccard coefficient and the Jaccard distance can be expressed as:

$$J(A, B) = \frac{|A \cap B|}{|A \cup B|} \tag{5.1}$$

$$d_j(A, B) = 1 - J(A, B) = \frac{|A \cup B| - |A \cap B|}{|A \cup B|} \tag{5.2}$$

where A and B are the two sample sets and currently are the quantum bits of two quantum registers, so 5.2 can be expressed as follows:

$$d_j = \frac{q + r}{p + q + r} \tag{5.3}$$

where p is the number of bits, that are equal to 1 for both registers, q is the number of bits, equal to 1 for the left-hand-side register (figuratively said) and 0 for the right-hand-side, and r is the opposite of q. It is easy to associate it with the *Cosine* similarity and detect this similarity - this statement will be mentioned below in Sect. 3. *Update* or minimize the average distance the means to be the centroids of the observations in the new clusters:

$$m_i^{(t+1)} = \frac{1}{\left| S_i^{(t)} \right|} \sum_{x_j \in S_j^{(t)}} x_j \tag{6}$$

Again, since this is a quantum approach, another operation, that is close to the bitwise operations, is needed. The *majority* vote, mentioned in Sect. 2, has exactly this purpose, so (6) will not be applicable in this form. The majority vote would be used to redefine the vector clustering. For each cluster S, every qubit vector x, belonging to it, is taken into consideration. Every qubit of an accumulator vector x is increased by 1 if the corresponding bits in x are 1. At the end, the majority rule is used to form the new centroid m': for each element x_i in x, if the majority of vectors voted for 1 as bit value in x_i, then m'_i takes 1, otherwise - takes 0. One can describe the algorithm in the following manner:

> *Descriptors for this cluster are taken in the following shape: [177, 23..., 234] – the length is 32;*
> *They are binarized as follows: [10110001, [00010111..., 11101010];*
> *The bits in each column are summarized through the whole vector of binary descriptors;*
> *If the values, corresponding to the given bits are more than N/2, where N is the length of the array of binary image descriptors, then 1 is stored as a bit in the newly created centroid.*
> *A transformation of the evaluated binary data array is applied for the creation of an array of byte, while transforming the binary bucket to their DEC number values.*

The whole clustering algorithm stops execution, when the assignments no longer change. The main idea is still to partition N observations into k clusters. If the measurement space is D dimensional the time complexity of the algorithm would be $\mathcal{O}(D)$, where each step takes time $\mathcal{O}(N^2 D)$. Having the observations above it would be expected, that the algorithm takes $\mathcal{O}(N\log(ND))$. Both times N is required at least once, because every vector is tested individually for the reassignment at each step. Based on the previous analysis in [3], the initialization procedure of the input vectors can be used as continuation. Thus, as observed in Fig. 1 the images are fed as an input and their ORB descriptor feature vectors are extracted: from the test image $\overrightarrow{x_0}$ and the vector from the training images - $\overrightarrow{x_j}$, where $j \in 1, 2, \ldots N$ (the number of the trained observations, e.g. descriptor feature vectors). D is the number of dimensions of the feature vector. The feature vectors must be stored in $|\alpha\rangle$ and $|\beta\rangle$ registers, respectively [3, 12]:

$$|\alpha\rangle = \frac{1}{\sqrt{D}} \sum_{i=1}^{D} |i\rangle \left(\sqrt{1 - x_{0i}^2} |0\rangle + x_{0i} |1\rangle \right) |0\rangle \tag{7.1}$$

$$|\beta\rangle = \frac{1}{\sqrt{N}} \sum_{j=1}^{N} |j\rangle \frac{1}{\sqrt{D}} \sum_{i=1}^{D} |i\rangle |0\rangle \left(\sqrt{1 - x_{ji}^2} |0\rangle + x_{ji} |1\rangle \right) \tag{7.2}$$

Continuing with the preparation of the states above, where the initial qubits $|0\rangle$ are needed and then the storage of the feature vectors, using (2). Since this operation has a long mathematical description, it will not have a focus-on here, and one may refer to [12] for additional clarification. The next important step is the distance measure. The observations above from (3.1, 3.2) and (5.1, 5.2) are combined and the definition of the similarity between two vectors is:

$$d(x_0, x_j) = P(|1\rangle) \tag{8}$$

It could be emphasized on, that x_0 and x_j are particular states of $|\alpha\rangle$ and $|\beta\rangle$, respectively. The smaller probability in (8) means the more similar are the two images. The resulted state is $|\psi\rangle$:

$$|\psi\rangle = \frac{1}{\sqrt{N}} \sum_{j=1}^{N} |j\rangle \left[\sqrt{1 - d(x_0, x_j)} |0\rangle + \sqrt{d(x_0, x_j)} |1\rangle \right] \tag{9}$$

The last remaining quantum evaluation is the *Grover*'s algorithm. It especially effective when a fast search is required [15]. Let's look into its steps:

1. Consider initial state: $|0\rangle^{\otimes N}$, where N is the number of inputs, e.g. the size of the database.
2. Apply Hadamard gate on the first N qubits to get a uniform superposition of all possible arguments.
3. Apply an oracle g, where g is a function that maps database entries to 0 or 1. $g(x) = 1$, if and only if x satisfies the search criterion $(x = \omega)$. A quantum black-box is provided and a subroutine in the form of a unitary operator U_ω is executed, where:

$$\left\{ \begin{array}{ll} U_\omega |x\rangle = |-x\rangle & \text{for } x = \omega, g(x) = 1 \\ U_\omega |x\rangle = |x\rangle & \text{for } x \neq \omega, g(x) = 0 \end{array} \right. \tag{10.1}$$

It continues with *Grover's diffusion operator* U_s:

$$U_s = 2|s\rangle\langle s| - I_N \tag{10.2}$$

where I is the identity matrix and $|s\rangle$ is the uniform superposition over all states:

$$|s\rangle = \frac{1}{\sqrt{N}}\sum_{x=0}^{N-1}|x\rangle$$

4. Subsequently (10.1) and (10.2) are executed $\sqrt{N}$ number of times. A schematic description of the algorithm can be seen below in Fig. 3:
5. Through an ancilla bit the measurement Ω is obtained and it results to eigenvalue λ_ω. Its probability will approach 1 for $N \gg 1$. ω is evaluated later from here.

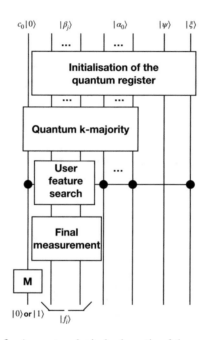

Fig. 3. A quantum logical schematic of the system.

4   Basis of Designing a Quantum Machine Learning System for Recommendation of Image Descriptors

As described above, all of the required nodes of building the system, that can be observed in Fig. 1. Also, the required operations for interpretation of the system's behavior in the quantum world is observed in detail. It is important to mention that the following schematics and algorithm explanation as non-existent in the real-world, because of the nature of the quantum hardware and its price. The pseudo-code for the algorithm is as follows ($\vec{x}$ denotes universally the descriptors):

1. Extract ORB binary image descriptors from the test image and the training images.
2. Randomly generate k binary centroids C
3. **while** C not changed **do:**
   > **for** $x \in \vec{x}$:
   >> $c_x \leftarrow argmin(d(c,x)), c \in C$
   >
   > **for** $c \in C$:
   >> **for** $x \in \vec{x}|_{c_x=c}$
   >>> v accumulates x votes
   >>
   >> $c' \leftarrow Majority(v)$,
   >> where c' is the new centroid

   end while
4. Execute *Grover's* algorithm over the clusters, using the *additional user descriptors (ORB)*
5. Provide results

Still, there are good quantum simulation examples. Given such an environment, a probable implementation of the recommendation system is expected, although it will not be able to prove by certain measurements the theoretical basis of this paper. An abstract overview of the technology is proposed with the intention to implement it later-on in a future work. The represented algorithm can also be designed as a quantum logical circuit. As already mentioned in the beginning of this paper, it follows the rules for designing a normal logical circuit. What is required here, is a set of quantum registers:

$c_0|0\rangle$ - auxiliary qubit, where the obtained probability distribution of the clustered elements can be amplified by Grover's algorithm to improve the probabilistic properties, if such input is provided by the user. If c_0 has measured probability of $|1\rangle$ one needs to reverse the computation of the *k-majority* step and retry with another value of k;

$|\beta_j\rangle$ - represents the states of the training images' feature vectors' registers;

$|\alpha_0\rangle$ - represents the states of the test image's feature vector's registers;

$|\psi\rangle$ - the state, which encodes the identifier of the recommended ORB descriptors;

$|\xi\rangle$ - the register, which represents user's additional input;

The end result is held in $|f_i\rangle$. This can be observed on the schematic below in Fig. 4:

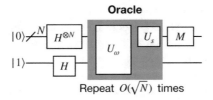

Fig. 4. An example quantum logical schematic for Grover's algorithm and iteration steps

5 Additional Complexity Analysis and Discussion

In order to prove the need of such algorithm transformation it is required to evaluate the theoretical performance and the possible limitation. Additionally, a comparison with the classical approach is presented below, as each phase is compared one-to-one. Let's go again through the main system blocks, where the initialization time and the read/write operation times are taken as $\mathcal{O}(1)$. This includes also the preparation stages. Another step that could be easily ignored in the given setup is the ORB descriptors extraction, because for the purposes of this paper, this algorithm hasn't been interpreted to a quantum one, as discussed above. Still, the complexity should be acknowledged, since it is $O(N)$, where N is the number of different descriptors.

The first big difference appears, when the clustering state is reached - this is actually the main reason to look towards the quantum representation. As mentioned in Sect. 2, the k-means has a computational complexity of $O(ndk)$, which even for small number of k or d is still a serious overhead. Execution of such algorithm on QRAM should happen with complexity $O(log(ndk))$. A major drawback in both cases of this algorithm is the value of k, because as per design it – the algorithm, assigns it near the number of classes. This is a good remark for future implementations considering Duerr's algorithm [16] for calculating the best fit of k. In [5] a mechanism is provided, using the *Silhouette* method for measuring the clustering quality and suggesting a better number of centroids. In both cases there could be an expensive operation, instead of risking the clustering quality with a randomized k cluster assignment- this should be analyzed further.

The last part of this recommendation system is the search operation, which classically happens in linear time for trivial applications, e.g. $\mathcal{O}(N)$, where N is the number of elements. Grover's algorithm will execute the same operation on a quantum computer with complexity $\mathcal{O}(\sqrt{N})$. For the purposes of this paper the *Graz-02* dataset is used [17]. It consists of 1096 images, split into three categories: bikes, cars, people. All of the image are colored, and have resolution of 640x480. Their ORB descriptor arrays are extracted, but they are kept as RGB, e.g. there's no need to *grayscale* them - this is because the target is to load the system as much as possible, and also – the clustering precision is being aimed at. Since there are three classes, $k = 3$ could be taken as initial value. This setup is used as the *training* environment. It is expected, that a *test* image is fed and a cluster is suggested, thus executing the *pattern recognition* part. One important remark would that roughly 140300000 bits for storing the data into the registers are required, which is fine for the classical case - this is around 20 MB of RAM.

Unfortunately, it becomes problematic if a quantum simulation on a classical PC is used, because 2^N bits would be required to hold N qubits, where N depends on the input parameters of the system. The tests are conducted on a powerful desktop machine, using Python 3.x and AMD Ryzen, 8-core CPU. Additionally, IBM's *qiskit* environment is used for the quantum representation of the algorithm - it provides abstractly simulated parts as quantum registers, quantum logic gates and operations. It is important to mention, that almost the whole RAM was depleted during the tests, but IBM provide a remote usage of their quantum chips if needed, which can be used in a future work. Although the application is multithreaded, when the tests are executed and the following results in interest of time are received, one can distinguish visually the performance of the quantum approach, as shown in Fig. 5.

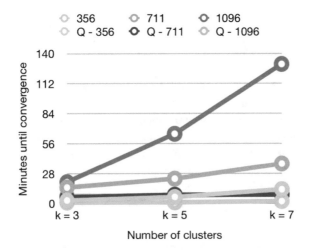

Fig. 5. Time dependency and usage analysis of a few scenarios for the system.

6 Conclusion

It must be emphasized on the fact, that the biggest part of these time metrics, is actually the *assignment* step of the algorithm. Both the Jaccard-Needham and the Majority vote steps are efficient and happen in a timely manner, as expected - they are around 20% of the values of observed time metrics. Another big challenge is the number of iterations of the clustering algorithm. On applications without complex inputs, the convergence usually doesn't happen after 50 iterations. Because of the characteristics of the binary vectors, it is almost impossible to expect a quick convergence and therefore, the quick-search through Grover's algorithm provides almost immediate result, as observed above. The precision of the algorithm is also high, because of the small number of classes and the high quality of the input images. When a known object is used as a test image, the result is correct over 90% of the time. Unknown objects, for example glasses, cannot be detected well, but if they are fed enough times to the system, the clusters will re-arrange and the class *glasses* will become available - this is the reason behind the unsupervised approach.

The current paper describes a novel approach towards recommendation systems in two main areas: improvements in the classical clustering technique for both speed-up and quality - clustering is not usually executed over binary feature vectors, so it may be considered, that this alone as an achievement; quantum interpretation of the classical approach in order for it to be transformed and run on a quantum computer. The ORB descriptors have been extracted from the Graz-02 dataset in order to prepare training data. The drawback is in its small number of classes, which leads to the impossibility of clustering and recommending more complicated objects with the current setup. Still, the algorithm performs well and, since it is an unsupervised approach, additional classes can be self-added later-on. Two unique steps have been used in the clustering's technique in order to deal with the binary data format: The distances are calculated through the Jaccard-Needham similarity measure, and the Majority votes measurement is used in order to replace the means calculation. The recommendation step is amplified through the usage of a search algorithm, which intends to lower the number of iterations and leads the technique to a faster convergence.

All of the mentioned steps above are also useful into translating this system into its quantum representation. Unfortunately, this implementation cannot be tested thorough, because of the nature of the quantum computing. Even running it on a simulator has proven as a major challenge, because the memory requirement raises to 2^N in order to bring an exponential speed-up. As a final remark, it is clear, that such an application is not useful in real-time scenarios, but provides promising results and easy implementation in the area of quantum computing.

References

1. NY Times about Google Images. https://bits.blogs.nytimes.com/2010/07/20/google-updates-its-image-search/?searchResultPosition=4. Accessed 21 May 2019
2. Ovhal, P.M., Phulpagar, B.D.: Plagiarized image detection system based on CBIR. Int. J. Emerg. Trends Technol. Comput. Sci. (IJETTCS) **4**(3), 8–12 (2015). ISSN 2278–6856.
3. Andreev, D.: Analysis of machine learning methods and algorithms in a quantum entanglement-based environment. In: Proceedings of 47th Sprint Conference of UMB, Borovets, April 2018, pp. 129–137
4. Karamanov, N., Andreev, D., Pfeifle, M., Bock, H., Otto, M., Schulze, M.: Map line interface for autonomous driving. In: 21st IEEE International Conference on Intelligent Transportation Systems, Proceedings of IEEE ITSC. IEEE Computer Society (2018). https://doi.org/10.1109/itsc.2018.8569378
5. Andreev, D., Petrakieva, S., Taralova, I., Qiao, Z.: Applying quantum machine learning approach for detecting chaotically generated fake usernames of accounts. In: 13th International Conference for Internet Technology and Secured Transactions (ICITST – 2018), Cambridge, United Kingdom, December 2018
6. Grana, C., Borghesani, D., Manfredi, M., Cucchiara, R.: A fast approach for integrating ORB descriptors in the bag of words model. In: SPIE - The International Society for Optical Engineering, February 2013. https://doi.org/10.1117/12.2008460
7. Bonstanci, E.: Is hamming distance the only way for matching binary image feature descriptors? IEEE Electron. Lett. **50**(1), 806–808 (2014). ISSN: 0013-5194
8. Sawerwain, M., Wroblewski, M.: Application of quantum k-NN and Grover's algorithms for recommendation big-data system. In: 39th International Conference of Information Systems Architecture and Technology (ISAT), pp. 235–244 (2018)
9. Chakraborty, S., Banerjee, S., Adhikari, S., Kumar, A.: Entanglement in the Grover's search algorithm. IMSC, 30 May 2013
10. Godfrin, C., Ferhat, A., Ballou, R., Klyatskaya, S., Ruben, M., Wemsdorfer, W., Balestro, F.: Operating quantum states in single magnetic molecules: implementation of Grover's quantum algorithm. Phys. Rev. Lett. (PRL) **119**, 187702 (2017)

11. Spectrum Article Page. https://spectrum.ieee.org/tech-talk/computing/hardware/20-entangled-qubits-brings-the-quantum-computer-closer. Accessed 08 Apr 2019
12. Dang, Y., Jiang, N., Hu, H., Ji, Z., Zhang, W.: Image classification based on quantum kNN algorithm. Quantum Inf. Process. **17**(9), 239 (2018)
13. Rublee, E., Rabaud, V., Konolige, K., Bradski, G.: ORB: and efficient alternative to SIFT or SURF. Willow Garage, California http://www.willowgarage.com
14. Luongo web-page for swap-test description. https://luongo.pro/2018/01/29/Swap-test-for-distances.html. Accessed 10 Apr 2019
15. Jiayu, Z., Junsuo, Z., Fanjiang, X., Haiying, H., Peng, Q.: Analysis and simulation of Grover's search algorithm. Int. J. Mach. Learn. Comput. **4**(1), 21–23 (2014)
16. Duerr, C., Heiligman, M., Hoyer, P., Mhalla, M.: Quantum query complexity of some graph problems. Lecture Notes in Computer Science, vol. 3142. Springer, Heidelberg (2004)
17. Graz-02 image database. https://lear.inrialpes.fr/people/marszalek/data/ig02. Accessed 20 Apr 2019

A Hybrid Approach in Future-Oriented Technology Assessment

Ewa Chodakowska[(✉)]

Bialystok University of Technology, 45A, Wiejska Street,
15-351 Bialystok, Poland
e.chodakowska@pb.edu.pl

Abstract. Technology Assessment has been a growing field of study for the few past decades. Intensive work on solving the problem of proper technology assessment has translated into the development, improvement or adjustment of the method and models used in technology evaluation projects. The article aims to present a new hybrid model that uses the Rough Sets approach and the DEA method to increase the objectivity in the selection of priority technologies in future-oriented technology assessment projects. Real-data application proved that this model: (i) reduces the number of considered assessment criteria by a few times without a significant change in technology rankings; (ii) gives individual objective weights to the criteria and allows highlighting the "strengths" of each technology; (iii) from the point of view of efficiency, considers the attractiveness of the development of each technology and the rationality of allocating resources required for the development; (iv) allows the inclusion of a possible contradiction among expert opinions.

Keywords: Future-Oriented Technology Assessment · Data Envelopment Analysis · Rough Sets · Model

1 Introduction

1.1 Future-Oriented Technology Assessment Methods and Models

Technologies have always played a driving and transformational role. The importance of technologies is undisputed for enterprises, where they constitute the basis of a long-term competitive advantage, but also for the economy and society, where they act as the main determinant of sustainable development. The recognition of the importance of technologies and the need to direct as well as the need to support sustainable innovations [47] gives rise to the propositions of various methods, models and tools required for their analysis and evaluation. The development of methods for technology evaluation started in the 1960s [63], but the issue is particularly important nowadays when the speed of technological change and the resulting uncertainty create new challenges for the management of technologies [40]. In the field of future technology management, three approaches have been developed: Technology Assessment (TA), Technology Foresight (TF) and Technology Forecasting, which together are included in the concept of Future-Oriented Technology Assessment (FTA) [5, 20, 43].

© Springer Nature Switzerland AG 2020
K. Arai et al. (Eds.): FTC 2019, AISC 1069, pp. 512–525, 2020.
https://doi.org/10.1007/978-3-030-32520-6_38

A wide set of methods that can be used in FTA includes both methods adapted from other areas as well as dedicated specifically for technology analysis. This set has been systematically supplemented with new or modified methods, adapted to specific conditions and design requirements, enabling the achievement of the assumed goals in technology assessment projects. Porter [50], Miles and Keenan [41], Popper and Korte [46], Popper [48, 49] are authors of the most popular classifications and evaluations of methods.

Although the set of potential methods of technology analysis is very large, in practice, technological research usually relies on several simplest methods [33, 49], and the most popular among them is the Delphi method [9, 17, 53].

The subject of developing innovative methods, models and tools is still valid because of the necessity to analyse technologies and the imperfection of methods currently used in the assessment process. The research problem in the article concerns the use of a hybrid model in the process of technology assessment to strengthen the objectivity. The proposed hybrid model uses methods from Rough Sets (RS) by Pawlak [45] and Data Envelopment Analysis (DEA) by Charnes, Cooper and Rhodes [8].

1.2   DEA and Rough Sets in Technology Analysis

The basic applications of Rough Sets are: (i) reducing and eliminating redundancies in data sets, (ii) discovering relationships between objects, attributes and generating decision rules, and (iii) assigning weights to attributes based on their significance. Thus, they are used as tools of data mining technology or, more broadly, knowledge technology. In these functions, RS are also used in the technology assessment. The selected examples of Rough Sets application are presented in Table 1.

The review of the works indicates the relatively rare use of Rough Sets in the Future-Oriented Technology Assessment. The quoted papers elaborate on assessments of products received from implemented technologies and organisations that use particular technologies or technology assessment from the perspective of a chosen aspect.

An analysis was made of the current application of the DEA in technology assessment. The use of DEA in the technology analysis so far mainly concerns the assessment of the eco-efficiency of selected technologies, which results from the possibility of including criteria whose value increase is undesirable and at the same time impossible to eliminate (waste, pollution, e.g. greenhouse gases etc.). Technology selection and technology forecasting are other identified applications. The mentioned areas of application with examples are listed in Table 2.

Various listed applications indicate the validity and suitability for the technology assessment of the DEA method and Rough Sets. In the literature, the integration of Rough Sets with the DEA method is relatively unpopular and rarely used. These tools were used, inter alia, in the following areas: forecasting business failures [60], supply chain efficiency [68], and the Japanese banking sector efficiency [59].

RS and DEA have a large potential that is not sufficiently used in technology management and assessment. In particular, the assessment of new or emerging technologies in the context of Future-Oriented Technology Assessment using the DEA and RS methods is hardly exploited. The aim of the article is to propose an innovative hybrid technology prioritisation model, in which the RS methods are used in the

Table 1. Rough Sets in technology assessment

Application	Authors (year)	Problem/subject of analysis
Reduction of assessment criteria	Lu, Huang and Wang (2007) [38]	The system for new technology assessment
	Tsai, Lai, Chang and Watada (2009) [64]	Dilemmas of R&D partnership in the Taiwanese engineering industry
	Wu and Lin (2012) [67]	Quality of e-learning
	Ciflikli and Kahya-Ozyirmidokuz (2012) [11]	Quality in the production of rugs
	Wang, Jia and Wang (2015) [66]	The efficiency of coal technologies
	Gao, Zhang, Wu, Yin and Lu (2017) [18]	Development of family farms
	Sharma, Dua, Singh, Kumar and Prakash (2018) [56]	Energy management systems
Decision rules induction	Tsai, Lai, Chang and Watada (2009) [64]	Dilemmas of R&D partnership in the Taiwanese engineering industry
	Wang, Chin and Tzeng (2010) [65]	R&D and innovation activity of high-tech enterprises
	Hemert and Nijkamp (2010) [22]	R&D and innovative activity of countries
	Jian, Liu and Liu (2010) [24]	The selection of a regional key technology
	Ciflikli and Kahya-Ozyirmidokuz (2012) [11]	Quality in the production of rugs
	Górny, Kluska-Nawarecka, Wilk-Kołodziejczyk and Regulski (2015) [19]	Heat treatment parameters
	Liang and Dijk (2016) [35]	Rainwater collection system
	Shiau and Chuen-Yu (2016) [57]	Impact of a wind farm
	Sharma, Dua, Singh, Kumar and Prakash (2018) [56]	Energy management systems
Determination of weights	Lu, Huang and Wang (2007) [38]	The system of new technology assessment
	Li, Wu and Zhang (2009) [34]	The implementation of lean production
	Lee, Lee, Seol and Park (2012) [30]	The assessment of the concept of services (video games)
	Luo and Hu (2015) [39]	The risk of technological innovations in agricultural cooperatives
	Lai, Liu and Georgiev (2016) [28]	The integration of low-carbon technology
	Bai and Sarkis (2017) [4]	A hypothetical production technology
	He, Pang, Zhang, Jiao and Chen (2018) [21]	The level of development of selected countries in the field of clean energy
Modelling of linguistic uncertainty	Zeng, Huang, Yang, Wang, Fu, Li and Li (2016) [70]	The management of a regional ecosystem

functions of tools (i) to limit the number of analysis criteria and (ii) to model inconsistencies of expert opinions, and the DEA method—for estimating optimal weights for selected criteria. The postulated solution significantly improves the technology assessment process and increases the objectivity of the results.

Table 2. DEA method in technology assessment

Application	Authors	Problem/subject of analysis
Sustainable development	Liu, Sun and Xu (2013) [37]	Eco-efficiency of water systems in China
	Sueyoshi and Goto (2014) [61]	Environmental impact of technological innovations in the Japanese industrial sector
	Shabani, Saen and Torabipour (2014) [55]	Environmentally friendly technologies of cooling towers in power plants
	Fan, Zhang, Zhang and Peng (2015) [14]	CO_2 utilisation technologies
	Kwon, Cho and Sohn (2017) [27]	Energy production from renewable sources in European countries, considering the level of CO_2 emissions
Technology selection	Cuhls and Kuwahara (1994) [12]	Results of the Japanese and German Delphi studies
	Lee, Lee, Seol and Park (2008) [31]	R&D priorities of the technology foresight project
	Saen (2009) [54]	Selection of robots (data: Khouja (1995) [26])
	Alinezhad, Makui, Kiani Mavi and Zohrehbandian (2011) [1]	Selection of robots (data: Karsak and Ahiska (2005) [25])
	Lee, Mogi and Hui (2013) [32]	R&D resources against high oil prices
	Amin and Emrouznejad (2013) [2]	Selection of robots (data: Karsak and Ahiska (2005) [25])
	Yu and Lee (2013) [69]	Emerging nanotechnologies
Technological forecasting	Lamb, Anderson and Daim (2010) [29]	R&D goals
	Inman, Anderson and Harmon (2006) [23]	Jet fighter aircraft
	Anderson, Daim and Kim (2008) [3]	Wireless communication

2   A Hybrid Model of Technology Assessment

2.1   Model Formulation

The use of RS elements in the hybrid model allows deleting the limitations of the application of DEA to data, strictly and uniquely defined. The hybrid model adopts the concept of rough variable ξ [36] to present inconsistencies in the assessments:

$$\xi = ([a,b],[c,d]), \text{ for } c \le a < b \le d \tag{1}$$

For the given rough variable ξ and $\alpha \in (0,1]$, the α-optimistic value and α-pessimistic value could be defined respectively [58, 62]:

$$\xi^{sup(\alpha)} = \begin{cases} (1-2\alpha)d + 2\alpha c, & \text{if } 0 \le \alpha \le \frac{d-b}{2(d-c)} \\ 2(1-\alpha)d + (2\alpha-1)c, & \text{if } \frac{2d-a-c}{2(d-c)} \le \alpha \le 1 \\ \frac{d(b-a)+b(d-c)-2\alpha(b-a)(d-c)}{(b-a)+(d-c)}, & \text{otherwise} \end{cases} \tag{2}$$

$$\xi^{inf(\alpha)} = \begin{cases} (1-2\alpha)c + 2\alpha d, & \text{if } 0 \le \alpha \le \frac{a-c}{2(d-c)} \\ 2(1-\alpha)c + (2\alpha-1)d, & \text{if } \frac{b+d-2c}{2(d-c)} \le \alpha \le 1. \\ \frac{c(b-a)+a(d-c)-2\alpha(b-a)(d-c)}{(b-a)+(d-c)}, & \text{otherwise} \end{cases} \tag{3}$$

Employing DEA in the hybrid model assumes an assessment based on the efficiency relation, which measures the ratio of the benefits (outputs) function to the expenditures (input) function characterising the technology. Technologies achieve the highest scores, which bring potentially highest benefits in relation to costs.

The article uses the following standard notation of DEA models:

$j = 1,2, \ldots, j_o, \ldots, n$—the number of technology,
$X_j = (x_{j1}, x_{j2}, \ldots, x_{jm}, \ldots, x_{jM})$—the input vector of technology j,
$Y_j = (y_{j1}, y_{j2}, \ldots, y_{js}, \ldots, y_{jS})$—the output vector of technology j,
M—the number of input variables,
S—the number of output variables,
λ_j—the vector of non-negative weights defining the intensity of the technology j in the optimal technology of the object being assessed.

The linear programming models of the SE-DEA using the concept of a rough variable and its optimistic and pessimistic values are [68]:

$$\min \theta^{sup(\alpha)} \tag{4}$$

$$\sum_{j=1, j \neq jo}^{n} \lambda_j x_{jm}^{sup(\alpha)} + \lambda_{jo} x_{jom}^{inf(\alpha)} \le \theta^{sup(\alpha)} x_{jom}^{inf(\alpha)}, \; m = 1, \ldots, M$$

$$\sum_{j=1, j \neq jo}^{n} \lambda_j y_{js}^{inf(\alpha)} + \lambda_{jo} y_{jos}^{sup(\alpha)} \ge y_{jos}^{sup(\alpha)}, \; s = 1, \ldots, S$$

$$\lambda_j \ge 0, j = 1, \ldots n$$

$$\min \theta^{inf(\alpha)} \tag{5}$$

$$\sum_{j=1, j \neq jo}^{n} \lambda_j x_{jm}^{inf(\alpha)} + \lambda_{jo} x_{jom}^{sup(\alpha)} \le \theta^{inf(\alpha)} x_{jom}^{sup(\alpha)}, \; m = 1, \ldots, M$$

$$\sum_{j=1, j\neq jo}^{n} \lambda_j y_{js}^{sup(\alpha)} + \lambda_{jo} y_{jos}^{inf(\alpha)} \geq y_{jos}^{inf(\alpha)}, \ s = 1, \ldots, S$$

$$\lambda_j \geq 0, j = 1, \ldots n$$

where:

$\left[x_{jm}^{sup(\alpha)}, x_{jm}^{inf(\alpha)} \right]$, $\left[y_{js}^{sup(\alpha)}, y_{js}^{inf(\alpha)} \right]$—the range of input and output values obtained for the rough variables describing a technology: $\left(\left[x_{mj}^{a}, x_{mj}^{b} \right], \left[x_{mj}^{c}, x_{mj}^{d} \right] \right)$, $\left(\left[y_{sj}^{a}, y_{sj}^{b} \right], \left[y_{sj}^{c}, y_{sj}^{d} \right] \right)$ for $0,5 \leq \alpha \leq 1$.

The use of the concept of rough variables in the DEA model implies the consideration of two variants of the assessment: optimistic and pessimistic, formulated for the lower and upper limits of the variable ranges. And the result is the range of efficiency indicators for the assumed level of α:

$$\left[\theta^{sup(\alpha)}, \theta^{inf(\alpha)} \right] \tag{6}$$

In the hybrid model, in addition to using the concept of a rough variable to formalise vague values, it is also proposed to use the RS method to reduce the number of criteria. In other words, replacing the original set of criteria with a selected reduct, i.e. an independent set of attributes that maintains the distinguishability.

The indication of reducts is related to the indiscernibility relations (IND) for each subset of attributes B defined as follows [44]:

$$IND(B) = \left\{ (x, x') \in U^2 : \bigwedge_{a \in B} a(x) = a(x') \right\} \tag{7}$$

where:

U—a non-empty, finite set of objects (the universe);
$(x, x') \in U^2$—objects in the universe;
$a(x), a(x')$—values of attribute a for x i x';
B—a subset of attributes.

Each minimal set of attributes C that maintains the indiscernibility relation computed on the set of attributes B such that: $IND(C) = IND(B)$ is called a reduct of B and is denoted as:

$$C = RED(B). \tag{8}$$

The core of B is the intersection of reducts:

$$CORE(B) = \cap RED(B). \tag{9}$$

The core contains the attributes from B, which are considered of the greatest importance for classification.

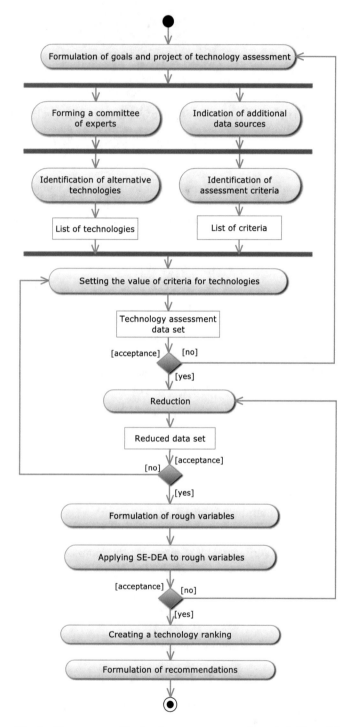

Fig. 1. The process of technology assessment using a hybrid model

2.2   The Hybrid Model in the Process of Technology Assessment

The presented process framework was developed based on [6, 7, 10]. The technology assessment using the hybrid model starts from the standard phase of objective setting, appointing a committee of experts/decision makers, identifying available alternative technologies and a set of criteria. In the next step, the criteria values for each technology are set. The use of a hybrid model is positioned in the reduction of the number of attributes in the technology assessment data set and the creation of the ranking by DEA. First, the data set is reduced, and for the reduced set, rough variables are formed, expressing imprecision/inconsistency of the assessment. Based on rough variables of selected assessment criteria, the SE-DEA model is used for classifying the technologies. The process ends with the construction of the ranking and recommendations. The proposed process in UML notation is shown in Fig. 1.

The presented technology assessment process fulfils the need for an objective multi-attribute assessment with the modelling of ambiguity. Earlier data reduction allows specifying the most important attributes of technology assessment that affect the results.

3   Empirical Application

3.1   Use of a Hybrid Model in Technology Assessment

Verification of the proposed hybrid model of technology assessment is presented using the example of data from the project 'NT FOR Podlaskie 2020' Regional Strategy of Nanotechnology Development (http://ntfp2020.pb.edu.pl/). The analysed data set consisted of 57 alternative technologies $T = \{T_1, T_2, \ldots, T_{57}\}$, 13 attractiveness criteria $B = \{A_1, A_2, \ldots, A_{13}\}$ and 8 feasibility criteria $B = \{F_1, F_2, \ldots, F_8\}$. The values of the technology assessment data set were expert responses on a scale of 1 to 5 [42]. The project researchers applied a method of key technologies, in which the basis for qualifying to the set of key technologies were above-average values of the average of the feasibility and attractiveness criteria weighted by the declared expert knowledge (also subjectively determined on a scale of 1–5).

The analysis of the data set with the proposed hybrid model started from the reduction in sub-sets of feasibility (inputs) and attractiveness (outputs) for the aggregated (average) values of the responses of experts weighted by the declared knowledge. For this purpose, ROSE 2 program written at the Poznan University of Technology was used [51, 52]. Three methods of discretisation of the criteria values were tested using the entropy minimisation algorithms implemented in ROSE 2 [15, 16]. The results are shown in Table 3.

The general principle of selecting the final reduct [13] says that the set should contain as few attributes as possible and should not eliminate the attributes assessed by the decision makers as the most important ones. In the provided example, the sets with the smallest number of elements were analysed, but at the same time consisting of the most frequently repeating attributes in the reducts obtained by different methods of discretisation. Finally, the following sets of criteria for attractiveness and feasibility after global discretisation were selected: $CORE(B) = \{A_1, A_2, A_3, A_9\}$ and

Table 3. Reducts and core—different methods of discretisation.

	CORE	RED
Feasibility criteria		
Local discretisation	{F1, F2, F5, F7}	{F1, F2, F5, F7, F8} {F1, F2, F3, F5, F7}
Local discretisation to the set {0,1,2}	{F1, F3, F5, F6, F7}	{F1, F3, F5, F6, F7}
Global discretisation	{F1, F2, F3, F7}	{F1, F2, F3, F7}
Attractiveness criteria		
Local discretisation	{A2}	14 different reducts
Local discretisation to the set {0,1,2}	–	104 different reducts
Global discretisation	{A1, A2, A3, A9}	{A1, A2, A3, A9}

$CORE(B) = \{F_1, F_2, F_3, F_7\}$. Next, rough variables of expert assessments for selected criteria using the nominal dispersion index h were formulated to account for their inconsistencies. Arbitrary $\pm20\%h$ and $\pm10\%h$ were assumed to design the lower and upper approximations given the relation "the less coherent answers, the wider the range." The technology scores were calculated using the SE-DEA model for rough variables and $\alpha = 0.6$ and $\alpha = 0.8$. The results obtained are shown in Fig. 2.

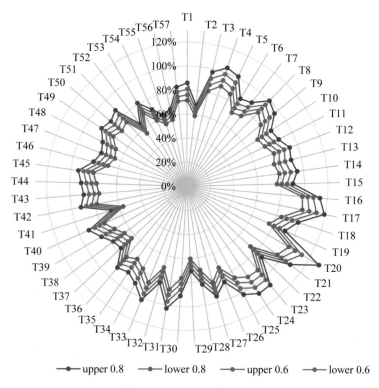

Fig. 2. Technologies assessment by the hybrid model

To sum up this example, the following benefits resulting from the use of the hybrid technology model can be highlighted:

1. The concept of reducts from the Rough Sets theory was used to make the number of considered criteria much smaller. In the 'NT FOR Podlaskie 2020', the original set of 21 attributes regarding attractiveness and feasibility was reduced to 8. Thus, the interpretation of the final technology assessment was simplified to analyse the values of only several variables. At the same time, as it was verified empirically, it did not significantly affect the technology rankings determined by the DEA. The Spearman's correlation between DEA rankings reduced and was 0.85 for the full data set.
2. The application of the DEA method gave objective weight to the technology assessment criteria. The individual adjustment of the weights allowed to emphasise the specific features of the technology, i.e. to assign criteria, in which the technology received the highest scores and, respectively, the highest possible weights in the case of attractiveness as well as the lowest possible scores for feasibility criteria. Thus, the arbitrariness of the considered criteria weights was eliminated.
3. The inclusion of the rough variable into the DEA method allowed for the possible contradiction between expert opinions. The result of the technology assessment in the form of ranges allowed to see discrepancies in technology assessments. This can be the basis for formulating alternative balanced recommendations and creating complementary scenarios.

Accordingly, the assumption can be made that the use of a hybrid prioritisation model justifies the choice of technology more rationally and objectively. In the 'NT FOR Podlaskie 2020' project, a hybrid model of technology assessment could have been implemented instead of the chosen variant of the key technology method to increase the quality and, thus, the confidence in the results of the selection of priority technologies.

4 Conclusion

The use of different technology assessment methods arises from the need to support the knowledge and competencies of decision-makers at the global, national, regional or institutional levels. The prospective management of technologies and the creation of a technological development strategy entail high risk due to its speculative nature of expert opinions but also due to a large number of aspects to be considered.

The hybrid model makes it possible to indicate the technologies that are worth supporting and investing, as they can translate into tangible economic success, achieved and maintained competitive advantage and/or sustainable social development. The advantages of a hybrid technology prioritisation model are:

i. The eliminated need for reaching an expert consensus on particularly valid areas of assessment or the weightings of criteria that would balance different opinions, values, and needs. The predefined wide range of criteria subjected to a mathematical procedure of estimating the significance and reduction. Weights of criteria are set via solving the linear programming task.

ii. The considered ratio of effectiveness and not only the attractiveness of development of each technology but also the rationality of the allocation of resources for development in relation to the alternatives considered.

iii. The formulated and modelled nonconformities of subjective expert evaluations.

The future development of the model is related to the additional analytical capabilities of DEA, which allow determining the most important attributes for the assessed technology that affect the results achieved and simulate the impact of changes in attribute values or defining groups of competitive technologies. Another possibility is to adapt the model to project specifics through the innovative formulation of rough variables.

Acknowledgment. The research was conducted within project G/WIZ/5/2018 financed from National Science Centre funds (DEC 2018/02/X/ST8/02000).

References

1. Alinezhad, A., Makui, A., Kiani Mavi, R., Zohrehbandian, M.: An MCDM-DEA approach for technology selection. J. Industr. Eng. Int. **7**(12), 32–38 (2011)
2. Amin, G.R., Emrouznejad, A.: A new DEA model for technology selection in the presence of ordinal data. Int. J. Adv. Manuf. Technol. **65**, 1567–1572 (2013)
3. Anderson, T.R., Daim, T.U., Kim, J.: Technology forecasting for wireless communication. Technovation **28**(9), 602–614 (2008)
4. Bai, C., Sarkis, J.: Improving green flexibility through advanced manufacturing technology investment: modeling the decision process. Int. J. Prod. Econ. **188**, 86–104 (2017)
5. Cagnin, C., Havas, A., Saritas, O.: Future-oriented technology analysis: its potential to address disruptive transformations. Technol. Forecast. Soc. Chang. **80**(3), 379–385 (2013)
6. Chan, F.T.S., Chan, H.K., Chan, M.H., Humphreys, P.K.: An integrated fuzzy approach for the selection of manufacturing technologies. Int. J. Adv. Manuf. Technol. **27**, 747–758 (2006)
7. Chan, F.T.S., Chan, M.H., Tang, N.K.H.: Evaluation methodologies for technology selection. J. Mater. Process. Technol. **107**(1–3), 330–337 (2000)
8. Charnes, A., Cooper, W.W., Rhodes, E.: Measuring the efficiency of decision-making units. Eur. J. Oper. Res. **2**(6), 429–444 (1978)
9. Choi, M., Choi, H.-L., Yang, H.: Procedural characteristics of the 4th Korean technology foresight. Foresight **16**(3), 198–209 (2014)
10. Chuu, S.-J.: Selecting the advanced manufacturing technology using fuzzy multiple attributes group decision making with multiple fuzzy information. Comput. Ind. Eng. **57**(3), 1033–1042 (2009)
11. Ciflikli, C., Kahya-Ozyirmidokuz, E.: Enhancing product quality of a process. Ind. Manage. Data Syst. **112**(8), 1181–1200 (2012)
12. Cuhls, K., Kuwahara, T.: Outlook for Japanese and German Future Technology - Comparing Technology Forecast Surveys. Technology, Innovation and Policy. Physica-Verlag, Heidelberg (1994)
13. Dimitras, A.I., Słowiński, R., Susmaga, R., Zopounidis, C.: Business failure prediction using rough sets. Eur. J. Oper. Res. **114**(2), 263–280 (1999)

14. Fan, J.-L., Zhang, X., Zhang, J., Peng, S.: Efficiency evaluation of CO2 utilization technologies in China: a super-efficiency DEA analysis based on expert survey. J. CO2 Utilization 11, 54–62 (2015)
15. Fayyad, U.M., Irani, K.B.: Multi-interval discretization of continuous-valued attributes for classification learning. In: Proceedings of the 13th International Joint Conference on Artificial Intelligence (IJCAI-1993) (1993)
16. Fayyad, U.M., Irani, K.B.: On the handling of continuous-valued attributes in decision tree generation. Mach. Learn. 8(1), 87–102 (1992)
17. Förster, B.: Technology foresight for sustainable production in the German automotive supplier industry. Technol. Forecast. Soc. Chang. 92, 237–248 (2015)
18. Gao, Y., Zhang, X., Wu, L., Yin, S., Lu, J.: Resource basis, ecosystem and growth of grain family farm in China: based on rough set theory and hierarchical linear model. Agric. Syst. 154, 157–167 (2017)
19. Górny, Z., Kluska-Nawarecka, S., Wilk-Kołodziejczyk, D., Regulski, K.: Methodology for the construction of a rule-based knowledge base enabling the selection of appropriate bronze heat treatment parameters using rough sets. Arch. Metall. Mater. 60(1), 309–312 (2015)
20. Halicka, K.: Innovative classification of methods of the future-oriented technology analysis. Technol. Econ. Dev. Econ. 22(4), 574–597 (2016)
21. He, Y., Pang, Y., Zhang, Q., Jiao, Z., Chen, Q.: Comprehensive evaluation of regional clean energy development levels based on principal component analysis and rough set theory. Renewable Energy 122, 643–653 (2018)
22. van Hemert, P., Nijkamp, P.: Knowledge investments, business R&D and innovativeness of countries: a qualitative meta-analytic comparison. Technol. Forecast. Soc. Chang. 77(3), 369–384 (2010)
23. Inman, O.L., Anderson, T.R., Harmon, R.R.: Predicting U.S. jet fighter aircraft introductions from 1944 to 1982: a dogfight between regression and TFDEA. Technol. Forecast. Soc. Chang. 73, 1178–1187 (2006)
24. Jian, L., Liu, S., Liu, Y.: The selection of regional key technology based on the hybrid model of grey fixed clustering and variable precision rough set. In: ISTASC 2010 Proceedings of the 10th WSEAS International Conference on Systems Theory and Scientific Computation, pp. 54–59 (2010)
25. Karsak, E.E., Ahiska, S.S.: Practical common weight multicriteria decision-making approach with an improved discriminating power for technology selection. Int. J. Prod. Res. 43(8), 1537–1554 (2005)
26. Khouja, M.: The use of data envelopment analysis for technology selection. Comput. Ind. Eng. 28(1), 123–132 (1995)
27. Kwon, D.S., Cho, J.H., Sohn, S.Y.: Comparison of technology efficiency for CO2 emissions reduction among European countries based on DEA with decomposed factors. J. Clean. Prod. 151, 109–120 (2017)
28. Lai, X., Liu, J.X., Georgiev, G.: Low carbon technology integration innovation assessment index review based on rough set theory - an evidence from construction industry in China. J. Clean. Prod. 126, 88–96 (2016)
29. Lamb, A., Anderson, T.R., Daim, T.U.: Difficulties in R&D target-setting addressed through technology forecasting using data envelopment analysis. In: Technology Management for Global Economic Growth, PICMET, pp. 1–9 (2010)
30. Lee, C., Lee, H., Seol, H., Park, Y.: Evaluation of new service concepts using rough set theory and group analytic hierarchy process. Expert Syst. Appl. 39, 3404–3412 (2012)
31. Lee, H., Lee, C., Seol, H., Park, Y.: On the R&D priority setting in technology foresight: a DEA and ANP approach. Int. J. Innov. Technol. Manage. 5(2), 201–219 (2008)

32. Lee, S.K., Mogi, G., Hui, K.S.: A fuzzy analytic hierarchy process (AHP)/data envelopment analysis (DEA) hybrid model for efficiently allocating energy R&D resources: in the case of energy technologies against high oil prices. Renew. Sustain. Energy Rev. **21**, 347–355 (2013)
33. Li, N., Chen, K., Kou, M.: Technology foresight in China: academic studies, governmental practices and policy applications. Technol. Forecast. Soc. Chang. **119**, 246–255 (2017)
34. Li, S., Wu, C., Zhang, H.: Key technology analysis of implementing lean production. In: IEEE 16th International Conference on Industrial Engineering and Engineering Management, vol. 1–2, pp. 1993–1996 (2009)
35. Liang, X., van Dijk, M.P.: Identification of decisive factors determining the continued use of rainwater harvesting systems for agriculture irrigation in Beijing. Water **8**(1), 7 (2016)
36. Liu, B.: Uncertain Theory: An Introduction to Its Axiomatic Foundation. Springer, Heidelberg (2004)
37. Liu, Y., Sun, C., Xu, S.: Eco-efficiency assessment of water systems in China. Water Resour. Manage **27**(14), 4927–4939 (2013)
38. Lu, W.-G., Huang, L.-C., Wang, J.-W.: The new technology evaluation based on rough-set theory. In: PICMET 2007 - 2007 Portland International Conference on Management of Engineering & Technology, pp. 883–886 (2007)
39. Luo, J.-L., Hu, Z.-H.: Risk paradigm and risk evaluation of farmers cooperatives' technology innovation. Econ. Model. **44**, 80–85 (2015)
40. Magruk, A.: Concept of uncertainty in relation to the foresight research. Eng. Manage. Prod. Serv. **9**(1), 46–55 (2017)
41. Miles, I., Keenan, M.: Overview of methods used in foresight. The Technology Foresight for Organisers Training Course, United Nations Industrial Development Organisation, Ankara (2003)
42. Nazarko, J., Magruk, A. (eds.): Kluczowe nanotechnologie w gospodarce Podlasia. Oficyna Wydawnicza Politechniki Białostockiej, Białystok (2013)
43. Nazarko, Ł.: Future-oriented technology assessment. In: 7th International Conference on Engineering, Project, and Production Management, Procedia Engineering, vol. 182, pp. 504–509 (2017)
44. Pawlak, Z., Skowron, A.: Rudiments of rough sets. Inf. Sci. **177**, 3–27 (2007)
45. Pawlak, Z.: Rough sets. Int. J. Inf. Comput. Sci. **11**, 344–356 (1982)
46. Popper, R., Korte, W.: XTREME EUFORIA: Combining Foresight Methods, EU-US Seminar: New Technology Foresight, Forecasting & Assessment Methods, Sewilla (2004)
47. Popper, R., Popper, M., Velasco, G.: Towards a more responsible sustainable innovation assessment and management culture in Europe. Eng. Manage. Prod. Serv. **9**(4), 7–20 (2017)
48. Popper, R.: Foresight methodology. In: Georghiou, L., Harper, J.C., Keenan, M., Miles, I., Popper, R. (eds.) The Handbook of Technology Foresight. Concepts and Practice. Edward Elgar Publishing Limited, Northampton (2008)
49. Popper, R.: How are foresight methods selected? Foresight **10**(6), 62–89 (2008)
50. Porter, A.L.: Technology assessment. Impact Assess. **13**(2), 135–151 (1995)
51. Predki, B., Słowiński, R., Stefanowski, J., Susmaga, R., Wilk, S.: ROSE - software implementation of the rough set theory. In: Polkowski, L., Skowron, A. (eds.) Rough Sets and Current Trends in Computing. Lecture Notes in Artificial Intelligence, vol. 1424, pp. 605–608. Springer-Verlag, Heidelberg (1998)
52. Predki, B., Wilk, S.: Rough set based data exploration using ROSE system. In: Ras, Z.W., Skowron, A. (eds.) Foundations of Intelligent Systems. Lecture Notes in Artificial Intelligence, vol. 1609, pp. 172–180. Springer-Verlag, Heidelberg (1999)
53. Proskuryakova, L.: Energy technology foresight in emerging economies. Technol. Forecast. Soc. Chang. **119**, 205–210 (2017)

54. Saen, R.F.: Technology selection in the presence of imprecise data, weight restrictions, and nondiscretionary factors. Int. J. Adv. Manuf. Technol. **41**(7–8), 827–838 (2009)
55. Shabani, A., Saen, R.F., Torabipour, S.M.R.: A new data envelopment analysis (DEA) model to select eco-efficient technologies in the presence of undesirable outputs. Clean Technol. Environ. Policy **16**(3), 513–525 (2014)
56. Sharma, S., Dua, A., Singh, M., Kumar, N., Prakash, S.: Fuzzy rough set based energy management system for self-sustainable smart city. Renew. Sustain. Energy Rev. **82**, 3633–3644 (2018)
57. Shiau, T.-A., Chuen-Yu, J.-K.: Developing an indicator system for measuring the social sustainability of offshore wind power farms. Sustainability **8**(5), 470 (2016)
58. Shiraz, R.K., Charles, V., Jalalzadeh, L.: Fuzzy rough DEA model: a possibility and expected value approaches. Expert Syst. Appl. **41**(2), 434–444 (2014)
59. Shiraz, R.K., Fukuyama, H., Tavana, M., Di Caprio, D.: An integrated data envelopment analysis and free disposal hull framework for cost-efficiency measurement using rough sets. Appl. Soft Comput. **46**, 204–219 (2016)
60. Shuai, J.J., Li, H.L.: Using rough set and worst practice DEA in business failure prediction. In: Ślęzak, D., Yao, J., Peters, J.F., Ziarko, W., Hu, X. (eds.) Rough Sets, Fuzzy Sets, Data Mining, and Granular Computing. RSFDGrC. Lecture Notes in Computer Science, vol. 3642, pp. 503–510. Springer, Heidelberg (2005)
61. Sueyoshi, T., Goto, M.: Environmental assessment for corporate sustainability by resource utilization and technology innovation: DEA radial measurement on Japanese industrial sectors. Energy Econ. **46**, 295–307 (2014)
62. Tohidi, G., Valizadeh, P.: A non-radial rough DEA model. Int. J. Math. Model. Comput. **1**(4), 257–261 (2011)
63. Tran, T.A., Daim, T.: A taxonomic review of methods and tools applied in technology assessment. Technol. Forecast. Soc. Chang. **75**(9), 1396–1405 (2008)
64. Tsai, Y.-H., Lai, W.-H., Chang, P.-L., Watada, J.: Dilemma of behavioral uncertainty of R&D alliance in Taiwan machinery industry. In: IEEE International Conference on Fuzzy Systems, vol. 1–3, pp. 439–1444 (2009)
65. Wang, C.-H., Chin, Y.-C., Tzeng, G.-H.: Mining the R&D innovation performance processes for high-tech firms based on rough set theory. Technovation **30**(7–8), 447–458 (2010)
66. Wang, X., Jia, F., Wang, Y.: Evaluation of clean coal technologies in China: based on rough set theory. Energy Environ. **26**(6–7), 985–995 (2015)
67. Wu, H.-Y., Lin, H.-Y.: A hybrid approach to develop an analytical model for enhancing the service quality of e learning. Comput. Educ. **58**(4), 1318–1338 (2012)
68. Xu, J., Li, B., Wu, D.: Rough data envelopment analysis and its application to supply chain performance evaluation. Int. J. Prod. Econ. **122**(2), 628–638 (2009)
69. Yu, P., Lee, J.H.: A hybrid approach using two-level SOM and combined AHP rating and AHP/DEA-AR method for selecting optimal promising emerging technology. Expert Syst. Appl. **40**, 300–314 (2013)
70. Zeng, X.T., Huang, G.H., Yang, X.L., Wang, X., Fu, H., Li, Y.P., Li, Z.: A developed fuzzy-stochastic optimization for coordinating human activity and eco-environmental protection in a regional wetland ecosystem under uncertainties. Ecol. Eng. **97**, 207–230 (2016)

Detecting Spam Tweets in Trending Topics Using Graph-Based Approach

Ramesh Paudel[(⊠)], Prajjwal Kandel, and William Eberle

Tennessee Technological University, Cookeville, TN 38501, USA
{rpaudel42,prkandel42}@students.tntech.edu, weberle@tntech.edu

Abstract. In recent years, social media has changed the way people communicate and share information. For example, when some important and noteworthy event occurs, many people like to "tweet" (Twitter) or post information, resulting in the event trending and becoming more popular. Unfortunately, spammers can exploit trending topics to spread spam more quickly and to a wider audience. Recently, researchers have applied various machine learning techniques on accounts and messages to detect spam on Twitter. However, the features of typical tweets can be easily fabricated by the spammers. In this work, we propose a graph-based approach that leverages the relationship between the named entities present in the content of the tweet and the document referenced by the URL mentioned in the tweet for detecting possible spam. It is our hypothesis that by combining multiple, heterogeneous information together into a single graph representation, we can discover unusual patterns in the data that reveal spammer activities - structural features that are difficult for spammers to fabricate. We will demonstrate the usefulness of this approach by collecting tweets and documents referenced by the URL in the tweet related to Twitter trending topics, and running graph-based anomaly detection algorithms on a graph representation of the data, in order to effectively detect anomalies on trending tweets.

Keywords: Twitter · Spam detection · Anomaly Detection · Graph-based anomaly

1 Introduction

Twitter, a popular social media, or micro-blogging, site, allows users to post information, updates, opinions, etc. using tweets. Given its wide-spread popularity for immediately sharing thoughts and ideas, adversaries try to manipulate the micro-blogging platform and propagate off-topic content for their selfish motives [1,2]. Compounding the issue, as the popularity increases around a certain event, more people tweet about the topic, thereby increasing its "trending" rate. Spammers then exploit these popular, trending Twitter topics to spread their own agendas by posting tweets containing keywords and hash-tags of the trending topic along with their misleading content. Ideally, one would like to be

© Springer Nature Switzerland AG 2020
K. Arai et al. (Eds.): FTC 2019, AISC 1069, pp. 526–546, 2020.
https://doi.org/10.1007/978-3-030-32520-6_39

able to identify anomalous tweets on a trending topic that have the potential to mislead the population, or even possibly cause further harm. Currently, Twitter allows users to report spam, and after an investigation, an account can be suspended. However, suspending a spam account is not an efficient technique to deal with spam related to trending topics because the suspension process is slow, and the trending topics usually last for only a few hours or a day at most [2]. Therefore, the focus of the *anomaly detection on trending topics* in this work is on the detection of tweets containing spam, instead of detecting spam accounts.

One of the more malicious activities involves a spammer who includes a URL in the tweet, leading the reader to a completely unrelated website. It is reported that 90% of anomalous tweets contain unrelated or misleading URLs [1,3]. People use shortened URLs or links in their tweet because of the limited number of characters (280) available in the tweet. Since it is common for tweets to include shortened text, so as to fit within the character limits, spammers can conceal their unrelated/malicious links with shortened URLs. Hence, the problem with a shortened URL is that users do not know what is the actual domain until the link is clicked. The existing approaches for spam detection on Twitter use various machine learning tools on user-based features (e.g. number of followers, number of tweets, age of the user account, number of tweets posted per day or per week, etc.) and content-based features (e.g. number of hashtags, mentions, URL, likes, etc.) [1,3–9]. Though user and content-based features can be extracted efficiently, an issue is that these features can also be fabricated easily by the spammer [2,10,11]. However, being able to hide an inconsistency between the topic of a tweet and the topic of the document referred by URLs in the tweet is much harder [2,12].

In this work, we propose an unsupervised, two-step, graph-based approach to detect anomalous tweets on trending topics. First, we extract named entities (like place, person, organization, product, event, or activity) present in the tweet and add them as key elements in the graph. As tweets on a certain topic share the contextual similarity, we believe they also share same/similar named entities. These named entities representing relevant/similar topics can have a relationship (e.g., shared ontology) amongst themselves, which we believe if represented properly, will provide broader insight on the overall context of the topic. As such, graphs can be a logical choice for representing these kinds of information where a node can represent a named entity and an edge can represent the relationships between them. Using a well-known graph-based tool like GBAD [13], we then discover the normal and anomalous behavior of a trending topic. Second, we propose adding hyperlinked document information because anomalies that could not be detected from tweets alone could be detected using both the document and tweets. It is our assumption that a better understanding of patterns and anomalies associated with entities like person, place, or activity, cannot be realized through a single information source, but better insight can be realized using multiple information sources simultaneously. For instance, one can discover interesting patterns of behavior about an individual through a single social media account, but better insight into their overall behavior can be realized by examining all of their social media actions simultaneously. Analyzing multiple information sources for anomaly detection on Twitter has been explored in the past. For example, the inconsistencies between the tweet and the document referred to by a URL in the tweet using cosine similarity [12] and a language model [2] were studied for potential anomaly detection. But, the cost for [2] is high as each tweet with a link is treated as a suspect and [12] need a predefined source of reliable information for each topic which makes these approaches less flexible in real-time trending topics.

Using the above mentioned 2-step approach, we aim to detect the following types of spam/anomalies in trending tweets that are consistent with the spam scenarios listed by Twitter [14].

1. *Keyword/Hashtag Hijacking*: Using popular keywords or hashtags to promote the tweet that are not related to the topic. This is done to promote anomalous tweets to a wider audience by hijacking popular hashtags and keywords.
2. *Bogus link*: Posting a URL that has nothing to do with the content of the tweet. This is done to generate more traffic to the website. Another scenario of bogus link is link piggybacking. For example, posting an auto redirecting URL that goes to legitimate website but only after visiting an illegitimate website. Another way is to post multiple links where one link can be a legitimate link while another can be a malicious or unrelated link. The motivation behind link piggybacking is to generate traffic to the illegitimate website by concealing the link inside a legitimate website. This can also be accomplished by using a tiny URL.

To verify our approach, we collect tweets (containing URLs) related to two separate (and very different) trending topics during the summer of 2018: FIFA World Cup and NATO Summit (Sect. 3). We then construct graphs using information from the tweet text and the document referred inside the tweet (Sect. 4), followed by using a graph-based anomaly detection tool (Sect. 5). We then compare the performance of our proposed approach with several existing approaches to show the effectiveness of a graph-based approach (Sect. 6).

2   Related Work

Anomaly detection in Twitter data can be about detecting anomalous/spam accounts or about detecting anomalous/spam tweets. Verma et al. [5] summarize spam account detection techniques in Twitter along with their analysis and comparison. Similarly, Tingmin et al. [15] present the detailed analysis, discussion and comparative studies of existing approaches on both spam accounts and spam tweets detection. Most of the approaches for detecting spam accounts use user-based and content-based features. Benevenuto et al. [1] use SVM on 62 sets of user-based and content-based features to classify spammer and non-spammer accounts. Soman and Murugappan [16] use fuzzy K-mean clustering to group the similar user profiles with the same trending topics, and an extreme learning machine (ELM) algorithm is applied to analyze the growing characteristics of spam with similar topics in Twitter from the clustering result.

Wu et al. [17] propose WordVector and deep-learning techniques to extract text-based features that are hard to fabricate by the spammer. These features are then fed into traditional classifiers. Boididou et al. [18] propose a semi-supervised approach based on bagging that uses different sets of tweet-based (TB) and user-based features (UB). Meda et al. [9] apply Random Forest by using only 5 features on the same dataset used by [1] and got comparable results to [1]. Ameen and Kaya [6] compare the performance of Naive Bayes, Random Forest, J48 and IBK classifiers to classify 1135 user accounts into spammer and non-spammer accounts. They report that Random Forest shows the best results among the four classifiers. Another approach that demonstrates the superiority of Random Forest over other machine learning algorithms (Naïve Bayesian, Support Vector Machine, and K-NN) to detect spam accounts is by McCord and Chuah [8].

Chao et al. [11] collected over 600 million public tweets, labelled around 6.5 million spam tweets, extracted 12 light weight statistical features, and conduct experiments on different machine learning algorithms simulating various scenarios. They did so to better understand the effectiveness and weakness of different algorithms for timely Twitter spam detection. Though user and content-based features can be extracted efficiently, an issue is that these features can also be fabricated easily by the spammer [2,10,11]. A more sophisticated approach by Wang [4] uses content-based along with graph-based features on a user's "follower" and "friend" relationships. These features are provided to a Bayesian classifier for classifying spam and non-spam accounts. Yang et al. [19] designed another sophisticated and robust approach by analyzing evasive techniques of a spammer. They propose the use of 10 detection features: 3 graph-based, 3 neighbor-based, 3 automation-based, and 1 timing-based feature as the input for several machine learning algorithms. Lee et al. [20] use social honeypots for harvesting deceptive spam accounts and perform statistical analysis on a spam account's properties to create spam classifiers to actively filter out existing and new spammers.

Besides spam account detection, there has been some research focusing on anomalous content. The author in [7,21] propose an anomalous tweet detection scheme by leveraging the features of embedded URL in the tweets. URL-based features show the discriminative power for classifying spam but these schemes can only detect anomalous tweets containing URLs and miss anomalous tweets containing only text or fabricated URLs [22]. Anantharam et al. [12] analyze tweet content along with documents referenced in the URL for assessing their relevance to an event/topic. Gupta et al. [23] propose the framework that takes the user-based, content-based and tweet-text features to classify the tweets using a Neural Network with accuracy up to 91.65%. Song et al. [10] use distance and connectivity between a message sender and a message receiver by constructing directed graphs based on the following and followed relations in Twitter to decide whether the message is spam or not. Martinez-Romo and Araujo [2] effectively classified trending tweets as spam or non-spam by exploiting the divergence between the statistical language models in the topic, the tweet, and the page linked from the tweet. This was based on the fact that if the tweet is spam, language models are likely to be different as the spammer usually tries to divert traffic to sites that have no semantic relation [2].

Though the use of graph-based features have better results in detecting spam accounts/tweets [4,10,19], little or no work has been done applying graph mining techniques to detect anomalous URL links in tweets. Graphs provide a powerful machinery for effectively capturing the long-range correlations among interdependent data objects/entities. Graph-based approaches have been successfully applied for anomaly detection in a wide array of applications [24]. Therefore, we plan to extend a graph-based approach for detecting spam tweets on trending topics. In this work, we will use the publicly-available graph-based anomaly detection (GBAD) tool [13] that has been used for detecting network intrusions [25,26], reporting anomalies in telecom data [27], discovering unusual elderly patient activities in a smart homes [28], detecting anomalous activity and potential fraud scenarios in medicare claim files [29], etc.

3 Data

The dataset used in this research consists of tweets and documents (primarily news stories) mentioned in the tweets. The detailed process of data collection and the description of data is presented in this section.

The data was collected using Twitter's standard search API. We collected tweets related to two trending topics, "FIFA World Cup" and "NATO Summit", during the summer of 2018. The results from Twitter's search API contains tweet text, Twitter handle name, any hashtags and URLs mentioned in the tweet, as well as all publicly available information about the user including their name. The data for tweets is a JSON dump of individual tweets.

We added a news filter onto the query to gather a sample of tweets containing links to news articles. We also added a filter for the English language so that we only get English tweets. We observed that the textual content of some tweets is the same as the title of the news (referred by the URL). Since our objective is to combine multiple sources of information together to learn new information, these particular tweets (containing the title of the news) does not provide any additional content (i.e. redundant). So, we used a python-based library called "difflib" to filter out tweets whose text content matches the title of the article referred by the URL in the tweet. To acquire the news information from the news URL, we used a python-based crawler called "Newspaper"[1]. Using Newspaper, we extracted the news title, body, summary, author name, published URL, domain name of the published website, and published date. Because of the challenges like dead URLs, URLs leading to multimedia content (videos or photos), non-English tweets, and links to non-English documents, the resulting dataset is smaller than we hoped. But, as we will show, it still demonstrates the effectiveness of our proposed approach in addressing the problem.

---

[1] https://github.com/codelucas/newspaper.

Table 1. Total number of tweet/news and anomalies in each topic

Trending topic	Total tweet/news	Anomalies		
		Keyword hijacking	Bogus link	Total
World Cup	1,463	2	20	22
NATO Summit	1,716	0	11	11

After data collection, we manually inspected the data for the amount of spam present in the dataset so that we can measure the performance of our approach. It should be noted that our approach is unsupervised and does not need any labelled data. We used the following criteria that are consistent with [1] to label the spam tweet.

- If the tweet has keywords related to the trending topic but the document referred by the URL does not have any.
- If the tweet has multiple links and if any of the links refer to a document not related to the trending topic.
- If the tweet has a URL that redirects to an unrelated website before redirecting to the related website. This usually occur when the tweet has a tiny URL.

Table 1 shows the number of tweets/news and anomalies in two datasets. The datasets are publicly available on https://rpaudel42.github.io/pages/dataset.html.

4   Methodology

The tweets on a certain topic should have contextual similarity. It is our assumption that this contextual similarity can be perceived by the presence of same/similar named entities in the tweet. One of the key ideas of our approach is to extract these named entities, in addition to user- and content-based features, and map their relationships using a graph. We also assume that two independent named entities are considered related if they share a common ontology. We believe these entities and their relationships (ontology) will provide valuable information about the context of the tweet. First, we will run anomaly detection on a graph using the information (named entity, user-based and content-based features) from the tweets. Second, we will run anomaly detection on a graph using information from both the tweets and the documents referred by the URL in the tweet. This will allow us to demonstrate that new anomalies can be discovered using information from multiple sources related to same entity (trending topic) which otherwise could not haven been detected using single source (tweet). The basic methodology aimed at discovering anomalies in trending Twitter topics consists of four key modules: Named Entity Extractor, Ontology Generator, Graph Parser and Graph-Based Anomaly Detection, as shown in Fig. 1. We will discuss each of these modules briefly in the following section.

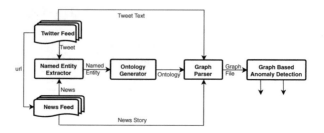

Fig. 1. Basic methodology used for experiment

4.1   Named Entity Extractor

Named entities are the real-world objects (either abstract or physical) like place, person, organization, company, date, time, etc. that can be denoted with a proper name. For our experiments, we focus particularly on three types of named entities: PERSON, ORGANIZATION and LOCATION. The python based Natural Language toolkit, NLTK [30], is used for extracting named entities from news and tweets. NLTK provides a classifier that has already been trained to recognize named entities. We collected the named entities by using the NLTK ne_chunk() function. Then, extracted entities are passed to the Ontology Generator for generating the ontology.

4.2   Ontology Generator

The primary use of an ontology in information systems is the description and structuring of shared knowledge [31]. Extracting an ontology for entities that share common structure provides added knowledge about those entities. We used DBpedia for generating the ontology. The DBpedia project has a large-scale knowledge base that extracts structured data from Wikipedia [32]. We created a python based script to query DBpedia by passing a named entity and its class (person, organization or location) to the API. The result is an ontology hierarchy that can be represented in graph form. A sample snapshot of the ontology structure for the class "Place" is given in Fig. 2. The leaf node in Fig. 2 (represented by a rectangle) are the Named Entities. For example, if the named entity is "Nashville", the result will be the hierarchy *[Place→ Populated Place→Settlement→City→Nashville]*. There can be ambiguity between entities that share common names but have no shared structure. For example, the entity "Apple" can be a technology company or a fruit. To avoid this ambiguity, we used "Apple" as the query string and "Organization" as the query class for generating an ontology for the company "Apple"; and "Apple" as the query string and "Fruit" as the query class for generating an ontology for the fruit "Apple". But if entities belong to the same ontology class (two people with same name, like John Doe the Businessman and John Doe the Athlete), the ambiguity still exists.

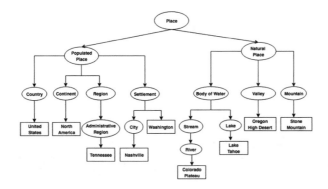

Fig. 2. Example of ontology for class place.

In this case, we have chosen to use the ontology of the first result returned by DBpedia API as that is the most likely match (i.e., most popular). The hierarchy thus generated is then passed to the Graph Parser for creating the graph used for our experiments.

4.3   Graph Parser

The desired input for our anomaly detection is a graph file. In order to create a graph input file, we created a Python-based parser that reads the data from the tweet and the news referred by the URL, and constructs the graph. The experiment is done on two graphs: a graph using information from just the tweets, and a graph using both news and tweet information. We named them *Tweet Graph* and *Mixed Graph* respectively. The basic layout of the *Mixed Graph* is shown in Fig. 3 and the process of creating the graph is described below.

In the graph, we include three content-based features (mentioned twitter handle, hashtag, and tweet posted date) and two user-based features (location and description). While we are extracting location and description as the two user-based features in the graph, we ignore other attributes like screen name, language, timezone, etc. because we cannot extract named entities (place, person, and organization) from these attributes. The Twitter account is represented by a *"handle"* node in the graph. If the Twitter account has a description, we extract the named entity (along with the ontology hierarchy) from the description and link them to the *"handle"* node with an edge *"is_about"*. Similarly, the location is linked to the *"handle"* node using an edge *"is_from"*. For example, if the handle has a location "New York" and a description "NYU Grad students, blogger, coffee lover", the Named Entity Extractor will extract 'NYU" and "Blogger" as the organization. This graph will therefore represent the person who is a New York University graduate student living in "New York" and loves to blog. An edge *"post"* between the Twitter handle and *"tweet"* represents that a handle posted the tweet. The entity *"hashtag"*, is linked to a tweet node with a *"has"* edge. The named entity (with their ontology hierarchy) extracted from the tweet

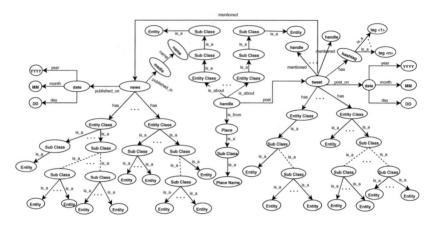

Fig. 3. Graph layout showing news-tweets features, named entities and their ontology. Dotted line represents multiple node or edge.

text are linked to the tweet node with a *"has"* edge. Tweet posted date is represented as an edge labelled *"post_on"* between *"tweet"* and *"date"* nodes. In addition, day, month, and year values are added as attributes for the date node.

The news is represented by a *"news"* node in the graph. The information from the news mentioned by the URL can be added to a *Tweet Graph* to create the *Mixed Graph*. This can be done by adding an edge labelled *"mentioned"* between *"handle"* and *"news"* nodes. Two features, published date and the publisher media, are added to the graph by representing them as a node in *Mixed Graph*. The News published date is represented as an edge labelled *"published_on"* between *"news"* and *"date"* nodes. Again, day, month and year values are added as attributes for the date node. The relationship between *"media"* and *"news"* is represented by an edge labelled *"published_in"*, and the media name is added as an attribute to "media". Named entities and their ontology are extracted from news text and added to the graph in similar way as it is done in a tweet text. It should be noted that Fig. 3 is just a visualization, as the actual graph input files are plain ASCII text files. The number of vertices and edges on the *Tweet Graph* and the *Mixed Graph* constructed for each of the datasets is shown in Table 2.

4.4   Graph-Based Anomaly Detection

In order to lay the foundation for this effort, we hypothesize that a real-world, meaningful definition of a graph-based anomaly is an unexpected deviation to a normative pattern, which is defined as follows:

Definition 1. *A labeled graph $G = (V, E, F)$, where V is the set of vertices (or nodes), E is the set of edges (or links) between the vertices, and the function F assigns a label to each of the elements in V and E.*

Table 2. Number of vertices and edges in graphs for both dataset

Graph name	Number of vertices	Number of edges
FIFA World Cup Dataset		
Tweet Graph	23,603	22,140
Mixed Graph	88,887	87,424
NATO Summit Dataset		
Tweet Graph	28,116	26,400
Mixed Graph	90,224	88,508

Definition 2. *A subgraph SA is an anomalous in graph G if* $(0 < d(SA, S) < TD)$ *and* $(P(SA|S) < TP)$*, where* $P(SA|S)$ *is the probability of an anomalous subgraph SA given the normative pattern S in G. TD bounds the maximum distance* (d) *an anomaly SA can be from the normative pattern S, and TP bounds the maximum probability of SA.*

Definition 3. *The score of an anomalous subgraph SA based on the normative subgraph S in graph G is* $d(SA, S) * P(SA|S)$*, where the smaller the score, the more anomalous the subgraph.*

The advantage of graph-based anomaly detection is that the relationships between entities can be analyzed for structural oddities in what could be a rich set of information, as opposed to just the entities' attributes. However, graph-based approaches have been prohibitive due to computational constraints since graph-based approaches typically perform subgraph isomorphism, a known NP-complete problem. Yet, in order to use graph-based anomaly detection techniques in a real-world environment, we need to take advantage of the structural/relational aspects found in dynamic, streaming data.

In order to test our approach, we will use the publicly available GBAD test suite, as defined by [13]. Using a greedy beam search and a minimum description length (MDL) heuristic, GBAD first discovers the "best" subgraph, or normative pattern, in an input graph. The MDL approach is used to determine the best subgraph(s) as the one that minimizes the following:

$$M(S, G) = DL(G|S) + DL(S)$$

where G is the entire graph, S is the subgraph, $DL(G|S)$ is the description length of G after compressing it using S, and $DL(S)$ is the description length of the subgraph. The complexity of finding the normative subgraph is constrained to be polynomial by employing a bounded search when comparing two graphs. Previous results have shown that a quadratic bound is sufficient to accurately compare graphs in a variety of domains [13]. GBAD can discover three general categories of anomalies in a graph; insertions, modifications and deletions. Insertions would constitute the presence of an unexpected vertex or edge. Modifications would consist of an unexpected label on a vertex or edge. Deletions would constitute the unexpected absence of a vertex or edge. For more details regarding the

GBAD algorithms, the reader can refer to [13]. In summary, the key to the GBAD approach is that anomalies are discovered based upon small deviations from the norm - not outliers, which are based upon significant statistical deviations from the norm.

Table 3. Number of anomalies detected using $TD = 0.35$ based on graphs and datasets

Dataset	Anomalies	Graphs	
		Tweet Graph	Mixed Graph
World Cup	Keyword hijacking	0 (2)	2 (2)
	Bogus link	5 (20)	20 (20)
	Total	**5 (22)**	**22 (22)**
NATO Summit	Keyword hijacking	–	–
	Bogus link	5(11)	11(11)
	Total	**5 (11)**	**11 (11)**

5   Experimentation and Results

The experimentation process involves running the GBAD tool[2] on the *Tweet Graph* and *Mixed Graph* constructed from both FIFA world cup and NATO summit datasets. GBAD uses a compression technique to discover the normative patterns in the graphs; the normative patterns are then used to identify the anomalous structures. In other words, GBAD analyzes the complete dataset through the lens of the selected normative patterns in order to discover the anomalies - and in our case, the spam. We select $TD = 0.35$ as our anomaly detection threshold for GBAD. Table 3 shows the number of anomalies detected by the proposed approach using the normative patterns shown in Figs. 4 and 7 for the World Cup and NATO datasets respectively. We now discuss the results of anomaly detection on each of the trending topics.

5.1   FIFA World Cup

The normative pattern reported by GBAD on the FIFA world cup dataset for *Tweet* and *Mixed Graph* is shown in Fig. 4. The normative pattern for the *Tweet Graph* (Fig. 4(a)) indicates that the tweet was posted by a Twitter handle on July 2018 and has a hashtag of "worldcup" and mentions another twitter handle. Using this normative pattern on *Tweet Graph*, we were able to discover only 5 anomalies (out of 22 anomalies) present on this dataset. The fewer number of anomalies is reported by the *Tweet Graph* because the graph constructed using just the text of the tweet is not able to provide enough context. Therefore, we will extend the context by applying extra information from the news associated

---

[2] www.gbad.info.

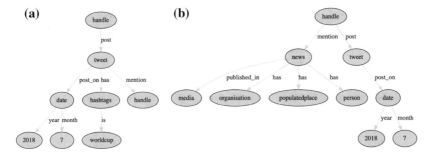

Fig. 4. Normative pattern on World Cup Dataset (a) Tweet Graph (b) Mixed Graph

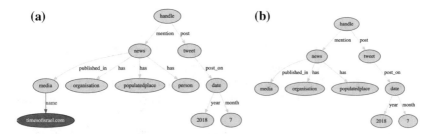

Fig. 5. Anomalous patterns reported in mixed graph (a) addition of node (b) deletion of node *person* in FIFA world cup dataset

Fig. 6. Example screen-shot of (a) news and (b) tweet showing keyword hijacking anomaly in world cup dataset

with the tweet in *Mixed Graph*. This will allow us to identify more informative normative patterns and anomalies.

The normative pattern for the *Mixed Graph* (Fig. 4(b)) indicates that a tweet posted by a Twitter handle in July of 2018 mentions news published in some media and has a person, an organization, and a populated place. This normative pattern provides extra information about the news besides tweet information. Using this normative pattern on the *Mixed Graph* we discovered several types of anomalies. For example, the anomaly represented by Fig. 5(a) has an extra node *"timesofisrael.com"* (shown as a red node) linked to a *"media"* node by an edge *"name"*. In this particular case, GBAD reported the anomaly because the tweet has the keywords "FIFA" and "WorldCup" but the link posted has news not related to the world cup - instead it was political news from Israel. This is an example of a keyword hijacking anomaly that we introduced earlier. The screen shot of the tweet and news related to this anomaly is shown in Fig. 6. Similarly, another instance of an anomaly (as shown in Fig. 5(b)) has a *"person"* node missing from the normative patterns. Upon inspection, we find that the tweet had two URLs; one was a legitimate URL that has real world cup news, while the other was a link to a website that was not related to the World Cup.

5.2   NATO Summit

Figure 7 shows the normative pattern discovered by GBAD on the *Tweet* and *Mixed Graph* for the NATO dataset. The normative pattern for the *Tweet Graph* (as shown in Fig. 7(a)) indicates that a twitter handle posted a tweet in July of 2018 that has a person, an organization, and another twitter handle mentioned. Using it for anomaly detection on the *Tweet Graph*, we are able to discover only 5 anomalies (out of a total of 11 anomalies) on the NATO dataset. The sub-graph representing these 5 anomalies present in *Tweet Graph* is shown in Fig. 8(a) where the label of the node *"person"* in terms of the normative pattern is modified into a label *"hashtags"* (represented by a red node).

However, the normative pattern discovered by GBAD on the *Mixed Graph* on the NATO dataset (as shown in Fig. 7(b)) provides extra information about news besides the information from the tweets. This normative pattern indicates that a tweet posted by a Twitter handle in July of 2018 mentions news published in some media that has a person "Donald Trump". Using this normative pattern for anomaly detection, we discover several anomalies. The anomalies represented by the anomalous sub-graph in Fig. 8(b) has nodes *"person"* and *"donald trump"* in the normative pattern modified to *"organization"* and *"company"* respectively (marked by the red nodes). Further inspecting the instances of the anomaly represented by this sub-graph, we discover that the tweet mentions two URLs. The second URL is the legitimate URL that refers to the news about the NATO summit, while the first URL links to a website called *"robinhood.com"* which talks about investing in the stock market. This is not related to the NATO summit. Hence, it is a bogus link (piggybacking) anomaly. The screen-shot of the tweet and the website referred by the URL is shown in Fig. 9.

Table 3 shows the numbers and types of anomalies detected by our proposed approach on both datasets. Running GBAD on just the *Tweet Graph*, we were able to detect 5 out of 22 anomalies, while all known 22 anomalies were discovered successfully using the *Mixed Graph* on the FIFA world cup dataset. Similarly, in the NATO summit dataset, we were able to detect 5 anomalies using the *Tweet Graph*, while all 11 anomalies were discovered using the *Mixed Graph*. More anomalies are discovered using *Mixed Graph* in both cases because we are able to extend the context by applying extra information from the news

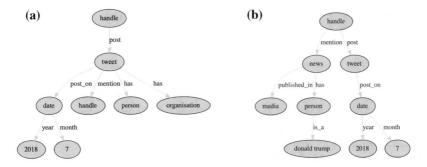

Fig. 7. Normative pattern on NATO Summit Dataset (a) Tweet Graph (b) Mixed Graph.

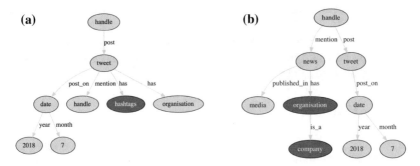

Fig. 8. Anomalous patterns on (a) Tweet Graph (b) Mixed Graph in NATO dataset.

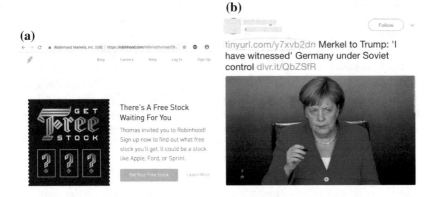

Fig. 9. Example screen-shot of (a) news and (b) tweet showing bogus link anomaly in NATO summit dataset.

associated with a tweet. The fact that we were able to detect both types of spam (keyword/hashtag hijacking and bogus link) using our approach supports our first hypothesis: "*Normal and anomalous behavior of a trending topic can be discovered using a graph-based tool on a graph representing named entities and their relationship"*. Though we were able to detect a few anomalies in the *Tweet Graph* for both datasets, we are clearly more successful using the *Mixed Graph*, which represents using multiple sources. This is because the graph constructed using just the text of the tweet was not able to provide enough context. There are cases where the content of the tweet (represented by *Tweet graph*) looks normal, but after adding the information from the news, we can tell that the tweet was, in fact, an anomaly. For example, the tweet shown in Fig. 9 looks legitimate (couldn't detect it as an anomaly in *Tweet Graph*), but after adding the information referred by the tiny URL we are able to identify that the tweet was spam (i.e. is detected as an anomaly in the *Mixed graph*). This demonstrates our second hypothesis: *"Adding information from the document referred by the URL together with information from the tweet into the graph, anomalies that could not be detected using just the information from tweet can be detected"*. Also, it is hard for the spammer to fabricate information in both the tweet and the document referred to by the URL in the tweet. The proposed graph-based approach exploits this type of dissimilarity in the content of the tweet and the document for anomaly detection.

6   Evaluation

In this section, we evaluate the performance of our graph-based approach. The dataset is highly imbalanced because spam tweets are very few in comparison to normal tweets. This challenges the interpretation of the standard evaluation metrics like accuracy or error rate [33–35]. For example, the FIFA world cup dataset has a class ratio of almost 99:1 and 99% accuracy is achievable by always predicting the majority class (i.e., normal tweets). Therefore, accuracy or error rate in this case does not provide adequate information on a classifier's functionality [33,35]. Other evaluation metrics that are frequently adopted in the research community to provide comprehensive assessments of imbalanced learning problems are precision, recall, and F1-score [33,35]. The focus of learning algorithms should be towards improving the recall, without sacrificing the precision. The F1-score incorporates both precision and recall, and the "goodness" of a learning algorithm for the minority class can be measured by the F1-score [35–37].

6.1   Baseline Approaches

In order to evaluate our approach, we compare our results against some well-known approaches that deal with the detection of spam tweets.

Benevenuto et al. [1] proposes an approach to detect spam tweets as well as the spam accounts using an SVM-based classifier. We compare our results with their spam tweet detection approach. The SVM classifier for spam tweet detection uses the following attributes for each tweet: number of words from a list of spam words, number of hashtags per words, number of URLs per words, number of words, number of characters in the text, number of characters that are numbers, number of URLs, number of hashtags, number of mentions, number of times the tweet has been replied, and if the tweet was posted as a reply.

Chen et al. [11] extract 12 user and content-based features and conduct experiments on different machine learning algorithms simulating various scenarios. They conclude random forest was the best approach. Account age, number of followers, followings, favorite users, list joined, and tweets posted are the user-based features. Similarly, the number of digits, characters, URLS, hashtags, mentions, and retweets on the tweet text are the content-based features. We compare the result of our approach against the performance of the random forest.

Anantharam et al. [12] proposes an approach to spot topically anomalous tweets in twitter streams by analyzing the content of the document pointed to by the URLs in the tweets in preference to their textual content. They manually identify reliable sources of information and compute the average cosine similarity Sim_{avg} amongst them. For every incoming tweet, they compute the cosine similarity between the document pointed to by the URL in the tweet and each of the trusted documents, then calculate the maximum cosine similarity Sim_{max}. If this maximum cosine similarity is less than the average cosine similarity among trusted documents (i.e. $Sim_{max} < Sim_{avg}$), the tweet is flagged as an anomaly.

Boididou et al. [18] proposes a semi-supervised framework that relies on two independent classification models built on the training data using tweet-based (TB) and user-based features (UB). Both models are built using a bagging technique. At prediction time, an agreement-based retraining strategy is employed (fusion) to combine the outputs of these two models (in a semi-supervised manner) which increases the generalization capabilities of the framework given tweets from a new unknown event. A corpus of labeled posts is necessary to build these classification models.

6.2   Discussion

We now present the performance evaluation of the proposed approach and the baseline approaches in terms of precision (P), recall (R), and F1-score. The performance score is shown in Table 4.

Using an anomaly detection threshold, $TD = 0.35$, without any other parameters, we are able to get a recall of 100% on both datasets. This indicates that the proposed graph-based approach can successfully detect spam tweets in trending topics. The higher recall is good for any anomaly detection problem, however, one also wants to achieve high precision or F1-score. The trade-off between precision and recall can be decided according to the need of our system. With some tradeoff on recall we are able to improve precision and F1-score by tuning GBAD

Table 4. Performance evaluation on both dataset

Approach	P	R	F1-score
World Cup Dataset			
Graph-based (non-parametric)	0.156	**1.0**	0.270
Graph-based (parametric)	0.516	**0.727**	**0.603**
Benevenuto et al. [1]	**0.799**	0.181	0.296
Chen et al. [11]	0.783	0.459	0.575
Anantharam et al. [12]	0.235	0.364	0.286
Boididou et al. [18]	0.692	0.409	0.514
NATO Summit Dataset			
Graph-based (non-parametric)	0.136	**1.0**	0.239
Graph-based (parametric)	**0.539**	**0.636**	**0.583**
Benevenuto et al. [1][a]	0.333	0.022	0.042
Chen et al. [11]	0.519	0.30	0.361
Anantharam et al. [12]	0.235	0.364	0.286
Boididou et al. [18]	0.375	0.272	0.315

[a]This approach was unable to classify spam on the NATO dataset, so we up-sampled spam tweets to make the spam to no-spam ratio $\approx$1:50

parameters[3]. Although the result of our parametric approach demonstrates the modest F1-score, the result can be easily modified based on our need (whether we need higher recall or higher precision). We present the result of both parametric as well as non-parametric approach in Table 4.

Our parametric approach has better performance in terms of recall and F1-score against all four baseline approaches on the FIFA World Cup dataset. Benevenuto et al. Chen et al. and Boididou et al. have better precision than our approach. However, the low recall and F1-score by these approaches indicates that our proposed graph-based approach has superior performance. Similarly, on the NATO summit dataset, our graph-based (parametric) approach has a better performance in terms of all metrics when comparing with all four baseline methods. Benevenuto et al. and Chen et al. use a machine learning approach that is more likely to be affected by the imbalanced nature of the data. Benevenuto et al. use a 1:3 ratio of spam to non-spam data in their original work while Chen et al. use 1:19. The dataset used in our experiment has a spam to non-spam ratio of $\approx$1:100. The low performance of these two approaches is because of their inflexibility in a highly imbalanced dataset. Also, Benevenuto et al. Chen et al. and Boididou et al. require a labeled dataset for training their classifier. However, our proposed graph-based approach is unsupervised and is not affected by the class imbalance problem. In fact, the performance of the proposed approach

---

[3] We use $maxAnomalousScore = 24$ for FIFA World Cup and $maxAnomalous Score = 52$ for NATO Summit datasets.

gets better in the data where the spam is rare (see definition of an anomaly in Sect. 4.4) which is usually the case in a Twitter trending topic. The approach by Anantharam et al. needs a predefined reliable information source for each topic, making it less flexible in real-time Twitter trends where new topics are evolving quickly. Also, the context generated (named entities and their relationships) from both the tweet and the document referred to by the URL in the tweet as used in *Mixed Graph* is hard to fabricate by the spammer. Furthermore, the better recall and F1-score in comparison to the existing approaches make our proposed unsupervised graph-based approach a solid approach for spam detection in a trending topic.

7   Conclusion and Future Work

In this work, we propose an unsupervised graph-based approach that leverages the relationship between the named entities present in the content of the tweet and the document referenced by the URL mentioned in the tweet for detecting possible spam. Our graph-based approach has superior performance in terms of recall and F1-score to that of existing approaches. Graphs provide a powerful machinery for effectively capturing the long-range correlations among interdependent data objects/entities. This interdependent nature of the data can be represented as an edge between each entity represented as nodes. We further claimed that a better understanding of patterns and anomalies associated with an entity like person, place, or activity, cannot be realized through a single information source, but better insight can be realized using multiple information sources simultaneously. We were able to discover new and unknown anomalies in the *Mixed Graph* which were not discovered in a *Tweet Graph*. We demonstrated this by collecting tweets relating to two different trending topics. Our approach focuses on the detection of spam tweets instead of the spam account. And the detection of the spam tweet can be useful for filtering spam in near real-time, while the detection of a spam account is about identifying such accounts retrospectively and blocking them so that they cannot spread future spam.

Although our research showed some good results, we were limited to static graphs. However, since the Twitter social graph can be extremely huge, graph construction and analysis can be time and resource consuming. If we think of a near real-time anomaly detection tool in social media or the web, it should be able to handle data that comes in streams. So, in the future we would like to focus on analyzing a near real-time feed by converting data streams into graph streams. Our experimentation was done on data sets of two trending topics, but we would like to extend it to more topics and test the robustness of the proposed approach. While our approach is not designed to discover anomalies in tweets that do not have URLs, discovering anomalies in tweets with URLs is key because it is reported that 90% of anomalous tweets contain unrelated or misleading URLs [1,3]. We would also like to use this approach for proactive criminal/terrorist activity detection by fusing multiple social media feeds. This is because we believe this anomaly detection technique will be able to track the change in behavior of an individual, particularly in terms of an individual's communications and transactions that may be represented in a graph.

References

1. Benevenuto, F., Magno, G., Rodrigues, T., Almeida, V.: Detecting spammers on twitter. In: Collaboration, Electronic Messaging, Anti-abuse and Spam Conference (CEAS), vol. 6, p. 12 (2010)
2. Martinez-Romo, J., Araujo, L.: Detecting malicious tweets in trending topics using a statistical analysis of language. Expert Syst. Appl. **40**(8), 2992–3000 (2013)
3. Gayo Avello, D., Brenes Martínez, D.J.: Overcoming spammers in twitter–a tale of five algorithms. In: Spanish Conference on Information Retrieval. CERI (2010)
4. Wang, A.H.: Don't follow me: spam detection in twitter. In: Proceedings of the 2010 International Conference on Security and Cryptography (SECRYPT), pp. 1–10. IEEE (2010)
5. Verma, M., Sofat, S.: Techniques to detect spammers in twitter–a survey. Int. J. Comput. Appl. **85**(10), 27–32 (2014)
6. Ameen, A.K., Kaya, B.: Detecting spammers in twitter network. Int. J. Appl. Math. Electron. Comput. **5**(4), 71–75 (2017)
7. Thomas, K., Grier, C., Ma, J., Paxson, V., Song, D.: Design and evaluation of a real-time URL spam filtering service. In: 2011 IEEE Symposium on Security and Privacy (SP), pp. 447–462. IEEE (2011)
8. Mccord, M., Chuah, M.: Spam detection on twitter using traditional classifiers. In: International Conference on Autonomic and Trusted Computing, pp. 175–186. Springer (2011)
9. Meda, C., Bisio, F., Gastaldo, P., Zunino, R.: A machine learning approach for twitter spammers detection. In: 2014 International Carnahan Conference on Security Technology (ICCST), pp. 1–6. IEEE (2014)
10. Song, J., Lee, S., Kim, J.: Spam filtering in twitter using sender-receiver relationship. In: International Workshop on Recent Advances in Intrusion Detection, pp. 301–317. Springer (2011)
11. Chen, C., Zhang, J., Chen, X., Xiang, Y., Zhou, W.: 6 million spam tweets: a large ground truth for timely twitter spam detection. In: 2015 IEEE International Conference on Communications (ICC), pp. 7065–7070. IEEE (2015)
12. Anantharam, P., Thirunarayan, K., Sheth, A.: Topical anomaly detection from twitter stream. In: Proceedings of the 4th Annual ACM Web Science Conference, pp. 11–14. ACM (2012)
13. Eberle, W., Holder, L.: Anomaly detection in data represented as graphs. Intell. Data Anal. **11**(6), 663–689 (2007)
14. Twitter: Report Spam on Twitter. https://help.twitter.com/en/safety-and-security/report-spam. Accessed 9 Oct 2018
15. Wu, T., Wen, S., Xiang, Y., Zhou, W.: Twitter spam detection: survey of new approaches and comparative study. Comput. Secur. **76**, 265–284 (2018)
16. Soman, S.J., Murugappan, S.: Detecting malicious tweets in trending topics using clustering and classification. In: 2014 International Conference on Recent Trends in Information Technology, pp. 1–6. IEEE (2014)

17. Wu, T., Liu, S., Zhang, J., Xiang, Y.: Twitter spam detection based on deep learning. In: Proceedings of the Australasian Computer Science Week Multiconference, p. 3. ACM (2017)
18. Boididou, C., Papadopoulos, S., Apostolidis, L., Kompatsiaris, Y.: Learning to detect misleading content on twitter. In: Proceedings of the 2017 ACM on International Conference on Multimedia Retrieval, pp. 278–286. ACM (2017)
19. Yang, C., Harkreader, R., Gu, G.: Empirical evaluation and new design for fighting evolving twitter spammers. IEEE Trans. Inf. Forensics Secur. **8**(8), 1280–1293 (2013)
20. Lee, K., Caverlee, J., Webb, S.: Uncovering social spammers: social honeypots+machine learning. In: Proceedings of the 33rd International ACM SIGIR Conference on Research and Development in Information Retrieval, pp. 435–442. ACM (2010)
21. Lee, S., Kim, J.: WarningBird: a near real-time detection system for suspicious URLs in twitter stream. IEEE Trans. Dependable Secure Comput. **10**(3), 183–195 (2013)
22. Egele, M., Stringhini, G., Kruegel, C., Vigna, G.: COMPA: detecting compromised accounts on social networks. In: NDSS (2013)
23. Gupta, H., Jamal, M.S., Madisetty, S., Desarkar, M.S.: A framework for real-time spam detection in twitter. In: 2018 10th International Conference on Communication Systems & Networks (COMSNETS), pp. 380–383. IEEE (2018)
24. Akoglu, L., Tong, H., Koutra, D.: Graph based anomaly detection and description: a survey. Data Min. Knowl. Disc. **29**(3), 626–688 (2015)
25. Noble, C.C., Cook, D.J.: Graph-based anomaly detection. In: Proceedings of the Ninth ACM SIGKDD International Conference on Knowledge Discovery and Data Mining, pp. 631–636. ACM (2003)
26. Paudel, R., Harlan, P., Eberle, W.: Detecting the onset of a network layer dos attack with a graph-based approach. In: FLAIRS Conference, pp. 38–43 (2019)
27. Chaparro, C., Eberle, W.: Detecting anomalies in mobile telecommunication networks using a graph based approach. In: FLAIRS Conference, pp. 410–415 (2015)
28. Paudel, R., Eberle, W., Holder, L.B.: Anomaly detection of elderly patient activities in smart homes using a graph-based approach. In: Proceedings of the 2018 International Conference on Data Science, pp. 163–169. CSREA (2018)
29. Paudel, R., Eberle, W., Talbert, D.: Detection of anomalous activity in diabetic patients using graph-based approach. In: FLAIRS Conference, pp. 423–428 (2017)
30. Bird, S., Klein, E., Loper, E.: Natural Language Processing with Python: Analyzing Text with the Natural Language Toolkit. O'Reilly Media Inc., Sebastopol (2009)
31. Jurisica, I., Mylopoulos, J., Yu, E.: Ontologies for knowledge management: an information systems perspective. Knowl. Inf. Syst. **6**(4), 380–401 (2004)
32. Lehmann, J., Isele, R., Jakob, M., Jentzsch, A., Kontokostas, D., Mendes, P.N., Hellmann, S., Morsey, M., Van Kleef, P., Auer, S., et al.: DBpedia-a large-scale, multilingual knowledge base extracted from wikipedia. Semant. Web **6**(2), 167–195 (2015)
33. Sun, Y., Wong, A.K., Kamel, M.S.: Classification of imbalanced data: a review. Int. J. Pattern Recogn. Artif. Intell. **23**(04), 687–719 (2009)
34. Krawczyk, B.: Learning from imbalanced data: open challenges and future directions. Prog. Artif. Intell. **5**(4), 221–232 (2016)
35. He, H., Garcia, E.A.: Learning from imbalanced data. IEEE Trans. Knowl. Data Eng. **21**(9), 1263–1284 (2008)

36. Chawla, N.V., Lazarevic, A., Hall, L.O., Bowyer, K.W.: SMOTEBoost: improving prediction of the minority class in boosting. In: European Conference on Principles of Data Mining and Knowledge Discovery, pp. 107–119. Springer (2003)
37. Guo, H., Viktor, H.L.: Learning from imbalanced data sets with boosting and data generation: the DataBoost-IM approach. ACM SIGKDD Explor. Newslett. **6**(1), 30–39 (2004)

Bottom-Up Strategy for Data Retrieval and Data Entry over Front-End Application Software

Cierto Trinidad Rusel$^{(\boxtimes)}$ ⬤ and Bernardo Tello Alcides ⬤

Universidad Nacional Hermilio Valdizán de Huánuco,
Pillco Marca, Huánuco, Peru
`rcierto@unheval.edu.pe`

Abstract. A few people implement "pattern" and "best practices" without analyzing its efficiency on their projects. Consequently, the goal in this article is to persuade software developers that it is worth to make an earnest effort to evaluate the use of best practices and software patterns. For such motive, in this study we took a concrete case system for geographical locations inputs through user interfaces. Then, we performed a comparative study on a conventional method with our approach, named "reverse logistic" to recover results about, by measuring the time that a user spends performing activities when entering data information into a system framework. Shockingly, we had a diminished of 59% within the sum of time went through in comparison to the time went through on the conventional strategy. This result lays a establish for feeding data from the typical final step and search based on string matching algorithms, speeding up the interaction between people and computer response.

Keywords: Data retrieval · UI · Human-machine interaction · Front-end · Back-end · Interaction design · Data entry

1 Introduction

The essential goals of user interface designing is to create user interfaces which are user-friendly, self-explanatory and effective [7, 13].

In our study, we have focused on productivity and efficiency with respect to time spent during data-entry.

Data-entry implies to user tasks involving both input of information data to a computer and computer answer to those inputs.

Inefficiencies caused by poorly designed data entry transactions are so visible that numerous users interface design in clerical jobs deal with data entry [2] and enquiries [6].

The greatest need, however, in current information systems is for enhancing the logic of data entry [4].

Hence, the study presented here deals with data entry algorithms, insofar as conceivable, in the want of attention to their hardware implementation.

© Springer Nature Switzerland AG 2020
K. Arai et al. (Eds.): FTC 2019, AISC 1069, pp. 547–555, 2020.
https://doi.org/10.1007/978-3-030-32520-6_40

Data information can be entered into a computer in an assortment collection of ways. The most effortless type of data entry consists merely of pointing at something. In more complex sorts of data entry is the control of the format of data inputs by the user. In general, in the industrial design [6] and in human–computer interaction, the aspiration is to allow effective operation. Human–Computer Interaction is the study of the way in which computer technology influences human work and activities [3].

For an effective operation and efficiency, in this study, we selected several websites using the coding system for geographical locations, called "Ubigeo".

The remaining sections in this article are organized as follows: In Sect. 2, we state a few assumptions to be true in this article. In Sect. 3, the case is discussed: Traditional method or top down approach for data recovery. Then we also measure the time used by the user when selecting data to input into a system.

Analyzing the nature of each website in connection to the data recovery, the processes through which we acquire data from servers by using queries, we found a common include top-down feature for choosing administrative subdivision: Regions, Provinces, and Districts. In Sect. 4, to test we have created a bottom-up approach.

In Sect. 5, the use of time measured in Sects. 3 and 4 is discussed to show comparability. Finally, in Sect. 6, we conclude.

2   Assumptions

In this section, we assumed the following propositions to be true statement for this study.

With the advent of robustness and reliability of computer power, more work is carried out by the front-end activities in web applications [13].

The only way a person interacts with a computer, tablet, smartphone or other electronic gadget is through user interface [8]. The blind assumption that introducing new techniques must be good. Algorithms for recovering data from the structured database that resembles Google search engines are easily implemented autocomplete in each section of "Ubigeo". This should start from the District, Province, Region and Country [1] and [5]. See Fig. 5.

3   Top-Down Approach for Data Retrieval: Traditional Method

Within the traditional approach, the "Ubigeo" begins from the first level of the geographic location database [11], begins with the country, followed by the Regions as the second level, Provinces as the third level and Districts as the fourth level respectively [2]. See Figs. 1, 2 and 3.

Figure 1 shows the typical data entry related to an identifiable natural person. Let's focus on the ID number, corresponds to level 1, the date of birth at level 2, name of the parents at levels 3 and 4, place of birth at level 5, Region, Province and District at levels 6, 7 and 8, respectively. In Fig. 2, we focus on the identifiable person to record the transaction on since starting point of the merchandise to be transported and in

INGRESE SUS DATOS:

Número de DNI:	
Fecha de Nacimiento:	Día ▼ Mes ▼ Año ▼
Nombre Padre:	
Nombre Madre:	
Lugar de Nacimiento:	Peru ▼
Departamento de nacimiento:	LIMA ▼
Provincia de nacimiento:	LIMA ▼
Distrito de nacimiento:	SAN ISIDRO ▼

Fig. 1. Traditional data entry for geographical locations

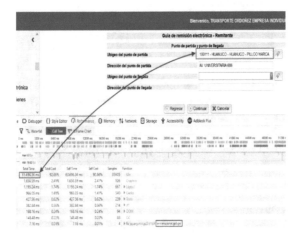

Fig. 2. Traditional data entry for geographical locations

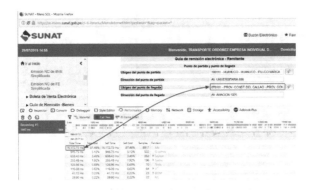

Fig. 3. Traditional data entry for geographical locations

Fig. 3, since the point of arrival of the merchandise to be transported. In both Figs. 2 and 3, the identifiable person uses the traditional approach of "Ubigeo", starting from the levels; Region, Province and District with a drop-down Combo-Box, where it allows the user to select and write the text of an item in the list that is displayed. Obviously, for each level of selection, there is a query to the database that retrieves a subset of place names and fills the combo box of the next level, after using said data recovery to complete the fields selected by the user, generating long time, expressed in milliseconds [14]. See Tables 1 and 2 and Fig. 4.

Table 1. Time spent for each level of geographic location using traditional method.

Geographical indications (Ubigeo)	Time with milliseconds
Country (País)	40 ms
Region (Departamento)	30 ms
Province (Provincia)	40 ms
District (Distrito)	21 ms
Total	**131 ms**

We have measured the number of milliseconds elapsed for each level of selection as shown in Tables 1 and 2, we show the descriptive statistics too. As it is shown we need on average 131 ms for our complete search. A bar graph is also provided below in Fig. 4.

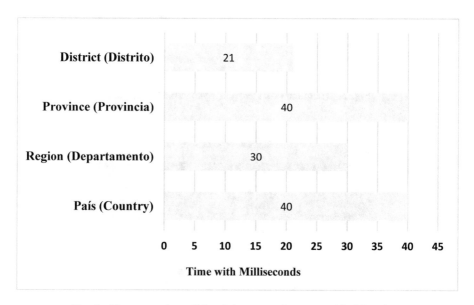

Fig. 4. Time spent in traditional data entry for geographical locations

Table 2. Descriptive statistics shown (see Table 1).

Time with milliseconds	
Mean	32,75
Standard error	4,571196051
Median	35
Mode	40
Standard deviation	9,142392101
Sample variance	83,58333333
Kurtosis	−1,741499331
Skewness	−0,768499333
Range	19
Minimum	21
Maximum	40
Sum	131
Count	4
Confidence level (95.0%)	14,54758598

4   Proposed Method: Reverse Logistic for Data Retrieval

In this section, we present a new way of thinking when developing software for data recovery: Starting from the root to the first level (District, Province, Region and Country) [2].

We advocate an upward approach when retrieving data from the geographic location database. For example writing from the root of the location. We only need any graphic control element that allows the user to enter text information into the text box, combined drop-down box. These will always be available in all programming languages and web design forms [2].

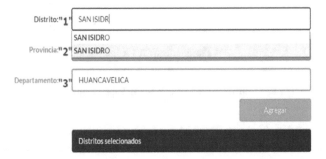

Fig. 5. Reverse logistic for data retrieval during data entry for geographical locations

Figure 5 shows the implementation of our approach. We start with the lowest level entry, in this case District (Distrito). The program displays immediate results below the edit area as we type the text. As we begin to type the search, it anticipates what we are looking for and begins to show results for our search [10].

Then, once we choose the item, the remaining level of search will immediately be populated with the correct data. Additionally, the underlying algorithm looks only for clues, to give us back a list of choices, from which we decide exactly what we want to select, for this search is string matching based algorithms [8].

In Table 3 and Fig. 6, we can see the time spent for each level of geographic location using reverse logistic method.

Table 3. Time spent for each level of geographic location using reverse logistic method.

Geographical indications (Ubigeo)	Time with milliseconds
Country (País)	24 ms
Region (Departamento)	06 ms
Province (Provincia)	06 ms
District (Distrito)	18 ms
Total	**54 ms**

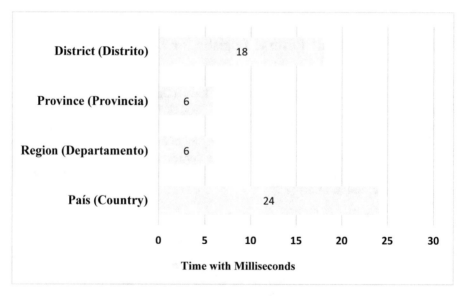

Fig. 6. Time spent reverse logistic for data retrieval during data entry for geographical locations.

In Sect. 3, inefficiencies emerge from running a query processes in each level of geographic location, yielding various separate sets of geography location [11].

5   Drawing a Comparison Between Traditional Method Data Retrieval and Reverse Logistic and Its Justification

In this section, we examine the result of the previous past two sections and clarify the timesaving result. First, from Table 1 we show the result of the traditional method of "Ubigeo" and in Table 3, we show the reduction of time with the method proposed in this article, where we find the real amount of the decrease in the time used in each case, reducing by 77 ms; then 131 ms was reduced to 54 ms. This number corresponds to 59% of the total time used in the traditional method of data entry and recovery. In the Eq. 1 [2], we can see the time reduction demonstration spent for each level of geographic location using reverse logistic method.

$$(0, 59)(131) \cong 77 \tag{1}$$

This clearly informs us that less time is spent on the front-end when a user's input data uses our reverse regression approach. We lower the time by 59%. In addition, when entering data into the system using our logistic reverse regression approach, the information appears instantaneously instead of slowly descending from the top of the hierarchical geographical locations. For many years, we have been using the traditional search for geographic location from top to bottom almost without thinking because it has become a habit, pattern or recommended practice at some time in the past. It has justification in the first computational capabilities when the computer was much slower than recent versions. In that sense the search for geographical location must be starting from the last level (District, Province, Region and Country). Where Table 4 shows the Descriptive Statistics (see the frequency Table 3).

Table 4. Descriptive statistics shown (see Table 3).

Time with milliseconds	
Mean	13,25
Standard error	4,422951503
Median	11,5
Mode	6
Standard deviation	8,845903006
Sample variance	78,25
Kurtosis	−3,114087109
Skewness	0,513224721
Range	18
Minimum	6
Maximum	24
Sum	53
Count	4
Confidence level (95.0%)	14,07580567

From this perspective, patterns and best practices can become obsolete or irrelevant after a time. It happens so faster with the increasingly rapid pace of technological development towards the end of the twentieth century and our present century.

To make stuff worse, developer had not learnt how to deal with these rapid changes in their undergraduate programs. We are certain with this result since it has a justification in Aristotle's Law of Identity. As within the standard terminology in database theory, each element is called entity. Thus, each geographical location is an entity. Therefore, it has a sustenance in Aristotle's Law of Identity where each entity exists as something and peculiar, particular or specific [2]. In our proposed approach we begin with the entity at the lowest level to recover information for the highest level.

6 Conclusions

By changing the logic of the data entry algorithm, we have reduced the time taken by the user to enter data into a database system. It can be deduced that they have revealed how patterns and best practices in software engineering can become obsolete over time due to rapid improvement in bandwidth and programming language performance [14].

Therefore, it is worth making a great effort to evaluate the use of best practices and software patterns in your projects before implementing them, according to the advent of an increasingly powerful computer. It allows us a new method or practice, to experience a change from what we are doing now to do something different, it may even be the reverse way.

The research focuses on the time efficiency aspect of the human-computer interaction in this study [13]. The other aspects, such as intuitive, easy to use, self-explanatory, require behavioral sciences, design, etc. which is past the reach of four knowledge (Philosophical, empirical, scientific and intuitive) taken as a reference [9].

Knowing the aforementioned area, can lead us to further examine efficiency and flexibility. The research aims to demonstrate that, if it is possible to reduce the response time, generating savings in transmission costs, computation costs [12], as shown (see Table 3) less processing time by the human operator, generating lower consumption of electrical energy in each computing device, avoiding global warming with the lower consumption of electrical energy and unnecessary computing costs.

Then we must ask ourselves, how should the bottom-up strategy be designed for data recovery input on the front-end application software? The answer to these questions enhances future longitudinal research.

References

1. Brin, S., Page, L.: The anatomy of a large-scale hypertextual web search engine. Comput. Netw. ISDN Syst. **30**(1–7), 107–117 (1998). https://doi.org/10.1016/S0169-7552 (98)00110-X
2. Trinidad, R.C., Tello, A.B.: Estrategia de recuperación de datos desde el último nivel reduciendo el tiempo de registro en el software. Rev. Bol. Redipe **7**(12), 201–206 (2018)
3. Dix, A., et al.: Human-Computer Interaction. Pearson, Harlow (2003)

4. Fender, J., Young, C.: Front-End Fundamentals. Leanpub (2014)
5. Grehan, M.: How Search Engines Work. Incisive Interactive Marketing LLC. 57 (2002)
6. Helfand, J.: Paul Rand: American Modernist. William Drenttel, New York (1998)
7. Johnson, J.: Designing with the Mind in Mind Simple Guide to Understanding User Interface Design Guidelines. Morgan Kaufmann Publishers Inc., San Francisco (2014)
8. Loudon, K.: Developing Large Web Applications: Producing Code That Can Grow and Thrive. O'Reilly Media Inc., Sebastopol (2010)
9. Meichtry, Y.J.: The nature of science and scientific knowledge: implications for a preservice elementary methods course. Sci. Educ. 8(3), 273–286 (1999)
10. Pantano, E., et al.: Internet Retailing and Future Perspectives. Taylor & Francis, Milton Park (2016)
11. Payne, L.D., et al.: A location-aware mobile system for on-site mapping and geographic data management. In: Proceedings of the 10th ACM Conference on SIG-Information Technology Education - SIGITE 2009, Fairfax, Virginia, USA, p. 166. ACM Press (2009). https://doi.org/10.1145/1631728.1631773
12. Sun, H.-M., et al.: An authentication scheme balancing authenticity and transmission for wireless sensor networks. In: 2010 International Computer Symposium (ICS2010), Tainan, Taiwan, pp. 222–227. IEEE (2010). https://doi.org/10.1109/COMPSYM.2010.5685515
13. Tidwell, J.: Designing Interfaces: Patterns for Effective Interaction Design. O'Reilly Media Inc., Sebastopol (2005)
14. Turban, E., et al.: Information Technology for Management: On-Demand Strategies for Performance, Growth and Sustainability. Wiley, Hoboken (2018)

A Tickless AMP Distributed Core-Based Microkernel for Big Data

Karim Sobh[✉] and Amr El-Kadi

American University in Cairo AUC, Cairo, Egypt
{kmsobh,elkadi}@aucegypt.edu
https://www.aucegypt.edu/

Abstract. Operating system kernels are designed to manage the underlying hardware resources of an environment and avail them to applications in a unified secure way. General purpose kernels perform average over different diversified application profiles and workloads. Their policies are not designed to be a perfect-fit for Big Data applications. In this paper, we present a special purpose tickless AMP distributed kernel designed with Big Data in mind. Design goals and decisions are presented together with the proposed kernel architecture and main subsystems. A prototype of the proposed kernel was used to conduct experiments for assessing the performance gain in Big Data applications in comparison with the open source Linux kernel. Experiment results show a gain in performance that ranges from 1.52 to 4.89 folds based on the target application profile and workload. The new proposed kernel was able to achieve 2.34 folds speed up over Linux/Hadoop using the TeraSort benchmark. Finally, a factorial ANOVA and regression analysis was conducted to assess the different factors affecting the gain in performance, and we were able to show and quantify the positive influential effect of our proposed new kernel on the performance.

Keywords: Asymmetric multiprocessing · Tickless kernels · Microkernel ·
Distributed kernels · Multi-core kernels · Big data · MapReduce ·
Inter-Processor Interrupts (IPI) · Network protocols

1 Introduction

Big data processing has increasing demands for consolidated processing power. Currently, the approach adopted in big data processing is distributed processing over clustered commodity hardware running network-based operating systems, such as Linux, FreeBSD, and other UNIX based systems. Such operating systems are general purpose systems designed to function with average performance across all applications' profiles [8]. Consequently, the performance is not optimal for big data processing due to the fact that buffer pools, file system, process management, and inter-process communication are not designed with big data in mind [7,38,44].

Most general-purpose operating systems adopt a Symmetric Multi-Processor (SMP) architecture, yet we will show that designing a special-purpose distributed Asymetric Multi-Processor (AMP) multi-core microkernel specialized in Big Data processing can

© Springer Nature Switzerland AG 2020
K. Arai et al. (Eds.): FTC 2019, AISC 1069, pp. 556–577, 2020.
https://doi.org/10.1007/978-3-030-32520-6_41

boost the performance and achieve more resource utilization across the whole processing environment. Moreover, virtualization support enhances scalability through virtual machine isolation, and the usage of such microkernel in conjunction with existing traditional kernels to serve tasks that it excels in [8, 16, 29].

In this paper, we present a new distributed multi-core microkernel designed specifically for Big Data. A set of experiments were performed to demonstrate gain in performance based on different workload profiles. Moreover, factorial ANOVA [21] and regression analysis are presented to quantify the effect of different experimental factors on the execution time, and predict execution time for setups that we could not apply experimentally.

In Sect. 2, we present the background followed by the related work in Sect. 3. The problem characterization and motivation are presented in Sect. 4. The details of our new microkernel are presented in Sect. 5. The experiments and their results are presented in Sect. 6 followed by analysis and discussion in Sect. 7. Finally, we conclude and present our future work in Sect. 8.

2   Background

Most modern general purpose operating systems are designed to achieve multi-programming, timesharing, and interactivity which were all inherited from CTSS, MULTIICS, Titan, and maturely reintroduced by Dennis Ritchie and Ken Thompson at Bell Labs in 1969 through their first UNIX version implementation. UNIX was initially designed to share a single CPU time among different processes through time slicing mechanisms. Moreover, a running process attempting to perform an I/O operation yields the CPU for another one ready to run, which cuts down the CPU idle time through interleaved I/O and processing [10, 32]. Despite this enhanced overall system throughput in interactive workloads, considerable performance degradation in Big Data long running batch tasks is experienced.

One way to achieve time sharing and interactivity in modern general purpose kernels is through interrupt source availed via hardware granting control to the kernel periodically, by switching from user-mode to kernel-mode, and allow switching to another process based on some scheduling logic. Examples of such hardware components are the Programmable Interrupt Controller (PIC), the Programmable Interval Timer (PIT), and the Advanced Programmable Interrupt Controller (APIC) available in Intel/AMD processors and supported by Intel and IBM PC [41].

The adopted interrupt timer-based model is very convenient to give the illusion of concurrent execution of processes in single CPU environments. When multi-core architectures started to prevail, general purpose kernels were extended to provide support through a Symmetric Multi-Processing (SMP) approach for simple and smooth migration with minimal modification to the target kernel architectural design. Each CPU runs the same kernel image with an extra extension module that allows coordination between different CPUs through Inter-Processor Interrupts (IPIs). Another alternative is Asymmetric Multi-Processing (AMP) kernels which assign CPUs different roles and tasks, yet they require major fundamental changes in the design and implementation of already existing kernels and hence were not widely adopted [8, 24, 25, 34].

3 Related Work

A Linux kernel extension implementing a tickless capability in idle time, called dynticks, is presented by Siddha et al. in [36] for better power utilization. A tickless kernel is a one that eliminates or at least reduces the number of interrupts generated by the hardware timer, and hence save the CPU cycles exhausted in servicing such timer interrupts. Dynticks maintains a quite kernel across all available cores and uses the advanced HPET timer to extend the sleep time of the CPUs while being idle. The HPET can be programmed as per CPU avoiding the broadcast to wake up all CPUs.

Linux kernels started supporting tickless operations as of version 3.10 [4]. There are two modes for tickless operations. The first mode will stop the timer on idle CPUs. The second mode is achieved by integrating dynticks [36] into the Linux kernel. Dynticks disables the timer on a specific core as long as there is a single process running. Practically, the timer cannot be fully disabled, but needs to fire at a much lower rate. The tickless kernel extension has no insight into the application logic and requires a lot of administration work to set processes affinity. Moreover, spawning light weight processes by data processing engines will bring the timer back. The tickless Linux extension was also integrated into the Linux Kernel for IBM Mainframe Z/OS which is deployed in logical partitions implementing application servers on Linux for System Z [27].

Yuan et al. presented GenreOS in [42] as an AMP kernel for multi-core systems. GenreOS partition cores into application cores, interrupt cores, and kernel cores. It implements the Linux Kernel interfaces for comparability although its main goals are performance and scalability. The main characteristics of GenreOS are running applications on dedicated cores with minimal interruption, kernel services implemented as a serial process to eliminate contention, and the use of a slim scheduler policy with low overhead. Experiments presented a considerable gain in performance over Linux with database and web server workloads. The distributed execution dimension was not introduced in GenreOS.

Considerable work has been done in the field of distributed microkernel operating systems, the most famous is the distributed operating system Amoeba that operate on a set of computing nodes to achieve a single system image of the collective resources of the nodes. Mach is another example of distributed microkernels, evolved from a monolithic one and developed for VAX running four CPUs multiprocessor [23,39].

L4 is a second generation high performance microkernel designed to work on the ix86 processor family and written entirely in assembly. L4 only handles processes preemption, without the need of threads or scheduling features, and runs I/O in user space. The GNU HURD has migrated from Mach to the L4 microkernel to eliminate heavyweight execution mechanisms [3,11,18].

4 Problem Characterization and Motivation

There are two main computing problems that act as a brick wall for the development and evolution of Big Data processing. The first problem is that Moores law [22] has lately been questioned and will most probably neither hold valid in the present nor the future due to physical limitations in the integrated circuits industry. We will not be able to achieve exponential growth of single processors by adding more silicon [9,35,40], and thus multi-core architectures are more likely to be adopted. The second problem is the nature and amount of data being collected and processed, which exceeds the storage and processing capacity of a single computer [13,37], even if it were a super computer, and hence scalability can only be achieved through distributed execution.

General purpose kernels are designed to provide best-effort performance for a wide range of application profiles. They do not have enough insight into the applications logic, needs, and requirements, rather they conduct default management mechanisms which result in performance loss and poor resource utilization [7,8,44]. The inefficiency of general purpose kernels in the domain of Big Data is due to the generality adopted in scheduling, file systems, memory management, buffer pools, and interprocess communication [38].

Two main reasons are the basis of this problem, namely SMP architecture adopted by general purpose kernels, and unneeded overhead due to generality. Using SMP timer-based kernels in Big Data is not the optimal approach to adopt. In the situation of having one CPU a timer is a must, but with the introduction of multi-core architectures not all CPUs need to have the timer enabled. Some processors can be assigned scheduling tasks, in which case the timer is essential, while other processors can be assigned data intensive tasks without interruption. In that case, data processing intensive cores can still be preempted via IPIs initiated by scheduler cores upon predefined task deadline expiration [5,31,34].

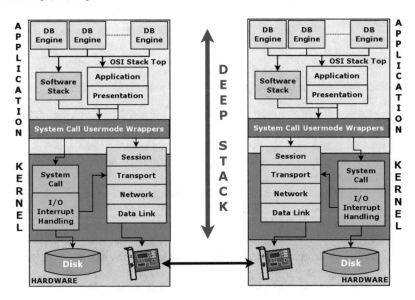

Fig. 1. Deep system stack

Another source of overhead is the deep stack of general purpose kernels and their system libraries, deployed to achieve application portability through respecting the

POSIX standard (Portable Operating System Interfaces for UNIX) as well as legacy systems. Such deep stack originated from the general purpose objective of such kernels. Figure 1 demonstrates a typical system wide stack of a conventional UNIX based environment. Two remote components of a distributed middleware attempting to exchange messages will traverse the whole system and network stack, going through different layers per single message exchange.

The motivation behind this work is to investigate outcomes of designing a special-purpose AMP distributed multi-core microkernel for Big Data processing. Our approach is to eliminate all possible overheads entailed by general-purpose kernels to cater for different application profiles and workloads, and direct our design goals and decisions to focus on precise well-known Big Data processing models. Moreover, designing special network protocols, that run side by side with the typical network layers, but have high priority for Big Data network operations.

5   Proposed Kernel

Our approach is greatly dominated by the idea that there should be more than one dimension, on the kernel design level, that needs to be tackled to gain a considerable boost in performance in Big Data processing. Focusing on one dimension will not allow us to fully achieve our target goal. In essence, our design goals and decisions aim at orchestrating different design aspects on various architectural fronts to work in harmony toward better performance in addition to being able to function reasonably with other applications.

5.1   Design Goals

The main design goals for the new proposed kernel can be enumerated as:

1. Better processing performance and eliminated overhead of servicing unnecessary interrupts.
2. Data processing tasks should execute with minimal interruption.
3. Single System Image (SSI), with a single CPU/Core as the building block of the target architecture rather than computer nodes.
4. Efficient utilization of system resources, and mainly (mainly CPUs/Cores).
5. Low latency I/O and network transfer.
6. Scalability through the ability to add more computer nodes (incorporating their CPUs/Cores).
7. Ability to assign workload to different CPUs/Cores, irrespective of their location (Location Transparency).
8. An extensible framework to add new kernel services without the need to modify the kernel architecture.

5.2   Design Decisions

The main design decisions we have adopted to achieve the above design goals are:

1. Each CPU/Core runs an independent copy of the kernel image using an AMP model.
2. Each CPU/Core has a single separate address space trying to achieve, as much as possible, a lock-free environment.
3. Each processing CPU/Core can have at most one single running process at any point in time with the possibility of multiple I/O preempted threads.
4. Each CPU/Core can be assigned a specific task and hence a role.
5. Available roles that can be assigned to CPUs/Cores are initially Management, Disk I/O, Network I/O, and Processing. New roles can be added in the future based on the needs.
6. A single CPU/Core can be assigned more than one task/role as long as they do not contradict.
7. Only management CPUs/Cores will have the timer enabled.
8. Within a single computing node there should be at least one CPU/Core assigned the management role and hence the timer will be enabled on that CPU/Core.
9. A CPU/Core should be assigned to each available physical network interface.
10. Multiple network interfaces can be assigned to the same CPU/Core.
11. CPUs/Cores communicate and synchronize over Inter-Processor Interrupts (IPIs) irrespective of their location, hence IPIs should be extended to work over the network and delivery guarantees need to be implemented.
12. Messages between different microkernel services are delivered over IPIs which encapsulate extensible messages that can be customized based on the service type.

5.3   Proposed Kernel Architecture

The architecture aims mainly at consolidating distributed compute nodes connected via a very high speed network and considering their collective cores as a pool of cores to be used in different tasks. The physical boundaries should not be perceived by the running applications through transparently utilizing the underlying network. The applications should utilize the underlying hardware transparently as a consolidated single computing power that has many CPUs/Cores and each of which has access to memory and hardware resources.

The proposed microkernel architecture is build on top of a set of subsystems that coherently collaborate together in a loosely coupled approach for maximum scalability. The different subsystems are an AMP microkernel framework, IPI over Ethernet (IPIoE) network protocol that incorporates a gossip protocol for resource updates, a BareMetal Operating System Markup Language (BOSML) to encapsulate microkernel messages, and a Fast Data Transfer Network Protocol (FDTNP) designed for distributed data processing.

AMP Microkernel Framework

The AMP microkernel composed of a number of initial federated services that can be extended. A proper set of services are started on target CPU/Core based on its role. Microkernel services communicate over extensible messages that are exchangeable between cores. Services can use each other and a considerable portion of them are implemented in the user-mode. The main supported services are registry service, memory service, gossip service, monitor service, network service, file system service, and shell service.

The **registry service** is run by management cores and is responsible for keeping record of the CPUs/Cores irrespective of their location, local or remote. Since it is handled by management cores, at least one registry service will be running per node. The registry service utilizes the gossip service to collect data about the environment resources. CPUs/Cores can join and leave dynamically at runtime. A very important duty of the registry service is to resolve CPUs/Cores addresses. Each CPU/Core is given a universal hash which resolves to a CORE ID and a Network MAC hardware address. A table is maintained by the registry service that contains the CORE IDs, their corresponding ID and MAC addresses, and their role.

Exactly one instance of the **memory service** should exist within a node. The memory service is responsible for maintaining a physical frames availability bitmap. The memory service can be invoked by different CPUs/Cores to allocate and deallocate physical frames based on memory needs by applications running on them. The actual page mapping is performed by the requesting core.

The **gossip service** is responsible for broadcasting CPU/Core information on the network to other compute nodes. It is also responsible for collecting gossip broadcast messages and availing them to the registry service to keep it up to date. The gossip service also broadcasts information about messages exchanged between different CPUs/-Cores that it is aware of. To eliminate network contention and overhead, not everything is being broadcast, rather only important events that inform registry services about the availability of resources; e.g. a processing core is assigned a task or finished a task. Gossip messages are encapsulated in a single network packet to avoid the need for flow control and a markup message format is adopted to allow for extensibility. The gossip service is run by a management core and utilizes the network service and IPIoE.

The **monitor service** is essentially handled by a management CPU/Core as it requires the timer to be enabled. One of the main design goals is that Big Data processing tasks should execute with minimal interruption, and hence processing CPU/Cores will essentially have the timer disabled. A problem is very likely to arise if something goes wrong while a data processing task is executing as there is no way that it can pre-empt itself. To countermeasure that, a Big Data processing task registers to the monitor service with some estimate deadline and some termination conditions. If the task fails to finish execution before the deadline its CPU/Core will be interrupted by the monitoring service using IPIoE. If the task needs more time, which can be detected through checking the termination conditions, it will be rescheduled to the same CPU/Core.

The **network service** runs on a network CPU/Core and is responsible for managing the network stack. The network service can be configured to manage the whole network protocol stack, or a subset of it. It is very important to be able to assign a specific protocol to a specific network service running on a specific network CPU/Core and connected to a specific dedicated network interface. For example, in large deployments isolating the IPIoE protocol on a separate network infrastructure will considerably enhance performance.

The **filesystem service** runs on a disk CPU/Core and is responsible for managing a number of filesystem partitions. A filesystem service implements a VFS-like architecture where different filesystems can be registered and managed by a filesystem service. Different filesystem services can be running on different disk CPUs/Cores within the same node to increase disk bandwidth. For example, assigning disk filesystem partitions of ATA controllers, primary and secondary, to different filesystem services allows concurrent I/O if each controller has a different DMA.

The **shell service** is a simple special network service that gives users access to a command line shell through which commands can be issued to start execution of applications.

IPI over Ethernet (IPIoE)

An IPI is an interrupt initiated by one Core/CPU to cause another one to execute an Interrupt Service Routine (ISR). IPIs are guaranteed to be delivered, but the recipient CPU/Core has no clue about the sender. Moreover, IPIs do not carry any data, and one way to associate data with it is through some form of shared memory.

The IPIoE is a subsystem that allows exchanging messages between different CPU/-Cores in a distributed environment. Unlike traditional IPIs which can only be used between cores that belong to the same node, IPIoE can cross the boundaries of the physical hardware. The IPIoE protocol is located in the OSI model just right above the data link layer side by side with the ARP protocol to have higher priority. As any network protocol, each IPIoE packet is composed of a header and a payload. The IPIoE protocol is designed to work on local network as addressing is done using the network MAC address coupled by the CPU/Core APIC Id fetched from the ACPI tables.

Upon a CPU/Core attempting to send an IPI, the IPIoE subsystem will encapsulate the IPI in an IPIoE network frame to be delivered over network in case of a remote recipient and over shared memory in case of a local one in a transparent manner. This allows encapsulating data, of maximum 1400 bytes, into the packet payload. Another benefit of such model is that we can encapsulate the sender information in the IPIoE packet header. The main constraint is that all IPIoE messages need to fit into an ethernet frame.

IPIoE provides an infrastructure for some synchronization models that can be utilized by distributed applications to implement dependability. For example, a Client/Server Request/Reply model can be achieved to implement remote procedure calls. Also a scatter/gather model can be implemented over IPIoE to achieve an MPI-like execution model. IPIoE is used by the gossip service to broadcast and collect data about distributed CPUs/Cores as well as available resources. Moreover, IPIoE is used to exchange messages between different microkernel services running on different CPUs/-Cores irrespective of their location.

BareMetal OS Markup Language (BOSML)

BareMetal Operating System Markup Language, BOSML, is an extensible markup language that is used for encoding messages superimposed over IPIoE. BOSML allows for maximum extensibility and interoperability. A new user-mode kernel service can be added and all what is needed is to define a BOSML schema as an interface for other services to access it.

Fast Data Transfer Network Protocol (FDTNP)

Big data processing needs a special protocol to transfer data between different compute nodes. A specialized data transfer protocol can enhance the performance through eliminating overhead resulting from generality of protocols like TCP and UDP. In Big Data applications, the workload profile we are catering for adopts a mechanism that follows the order of sending data, processing it, and sending it back to store results. A major feature that we do not really need is synchronicity. Also the size of the data to be sent within a communication session is known beforehand. Moreover, we assume in our architectural design the availability of a high-speed underlying local network with minimal packet loss. A network protocol that can utilize such opportunities will try to send as much data as possible and group acknowledgments in one single packet. The FDTNP allows the establishment of a communication channel between two cores and is designed to work on a local network with addressing built on top of the hardware MAC coupled with the APIC ID of the communicating cores.

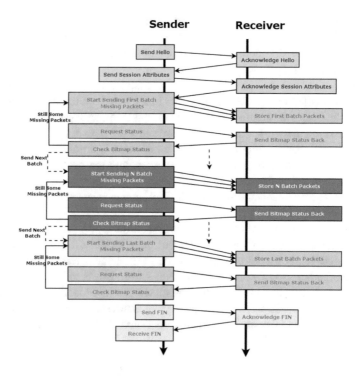

Fig. 2. FDTNP workflow

Figure 2 demonstrates the FDTNP protocol designed such that a sender establishes a connection session with a receiver using 3-way handshake through which agreements between the two parties are established. The size of the data to be transmitted and the batch size in number of packets are predetermined by the sender and sent during the handshake phase. The sender keeps on sending packets with sequence numbers to complete a batch before it sends the receiver asking about the status. When the receiver receives the status packet it replies with one packet that contains a bit map indicating which packets were received and which were lost. The sender will keep on sending the missing packets until it receives a short message from the receiver in response to a status message indicating the full reception of the current batch. Subsequent batches are transmitted until all the data is transferred after which the connection is closed. A 2-way connection termination mechanism is used if any party would like to terminate a connection session before the whole data is transferred.

To achieve very high priority in processing FDTNP packets, FDTNP is located just right above the data link layer side by side with the ARP and IPIoE protocols. FDTNP packets are composed of a header and a payload. The header is guaranteed not to exceed 100 bytes leaving 1400 bytes to be used as a payload. To be able to acknowledge a batch with a single packet, a batch cannot contain packets more than what can be represented by the bitmaps that a single packet can store, which is 11200 packets.

Overall Architecture Workflow
Figure 3 demonstrates the overall architecture of the proposed distributed AMP kernel. Each node contains a number of CPU/Cores, assigned different roles. Essentially, management cores have the timer enabled. Based on a CPU/Core role, different services are started on it. Services running on different CPUs/Cores exchange messages using IPIoE. Services running on CPUs/Cores located in the same node exchange IPIoE over shared memory. On the other hand, remote cores communicate over ethernet frames using the distributed IPIoE infrastructure.

The Detailed View box in Fig. 3 demonstrates more details about the internal kernel architecture within a single core. Each core has its own single page table achieving single address space per CPU/Core which will eliminate TLB flushing. The gossip service communicates with the IPIoE through local microkernel message. The IPIoE and the FDNTP services access their corresponding network layers directly bypassing the OSI network stack to give IPIoE and FDNTP a higher priority.

5.4   Kernel Implementation

We built a prototype of the proposed kernel based on x86_64 baremetal implementation that is built from scratch to optimize all the operations in favor of Big Data processing. We used the APIC high precision timer and APIC/IO to be able to route I/O interrupts to different CPUs/Cores based on their target roles. Each CPU/Core has a single PML4 page table with the kernel space being mapped with ring-0 access protection.

A full network stack is implemented supporting ARP, ICMP, IP, TCP, and UDP network protocols. Moreover, a prototype version of the IPIoE is implemented with features needed to run our target experiments, as well as a full implementation of FDNTP. The kernel prototype supports ATA interfaces as well as Intel e1000 network drivers. Moreover, a simple filesystem is implemented to store data within a VFS-like filesystem hierarchy.

A set of system calls are implemented to enable a running application to read and write data from disk and network from the userspace. The system call wrappers are designed to run most of its logic in user-mode through implementing most of the kernel policies in user-mode, and only switch to kernel-mode when needed; to access the hardware or to send microkernel messages over IPIoE.

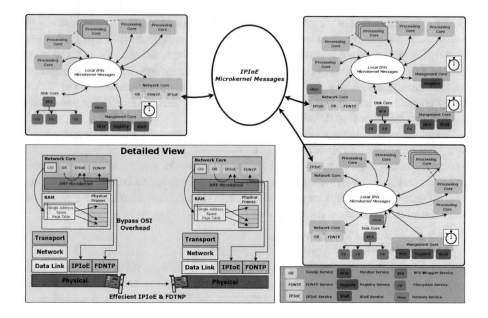

Fig. 3. Overall architecture workflow

When the kernel boots up the bootstrap processor (BSP), it reads configuration files and starts up different application cores (AP) from a special trampoline that assigns each core its role based on the target configuration. The AP CPUs/Cores register their information to the registry service when they are fully started.

The kernel prototype is designed to run from within a virtual machine as we did not want to invest much time trying to support diversified hardware through writing a lot of device drivers.

6   Experiments

In this section, we present the results of three experiments that were designed to compare our proposed kernel against the open source Linux kernel and to estimate the boundaries and range of performance gain among different workloads. The first two experiments target measuring the performance gain of applications that can run on an isolated single CPU/Core. This allows us to measure the speed gained by having a controlled tickless kernel to be able to estimate the maximum and minimum possible gain in performance. The third experiment is designed to measure the performance gain in a real distributed networked Big Data application to test the applicability of the proposed kernel in a real life application.

6.1   Word Count

The first experiment is the simplest one and it shows the best performance gain possible. A simple C program is used to count the words in a text file. Words are defined as tokens separated by white spaces. The program first allocates memory to load the whole content of the target file into. A tight loop scans the buffer, byte by byte, incrementing a counter upon encountering a white space. The program is designed to isolate the I/O and memory allocation from processing of the memory buffer.

Listing 1.1 shows the source code of the word count application used in both environments. Basically we were keen to only measure the execution time of the loop from line 19–21 on Linux compared to our proposed kernel using different data sizes. We used the taskset command in case of Linux to attach the program to a specific CPU/-Core. Figure 4 shows the gain in performance over different file sizes ranging from 1 GB to 24 GB. An average of 4.89 folds was achieved by the tickless kernel over Linux.

```
1  int main(int argc, char **argv) {
2     if ( argc != 2 ) {
3         printf ("usage: %s <path to file>\n",argv[0]);
4         exit(1);
5     }
6     FILE * f = fopen (argv[1],"rb");
7     if ( f != NULL) {
8         fseek(f,0,2);
9         uint64_t fsize = ftell(f);
10        fseek(f,0,0);
11        char * buffer = (char *)calloc(fsize+10,sizeof(char))
               ;
12        fread(buffer,1,fsize,f);
13        fclose (f);
14        uint64_t word_count = 0;
15        time_t t1 = time(NULL);
16        uint64_t i = 0;
17        for ( ; buffer[i] != 0 ; i ++)
18            if ( buffer[i] == ' ' ) word_count ++;
19        time_t t2 = time(NULL);
20        free(buffer);
21        printf ("counted: %lu/%lu words in %lu\n",word_count,
               i,t2-t1);
22    }
23    else printf("Error opening text file\n");
24    return 0;
25 }
```

Listing 1.1. Word Count

This is the maximum speed up that we were able to achieve in all experiments and we attribute that to the fact that the workload is sequential and memory is being read spatially in a cache friendly way that allows readahead to maximize performance by reducing CPU/Core stalls that usually results from cache misses. Moreover, the word count application has minimal memory updates and writes which eliminates any memory contentions and cache updates. Finally, we have excluded all I/O and ring- levels switches from our measurements by performing all I/O at once before starting the processing.

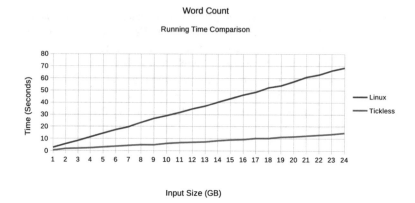

Fig. 4. Word count results

6.2   Quick Sort

The second experiment, quick sort, is also chosen to run on an isolated CPU/Core and was customized to sort data generated by TeraGen which is part of the TeraSort benchmark [19,20]. The sort algorithm is written completely in C++ without any dependability on the C++11 STD library. The program was designed in a way to isolate the I/O from the core sort processing. It is designed in a way that was used later to build the TeraSort distributed application for our AMP tickless kernel presented in the next section.

We executed the same C++ quick sort program on both environments on a single CPU/Core with different data sizes to test the gain in performance. The sizes used were 2, 4, 8, 12, 16, 20, and 24 GB. The time recorded accounts only for the sort processing excluding the I/O operations for reading the unsorted data and writing it back after sorting. Figure 5 presents the results of the experiment comparing the two environments and showing a gain in performance of average 1.52 folds.

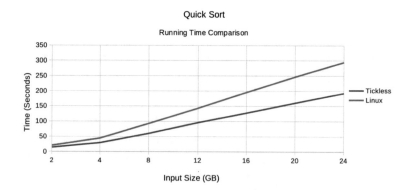

Fig. 5. Quick sort results

The gain in performance is much less than the word count application. We attribute this decrease in performance gain to the aggressive memory access of the quick sort algorithm represented in comparing data from different locations in memory, which results in many cache misses. Moreover, a lot of swaps are performed as part of how the algorithm works which again pressures the cache to perform updates and write backs. This results in considerable amount of CPU/Core stalls that are constant between the two environments.

6.3   TeraSort Benchmark

The third experiment is designed to test the end-to-end run time of a real distributed application, including disk I/O and network transfer. In this experiment we will compare the Terasort algorithm run time over the traditional Hadoop Big Data platform and our proposed AMP tickless kernel deployed over a distributed environment. We used a deployment environment of 4 hardware nodes as per the table in Fig. 6.

Processor	Model	Cache (L2)	Cores/Threads	RAM	VMs
iCore-7	i7-3770 3.40GHz	8 MB	4/8	24 GB (6x4 DDR3 1600 MHz)	5
iCore-7	i7-3770 3.40GHz	8 MB	4/8	24 GB (6x4 DDR3 1600 MHz)	5
iCore-7	i7-3770 3.40GHz	8 MB	4/8	24 GB (6x4 DDR3 1600 MHz)	5
Xeon	E5410 2.33GHz	6 MB	8/8	32 GB (8x4 DDR2 1333 MHz)	6

Fig. 6. TeraSort environment configuration

The whole environment has a total of 24 physical cores, 16 of which supports hyper-threading resulting in 32 virtual core threads, and a total memory of 104 GB of RAM. We used the VirtualBox hypervisor to instantiate 21 virtual machines, each of which has 4 cores and 4 GB of RAM. The total number of virtual cores perceived by the virtual machines is 84 cores, which is much more than the number of physical CPUs/Cores.

We have set up 2 different environments with the same configuration as above. The first environment runs the Linux Debian Distribution with Hadoop and HDFS [2,14,26,43] installed and configured to run Hadoop MapReduce [6] implementation of TeraSort. The Second environment runs our proposed AMP distributed tickless kernel. Each virtual machine has a dedicated management core with the timer enabled, a dedicated core for networking and disk I/O, and 2 worker cores, which sums up to a total of 40 workers. A C++ implementation of TeraSort that integrates the quick sort algorithm presented in subsection 7.2 is used to run the TeraSort benchmark.

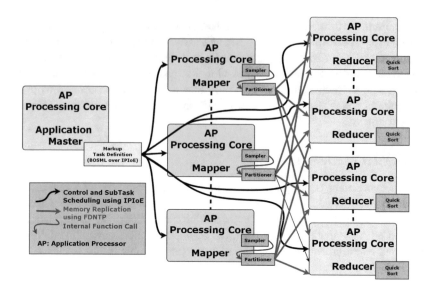

Fig. 7. AMP tickless Kernel teraSort environment

Figure 7 demonstrates our implementation of the TeraSort algorithm which is built up of three components: Application Master, Mapper, and Reducer. There is only one instance of the application master and it runs on a dedicated worker core. A number of mappers and reducers can be launched, each on a dedicated worker core, based on the sort task configuration. All application components communicate and synchronize over IPIoE. In Hadoop there is no control over the number of mappers as it is dependent on where a target input file distributed blocks are stored, but we can configure a TeraSort task to use a specific number of reducers which greatly affects the execution time. In our proposed kernel environment, we fixed the number of mappers to 20 which will reserve 20 cores for that, and we will change the number of reducers in different experiment runs.

In our proposed kernel implementation of TeraSort the application master starts first, issues IPIoEs to start up mappers, assign them their jobs, and provide them with the intended number of reducers so they can split the data accordingly. Each mapper loads its data from the disk storage and applies sampling techniques to identify data partitions. When the mappers finish their tasks they notify the application master using IPIoE messages to start up the reducers and hence the shuffling phase, data partitioning and network transfer over FDNTP, is started to assign each reducer a subset of the data. Each reducer launches its quick sort engine to sort its subset of the data and store its output on disk. The sampling phase carried out by the mappers ensures that the partitions generated by different reducers are organizationally sorted.

We have used TeraGen to generate 9 datasets with different sizes: 2, 4, 8, 12, 16, 24, 28 and 32 GB. We conducted different runs with different data sizes and different number of reducers: 10, 12, 14, 16, 18, and 20 reducers. Some limitations restricted running experiments with all combinations of sizes and reducers; for example, due to memory limitations we could not run the sort on data larger than 12 GB with 10 reducers, and we could not sort 32 GB except with 20 reducers. Figure 8 shows the gain in the execution time by the proposed kernel over Linux/Hadoop using different reducers and input sizes.

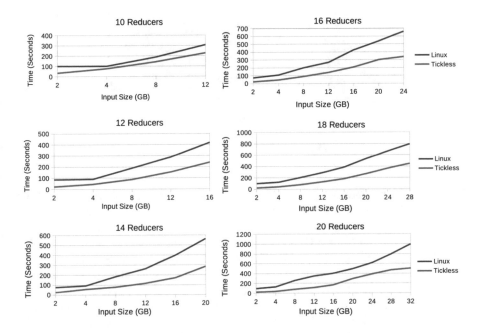

Fig. 8. AMP tickless Kernel vs Linux runtime (Reducers/Input Size)

Figure 9 shows the speed up percentage of the AMP Tickless kernel over Linux/Hadoop for different reducers and input sizes. The performance gain increased as we increase the number of reducers within the same input size. Increasing the input size using the same number of reducers has a negative effect on the performance gain as the physical hardware resources are fixed across all experiment runs. The average speed up of the AMP Tickless Kernel over Linux/Hadoop is 234%.

We were able to isolate and measure the shuffling phase in our TeraSort implementation, yet within the shuffling phase we could not isolate the network transfer time and data partitioning. Figure 10 shows the shuffling speed with respect to the number of reducers used. Up to 18 reducers, the speed increased proportionally with the number of reducers. Speed starts to decrease after that because of the network congestion resulting from communication between larger numbers of reducers resulting in higher packet loss rate. We also attribute this to the target underlying infrastructure used, where 5 virtual machines are running per physical node on average, and bridge over, and sharing, the same physical network interface.

Finally, we have conducted a factorial ANOVA and regression analysis. Our investigated factors are the input size, number of reducers, and the environment; AMP Tickless

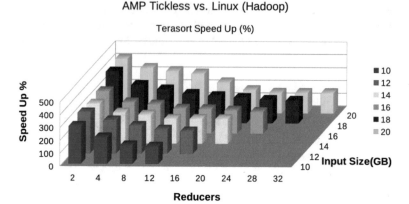

Fig. 9. AMP tickless Kernel speed up

Kernel and Linux/Hadoop. The yield is basically the end-to-end execution time of the overall TeraSort run. Each run in our experiment has fixed factor values and is executed 3 times to collect 3 readings. Figure 11 presents the ANOVA and the regression analysis results. The Normal Q-Q graph demonstrates the normality of the data which is further verified using Anderson-Darling and Shapio-Wilk normality tests comparing the P-Values with significance level of 0.01. The ANOVA results table shows that all the target factors are significant as well as the interaction between the environment and the reducers, and the interaction between the input size and the reducers. More importantly, the mean squares pie plot shows the magnitude of the influence of each factor. Finally, a regression analysis is presented with a very high R-Squared and adjusted R-Squared values which emphasizes that the model can be used to predict results for factor values outside the experiment range.

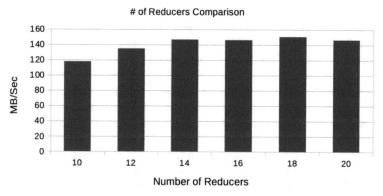

Fig. 10. Shuffling speed

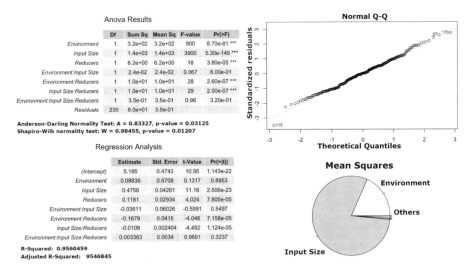

Fig. 11. ANOVA results and regression analysis

Figure 12 demonstrates the factors interaction using 2-D and 3-D graphs. It is evident from the graphs that the environment has a significant effect on the execution time which emphasizes the gain in performance by the AMP tickless kernel over Linux/Hadoop.

7   Analysis and Discussion

It is evident from the experiment results that building a special purpose kernel for Big Data processing achieves considerable performance gain. Taking into account the network transfer and optimizing it to suit distributed Big Data processing has a considerable contribution to the performance gain we observed in the TeraSort experiment over the single processor quick sort experiment.

It is worth mentioning that we were forced to oversubscribe the CPUs/Cores in our test environment because of the limited availability of HW resources. A resource contention is realized over the physical CPUs/Cores assigned to different virtual machines. We anticipate that if the experiments are repeated with dedicated CPU/Cores for each virtual machine a better performance gain can be achieved, which we believe needs to be verified experimentally.

The ANOVA analysis shows that the input size has a huge significant magnitude over the execution time. This might give the illusion of poor scalability, yet the fixed physical resources used explains the inflated impact of the input size. Increasing the virtual resources pressures the underlying fixed physical resources through the aggressive multiplexing of different virtual machines over the limited CPUs/Cores available. We believe that increasing the physical hardware resources in proportion to the workload will keep the proportional gain in performance constant. A larger hardware environment than the one we used will be needed to verify this aspect experimentally. In any

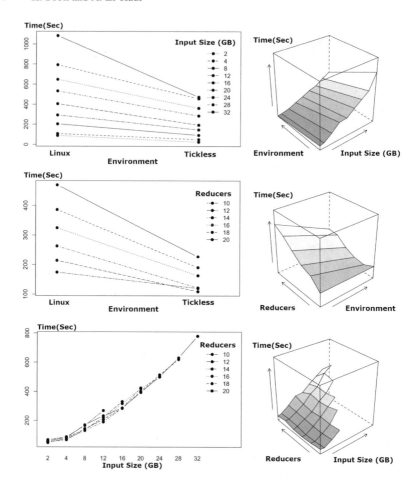

Fig. 12. Factors interaction models

case, the underlying physical environment is common in both cases, AMP Tickless and Linux/Hadoop.

It is evident that the added tickless feature contributed largely to the gain in performance. The major gain of the tickless approach, other than dedicating CPUs/Cores for processing tasks, is the realization of a single address space per CPU/Core which saves up a lot of TLB flushing. Moreover, soft irqs are being handled by the management cores which relieves the processing cores from having to check and handle them.

One major aspect of the success of the proposed AMP tickless kernel is the transparent interaction of the different CPUs/Cores irrespective of their location through the transparent IPIoE messaging mechanism. The TeraSort application was written in a way that the user level logic does not perceive the distributed nature of the underlying environment which realizes a Single System Image (SSI) allowing easy and reliable scalability without any changes to the application logic.

A very valid question is why an already existing kernel was not used and try to modify it in a way that would achieve all of the above. The main problem based on [15,17,27,28,30,33,36] is that by disabling the timers a lot of prerequisites are no longer available. For example, the I/O subsystem in the Linux kernel for example is based on soft_irqs, work queues, and deferred execution of I/O responses; basically you can not do I/O while handling an I/O. Essentially, with no interrupts enabled, there is no irq_exit, and hence there is no soft_irqs. soft_irqs are essential for handling pending software interrupts that sometimes need to be handled on more than one core in an SMP environment. In addition to the challenging time handling, interrupt rescheduling and APIC/IO routing, availing the preconditions to sustain tickless operations is relatively very difficult to achieve. In a general purpose kernel serving a distributed complex application, like TeraSort, it is challenging to ensure that each CPU/Core does not have more than one task at a time.

8   Conclusion and Future Work

In this paper we presented a new special purpose kernel for Big Data processing. Our approach was to tackle the problem of Big Data processing from different angles to be able to maximize the performance gain. A set of design goals and decisions were presented stemming from our belief in distributed AMP kernels where CPUs/Cores need to be dedicated for data processing, and isolating I/O and management tasks on special CPUs/Cores. A new AMP distributed kernel architecture was presented with its different components and a prototype of a subset of the features was presented. Experiments show that our approach achieved considerable gain in performance in different application profiles and workloads. The magnitude of the performance gain was measured to range from 1.52 to 4.89 folds based on the target application, size of data, and dedicated resources available. A considerable gain in performance of average 2.34 folds was achieved in the distributed TeraSort benchmark over Linux/Hadoop.

Our future work will focus on completing all features of the AMP kernel. We are planning to port most of the standard C library instead of the limited functions that we needed to run the experiments. Trying to make our kernel POSIX complaint as much as possible. We believe that we need to run more experiments on larger pools of hardware resources to test the scalability of our AMP distributed tickless kernel. A major dimension that we are planning to dedicate a lot of research for is the security of the communication over IPIoE and FDTNP, as well as securing the microkernel policy that runs in the user mode. Finally, we will concentrate on a real life application to show the real benefits of the new kernel, and currently we are investigating running a bioinformatic distributed protein sequence aligner on the new kernel to measure the gain in performance over the traditional classical Blast [1, 12] sequence aligner.

References

1. Altschul, S.F., Gish, W., Miller, W., Myers, E.W., Lipman, D.J.: Basic local alignment search tool. J. Mol. Biol. **215**(3), 403–410 (1990)
2. Borthakur, D.: HDFS architecture guide (2008)

3. Brinkmann, M.: Porting the GNU hurd to the L4 microkernel. Free Software Foundation Inc. (2003)
4. Corbet, J.: (Nearly) full tickless operation in 3.10, May 2013. https://lwn.net/Articles/549580/. Accessed on 03 Feb 2019
5. De Witte, N., Vincke, R., Landschoot, S.V., Steegmans, E., Boydens, J.: Comparing dual-core SMP/AMP performance on a telecom architecture (2013)
6. Dean, J., Ghemawat, S.: Mapreduce: simplified data processing on large clusters. In: OSDI 2004: Sixth Symposium on Operating System Design and Implementation, San Francisco, CA, pp. 137–150 (2004)
7. Engler, D.R., Kaashoek, M.F.: Exterminate all operating system abstractions. In: Proceedings 5th Workshop on Hot Topics in Operating Systems (HotOS-V), May 1995, pp. 78–83 (1995)
8. Giceva, J., Zellweger, G., Alonso, G., Rosco, T.: Customized OS support for data-processing. In: Proceedings of the 12th International Workshop on Data Management on New Hardware, DaMoN 2016, pp. 1–6. ACM, New York (2016)
9. Guarnieri, M.: The unreasonable accuracy of moore's law [historical]. IEEE Ind. Electron. Mag. **10**(1), 40–43 (2016)
10. Brinch Hansen, P. (ed.): Classic Operating Systems: From Batch Processing to Distributed Systems, 1st edn. Springer Publishing Company, New York (2010)
11. Heiser, G., Elphinstone, K.: L4 microkernels: the lessons from 20 years of research and deployment. ACM Trans. Comput. Syst. **34**(1), 1–29 (2016)
12. Hunter, C.I., Mitchell, A.L., Jones, P.H., McAnulla, C., Pesseat, S., Scheremetjew, M., Hunter, S.: Metagenomic analysis: the challenge of the data bonanza. Briefings Bioinform. **13**(6), 743–746 (2012)
13. Ionescu, B., Ionescu, D., Gadea, C., Solomon, B., Trifan, M.: An architecture and methods for big data analysis. In: Valentina, E.B., Lakhmi, C.J., Branko, K. (eds.) Soft Computing Applications, pp. 491–514. Springer International Publishing, Cham (2016)
14. Jlassi, A., Martineau, P.: Benchmarking hadoop performance in the cloud - an in depth study of resource management and energy consumption, pp. 192–201 (2016)
15. Khan, N., Yaqoob, I., Abaker, I., Hashem, I.A.T., Zakira, I., Kamaleldin, W.M.A., Alam, M.A., Shiraz, M., Gani, A.: Big data: survey, technologies, opportunities, and challenges. Sci. World J. (2014)
16. Kiefer, T., Schlegel, B., Lehner, W.: Experimental evaluation of NUMA effects on database management systems. In: BTW (2013)
17. Klein, J., Gorton, I.: Runtime performance challenges in big data systems. In: Proceedings of the 2015 Workshop on Challenges in Performance Methods for Software Development, WOSP 2015, pp. 17–22. ACM, New York (2015)
18. Le Mignot, G.: The GNU hurd. In: Extended Abstract of Talk at Libre Software Meeting, Dijon, France (2005)
19. Li, S., Supittayapornpong, S., Maddah-Ali, M.A., Avestimehr, A.S.: Coded terasort. CoRR, abs/1702.04850 (2017)
20. MAPR. Terasort benchmark comparison for yarn. https://mapr.com/resources/terasort-benchmark-comparison-yarn/. Accessed on 31 Jan 2019
21. Montgomery, D.C.: Design and Analysis of Experiments, 8th edn. Wiley, New York (2012)
22. Moore, G.E.: Cramming more components onto integrated circuits, reprinted from electronics, vol. 38, no. 8, 19 April 1965, pp. 114. IEEE Solid-State Circ. Soc. Newsl. 11(3), 33–35 (2006)
23. Mullender, S.J., van Rossum, G., Tananbaum, A.S., van Renesse, R., van Staveren, H.: Amoeba: a distributed operating system for the 1990s. Computer **23**(5), 44–53 (1990). https://doi.org/10.1109/2.53354

24. Murata, Y., Kanda, W., Hanaoka, K., Ishikawa, H., Nakajima, T.: A study on asymmetric operating systems on symmetric multiprocessors. In: Proceedings of the 2007 International Conference on Embedded and Ubiquitous Computing, EUC 2007, pp. 182–195. Springer-Verlag, Heidelberg (2007)

25. Narayanaswamy, G., Balaji, P., Feng, W.: An analysis of 10-gigabit ethernet protocol stacks in multicore environments. In: 15th Annual IEEE Symposium on High-Performance Inter-connects (HOTI 2007), Aug 2007, pp. 109–116 (2007)

26. O'Malley, O.: Terabyte sort on apache hadoop (2008)

27. Parziale, L.: Sap on db2 9 for z/os: Implementing application servers on Linux for system z. IBM, International Technical Support Organization (2005)

28. Pavlo, A., Paulson, E., Rasin, A., Daniel, J.A., David, J.D., Samuel, M., Michael, S.: A comparison of approaches to large-scale data analysis. In: Proceedings of the 2009 ACM SIGMOD International Conference on Management of Data, SIGMOD 2009, pp. 165–178. ACM, New York (2009)

29. Psaroudakis, I., Scheuer, T., May, N., Sellami, A., Ailamaki, A.: Scaling up concurrent main-memory column-store scans: towards adaptive NUMA-aware data and task placement. In: The Proceedings of the VLDB Endowment, vol. 8, no. 12, pp. 1442–1453 (2015)

30. Rabl, T., Jacobsen, H.-A., Mankovskii, S.: Big data challenges in application performance management. In: XLDB 2011, Middleware Systems Research Group. http://msrg.org

31. Regehr, J., Stankovic, J.A.: Augmented CPU reservations: towards predictable execution on general-purpose operating systems. In: Proceedings Seventh IEEE Real-Time Technology and Applications Symposium, May 2001, pp. 141–148 (2001)

32. Ritchie, D.M., Thompson, K.: The unix time-sharing system. Commun. ACM **17**(7), 365–375 (1974)

33. Santosa, M.: Choosing the right timer interrupt frequency on Linux. Skripsi Progr. Stud. Sist. Inf. (2010)

34. Scogland, T., Balaji, P., Feng, W., Narayanaswamy, G.: Asymmetric interactions in symmetric multi-core systems: analysis, enhancements and evaluation. In: SC 2008: Proceedings of the 2008 ACM/IEEE Conference on Supercomputing, November 2008, pp. 1–12 (2008)

35. Shalf, J.M., Leland, R.: Computing beyond Moore's law. Computer **48**(12), 14–23 (2015)

36. Siddha, S., Pallipadi, V.: AVD Ven. getting maximum mileage out of tickless. In: Proceedings of the Linux Symposium, vol. 2, pp. 201–207. Citeseer (2007)

37. Slagter, K., Hsu, C.-H., Chung, Y.-C.: An adaptive and memory efficient sampling mechanism for partitioning in mapreduce. Int. J. Parallel Prog. **43**(3), 489–507 (2015)

38. Stonebraker, M.: Operating system support for database management. Commun. ACM **24**(7), 412–418 (1981)

39. Andrew, S.T.: Distributed Operating Systems. Prentice-Hall Inc., Upper Saddle River (1995)

40. Theis, T.N., Wong, H.P.: The end of Moore's law: a new beginning for information technology. Comput. Sci. Eng. **19**(2), 41–50 (2017)

41. Walter, U., Oberle, V.: μ-second precision timer support for the Linux Kernel (2001)

42. Yuan, Q., Zhao, J., Chen, M., Sun, N.: Generos: an asymmetric operating system kernel for multi-core systems. In: 2010 IEEE International Symposium on Parallel Distributed Processing (IPDPS), April 2010, pp. 1–10 (2010)

43. Zaharia, M., Chowdhury, M., Das, T., Dave, A., Ma, J., McCauley, M., Franklin, M., Shenker, S., Stoica, I.: Fast and interactive analytics over hadoop data with spark. In: USENIX login, vol. 37, pp. 45–51 (2012)

44. Tamer Özsu, M., Valduriez, P.: Distributed data management: unsolved problems and new issues. In: Casavant, T.L. Singhal, M. (eds.) IEEE Computer Readings in Distributed Computing Systems, pp. 512–544. Society Press (1994)

K-means Principal Geodesic Analysis
on Riemannian Manifolds

Youshan Zhang[(✉)]

Computer Science and Engineering, Lehigh University, Bethlehem, PA 18015, USA
yoz217@lehigh.edu

Abstract. Principal geodesic analysis (PGA) has been proposed for data dimensionality reduction on manifolds. However, a single PGA model is limited to data with a single modality. In this paper, we are the first to propose a generalized K-means-PGA model on manifolds. This model can analyze multi-mode nonlinear data, which applies to more manifolds (Sphere, Kendall's shape and Grassmannian). To show the applicability of our model, we apply our model to spherical, Kendall's shape and Grassmannian manifolds. Our K-means-PGA model offers correct geometry geodesic recovery than K-means-PCA model. We also show shape variations from different clusters of human corpus callosum and mandible data. To demonstrate the efficiency of our model, we perform face recognition using ORL dataset. Our model has a higher accuracy (99.5%) than K-means-PCA model (95%) in Euclidean space.

Keywords: K-means · Principal Geodesic Analysis · Manifolds

1 Introduction

Principal component analysis (PCA) has been widely used in analyzing high-dimensional data. It also has other applications. For example, object recognition has attracted more attention in the field of pattern recognition and artificial intelligence. Many literature addressed this issue that included steps of object prepossessing, feature extraction and recognition [1–3]. The first appearance-based method applied the PCA model to recognize images [1]. Some techniques added additionally incorporate facial shape, which included morphable model and active appearance model [2,3]. Other models considered all images as a Lambertian object, and recognized images from orthonormal basis [4,5]. A statistical model was proposed to show variations of normal surface direction, which used PCA to transfer surfaces norm to the tangent space [6,7]. However, PCA has difficulties in analyzing nonlinear data. Nowadays, the convolutional neural network has been widely used in object recognition. However, it always needs a sufficient number of training data to get a good classifier [8,9]. Principal geodesic analysis (PGA) has been proposed to analyze the nonlinear data on Riemannian manifolds [10], and it also can be used to analyze facial shape variations [11–13]. However, previous work did not provide a general model for all manifolds,

© Springer Nature Switzerland AG 2020
K. Arai et al. (Eds.): FTC 2019, AISC 1069, pp. 578–589, 2020.
https://doi.org/10.1007/978-3-030-32520-6_42

which is only applicable to spherical manifold. PGA model also cannot deal with multi-cluster data, but there are always different modes of data.

In this paper, we aim to explore whether K-means-PGA model is applicable to general manifolds for shape analysis and whether our model can achieve a higher accuracy than K-means-PCA model for face recognition. Our contributions are in two-fold:

(1). A general K-means-PGA model for more manifolds.
(2). The detailed calculations of K-means-PGA model follow the geometry of data. Also, it has a higher accuracy than the cases in Euclidean space.

We first define a more general statistical model, which is applicable to more manifolds. We then draw the concept of K-means-PCA in Euclidean space and define a general objective function of K-means-PGA model on manifolds. We also introduce the detailed computations of the exponential map and logarithmic map on three manifolds. To show the applicability of our model, we apply our model to the spherical and Kendall's shape manifolds. Our K-means-PGA model offers correct geometry geodesic recovery than K-means-PCA since K-means-PGA allows the model to capture the specific geometry of data. In addition, we show shape variations from different clusters of human corpus callosum and mandible data. To demonstrate the efficiency of our model, we perform the face recognition using ORL data set. We further explore whether the dimensionality reduction can affect the recognition rate of our model. We compare our method with K-means-PCA model, empirical results show that the recognition rate can achieve 99.5% with a reduced dimensionality as low as 67 (vs. original 360).

2   K-means Principal Component Analysis

In this section, we briefly review the K-means principal component analysis model in the Euclidean space.

2.1   K-means on Euclidean Space

The purpose of the K-means algorithm is to partition the given data points into a predefined amount of k clusters. K-means algorithm aims to optimize the objective function of Eq. 1. The algorithm is described in Algorithm 1, it starts with a random set of k center-points μ_k. In the update step, all data x are assigned to their nearest center-point in Eq. 2. If multiple centers have the same distance to the data, we will randomly choose one.

$$
J = \sum_{n=1}^{N} \sum_{k=1}^{K} r_{nk} ||x_n - \mu_k||^2, r_{nk} = \begin{cases} 1 & x_n \in S_k \\ 0 & \text{otherwise} \end{cases}, \tag{1}
$$

where x is the observed data, μ_k is the k centers. K is the number of clusters, and S_k is one cluster data.

$$
S_k = \{x_n : \left\| x_n - \mu_k \right\|^2 \leq \left\| x_n - \mu_j \right\|^2, j \neq k\}, \quad \mu_k^{'} = \frac{1}{|S_k|} \sum_{x_j' \in S_k} x_j', \tag{2}
$$

We then recalculate the mean of assigned observations to original center-points μ'_k, where $|S_k|$ is the number of cluster S_k.

Algorithm 1. K-means on Euclidean space

Input: $x_1, \cdots, x_N \in X_{i=1}^N$, and the number of clusters k
Output: k clusters data X_k^K
  1: Randomly choose k centroids from data set X
  2: Compute distances from each data to each centroid, and assign each data to its closest centroid
  3: Replace each centroid by the mean of partitioned data in step 2
  4: Repeat step 2 and 3 until all k centroids are chosen

2.2 K-means-PCA Algorithm

After calculating k clusters data from Algorithm 1, we then obtain the K-means-PCA algorithm as following:

Algorithm 2. K-means-PCA

Input: k clusteres data X_k^K from Alg. 1
Output: Eigenvectors ve_k and eigenvalues λ_k of each cluster
  1: **For** $k = 1$ to K
  2:   $\mu_k = \frac{1}{N_k}\sum_{i=1}^{N_k} x_{ik}$
  3:   $S = \frac{1}{N_k}\sum_{i=1}^{N_k}(x_{ik} - \mu_k)(x_{ik} - \mu_k)^T$
  4:   ve_k, λ_k = eigenvectors/eigenvalues of S // using singular value decomposition
  5: **end**

---

3 K-means Principal Geodesic Analysis

3.1 Riemannian Geometry

In this section, we first review three basic concepts (geodesic, exponential and logarithmic map) of Riemannian geometry (more details are provided by others [14–16]).

Geodesic. Let (M, g) be a Riemannian manifold, where g is a Riemannian metric on the manifold M. A curve $\gamma(t) : [0, 1] \rightarrow M$ and let $\gamma'(t) = d\gamma/dt$ be its velocity. The operation $D \cdot /dt$ is called a *covariant derivative* (also called a connection on M), which is denoted as $\nabla_{\gamma'(t)}$ or $\nabla_{\gamma'}$. A vector field $V(t)$ along γ is parallel if $\frac{DV(t)}{dt} = \nabla_{\gamma'}V = 0$. We call γ a geodesic if $\gamma'(t)$ is parallel along γ, that is: $\gamma'' = \frac{D\gamma'}{dt} = \nabla_{\gamma'}\gamma' = 0$, which means the acceleration vector (directional derivative) γ'' is normal to $T_{\gamma(t)}M$ (the tangent space of M at $\gamma(t)$). A geodesic is also a curve $\gamma \in M$ that locally minimizes $E(\gamma) = \int_0^1 ||\gamma'(t)||^2 dt$. Here $|| \cdot ||$ is called *Riemannian norm*, for any points $p \in M$, and $v \in T_pM$, $||v||$ is defined by: $||v|| = \sqrt{g_p(v,v)}$. $g_p(u,v)$ is called *Riemannian inner product* of two tangent vectors $u, v \in T_pM$, which can also be denoted by $\langle u, v \rangle_p$ or simply $\langle u, v \rangle$. The norm of velocity in a geodesic γ is constant, that is: $||\gamma'(t)|| = c$ [15]. Note that geodesics are straight lines in Euclidean space ($\mathbb{R}^n$).

Exponential Map. For any point $p \in M$ and its tangent vector v, let $\mathcal{D}(p)$ be the open subset of T_pM defined by: $\mathcal{D}(p) = \{v \in T_pM | \gamma(1)$ is defined, where γ is the unique geodesic with initial conditions $\gamma(0) = p$ and $\gamma'(0) = v$. The *exponential map* is the map $\text{Exp}_p : \mathcal{D}(p) \to M$ defined by: $\text{Exp}_p(vt_{=1}) = \gamma(1)$, which means the exponential map returns the points at $\gamma(1)$ when $t = 1$. If $\omega \in \mathcal{D}(p)$, then the line segment $\{t\omega | 0 \leq t \leq 1\}$ is constrained to be in $\mathcal{D}(p)$. We then define: $\text{Exp}_p(vt) = \gamma(t)$, where $0 \leq t \leq 1$.

Logarithmic Map. Given two points p and $p' \in M$, the *logarithmic map* takes the point pair (p, p') and maps them into the tangent space T_pM, and it is an inverse of the exponential map: $\text{Log}(p, p') \to T_pM$. $\text{Log}(p, p')$ can also be denoted as: $\text{Log}_p p'$. Because Log is an inverse of the exponential map, we can also write: $p' = \text{Exp}(p, \text{Log}(p, p'))$. The *Riemannian distance* is defined as $d(p, p') = ||\text{Log}_p(p')||$. In Euclidean space ($\mathbb{R}^n$), the logarithmic map is the subtraction operation: $\text{Log}_p(p') = p' - p$ [17].

3.2 K-means on Manifold

We define a general objective function J for K-means algorithm on manifolds.

$$J = \sum_{n=1}^{N}\sum_{k=1}^{K} ||\text{Log}_{x_n}(\mu_k)||^2 = \sum_{n=1}^{N}\sum_{k=1}^{K} d(x_n, \mu_k)^2 \tag{3}$$

Alternatively, we aim to minimize the following energy function:

$$E = \arg\min_J \sum_{n=1}^{N}\sum_{k=1}^{K} d(x_n, \mu_k)^2, \tag{4}$$

where μ_k is the intrinsic mean, which is calculated in Algorithm 4, $|S_k|$ is the number of cluster S_k. By using Algorithm 3, we obtain the k clusters data on manifolds.

3.3 Principal Geodesic Analysis (PGA)

Similar to the PCA model, PGA was proposed by Fletcher (2004), it aims to reduce the dimensionality of nonlinear data [10]. Differ from the Algorithm 2, we need to calculate the intrinsic mean of data. We then perform the PGA in Algorithm 4.

Algorithm 3. K-means on manifold

Input: $x_1, \cdots, x_N \in X_{i=1}^N$, number of clusters k
Output: k clusters data X_k^K
1: Randomly choose k centroids from data set X.
2: Compute the **Riemannian distances** (using Log map) from each data to each centroid, and assign each data to its closest centroid
3: Replace each centroid by the **intrinsic mean** of partitioned data points in step 2
4: Repeat step 2 and 3 until all k centroids are chosen

Algorithm 4. Principal Geodesic Analysis

Input: $x_1, \cdots, x_N \in M$
Output: Eigenvectors ve and eigenvalues λ of input data
1: $\mu_0 = x_1$
2: **Do**
   $\Delta\mu = \frac{\tau}{N} \sum_{i=1}^N \text{Log}_{\mu_j x_i}$
3: $\mu_{j+1} = \text{Exp}_{\mu_j}(\Delta\mu)$
4: **While** $||\Delta\mu|| > \epsilon$ // calculate intrinsic mean of $\{x_i\}$
5: $x_i' = \text{Log}_\mu(x_i)$
6: $S = \frac{1}{N} \sum_{i=1}^N x_i' x_i'^T$
7: $ve_k, \lambda_k = $ eigenvectors/eigenvalues of S

3.4   K-means-PGA Algorithm

After calculating the data of each cluster from Algorithm 3, and we can combine it with Algorithm 4 to obtain the K-means-PGA algorithm as following:

Algorithm 5. K-means-PGA

Input: k clusters data X_k^K from Alg. 4
Output: Eigenvectors ve_k and eigenvalues λ_k of each cluster
1: **For** $k = 1$ to K
2:   $\mu_k = $ intrinsic mean of $\{x_{ik}\}$
3:   $x_{ik}' = \text{Log}_\mu(x_{ik})$
4:   $S = \frac{1}{N} \sum_{i=1}^N x_{ik}' x_{ik}'^T$
5:   $ve_k, \lambda_k = $ eigenvectors/eigenvalues of S
6: **end**

3.5   Computation of Exp and Log Map Manifolds

Sphere Manifold. One of the well-known spherical manifolds is the 3D sphere (2D surface embedding in 3D space). Let r be the radius of the sphere, u the azimuth angle and v the zenith angle. Then any points on 3D sphere can be expressed by: $X = (r\sin(u)\sin(v), r\cos(u)\sin(v), r\cos(v))$. The generalized $n-1$ dimensional hyper-sphere embedded in $\mathbb{R}^n$ Euclidean space $(x_1, x_2, \cdots, x_n)$ has the constraint of: $\sum_{i=1}^n x_i^2 = r^2$, here r is the radius of such a hyper-sphere, we set $r = 1$. Let p and p' be such points on on a sphere embedded in $\mathbb{R}^n$, and let v be a tangent vector at p.

The Log map between two points p, p' on the sphere can be computed as follows:

$$v = \text{Log}(p, p') = \frac{\theta \cdot L}{||L||}, \quad \theta = \arccos(\langle p, p' \rangle), \tag{5}$$
$$L = (p' - p \cdot \langle p, p' \rangle)$$

where $p \cdot \langle p, p' \rangle$ denotes the projection of the vector p' onto p. $||L||$ is the *Riemannian norm* as defined in Sect. 3.1.

Given base point p, and its estimated tangent vector v from Eq. 5 and t, we can compute the Exp map as:

$$\text{Exp}(p, vt) = \cos \theta \cdot p + \frac{\sin \theta}{\theta} \cdot vt, \quad \theta = ||vt||. \tag{6}$$

This explanation is based on that of Wilson and Hancock (2010); see them for details of Log map and Exp map on the Sphere manifold.

Kendall's Shape Manifold. A more complex manifold was studied by David G. Kendall [18]. Kendall's shape can provide a geometric setting for analyzing arbitrary sets of landmarks. The landmark points in Kendall's space are a collection of points in Euclidean space. Before calculating the Log and Exp maps, we use some transformations (scale and rotation) to get the pre-shape of data.

Given two shapes $p, p' \in V$ with $d \times n$ matrix (d is the dimension of the shape, n is the number of points), to construct Kendall's shape space, first, we remove the translation and scale of the shape. To get the pre-shape of an object, we eliminate the centroid (subtract the row means from each row) and scale it into unit norm (divide by the Frobenius norm). Then, we remove the rotation of the shape using Orthogonal Procrustes Analysis (OPA) [19]. OPA solves the problem of finding the rotation R^* that minimizes distance between p and p'.

$$R^* = \arg \min_{R \in SO(d)} ||Rp - p'||,$$

where SO means a special orthogonal group. OPA performs singular value decomposition of $p \cdot p'^T$; let $[U, S, V] = \text{SVD}(p \cdot p'^T)$, then the $R* = UV^T$.

The Log map between two shapes p, p' of Kendall's shape manifold is also given by finding the rotation between p and p' first. To find the rotation of p', we calculate the singular value decomposition of $p \cdot p^T$; then we find the rotation of p' according to $p'_{(Rot)} = \text{rotation} \cdot p'$. Then the Log map is given by:

$$v = \text{Log}(p, p') = \frac{\theta \cdot L}{||L||}, \quad \theta = \arccos(\langle p, p'_{(Rot)} \rangle), \tag{7}$$
$$L = (X_{T(Rot)} - p \cdot \langle p, p'_{(Rot)} \rangle),$$

where $\langle p, p'_{(Rot)} \rangle$ denotes the projection of the vector p' onto p.

The Exp on Kendall's shape is the same as sphere manifold (Eq. 6). Refer to Kendall (1984) to see the details of Log map and Exp map Kendall's manifold.

Grassmannian Manifold: Before we introduce the Exp and Log map of Grassmannian manifold, we should get the subspace of the data, since Grassmannian manifold $\mathbb{G}_{N,d}$ is defined as $d-$dimension subspace. Therefore, it is necessary to use Principal Geodesic Analysis to get the subspace (submanifold) of Grassmannian manifold. Algorithm 4 describes how to calculate the intrinsic mean and PGA reduction of Grassmannian manifold. Please refer to [10] for more details. Suppose that we get the subspace from PGA of the data (that is p and p'), again to get the sampling from p to p', we need to calculate the Log map and Exp map. Differ from [20], and we use the following calculations:

Then the Log map is given by:

$$X = (I - p \cdot p^T) \cdot p' \cdot (p^T \cdot p')^{-1}, \quad [U, s, V] = \mathrm{SVD}(X),$$
$$v = \mathrm{Log}(p, p') = U \cdot \theta \cdot V^T, \quad \theta = \arctan s, \tag{8}$$

where I is the identify matrix, and θ is the principal angles between p and p'.

The exponential map is computed by base point p, and the estimated initial velocity v, that is,

$$[U, \theta t, V] = \mathrm{SVD}(vt), \quad \mathrm{Exp}(p, vt) = p \cdot V \cdot \cos(\theta t) + U \cdot \sin(\theta t) \cdot V^T, \tag{9}$$

Refer to [21] and [22] to see the details of geometry on Grassmannian manifold.[1]

4   Experiments

In this section, we demonstrate the effectiveness of our K-means-PGA model by using both synthetic data and real data.

4.1   Sphere

To validate our model in sphere manifold, we randomly simulate 764 data points on a unit sphere S^2 with three clusters. We can visualize the true geodesic from the Fig. 1. The blue line is the estimated geodesic of K-means-PGA model, while the yellow line is true geodesic. We can find that estimated geodesics are almost overlaying with true geodesic, which demonstrates the correctness of our model. We also compare the estimated results of our model with K-means-PCA in the Euclidean space. As shown in Fig. 1c, estimated geodesics of K-means-PGA are curves on the sphere, but the estimated geodesic of K-means-PCA is straight lines, which cannot recover true geodesics on the sphere. Figure 1b illustrates the limitation of PGA model since it cannot estimate geodesics of multimode data since only one geodesic can be estimated. This result also illustrates the advantage of K-means-PGA model than the traditional PGA model.

---

[1] Source code is available at: https://github.com/heaventian93/Kmeans-Principal-Geodesic-Analysis.

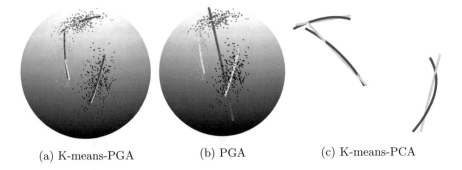

<table>
<tr><td>(a) K-means-PGA</td><td>(b) PGA</td><td>(c) K-means-PCA</td></tr>
</table>

Fig. 1. (a) The estimated geodesic using K-means-PGA. The blue line is the estimated geodesic, and the yellow line is the ground truth. (b) The estimated geodesic (magenta color) of PGA model, which is the limitation of PGA model since it only shows one curve. (c) The comparison of estimated geodesics of K-means-PGA and K-means-PCA. The blue line is the estimated geodesic in (a). Cyan lines are estimated geodesics using K-means-PCA, which are straight lines.

(a) Cluster one   (b) Cluster two

Fig. 2. The shape variations of two clusters corpus callosum using K-means-PGA along the first geodesic with $[-\sqrt{\lambda}, \sqrt{\lambda}]$.

4.2 Kendall's Shape

Corpus Callosum Data. To show K-means-PGA model for demonstrating classification for the real shape changes, we apply it to corpus callosum data. The corpus callosum data are derived from MRI scans of human brains. The data is from the ABIDE database,[2] which is used in Hiess et al., (2015). The valid data contains 1100 patients; we extract the boundary of each corpus callosum shape, and each shape contained 2×142 points. As shown in Fig. 2, it demonstrates corpus callosum shape variations of two different clusters, which are generated from the points along the estimated principal geodesics: $\mathrm{Exp}(\mu, \alpha_i v e_i)$, where $\alpha_i = [-\sqrt{\lambda}, \sqrt{\lambda}]$, for $i = 1$, which means the largest the eigenvalue. It shows that the anterior, mid-caudate and posterior of corpus callosum from the unhealthy group (Fig. 2a) is significantly larger than the healthy group (Fig. 2b).

Mandible Shape. The mandible data is extracted from a collection of CT scans of human mandibles. It contains 77 subjects and the age is from 0 to 19 years.

---

[2] https://sites.google.com/site/hpardoe/cc_abide.

We sample 2×400 points on the boundaries. Figure 3b shows examples of 3D raw data of mandible and 2D shape data. Shape variations of mandible data from male (Fig. 4a) and female group (Fig. 4b) are shown in Fig. 4. In general, male mandibles have larger shape variations in the temporal crest, middle part and the base than that in female mandibles, which is consistent with previous studies [23].

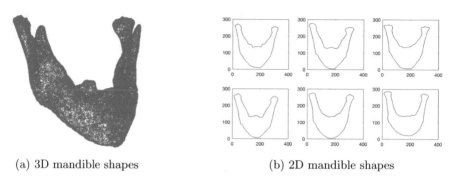

(a) 3D mandible shapes   (b) 2D mandible shapes

Fig. 3. Example of 3D and 2D mandible shapes.

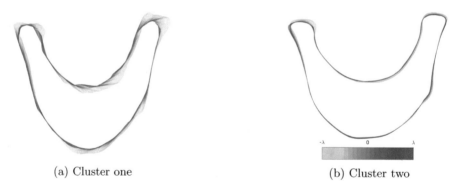

(a) Cluster one   (b) Cluster two

Fig. 4. The shape variations of two clusters mandible data using K-means-PGA along the first geodesic with $[-\sqrt{\lambda}, \sqrt{\lambda}]$.

4.3 Grassmannian Manifold

We use the ORL face database for examining face recognition task. This dataset is composed of 400 images with the size of 112×92. There are 40 persons and ten images for each person [24]. For the experiments, we use the ten-folds-cross-validation, we chose 90% of images of each category as training data and the rest as test data. As shown in Fig. 5, we observe eight eigenface examples which are chosen from the first component of the K-means-PGA model. From Fig. 6, the face recognition percentage of K-means-PGA model (99.5%) is significantly higher than K-means-PCA model (95%), which illustrates the superiority of our model. The K-means-PGA model reduces the dimensionality of input images and only 67 dimensionalities out of 360 can extract enough number of features for face recognition.

Fig. 5. The first eigenvector of each cluster, which shows the most important feature of each category.

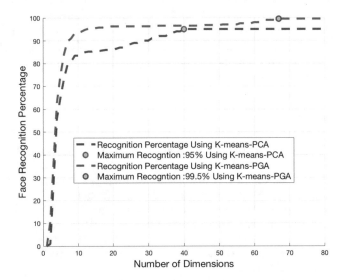

Fig. 6. The comparison of face recognition percentage of K-means-PGA and K-means-PCA with different number of dimensionalities.

5   Discussion

The experiments for sphere and Kendall's shape demonstrate that K-means-PGA model is suitable for shape analysis, especially for nonlinear data. Also, our model achieves a higher facial recognition accuracy than K-means-PCA model since the K-means-PGA model can capture the intrinsic geometry of data. To choose a proper manifold for different datasets, we suggest that sphere manifold is suitable for data, which lies on hypersphere; Kendall's manifold is the best choice if data are 2D landmark shapes, and Grassmannian manifold will have a good performance if data should be prepossessed in the subspace.

However, there are two major limitations of our model. First, the computation time is $\mathcal{O}(i \times k \times n \times d)$, i is the number of iteration, k is the number of clusters, n is the number of samples, and d is the dimensionality of each sample feature. If $i \times k \times n \times d$ is bigger, our model will consume longer computation time. Secondly, the number of cluster k cannot be automatically determined. In our experiment, we determine the number of clusters k using the minimal error, which defined in Eq. 4.

6   Conclusion

In this paper, we first define a general model for K-means-PGA model on different Riemannian manifolds, and we give closed-form expressions of Log map and Exp map on three manifolds. We conduct four experiments on three manifolds, and experimental results show that K-means-PGA model is able to analyze shape variations in multimodal data, and the accuracy of K-means-PGA model is relatively higher than the model in Euclidean space. For future work, we will apply our model in more manifolds, and analyze detailed changes of data (e.g., gesture recognition in images).

References

1. Turk, M., Pentland, A.: Eigenfaces for recognition. J. Cogn. Neurosci. **3**(1), 71–86 (1991)
2. Timothy, F.C., Gareth, J.E., Christopher, J.T.: Active appearance models. IEEE Trans. Pattern Anal. Mach. Intell. **23**(6), 681–685 (2001)
3. Blanz, V., Vetter, T.: Face recognition based on fitting a 3D morphable model. IEEE Trans. Pattern Anal. Mach. Intell. **25**(9), 1063–1074 (2003)
4. Basri, R., Jacobs, D.: Lambertian reflectance and linear subspaces. In: Proceedings Eighth IEEE International Conference on Computer Vision 2001. ICCV 2001, vol. 2, pp. 383–390. IEEE (2001)
5. Peter, N.B., David, J.K.: What is the set of images of an object under all possible illumination conditions? Int. J. Comput. Vision **28**(3), 245–260 (1998)
6. William, A.P.S., Edwin, R.H.: Recovering facial shape using a statistical model of surface normal direction. IEEE Trans. Pattern Anal. Mach. Intell. **28**(12), 1914–1930 (2006)
7. William, A.P.S., Edwin, R.H.: Face recognition using 2.5D shape information. In: 2006 IEEE Computer Society Conference on Computer Vision and Pattern Recognition, vol. 2, pp. 1407–1414. IEEE (2006)
8. Yin, X., Liu, X.: Multi-task convolutional neural network for pose-invariant face recognition. IEEE Trans. Image Process. **27**(2), 964–975 (2018)
9. Ding, C., Tao, D.: Trunk-branch ensemble convolutional neural networks for video-based face recognition. IEEE Trans. Pattern Anal. Mach. Intell. **40**(4), 1002–1014 (2018)

10. Fletcher, P.T., Lu, C., Pizer, S.M., Joshi, S.: Principal geodesic analysis for the study of nonlinear statistics of shape. IEEE Trans. Med. Imaging **23**(8), 995–1005 (2004)
11. Dickens, M.P., Smith, W.A.P., Wu, J., Hancock, E.R.: Face recognition using principal geodesic analysis and manifold learning. In: Iberian Conference on Pattern Recognition and Image Analysis, pp. 426–434. Springer (2007)
12. Pennec, X.: Probabilities and statistics on Riemannian manifolds: a geometric approach. PhD thesis, INRIA (2004)
13. Jing, W., William, A.P.S., Edwin, R.H.: Weighted principal geodesic analysis for facial gender classification. In: Iberoamerican Congress on Pattern Recognition, pp. 331–339. Springer (2007)
14. Perdigão do Carmo, M.: Riemannian Geometry. Birkhauser (1992)
15. Gallier, J.: Notes on differential geometry and lie groups (2012)
16. Pennec, X.: Statistical computing on manifolds: from Riemannian geometry to computational anatomy. In: Emerging Trends in Visual Computing, pp. 347–386. Springer (2009)
17. Zhang, M., Fletcher, P.T.: Probabilistic principal geodesic analysis. In: Advances in Neural Information Processing Systems, pp. 1178–1186 (2013)
18. David, G.K.: Shape manifolds, procrustean metrics, and complex projective spaces. Bull. Lond. Math. Soc. **16**(2), 81–121 (1984)
19. John, C.G.: Generalized procrustes analysis. Psychometrika **40**(1), 33–51 (1975)
20. Kyle, A.G., Anuj, S., Xiuwen, L., Paul Van, D.: Efficient algorithms for inferences on Grassmann manifolds. In: 2003 IEEE Workshop on Statistical Signal Processing, pp. 315–318. IEEE (2003)
21. Edelman, A., Arias, T.A., Smith, S.T.: The geometry of algorithms with orthogonality constraints. SIAM J. Matrix Anal. Appl. **20**(2), 303–353 (1998)
22. Absil, P.-A., Mahony, R., Sepulchre, R.: Riemannian geometry of Grassmann manifolds with a view on algorithmic computation. Acta Applicandae Mathematica **80**(2), 199–220 (2004)
23. Moo, K.C., Anqi, Q., Seongho, S., Houri, K.V.: Unified heat Kernel regression for diffusion, Kernel smoothing and wavelets on manifolds and its application to mandible growth modeling in CT images. Med. Image Anal. **22**(1), 63–76 (2015)
24. Guodong, G., Stan, Z.L., Kapluk, C.: Face recognition by support vector machines. In: Proceedings Fourth IEEE International Conference on Automatic Face and Gesture Recognition 2000, pp. 196–201. IEEE (2000)

Evaluation of Missing Data Imputation Methods for an Enhanced Distributed PV Generation Prediction

Aditya Sundararajan and Arif I. Sarwat[✉]

Florida International University, Miami, FL, USA
asarwat@fiu.edu

Abstract. To effectively predict generation of distributed photovoltaic (PV) systems, three parameters are critical: irradiance, ambient temperature, and module temperature. However, their completeness cannot be guaranteed because of issues in data acquisition. Many methods in literature address missingness, but their applicability varies with missingness mechanism. Exploration of methods to impute missing data in PV systems is lacking. This paper conducts statistical analyses to understand missingness mechanism in data of a real grid-tied 1.4MW PV system at Miami, and compares the imputation performance of different methods: random imputation, multiple imputation using expectation-maximization, kNN, and random forests, using error metrics and size effect measures. Imputed values are used in a multilayer perceptron to predict and compare PV generation with observed values. Results show that values imputed using kNN and random forests have the least differences in proportions and help utilities make more accurate prediction of generation for distribution planning.

Keywords: Distributed PV · Missing data · Data processing · Imputation methods · PV Generation Prediction

1 Introduction

In future high penetration scenarios of distributed photovoltaic (PV) systems, most power system applications such as state estimation, optimal power flow, economic dispatch, and unit commitment would rely on accurate prediction of PV generation [1,2]. These predictive models require a significant wealth of complete, good quality historical observations on three primary features: irradiance, ambient temperature, and module temperature, for training and validation [3]. Completeness, also called coverage, implies that the data is devoid of gaps between values that can be intuitively detected, such as unexpected jumps in time-series values, unlabeled exceptions in categorical values, and discrepancy between expected and observed number of assets [4].

Many methods exist in the literature to impute missing values, but their performance varies with the type of data, the application domain, and the mechanism

© Springer Nature Switzerland AG 2020
K. Arai et al. (Eds.): FTC 2019, AISC 1069, pp. 590–609, 2020.
https://doi.org/10.1007/978-3-030-32520-6_43

of missingness. Many application domains have been the focus of different works in the literature: the advanced metering infrastructure and smart meters [5–7], electric vehicle charging [8], low-voltage distribution networks [9,10], and load forecasting [11]. Given the low penetration of existing distributed PV systems, much research has not been done to find suitable methods to impute missing values that are typically seen in datasets of irradiance, ambient temperature, and module temperature that are caused by a loss of communications, poor data acquisition, erroneous calibration, system failure or memory corruption, and external factors such as inclement weather [12,13]. The utilities primarily rely on elementary methods such as listwise record deletion, mean, zero, and median value substitutions to handle missingness. These methods reduce data quality, cause information loss, and introduce bias into the existing data. In future high penetration scenarios, however, missingness in these parameters significantly impact the training and generalization of the self-adaptive models used to predict PV generation [14,15]. Further, identifying missingness in data is a first step to detect potentially fraudulent or malicious records in a typical utility data flow pipeline [16–18]. Hence, there is a need to first explore the likely missingness mechanism in these data and determine suitable methods to impute missing values without jeopardizing the data's inherent properties. Bridging this need also forms the goal of this paper.

Key contributions of this work are that it: **(1)** provides a robust methodology that can be adopted by utilities and data analysts to reliably impute missing values of irradiance, ambient temperature, and module temperature for a more accurate prediction of distributed PV generation (Sect. 2); **(2)** is the first work to statistically analyze the missingness mechanism for the application domain of distributed PV generation prediction (Sect. 2.2); **(3)** is the first work to apply non-elementary methods to impute missing values of irradiance, ambient temperature, and module temperature (Sects. 3 and 4); and **(4)** brings the utilities closer to achieving operational visibility into remote distributed PV systems.

Imputation methods in literature treat missingness as a big data challenge but do not characterize the missingness mechanism first [19]. This is important because most imputation methods assume the data contains values missing at random (MAR), whereas the methods suitable to impute values for a particular type of data with a likely missingness mechanism might not be applicable to a dataset of another domain. A majority of smart grid data is structured [20]. Many methods to handle missingness in structured data exist. Single imputations such as mean-value or median-value, multiple imputation, and self-adaptive models such as k-nearest neighbors (kNN), multilayer perceptron (MLP), and self-organizing maps were applied to estimate missing values and offer prognosis for cancer patients [21]. However, the study did not disclose the missingness mechanism. The necessity of imputation and an analysis of when to use it was conducted by [22], who also prescribed a model-based method using conditional density to estimate missing values. The work highlighted that methods imputing values using relationships between observable variables are useful. The work in [23] compared 6 known imputation methods (kNN, fuzzy k-means, singular value decomposition, Bayesian principal component analysis, and multiple imputation by chained equations) for different categorical datasets and concluded, using metrics such as root mean square error (RMSE) and classification errors, that the Bayesian principal component analysis fared better. The work in [24] highlighted the significance of imputation methods to the PV systems domain and applied different methods to impute missing values of irradiance and power. However, the methods did not include more powerful machine learning models, and not much information was provided about the missingness mechanism. Surveys bring to light future challenges in querying and indexing data with missing values, developing missing data handling systems that work across domains and handle all mechanisms, and finding a trade-off between runtime and accuracy of imputation [25–28].

2   The System of Study

Figure 1 illustrates, for the application domain of PV generation prediction, the methodology to impute missing values in structured, time-series interval data of irradiance, ambient temperature, and module temperature collected in real-time from the 1.4 MW grid-tied PV system at Miami, Florida. Following the customary cleaning and exploratory analysis, the likely missingness mechanism is understood through statistical tests. Depending on the mechanism, different solutions are proposed. Imputation methods can be applied only when the mechanism is of MAR, in which case different error and size of effect measures are used to compare imputation performance. To further identify the method(s) that work(s) best for these types of data, the values imputed using different methods are fed into an MLP model that predicts PV generation. Residual errors between predicted and observed generation serve as strong indicators of the influence of imputation on data integrity. The PV system has 4,480 modules and 46 smart string inverters that can either be connected to consumer loads or fed directly into the grid. Further description of the PV system and its characteristics can be found in the authors' previous works [29,30]. The dataset consists of four

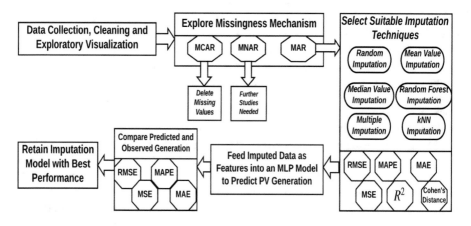

Fig. 1. Proposed methodology to determine a suitable imputation technique.

attributes: irradiance (measured in W/m^2, the amount of solar power incident per square meter of PV modules), ambient temperature (measured in °F), module temperature (also measured in °F, temperature of PV modules), and PV generation (measured in kW, the aggregate amount of power generated by the modules). All these values are recorded in real-time for one year in 15 minute intervals.

2.1   Exploratory Visualization and Analysis

Figure 2a, 2b, and 2c illustrate the explorations done to understand the non-monotone missing patterns in the observations, which amount to 20% of the total dataset [31–33]. Figure 2a shows the number of missing values grouped by attributes. It also shows the distribution of missing observations, where grey cells denote present values and black cells the missing values. For each attribute, the distribution of missing values across each record is shown in Fig. 2b in red. Further, the values color-coded in lighter shades of grayscale denote low magnitudes and those with higher magnitudes are coded in darker shades. The patterns and combinations of missingness across different cases of the dataset are illustrated by Fig. 2c, where there are 1312, 1308, 1223, and 1215 missing values for ambient temperature, module temperature, irradiance, and PV generation, respectively. There are 79 cases where both ambient and module temperature values are missing together, 79 where module temperature and PV generation are both missing, and so on. This provides a comprehensive view of the missing patterns by cases for the entire dataset. It is to be noted that these missing patterns were not synthetically introduced or assumed but were observed in the real, captured data.

2.2   Exploring the Mechanism of Missingness

Missingness can be introduced in the data by three mechanisms: missing completely at random (MCAR), MAR, and missing not at random (MNAR) [34,35]. Understanding the nature of missingness is important prior to applying imputation because most of the available methods for imputation assume MAR missingness [36].

MCAR: Missing values have no relationship with other values either missing or present. This can be treated as ignorable since the information about missing values need not be modeled, and is ideal for listwise deletion since it does not create bias.

MAR: Likelihood of values being missing can be modeled using just the observed values in a dataset, since they do not depend on other missing observations.

MNAR: Missingness is neither MCAR nor MAR but is dependent on missing values alone; estimating missing values requires a knowledge of how the missingness occurred.

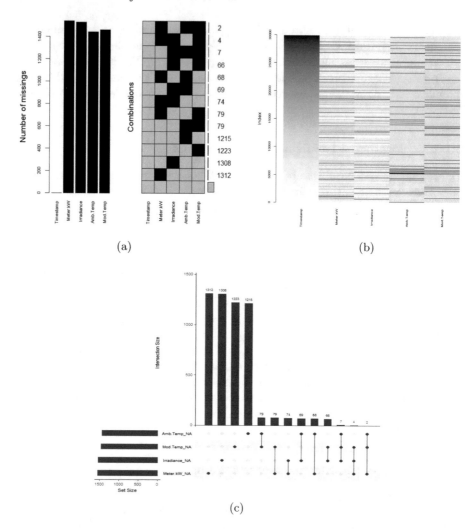

Fig. 2. Exploratory analysis of missing values: (a) Number of missing values and distribution of combinations of missing values grouped by attributes, (b) A matrix plot showing missing values in red and observed values in grayscale, and (c) Different groups of missing patterns seen across the dataset.

The test proposed in [37] is used to test for MCAR missingness. The dataset with missing values is split into three batches to reduce the amount of memory required for processing, where the first two batches comprise 10,000 cases each, and the third has 9,831 cases. The results are shown in Fig. 3. In all three runs, the test first imputed values using the method prescribed in [38] that considers a case-based independence of values with linearly related variables, and assumes their cumulative distribution functions to be continuous. For each batch, the approach applies a modified Hawkins test on the imputed set that

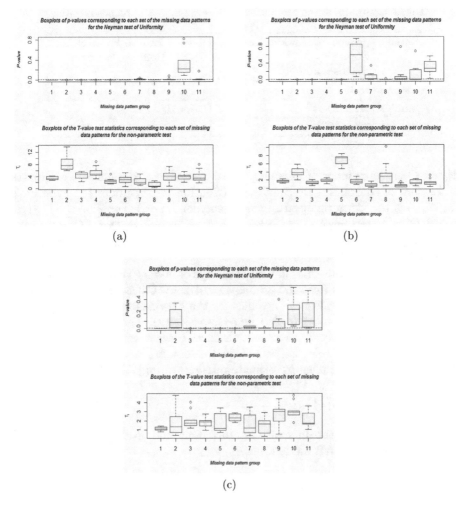

Fig. 3. Test of the given dataset for likelihood of MCAR missingness, with dataset separated into three batches, (a), (b), and (c), by visualizing boxplots for p and t-values using Neyman's test of uniformity and non-parametric test, respectively.

uses Neyman's test for uniformity shown in Eq. (1) [39], the rejection of which (at large values of N_{ij}) prompts a non-parametric test. Rejection of the second test implies rejecting the null hypothesis ($\mathcal{H}_0$) that data is MCAR.

$$N_{ij} = \sum_{j=1}^{4} [\frac{1}{\sqrt{n_i}} \sum_{l=1}^{n_i} \mathcal{F}_j(\mathcal{X}_{il})], \forall i \in [1, G], i \in \mathbb{Z}^+ \tag{1}$$

Table 1. Results of the test for MCAR, normality, and homoscedasticity

Batch	p-value for Hawkins Test of Normality and Homoscedasticity	p-value for Non-Parametric Test of Homoscedasticity
Batch 1	≈ 0	5.2×10^{-11}
Batch 2	1.1×10^{-105}	9.5×10^{-6}
Batch 3	3.5×10^{-175}	2.2×10^{-6}

where, G is the number of missing pattern groups, j is the number of degrees of freedom of the central chi-squared distribution for N_j under the hypothesis $\mathcal{H}_0$, $\mathcal{F}_j(\cdot)$ is the jth normalized Legendre polynomial orthogonal on the interval $[0, 1]$ [40], n_i is the number of cases of missing values such that $\sum_{i=1}^{G} n_i$ is the total number of missing values in the data's batch, and $\mathcal{X}_{il}$ is the lth case of observations in the ith group.

Results in Fig. 3 show that for the 11 groups of missing patterns, the p-values vary, with their values for some groups almost zero. Considering a significance level of 0.05, Table 1 shows the p-values for the Hawkins test of normality and homogeneity of covariances (called homoscedasticity) and the non-parametric test for homoscedasticity. The low p-values in all cases reject the null hypotheses for both tests, thereby rejecting MCAR [41]. Assuming MAR, different imputation methods can be applied to estimate missing values [42].

3   Imputation Methods

This section describes the different imputation methods evaluated in this paper, all of which assume MAR mechanism.

3.1   Random Imputation

The rationale behind random imputation is that the missing values of an attribute could be appropriated by replacing them with randomly drawn values from a sample of observed values of the same attribute. For a given missing value $\mathcal{X}_{il}$ of the lth case in the ith group for an attribute Y in the dataset, such that $i \in [1, G], \forall i \in \mathbb{Z}^+$ and X_{obs} is the set of observed values of Y, $\hat{\mathcal{X}}_{il}$ imputed by random imputation is given by:

$$\hat{\mathcal{X}}_{il} = f(X_{obs}, \sum_{i=1}^{G} n_{iY}) \tag{2}$$

where, $f(\cdot)$ is the random sampling function that draws samples from the given X_{obs} with repetition, and $\sum_{i=1}^{G} n_{iY}$ denotes the total number of missing values for the attribute Y. Random imputation does not explore the relationship between variables [42]. However, it could be combined with regression model fitting for missing value prediction.

3.2   Mean and Median-Value Imputations

These two single imputation methods use the mean and median, respectively, of an attribute as the substitution value for all missing values. For a given attribute Y comprising $\mathcal{X}_{il}$ missing values, the mean-value imputation is given by:

$$\hat{\mathcal{X}}_{il} = \bar{X}_{obs}, \forall i \in [1, G] \; for \; Y, \; \mathbb{Z}^+ \tag{3}$$

Mean-value imputation is computationally efficient and preserves the mean of the dataset, but biases the standard error and does not preserve the relationship among variables of the data [43,44]. Therefore, applying correlation and regression for original and imputed datasets show different results. Median-value imputation is given by:

$$\hat{\mathcal{X}}_{il} = f(X_{obs}) \forall i \in [1, G] \; for \; Y, \; \mathbb{Z}^+ \tag{4}$$

Here, the function $f(\cdot)$ calculates the median of its argument vector. Similar to mean-value imputation, this method preserves the median but suffers from similar pitfalls. Multiple imputation is a better alternative.

3.3   Multiple Imputation

Multiple imputation performs m iterations of imputations on a dataset containing missing values to generate m corresponding complete datasets, one of which can be selected based on user-specified criteria. The missing values are filled using the distribution of imputations that account for the uncertainty in missing data. In addition to assuming the mechanism is MAR, this method assumes $\mathcal{X} = \mathcal{N}(\mu, C)$, that the dataset $\mathcal{X}$ is multivariate normal with μ as the mean vector and C as the covariance matrix [45]. The method applied in this paper uses expectation-maximization with bootstrapping (EMB) to perform the imputation. By the definition of MAR, the approach concerns only with the set of complete parameters, $\Theta = (\mu, C)$, to impute missing values. The likelihood of observed data, considering the inference on complete-values parameters is [46,47]:

$$\mathcal{L}(\Theta|X_{obs}) \propto \int p(\mathcal{X}|\Theta) dX_{miss} \tag{5}$$

where, $\mathcal{X}_{miss} = \{\mathcal{X}_{il}\}$ across l cases and i groups. To reduce the computational complexity involved in drawing from the posterior probability $p(\mathcal{X}_{obs}|\Theta)$, bootstrapping with EM is used, that also accounts for the uncertainty in estimation. Let $\mathbf{M}$ denote a matrix of dimensions the same as $\mathcal{X}$ such that a cell in $\mathbf{M} = 1$ if the corresponding cell in $\mathcal{X} \in X_{miss}$, and 0 otherwise. Then, specifically during the expectation step of EMB, the imputed value $\hat{\mathcal{X}}_{miss}$ is given by [45,47]:

$$\hat{\mathcal{X}}_{miss} = X_{obs} + \mathbf{M} \times (X_{obs}\Theta\{1 - M\}^{iter}) \tag{6}$$

where, $iter$ denotes the imputation iteration index. This method preserves the relationship between variables and has been shown to perform better than single imputation methods across different domains.

3.4 K-Nearest Neighbors (kNN)

The characteristics of the observed values are used to impute the missing values in this method, which makes the imputed values naturally occurring, and the method capable of preserving the original properties and structure of the dataset. In this method, the missing cell is assigned a value derived from applying the weighted mean of the values obtained from k neighbors with observed values that are nearest to that missing cell. The assignment of weights is proportional to the Minkowski norm distance, d_{MN}, between that neighboring cell and the missing cell in question [48]. The distance of order p for the kth neighbor is calculated by:

$$d_{MN} = (\sum_{i=1}^{n} |X_{obs}^k - X_{miss}|)^{1/p} \tag{7}$$

where, $n=\sum_{i=1}^{G} n_i$, $p=2$ (Euclidean distance), and K denotes the total number of nearest neighbors. The weight for a given X_{obs} is assigned as $e^{-d_{MN}}$ such that:

$$\hat{\mathcal{X}}_{il} = \frac{\sum_{k=1}^{K}(e^{-d_{MN}} \times X_{obs}^k)}{k} \tag{8}$$

The number of nearest neighbors is selected by observing the variation of RMSE across different number of neighbors is shown in Fig. 4 for each attribute. The trend of the variations show that for all three attributes, the RMSE was lowest when the number of nearest neighbors, $k=2$. Hence, the imputation is run with this parameter.

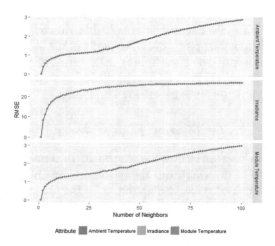

Fig. 4. Variation of RMSE with number of nearest neighbors.

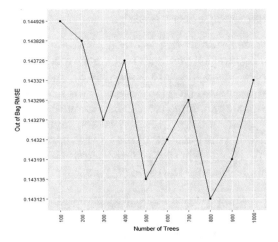

Fig. 5. Out of bag error (nRMSE) versus number of trees in the forest.

3.5 Random Forest

A random forest is an ensemble of multiple decision trees, and it is designed to consider imputation as a regression problem. The number of trees to grow in the random forest is selected by observing the variation of RMSE ($nRMSE$) shown in Eq. (9) across different number of trees in the random forest as illustrated by Fig. 5. In the equation, $Var(\cdot)$ denotes the function to calculate variance of $(\cdot)$. There is a fluctuating trend in RMSE values, but it is the lowest when the number of trees to grow is 800. Hence, the imputation is run with this parameter. The method splits a given dataset $\mathcal{X}$ attribute-wise into Y_{obs} denoting the attribute(s) with observed values, and Y_{miss} denoting the attribute(s) with missing values. Further, X_{obs} and X_{miss} (defined here as the observations corresponding to Y_{obs} and Y_{miss} unlike previously used in this paper).

A mean-value imputation is first conducted to fill the missing values in $\mathcal{X}$ with an initial set of values. Then, a training set comprising X_{obs} inputs and Y_{obs} target is fed into the random forest model with 800 trees. The trained model is then used to predict Y_{miss} using X_{miss}. Over multiple iterations, the values are fine-tuned until the difference between imputed values over two subsequent iterations, calculated using Eq. (9) stops decreasing [49].

$$s = \frac{(\hat{\mathcal{X}}_{iter} - \hat{\mathcal{X}}_{iter-1})^2}{(\hat{\mathcal{X}}_{iter})^2}, \quad nRMSE = \sqrt{\frac{(\mathcal{X} - \hat{\mathcal{X}})^2}{Var(\mathcal{X})}} \qquad (9)$$

4 Discussion of Results

This section discusses the results obtained by applying different imputation methods.

4.1   Performance Comparison Metrics

Five standard metrics available in the literature are used here to compare the performance of the imputation methods: root mean squared error (RMSE), mean squared error (MSE), mean absolute error (MAE), mean absolute percentage error (MAPE), and R-squared (R^2) [50]. Consider the following for each attribute in the dataset. Let the original values of the attribute be $\mathcal{X}=\{X_{il}, X_{obs}\}$ such that $i \in [1, G], \forall i \in \mathbb{Z}^+$ and $l \in [1, n_i], \forall l \in \mathbb{Z}^+$, and X_{obs} is the number of complete cases. Also let the same attribute of the dataset with missing values imputed be $\hat{\mathcal{X}}=\{\hat{X}_{il}, X_{obs}\}$, where $\hat{X}_{il}$ is the imputed value of the lth missing case in the ith group. If $\bar{\cdot}$ denotes the mean of $\cdot$, the metrics can then be calculated using Eq. (11).

The statistical tests of hypotheses can be supplemented using Cohen's distance test, also called h-test, that qualifies as a measure of effect size. When measured between two sample distributions, $\mathcal{X}$ and $\hat{\mathcal{X}}$, the distance measure could be used to explore the difference in proportions using the standard rule of

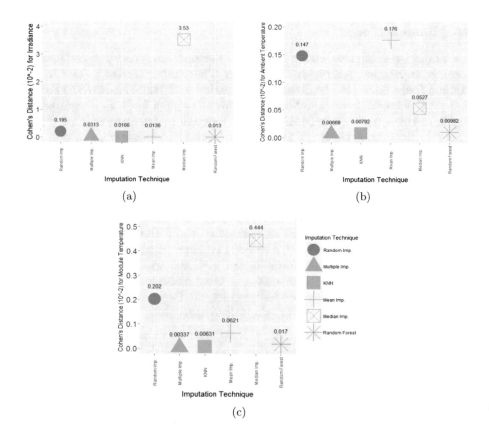

Fig. 6. Cohen's distance to measure similarity between original and imputed datasets for: (a) Irradiance, (b) Ambient Temperature, and (c) Module Temperature.

thumb: $h = 0.2$ implies a small difference, $h = 0.5$ implies a medium difference, and $h = 0.8$ means a large difference. The Cohen's distance, calculated using the below set of equations, is shown for irradiance, ambient temperature, and module temperature in Fig. 6a, 6b, and 6c, respectively.

$$\dot{x} = JH'(x)$$
$$x(0) = x(T) \tag{10}$$

$$RMSE = \sqrt{\frac{\sum_{i=1}^{G} \frac{\sum_{l=1}^{n_i}(\hat{\mathcal{X}}_{il} - \mathcal{X}_{il})^2}{n_i}}{G}},$$

$$MSE = (RMSE)^2, \quad MAE = \frac{\sum_{i=1}^{G} \frac{\sum_{l=1}^{n_i}|\hat{\mathcal{X}}_{il} - \mathcal{X}_{il}|}{n_i}}{G},$$

$$MAPE = \frac{\sum_{i=1}^{G} \frac{\sum_{l=1}^{n_i}|\frac{\hat{\mathcal{X}}_{il} - \mathcal{X}_{il}}{\mathcal{X}_{il}}|}{n_i}}{G} \times 100,$$

$$R^2 = \frac{1 - \sum_{i=1}^{G}\sum_{l=1}^{n_i}(\mathcal{X}_{il} - \hat{\mathcal{X}}_{il})^2}{1 - \sum_{i=1}^{G}\sum_{l=1}^{n_i}(\mathcal{X}_{il} - \bar{\mathcal{X}}_{il})^2} \tag{11}$$

$$h = \frac{M_{\mathcal{X}} - M_{\hat{\mathcal{X}}}}{SD_{\mathcal{X}\hat{\mathcal{X}}}}, \quad SD_{\mathcal{X}\hat{\mathcal{X}}} = \sqrt{\frac{SD_{\mathcal{X}}^2 + SD_{\hat{\mathcal{X}}}^2}{2}} \tag{12}$$

where, $M_{\mathcal{X}}$ and $M_{\hat{\mathcal{X}}}$ are the sample means of $\mathcal{X}$ and $\hat{\mathcal{X}}$, respectively; $SD_{\mathcal{X}}$ and $SD_{\hat{\mathcal{X}}}$ the respective standard deviations.

4.2   Analysis of Results

Table 2 summarizes different statistical parameters retrieved from analyzing the probability density functions of the original data that contains missing values, and the different imputed datasets. Using the models described in Sect. 3, irradiance, ambient temperature, and module temperature were imputed. A mark of good performance is when the imputation does not significantly alter the data's statistical properties such as mean, median, standard deviation, skewness, and kurtosis. In Table 2, the data marked as "Original Data" refers to the dataset with missing values. Properties of the original and imputed data should match if the imputation method performed well. Accordingly, by this comparison, it is seen that for irradiance, kNN, random forest, and multiple imputation methods performed well for most of the aforementioned properties. This observation also corroborates for the other two attributes, where closeness of the imputed values to the original values is mostly within the acceptable error limits of ± 0.1.

The residuals for imputed values are shown in Table 3, where for irradiance, kNN, random forest, and multiple imputation methods have lower RMSE, MSE, and MAE scores than other methods. A higher R^2 statistic is observed for

Table 2. Statistical parameters for original and imputed datasets

Parameter	Value	Original data	Random	Multiple	kNN	Mean value	Median value	Random forest
Mean	Irradiance	193.51	194	193.59	193.55	193.54	183.89	193.471
	Ambient Temp	79.62	79.65	79.62	79.62	79.66	79.63	79.62
	Module Temp	80.6	80.65	80.6	80.6	80.62	80.49	80.61
Median	Irradiance	4.89	5.48	10.75	4.99	22.35	5.03	5.292
	Ambient Temp	79.05	79.09	79.06	79.056	79.56	79.07	79.065
	Module Temp	78.05	78.06	78.05	78.045	78.82	78.04	78.04
Standard Deviation	Irradiance	274.14	274.35	274.25	274.14	267.1	270.31	273.904
	Ambient Temp	23.25	23.23	23.24	23.249	22.69	22.69	23.246
	Module Temp	25.16	25.26	25.14	25.15	24.58	24.58	25.152
Skewness	Irradiance	1.29	1.287	1.288	1.289	1.326	1.369	1.289
	Ambient Temp	6.41	6.43	6.38	6.412	6.605	6.609	6.415
	Module Temp	4.89	4.89	4.88	4.888	5.012	5.023	4.888
Kurtosis	Irradiance	3.48	3.474	3.494	3.479	3.67	3.704	3.476
	Ambient Temp	62.861	62.97	62.49	62.86	66.16	66.19	62.905
	Module Temp	43.69	43.54	43.69	43.71	45.82	45.88	43.7

these methods. The MAPE scores for irradiance were not calculable because it uses the observed values of irradiance in the denominator, which were zero for specific instances of time (when there was no sunlight). A similar trend in the variation of error metrics is observed for the other two attributes as well. Among kNN, random forest, and multiple imputation, however, kNN does the best, followed closely by random forest, and later by multiple imputation. This sequence corroborates with the results of Table 2. For each of the three attributes, five statistical properties have been tabulated, making a total of 15 parameters. The values imputed using kNN had 13 instances where the original and imputed values were similar with an error margin of ± 0.1, followed closely by random forest with 12 instances within the margin, then by multiple imputation with 10 instances within the margin.

The Cohen's distance values computed using Eq. (12) for the three attributes is shown in Fig. 6. For irradiance, all methods except median value imputation have values less than 0.2, but random forest has the lowest difference in proportions.

Table 3. Average values of error metrics for the different imputation methods

Imputed variable	Imputation Method	RMSE	MSE	MAPE	R^2	MAE
Irradiance	Random	212.05	44964	NaN	0.49	81.37
	Multiple	48.75	2385.41	NaN	0.968	17.43
	kNN	8.09	65.54	NaN	0.999	2.06
	Mean value	149.76	22427	NaN	0.698	67.499
	Median value	181.92	33096	NaN	0.608	57.573
	Random forest	29.01	841.77	NaN	0.989	7.184
Ambient temperature	Random	17.68	312	34.7	0.483	4.194
	Multiple	6.12	37.53	6.738	0.928	1.422
	kNN	0.392	0.163	0.002	0.999	0.163
	Mean value	12.23	149.63	34.74	0.707	2.59
	Median value	12.22	149.45	34.48	0.707	2.587
	Random forest	1.417	2.008	0.006	0.996	0.444
Module temperature	Random	19.42	377	23.48	0.487	5.668
	Multiple	6.095	37.242	5.877	0.939	1.448
	kNN	0.508	0.258	0.002	0.999	0.163
	Mean value	12.92	167	23.011	0.723	3.61
	Median value	12.96	168	22.27	0.722	3.537
	Random forest	1.729	2.992	0.007	0.995	0.547

For ambient and module temperatures, kNN, multiple imputation, and random forest tend to have the smallest of all differences in proportions. While all methods had a difference less than 0.2 for ambient temperature, random imputation and median value imputation had Cohen's values above 0.2 and less than 0.5 for module temperature, implying a medium difference. These results show that for the application of estimating missing values in PV generation data, kNN, random forests, and multiple imputation using expectation maximization could be the most applicable methods. This conclusion is further verified by feeding imputed values of irradiance, ambient temperature, and module temperature as inputs into a 5-layer MLP model to predict PV generation.

4.3   Estimating PV Generation

The MLP model used in this paper was designed to predict PV generation using irradiance, ambient temperature, and module temperature as features. The model's architecture has been designed on a trial and error basis with a combination of hyperparameters that yields the most accurate prediction. In the future, the tuning of these model hyperparameters could be explored. comprises 1 input layer with 3 units, each representing a feature, 3 hidden layers with 25, 11, and 6 units, and an output layer with 1 unit whose output is the target, PV generation. Each layer, l, has a specific activation function, with weights initialized to samples drawn from Xavier uniform distribution [51] for the first hidden layer ($l=2$) and uniform distribution [52] for subsequent layers ($l=3,4,5$): tanh activation [53] for $l=2$, sigmoid [54] for $l=3$, tanh for $l=4$, and softplus [55] for $l=5$. All datasets (original and imputed) are of the same length, and are split to have 85% of the values for training and 15% for validation and testing. MSE is used as the loss function, defined as $\mathcal{L}(\hat{y}^{(l)}, y^{(l)}) = \frac{1}{m}\sum_{k=1}^{m}(\hat{y}_k^{(l)} - y_k^{(l)})^2$, where l is the index of MLP layers, m is the number of samples, $y = \hat{\mathcal{X}}_{observed}$ and $\hat{y} = \hat{\mathcal{X}}_{predicted}$. To train the model, backpropagation was used, with an improvement to the traditional stochastic gradient descent by calculating individual adaptive learning rates from the first and second moment estimates of the gradients for each parameter [56]. Validation was conducted using holdout method, with the testing dataset reused for validation. Further details about the model is beyond the scope of this paper.

In Table 4, the error metrics that measure MLP's performance in PV generation prediction are summarized. These metrics are calculated using the same equations as in Eq. (11), but here, y and $\hat{y}$ replace $\mathcal{X}$ and $\hat{\mathcal{X}}$. Imputed datasets are fed to predict PV generation and the model's performance is analyzed by comparing with the corresponding observed generation values. This serves as another way to evaluate the imputation performance because the same MLP model is able to train on imputed values and understand the nonlinear relationships between the features and target, thereby predicting generation that matches closely with observed values. Among the imputed datasets, kNN and

Table 4. MLP performance on different imputed datasets.

Imputation method	Phase	RMSE	MSE	MAE
Random	Training	11.954	1.429	5.948
	Validation	11.269	1.27	4.919
Multiple	Training	2.994	0.091	1.528
	Validation	1.423	0.008	0.832
kNN	Training	5.083	0.258	2.678
	Validation	2.029	0.041	1.358
Mean value	Training	8.628	0.744	4.672
	Validation	6.319	0.339	3.284
Median value	Training	9.278	0.861	4.138
	Validation	6.778	0.459	2.575
Random forest	Training	4.913	0.241	2.486
	Validation	1.929	0.037	1.297

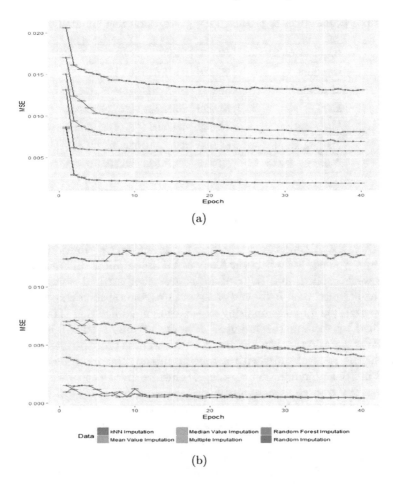

Fig. 7. Variation of MSE values across 40 epochs during: (a) MLP model training and (b) MLP model validation.

random forest have low error values. This is also supported by observing how MSE varies across the 40 epochs of training (Fig. 7a) and validation (Fig. 7b) of the MLP model. Specifically, it can be seen that during both these phases, the only imputed datasets that have a similar trend in values of MSE vs number of epochs are kNN and random forest. Table 5 shows the error metrics that now measure the difference in values between the PV generation values predicted using different imputation methods. Again, the values of RMSE, MSE, MAE, and MAPE provide reasonable credibility that among all methods, kNN and random forest performed consistently better, followed by multiple imputation.

Table 5. Residuals of predicted PV generation using imputed datasets

Imputation method	RMSE	MSE	MAE	MAPE
Random	97.286	9464.623	51.18	7.176
Multiple	52.071	2711.403	25.408	0.68
kNN	51.229	2624.427	26.786	0.964
Mean value	68.265	4660.081	38.427	3.898
Median value	65.963	4351.110	34.369	2.418
Random forest	50.235	2523.533	25.395	0.982

5   Conclusion and Future Work

In this paper, the missingness mechanism for data from distribution grid-tied PV systems was studied. Following the likely MAR assumption, different imputation methods were applied, and their performances compared. Results showed that among the six methods, kNN and random forests performed the best, followed closely by multiple imputation using expectation maximization. The results were also verified by feeding the imputed datasets as inputs to a 5-layer MLP to predict PV generation. A key limitation of this work is that the methods working best with this type of data with a specific missingness mechanism might not apply to another domain. Hence, the conclusions from this work can be used by utilities, PV system owners, installers, and analysts who work with the data from the systems for higher-level analytics. The future work will explore the possibilities of using one measurement to recreate, synthesize, or estimate the other parameters within a dataset. Performance analysis of imputed datasets by predicting PV generation was considered in this work, but the prediction model could be fine-tuned by considering more input variables such as wind velocity, cloud coverage, and more.

Acknowledgments. The work published is a result of the research sponsored by the National Science Foundation (NSF) CNS division under the award 1553494.

References

1. Sundararajan, A., Olowu, T.O., Wei, L., Rahman, S., Sarwat, A.I.: A case study on the effects of partial solar eclipse on distributed photovoltaic systems and management areas. IET Smart Grid (2019)
2. Peterson, Z., Coddington, M., Ding, F., Sigrin, B., Saleem, D., Horowitz, K., et al.: An overview of distributed energy resource (DER) interconnection: current practices and emerging solutions. NREL Tech. rep. (number NREL/TP-6A20-72102), April 2019. https://www.nrel.gov/docs/fy19osti/72102.pdf
3. Sarwat, A.I., Amini, M., Domijan, A., Damjanovic, A., Kaleem, F.: Weather-based interruption prediction in the smart grid utilizing chronological data. J. Mod. Power Syst. Clean Energy **2**, 308–315 (2015)

4. Sundararajan, A., Khan, T., Moghadasi, A., Sarwat, A.I.: A survey on synchrophasor data quality and cybersecurity challenges, and evaluation of their interdependencies. J. Mod. Power Sys. Clean Energy, 1–19 (2018)

5. Jeng, R.S., Kuo, C.Y., Ho, Y.H., Lee, M.F., Tseng, L.W., Fu, C.L., et al.: Missing data handling for meter data management system. In: Proceedings of the Fourth International Conference on Future Energy Systems. e-Energy 2013, pp. 275–276. ACM, New York (2013). https://doi.org/10.1145/2487166.2487204

6. Peppanen, J., Zhang, X., Grijalva, S., Reno, M.J.: Handling bad or missing smart meter data through advanced data imputation. In: 2016 IEEE Power Energy Society Innovative Smart Grid Technologies Conference (ISGT), pp. 1–5 (2016)

7. Kodaira, D., Han, S.: Topology-based estimation of missing smart meter readings. MDPI Energies 11(224), 1–18 (2018)

8. Majidpour, M., Chu, P., Gadh, R., Pota, H.R.: Incomplete data in smart grid: treatment of missing values in electric vehicle charging data. In: 2014 International Conference on Connected Vehicles and Expo (ICCVE), pp. 1041–1042 (2014)

9. Genes, C., Esnaola, I., Perlaza, S.M., Ochoa, L.F., Coca, D.: Robust recovery of missing data in electricity distribution systems. IEEE Trans. Smart Grid, 1 (2018)

10. Olowu, T.O., Jafari, M., Sarwat, A.I.: A multi-objective optimization technique for Volt-Var control with high PV penetration using genetic algorithm. In: 2018 North American Power Symposium (NAPS), pp. 1–6 (2018)

11. Jurado, S., Nebot, A., Mugica, F., Mihaylov, M.: FIR forecasting strategies able to cope with missing data: a smart grid application. Appl. Soft Comput. 51, 225–238 (2017)

12. Khalid, A., Sundararajan, A., Sarwat, A.I.: A multi-step predictive model to estimate Li-Ion state of charge for higher C-rates. In: 2019 IEEE International Conference on Environment and Electrical Engineering and 2019 IEEE Industrial and Commercial Power Systems Europe (EEEIC / I&CPS Europe) (2019)

13. Olowu, T.O., Sundararajan, A., Moghaddami, M., Sarwat, A.: Fleet aggregation of photovoltaic systems: a survey and case study. In: 2019 IEEE Power Energy Society Innovative Smart Grid Technologies Conference (ISGT) (2019)

14. Sundararajan, A., Chavan, A., Saleem, D., Sarwat, A.I.: A survey of protocol-level challenges and solutions for distributed energy resource cyber-physical security. MDPI Energies 9, 2360 (2018)

15. Parvez, I., Sarwat, A.I., Wei, L., Sundararajan, A.: Securing metering infrastructure of smart grid: a machine learning and localization based key management approach. Energies, 9(9) (2016). https://www.mdpi.com/1996-1073/9/9/691

16. Sundararajan, A., Wei, L., Khan, T., Sarwat, A.I., Rodrigo, D.: A tri-modular framework to minimize smart grid cyber-attack cognitive gap in utility control centers. In: 2018 Resilience Week (RWS), pp. 117–123 (2018)

17. Sundararajan, A., Sarwat, A.I., Pons, A.: A survey on modality characteristics, performance evaluation metrics, and security for traditional and wearable biometric systems. ACM Comput. Surv. 52(2), 1–35 (2019)

18. Wei, L., Sundararajan, A., Sarwat, A.I., Biswas, S., Ibrahim, E.: A distributed intelligent framework for electricity theft detection using Benford's law and stackelberg game. In: Resilience Week. pp. 5–11 (2017)

19. Zhang, Y., Huang, T., Bompard, E.F.: Big data analytics in smart grids: a review. Energy Inf. 1(1), 8 (2018). https://doi.org/10.1186/s42162-018-0007-5

20. Olowu, T.O., Sundararajan, A., Moghaddami, M., Sarwat, A.I.: Future challenges and mitigation methods for high photovoltaic penetration: a survey. Electr. Power Syst. Res. (2018)

21. Jerez, J.M., Molina, I., Garcia-Laencina, P.J., Alba, E., Ribelles, N., Martin, M., et al.: Missing data imputation using statistical and machine learning methods in a real breast cancer problem. Artif. Intell. Med. **50**, 105–115 (2010)
22. Mittag, N.: Imputations: benefits, risks and a method for missing data (2013). http://home.cerge-ei.cz/mittag/papers/Imputations.pdf
23. Schmitt, P., Mandel, J., Guedj, M.: A comparison of six methods for missing data imputation. J. Biometrics Biostatistics **6**(1), 1–6 (2015)
24. Livera, A., Phinikarides, A., Makrides, G., Georghiou, G.E.: Impact of missing data on the estimation of photovoltaic system degradation rate. In: 2017 IEEE 44th Photovoltaic Specialist Conference (PVSC), pp. 1954–1958 (2017)
25. Miao, X., Gao, Y., Guo, S., Liu, W.: Incomplete data management: a survey. Front. Comput. Sci. **12**(1), 4–25 (2018)
26. Kang, H.: The prevention and handling of the missing data. Korean Soc. Anesthesiologists **64**(5), 402–406 (2013)
27. Addo, E.D.: Performance comparison of imputation algorithms on missing at random data. Master's Thesis submitted to East Tennessee State University, pp. 1–129 (2018). https://dc.etsu.edu/cgi/viewcontent.cgi?article=4839&context=etd
28. Horton, N.J., Kleinman, K.P.: Much ado about nothing: a comparison of missing data methods and software to fit incomplete data regression models. Am. Stat. **1**, 79–90 (2011)
29. Sundararajan, A., Sarwat, A.I.: Roadmap to prepare distribution grid-tied photovoltaic site data for performance monitoring. In: International Conference on Big Data, IoT and Data Science (BID), pp. 101–115 (2017)
30. Anzalchi, A., Sundararajan, A., Moghadasi, A., Sarwat A. power quality and voltage profile analyses of high penetration grid-tied photovoltaics: a case study. In: 2017 IEEE Industry Applications Society Annual Meeting, pp. 1–8 (2017)
31. Zhang, Z.: Missing data exploration: highlighting graphical presentation of missing pattern. Ann. Transl. Med. **3**(22), 356–362 (2015)
32. Kabacoff, R.I.: Advanced Methods for missing data (2015). https://rstudio-pubs-static.s3.amazonaws.com/4625_fa990d611f024ea69e7e2b10dd228fe7.html
33. Cheng, X., Cook, D., Hofmann, H.: Visually exploring missing values in multivariable data using a graphical user interface. J. Stat. Softw. **68**(6), 1–23 (2015)
34. Rubin, D.B.: Inference and missing data. Biometrika **63**(3), 581–592 (1976)
35. Little, R.J.A., Rubin, D.B.: Statistical Analysis with Missing Data. 2nd edn. (2002). https://www.wiley.com/en-us/Statistical+Analysis+with+Missing+Data%2C+2nd+Edition-p-9780471183860
36. Garson, G.D.: Missing values analysis & data imputation. In: Statistical Associates Blue Book Series (2015)
37. Jamshidian, M., Jalal, S.: Tests of Homoscedasticity, Normality, and Missing Completely at Random for Incomplete Multivariate Data. J. Psychometrika **75**(4), 649–674 (2010)
38. Srivastava, M.S., Dolatabadi, M.: Multiple imputation and other resampling schemes for imputing missing observations. J. Multivar. Anal. **100**(9), 1919–1937 (2009)
39. Rayner, G.D., Rayner, J.C.W.: Power of the Neyman smooth tests for the uniform distribution. J. Appl. Math. Decis. Sci. **5**(3), 181–191 (2001)
40. Yu, B.P., Lemeshko, B.: A review of the properties of tests for uniformity. In: 12th International Conference on Actual Problems of Electronics Instrument Engineering (APEIE), vol. 1 (2014)

41. Jamshidian, M., Jalal, S., Jansen, C.: MissMech: an R package for testing homoscedasticity, multivariate normality, and missing completely at random (MCAR). J. Stat. Softw. **56**(6), 1–31 (2014)

42. Gelman, A.: Missing-data imputation. In: Data Analysis Using Regression and Multilevel/Hierarchical Models, pp. 529–544 (2006)

43. Zhang, Z.: Missing data imputation: focusing on single imputation. Ann. Transl. Med. **1**, 1–8 (2016)

44. Lodder, P.: To Impute or not Impute: That's the Question. Paper Methodological Advice, University of Amsterdam, pp. 1–7. http://www.paultwin.com/wp-content/uploads/Lodder_1140873_Paper_Imputation.pdf

45. Dempster, A.P., Laird, M.M., Rubin, D.B.: Maximum likelihood from incomplete data via the EM algorithm. J. Roy. Stat. Soc. B. **39**(1), 1–38 (1977). www.jstor.org/stable/2984875

46. Honaker, J., King, G.: What to do about missing values in time-series cross-section data. Am. J. Polit. Sci. **2**, 561–581 (2010)

47. Honaker, J., King, G., Blackwell, M.: Amelia II: a program for missing data. J. Stat. Softw. **7**, 1–47 (2011). https://gking.harvard.edu/files/gking/files/amelia_jss.pdf

48. Beretta, L., Santaniello. A.: Nearest neighbor imputation algorithms: a critical evaluation. In: 5th Translational Bioinformatics Conference (TBC), no. 3, pp. 198–208 (2015)

49. Stekhoven, D.J., Buhlmann, P.: MissForest-non-parametric missing value imputation for mixed-type data. J. Bioinform. **1**, 112–118 (2012)

50. Khalid, A., Sundararajan, A., Acharya, I., Sarwat, A.I.: Prediction of Li-Ion battery state of charge using multilayer perceptron and long short-term memory models. In: 2019 IEEE Transportation Electrification Conference (ITEC) (2019)

51. Glorot, X., Bengio, Y.: Understanding the difficulty of training deep feedforward neural networks. In: 13th International Conference on Artificial Intelligence and Statistics (AISTATS), pp. 1–8 (2010)

52. Nguyen, D., Widrow, B.: Improving the learning speed of 2-layer neural networks by choosing initial values of the adaptive weights. In: 1990 IJCNN International Joint Conference on Neural Networks, vol. 3, pp. 21–26 (1990)

53. Kalman, B.L., Kwasny, S.C.: Why tanh: choosing a sigmoidal function. In: Proceedings of IJCNN International Joint Conference on Neural Networks. vol. 4, pp. 578–581 (1992)

54. Karlik, B., Olgac, A.V.: Performance analysis of various activation functions in generalized MLP architectures of neural networks. Int. J. Artif. Intell. Expert Syst. **4**, 111–122 (2011)

55. Zheng, H., Yang, Z., Liu, W., Liang, J., Li, Y.: Improving deep neural networks using softplus units. In: 2015 International Joint Conference on Neural Networks (IJCNN), pp. 1–4 (2015)

56. Kingma, D.P., Ba, J.L.: Adam: a method for Stochastic Optimization. In: International Conference on Machine Learning, pp. 1–15 (2015)

Document Clustering by Relevant Terms: An Approach

Cecilia Reyes-Peña, Mireya Tovar Vidal$^{(\boxtimes)}$, and José de Jesús Lavalle Martínez

Faculty of Computer Science, Benemérita Universidad Autónoma de Puebla,
14 sur y Av. San Claudio, C.U., Puebla, Puebla, Mexico
reyesp.cecilia@gmail.com, mtovar@cs.buap.mx, jlavallenator@gmail.com

Abstract. In this work, a document clustering based on relevant terms into an untagged medical text corpus approach is presented. To achieve this, to create a list of documents containing each word is necessary. Then, for relevant term extraction, the frequency of each term is obtained in order to compute the word weight into the corpus and into each document. Finally, the clusters are built by mapping using main concepts from an ontology and the relevant terms (only subjects), assuming that if two words appear in the same documents these words are related. The obtained clusters have a category corresponding to ontology concepts, and they are measured with cluster from K-Means (assuming the k-Means cluster were well formed) using the Overlap Coefficient and obtaining 70% of similarity among the clusters.

Keywords: Documents clustering · Relevant terms · Medical corpus

1 Introduction

Nowadays, onto the web the information does not have a general structure that indicates the specific domain to which it belongs. Sometimes, there is grouped information (corpus) about general domains which facilitate its use, but at the same time, the information does not have indicators for specifying the main categories into the domain.

In natural language processing, the use of relevant terms is very important for identifying the main content of a document in a synthesized way; in general, the relevant terms represent the main idea of a text. The relevant terms can be represented by one word (keyword) or more words (keyphrase), the keywords extraction task is simpler than keyphrases extraction, because the use of keyphrases involves problems associated to synonym and polysemy [1]. The relevant term identification is a powerful tool that allows to extract from a text some features in order to simplify the document clustering. Document clustering is a task belonging to unsupervised learning that does not require train data and does not have a defined label for the obtained cluster [2]; this task separates text documents into clusters according their informative features; the type of features depends on the selected methods for information processing [3]. The clustering

© Springer Nature Switzerland AG 2020
K. Arai et al. (Eds.): FTC 2019, AISC 1069, pp. 610–617, 2020.
https://doi.org/10.1007/978-3-030-32520-6_44

into the medical area permits to manage documents from the similarity between them, facilitating the information search around a specific topic.

In this work, a document clustering based on relevant terms into a medical document corpus approach is presented. For this it is necessary to create a list of documents which contain each word; then, for relevant term extraction each frequency was obtained for calculating the word weight into the corpus. Finally, the clusters are built by mapping using main concepts from an ontology and the relevant terms, assuming that if two words exist in the same documents these words are related. The medical documents were obtained from MEDLINE [4] repository, which contains material related to biomedicine and health care as documents, short texts, and books [5]. The results are compared whit cluster from K-Means.

This document is divided into five sections: Sect. 2 contains the related works about the relevant term extraction and document clustering; Sect. 3 describes the method used in this work, Sect. 4 shows the obtained results, and Sect. 5 contains conclusions and future work.

2   Related Works

There are many strategies for relevant term extraction, which consider different approaches as linguistic (using linguistic features of the words, sentences and documents), statistical (independent of the language), based on relevant term location in the document and term weight among others [1]. Habibi et al. [6] present an approach for document recommendation in conversations by keyword extraction and clustering; a model topics in conversation based in probabilistic latent semantic is using for keyword extraction and the keyword clustering uses a ranking score for each word according a main topic.

Concerning document clustering, Steinbach et al. [7] presented a comparison between two algorithms for document clustering: agglomerative hierarchical clustering and K-Means (agglomerative and traditional). In the same context, Balabantaray et al. [8] made a comparison between K-Means and K-Medoids for document clustering; they converted the documents to vectors by tokenization, stop-words removal, weighted using Term Frequency- Inverse Document Frequency (TF-IDF) and a matrix using weight term in each document, which then would be used as input for both algorithms.

Pinto and Rosso [5] propose a short text clustering using around 900 abstracts about cancer from MEDLINE by Expectation Maximization and K-means using categories as: blood, bone, brain, genitals, liver, therapy among others; the results were evaluated by F-measure. There is a proposal to create an algorithm for document clustering that allow the overlap among the cluster starting a retrieval information system in order to satisfy user profile identification task using tagged list for 3-grams and TF-IDF weight [9]. Jensi et al. [2] propose some steps for document clustering: text document collections preprocessing, document-term matrix construction, dimension reduction, text document clustering and optimization.

Other approach for document clustering is the proposed by Sunghae Jun et al. [10], starting with a document-term matrix, assigned a score based on frequencies to reduce the number of variables, then they mapped the data using Gaussian kernel and determining the number of cluster by Silhouette measure; finally the clusters are built using K-Means.

3   Approach

The presented approach has two main stages: Pre-processing Text and Clustering (Fig. 1). The first stage processes the corpus to get the relevant terms, and using set theory, a relationship percentage can be assigned to them. At the second stage, the relevant terms are mapped whit main concepts from Disease Ontology (DO) [11] to assign a category and group with the index list.

3.1   Pre-processing Text

The first stage in this task is the morphological and syntactic tagging for all words to identify the keywords and the keyphrases considering the noun locations. Then the index lists are created, for this, a number for each document is assigned (indexing) and each word is stored with the index of the document in which it appears (see Fig. 2).

The next process is about the relevant term extraction, those terms having highest weight values must be appropriated for representing a category and considered as *relevant terms*. To assign a weight value, a simple ranking technique is used. The simple ranking technique calculates the frequency of each term on each document and the relevance of each term in the corpus. Then, using those values, the final weight is obtained [12]. To decide if a term can be a category, each term is tagged, based on its lexical properties, and only the nouns are selected.

3.2   Clustering by Relevant Terms

For document clustering, the amount of documents where two relevant terms appear is considered as an indicator of a relationship between the relevant terms. To define the relationship grade between the terms using the set theory, the similarity is calculated using the average of the Jaccard (Eq. 1), Sorensen-Dice (Eq. 2), Overlap (Eq. 3) and Tversky (Eq. 4) coefficients; where A and B are document index sets of two relevant terms, respectively.

$$Jaccard = \frac{|A \cap B|}{|A \cup B|} \tag{1}$$

$$Sonrense - Dice = 2 * \frac{|A \cap B|}{|A| + |B|} \tag{2}$$

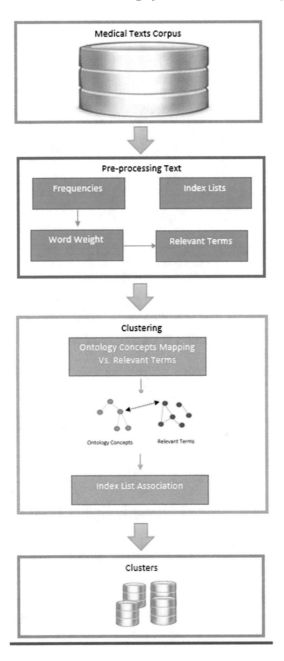

Fig. 1. Proposed approach.

DOCUMENT NUMBER	CONTENT
1	Adverse reactions to antimicrobial agents consist of toxics effects mediated ...
2	Records of 75 patients with syringomyellia are review. Their clinical course fell into one of three main ...
3	Eleven patients with established Hodgkin's disease were treated with vinblastine sulfate ...
4	During an outbreak of infectious hepatitis at a housing development, Coxsackie a10 virus ...
...	...

Term	Document Index
Patient	1, 2, 4, ...
Records	2, 5, 7, ...
Hodgkin's disease	3, 9, 11, ...
Toxic effects	1, 15, 66, ...
Cancer	6, 10, 17, ...
Pregnancy	7, 22, 41, ...
Complications	8, 56, 72, ...
Treatment	6, 11, 89, ...
Virus	4, 19, 21, ...
Clinical course	10, 21, 34, ...
Child	21, 45, 78, ...
Mental health	99, 101, 512, ...
...	...

Fig. 2. Index lists.

$$Overlap = \frac{|A \cap B|}{min(|A|, |B|)} \tag{3}$$

$$Tversky = \frac{|A \cap B|}{|A \cup B| + \alpha|A - B| + \beta|B - A|} \tag{4}$$

The categories used to group the documents are taken from DO. Figure 3 shows the DO main concepts inside the box which are mapped to relevant terms and their relationships with other relevant terms in order to add the document index list of each term.

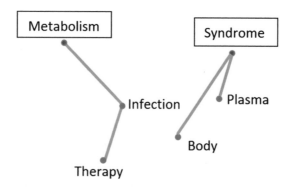

Fig. 3. Relevant terms relationships with ontology concepts.

4   Results

In this method, 50,433 short texts were used, those were extracted from MED-LINE without a particular category, the obtained terms are 281053, and only 500 of them were considered as relevant terms. The total of relationships among the relevant terms is 124750 of which 2722 indicate a null relation (grade of relation equal to 0) and 122028 indicate a grade of relation (bigger than 0 and smaller or equal than 1).

The total of concepts taken from the DO was 190, of which only five have tied with the document index list. The most results are shown in the next table (see Table 1).

Table 1. Document clustering by ontology concepts.

Concept	Number of documents	Related terms
Cancer	49667	Study, Skin, Stage, Leukemia, Cycle, Bone, Laboratory, Hormone, Experience, Lack, Risk
Syndrome	49667	Agent, Mass, Cytoplasmic, Sex, Child, Human, DNA, Diagnosis
Metabolism	49644	Excretion, Cholesterol, Energy, Fat, Activity, Treatment, Acid, Protein, Gene
Mental Health	2226	Experimental
Cellular proliferation	1211	Intracellular, Proliferation

In this work, k-Means is taken as baseline for measuring the cluster construction, for this, each document of MEDLINE corpus was converted in a vector (*Word2Vec*) and we used a $k = 5$. In Table 2, it can see the amount of document and relevant terms per cluster obtained by k-Means; the cluster relevant terms show a difficulty for assigning a category that is solved by the proposed work.

Table 2. Document clustering by K-Means.

Cluster	Number of documents	Related terms
Cluster1	27720	Cell, Blood, Protein, Tumor, Cancer, Virus, DNA
Cluster2	11035	Activity, Cell, Protein, Treatment, Gene, Plasma
Cluster3	3413	DNA, Virus, Presence, RNA, Apoptosis, Membrane
Cluster4	1693	Group, Protein, Treatment, Effect, DNA, Disease, Plasma, Factor
Cluster5	6571	Cell, DNA, Gene, Disease, Age, Blood, PH

In order to evaluate the cluster, the resulted clusters were compared with the obtained clusters by the ontology concepts using the Overlap Coefficient (see Table 3).

Table 3. Similarity among clusters.

K-Means clusters	Ontology concept clusters	Overlap coefficient
Cluster1	Mental Health	0.5219
Cluster2	Syndrome	0.9982
Cluster3	Metabolism	0.9967
Cluster4	Cellular Proliferation	0.0316
Cluster5	Cancer	0.9878

k-Means is a popular clustering algorithm with good results, so that we can assume the high similarity values in the three clusters indicate the clusters are well formed. However, there is a cluster with similarity less than 0.1, this could be caused by the overlap into the proposed work clusters.

5   Conclusions and Future Work

This work presents a document clustering approach based on relevant terms and their relations using the same concepts of DO as categories, this causes that all proposed categories are focused to disease classification and does not consider other topics related to the medical area. Considering the obtained results, it is concluded that the document clustering is difficult task when the domains in a corpus are very close, i.e., all the documents belong the same general domain that can be divided into topics more specific given that only around the 2.5% of the ontology concepts were related with the relevant terms in spite of the corpus and the ontology are in the same domain. About the cluster evaluation, the proposed work has around 70% of similarity respect k-Means clustering, that was obtained by the average of cluster similarity; this percentage is expected to rise when the overlap among the clusters be deleted. However, the proposed work has the advantage about to assign a category for the clusters.

In future work, others weighted strategies will be implemented in order to provide new relevant terms which can be integrated into the DO, to identify relevant terms that can strengthen the relevant term selection and develop algorithms to delete the overlap into the clusters and create rules for automatic population of documents, using the ontology as the main structure.

Acknowledgment. This work is supported by the Sectoral Research Fund for Education with the CONACyT project 257357, and partially supported by the VIEP-BUAP project.

References

1. Siddiqi, S., Sharan, A.: Keyword and keyphrase extraction techniques: a literature review. Int. J. Comput. Appl. **109**(2), 18–23 (2015)

2. Jensi, R., Wiselin, J.G.: A survey on optimization approaches to text document clustering. Int. J. Comput. Sci. Appl. **3**, 31–44 (2013)
3. Abualigah, L.M., Khader, A.T., Al-Betar, M.A., Alomari, O.A.: Text feature selection with a robust weight scheme and dynamic dimension reduction to text document clustering. Expert Syst. Appl. **84**, 24–36 (2017)
4. Medline. https://www.nlm.nih.gov/bsd/pmresources.html. Accessed 02 Aug 2019
5. Pinto, D., Rosso, P.: KnCr: a short-text narrow-domain sub-corpus of medline. In: Proceedings of TLH-ENC 2006, pp. 266–269 (2006)
6. Habibi, M., Popescu-Belis, A.: Keyword extraction and clustering for document recommendation in conversations. IEEE/ACM Trans. Audio Speech Lang. Process. **23**, 746–759 (2015)
7. Steinbach, M., Karypis, G., Kumar, V., et al.: A comparison of document clustering techniques. In: KDD Workshop on Text Mining, Boston, vol. 400, pp. 525–526 (2000)
8. Balabantaray, R.C., Sarma, C., Jha, M.: Document clustering using k-means and k-medoids. CoRR, abs/1502.07938 (2015)
9. Beltrán, B., Ayala, D.V., Pinto, D., Martínez, R.: Towards the construction of a clustering algorithm with overlap directed by query. Res. Comput. Sci. **145**, 97–105 (2017)
10. Jun, S., Park, S.-S., Jang, D.-S.: Document clustering method using dimension reduction and support vector clustering to overcome sparseness. Expert Syst. Appl. **41**(7), 3204–3212 (2014)
11. Disease ontology. https://www.disease-ontology.org. Accessed 02 Aug 2019
12. Reyes-Peña, C., Pinto-Avendaño, D., Vilariño Ayala, D.: Emotion classification of twitter data using an approach based on ranking. Res. Comput. Sci. **147**(11), 45–52 (2018)

Smart Consumption Study Using Data Analysis and IoT in Residential Level Refrigeration Machines

X. de L. Luiguy[(⊠)], H. O. A. Clayton, and H M. S. Sergio

Estácio de Belém, CEP 66, Belém, Pará 812-480, Brazil
lima.1x10@gmail.com

Abstract. The present article pretends to dissert about the theme: "Case study using Internet of Things", with the objective to define, elucidate and classify the IOT's functions, directly impacts and non-directly, which this research would have on medium and long term. This work pretends to explain, through correlated works, or with directly relation to theme, doing an analytic approach and true about the subject and comparison of data together the Evolution of technology linked to the use of IOT for the building automation. The data exposed by this article will be acquired from a prototype created by the authors with the objective to analyze the energy efficiency inside a controlled room of a building. The data will be collected and will be analyzed to the making of energetic profile referring to that building and of this analysis are created proposals to increase the energetic efficiency based on that profile.

Keywords: Internet of Things · Energy efficiency · Building's smartification · Power saving

1 Introduction

The buildings represent about 30% of the total energy consumption of the planet and are part of the sectors with the largest participation in the world energy matrix. Soon Brazil will account for more than 30% of total energy consumption as the processes of urbanization, middle class growth and modeling of buildings accelerate. Therefore, the services, commercial and public sectors, as well as residential, are essential for any energy efficiency policy.

In Brazil, under the National Policy for the Conservation of the Rational Use of Energy, the Ministry of Mines and Energy approved the technical regulation for the energy efficiency of commercial Buildings and public Services in 2009. In the year 2010, the analogous regulation for residential buildings entered into force. Both regulations establish the requirements for evaluation and classification of Energy Efficiency in Buildings, and they were made compulsory after 5 years of publication [1].

According to the 10-year plan for energy expansion, we will have the residential energy consumption of approximately 166,888 GWh/year by 2020 in the national territory, if we add a commercial area of 123,788 GWh/year, we will have a total of 290,676 GWh/year, which may be higher than the industrial area in 2020 [5, 8].

© Springer Nature Switzerland AG 2020
K. Arai et al. (Eds.): FTC 2019, AISC 1069, pp. 618–630, 2020.
https://doi.org/10.1007/978-3-030-32520-6_45

Well, according to the Folha de São Paulo newspaper. Brazil in the year 2014 wasted more than 10% of its electricity, totaling around R $12.64 billion. The half being residential consumers [2]. "Of all the energy consumed in Brazil during 2014, more than 10% were lost at random, shows a survey of ABESCO, the association that brings together service conservation companies" [6].

With all these questions being, how will be the use of energy resources in the face of all future demands and expectations of growth and, will the final consumer, mainly the residential profile, be prepared for the rational and efficient management of the energy use made available to him by the energy concessionaire?

For this reason, the primary objective of this work is to show that it is possible to construct a cheap solution for specific equipment that may be villains in energy expenditure, without directly impacting user's comfort, but qualifying the use of energy in a rational and conscious way.

2   Theorical Aspects

In order to place energy efficiency in the current scenario from the industrial revolution to the so-called sustainable development, it is necessary to first understand how energy is embedded in the same scenario and what kind of role, actions and activities related to the production and management of energy must be fulfilled to accelerate the path to sustainability. Since the concept of sustainable development is of interest and global application, it is necessary to approach energy internationally, including its relations with fundamental aspects of the construction of development and sustainability, and with socio-environmental aspects, which involve several issues of great importance, such as equity. Added to the great difficulties of consensus among nations, today, this issue becomes crucial and a priority among the great debates in progress around the world.

In the timeline, the global discussion of the sustainable development model began at the 1972 United Nations Conference on the Human Environment and continues today in an increasingly broad and participatory scenario, catalyzed by the globalization process, which is also a challenge to sustainable development. A discussion whose course, full of comings and goings, will not be deepened in this article, mainly for reasons of objectivity, although it is contained in several parts of the work and in much of the bibliography suggested, because of the strong impacts of energy and energy efficiency in the construction of a model of sustainable development, as will be recognized throughout the work.

In view of the current situation faced by humanity, energy is seen as a basic need necessary for the integration of the human being into development. This is because, among other things, it provides opportunities and a greater variety of alternatives for both the community and the individuals. The supply of energy in its various forms, with acceptable cost and guaranteed reliability, is a fundamental requirement for a region's economy to be fully developed and for each individual and the community to have adequate access to various essential services for increasing quality of life, such as education, sanitation and health. Thus, it also becomes the basic condition of citizenship.

Some facts do not have a certain/correct date, others have approximate date, and some have precise dates. The emergence of worldwide concern with the issue of energy efficiency in buildings is accurate: October 17, 1973, the date known as the first oil shock. That day, OPEC's major producers reduced the extraction of oil, pushing the price of the barrel from \$2.90 to \$11.65 in just 90 days, meaning that the value of the barrel was multiplied by a factor very close to four. Figure 1, located below, illustrates the impact of this measure on the historical series 1861–2001.

The increase in the price of the barrel from about 3 to about 12 dollars in such a short period caused an immediate crisis in the buildings sector. Until that date, there was no global concern about the energy issue in the transport sector, nor on the building sector. In the United States, for example, large buildings in the commercial sector consumed up to 100 kWh/m^2 per month, or 8–10 times higher than those currently practiced [1]. Power consumption was not a major concern in the United States, nor in the rest of the developed or developing world. There were no public or government policies that interfered with the issue in order to discipline consumption in commercial, service or residential buildings. In this respect, the oil shock triggered a new process in the history of buildings on the international scene: the large-scale application of regulations with the force of law aimed at reducing energy consumption in buildings and incentive policies with the same objective.

Specifically after the 1973 Gulf War crisis, which had a real impact in the first three months of 1974, another important step in addition to the creation of the International Energy Agency (IEA) was given in the specific field of buildings: the development of the first regulations with restrictions on energy consumption, supported by the force of the law and known as energy regulations. Regarding energy efficiency, before the implementation of these regulations, the building sector depended, worldwide, on the desire of the architect or entrepreneur to adopt measures that would bring better efficiency. Such measures were generally considered to be passive or active technologies, or both, but implanted by voluntary decisions of the designers or owners. As technical support and foundation, many countries already had technical books supporting the architect in these areas. Such publications were developed by governmental technology institutes or civil society organizations [1].

Consequently, the average change in consumption in the period between 1974 and 1985 was 0%, that is, there was no growth in consumption, although GDP and quality of life increased (Gadget and Rosenfeld, 1991, p. 25) (Fig. 1). From 2010, as shown in Fig. 1, the North American consumption may fall into a rather optimistic scenario, however for this to happen, more aggressive policies should be taken, not only for the transport sector but also for the buildings. Most likely, the current trend is continuous, given that aggressive policies are not implemented from one moment to the next and demand acceptance from the part of society. Neither the United States nor France, or the other European countries in which energy regulations were implemented, was there a decline in the quality of the architecture or a reduction in the architects' creation possibilities. The implemented standards were not intended to provide descriptive revenue, but rather to provide the conditions for architects to adapt the most convenient building materials to the local climate and the most aggressive weather conditions, opting for example to isolate the building in order to avoid losses [1].

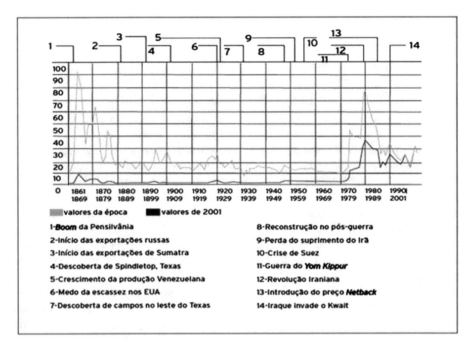

Fig. 1. Change in the average value of the barrel of oil over time

Since 1974, several countries have adopted policies for the implementation of energy regulations for buildings. The European Union, for example, has instituted a policy in this respect, which has been adopted by the member countries in different periods, giving significant effects of reducing consumption and raising the quality of habitability of buildings.

The crisis in the Brazilian electricity sector and the consequent rationing, in late 2001 and early 2002, brought to the attention and discussion of the population the issues of energy conservation and energy efficiency. What was seen, at that moment, was a very positive response from the population, mainly regarding the change in their consumption habits, besides tests of the adoption of efficiency policies, such as the distribution of more economical lamps, the facilitation of financing of more efficient equipment, etc. [1].

The occurrence of such actions was not, however, new for those who work in the sector, since, for some time now, institutions, whether governmental or not, have acted accordingly. What was new was the good response of the population and the acceleration of the process, whose speed, under normal conditions, had left much to be desired. Today, eight years later, unfortunately, what can be noticed is almost the return to pre-rationing conditions. This demonstrates that the good response of the population occurred due to the economic penalties imposed in that situation and missed the great opportunity to continue with actions aimed at changing habits [1].

In this sense, it is important to emphasize some general problems that act by inhibiting the more firm and positive adoption of measures of conservation of electric energy, among which the following stand out: The technology-based gap - Brazil already commercializes several efficient technologies, still small and not accessible to the majority of the population in function of the costs incurred. Some technologies include imported components. In this case, it is necessary to implement measures that reduce import duties completely or partially. On the other hand, some technologies do not meet international standards, which detract from exports. However, these technologies must be very well evaluated, because the Brazilian operational and climatic conditions are different from the countries of export, our pattern may not be the same.

However, there is still the problem of high interest rates, which hampers the economic viability of investments in conservation projects. Lack of information - lack of information on efficient market, as well as in relation to conservation measures, can be pointed out as a strong barrier to the implementation of conservation projects. It has been sought to solve this type of problem through information programs based on the production and distribution of manuals and other educational materials, linked advertisements on TV, radio, etc. but this can still be improved, especially with regard to habits of conservation, both by the low-income population and by the high-income population.

Looking at the Brazilian scenario, according to the report of the Ministry of Mines and Energy (MME), of the international ranking of energy and socioeconomics, data for the year 2014 and 2015, we have Brazil occupying the 5th position in world population ranking, 7th place in energy availability and GDP, 9th in energy consumption, but 71st in per capita energy. According to the National Planning of energy development, PDE, in 2030 Brazil will have a consumption of up to 35.8% of the national energy matrix, including the commercial and industrial sectors. If we consider the public consumption of electric energy, included in this report, this number rises to up to 44.8% of consumption in the national energy matrix [2].

If there was a comparison of the consumption of the national energy matrix with the industrial sector vs. the public, commercial and residential sectors, we will have 49.2% of the industrial sector, compared to 44.8% of the other. Following this rationing line, the relevance of having a means of control and efficiency in the use of these consumption profiles (residential, commercial and public). It could impact practically 50% of the total energy consumption generated by 2030 [2].

According to ABESCO, the sum of energy waste from the residential, industrial and commercial sectors from 2014 to 2016 was 1,402,206 GWh, taking into account the same data, it was estimated that there was a loss, taking into consideration only residential and commercial sectors, about 664.66 GWh, thus accounting for about 47% of the total waste of energy in the study years [3].

Despite the awareness campaigns for the rational use of energy, the commitment of the national industry to manufacture more efficient and effective equipment, there is still a big gap in the control and measurement of energy waste.

For this reason, the advancement of technology has proved to be a great ally for bridging the barriers described above. The IOT, "Internet of Things", has grown more and more popularized in several sectors, but mainly as a tool of residential automation [3, 4].

This leads us to a possibility of using IOT for a quantitative and qualitative analysis of energy use, from rationalization to rationing. A low-cost output can inform the user of their amount of waste and can act as an automation agent for residential or commercial environments [4].

3 Methodology

Some hardware have been implemented in the industry, since the first industrial network implementation, which are the traditional DCS, as there has been evolution of electronics since the 70's, new technologies have been developed throughout history to control the industrial process, focused mainly on the means of production, process automation and energy efficiency. Since the industrial sector moves the country and is undoubtedly the sector that consumes the most electricity.

However, out of the industrial paradigm, this evolution did not follow in an equal way when we look at the final consumer. This is recognized even through the electric power system, where the generation and transmission receive greater attention in levels of publication and technology, since the distribution and consumption, are the bottleneck of the concessionaires of energy and a great standoff for the advance of the technology of Smart Grid.

From the consumer's point of view, the technology for control and science of consumption of a specific equipment is usually measured through the labeling made and regulated through PRODIST.

However, the way to report consumption through labeling is extremely flawed, since an equipment can, after some time, have changes in its electrical variables. Still, there is no customer aware of how much an equipment consumes and how much it could optimize this consumption without directly impacting your comfort.

Only then do we apply the formula (1), with a new profile, reducing the number of hours of use of the equipment, compensating it with the appropriate offset temperature of the machine, so that the refrigeration machine does not have to start all the time and decrease the refrigeration cycle.

$$\frac{.h'.d'}{1000} = \text{Kwh/d} \tag{1}$$

At where:
Pw: Active power consumed by the machine
h': is the number of hours defined by the profile
d': is the number of days of machine operation. Kwh/d: unit of energy consumed as a function of days

Based on formula (1), we were able to calculate the value in kwh/day of machine consumption, before and after applying the tillering process, to calculate the consumption value before the process and after, thus having parameters to analyze if there was a reduction of cost due to the reduction of the time of use, for the new optimized use time.

Being this time of use changed by user's will, or by control of the module through the presence sensor and hours of use in "pre-set", therefore defining schedules of use so that the user does not occur memory lapse switch off the appliance, or excessive use of refrigeration.

The prototype has its operation illustrated by Fig. 2, it consists of two microcontrollers, which communicate with each other through UART protocol, where the ATMEGA 328, better known as Arduino, is responsible for the interface between microcontroller and sensors, and the ESP8266 is responsible for receiving data previously processed by ATMEGA, making a "POST" type request to the server. The server, in turn, will have some prerequisite items, a MYSQL database, Apache2 server, PHP7.X and network, intranet or internet access.

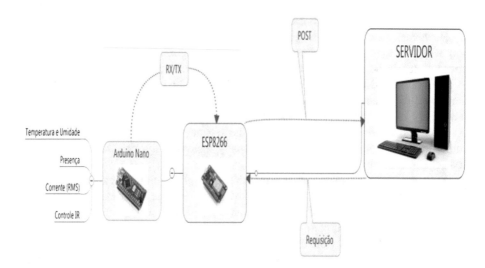

Fig. 2. Communication architecture and analysis

Using the architecture of Fig. 2, we can collect real-time data from the sensors, analyze the data collected and stored in the database, and then make up the area of interest as well as the total energy involved, the main form of energy measurement is through the non-invasive magnetic flux sensor. This sensor works as a current transformer, reacting to the magnetic flux of a conductor in its primary, having a RTC, of 1: 2000, it decreases the input signal 2000 times, making it feasible to acquire a signal of up to 400 amps, depending on the data acquisition board attached to this sensor.

For this case study the current sensors (CTs), which will be used, are those of 20 and 100 A, since the currents of interest are limited to this interval (Fig. 3).

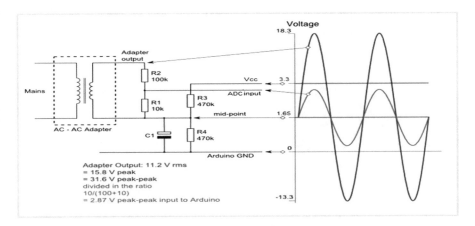

Fig. 3. Example of current sensor operation

Module is also equipped with a PIR sensor and an IR LED. The PIR sensor will send a presence signal if the user is in the area of measurement node, and the IR will convert the control signal from the air conditioner machine to the infrared light signal, allowing the module to change the temperature, turn it on or off according to the user's need and configuration.

In addition, the existence of a DHT11 sensor with a 0.5% temperature reading error causes an external comparison between the chosen temperature and the temperature read in the environment by the measuring node. Thus, the measuring node is a second temperature reference "sensed" by the environment.

The system voltage drops, the voltages of interest are 127 and 220 volts, values adopted by the CELPA energy concessionaire, were not considered for the purpose of analysis where the research was carried out.

The main data of interest will be temperature, humidity and current, since the focus of this article is to work on the great "villain" of waste, which is, within a context of predial cooling [7, 10].

Therefore, sampling was carried out in a closed and controlled environment, located in Icoaraci, for the data collection and profile analysis, in this experiment 1800 samples were collected and sent every 10 s to the server. An average of 10 samples per minute was carried out, with a mean of sampling per minute, making the sensor sensitivity more compatible with the variation characteristics of the cooling machine system. Figure 4, contained in Annex A, illustrates a portion of the samples made and stored in the MYSQL database.

The data collected were available in [10], through a cloud for better visualization of this collection, using this first sampling it is possible to realize with an auxiliary software for plotting graphs, the behavior of the input and output variables in question, which are respectively, humidity, temperature and current. The software used was MATLAB, due to the range of mathematical tools that can be used with this software. An example graph is shown in Fig. 5, where the temperature is blue and the current is orange.

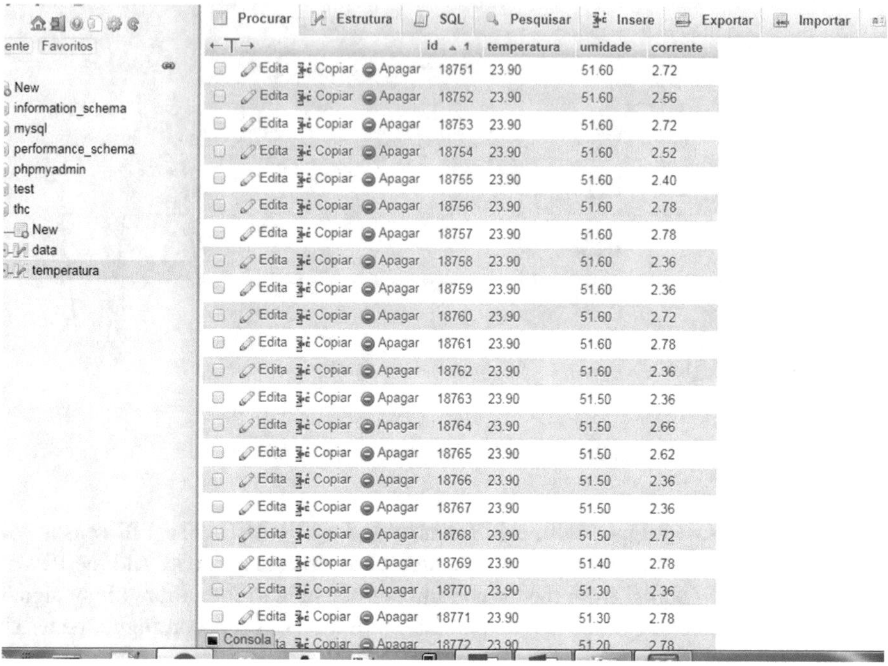

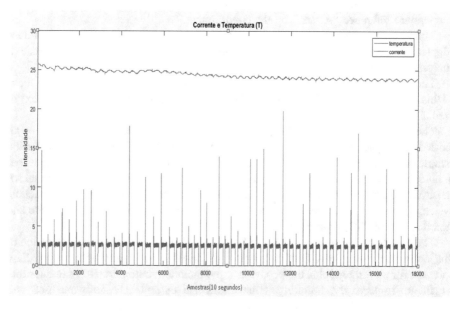

Fig. 4. Database MYSQL with data extracted from profile

Fig. 5. Temperature and current vs samples

On the basis of Fig. 5, it is clear the moments in which the refrigerating machine in question, a Springer air conditioner of 12000 BTUS, enters working period, that is, it leaves the inertia, and "shoots", entering the temperature maintenance period only by circulating ambient air through the evaporator of the machine. From the data of Fig. 5 we obtained Table 1:

Table 1. Collected data

Collected data maximum and medium	
Maximum current	19.4 A
Average current	2.76 A
Maximum temperature	25.5 °C
Average temperature	23.9 °C

According to Table 1 extracted from the graph of Fig. 5 we can conclude that the machine varied from 1.6 °C from the starting point of 25.5 °C to 23.9 °C, and the minimum temperature reached was 23.2°, these temperatures were measured by the measuring node that is 1.8 meters away from the evaporator of the refrigeration machine, the cooling conditions of the machine control were 22 °C, therefore we have a difference of approximately 2 °C between the values of the physical control machine to the measuring node.

Through the data of Table 1, y preserving the same operating profile previously collected by the measuring node as shown in Fig. 5, the values of the "off-set" for the operating limits of the test profile were defined, but with the value reduced by 2 °C, since it was the error calculated between the node measurement values and the values depicted in the physical control of the machine.

The value chosen to be inserted into the memory of the module preserving the conditions portrayed in the previous item was 22 °C, this value was inserted through the copy of the hexadecimal code of the control that is emitted in the form of light, more specifically in the form of infrared signal by the physical control of the machine. For the copy, it was necessary to use an IR light sensitive diode component, HS0038, where it does through a library specific to the Arduino micro controller, the transcription of the light signal to hexadecimal.

Table 2. Data collected on test profile

Collected data maximum and medium	
Maximum current	18.9 A
Average current	2.68 A
Maximum temperature	25.6 °C
Average temperature	23.7 °C

The result of the test profile generated Tables 2 and 3. Table 2 has values similar to Table 1, since the operating conditions were maintained, except for the number of hours of machine operation that can be seen in comparative (Table 3).

Table 3. Comparison of data from Tables 1 and 2

Collected data maximum and medium		
Maximum current	19.4 A	18.9 A
Average current	2.76 A	2.68 A
Maximum temperature	25.5 °C	25.6 °C
Average temperature	23.9 °C	23.7 °C
Number of total utilization hours	6 h	5 h and 10 min

The data of Table 3 are results of the maintenance of the profile of the operation and collection, however with the change of the factor of time of use gives machine of 30 min less in the timer of the module and rule of starting the machine from the presence of the user detected by the PIR sensor and from the preset time which was 23 o'clock GMT-3, as a consequence of this the refrigeration machine was only switched on if the user was present in the space in question and if the configured time were in agreement with the conditions for the cooling head.

Because of this, we had a reduction of 50 total minutes in the mean of collection days that are available in [10]. Therefore, if we take the data of Table 3 considering voltage of the power supply network as fixed in 220 volts we will have:

$$\frac{220 * 2,76 * 6 * 1}{1000} = 3.643 \, \text{Kwh/d} \tag{2}$$

$$\frac{220 * 2,68 * 5,166 * 1}{1000} = 3.045 \, \text{Kwh/d} \tag{3}$$

The result obtained by Eq. (1), applied in (2) and (3), where in (2) no operating rule was used and in (3) the rules defined in the item above for Table 2 were used. We have that the result in (3) is 16.41% lower than in (2), thus generating an economy for the end user based on the time usage factor.

3.1 Correlated Works Comparison

Based on the related works, we can see the evolution of the internet of things and residential automation, now when we compare the works of [11–13]. Where the work of [11] still implements residential automation, dependent on a computer, since in 2014, micro-computers such as raspberry pi, seen in [12], still had high prices and incomplete configurations in 2014. The proposal of [11] is already outdated, since its hardware architecture is also. However the structure of communication and organization of this work, still remains alive, as can be seen in this article and in the work of [12] and [13].

The work of [12] and [13], where 12 describes a system only theoretical, but viable, has the same function and flexibility as this article, with the differential, being the differential of this work at a low implantation cost.

The work of [13] is very similar to this article from hardware structure to communication structure; however, this paper proposes a structure of simulation and cloning of the control of the refrigeration machine, which is non-existent in [13].

In addition to this, all the communication infrastructure used in [13] was done through an auxiliary software called "Blynk", limiting the work the same limitations and problems that the software may have. Already this infrastructure made for this paper, has its intranet base, can be flexible for each residence and works even without the internet service.

4 Conclusion

Weighing the theoretical and practical pocket ark behind residential automation and its benefits when well applied, in each case, investment in intelligent control and support in decision-making [9] as well as rational and appropriate use of the available energy resources, aiming at environmental conservation and distancing from waste.

Consequently, investment in control and automation within the residential environment becomes extremely beneficial as a tool for waste control and more efficient use from the economic point of view.

Therefore, the development of projects aimed at waste control, conscious use of energy resources and environmental awareness are of great importance not restricted only to the residential environment, but all activity that uses electric energy as a primary source.

As can be seen in this work considering the data presented together with the methodology, the proposed architecture was efficient and effective. Resulting in an economy for this case study of 16.41% for the refrigeration machine which this project was applied.

Some improvements must be made in future work, such improvements can be both software level, such as: creation of a more efficient and automatic algorithm for decision making, survey of new curves in different operating situations, calculation of temperature transition inserted in the algorithm to prevent early machine firing and a graphical web interface for better user viewing.

Other improvements can be made at the hardware level such as: Use of more accurate sensors and more quantity per measurement node, existence of more than one measurement point, communication channel between these measuring nodes and printed circuit board specific for module with protection against network surges.

With everything, it is concluded through this article that there was success in the proposed case study choosing a viable and low-cost solution, since the average cost of the module was $50,00 US Dollars. Having a much lower cost than the automation systems available in the market and with a differentiated architecture and specific sensors for the control and analysis gives cooling machine.

References

1. Marcelo, A., Lineus, B.: Eficiência energética em edifícios, 1st edn. Booker, Brazil (2015)
2. Plano Decenal de expansão de energia de 2020. Accessed 7 Sep 2018
3. ABESCO. http://www1.folha.uol.com.br/mercado/2015/02/1586778desperdicioconsome10-da-energia-eletrica-no-pais-diz-associacao.shtml, jornal folha de são paulo, 15 de fevereiro de 2015. Accessed 08 Apr 2018
4. Antonio, A., Roderval. M., Vilson. G.: A internet das coisas aplicada ao conceito de eficiência energética: uma analise quantitativo- qualitativo do estado do estado da literatura, UFPR papers, vol. 5, no. 2 (2016)
5. Planejamento de eficiência energética, ministério de minas e energia. http://www.mme.gov.br/documents/10584/1432134/plano+nacional+efici%c3%aancia+energ%c3%a9tica+%28pdf%29/74cc9843-cda5-4427-b623-b8d094ebf863. Accessed 29 Jun 2018
6. Referência ABESCO, dados divulgados em periódico. http://conexaomineral.com.br/noticia/472/brasil-teve-desperdicio-alarmante-de-energia-nos-ultimos-tres-anos.html#foto. Accessed 05 Jun 2018
7. Vilões do desperdicio. https://www.ecycle.com.br/component/content/article/67-dia-a-dia/2918-os-quatro-maiores-viloes-da-conta-de-luz-da-sua-casa.html. Accessed 05 Jun 2018
8. Prodist 5. http://www.aneel.gov.br/prodist. Accessed 09 Jul 2018
9. Sergio, H.: estratégias de planejamento para otimização do consumo residencial de energia elétrica: uma abordagem baseada em smart home e sistemas fuzzy. UFPA (2017)
10. Dados de coleta de central 12000 btus. https://drive.google.com/drive/folders/1atakt4qvaqtwio5ip3fbhcpgkhzvdwyb. Accessed 15 Aug 2018
11. Patil, V., Bade, K., Bendale, Y.: An optimized use of smart air conditioner using cognitive IOT. IJCSMA (2018)
12. Jordi, S., David, P., Angelos, A., Christos, V.: Smart HVAC control in IoT: energy consumption minimization with user comfort constraints. Sci. World J. **2014**, 11 (2014)
13. Mehmet, T., Hayrettin, G.: An Internet of Things based air conditioning and lighting control system for smart home. Am. Sci. Res. J. Eng. Technol. Sci. **50**(1), 181–189 (2018)

Factors Affecting Consumer-to-Consumer Sales Volume in e-Commerce

Moutaz Haddara[1,2(✉)] and Xin Ye[3]

[1] Kristiania University College, 0186 Oslo, Norway
moutaz.haddara@kristiania.no
[2] Graduate School of Chinese Academy of Social Sciences,
Beijing 102411, China
[3] TIAS School for Business & Society, 3511 RC Utrecht,
Holland, The Netherlands
roxy-ye@qq.com

Abstract. Evaluation matrices in online platforms are key parts in Consumer-to-consumer (C2C) marketplaces. Among the C2C platform providers, Taobao plays an important role in the Chinese virtual market. In this paper, the authors aim at identifying the various indexes for the evaluation matrices system and investigate the influential factors on the performance of the merchants on the platform, which are generally measured by the sales volume. Apple's iPhone 7 plus is selected as the sample product in this study. Besides the sellers of this product, sales data and sellers' indexes are collected and analyzed via various statistical techniques. Our results show that accumulated credit, consumer favorable rate, matching score and consumer service rate have a positive impact on sales. Accumulated credit has the strongest influence on sales, and service score has a negative influence on sales. Finally, our recommendations for C2C sellers are provided according to the findings.

Keywords: C2C · Evaluation matrices · Indexes · e-Commerce · iPhone · Taobao

1 Introduction

With the emergence of the Internet, e-Commerce has grown rapidly. e-Commerce users have reached 361 million in 2014 [1]. e-Commerce has become one of the most important economic engines for China's trade market. There are four main types of e-Commerce business models, Business-to-Business, Business-to-Consumer, Consumer-to-Business and Consumer-to-Consumer (C2C). In China's domestic market, the most outstanding C2C platform operator is Taobao. In 2014, Taobao's brand penetration rate was 87%, leading an irreplaceable position for competitors. Taobao, was created by the business-to-business company Alibaba Group, as a sub-branch platform for domestic C2C. As of Jan 2015, Taobao's market share was 7% and it occupies the first place in C2C retail business [1]. Unlike traditional businesses, C2C purchasing behavior is different from that of traditional business [2]. Online searching functions and the evaluation panel enable consumers to approach products more dynamically. The

© Springer Nature Switzerland AG 2020
K. Arai et al. (Eds.): FTC 2019, AISC 1069, pp. 631–643, 2020.
https://doi.org/10.1007/978-3-030-32520-6_46

browsing and purchasing behaviors are more likely to depend on the platform's evaluation matrices [3]. The evaluation matrices are comprehensive estimation frameworks, which assess the sellers' product and service quality via consumers' reviews. They include indexes such as accumulated credit, favorable review, service attitude, etc. Customers use the filter function to select sellers or products according to their preference; the filter can be a high score of delivery or a good service attitude. This search behavior can lead to different purchasing decisions. The current evaluation matric on Taobao is a combination of numerous variables. Although Taobao has improved its classification, there are at least ten variables in the evaluation matrices. The matrices have three main big categories and some minor categories. Existing literature has done some research about part of the factors' influence on sales. But due to different researcher's definition of the same variables and classification criteria, the definition of the variables varies and is vague. The aim of this study is to examine the factors affecting sales under a C2C environment within the Taobao e-market place.

The remainder of the paper is organized as follows. The research background and research hypotheses are presented in Sect. 2. Section 3 provides an overview on the research methodology adopted in this paper, and the data analysis process. Our results and findings are presented in Sect. 4. Section 5 presents a discussion on the findings, followed by conclusions in Sect. 6.

2   Research Background

In Dec 1997, the International Business Commerce defined electronic-commerce as: the transaction happens by an electronic channel, not by face-to-face trade or any other form [4]. Because of the virtual transaction environment, trust is challenging in a C2C context. Author in [5] claim that all the risks happen in C2C business will cause severe trust issues. Jones and Leonard [6] pointed out that two factors can influence a customer's trust. One derives internally and is a natural propensity to trust. Another is external; it includes trust of buyers/sellers and third-party recognition [6]. Author in [7] suggest that C2C customers can't easily decide to select a certain product or a certain seller. Before their purchases, they will refer to some clues like price, seller credit and seller review record. Author in [8] assert that Taobao dominates China's C2C market and that the crisis of faith is serious, as individual sellers are making the best to pursue short-term interest. This hidden rule tends to be increasingly worsening. Thus, online shopping faces severe risk, and it is affecting the sustainable development of C2C businesses. Fraud is also an outstanding problem in C2C business. C2C business has a low entry barrier, however, it usually lacks enough information sharing and transparency [9]. This may lead to severe fraudulent activities. In an investigation of fraud's impact on credit systems, it was reported that the refund process and transaction scale are important factors [10]. Thus, it is essential to include them in C2C evaluation matrices. Author in [11] found that there are three factors affecting fraud, namely, product quality, service and logistics speed. Trust and price are important issues in C2C marketplaces [3]. Thus, it is believed that evaluation matrices systems help to solve trust issues [10]. Customers use the evaluation system to get a proof of the product features, service and product quality [11].

2.1   Sales on Taobao

On Taobao, there are several indexes that can be used to demonstrate sales. Aggregate shop sales, single product sales, amount of payment and the number of completed transactions. All of the indexes are calculated in a period of the latest (one) month. *Single product sales* is an index only used for upmarket shopping mall "TMall's" products. It shows the purchase records of a product. This number may not be the most accurate of purchase records, because it is measured after the purchase process is completed, not confirming if the package was received. It is similar as the number of payments of regular sellers discussed later on in this paper.

2.2   Price on Taobao

There are three price indexes on Taobao. Product price, delivery fee and discount [12]. Some products include delivery fees in the price. The iPhone 7 plus is an expensive product, most of the sellers don't charge for extra delivery fees, thus the price in this study will be product price. Existing research suggest that price is a significant factor for sales [12].

2.3   Evaluation Matrices on Taobao

Taobao's evaluation matrices work together with the participation of two parties. One is a buyer, another one is a seller. Both of them can give feedback on each other after a transaction is completed. Buyers can update review score for the purchasing and their user experience. Later, sellers receive buyers' feedback and give a feedback to buyers. Review from buyers to sellers is the impact factor in this research study.

Dynamic score is one panel of the evaluation matrices introduced by Taobao since Feb 2009 [12]. It includes three indexes, matching degree, service attitude and delivery quality. There are five scales, from the lowest to the highest, 1 to 5. Because of the fierce competition on the platform, sellers adopt various kinds of bonuses to get high score from buyers. It is common to find that the score is above 4.5. Existing literature indicates that dynamic score doesn't have a significant impact on sales [12].

In the matrices system, there are four indexes. Refund processing speed, refund rate, dispute rate and the number of penalties. In order to maintain a good reputation, sellers will barely go into dispute or violate rules. Therefore, the last two indexes are usually zero. Refund processing speed is the average time took a seller to complete an entire refund procedure in the latest 30 days. Refund rate is the ratio of completed refunds to the total transactions in the latest 30 days.

Another metric on Taobao is the credit score. Author in [13] established a comprehensive mathematical model to evaluate seller credit on the basis of grey theory evaluation method. This model is expected to reflect the fraudulent behavior of sellers during transactions.

2.4   Factors Affecting Sales on Taobao

Several studies focused on identifying the factors that affect C2C sales as presented above. In order to make a clear identification of the current evaluation matrices of Taobao and conduct an overall investigation on the sales influencing factors, this study conducts its own unique exploration and will raise the following research question:

- What are the factors affecting C2C sales based on Taobao platform?

To answer the main research question, the following sub questions were also explored.

- Q1: What are the matrices of Taobao's evaluation system?
- Q2: How to identify the current indexes on Taobao?
- Q3: Whether there are relationships between these indexes and sales on Taobao? How strong are they?
- Q4: Whether there is any moderating effect?

3   Research Hypotheses

In this paper, the authors have chosen one fixed product, the Apple's iPhone 7 plus, as the sample object. In Taobao, due to the fierce competition and serious shop violation handling, sellers will not take a risk to offer fake products, especially with a product at a high price, such as the iPhone 7 plus. Therefore, quality is guaranteed and no big difference between various sellers. In addition, the Apple corporation is directly responsible for the warranty service of their products during the maintenance period. Sellers are not responsible for the quality of the product; they will not offer an after-service for repairs. Therefore, warranty and quality are not considered as factors in this study.

- Hypothesis 1: Price has a negative impact on sales.
- Hypothesis 2: Customer favorable rate has a positive impact on sales.
- Hypothesis 3: Dynamic score has a moderating effect on the relationship between accumulated credit and sales.

4   Research Methodology

This study adopted a deductive research approach and a quantitative methodology [14]. Starting from literature review, to formulating research questions and hypotheses that are derived from previous studies. These hypotheses are then tested by quantitative methods of analysis. In this research study, we have chosen Apple's iPhone 7 plus 32 GB Golden as our constant research object and variable. SPSS was used for the data analysis. As first steps, we have conducted a normality of data test, and multi-collinearity, especially for the aggregated dynamic scores and the three dimensions, matching degree, service attitude and delivery quality. From the preliminary collected data, the range of some variables was quite huge. Therefore, the test has adopted a log transformation to reduce the distance between the variables. Later, a regression analysis was conducted, and for testing the moderation, a moderating test was steered.

4.1 Data Sources

The selection of the right index is a complex job. It should reflect the internal structure and external status of the system. Therefore, the selection process has followed some basic rules:

- First are comprehensiveness and integrity. On-line shopping is an integral process. The system should be analyzed and designed based on the whole process.
- Second is consistency. The object of the design should serve the research object. All of the indexes must have relevance to sales.
- Thirdly is the scientific nature of it. There should not be any missing indexes or unreliable phenomenon presented.
- Fourth is logic. The system should follow a progressive relationship and delve deeply into the process from macroscopic scale to a microscopic scale.
- Finally, is independence, as to cover important indexes and avoid duplication, as it is essential to keep independence between indexes.

The adequate method of index selection was identified through our literature review. Previous studies provided a good reference of the essential indexes. For example, [15] found that customers will most likely purchase a product when they are sure that the sellers can assure that the package will be delivered safely and on time. He argues that delivery has a significant impact on increasing visitors volume. Author in [16] pointed out that a negative review will increase the degree that buyers feel about the risks and it will prevent them from completing a purchase. Author in [12] reported that product description as service score plays a positive role in promoting visit volume. Figure 1 below provides an overview of the indexes selected in this study.

Category	Index	Unit	Value range
Dynamic Scoring	Aggregate Dynamic Score	Point	0-5
	Matching Score	Point	0-5
	Service Attitude Score	Point	0-5
	Delivery quality Score	Point	0-5
Service Scoring	Refund Processing Time	Day	$[0,+\infty)$
	Refund Rate	%	0-1
Credit Scoring	Favorable Rate	%	0-1
	Accumulate Credit	Point	$[0,+\infty)$
Price	Price	RMB	$[0,+\infty)$
Sales	Sales	Piece	$[0,+\infty)$

Fig. 1. Indexes selected

4.2   Data Collection

The data collected are all secondary data and have been collected directly from Tao-bao's website. The scrapped data was all related to Apple 7 plus mobile phone, color as 'Golden', and 32 GB as its capacity. The collected data contained several variables including the shop profile and other indexes such as the dynamic score. The sample strategy is purposive sampling, indexes with missing values or those that can't meet the research requirements such as outliers are removed directly during the collection process. At the end, 30 samples have been collected.

4.3   Sampling Strategy

There are numerous operational methods on the platform. For instance, sellers who pay higher service fees can register in an upmarket sub-branch platform called "TMALL". Joining the program enables sellers to get rid of the credit evaluation panel on their product's page. Therefore, random sample selection is not practical. This research has adopted a purposive sampling method to collect products from regular small sellers to guarantee that there are no missing values for variables. Figure 2 provides a sample of the raw data collected.

	Retail Name	Website	Price(RMB	Sales	Accumulat	Agr	Dynan	Matching	Service att	Delivery	Customer	Customer	Favorable rate
1													
2	Lixiaomai	https://rate	5330	78	32411	4.9	4.9	4.9	4.9	0.8	0.00802	0.9997
3	Zuidizuicheng	https://rate	5298	78	189229	4.9	4.8	4.8	4.9	1.06	0.001618	0.9987
4	Mactuancom	https://rate	5650	173	22835	4.9	4.9	4.9	4.9	0.64	0.0655	0.9994
5	Qitai	https://rate	4498	942	199695	4.8	4.8	4.8	4.8	1.11	0.2498	0.9948
6	Liubingtuan	https://rate	4420	775	113065	4.8	4.8	4.8	4.8	0.74	0.2215	0.996
7	shoujitaotao	https://rate	4359	51	5803	4.8	4.8	4.8	4.8	2.61	0.117	0.9947
8	jinjishumahar	https://rate	4040	225	10580	4.9	4.9	4.9	4.9	0.79	0.1749	0.997
9	ronghua	https://rate	3999	133	15685	4.7	4.8	4.7	4.7	0.6	0.151	0.9939
10	qiqi	https://rate	4550	7	12935	4.7	4.8	4.8	4.8	0.96	0.076	0.9729
11	laoqi	https://rate	5650	89	81652	4.9	4.9	4.9	4.9	0.45	0.1013	0.9845
12	qunziyefengk	https://rate	5760	230	63092	4.8	4.8	4.9	4.9	0.6	0.16	0.9955
13	jinyi	https://rate	5460	39	17117	4.8	4.8	4.8	4.8	0.83	0.1186	0.9848
14	hengshi	https://rate	5280	20	7140	4.8	4.8	4.9	4.8	1	0.1021	0.9973

Fig. 2.   Raw data

4.4   Data Analysis Method

As the data needs to be statistically significant, a series of checks have been conducted. Usually data cleaning is the first step in analysis; however, this process was ignored because the data is collected through purposive sampling and was highly structured. The Normality test is to make sure the dataset is statistically significant. Most statistical tests reply upon the assumption that data is normally distributed, however, the results can be weak if the sample size is small. It is accepted that a research is done on the assumption of normality [17]. In this research, Skewness and Kurtosis statistical approaches were applied. Skewness reflects the symmetry of the distribution. On the other hand, Kurtosis reflects the peak of the distribution. However, it's not enough to judge the normality of a distribution simply by Skewness and Kurtosis. Therefore, the Kolmogorov-Smirnov and the Shapiro-Wilk tests were also conducted to support the tests. By comparing the results of both tests, the findings can be more informative and reliable.

4.5 Multivariable Linear Regression Model

A Multiple Linear Regression model is used to examine the relationship between two or more explanatory variables and a response variable by fitting a linear equation to observed data [18]. It is a statistical approach to describe the simultaneous associations of several variables with one continuous outcome [18]. One popular equation of the model is $Y = \beta 0 + \beta 1 X1 + \beta 2 X2 + \ldots + \beta k Xk + \varepsilon$, for $k = 1, 2, \ldots n$.

4.6 Moderating Effects

If the relationship between variable X and variable Y is the function of variable M, M is the moderator. The relationship between X and Y is influenced by M [18]. Moderator can be both quantitative and qualitative variable, it explains the direction and the degree of the relationship. One popular moderating model is $Y = aX + bM + cXM + e$, which can be transformed into $Y = bM + (a + cM)X + e$.

For a fixed M, this is a linear regression of X to Y. $a + cM$ is the regression coefficient of the relation between those two, e measures the degree of Moderating Effect.

5 Findings

5.1 Descriptive Statistics

This study obtained 30 samples via purposive sampling collection strategy. During the collection process, data with missing values and outliers were automatically abandoned. Therefore, this dataset has no missing values. Figure 3 reveals the descriptive statistics of the raw data and displays the central tendency of the data. Total distance of price, sales and accumulated credit are extremely higher than the rest of variables. This is reasonable, as a popular shop's accumulated credit may be as large as 300 times of another. While dynamic score and other scores on the platform have an upper limit of 5. The total distance of the dynamic score, matching score, service score and delivery score is 0.2, which is very low. Different magnitude of variables makes it hard to compare with each other. Therefore, this data will be logged and transformed to reduce the magnitude. Figure 4 shows the descriptive statics after log transformation. The distance between different variables is improved.

Before adopting the dataset to do further quantitative statistics, it is better to run a normality test to assure the significance of statistics [18]. This can be done by checking the value of Skewness and Kurtosis or conduct the Kolmogorov-Smirnov test (K-S) and Shapiro-Wilk (S-W) test [19]. The value of the Skewness and Kurtosis assessments are demonstrated in Fig. 5. With the exception of customer service rate and customer favorable rate, most of the variables' Skewness and Kurtosis values stay close to -1 or 1. This result shows that the data is not normal, but it doesn't mean that the parametric test is invalid. The research can still be carried out under the assumption of normality. Normality is a matter of degrees, not a cut-off point [18]. Therefore, there is no need to stop the research or transform the data into a more normal distributed pattern of data. Besides, it is imprudent to make a conclusion simply using the Skewness and Kurtosis test. A Shapiro-Wilk test is added to be more objective. Rather than Kolmogorov-Smirnov test, S-W test was selected because of the sample size. If the sample size is above 500, it's better to use the Kolmogorov-Smirnov test. If not, it's better to choose the Shapiro-Wilk test to undertake the Normality test [19]. Figure 6 reveals that not all the significance of the variables is above 0.05. Sample size has a negative impact on the normality test, if the sample is smaller than 50, and some of the outcomes can be ignored [18].

	N	Minimum	Maximum	Mean	Std. Deviation	Skewness		Kurtosis	
	Statistic	Statistic	Statistic	Statistic	Statistic	Statistic	Std. Error	Statistic	Std. Error
Price(RMB)	30	3999.0	5760.0	5001.367	537.8014	-.311	.427	-1.239	.833
Sales	30	6.0	942.0	169.867	216.2900	2.427	.427	6.355	.833
Accumulate Credit	30	1267.0	3446601.0	174938.533	624038.8586	5.312	.427	28.714	.833
Agr Dynamic Score	30	4.7	4.9	4.830	.0702	-.499	.427	-.781	.833
Matching	30	4.7	4.9	4.833	.0606	-.294	.427	-.550	.833
Service attitude	30	4.7	4.9	4.850	.0572	-.591	.427	-.620	.833
Delivery	30	4.7	4.9	4.843	.0626	-.635	.427	-.453	.833
Customer service processing time	30	.20	2.61	.8930	.49238	1.832	.427	4.478	.833
Customer service rate	30	.001618	.255600	.14004793	.064260002	-.079	.427	-.085	.833
Favorable rate	30	.9729	.9998	.994930	.0056108	-2.544	.427	7.852	.833
Valid N (listwise)	30								

Fig. 3. Descriptive statistics

	N	Minimum	Maximum	Mean	Std. Deviation	Skewness		Kurtosis	
	Statistic	Statistic	Statistic	Statistic	Statistic	Statistic	Std. Error	Statistic	Std. Error
Adj.Price	30	3.601951404	3.760422483	3.696585229	.0478072832	-.431	.427	-1.090	.833
Adj.Sales	30	.7781512504	2.974050903	1.908430230	.5933759710	-.333	.427	-.567	.833
Adj. Acc Credit	30	3.102776615	6.537391010	4.434781127	.7637944602	.570	.427	.571	.833
Adj. Dynamic score	30	.6720978579	.6901960800	.6839025519	.0063365045	-.521	.427	-.740	.833
Adj.Matching	30	.6720978579	.6901960800	.6842136158	.0054596395	-.320	.427	-.486	.833
Adj.Service	30	.6720978579	.6901960800	.6857123741	.0051424427	-.609	.427	-.546	.833
Adj.Delivery	30	.6720978579	.6901960800	.6851091001	.0056361824	-.657	.427	-.383	.833
Adj.Cumstomer service processing time	30	-.698970004	.4166405073	-.104059764	.2228193963	-.133	.427	1.206	.833
Adj.Customer service rate	30	-2.79102148	-.592439151	-.959241990	.4422452712	-3.148	.427	11.093	.833
Adj.Customer Favorable rate	30	-.011931797	-.000086868	-.002214210	.0024658331	-2.567	.427	7.987	.833
Valid N (listwise)	30								

Fig. 4. Descriptive statistics after log transformation

A correlation analysis within the explaining variables was conducted to select the final variables that will be used in the modeling process. The more the correlation coefficient is closer to 1 or −1, the stronger the correlation shall be. If the value is above 0.8, it means there is a strong correlation. Figure 7 presents that result. The dynamic sore and matching score's coefficient is 0.889, with a significance level of 0.01. The service score and delivery's coefficient is 0.818, accompanying with the same significance level.

	N	Minimum	Maximum	Mean	Std. Deviation	Skewness		Kurtosis	
	Statistic	Statistic	Statistic	Statistic	Statistic	Statistic	Std. Error	Statistic	Std. Error
Adj.Price	30	3.601951404	3.760422483	3.696585229	.0478072832	-.431	.427	-1.090	.833
Adj.Sales	30	.7781512504	2.974050903	1.908430230	.5933759710	-.333	.427	-.567	.833
Adj. Acc Credit	30	3.102776615	6.537391010	4.434781127	.7637944602	.570	.427	.571	.833
Adj. Dynamic score	30	.6720978579	.6901960800	.6839025519	.0063365045	-.521	.427	-.740	.833
Adj.Matching	30	.6720978579	.6901960800	.6842136158	.0054596395	-.320	.427	-.486	.833
Adj.Service	30	.6720978579	.6901960800	.6857123741	.0051424427	-.609	.427	-.546	.833
Adj.Delivery	30	.6720978579	.6901960800	.6851091001	.0056361824	-.657	.427	-.383	.833
Adj.Cumstomer service processing time	30	-.698970004	.4166405073	-.104059764	.2228193963	-.133	.427	1.206	.833
Adj.Customer service rate	30	-2.79102148	-.592439151	-.959241990	.4422452712	-3.148	.427	11.093	.833
Adj.Customer Favorable rate	30	-.011931797	-.000086868	-.002214210	.0024658331	-2.567	.427	7.987	.833
Valid N (listwise)	30								

Tests of Normality

	Kolmogorov-Smirnov[a]			Shapiro-Wilk		
	Statistic	df	Sig.	Statistic	df	Sig.
Adj.Price	.200	30	.004	.913	30	.018
Adj.Sales	.104	30	.200*	.965	30	.421
Adj. Acc Credit	.104	30	.200*	.968	30	.482
Adj. Dynamic score	.273	30	.000	.780	30	.000
Adj.Matching	.307	30	.000	.754	30	.000
Adj.Service	.342	30	.000	.710	30	.000
Adj.Delivery	.317	30	.000	.742	30	.000
Adj.Cumstomer service processing time	.108	30	.200*	.976	30	.698
Adj.Customer service rate	.268	30	.000	.623	30	.000
Adj.Customer Favorable rate	.241	30	.000	.720	30	.000

*. This is a lower bound of the true significance.

a. Lilliefors Significance Correction

Fig. 5. Results of skewness & kurtosis and test of normality

Correlations

		Adj.Price	Adj.Sales	Adj. Acc Credit	Adj. Dynamic score	Adj.Matching	Adj.Service	Adj.Delivery	Adj.Cumstomer service processing time	Adj.Customer service rate	Adj.Customer Favorable rate
Adj.Price	Pearson Correlation										
	Sig. (1-tailed)										
Adj.Sales	Pearson Correlation	-.299									
	Sig. (1-tailed)	.054									
Adj. Acc Credit	Pearson Correlation	.062	.525**								
	Sig. (1-tailed)	.373	.001								
Adj. Dynamic score	Pearson Correlation	.379*	.021	-.119							
	Sig. (1-tailed)	.019	.456	.266							
Adj.Matching	Pearson Correlation	.247	-.057	-.243	.889**						
	Sig. (1-tailed)	.095	.383	.098	.000						
Adj.Service	Pearson Correlation	.469**	-.219	-.098	.729**	.692**					
	Sig. (1-tailed)	.005	.122	.303	.000	.000					
Adj.Delivery	Pearson Correlation	.358*	-.058	.126	.712**	.602**	.818**				
	Sig. (1-tailed)	.026	.381	.253	.000	.000	.000				
Adj.Cumstomer service processing time	Pearson Correlation	-.477**	.233	.277	-.315*	-.406*	-.356*	-.260			
	Sig. (1-tailed)	.004	.107	.069	.045	.013	.027	.083			
Adj.Customer service rate	Pearson Correlation	-.271	.176	-.129	-.354*	-.179	-.112	-.363*	-.043		
	Sig. (1-tailed)	.073	.176	.249	.028	.172	.278	.024	.411		
Adj.Customer Favorable rate	Pearson Correlation	.040	.351*	.139	.284	.054	.197	.166	-.003	-.066	
	Sig. (1-tailed)	.417	.029	.232	.064	.388	.148	.191	.494	.365	

*. Correlation is significant at the 0.05 level (1-tailed).

**. Correlation is significant at the 0.01 level (1-tailed).

Fig. 6. Correlations

5.2   Multiple Linear Regression Results

After inputting all of the 10 variables into the linear regression function, the regression analysis was conducted. Figure 8 shows the results of the analysis. At a level of 0.1 significance, only accredited credit (P = 0.003), service (P = 0.099) and customer service rate(P = 0.043) are above 0.1, which is significant. Price's significance level reveals it may not have an impact on sales in our data sample. Besides, the collinearity statistics shows dynamic score's VIF value is above 10 (VIF = 12), which means there is a strong multicollinearity between the independent variables [18]. Dynamic score is an aggregated index made from matching service and delivery, which may lead to this result. To reduce the multicollinearity impact, dynamic score was removed from the next step to run the regression analysis. To adjust the model, the dynamic score variable was also removed, and Fig. 9 shows an enhanced result, as the multicollinearity effect was reduced. Some of the variables' significance level was improved. At a level of 10% significance, accumulated credit, matching score, service score, customer service rate and customer favorable rate have impact on sales. However, there are still three insignificant factors. The level of significance of all the variables seem to be adequate when compared to the first round of analysis, as all of them are under 5%. All the VIF values are around 2, there is no multicollinearity. The model's R and R2 values are high (R = 0.76, R2 = 0.578), which means that this regression equation can explain the sample in a good proportion [19]. In general, both the equation and the variables' parameters are deemed adequate. The equation of the linear regression model is:

$$sales = 0.58 * accumulated\ credit + 0.37 * customer\ favorable\ rate + 0.49$$
$$* matching + 0.30 * customer\ service\ rate - 0.54 * service$$

This infers that customers' favorable rate has a positive impact on sales. Every 1 point of accumulated credit increases, the sales will correspondently increase 0.37 points. Hypothesis 2 is resulted to be true. matching score's impact can't be ignored with a good P value and β value (P = .017, β = .49). Customer service rate plays an import role (P = .037, β = .303). Service score exceptionally reveals a negative impact on sales. The level of significance is good (P = .008), while the regression coefficient is negative 0.54.

Coefficients[a]

Model		Unstandardized Coefficients		Standardized Coefficients	t	Sig.	Collinearity Statistics	
		B	Std. Error	Beta			Tolerance	VIF
1	(Constant)	11.220	15.074		.744	.465		
	Adj.Price	-2.812	2.627	-.227	-1.071	.297	.399	2.504
	Adj. Acc Credit	.432	.127	.555	3.387	.003	.665	1.504
	Adj. Dynamic score	51.800	43.470	.553	1.192	.247	.083	12.047
	Adj.Matching	3.256	45.889	.030	.071	.944	.100	9.967
	Adj.Service	-67.328	38.872	-.583	-1.732	.099	.158	6.345
	Adj.Delivery	12.119	33.349	.115	.363	.720	.178	5.610
	Adj.Cumstomer service processing time	-.009	.501	-.003	-.018	.986	.505	1.981
	Adj.Customer service rate	.509	.236	.380	2.160	.043	.579	1.727
	Adj.Customer Favorable rate	58.897	41.769	.245	1.410	.174	.594	1.684

a. Dependent Variable: Adj.Sales

Fig. 7.   Linear regression results

Coefficients[a]

Model		Unstandardized Coefficients		Standardized Coefficients	t	Sig.	Collinearity Statistics	
		B	Std. Error	Beta			Tolerance	VIF
1	(Constant)	4.861	14.238		.341	.736		
	Adj.Price	-1.520	2.416	-.122	-.629	.536	.481	2.078
	Adj. Acc Credit	.408	.127	.525	3.209	.004	.681	1.467
	Adj.Matching	50.902	22.743	.468	2.238	.036	.417	2.400
	Adj.Service	-74.355	38.805	-.644	-1.916	.069	.161	6.199
	Adj.Delivery	25.786	31.626	.245	.815	.424	.202	4.946
	Adj.Cumstomer service processing time	.185	.479	.069	.386	.704	.564	1.773
	Adj.Customer service rate	.454	.233	.338	1.944	.065	.603	1.659
	Adj.Customer Favorable rate	88.155	34.127	.366	2.583	.017	.907	1.102

a. Dependent Variable: Adj.Sales

Fig. 8. Linear regression without dynamic score

Coefficients[a]

Model		Unstandardized Coefficients		Standardized Coefficients	t	Sig.	Collinearity Statistics	
		B	Std. Error	Beta			Tolerance	VIF
1	(Constant)	6.770	11.719		.578	.569		
	Adj. Acc Credit	.449	.109	.579	4.105	.000	.885	1.130
	Adj.Matching	53.684	21.018	.494	2.554	.017	.470	2.127
	Adj.Service	-62.706	21.811	-.543	-2.875	.008	.492	2.032
	Adj.Customer service rate	.406	.184	.303	2.206	.037	.934	1.071
	Adj.Customer Favorable rate	89.215	33.119	.371	2.694	.013	.928	1.077

Fig. 9. Linear regression with 5 variables

6   Implications and Discussion

Hypothesis 1: Price has a negative correlation with sales *(Not supported).*
The results reveal a total reverse conclusion. In the first two rounds of the regression analysis, price's significance level has never passed the requirement and remains insignificant. Several reasons may cause this. One is that the sales calculation is not scientifically precise. Some shops put all the variations of one product in one product page, for instance, iPhone 7 Plus with and without contract use the same sales collection channel, therefore, the sales level is an aggregate total of all types' sales. The precise sales of a certain type of one product can't be obtained from some product pages on Taobao. Another reason could be related to the weakness of the sample strategy.

Hypothesis 2: Customer favorable rate has a positive impact on sales *(Supported)*
This is supported, and the outcome is applicable in this case. The marginal effect of customer favorable rate is 0.37, and it shows a major positive impact on sales. This could be because the limitation of physical contact between buyer and products urges a new channel to fill the bridge of untouched products and customers. The user-friendly Reviewing panel of Taobao enables buyers to have a closer feeling of the products via product photo show and providing experience feedback.

Hypothesis 3: Dynamic score has a moderating effect on the relationship between accumulated credit and sales *(Not supported)*
The hypothesis is not supported. The moderating test outcome reveals that both the significance level of the changed model and the multiplicative variable CmD's regression coefficient is above the upper limit. Dynamic score is not a moderator between the relationship of accumulated credit and sales.

7   Conclusions

Our results show that the most important factors are: accumulated credit, customer favorable rate, matching score, customer service rate, and service score. Except for the service score, all the other factors have a positive impact on sales. The marginal effect of accumulated credit is the highest, 0.58. The marginal effect of customer favorable rate is 0.37. The marginal effect of matching score is 0.49. The marginal effect of customer service rate is 0.3. The marginal effect of service score is −0.54. This research findings provide a guide for C2C business management. Individual sellers can find a concise combination of the impact factors and apply them to the relevant investment as urgent improvements. For instance, setting up a new credit increase strategy by ensuring product quality, develop a visit volume project by improving the layout of the homepage or upgrade service level via a service staff training program.

This research enriches the literature on sales in a customer-to-customer business context and fills in the blanks among comprehensive factors. Extant literature indicates that customer-to-customer sales are affected by credit, product description, service level, reviews and logistics quality, but there is not a comprehensive research on a wider range of factors.

References

1. Shi, W.J.: Impact factors of e-commerce sales using SEM model. Chin. Acad. J. Electron. Publishing House, 10–15 (2015)
2. Yeah, A.Z.: The factors of affecting the sales of taobao store based on logistic regression. J. Taiyuan Norm. Univ. **64**, 39–41 (2016)
3. Greenberg, P.: CRM at the Speed of Light: Social CRM Strategies, Tools, and Techniques. McGraw-Hill (2010)
4. Che, C.: e-Commerce credit based on B2B model. http://kns.cnki.net/KCMS/detail/detail.aspx?dbcode=CMFD&dbname=CMFD0506&filename=2005043935.nh&v=MTU4Mjdm YitWdkZ5L2hXN3pQVjEyN0c3TzhIZGpQcXBFYlBJUjhlWDFMdXhZUzdEaDFFUM3 FUcldNMUZyQ1VSTDI=
5. Hu, X., Lin, Z., Whinston, A.B., Zhang, H.: Hope or hype: on the viability of escrow services as trusted third parties in online auction environments. Inf. Syst. Res. **15**, 236–249 (2004)
6. Jones, K., Leonard, L.N.: Trust in consumer-to-consumer electronic commerce. Inf. Manage. **45**, 88–95 (2008)
7. Roest, H., Rindfleisch, A.: The influence of quality cues and typicality cues on restaurant purchase intention. J. Retail. Consum. Serv. **17**, 10–18 (2010)

8. Epstein, G., Gady, W.: Alibaba's Jack Ma fights to win back trust. Forbes. Accessed from 23 Sep 2012 (2011)
9. Albert, M.R.: E-buyer beware: why online auction fraud should be regulated. Am. Bus. LJ. **39**, 575 (2001)
10. Nan, R., Xin, W.: Research on the improvement of C2C e-commerce credit evaluation system. Value Eng. 17 (2011)
11. Zhang, Y., Bian, J., Zhu, W.: Trust fraud: a crucial challenge for China's e-commerce market. Electron. Commer. Res. Appl. **12**, 299–308 (2013)
12. Chne, Y.J.: The relationship model between favorable review, credit, price and sales. In: Presented at the Chinese Statistical Education Conference (2015)
13. Zhang, Q., Tsao, Y.-C., Chen, T.-H.: Economic order quantity under advance payment. Appl. Math. Model. **38**, 5910–5921 (2014)
14. Bryman, A.: Social research methods. OUP Oxford (2012)
15. Palvia, P.: The role of trust in e-commerce relational exchange: a unified model. Inf. Manage. **46**, 213–220 (2009)
16. Hsieh, M.-T., Tsao, W.-C.: Reducing perceived online shopping risk to enhance loyalty: a website quality perspective. J. Risk Res. **17**, 241–261 (2014)
17. Saunder, M., Lewis, P., Thornhill, A.: Research Methods for Business Students, 6th edn. Always learning, London (2012)
18. Eberly, L.E.: Multiple linear regression. In: Topics in Biostatistics. pp. 165–187. Springer, New York (2007)
19. Thode, H.C.: Testing Fornormality. CRC Press, Boca Raton (2002)

A Cutting-Edge Unified and Stable Rule Design Pattern

Mohamed E. Fayad[1(✉)], Gaurav Kuppa[1], Siddharth Jindal[1],
and David Hamu[2]

[1] San Jose State University, San Jose, CA, USA
{m.fayad,gaurav.kuppa}@sjsu.edu,
siddharthajindall@gmail.com
[2] Liberty Consulting, Phoenix, AZ, USA
dave.hamu@gmail.com

Abstract. Often, changing market dynamics require business applications to quickly and efficiently adapt to the needs of the ensuing business environment. Business Rules excel in delivering software solutions that are implicitly adaptable to changing business requirements; thus they can prove to be an effective tool to provide necessary flexibility and control for rapidly deploying changes across a wide array of business operations. When a proper design is employed, business rules provide a robust and capable way of enabling enterprise software that adapts to changing business needs. In other words, business rules find varied applications and ways of use, for example, managing a pending problem, using it as production rules and for facilitating collaboration between various systems, etc. However, despite a plethora of tools and technologies available, most organizations still find it difficult to define or model workable business rules explicitly. Furthermore, from a macroscopic point of view, rules are important and inseparable artifacts in governing a software application to make it comply with the system goals. In this paper, the current ways to manage business rules along with their pitfalls are discussed, and therefore, a new approach for developing rule-based business architectures is proposed. The proposed approach allows for both managing and reusing business rules.

Keywords: Software reuse · Stable design patterns · Software Stability Model · Business Rules · Knowledge map · Adaptability

1 Introduction

Modern business applications require a system that has the ability to efficiently and effectively manage its processes to align itself with the business goals. Pertaining to specific needs and requirements, a number of tools to generate and manage business rules currently exist in the market. However, existing rule generating solutions possess formidable limitations, including:

(1) Very high development and maintenance cost.
(2) Prohibitive demands for scarce technical expertise.

© Springer Nature Switzerland AG 2020
K. Arai et al. (Eds.): FTC 2019, AISC 1069, pp. 644–653, 2020.
https://doi.org/10.1007/978-3-030-32520-6_47

(3) Inadequate expressive semantics for modeling very complex solutions.

(4) Lack of administrative features required to manage complex rule-base deployments and versioning. majority of these issues center around the limits of conventional modeling approaches which contend with tangible rather than changing aspects of a software system. A flexible approach to system modeling is paramount for a successful business rules-based solution.

Some software engineer may counter that rules engines are entirely unnecessary since a programming language is in and of itself a rules engine. This perspective is naïve and is equivalent to asserting that a plumbing system is but an assortment of pipes and fittings.

A rule, in general, is defined either to make an orderly system or to obtain uniformity for reaching the target. It is a verdict that must be followed while carrying out the set of operations for which it is specified. In essence, a rule finds application in almost every domain because it is the basic building block behind managing and controlling a system. Therefore, it is imperative to understand how rules can be generalized and reused to suit any number of applications. According to Software Stability Model (SSM), a rule can be classified as a Business Object (BO) as it has a specific duration and is subject to changes over a period of time [7–9]. AnyRule Stable Design Pattern [2–6] illustrated in this paper, identifies the core knowledge behind a rule, related to its Enduring Business Theme (EBT) which is *Governing*. This clear separation of knowledge from problem-specific artifacts makes it stable and reusable over an unlimited number of applications [11].

In short, we assert that business rules-based software modeling is a wholly unique process unto itself. The sort of generalizations which must be employed to successfully model a dynamic, flexible and adaptable system of rules for a given domain is far more specialized than one will find in a straightforward deterministic algorithm designed for a solution that is not designed to be changed. Furthermore, the modeling approach and methodology is different than what is employed when building software with an object-oriented language. Foremost, the understanding of the solution domain must be far deeper than the analysis that s required for a non-rule-based application. Software Stability Method is a specialized approach to software engineering that targets those essential aspects of a solution that are enduring (and therefore stable over time) and distinguishes those aspects of a model from those aspects of a solution that are dynamic and changing over time.

Section 2 of this paper discusses the problem associated with the existing ways of modeling systems based on traditional methods and elaborates this with the help of a few sample scenarios. It also touches upon the issue of current systems having failure rates and how well-defined rule-based systems can be a solution to this problem. In the latter half of this section, the new Stable Rule Pattern is discussed illustrating its reuse potential. In Sect. 3, we will discuss a related rule-based traditional pattern and measure it against the new stable design pattern: this evaluation, both qualitative and quantitative criteria. Section 4 will then provide a conclusion about this work, followed by references to end the paper.

2 The Study

2.1 Problem

In the Information Technology (IT) sector, various studies have shown high failure rates for the projects due to reasons ranging from confusing or changing requirements to poor design and inappropriate alignment of business processes with the organization's goals and objectives. According to an estimate by the US Government Accounting Office (GAO), it was found that out of a total of 840 federal projects, approximately 49% were either poorly planned or performing poorly. In 2014, a market survey conducted by Project Management Institute showed only 42% of organizations having a high alignment of projects to organizational strategy. Furthermore, according to a worldwide estimate, organizations and governments will spend approximately $1 trillion on IT services and infrastructure in the next few years. Most likely, 5 to 15% of the initiated projects will be found inadequate and subsequently abandoned while many more will be over-budget or over-time and may even need massive reworking. In other words, only a few of these IT projects will truly succeed.

The reasons behind the failure of new business initiatives or improvement of an existing business process are often complex. But it often comes down to the fact that the underlying problems the organization was facing were poorly understood or the process improvement initiatives were poorly attuned to solving the real needs of the organization [1]. This is largely due to the fact that these initiatives were not properly governed or elicited, owing to a lack of clearly defined and stable rules. Most organizations have a hard time identifying and articulating these rules because the rules and logic are scattered across the organization and owned by several stakeholders who may not possess the complete visibility of the interconnected nature of the problems that they are trying to solve. This makes the business applications susceptible to failures as the business itself may lack the agility and domain knowledge needed for adapting to rapidly changing business situations.

2.2 Discussion

Traditionally, a rule is understood in the limited context of a term or a statement which is accepted as true and used as a basis for reasoning or conduct. However, rules are the fundamental unit that governs an application in almost every domain irrespective of the problem concerned. Figure 1 depicts the scenario of a typical Loan Application process.

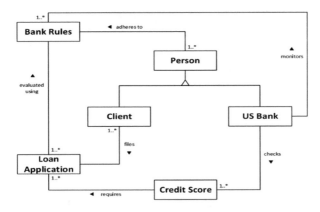

Fig. 1. Traditional Loan Application process

In this example, a Loan Application is submitted to a Bank, which is then evaluated as per the Bank Rules. However, any change in the application requirements or components will lead to the re-designing of the complete application architecture. For example, another application in similar context can be a sports application. In such a scenario, a Coach may record data and select players on the basis of their performance in some Training Match and even finalize their Playing Positions. But if we compare it with the earlier model, we can find striking similarities in the underlying context of the two models. Figure 2 gives an illustration of how another application deals with problem-specific and distinct classes but in a similar context, i.e. rule-based evaluation of players for selection into the final team.

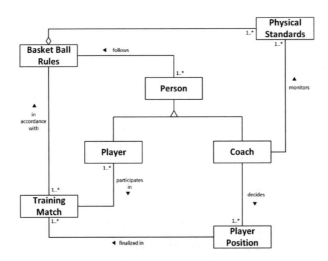

Fig. 2. Traditional model for a sport application

Let us consider one more example of an application dealing with rules in another way. Figure 3 shows an application scenario for a Conference Enrollment system. This model again consists of completely different classes if compared with the earlier two, but as can be seen, the underlying concept is about dealing with a set of rules to enroll for a conference.

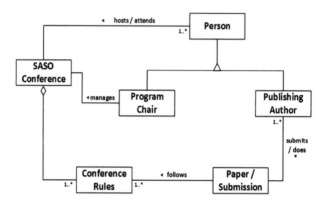

Fig. 3. Traditional conference enrollment model

All the afore-mentioned application scenarios highlight an important and glaring limitation with the traditional modeling approach [10], and that is, *Lack of Stability*. In traditional UML, a rule pattern that is self-adaptive to ever occurring changes and self-organizing to the problem at hand does not exist. Because of this, we need to rethink the problem every time we are required to model an application dealing with rules. This research demonstrates <u>AnyRule</u> as a Stable Design Pattern

2.3 Results

(a) The Stability Approach

As discussed earlier, a rule is the basic building block behind managing and controlling a system. A rule usually consists of a set of conditions and acts as a standard for different activities. It also provides detailed direction on how to convert a given strategy into actions. However, for the rules to have any value, they are required to be collected and stored in a manner that is consistent within the framework of the business requirements. Well-organized and well-structured rules become vital information for defining processes and executing actions. Given below are the stable representations of the application scenarios discussed earlier in this column. These applications are re-modeled using the newly proposed AnyRule Stable Design Pattern [5, 11]:

I. *Rules in Banking:* A bank, for example, Bank of America (AnyParty) defines the minimum criteria like a credit score (AnyConstraint), for or above which only they accept the credit card applications (AnyEntity). For this purpose, they hire professionals like underwriters (AnyParty), who evaluate each credit application

(AnyEntity) based on the applicant's credit score (AnyConstraint). However, to administer these requirements (AnyConstraint), they are governed by a set of Business Rules (AnyType) which are usually statements (AnyRule) consisting of conditions which help them evaluate each application without any bias or prejudice. These rules are generally recorded on rule manuals (AnyMedia) which are a part of the Business Requirement Document (see Fig. 4).

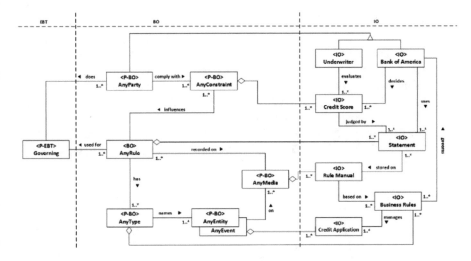

Fig. 4. Stable pattern: Loan Application process

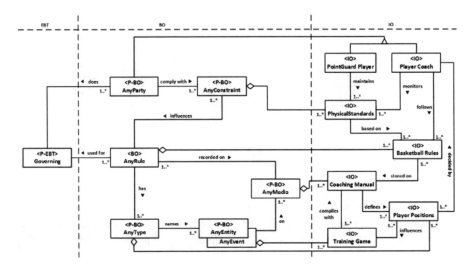

Fig. 5. Stable model for a sport application

II. *Rules in Sports:* During the trial and selection of basketball players (AnyParty), a coach (AnyParty) defines the minimum physical standards (AnyConstraint) for selection, based on which the players appear for trials. For selecting the best players, the coach (AnyParty) uses a defined set of rules (AnyRule) as recorded on the coaching manual (AnyMedia). The selection of the players has to be carried in accordance with the standard international basketball rules (AnyType). The players (AnyParty) performing best in practice matches (AnyEvent) and meeting the minimum physical requirements (AnyConstraint) are shortlisted for selection (see Fig. 5).

III. *Rule in Conference Enrollment:* Usually in technical conferences like the Self-Adaptive and Self-Organizing (SASO) Systems (AnyEvent) which is being held at Massachusetts Institute of Technology (MIT) in 2015, there are certain specific Conference Rules (AnyRule) that are needed to be followed by the Authors (AnyParty) to enroll and submit their work. These rules are often specified and approved by the Program Chair (AnyParty) and are also responsible for managing the conference. The authors can complete their Enrollment (AnyType) by completing an Electronic Form (AnyMedia), which also highlights the Submission Guidelines (AnyConstraint) to be followed (see Fig. 6).

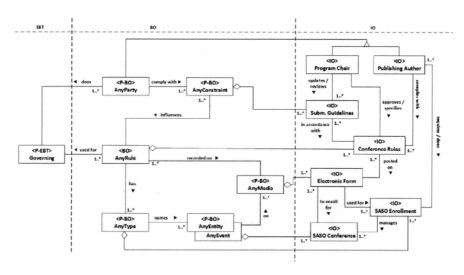

Fig. 6. Stable conference enrollment model

From the different applications of the AnyRule pattern discussed above, it is quite evident that the proposed pattern can be utilized to generate problem-specific solutions for different systems in a similar context. This capability gives the AnyRule Stable Design Pattern the required flexibility of unlimited reuse and adaptability to varying requirements [12–14].

3 Related Pattern and Measurability

3.1 Related Pattern

The pattern given below gives an abstract view of how the rules have been modeled traditionally. This pattern consists of a number of dependencies in the form of aggregation and inheritance. The model also includes tangible classes like the *client*, *GUI, Property List,* etc. (see Fig. 7).

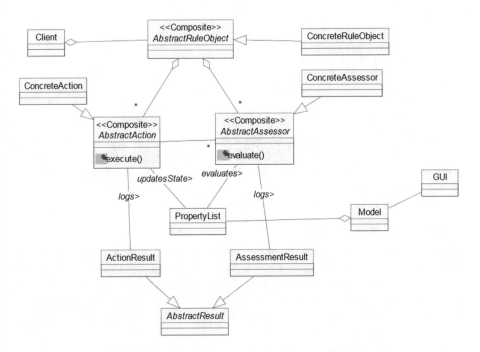

Fig. 7. Traditional compound rule object model [15].

3.2 Measurability

Table 1. Measurability study

Feature	TM	SSM
Number of tangible classes	4	0
Number of inheritances	5	0
Number of attributes per class	5–7	3
Number of operations per class	1–5	2
Number of applications	1	Unlimited

1. **Quantitative Measurability:** The total number of methods in any system can be calculated using the formula:

$T = C * M$; where,
   T = total number of operations
   C = total number of classes
   M = number of methods per class

(a) **Traditional Rule Model**

$C = 13$
$M = 3$
$T = 13 * 3 = 39$

(b) **Stable Rule Model**

$C = 8$
$M = 2$
$T = 8 * 2 = 16$

Stable Rule model is more applicable and accurate compared to traditional model since the level of complexity is way less as evident by the above calculations. The stable model based on three level architecture acquires a more detailed understanding of the problem requirements with lesser complexity. The Traditional Rule Model includes tangible classes, which make it vulnerable in the event of any change.

2. **Qualitative Measurability**

Whereas a traditional rule model can only be used for a particular application, the stable rule model is usable in multiple applications. When requirements change, the traditional rule model becomes obsolete or requires massive changes. On the other hand, stable rule model can be used again and again by simply adjusting it to the application scenario. This is further emphasized by the following quality measure.

$R = T_c - T_n$; where,
R = Reusability Factor
T_c = Total Number of Classes
T_n = Number of classes not reused

(a) **Traditional Rule Model**

Total Number of Classes = 13
Number of classes not reused = 13
Reusability of classes = 13 – 13 = 0

(b) **Stable Rule Model**

Total Number of Classes = 8
Number of classes not reused = 0
Reusability of classes = 8 − 0 = 8

From the above calculations, it is clear that stability model has much more reusable classes than the traditional model, and hence it has wide applicability (Table 1).

4 Conclusion

This paper represents the core knowledge about rules and proposes the AnyRule Stable Design Pattern. The usefulness and varied applicability of the pattern can be illustrated by the fact that it is based on the core knowledge underlying any type of rules which makes it reusable in unlimited applications and as much as required. Another goal of our research has been to contribute in the development of a stable and comprehensive Stable Business Rule Engine (SBRE) where both stable classes and application classes are designed separately from each other and whose composition is formally supported to ensure correctness and exactness. This separation of concerns allows for reusability and enables the building of Stable Business Rules that are adaptable and extendable to an unlimited number of applications. This engine will be accessible online by any business, thereby removing huge costs of ownership and maintenance by only paying a minimal license fee. Using the system, any user can set up the rules as per his or her requirements through direct interaction with the portal simply by choosing required features or answering a few questions.

References

1. Griss, M.L.: Software reuse: from the library to the factory. IBM Syst. J. **32**(4), 1–23 (1993)
2. Griss, M.L., Wentzel, K.D.: Hybrid domain-specific kits for a flexible software factory. In: Proceedings: SAC 1994, Phoenix, Arizona, March 1994
3. Gaffney, J.E., Cruickshank, R.D.: A general economics model of software reuse. In: Proceedings: 14th ICSE, Melbourne Australia, May 1992
4. Mahdy, A., Fayad, M.E., Hamza, H., Tugnawat, P.: Stable and reusable model-based architectures. In: ECOOP 2002, Workshop on Model-Based Software Reuse, Malaga, Spain, June 2002
5. Hamza, H., Fayad, M.E.: Model-based software reuse using stable analysis patterns. In: ECOOP 2002, Workshop on Model-based Software Reuse, Malaga, Spain, June 2002
6. Fayad, M.E.: Stable Design Patterns for Software and Systems. Auerbach Publications, Boca Raton, FL (2015)
7. Fayad, M.E., Altman, A.: Introduction to software stability. Commun. ACM **44**(9) (2001)
8. Fayad, M.E.: Accomplishing software stability. Commun. ACM **45**(1) (2002a)
9. Fayad, M.E.: How to deal with software stability. Commun. ACM **45**(4) (2002b)
10. Ross, R.G.: Business Rule Concepts: Getting to the Point of Knowledge, 4th edn. Business Rule Solutions, LLC, April 2013
11. Jindal, S.: Stable Business Rule Standards. Masters Thesis, San Jose State University (2015)
12. Fayad, M.E., Sanchez, H.A., Hegde, S.G.K., Basia, A., Vakil, A.: Software Patterns, Knowledge Maps, and Domain Analysis. Auerbach Publications, Boca Raton (2014)
13. Fayad, M.E., Cline, M.: Aspects of software adaptability. Commun. ACM **39**(10), 58–59 (1996)
14. Muehlen, Z., Michael, M.I., Kittel, K.: Towards integrated modeling of business processes and business rules. In: ACIS 2008 Proceedings, vol. 108 (2008). [Arch]
15. Arsanjani, A.: Rule object 2001: a pattern language for adaptive and scalable business rule construction. In: PLoP 2001 Conference on Business Rule Construction. National EAD Center of Competency, p. 12. IBM, Raleigh (2001)

3D Distance Transformations with Feature Subtraction

Mike Janzen and Sudhanshu Kumar Semwal$^{(\boxtimes)}$

Department of Computer Science, University of Colorado, Colorado Springs,
CO, USA
{mjanzen, ssemwa}@uccs.edu

Abstract. This paper presents a technique to implement three-dimensional distance transformation based on the city-block algorithm and selective removal of cells of interest using both serial and parallel techniques. We call this *feature subtraction*. One of the main benefits of feature subtraction is that the volume data could be representing as union of these subtracted features (e.g. spheres), replacing the cells in the volume data by a set of these disjointed spheres. The algorithm is implemented in the specialized graphics programming language called *Processing Language*.

Keywords: 3D distance transformation · Sphere packing · Medical applications

1 Introduction

Distance transformation is a well-known concept in the area of computer graphics and image processing. It is multidimensional and can be used as a filtering technique for 2D images as well as 3D volume data. The focus here will be on three dimensions by adapting a 2D serial implementation. Additionally, we present a unique parallel approach to solving this problem. When performing a distance transformation, the general approach is to compute the distance from each voxel of interest, or feature voxel, to the nearest unoccupied, or non-feature, voxel. The goal is similar in our 3D implementation as well: to create a three-dimensional volume grid of data and perform distance computations on each voxel. Each of these voxels will be assigned a distance value corresponding to the nearest non-feature voxel that can be reached from its grid location. After producing an algorithmically generated cubic set of voxel data containing distance calculations, we will demonstrate the algorithm by selectively removing a feature voxel and all neighboring voxels that can be reached in a set number of steps. Thus, the region of interest can be visualized as a set of these well-defined, distance transformation-based regions. As a real-world example to aid visualization, it is helpful to imagine a cluster of human cells where some are healthy, and some are cancerous. We consider the healthy cells to be non-feature cells, while the cancer cells are features. With this terminology, it becomes clear that we would seek to

© Springer Nature Switzerland AG 2020
K. Arai et al. (Eds.): FTC 2019, AISC 1069, pp. 654–662, 2020.
https://doi.org/10.1007/978-3-030-32520-6_48

remove the features (cancer), while preserving the non-feature cells (healthy cells). To see this algorithm in action in two dimensions, let us examine three Figures that show an initial dataset (Fig. 1), the distances after the forward mask is applied (Fig. 2), and the final data after the backward mask is applied (Fig. 3). A value of 0 means a healthy, non-feature cell, and a value of ∞ means that there is a feature or cancer cell at that point. These values were generated directly from the algorithm that was implemented. These tables clearly illustrate the interaction between the two masks. For instance, after the application of the forward mask the values around the healthy cell in the lower right corner are clearly wrong (a value of 5 when there is a non-feature cell directly next to it with a value of 0). Additionally, there are still cells with no distance information because the forward mask was unable to identify the location of any non-feature cells. Forward and backward masks for Distance Transformation has been discussed elsewhere [8, 13, 14], here we discuss the resultant 2D cells where we start with features of interest with value zero and rest with values infinite. Once the forward and backward mask is applied, all cells will have been updated with the correct distance information (Fig. 3). As mentioned above, the backward mask can only compare right-hand neighbors when being applied to the bottom row. A cascading effect results from this one correction since it allows the other cells whose values were incorrect to be recomputed and corrected. The resultant Table 3 shows a bottom left pixel with a value of 5 which indicate that feature pixels (with values) are approximately five radius distance away from the center of the pixel marked 5.

∞	∞	∞	∞	∞	∞	0
∞	0	∞	∞	∞	∞	∞
∞	0	∞	∞	∞	∞	∞
∞	∞	∞	0	∞	∞	∞
∞	∞	∞	∞	∞	∞	∞
∞	∞	∞	∞	∞	∞	∞
∞	∞	∞	∞	∞	∞	0

Fig. 1. Initial dataset

∞	∞	∞	∞	∞	∞	0
∞	0	1	2	3	4	1
∞	0	1	2	3	4	2
∞	1	2	0	1	2	3
∞	2	3	1	2	3	4
∞	3	4	2	3	4	5
∞	4	5	3	4	5	0

Fig. 2. Dataset after forward mask

2	1	2	3	2	1	0
1	0	1	2	3	2	1
1	0	1	1	2	3	2
2	1	1	0	1	2	3
3	2	2	1	2	3	2
4	3	3	2	3	2	1
5	4	4	3	2	1	0

Fig. 3. Dataset after forward mask

2   Previous Research and Significance of Proposed Work on Feature Subtraction

Volume rendering techniques have been popular since 1990s [1–4, 6] and can be used to render one frame of volume data [5]. Volume data is data in 3D grid format representing stacks of CT/MRI/Photographs stacked as 3D grid in three dimensions. Recent implementations of distance transformation have been shown to be effective in isolating circles from 2D images [11–14]. Our work has the same lineage as Speher packing or packing shapes inside a defined 2 or 3D area [7]. Here we are extending our previous work [13] to three-dimensional data by simulating a 3D grid with initialized with zeros those grid cells representing the region of interest, and then using the 3D Distance transformations to isolate spherical shapes from the 3D-data and implement that to arbitrary details and complexity. We call this *feature subtraction* below. One of the main benefits of feature subtraction is that the volume data could be representing as union of these subtracted features (e.g. disjoint spheres), thus volume data can be represented as set of these disjointed spheres.

3   Feature Subtraction

The program first launched a window, such as Fig. 4. Figure 5 demonstrates a possible series of steps. After launching the program, the user would click the drawing area, move the mouse to position the image as shown, then click to lock that view. Next, the user would click *Remove Cell* to remove the cell (and neighbors) with the largest distance value. Clicking the *Toggle Removals* changes the image to display removed cells in orange and clicking *Toggle Spheres* will display a sphere that encompasses the removed cells.

Proceeding from left to right in Fig. 5, we see the initial data, the same data rotated and translated, the data after removing the largest distance item, removed cells displayed (in orange), and the sphere displayed.

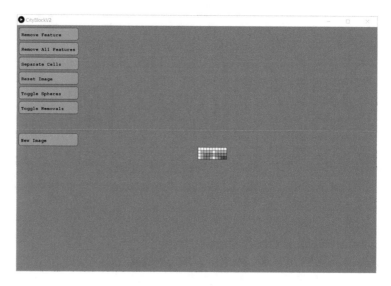

Fig. 4. Initial program window

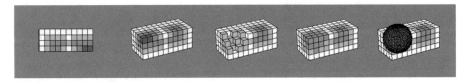

Fig. 5. Five Basic stages of processing

Next, we demonstrate the use of separation and spheres on the same data as shown in Fig. 4. In the upper left corner of Fig. 6, we see the same cell contents as the fourth image in Fig. 5; however this image was generated by clicking *Separate Cells* once. The image in the upper right of Fig. 6 shows the result of clicking *Separate Cells* a second time, and the remaining image shows the usage of *Toggle Spheres* in conjunction with the separation.

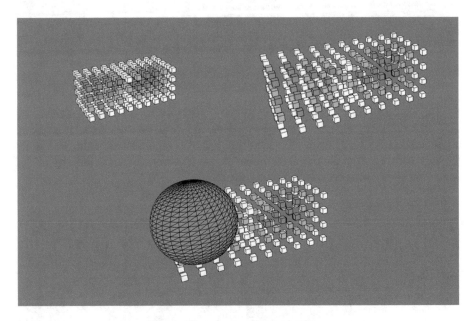

Fig. 6. Use of separation and spheres

A logical next step would be to remove a second feature. In Fig. 7, we show three different views of the same data. In the upper left, the removed cells are not displayed, in the upper right the removed cells are indicated by the orange cells, and on the bottom there are spheres indicating the removed cells.

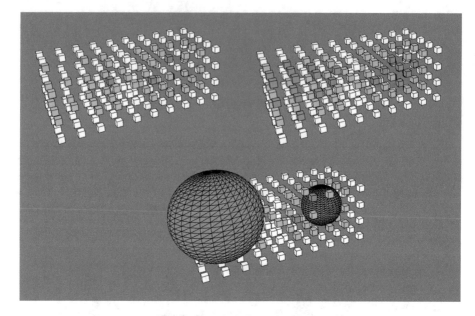

Fig. 7. Removal of second feature

The user may now wish to remove all the features, either by clicking *Remove All Cells* or by successive clicks of *Remove Feature*. In Fig. 8, we show three views of the original data from Fig. 5 with all feature cells removed. From left to right, we display the data with the cells entirely removed, with the removed cells indicated by orange cells, and finally with spheres indicating the removal.

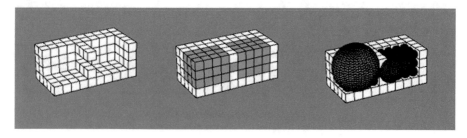

Fig. 8. All features removed

In Fig. 9, we demonstrate sphere interference and the ability to reset the image. First, we show the final portion of Fig. 8 but with separation factor increased. The second portion of the image shows the data with rotation and translation that clearly displays the interference of the two left-most spheres. Lastly, we show the result of clicking *Reset Image*. All spheres are now cleared, the view orientation is unchanged, and the original data has been restored.

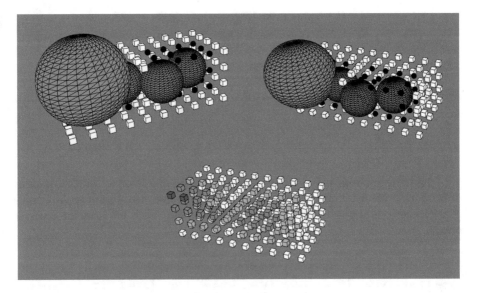

Fig. 9. Reset image

4 Results

Timing data was captured on a Microsoft Surface Book with 16.0 GB of RAM, an SSD, and an Intel Core i7-6600U CPU running at 2.60 GHz. The operating system is Microsoft Windows 64-bit with all current updates installed. The version of Java that was used is JDK 1.8.0_121, and we are using *Processing* version 3.3.6 [9, 10].

Serial Execution: Performing a distance transformation in three dimensions and removing the largest value feature are the main purposes of this program. The distance transformation and removal can be applied to data from any source that is in a compatible format. Therefore, the means by which we generated growths, displayed the data, and the provided UI exist mainly to demonstrate the distance transformation and feature removal.

To prove that this our solution is viable for other applications, we performed timing analysis on these aspects. We performed tests using cubic sizes of $10 \times 10 \times 10$, $25 \times 25 \times 25$, $50 \times 50 \times 50$, $100 \times 100 \times 100$, and $200 \times 200 \times 200$. For each size, we collected the following timing data and averaged them over 10 runs: growth generation, forward mask distance transformation, backward mask distance transformation, removal of largest feature and neighbors, and time to draw the image. This data is shown in Table 1.

Table 1. Serial timing information

Cubic size	Total cells	Mutant cells	Growth time (ms)	Fwd DT (ms)	Bkwd DT (ms)	Orphan check (ms)	Total DT (ms)	Remove largest (ms)	Remove radius (cells)	Draw time (ms)
10	1,000	300	0.70	0.50	0.20	0.10	0.70	1.50	4.00	260.20
25	15,625	4.687	1.55	1.91	1.73	0.36	3.64	5.27	10.82	427.45
50	125,000	37,500	2.80	10.30	11.10	2.40	21.40	25.80	23.50	1,405.50
100	1 m	300,000	11.40	28.30	29.30	5.30	57.60	162.70	45.70	5,295.10
200	8 m	2.4 m	41.00	117.80	103.40	15.10	221.20	336.40	94.10	37.226.50

With timing data captured, we analyzed how the timing changed with the number of cells. The cube with dimension 10 is considered the baseline. Each row in Table 2 contains data that shows how much of an increase these values are from the previous row.

Table 2. Serial execution time increases

Cubic size	Total cells Inc	Mutant cells Inc	Growth time Inc	Fwd DT Inc	Bkwd DT Inc	Orphan check Inc	Total DT Inc	Remove largest Inc	Remove radius Inc	Draw time Inc
10	1	1	1	1	1	1	1	1	1	1
25	15.6x	15.6x	2.21x	3.82x	8.64x	3.64x	5.19x	3.52x	2.70x	1.64x
50	8x	8x	1.81x	5.40x	6.43x	6.60x	5.89x	4.89x	2.17x	3.29x
100	8x	8x	4.07x	2.75x	2.64x	2.21x	2.69x	6.31x	1.94x	3.77x
200	8x	8x	3.60x	4.16x	3.53x	2.85x	3.84x	2.07x	2.06x	7.03x

What we see in these two tables is that the various stages of the algorithm vary linearly with the size of the data, thus giving the overall program a complexity of $O(n)$ where n is the number of pixels. We also see that the biggest detriment to the overall responsiveness of the program is the draw routine that is part of *Processing*. Even with 8 million cells, removing the largest feature (which includes the removal, resetting the values of all the remaining cells, and recomputing all the distances) is accomplished in an average of just over three tenths of a second.

Parallel Distance Transformation: Due to the overhead involved in parallelization, we only looked at the performance of large datasets and focused on simply examining the distance transformations since that is the only place where parallelism was implemented. Even with a simplistic partitioning scheme, the results are clearly in favor of parallelism for large datasets.

Cubic datasets of size 500, 700, and 900 were executed 10 times for both the serial and parallel versions of the program. The times were then totaled and compared to each other. For summary values, we averaged all the runs and computed the time decrease. For this data, we saw an execution time decrease of between 47% and 49% on average. Given the relatively simplistic partitioning scheme, it is believed that these results could be improved by fine tuning the number of slices given to a particular process. Regardless, it is clear that parallel programming is a powerful tool for solving this problem. Shown below (Table 3) are the average execution times for the serial and parallel versions of the program on a given dataset. The last column is the percentage decrease in execution time for the parallel implementation.

Table 3. Serial execution time vs parallel execution time

Cubic size	Serial forward DT average (ms)	Serial backward DT average (ms)	Serial total DT average (ms)	Parallel horiz DT average (ms)	Parallel vertical DT average (ms)	Parallel total DT average (ms)	Parallel execution decrease average
500	1,123	1,174	2,297	724	476	1,200	47.74%
700	3,042	3,349	6,391	2,003	1,300	3,303	48.31%
900	6,732	7,599	14,331	4,565	2,699	7,264	49.32%

5   Conclusion and Future Work

The biggest area that would benefit from additional effort would be implementing parallelism in a few crucial areas as well as refining the parallelization that has been demonstrated. When working with the 3D arrays, there are places where it would be possible to split the work among different processes or implement them on a GPU.

Since the draw() routine is the slowest part of the entire program, this is a good place to examine for further improvement. The slowness is due to the extensive work that the system must perform to create the voxels, space them appropriately, determine whether to add spheres and removed cells, and so on. Since these computations happen across the entire dataset, they can occur millions of times per call. Additionally, draw() is also called automatically by the system during rotations, separations, or changes in the data resulting in a substantial number of calls. One way to speed up this process could involve selectively processing only the cells which are visible to the user. This should be further investigated.

The city-block algorithm itself is strictly serial in terms of a given slice. It is not possible, for example, to split the forward and backward mask applications among processes, nor can a two-dimensional slice be split into multiple pieces. However, we have shown that it is possible to implement a parallel algorithm where each process is responsible for computing a given number of slices. Our implementation is only intended as a demonstration and could be improved by designing a more robust partitioning scheme. For example, choosing to divide the data among 10 processes was arbitrary and research could determine that there is an optimal number of slices per processor that produce the fastest execution. An additional benefit of improving the city-block algorithm is that it leads to benefits in the cell removal because the transformations must be performed after a removal occurs. The removal of a given cell provides several opportunities for parallelism, although speed increases may be minimal.

Acknowledgments. We want to acknowledge the role Slicer3D Community has played in our research directions. Although we have implemented the code using Processing Language, we want to implement the Distance Transformations algorithms using 3D Slicer platform in future. The second author of this paper is also grateful to Dr. Arcady Godin who introduced Dr. Leonid Perlovsky, and Dr. Perlovsky's work on Dynamic Logic, although not discussed in paper, was the motivation to start our work in 3D medical data visualization.

References

1. Lorenson, W.E., Cline, H.E.: Marching cubes: a high resolution 3D surface reconstruction algorithm. Comput. Graph. **21**, 163–169 (1987). SIGGRAPH 1987 Proc.
2. Levoy, M.: Display of surfaces from volume data. IEEE Comput. Graph. Appl. **8**, 29–37 (1988)
3. Upson, C., Keeler, M.: V-BUFFER: visible volume rendering. Comput. Graph. **22**, 59–64 (1988). SIGGRAPH 1988 Proc.
4. Drebin, R., Carpenter, L., Hanrahan, P.: Volume rendering. Comput. Graph. **22**, 65–74 (1988). SIGGRAPH 1988 Proc.
5. Spitzer, V.M., Whitlock, D.G.: High resolution electronic imaging of the human body. Biol. Photogr. **60**, 167–172 (1992)
6. Buchanan, D., Semwal, S.K.: A front to back technique for volume rendering. In: Computer Graphics International, Computer Graphics Around the World, Singapore, pp. 149–174. Springer-Verlag (1990)
7. Rogers, C.A.: Packing and Covering. Cambridge University Press, Cambridge (1964)
8. Borgefors, G.: Distance transformation in digital images. Comput. Vis. Graph. Image Process **34**, 344–371 (1986)
9. https://www.processing.org
10. https://github.com/processing
11. Junli, L, Xiuying, W.: A fast 3D euclidean distance transformation. In: 2013 6th International Congress on Image and Signal Processing (CISP 2013)
12. Bailey, D.G.: An Efficient Euclidean Distance Transformation
13. Janzen, M.: 3D Distance Transformation with Feature Subtraction, MS Project report. Advisor: Dr. SK Semwal, Department of Computer Science, University of Colorado, Colorado Springs, pp. 1–49 (2018)
14. Semwal, S.K., Janzen, M., Promersberger, J., Perlovsky, L.: Towards approximate sphere packing solutions using distance transformations and dynamic logic. In: Arai, K., Kapoor, S. (eds.) CVC 2019, AISC 944, pp. 1–14 (2020). https://doi.org/10.1007/978-3-030-17798-0-31

Space-Filling Curve: A Robust Data Mining Tool

Valentin Owczarek$^{(\boxtimes)}$, Patrick Franco, and Rémy Mullot

La Rochelle Université, Laboratoire Informatique, Image et Interaction (L3i),
La Rochelle, France
`valentin.owczarek1@univ-lr.fr`

Abstract. Due to the development of internet and the intensive social network communications, the number of data grows exponentially in our society. In response, we need tools to discover structures in multidimensional data. In that context, dimensionality reduction techniques are useful because they make it possible to visualize high dimension phenomena in low dimensional space. Space-filling curves is an alternative to regular techniques, for example, principal component analysis (PCA). One interesting aspect of this alternative is the computing time required (less than half a second where PCA spends seconds). Moreover with the algorithms provide results are comparable with PCA in term of data visualization. Intensive experiments are led to characterize this new alternative on several dataset covering complex data behaviors.

Keywords: Dimension reduction · Data visualization · Space-filing curves

1 Introduction

Due to the development of internet, to the intensive social network communications and the birth of internet of things (connected devices, home appliance), the number of data grows exponentially in our society. In response, we need tools to help us to "make data speak". Precisely, approaches and methodology are needed to discover structure in multidimensional data, to identify clusters, to detect outliers and to fit predicting model. In that context, dimensionality reduction techniques are useful, because they make it possible to visualize high dimension phenomena in low dimensional space. That is why in many domains, the reduction of dimensionality is considered as a preprocessing step (the first step) to explore multidimensional data. Recently in [1], Castro and Burns show the utility of space-filling curve (SFC) to reduce dimensions. In Particular, the capabilities of Hilbert curve to online data visualization is put forward compare to PCA [10,14], a standard method extensively used. One interesting aspect with this kind of curve is the time computing required to map a D-Dimensional data point to an 1-D index. With the algorithm provided [1], the mapping of one point (to $1-D$, $2-D$, or $3-D$) runs in a linear time. That is a sought property

© Springer Nature Switzerland AG 2020
K. Arai et al. (Eds.): FTC 2019, AISC 1069, pp. 663–675, 2020.
https://doi.org/10.1007/978-3-030-32520-6_49

because it is not highly sensitive to the number of data points to be processed (size of the data set) when every day the number of data increase. Therefore, using the Hilbert curve seems to be well adapted for online visualization purpose.

In other side, in [3], new space-filling curves are emerging. It has been proven that selecting alternative basic pattern (the SFC at the first order) instead of the classical one - issued from Reflected Binary Gray code - can lead to design curves with comparable (and sometimes better) locality preserving than the Hilbert curve, which so far is the reference.

In this article, the work initiate in [1] is extend to new competitive space-filling curves. High locality preserving curves are selected (including Hilbert). Original algorithms working with new curves are provided for mapping and reverse mapping (an index to a point and vice-versa). Discussion about resolution constraint (in the case of D-Dimensional → 2D, 3D) is provided in order to guide the user to make right choice on the data point resolution (order of curve). With SFC, mapping points to 1-D indexes is usual, but getting an 1-D index and turn back to a 2-D or 3-D points is more original. Clarification and details on this stage are brought out.

Furthermore, in order to get more consistent benchmark results, new data set are added. For example, Blob data-set, Tic-Tac-Toe and Sphere are integrated to cover more complex data behavior (linear and non-linear). The quality of dimensionality reduction is measured trough standard criteria: the topology preservation [5] and the Sammon Stress [12]. Comparative results show that new SFC can outperformed Hilbert curve and for some data-set, giving comparable interpretable results than PCA. Being able to quickly applied dimensionality reduction with interpretable results than standard technique seems interesting for data analysis point of view. Especially during the discovery of the data set, i.e. when a priori the characteristics of the variables and statistical relationship between them is not known.

2   Standard Dimension Reduction Techniques

2.1   Classical PCA

Principal component analysis [10, 14] is a linear projection of the data in a new reference system to maximize variance. Given a set of points $\mathbf{X}$ lying in $\mathbb{R}^D$ and the mean vector $\overline{\mathbf{X}}$. PCA find a new representation $\mathbf{Y}$ of $\mathbf{X}$ where $\mathbf{Y}$ lies in dimension D', with $D' \leq D$, such as:

$$\mathbf{Y} = W^t(\mathbf{X} - \overline{\mathbf{X}}) \tag{1}$$

where W is a $D \times D'$ matrix formed by the D' first eigenvectors corresponding to the highest eigenvalues of the covariance matrix computed on the data $\mathbf{X}$.

3   Space-Filling Curves Applied to Dimension Reduction

3.1   Space-Filling Curve: An Overview

Space-filling curves are continuous non-differentiable function. The first curve was discovered by Peano [9] in 1890. The function defines a bijection between a

D-dimensional grid and their $1 - D$ indexes I on the curve. The property behind theses functions is the locality-preserving level [7], the fact that if two points in the high dimensional space are close then their indexes are close too. In other words, the curve achieving the best locality-preserving level is the curve with less topological break.

One of the best curve according to the literature is the Hilbert curve [4] and the multidimensional extension based on the Reflected Binary Gray code (RBG).

Let's denote by S_n^D the D-dimensional space-filling curve at order n and the inverse function $\bar{S}_n^D$. Then if $\mathbf{X}_i$ and $\mathbf{X}_j$ are two close D dimensional point:

$$S_n^D(\mathbf{X}_i) = I_i, \ \bar{S}_n^D(I_i) = \mathbf{X}_i$$
$$S_n^D(\mathbf{X}_j) = I_j, \ \bar{S}_n^D(I_j) = \mathbf{X}_j \tag{2}$$
$$d(\mathbf{X}_i, \mathbf{X}_j) = \varepsilon_D, \ d(I_i, I_j) = \varepsilon_1$$

with I_i, I_j indexes and ε_D, ε_1 small.

A standard numerical characterization of the locality-preserving level of a space-filling curve can be achieved with the parameterized Faloutsos measure [2]:

$$L_{\mathbf{r}}(\bar{S}_1^D) = \frac{1}{2^D} \sum_{k,l\in[2^D], \ k<l, \ d(k,l)\leq\mathbf{r}} \max\{d(\bar{S}_1^D(k), \bar{S}_1^D(l))\} \tag{3}$$

with d a distance function. $L_1(\bar{S}_1)$, is the level of locality preserving reach by a curve S at radius $r = 1$. The Faloutsos criteria verifies that two close indexes, k and l, correspond to points, $\bar{S}_1^P(k)$ and $\bar{S}_1^P(l)$, respectively close in space. The more $L_{\mathbf{r}}(\bar{S}_1)$ tends to 1, the better is the level of locality preserving.

In [1], the preserving-locality level and the bijection properties of the Hilbert-like curve is used to map data points from an arbitrary D-dimensional space to a lower dimensional space D' (2 or 3) for data visualization purpose. The dimension reduction is performed with two steps:

1. Map all data points $\mathbf{X}$ to their indexes I using the function S_n^D.
2. Map back the indexes to $\mathbf{Y}$ in D' using the inverse space-filling curve function $\bar{S}_{n'}^{D'}$.

The space-filling curve mapping algorithm is a variant of the algorithm provided in [6], using the RBG pattern. Nevertheless recently, in [3,8], new algorithms emerge with better or equal performances in term of locality preserving.

The contribution in [8] opens new kind of algorithms based on the inheritance between the $n - 1$ and n order curve. Moreover, it has been proven experimentally that the locality-preserving level of the pattern induces a curve with less topological break [3].

In Table 1, the RBG pattern and an adjacency-based pattern $\mathcal{P}_{RBG}^*$ are shown, the adjacency-based pattern reaches a better locality-preserving level.

Table 1. p^*_{RBG}: The pattern achieving better locality preservation (through Faloutsos criteria) instead of classical RBG often used to build the Hilbert curve.

Pattern name	Faloutsos score					
	L_3	L_4	L_5	L_6	L_7	L_8
RBG	2.875	2.875	3	3.75	3.75	3.75
$\mathcal{P}^*_{RBG}$	2.875	2.875	3	**3.5625**	**3.5625**	3.75

p^*_{RBG} Pattern point list

0 1 1 0 0 1 1 0 0 1 1 0 0 0 1 1 1 1 1 0 0 0 0 0 0 0 0 1 1 1 1

0 0 0 0 0 0 0 0 0 0 0 0 0 0 0 0 1 1 1 1 1 1 1 1 1 1 1 1 1 1 1

0 0 0 0 1 1 1 1 1 1 1 1 0 0 0 0 0 0 1 1 1 0 0 1 1 0 0 1 1 0 0 1

0 0 0 0 0 0 0 0 1 1 1 1 1 1 1 1 1 1 1 1 1 1 0 0 0 0 0 0 0 0

0 0 1 1 1 1 0 0 0 0 1 1 1 0 0 1 1 0 0 1 1 1 0 0 0 0 1 1 1 1 0 0

The pattern is derived to a high order curve using a non regular reflection and rotation system i.e.

$$\mathfrak{Iso} = \mathfrak{Ref}_A \circ \mathfrak{Rot}_f$$

$$\mathcal{P}'_i(k) = \mathfrak{Ref}_A(\mathcal{P}_i(k)) = \begin{cases} 1 - \mathcal{P}_i(k) & \text{if } i \in A \\ \mathcal{P}_i(k) & \text{if } i \notin A \end{cases} \qquad (4)$$

$$\mathcal{P}' = \mathfrak{Rot}_f(\mathcal{P}) = \{\mathcal{P}(f(0)), ..., \mathcal{P}(f(D-1))\}$$

In the next section, the space-filling curve function initialized by a pattern $\mathcal{P}$ and an order n will be denoted by $S^{\mathcal{P}}_n$ and its inverse function by $\bar{S}^{\mathcal{P}}_n$. The pattern $\mathcal{P}$ gives the information on dimension D used in the transformation.

3.2   Dimension Reduction via New Space-Filling Curve: Proposition of Mapping Algorithm

This section is a didactic example of the space-filling curve algorithm allowing its initialization by any adjacency-based patterns exploiting the contribution in [8]. The version presented maps points $\mathbf{X} \in [0, 2^n - 1]^D$ to the indexes $\mathbf{I} \in [0, 2^{nD} - 1]$. The pattern in the example is $\mathcal{P}_{RBG}$, the RBG-based pattern in $D = 2$, i.e.

$$\mathcal{P}_{RBG} = \begin{pmatrix} 0 & 0 & 1 & 1 \\ 0 & 1 & 1 & 0 \end{pmatrix}$$

The attached isometry system $\mathfrak{Iso}_{RBG}$ for $D = 2$ is:

$$\mathfrak{Iso}_{RBG} = \begin{pmatrix} \mathfrak{Ref} & \emptyset & \emptyset & \emptyset & 1 \\ \mathfrak{Rot} & 1 & 0 & \emptyset & \emptyset & 1 & 0 \end{pmatrix}$$

where $\emptyset$ mean that there is no rotation or reflection needed.

By notation, the pattern and the isometry system are zero based array: the first point of the RBG pattern in $D = 2$ is, $\mathcal{P}_{RBG} = (0,0)$ and the associate isometry is $\mathfrak{Iso}_{RBG}^{0} = \begin{pmatrix} \mathfrak{Ref} & \emptyset \\ \mathfrak{Rot} & 1\ 0 \end{pmatrix}$.

From a Point to an Index: Let us define a point $x = (3, 2, 2)$, the question is how to compute $S_2^{\mathcal{P}_{RBG}}(x)$?

---

1. Initialize: $\mathcal{P} = \mathcal{P}_{RBG} = \begin{pmatrix} 0\ 1\ 1\ 0\ 0\ 1\ 1\ 0 \\ 0\ 0\ 1\ 1\ 1\ 1\ 0\ 0 \\ 0\ 0\ 0\ 0\ 1\ 1\ 1\ 1 \end{pmatrix}$,

   $n = 2,\ D = 3$
2. Get $|x|_2$ the binary representation of x with n digits:
   $|x|_2 = (11, 10, 10)$
3. Form pp_1 with the most significant digit of $|x|$: $pp_1 = (1, 1, 1)$
4. Calculate ip_1 the index of pp_1 on $\mathcal{P} : ip_1 = 5$
5. Apply the isometry $\mathfrak{Iso}_{RBG}^{ip_1} = \begin{pmatrix} \mathfrak{Ref} & 2\ 0 \\ \mathfrak{Rot} & 1\ 2\ 0 \end{pmatrix}$ to $\mathcal{P}_{RBG}$:

   $\mathcal{P} = \mathfrak{Iso}_{RBG}^{ip_1}(\mathcal{P}_{RBG}) = \begin{pmatrix} 0\ 0\ 1\ 1\ 1\ 1\ 0\ 0 \\ 1\ 1\ 1\ 1\ 0\ 0\ 0\ 0 \\ 1\ 0\ 0\ 1\ 1\ 0\ 0\ 1 \end{pmatrix}$
6. Form pp_2 with the second most significant digit of $|x|_2$: $pp_2 = (1, 0, 0)$
7. Calculate ip_2 the index of pp_2 on $\mathcal{P} : ip_2 = 5$
8. Deduce $\mathcal{F}_3^{\mathcal{P}_{RBG}}(x)$ from the ip_k index:
   $\mathcal{F}_2^{\mathcal{P}_{RBG}}(x) = \sum_{k=1}^{n} ip_k 2^{D(n-k)} = (5 \cdot 8) + 5 = 45$

Example of computation of an index from a point, with the space-filling function: $S_2^{\mathcal{P}_{RBG}}$. The 3-D point $x = (3, 2, 2)$ has the index $i = 45$ when the space is filled via the curve $S_2^{\mathcal{P}_{RBG}}$ at order $n = 2$, built from the $\mathcal{P}_{RBG}$ pattern generator using $\mathfrak{Iso}_{RBG}$. The pattern is transformed by the use of Eq. 4. The reverse mapping (index to point) is always possible and use the same isometry $\mathfrak{Iso}_{RBG}$.

Glossary

- D is the dimension of the space-filling curve.
- n is the order of the space filling curve.
- $\mathcal{P}$ is a pattern, a space-filling curve at $n = 1$, the seed of the space-filling curve algorithm.
- $S_n^{\mathcal{P}}(x)$, $\bar{S}_n^{\mathcal{P}}(i)$ are respectively the point to index and index to point space-filling curve functions, according to [8].
- x, $|x|_b$ are respectively a data point and the representation of the point in base b.
- pp_k is a point on the pattern.
- ip_k is an index on the pattern.

3.3   Application to Dimension Reduction

In [1], Space-filling curves are the key of a new dimension reduction method. The main difference between this technique and the other introduced in this paper is the non usage of the data. The idea is to project the entire D-cube induced by the space-filling curve S_n^D to the D'-cube defined by $S_n'^{D'}$ by computing the associate indexes I. There is no formal description of the technique in [1], the next section is a possible one.

Let $\mathbf{X}_i$ be a point in the D-dimensional space then $\mathbf{Y}_i$ the projection by $S_D^{\mathcal{P}}$ and $S_{D'}^{\mathcal{P}'}$ is:

$$\mathbf{Y}_i = \bar{S}_{n'}^{\mathcal{P}'}(S_n^{\mathcal{P}}(\mathbf{X}_i)) \tag{5}$$

with $\mathcal{P}$ and $\mathcal{P}'$ are two patterns lying, respectively in dimension D and D' with $D > D'$

Nevertheless, the algorithms in Sect. 3.2 map points from $[0, 2^n]^D$ to $[0, 2^{nD}]$. The data points must be transformed, we propose to rewrite $\mathbf{X}$ as:

$$\mathbf{X}_i^t = \mathbf{X}_i^t - \min(\mathbf{X}_i^t) \tag{6}$$

This transformation translates all the point in the interval $[0, 2^{nD}]$.

The orders of the two curves are free parameters, but user must respect the following inequality to avoid any collision i.e. two different points in high dimension with the same coordinate in the projected space:

$$2^{Dn} \leq 2^{D'n'} \tag{7}$$

4   Experiments with Classical and Space-Filling Curves Dimension Reduction Techniques

4.1   Experimental Conditions: Measure

In order to compare the different dimension reduction algorithms, two measure are employed: the topology preservation measure and the Sammon stress.

Topology Preservation Measure: The measure is proposed in [5], it qualifies the preservation of nearest neighbor between the high dimensional point $\mathbf{X}$ and the projected one $\mathbf{Y}$. The measure is based on a credit assignment scheme:

$$qm_{ji} = \begin{cases} 3 & NNX(j,i) = NNY(j,i) \\ 2 & NNX(j,i) = NNY(j,t) \ t \in [1,n], t \neq i \\ 1 & NNX(j,i) = NNY(j,t) \ t \in]n,k], n < k \\ 0 & else \end{cases} \tag{8}$$

where $NNX(j,i)$ and $NNY(j,i)$ correspond to the ith neighbor of respectively the point $\mathbf{X}_j$ and $\mathbf{Y}_j$. The definition of the nearest neighbor need a distance function, we used the euclidean distance.

The measure is defined as the normalized sum of the $qm_{i,j}$.

$$Tpm = \frac{1}{3nN} \sum_{j=1}^{N} \sum_{i=1}^{n} qm_{ji} \qquad (9)$$

with n and k initialized to 4 and 10.
The measure indicate a perfect mapping when $Tpm = 1$.

Sammon Stress: The measure is extracted from the Sammon mapping [1, 12] and can be interpreted as a distance error function. E_{SS} is equal to zero if all distances between original high dimensional points are respected in the low projected space. The original measure is sensitive to rescaling. In order to compare different technique which may rescaled data differently, the β term is added.
The measure is then equal to:

$$E_{SS} = \frac{1}{\sum\limits_{i=1}^{n-1} \sum\limits_{j=i+1}^{n} D(\mathbf{X}_i, \mathbf{X}_j)} \sum_{i=1}^{n-1} \sum_{j=i+1}^{n} \frac{(D(\mathbf{X}_i, \mathbf{X}_j) - \beta D(\mathbf{Y}_i, \mathbf{Y}_j))^2}{D(\mathbf{X}_i, \mathbf{X}_j)} \qquad (10)$$

with β equals to:

$$\beta = \frac{\sum\limits_{i=1}^{n-1} \sum\limits_{j=i+1}^{n} D(\mathbf{Y}_i, \mathbf{Y}_j)}{\sum\limits_{i=1}^{n-1} \sum\limits_{j=i+1}^{n} \frac{D(\mathbf{Y}_i, \mathbf{Y}_j)^2}{D(\mathbf{X}_i, \mathbf{X}_j)}} \qquad (11)$$

The distance function D is commonly the euclidean function.
The measure indicates a perfect distance preservation when $E_{SS} = 0$.

4.2 Experimental Conditions

The previously described dimension reduction techniques: Principal component analysis (PCA) and Space-Filling curve dimension reduction (SFC) are tested on six different datasets. The main characteristics of each dataset are listed in Table 2. Most of the datasets are extracted from OpenML [13] and other are issued from the Python version of Scikit-learn [11].

The presented results are the best scoring of each techniques. For the datasets: Forest Covertype and Letter Image Recognition Data, the result is the average score on five samples of 2000 attributes.

For the dimension reduction based on space-filling curves, the pattern used to map data points in high dimension is specified for each experiment. Hilbert corresponds to the RBG pattern, $(\mathcal{P}_{RBG}^*)$ is a pattern achieving better performance than the RBG one according to the parameterized Faloutsos measure (Eq. 3). Parameters n and n' are chosen to match the observation of Sect. 3.3 (Eq. 7).

Table 2. Dataset use for the experiments.

Dataset name	Dimension	Number of attributes	Number of classes
Concentric sphere	3	1000	2
Iris database	4	150	3
Blob	5	500	3
Tic-Tac-Toe	9	958	2
Forest covertype	12	581012	7
Letter image recognition data	16	20000	26

4.3   Results of the Several Experiences

Table 3. Results for the six datasets on two different measures. The projected dimension is $D' = 2$. For each measure, a bold font style indicate the best score.

Dataset	Method	Sammon stress	Topology preservation
Concentric sphere (3-D)	PCA	**0.0558**	**0.4624**
	Hilbert	0.2739	0.3650
Iris dataset (4-D)	PCA	**0.0063**	**0.5844**
	Hilbert	0.3212	0.4022
Blob (5-D)	PCA	**0.0237**	0.1791
	Hilbert	0.2579	0.2195
	SFC $(\mathcal{P}^*_{RBG})$	0.2439	**0.2291**
Tic-Tac-Toe (9-D)	PCA	0.1488	0.0328
	Hilbert	**0.1294**	0.1710
	SFC $(\mathcal{P}_A)$	0.1926	0.1709
	SFC $(\mathcal{P}_B)$	0.2207	**0.1745**
Forest covertype[a] (12-D)	PCA	**0.0052**	**0.3509**
	Hilbert	0.7578	0.2625
Letter image recognition data[a] (16-D)	PCA	**0.1110**	0.1167
	Hilbert	0.3450	**0.2277**

[a] Average score on 5 samples of 2000 attributes

The results for the two techniques (PCA and SFC) can be seen in Table 3, in the experiment concluded the dimension reduction goal is $D' = 2$. The time needed to reduce the dimensionality with SFC is always lower than PCA. On average it takes one second for PCA and less than half a second for SFC.

PCA reached every best Sammon Stress scores for each dataset except for Tic-Tac-Toe. For the Topology Preservation measure, SFC reached best score for three over six datasets. This result is expected since contrary to SFC, PCA is the a linear transformation (i.e. rotation) which do not change the distance between original $\mathbf{X}$ data point and projected one $\mathbf{Y}$. Nevertheless SFC tend to maximize the preservation of topology (cf. Sect. 3.1), explaining the results achieved for the Topology Preservation measure.

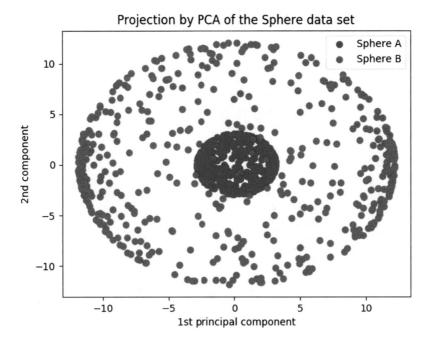

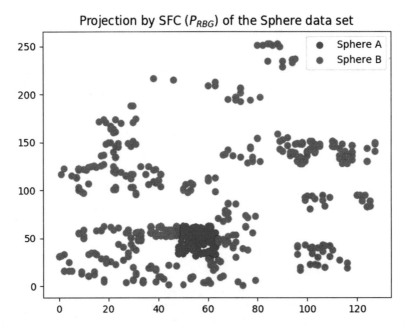

Fig. 1. Comparison of projection by SFC and PCA reduction technique on the Sphere data set. PCA failed to perfectly separate the two spheres, where SFC obtains a clear separation.

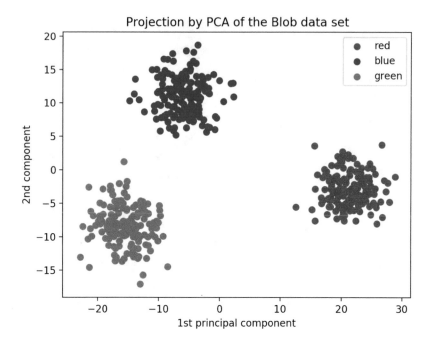

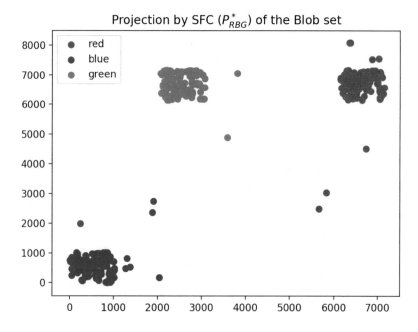

Fig. 2. Comparison of projection by SFC and PCA reduction techniques on the Blob data set. PCA creates well separated clusters on Gaussian distribution. SFC topology preservation properties are able to create several clusters by class.

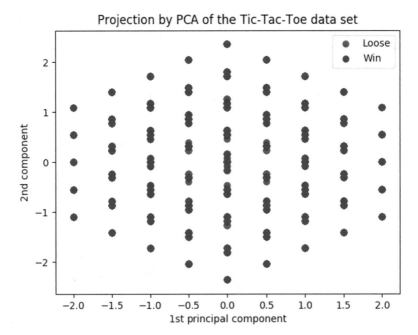

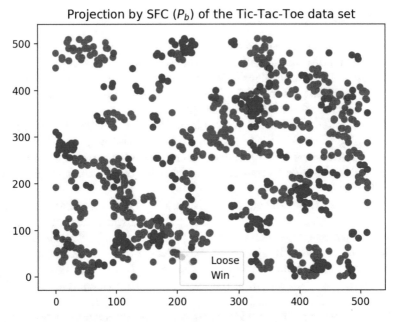

Fig. 3. Comparison of projection by SFC and PCA reduction techniques on the Tic-Tac-Toe dataset. Due to the data set properties, PCA cannot separate the two class (Loose and Win) and it creates a superposition of points, whereas SFC reducts the data set to an understandable projection where each point is distinct.

By only interpreting the theoretical measure results from Table 3, we can affirm that SFC is less effective than PCA. But the two measure used in this article can be criticized by observing the projection presented in Figs. 1, 2 and 3 representing, respectively the dimension reduction from D to $D' = 2$ for the Concentric Sphere, Blob and Tic-Tac-Toe dataset.

Fig. 1 shows the differences between the two dimension reduction techniques on the Concentric Sphere dataset, where PCA is not able to clearly separate the two spheres, SFC create a cluster with the Sphere A surround by Sphere B, a layout where a classification algorithm could obtain a perfect score.

PCA and SFC present quite similar results in term of visualization for the Blob dataset, Fig. 2. Reader may prefer the SFC visualization over the PCA one, in regard of the compactness of the clusters. Nonetheless, with both techniques we are able to determine the number of clusters in the original dimension $(D = 5)$.

For the Tic-Tac-Toe dataset, Fig. 3, the dimension reduction technique showed completely different visualization. PCA preserves the inner topology of the dataset, a grid, where we cannot identify clusters of the same class. In contrast, SFC is able to create several clusters. The projection informs us about the high dimension topology: data from the same class are close to one another. With this information, users can decide for example to use a nearest neighbour based clustering algorithm.

5   Conclusion

The proposition follows the work initiated in [1], using space-filling curve to create a new kind of dimension reduction technique. The approach is based on a new space-filling curve algorithm [8] allowing the use of new patterns (the space-filling curve at order $n = 1$). The new patterns reach comparable locality-preserving level than the Hilbert curve [3].

The methodology is intensively explained in Sect. 3, from the pattern to the curve mapping ($[0, 2^n - 1]^D \rightarrow [0, 2^{nD} - 1]$) and inverse mapping ($[0, 2^{nD} - 1] \rightarrow [0, 2^n - 1]^D$). Moreover a discussion about resolution constraint is established in order to guide the user to make the right choice on the data point resolution (order of curve).

Experiments are led to compare PCA and SFC dimension reduction method on six different datasets covering complex data behavior (linear and non-linear). In the light of these experiments, we can conclude that the dimension reduction method based on space-filling curve is a challenger of PCA. Both techniques are able to create visualization in low dimension space ($D' = 2$) of phenomena involving in high dimension, given information required to tune complex algorithms of data-mining. Nonetheless space-filling curve time complexity is not comparable with PCA, on average it takes less than half a second against seconds for PCA to compute the dimension reduction projection of the same dataset. Moreover the use of high locality preserving level patterns implies better overall results.

The SFC dimension reduction technique is used in this paper to visualize data set in $2 - D$, but dimension reduction technique are used in different application as preprocessing step. The dimension of the input data is reduct in order to apply for example machine-learning algorithm. A study of the differences between PCA and SFC dimension reduction technique as preprocessing would be interesting considering the low computing time cost.

References

1. Castro, J., Burns, S.: Online data visualization of multidimensional databases using the Hilbert space-filling curve. In: Pixelization Paradigm, pp. 92–109. Springer, Heidelberg (2007)
2. Faloutsos, C., Roseman, S.: Fractals for secondary key retrieval. In: Proceedings of the Eighth ACM SIGACT-SIGMOD-SIGART Symposium on Principles of Database Systems, PODS 1989, pp. 247–252. ACM (1989)
3. Franco, P., Nguyen, G., Mullot, R., Ogier, J.M.: Alternative patterns of the multidimensional Hilbert curve. Multimedia Tools Appl. **77**(7), 8419–8440 (2018)
4. Hilbert, D.: Ueber die stetige Abbildung einer Line auf ein Flächenstück. Math. Ann. **38**(3), 459–460 (1891)
5. Konig, A.: Interactive visualization and analysis of hierarchical neural projections for data mining. IEEE Trans. Neural Netw. **11**(3), 615–624 (2000)
6. Lawder, J., King, P.: Using state diagrams for Hilbert curve mappings. Int. J. Comput. Math. **78**(3), 327–342 (2001)
7. Moon, B., Jagadish, H.V., Faloutsos, C., Saltz, J.H.: Analysis of the clustering properties of the Hilbert space-filling curve. IEEE Trans. Knowl. Data Eng. **13**(1), 124–141 (2001)
8. Nguyen, G.: Courbes remplissant l'espace et leur application en traitement d'images. Ph.D. thesis, Université de La Rochelle (2013)
9. Peano, G.: Sur une courbe, qui remplit toute une aire plane. Math. Ann. **36**, 157–160 (1890)
10. Pearson, K.: On lines and planes of closest fit to systems of points in space. Philos. Mag. Series 6 **2**(11), 559–572 (1901)
11. Pedregosa, F., Varoquaux, G., Gramfort, A., Michel, V., Thirion, B., Grisel, O., Blondel, M., Prettenhofer, P., Weiss, R., Dubourg, V., Vanderplas, J., Passos, A., Cournapeau, D., Brucher, M., Perrot, M., Duchesnay, E.: Scikit-learn: machine learning in python. J. Mach. Learn. Res. **12**, 2825–2830 (2011)
12. Sammon, J.W.: A nonlinear mapping for data structure analysis. IEEE Trans. Comput. **C-18**(5), 401–409 (1969)
13. Vanschoren, J., van Rijn, J.N., Bischl, B., Torgo, L.: OpenML: networked science in machine learning. SIGKDD Explor Newsl. **15**(2), 49–60 (2014)
14. Wold, S., Esbensen, K., Geladi, P.: Principal component analysis. Chemometr. Intell. Lab. Syst. **2**(1–3), 37–52 (1987)

A Local-Network Guided Linear Discriminant Analysis for Classifying Lung Cancer Subtypes using Individual Genome-Wide Methylation Profiles

Yanming Li[✉]

University of Michigan, Ann Arbor, MI 48109, USA
liyanmin@umich.edu

Abstract. Accurate and efficient prediction of lung cancer subtypes is clinically important for early diagnosis and prevention. Predictions can be made using individual genomic profiles and other patient-level covariates, such as smoking status. With the ultrahigh-dimensional genomic profiles, the most predictive biomarkers need to be first selected. Most of the current machine learning techniques only select biomarkers that are strongly correlated with the outcome disease. However, many biomarkers, even though have marginally weak correlations with the outcome disease, may execute a strong predictive effect on the disease status. In this paper, we employee an ultrahigh-dimensional classification method, which incorporates the weak signals into predictions, to predict lung cancer subtypes using individual genome-wide DNA methylation profiles. The results show that the prediction accuracy is significantly improved when the predictive weak signals are included. Our approach also detects the predictive local gene networks along with the weak signal detection. The local gene networks detected may shed lights on the cancer developing and progression mechanisms.

Keywords: Cancer subtype prediction · Linear discriminant analysis · Local gene network · Ultrahigh-dimensionality

1 Introduction

In cancer-genomic studies, individual-level genomic profiles are often used to predict cancer status or cancer subtypes. With the advances of modern high-throughput technologies, the numbers of genetic biomarkers in individual genomic profiles are growing at a unprecedented speed. They are nowadays usually of exponential orders of the sample sizes. As in our motivating Boston Lung Cancer Study Cohort (BLCSC) datasets, genome-wide DNA methylation profiles are essayed for 153 lung cancer patients at more than 400,000 cytosine-guanine dinucleotide (CpG) sites. The patients are of three major lung cancer types. There are (i) 98 lung adenocarcinoma (LUAD) patients, which consist of 77 adenocarcinoma (AD) and 21 bronchioloalveolar carcinoma (BAC-AD) patients,

© Springer Nature Switzerland AG 2020
K. Arai et al. (Eds.): FTC 2019, AISC 1069, pp. 676–687, 2020.
https://doi.org/10.1007/978-3-030-32520-6_50

(ii) 45 lung squamous cells carcinoma (LUSC) patients and (iii) 10 patients of other types, which include subtypes such as large cell carcinoma, small cell carcinoma and carcinoid.

The three major lung cancer subtypes in the BLCSC study: adenocarcinoma, squamous cells carcinoma and large cell carcinoma account for 40%, 25–30% and 10–15% of the lung cancer cases, respectively. Adenocarcinoma arises from small airway epithelial, type II alveolar cells, which secrete mucus and other substances. It tends to occur in the periphery of the lung and tends to grow slower and more easy to be found before it has spread outside of the lungs. Squamous cells carcinoma arises from early versions of squamous cells and the airway epithelial cells in the bronchial tubes in the center of the lungs. It is more strongly correlated with cigarette smoking compared to the other lung cancer subtypes. Large cell carcinoma often begins in the central parts of the lungs, sometimes into nearby lymph nodes and the chest wall. It has been recognized that the genome-wide DNA methylation patterns undergo substantial changes across different lung cancer subtypes [16]. Therefore, identifying the CpG biomarkers that are most relevant to the etiologies of different lung cancer subtypes and using them for accurate prediction are clinically important for prognosis and prevention.

Technical Significance. In ultrahigh-dimensional settings, using all the CpG markers will lead to poor prediction results [2]. In order to achieve an accurate classification, it is important to select only the most predictive biomarkers. In cancer-genomic studies, most of the approaches select predictive biomarkers based on their marginal effects. This way, only the biomarkers strongly correlated with the disease outcome will be included in the prediction. Recent studies [9,10] show that many of the marginally weak biomarkers may execute a strong predictive effect on the outcome disease through their connections to the marginal strong markers. This makes a better sense from the system biology perspective, as the biomarkers are not acting on the outcome disease separately, but are working together as a whole complex system. Therefore, the connections between markers play a key role in the outcome prediction. In this paper, we tailor a recently developed, ultrahigh-dimensional classification method – mLDA [9], which was particularly designed for detecting marginally weak signals, to the whole genome-scale association studies, and use it to predict the lung cancer subtypes. As will be shown in our analysis results, the prediction accuracy is significantly improved by incorporating the weak signals.

Clinical Relevance. Our approach may impact the clinical practice of lung cancer as it significantly increases the lung cancer subtype prediction accuracy. It may also be applied in the fields such as artificial intelligence (AI) and remote diagnoses. In certain unforeseeable extreme circumstances, which forbid the on-site presence of human doctors, our approach can still provide in-time and low-cost diagnosis and prevention.

Our approach can also learn the local network structures between the predictive CpG markers at the same time. DNA methylation is a key determinant of gene expression [5]. It generally leads to gene silencing when the CpG sites residing in gene promoter regions get methylated. It is usually positively correlated with gene expression when the CpG sites are in gene body regions [11]. Therefore, the local network structures of the CpG markers reflect the gene networks they resided in. Uncovering such local gene networks renders important biological interpretation for the selected markers, as such biological networks can help with a more insightful understanding of the cancer developing and progression mechanisms.

The rest of the paper is organized as the follows. Section 2 introduces the BLCSC study and its methylation dataset. In Sect. 3, we briefly review the mLDA method, describe how we tailored it to fit the genome-wide methylation analysis by implementing chromosome-wise parallel computing and adjusting for other covariates such as the smoking status and age. Section 4 gives the analysis results and biological interpretation for our findings. The paper is concluded by Sect. 5, where a summary is provided along with a brief discussion of relevant issues.

2   The Boston Lung Cancer Study Cohort

The Boston Lung Cancer Study (BLCS) is a cancer epidemiology cohort (CEC) of 11,164 lung cancer cases enrolled at Massachusetts General Hospital, the Dana-Farber Cancer Institute, and the Brigham and Women's Hospital since 1992. The BLCSC has collected detailed demographic, smoking, occupational, and diet information, in addition to pathology, treatments, oncogenic (somatic driver) mutation status, and biosamples. The BLCS biorepository includes serum, white blood cells, DNA, and 2,000 tumor and surrounding tissues. The follow-up of the BLCSC has been high, approximately 95% for the cumulative follow-up, with nearly complete ascertainment of deaths using the National Death Index and other resources. By identifying the relevant cases and providing critical archived tumor tissues, through the collaboration within the Dana-Farber/Harvard Cancer Center (DF/HCC) Lung Cancer Program, the BLCSC supported the first discovery of the association between *EGFR* mutations and response to therapy with *EGFR-TKIs* in 2004, starting the era of targeted therapy.

The BLCSC methylation dataset currently collects genome-wide methylation profiles on more than 400,000 CpG sites, assayed using the Infinium Human-Methylation450 beadarray chips, along with other clinical and demographic indices, such as gender, age and smoking status in term of package per year, etc. for 153 lung cancer patients, consisting of 98 LUAD patients, 45 LUSC patients and 10 patients of other lung cancer types including large cell carcinoma, small cell carcinoma, carcinoid, etc.

2.1   Data Processing

After removing the CpG sites with missing values, there are 409,111 CpG sites left across the whole genome. The CpG sites numbers on each of the 22 automosomes and 2 sex chromosomes are given in Table 1. Since the mLDA method requires the input features to be from normal distribution, we did an exploratory analysis on each of the features. No severe deviation from the normal assumption was found. We also standardized each input feature by dividing its sample standard deviation to ensure that the variations of different features are at the same scale.

Table 1. Number of CpG sites on each chromosome in the BLCSC methylation dataset

chr 1	chr 2	chr 3	chr 4	chr 5	chr 6	chr 7	chr 8
39,813	29,650	21,212	17,112	20,828	30,305	25,057	17,902
chr 9	chr 10	chr 11	chr 12	chr 13	chr 14	chr 15	chr 16
8,332	20,795	24,520	20,487	10,251	12,660	12,933	17,940
chr 17	chr 18	chr 19	chr 20	chr 21	chr 22	chr X	chr Y
23,263	5,104	21,253	9,057	3,606	7,176	9,772	83

3   The mLDA Method and Modifications

The mLDA method was first introduced in [9] for multiclass linear discriminant analysis with ultrahigh-dimensional features. The mLDA method can detect predictive marginally weak signals and integrate them into outcome classifications. It also explores the local network structures between the selected features along the selection procedure. For the completeness of this paper, we present the major algorithmic steps of mLDA in the flowchart in Fig. 1.

We tailored the mLDA procedure particularly for analyzing the BLCSC genome-wide methylation data in the following two ways:

1. We parallalized the mLDA procedure so that the analysis can be conducted on the 24 chromosomes in a parallel way.
2. We modified the mLDA source codes to allow adjusting for covariates. Here we assume that the number of the adjusted covariates is much less than the sample size, the adjusted covariates are independent to the CpG markers and are jointly follow a multivariate normal distribution. The adjusted covariates will be always included into the classification.

4   BLCSC Data Analysis

We applied the mLDA classifier, along with some other conventional ultrahigh-dimensional classification methods designed only for detecting marginally strong signals, on the 24 chromosomes and then the selected predictive CpG sites across the whole genome were put together to predict the lung cancer subtypes.

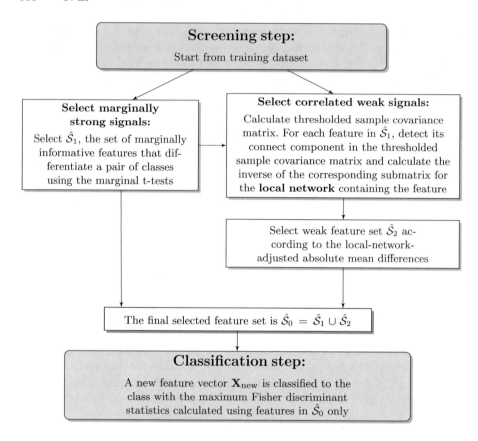

Fig. 1. Flowchart of mLDA procedure.

We included the "pack-year" (package-per-year) and "age" as the adjusted covariates. Denoted by $\mathbf{Z}$ the covariate matrix of "pack-year" and "age" in the training data and we reformatted the classification rule of mLDA as

$$
\hat{Y}_{\mathrm{new}} = arg\,max_{1 \leq k \leq K} \left\{ (\mathbf{Z}_{\mathrm{new}} - \hat{\boldsymbol{\mu}}_Z^{[k]}/2)' \hat{\boldsymbol{\Sigma}}_Z^{-1} \hat{\boldsymbol{\mu}}_Z^{[k]} \right.
$$
$$
\left. + \sum_{c=1}^{24} \sum_{w=1}^{B_c} (\mathbf{X}_{cw,\mathrm{new}} - \hat{\boldsymbol{\mu}}_{cw}^{[k]}/2)' \hat{\boldsymbol{\Sigma}}_{cw}^{-1} \hat{\boldsymbol{\mu}}_{cw}^{[k]} \right\}, \tag{1}
$$

where $\mathbf{Z}_{\mathrm{new}}$ is the covariate vector of "pack-year" and "age" for the new sample, $\hat{\boldsymbol{\mu}}_Z^{[k]} = \sum_{i:Y_i=k} \mathbf{Z}/n_k$ is the covariate mean vector for the k-th class evaluated from the training data, $\hat{\boldsymbol{\Sigma}} = cov(\mathbf{Z})$, c indexes the 24 chromosomes with the X chromosome indexed by 23 and Y chromosome by 24, w indexes the local networks consisting of predictive markers detected on each chromosomes, $\mathbf{X}_{cw,\mathrm{new}}$ is the vector of CpG markers within the w-th the local network detected on chromosome c, $\hat{\boldsymbol{\mu}}_{cw}^{[k]} = \sum_{i:Y_i=k} \mathbf{X}_{cw}/n_k$ and $\hat{\boldsymbol{\Sigma}}_{cw} = cov(\mathbf{X}_{cw})$ are the k-th class sample mean and covariance of CpG markers in the w-th the local network

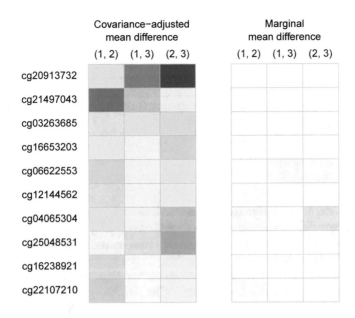

Fig. 2. Heatmap of ten example marginally weak CpG markers from the BLCSC methylation data. The right panel is the marginal mean differences between each class pair. The left panel is the local-network-adjusted mean differences between each class pair. Yellow is for positive values and blue is for negative values. The darker the color, the greater the absolute values are. The class pairs are: (1,2)=(LUAD, LUSC), (1,3)=(LUAD, others), (2,3)=(LUSC, others).

detected on chromosome c from the training data, respectively. For the MS method [2], we take each $\hat{\boldsymbol{\Sigma}}_{cw}$ to be a diagonal matrix. For the SIS method [3], we used pairwise logistic regressions with the two adjusted covariates.

To avoid over-fitting, we adopted a five-fold cross-validation procedure on the 153 subjects. Specifically, we divided the data into five fold with about the same sample sizes for each class in each fold. We selected predictive CpG sites on four folds combined as the training dataset and then used the selected markers to predict the lung cancer subtypes on the left-out fifth fold as the test dataset. We repeated such a procedure till each fold has been used as the test dataset once and computed the overall classification errors by summing the classification errors over all five test datasets. For mLDA, we used a fix thresholding parameter $\alpha = 0.5$ for the correlations between markers. We used a chromosome-specific thresholding parameter τ such that only the top 5 strongest marginally strong markers are selected on each chromosome and a chromosome-specific thresholding parameter ν determined by applying mLDA on the validation dataset, which is divided from the training dataset with about one third of its samples. Table 2 lists the classification errors from the different methods. The mLDA gave in total 4 misclassifications among the 153 patients. On the other hand, the marginal approaches gave much more misclassifications.

Table 2. Number of mis-classifications from analyzing the BLCSC methylation data using different methods

	mLDA	MS	SIS
LUAD	1	6	6
LUSC	2	13	12
Others	1	4	1
Total	4	23	19

- MS: [2].
- SIS: [3].

Figure 2 depicts the marginal mean differences and the local-network-adjusted mean differences for some selected marginally weak CpG markers. It can be seen that the predictive effects of these marginally weak features are manifested by integrating their local network structures.

Table 3 lists some of the selected CpG sites sitting in genes that had been identified to be associated with lung cancer in previous genome-wide association studies (GWAS). It is worthy pointing out that all these CpG sites are selected as marginally weak markers. With a small total sample size of 153, these markers are more likely to be missed in a GWAS. However, mLDA is able to pick them up by incorporating their connections with some other marginally strong markers.

Table 3. Selected CpG markers in lung cancer related genes reported in previous GWAS studies

CpG site name	chr	pos	gene	type	Literature GWASs
cg17319788	7	55121401	*EGFR*	MW	[1,24];
cg18394848	12	25403102	*KRAS*	MW	[7,12];
cg02842899	17	7589491	*TP53*	MW	[13,14];

- chr: chromosome.
- pos: base pair position.
- gene: the unique gene that the CpG site sit in; "inter-gene" means that the CpG site is in inter-gene region.
- MS: Marginally strong signal.
- MW: Marginally weak signal.

Table 4 lists the top selected CpG markers across all 24 chromosomes. The markers are ranked based on their maximum absolute local-network-adjusted mean difference values across the three class pairs in a decreasing order.

The CpG site cg21497043, locating within 5'UTR of the 1st exon in gene *GLS* on chromosome 2, is selected as a marginally weak marker differentiating both the LUAD-others and the LUSC-others class pairs. Its absolute marginal mean differences between class pairs LUAD-others and LUSC-others are 0.16 and 0.02, respectively. Its absolute local-network-adjusted mean differences between the two class pairs are 6.52 and 5.99, respectively (also see Fig. 2). Author in [19] found that *GLS1*, an isoform of *GLS*, to be a key enzyme for the dependency of the glutanamine (Gln) pathway, which plays an important role in sustaining lung cancer cell proliferation. [26] found that the *GLS* gene provides glutamine-derived carbon and nitrogen to pathways that support proliferation of non-small cell lung cancer cells. The local gene network of genes harboring cg21497043 and its connected predictive CpG sites is depicted in the upper panel of Fig. 3.

The CpG site cg18394848, locating within the 5'UTR region of gene *KRAS* on chromosome 12, also selected as a marginally weak marker, differentiates the LUSC-others classes. Its absolute marginal mean difference between the class pair is 0.04 and its absolute local-network-adjusted mean differences between the class pair is 5.51. The gene *KRAS* has been identified to be associated with lung cancer in genome-wide association studies (GWAS) such as [4,7,12,18,25]. Its positive correlation with *PFKM* support the findings that multiple glycolytic genes, including phosphofructokinases (*PFKL* and *PFKM*), are upregulated by the *KRAS* metabolic reprogramming [6]. The local gene network of genes harboring cg18394848 and its connected predicitve CpG sites is given in the lower panel of Fig. 3.

The CpG site cg07607126, sitting in gene *EBF1* on chromosome 5, is selected as a marginally weak marker. It differentiates class pairs LUAD-others and LUSC-others with absolute marginal mean differences between the two class pairs being 0.44 and 0.82, respectively, and absolute local-network-adjusted mean differences between the two class pairs being 5.41 and 5.25. *EBF1* gene plays an important role in the expression of *Pax5* gene, which is a key factor for the production of antibody-secreting cells [8]. The emerging roles of the *EBF* family genes on suppressing the lung cancer tumor cells are reported in [8] and [17].

It is extremely interesting that three CpG sites among the top selected markers are on the X chromosome. The effects of X chromosome genes on lung cancer were barely studied in the literature [20]. The *KLHL13* gene located in Xq24 has been reported to be associated with various types of cancers [15,21,22,27]. *MORC4* was found to associate with inflammatory bowel disease [23]. Further investigations are needed for associations between these X-chromosomal genes and lung cancer subtypes.

5 Discussion

We introduce a novel approach for detecting marginally weak signals in classification of lung cancer subtypes using individual genome-wide methylation profiles. The results show that marginally weak signals could be predictive to the outcomes through their connections to the marginally strong signals. The local gene network structures of those predictive CpG markers are inferred along with the variable selection procedure. By incorporating the weak signals, the outcome prediction performance can be significantly improved, especially when the number of the marginally weak signals is large. The procedure used here can be applied to prediction of other cancers or disease outcomes as well.

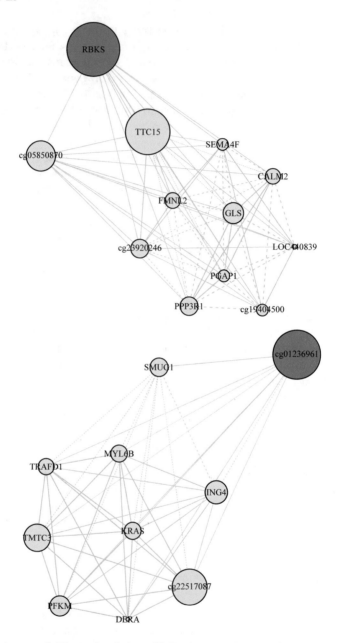

Fig. 3. Upper panel: Example of the *"GLS"*-signatured local gene networks on chromosome 2. The nodes represent the unique genes the CpG sites sit in. If the CpG is not located in any genes, the node is named after the CpG site name. Red nodes: genes for marginally strong CpG markers; Yellow nodes: genes for marginally weak CpG markers. Positive partial correlations are presented by solid-line edges. Negative partial correlations are presented by dotted-line edges. The node sizes are proportional to the absolute marginal mean differences between classes LUAD and others. Lower panel: Example of the *"KRAS"*-signatured local gene networks on chromosome 12. The node sizes are proportional to the absolute marginal mean differences between classes LUSC and others.

Table 4. Top selected CpG markers

CpG name	chr	pos	gene	Discriminant class pair	type	Correlated MS marker	Absolute marginal mean differences	mLDA importance score
cg16653203	2	47403915	CALM2	(2,3); (1,3)	MW	cg02599006	(2,3)=0.04; (1,3)=0.12	(2,3)=8.35; (1,3)=3.63
cg22517087	12	49208544	Inter-gene	(2,3)	MW	cg01236961	(2,3)=0.08	(2,3)=6.66
cg21284493	X	53123624	Inter-gene	(1,3)	MW	cg12485020	(1,3)=0.29	(1,3)=6.65
cg21497043	2	191745649	GLS	(1,3); (2,3)	MW	cg21497043	(1,3)=0.16; (2,3)=0.02	(1,3)=6.52; (2,3)=5.99
cg21626372	12	88536536	TMTC3	(2,3)	MW	cg01236961	(2,3)=0.01	(2,3)=5.87
cg00415263	7	26331488	SNX10	(2,3); (1,3)	MW	cg25748521	(2,3)=0.74; (1,3)=0.07	(2,3)=5.66; (1,3)=3.24
cg18394848	12	25403102	KRAS	(2,3)	MW	cg01236961	(2,3)=0.04	(2,3)=5.51
cg07607126	5	158527469	EBF1	(1,3); (2,3)	MW	cg23457837	(1,3)=0.44; (2,3)=0.82	(1,3)=5.41; (2,3)=5.25
cg13024202	X	117250753	KLHL13	(2,3)	MW	cg13113011	(2,3)=0.30	(2,3)=5.16
cg14479037	X	106242642	MORC4	(1,3)	MW	cg12485020	(1,3)=0.38	(1,3)=4.91
cg13799302	1	60392392	CYP2J2	(2,3); (1,3)	MS	–	(2,3)=2.61; (1,3)=2.73	(2,3)=3.50; (1,3)=3.17

- Selected CpG markers are ranked by its maximum absolute local-network-adjusted mean difference between a pair of classes in a decreasing order.
- chr: chromosome.
- pos: base pair position.
- gene: the unique gene that the CpG site sit in; "inter-gene" means that the CpG site is in inter-gene region.
- MS: Marginally strong signal.
- MW: Marginally weak signal.

The method can be extended to accommodate non-Gaussian distributed predictors. Thus it can be applied to classification problems with ultrahigh-dimensional categorical predictors, such as genome-wide single-nucleotide polymorphism (SNP) genotype data.

The local gene networks can be viewed as the gene co-regulation networks, which encode the structures of the partial correlations between genes. In our approach, the covariance matrix is assumed to be the same across different classes. When the covariance structures are heterogeneous across classes, extensions to nonlinear classification may be more appropriate and warrant future research.

Other than the correlation structures, other genome structures, such as the gene/pathway grouping structures are also worth to be incorporated in detection of weak signals. Some marginally weak signals may not be detectable by incorporating the correlation structures, such as a group of weakly correlated weak signals with a strong joint grouped effect. In such cases, it is worth investigating how to integrate different genome structures into weak signal detection for improving outcome classification or prediction. We will pursue this direction in future research.

References

1. Bell, D.W., et al.: Inherited susceptibility to lung cancer may be associated with the T790M drug resistance mutation in EGFR. Nat. Genet. **37**, 1315–1316 (2005)
2. Fan, J., Fan, Y.: High-dimensional classification using features annealed independence rules. Ann. Statist. **36**, 2605–2637 (2008)
3. Fan, J., Lv, J.: Sure independence screening for ultrahigh dimensional feature space (with discussion). J. R. Statist. Soc. B. **70**, 849–911 (2008)
4. Ferrer, I., Zugazagoitia, J., Herbertz, S., John, W., Paz-Ares, L., Schmid-Bindert, G.: KRAS-mutant non-small cell lung cancer: from biology to therapy. Cancer Biol Ther. **124**, 53–64 (2018)
5. Johansson, A., Flanagan, J.M.: Epigenome-wide association studies for breast cancer risk and risk factors. Trends Cancer Res. **12**, 19–28 (2017)
6. Kerr, E.M., Martins, C.P.: Metabolic rewiring in mutant KRAS lung cancer. FEBS J. **285**(1), 28–41 (2018). https://doi.org/10.1111/febs.14125
7. Kim, J., et al.: XPO1-dependent nuclear export is a druggable vulnerability in KRAS-mutant lung cancer. Nature **538**(7623), 114–117 (2016)
8. Liao, D.: Emerging roles of the EBF family of transcription factors in tumor suppression. Mol. Cancer Res. **7**(12), 1893–1901 (2009). https://doi.org/10.1158/1541-7786.MCR-09-0229
9. Li, Y., Hong, H.G., Li, Y.: Multiclass linear discriminant analysis with ultrahigh-dimensional features. Biometrics (2019). https://doi.org/10.1111/biom.13065
10. Li, Y., Hong, H.G., Ahmed, S.E., Li, Y.: Weak signals in high-dimensional regression: detection, estimation and prediction. Appl. Stoch. Models Bus. Ind. **35**, 283–298 (2019). https://doi.org/10.1002/asmb.2340
11. Lou, S., et al.: Whole-genome bisulfite sequencing of multiple individuals reveals complementary roles of promoter and gene body methylation in transcriptional regulation. Genome Biol. **15**(7), 408 (2014)

12. McKay, J.D., et al.: Large-scale association analysis identifies new lung cancer susceptibility LOCI and heterogeneity in genetic susceptibility across histological subtypes. Nature Genet. **49**, 1126–1132 (2017)

13. Mechanic, L.E., Bowman, E.D., Welsh, J.A., Khan, M.A., Hagiwara, N., Enewold, L., Shields, P.G., Burdette, L., Chanock, S., Harris, C.C.: Common genetic variation in TP53 is associated with lung cancer risk and prognosis in African Americans and somatic mutations in lung tumors. Cancer Epidemiol. Biomark. Prev. **16**(2), 214–222 (2007)

14. Mogi, A., Kuwano, H.: TP53 mutations in nonsmall cell lung cancer. J Biomed. Biotechnol. **2011**, 583929 (2011). https://doi.org/10.1155/2011/583929

15. Parsons, D.W., et al.: An integrated genomic analysis of human glioblastoma multiforme. Science **321**(5897), 1807–1812 (2008)

16. Pfeifer, G.P., Rauch, T.A.: DNA methylation patterns in lung carcinomas. Semin. Cancer Biol. **19**(3), 181–187 (2009). https://doi.org/10.1016/j.semcancer.2009.02.008

17. Rollin, J., Blechet, C., Regina, S., Tenenhaus, A., Guyetant, S., Gidrol, X.: The intracellular localization of ID2 expression has a predictive value in non small cell lung cancer. PLoS ONE **4**(1), e4158 (2009). https://doi.org/10.1371/journal.pone.0004158

18. Román, M., Baraibar, I., López, I., Nadal, E., Rolfo, C., Vicent, S., Gil-Bazo, I.: KRAS oncogene in non-small cell lung cancer: clinical perspectives on the treatment of an old target. Mol. Cancer **17**(1), 33 (2012). https://doi.org/10.1186/s12943-018-0789-x

19. Pieter, A., van den Heuvel, J., Jing, J., Wooster, R.F., Bachman, K.E.: Analysis of glutamine dependency in non-small cell lung cancer GLS1 splice variant GAC is essential for cancer cell growth. Cancer Biol. Ther. **13**(12), 1185–1194 (2012)

20. Schinstine, M., Filie, A.C., Torres-Cabala, C., Abati, A., Linehan, W.M., Merino, M.: Fine-needle aspiration of a Xp11.2 translocation/TFE3 fusion renal cell carcinoma metastatic to the lung: report of a case and review of the literature. Diagn. Cytopathol. **34**(11), 751–756 (2006)

21. Shah, S.P., et al.: Frequent mutations of genes encoding ubiquitin-mediated proteolysis pathway components in clear cell renal cell carcinoma. Nature genet. **44**(1), 17–19 (2012)

22. Sjöblom, T., et al.: The consensus coding sequences of human breast and colorectal cancers. Science **314**(5797), 268–274 (2006)

23. Söderman, S.P., et al.: Analysis of single nucleotide polymorphisms in the region of CLDN2-MORC4 in relation to inflammatory bowel disease. World J. Gastroenterol. **19**(30), 4935–4943 (2013)

24. Timofeeva, M.N., et al.: Influence of common genetic variation on lung cancer risk: meta-analysis of 14,900 cases and 29,485 controls. Hum. Mol. Genet. **21**, 4980–4995 (2012)

25. To, M.D., Wong, C.E., Karnezis, A.N., et al.: KRAS regulatory elements and exon 4A determine mutation specificity in lung cancer. Nature Genet. **40**, 1240–1244 (2008)

26. Ulanet, D.B., et al.: Mesenchymal phenotype predisposes lung cancer cells to impaired proliferation and redox stress in response to glutaminase inhibition. PLoS ONE **9**(12), e115144 (2014). https://doi.org/10.1371/journal.pone.0115144

27. Varela, I., et al.: Exome sequencing identifies frequent mutation of the SWI/SNF complex gene PBRM1 in renal carcinoma. Nature **469**(7331), 539–542 (2011)

Multiple Global Community Detection in Signed Graphs

Ehsan Zahedinejad[1], Daniel Crawford[1], Clemens Adolphs[1],
and Jaspreet S. Oberoi[1,2(✉)]

[1] 1QB Information Technologies, 200-1285 West Pender Street,
Vancouver, BC V6E 4B1, Canada
jaspreet.oberoi@1qbit.com
[2] School of Engineering Science, Simon Fraser University,
8888 University Drive, Burnaby, BC V5A 1S6, Canada

Abstract. Signed graphs serve as a primary tool for modelling social networks. They can represent relationships between individuals (i.e., nodes) with the use of signed edges. Finding communities in a signed graph is of great importance in many areas, for example, targeted advertisement. We propose an algorithm to detect multiple communities in a signed graph. Our method reduces the multi-community detection problem to a quadratic binary unconstrained optimization problem and uses state-of-the-art quantum or classical optimizers to find an optimal assignment of each individual to a specific community.

Keywords: Graph clustering · Modularity · Frustration · Community detection · Quantum annealing · Discrete optimization

1 Introduction

Signed graphs (SG) are ubiquitous in social networks [1–3]. They can encode the perception and attitude between individuals via signed links, where a positive link between two nodes can indicate friendship and trust, while a negative link denotes animosity and distrust [4]. Thus far, there has been impressive progress towards the development of methods for exploring tasks within SGs [5–7]. With the continuous rapid yearly growth of social media users, there is an immediate need for reliable and effective approaches to the modelling of social networks.

There exists a range of interesting problems within the SG domain, including link prediction [6,8], network evolution [7], node classification [9], and community detection. In this work, we focus on developing a multi-community detection algorithm [10–14]. The principal idea in community detection is to divide an SG into clusters such that the representations of users within a cluster are densely connected by positive links whereas those that belong to different clusters are connected by negative links. Community detection has numerous applications in various areas, including in medical science [15,16], telecommunications [17], the detection of terrorist groups [18], and information diffusion processes [19]. The

© Springer Nature Switzerland AG 2020
K. Arai et al. (Eds.): FTC 2019, AISC 1069, pp. 688–707, 2020.
https://doi.org/10.1007/978-3-030-32520-6_51

wide applicability of community detection makes it an important topic of study, and emphasizes the need to devise faster and more-effective approaches in its implementation.

Community detection research can be divided into four categories [10], that is, clustering based, mixture-model based, dynamic-model based, and modularity based. Our approach is modularity based: we maximize modularity (i.e., the number of edges that fall within clusters minus the expected number of edges within those clusters in an equivalent network wcih edges placed at random) and minimize frustration [11] (i.e., the number of negative edges within communities plus the number of positive links between communities) to discover communities in an SG.

Over the last decade, there has been a large body of work that has used modularity or a variant of it as a metric for detecting communities in an SG. For instance, the authors of [14] find communities by minimizing the frustration and those of [20] propose a community detection framework called SN-MOGA, using a non-dominated sorting genetic algorithm [12, 13] to simultaneously minimize frustration and maximize signed modularity. Authors in [21] investigate the role of negative links in SGs that use frustration as a metric.

We formulate the multi-community detection problem as a quadratic unconstrained binary optimization (QUBO) problem, the optimal solution of which corresponds to the solution of the multi-community detection problem. Our approach has several advantages over existing community detection algorithms. Unlike other approaches, our approach does not require an input from the user predefining the number of communities to be detected. It requires only an upper-bound on the number of communities in order to find the optimal number of communities. Also, in the case of finding more than two communities, our approach does not recursively divide the graph into two parts [14], creating artificial local boundaries between communities, but instead preserves the global structure of the SG. Lastly, our approach is applicable to any community detection metric which can be formulated as a QUBO problem.

In this work, we use two solvers to address a QUBO problem, a classical algorithm called parallel tempering Monte Carlo with isoenergetic cluster moves (PTICM) [22, 23], also known as "borealis" [24, 25], and a quantum annealer (the D-Wave 2000Q [26]). We have chosen PTICM because of its proven superiority over other QUBO solvers [24, 25], while experiments with the D-Wave device have helped us to understand that our approach can immediately benefit from advancements in quantum computing [27].

The current quantum annealer's processor has a small number of qubits, limiting the size of benchmark datasets we can use to test our algorithm. We have considered two approaches to remedy this constraint. First, trivially enough, we limit the maximum size of the selected dataset by choosing an SG with less than 64 nodes (this results in a QUBO problem with 256 binary variables). Second, we implement block coordinate descent (BCD) to enable our approach to find communities in larger SGs [28, 29]. BCD works by iteratively solving subproblems while keeping the rest of the variables fixed. The performance of BCD on several QUBO problems is reported in [29].

The main focus of this work is to propose a new algorithmic approach for finding multiple communities in an SG. As such, we do not intend to compare and benchmark the performance of different QUBO solvers for solving this problem. The paper is structured as follows. We present terminology and notation in Sect. 2. We then explain the concept of structural balance its relation to community detection Sect. 3. We give our proposed method for multi-community detection in Sect. 4. We explain our approach in comparing the performance of the proposed algorithms and our choice of benchmarking datasets in Sect. 5. We report our results in Sect. 6 and discuss them in Sect. 7. We conclude and suggest future research directions in Sect. 8.

2   Notation

In this section, we present the notation we use throughout this work. In our terminology, $\mathbf{G}(\mathrm{V}, \mathrm{E})$ denotes an SG where V is the set of vertices and $\mathrm{E} \subset \mathrm{V} \times \mathrm{V}$ denotes the set of edges that are present in the SG. The adjacency matrix of $\mathbf{G}$ is represented by A, where each element of this matrix is $+1$ (-1) when there is a positive (negative) relation between two nodes, and zero otherwise. We also define A' to be the positive adjacency matrix whose elements are 1s if there is a link between two nodes, and zero otherwise. Following the notation in [14], we define the elements of the positive (P) and negative (N) matrices as

$$\mathrm{P}_{ij} = \frac{A_{ij} + A'_{ij}}{2} \quad \text{and} \quad \mathrm{N}_{ij} = \frac{A'_{ij} - A_{ij}}{2}, \tag{1}$$

where the A_{ij} (A'_{ij}) are the elements of the adjacency (positive adjacency) matrix and $\{i, j\} \in \{\mathbf{V}\}$. The total number of nodes is n and the total number of edges is m. The number of non-zero entries in A, P, and N are denoted by $2 \times m$, $2 \times m_p$, and $2 \times m_n$, respectively. We refer to the positive degree of vertex i as p_i, and its corresponding negative degree as n_i. The degree of the vertex i is given by $d_i = p_i + n_i$. We denote a non-empty set of vertices by $\mathscr{C}$ and call it a community cluster.

Our algorithm divides a given community cluster into k communities $\mathscr{C}_1, \mathscr{C}_1, \mathscr{C}_2, \cdots, \mathscr{C}_k$ by optimizing the frustration or modularity as the objective function. Here, we assume that each $\mathscr{C}_l$ $(l \in \{1, 2, \ldots, k\})$ is a non-empty set of nodes, and that each node belongs to only one cluster (i.e., there is no overlap between clusters).

When solving for two clusters, we label each node with s_i $(i \in \{1, \ldots, n\})$, which as a solution takes the value $+1$ (-1) if it falls in the first (second) cluster. We denote $\mathbf{s}$ as the n-dimensional configuration vector and define it as

$$\mathbf{s} = [s_1, s_2, \ldots, s_n]. \tag{2}$$

For multi-cluster detection, we use one-hot encoding to label each node i with one of the k clusters. In particular, we denote the label of a node i by $\mathbf{s}_i$

and define it as

$$\mathbf{s}_i = [s_{i1}, s_{i2}, \ldots, s_{ik}], \tag{3}$$

where s_{ic} ($c \in \{1, 2, \cdots, k\}$) is 1 if node i belongs to the c-th cluster, and zero otherwise. Similar to (2), we define the $(k \times n)$-dimensional configuration matrix **S** as

$$\mathbf{S} = [\mathbf{s}_1^T, \mathbf{s}_2^T, \cdots, \mathbf{s}_n^T]. \tag{4}$$

3   Structural Balance

The notion of structural balance was first introduced in [4] to analyze the interaction between pairs of users whose relationships were expressed in terms of being either friends or foes. Later, an SG approach was taken in [30,31] in order to model the social structure between users. The concept of structural balance can perhaps most easily be explained for a simple graph comprising three nodes and then generalized to larger graphs. In what follows, we explain why minimizing the frustration or maximizing the modularity can be appropriate measures to take to find communities in SGs.

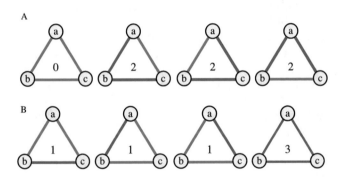

Fig. 1. Examples of different configurations of a signed complete graph with three nodes. Each circle represents a node. Nodes are connected via either a negative (a solid line in red) or positive (a solid line in blue) relationship. The numbers inside each graph indicate the total number of negative links in each configuration. The theory of strong social networks defines those configurations with an even (odd) number of negatives edges as structurally stable (unstable).

In Fig. 1, we give an example of an SG with three nodes {a, b, c}, where the relation between each pair of nodes can be negative or positive. Assuming the graph is complete (i.e., all nodes are connected to each other by negative or positive links), there exist only four possibilities in general that nodes can be interconnected (i.e., it is a triad). We label each configuration using the total number of negative edges present in that configuration (see Fig. 1).

The theory of strong social networks defines those configurations that have an even number of negative links as stable [32]. Examples of these stable configurations in Fig. 1 are the triad with zero negative links (representing mutual friends) and those with two negative links (representing a pair of friends with a common enemy). The other two sets of configurations with an odd number of total negative links are unstable, that is, the configurations with one negative tie (representing a pair of enemies with a common friend) and the configuration with three negative ties (representing three mutual enemies). We highlight that there is a generalized structural balance theory proposed in [33] in which configurations with a total of three negative links are also considered to be stable. Consistent with much other community detection research, we consider here such configurations to be unstable.

We can easily generalize the above discussion to larger graphs by checking that all the fully connected subgraphs of size three in a complete SG are structurally stable. In other words, we call a complete SG *structurally stable* if each 3-clique in SG is structurally stable [14]. In the case of an incomplete SG, we call it *balanced* when it is possible to assign ± 1 signs to all missing entries in the adjacency matrix, such that the resulting complete network is balanced.

In real-world applications involving SGs, it is often challenging to enforce the notion of structural balance. Instead, the easier concept of clusterizable graphs [33] is used throughout the literature mostly for solving community detection problems. In a clusterizable graph, nodes within clusters are connected by only positive links, whereas negative links connect nodes between clusters. Since minimizing the frustration or maximizing the modularity are means for finding clusters in a given network, we use these measures as metrics to find clusters with a high degree of structural balance. In the next section, we give a technical explanation of how frustration or modularity can be used to discover multiple communities in SGs.

4   Methods

In this section, we explain our approach to finding multiple communities in social networks by minimizing the frustration or maximizing the modularity within a network. For each of the frustration- or modularity-based approaches, we start with the simpler task of finding two communities in a given SG. We then use the two-community formalism as a building block of our k-community detection algorithm.

4.1   Frustration

Two-Community Detection. Based on the definition of network frustration, the overall frustration of a given SG is equal to the total number of negatives edges of that network, because all the positive edges lie within a single community (i.e., the entire network). Community detection based on the minimization of frustration considers the possibility of any clustering division that leads to a lower frustration contributed by each of the proposed clusters, when compared to the frustration of the full graph considered as a single community. Therefore, our goal is to assign a label s_i to each node $i \in \mathbf{V}$ such that the resultant assignment lowers (possibly to a minimum value) the frustration within the SG to a value smaller than the frustration of the entire network when considered to be a single community. After such an assignment, each node with a label of $+1$ (-1) will belong to the community $\mathscr{C}_1$ $(\mathscr{C}_2)$.

There are two cases that we should consider in order to formulate a measure of frustration. First, any two nodes that are connected by a positive link but are assigned to different communities should increase the frustration. Second, any two nodes which are connected by a negative link but are assigned to the same community should increase the frustration. We can mathematically express these two cases using the following formula [14] for the frustration $\mathscr{F}$:

$$\mathscr{F} = \sum_{i,j \in \mathscr{C}} A_{ij} - \mathbf{s}\mathbf{A}\mathbf{s}^T. \tag{5}$$

It is easy to verify that a node pair (i,j) increases the value of $\mathscr{F}$ by 1 if and only if they form a frustrated pair. The solution to the two-community detection problem is then given by the assignment $\mathbf{s}^*$ that minimizes $\mathscr{F}$. The minimum frustration will be zero in the case of a fully structurally balanced SG and positive in the case of a partially structurally balanced SG. We now explain our approach to generalizing this two-community detection algorithm into a k-community detection approach.

Multi-community Detection. As mentioned earlier, we use one-hot encoding to label each node i with one of the k communities. We require that clusters be non-overlapping, that is, each node will be assigned to exactly one cluster. Therefore, we have the following constraint on the label of a given node i,

$$\|\mathbf{s}_i\| = 1, \tag{6}$$

where $\|\cdot\|$ is L_1-norm operator. From (3) and (6), it follows that if the two nodes i, j belong to the same community, we have

$$\mathbf{s}_i \mathbf{s}_j^T = 1, \tag{7}$$

and zero otherwise.

Given (3–5) and (7), we can readily generalize the two-community frustration metric (5) into the k-community frustration metric $\mathscr{F}^k$:

$$\mathscr{F}^k = \sum_{i,j \in \mathscr{C}} A_{ij} - A_{ij} \mathbf{s}_i \mathbf{s}_j^T. \tag{8}$$

As we want the detected communities to be non-overlapping, we need to restrict the minimization of (8) to those configurations $\mathbf{s}$ that are feasible, that is, those that satisfy the constraint that $\|\mathbf{s_i}\| = 1$. Since a QUBO problem is, by definition, unconstrained, we need to transform this constrained problem into an unconstrained problem. This is done via a penalty method, using the penalty term

$$\mathscr{P} = M \sum_i (1 - \|\mathbf{s}_i\|)^2 \,. \tag{9}$$

Instead of minimizing the unconstrained objective function (8), we then minimize the penalized objective function $\mathcal{F}^k_{\mathscr{P}}$:

$$\mathcal{F}^k_{\mathscr{P}} = \mathcal{F}^k + \mathscr{P} \,. \tag{10}$$

It is easy to verify that the term inside $\mathscr{P}$ for index i is 0 if and only if $\|\mathbf{s_i}\| = 1$, and greater than 1 otherwise. This means that evaluating $\mathscr{P}$ on a feasible solution will yield a value of 0, and evaluating it on an infeasible solution will yield a value of *at least* M, which is chosen to be a sufficiently large positive number. Details on the selection of M are given in a later section. The effect of this term is to push the objective value of all infeasible solutions higher by at least M while leaving the objective value of a feasible solution unchanged. Thus, if a sufficiently large M is chosen, the optimal solution to $\mathcal{F}^k_{\mathscr{P}}$ is guaranteed to be feasible.

To conclude, in (10), we have transformed the k-community detection problem into a QUBO problem. An optimal solution $\mathbf{S}^*$, corresponding to the minimum value of $\mathcal{F}^k_{\mathscr{P}}$, will assign each node i to exactly one of k communities.

4.2 Modularity

In this section, we introduce modularity as another methodology for finding multiple communities in a given SG. For unsigned networks, *modularity* is defined as the difference between the number of edges that fall within a community and the number of edges in an equivalent network (i.e., a network with the same number of nodes) when permuted at random [34]. In other words, modularity quantifies a "surprise" measure which explains the statistically surprising configuration of the edges within the community. Maximizing modularity is then equivalent to having a higher expectation of finding edges within communities compared to doing so by random chance.

In what follows, we first give the formulation for the unsigned and signed networks when the underlying task is to find two communities within a network. We then describe our approach for multi-community detection.

Two-Community Detection—Unsigned Graphs. The notion of modularity has been largely used for detecting communities within unsigned networks (see [35]). Let us first consider the two-community detection problem. Without going into details, we use the approach from [35] and write the modularity, $\mathcal{M}^u$, (up to a multiplicative constant) as

$$\mathcal{M}^u = \mathbf{s}\mathbf{B}^u\mathbf{s}^\mathbf{T} \,, \tag{11}$$

where we have defined the real symmetric matrix $\mathbf{B}^{\mathrm{u}}$ as the modularity matrix with the elements

$$B_{ij}^{\mathrm{u}} = A_{ij} - \frac{d_i d_j}{2m},$$ (12)

where superscript u in (11) and (12) refers to the unsigned graph. In (12), the term $\frac{d_i d_j}{2m}$ is the expected number of edges between nodes i and j, and all the other symbols have their usual meanings. Given an optimal configuration $\mathbf{s}^*$ which maximizes (11), we can assign each node to one of the two communities $\mathscr{C}_1$ and $\mathscr{C}_2$. We next use the approach from [14] to explain how the modularity-based community detection method for unsigned graphs can be expanded into the community detection of SGs.

Two-Community Detection—Signed Graphs. In the case of a signed network, we need to reformulate (11) and (12) to include the effect of positive and negative edges. Assuming that our task is to cluster nodes in a given community cluster $\mathscr{C}$ into two clusters $\mathscr{C}_1$ and $\mathscr{C}_2$, we rewrite (11) and (12) into the modularity relation, $\mathcal{M}$, for SGs:

$$\begin{aligned}
\mathcal{M} = &\sum_{i,j\in\mathscr{C}_1} (P_{ij} - \frac{d_{p_i}d_{p_j}}{2m_p}) + \sum_{i,j\in\mathscr{C}_2} (P_{ij} - \frac{d_{p_i}d_{p_j}}{2m_p}) \\
&+ \sum_{i\in\mathscr{C}_1, j\in\mathscr{C}_2} (N_{ij} - \frac{d_{n_i}d_{n_j}}{2m_n}) + \sum_{i\in\mathscr{C}_2, j\in\mathscr{C}_1} (N_{ij} - \frac{d_{n_i}d_{n_j}}{2m_n}).
\end{aligned}$$ (13)

Focusing on the right-hand side of (13), we can merge the first two summation terms into a sum over all nodes by multiplying each of the summation terms by

$$\frac{1}{2}(1 + s_i s_j)$$ (14)

and merge the last two summation terms by multiplying each summation term by

$$\frac{1}{2}(1 - s_i s_j).$$ (15)

Then, using (14)–(15), we can rewrite (13) (up to some constant terms) as:

$$\mathcal{M} = \sum_{i,j\in\mathscr{C}} \left(P_{i,j} - N_{i,j} + \frac{d_{n_i}d_{n_j}}{2m_n} - \frac{d_{p_i}d_{p_j}}{2m_p} \right) s_i s_j.$$ (16)

Finally, it is straightforward to show that (16) can be written in matrix form as

$$\mathcal{M} = \mathbf{s}\mathbf{B}\mathbf{s}^{\mathrm{T}},$$ (17)

where $\mathbf{B}$ is called the "signed modularity matrix" and, given any two nodes $\{i,j\} \in \mathbf{V}$, B_{ij} is given by

$$B_{ij} = A_{ij} + \frac{d_{n_i}d_{n_j}}{2m_n} - \frac{d_{p_i}d_{p_j}}{2m_p}.$$ (18)

All symbols in (13) and (17–18) have their usual meanings. From (16) to (17), we used the relations in (1) such that $A_{i,j} = P_{i,j} - N_{i,j}$. An optimal configuration $\mathbf{s}^*$ which maximizes (17) will assign each node to one of the two communities $\mathscr{C}_1$ and $\mathscr{C}_2$.

k-Community Detection. We now generalize the idea of two-community detection to multi-community detection. We explain the core idea for SGs, but note that the method can be easily generalized to the case of unsigned networks.

Our first step in formulating the k-community detection algorithm is to generalize (13) for k communities as follows:

$$\mathcal{M}^k = \sum_{c=1}^{k} \sum_{i,j \in \mathscr{C}_c} \left(P_{ij} - \frac{d_{p_i} d_{p_j}}{2m_p} \right) + \sum_{c_1 \neq c_2} \sum_{\substack{i \in \mathscr{C}_{c_1}, \\ j \in \mathscr{C}_{c_2}}} \left(N_{ij} - \frac{d_{n_i} d_{n_j}}{2m_n} \right). \quad (19)$$

Following the same approach we took to derive (17), we can combine the first k summation terms on the right-hand side of (19) into a sum over all nodes by multiplying each of the summations by

$$\mathbf{s}^T \mathbf{s} \quad (20)$$

and merge the last $k(k-1)$ summation terms by multiplying each summation by

$$1 - \mathbf{s}^T \mathbf{s}. \quad (21)$$

Using (20)–(21) and constraint (9) (for the case of maximization we consider the negative value of $\mathscr{P}$) and the same one-hot encoding approach (3), we can generalize (17) to a k-community formulation and write the penalized modularity $\mathcal{M}_{\mathscr{P}}^k$ as

$$\mathcal{M}_{\mathscr{P}}^k = \sum_{ij \in \mathscr{C}} B_{ij} \mathbf{s}_i \mathbf{s}_j^T - M \sum_i (1 - \|\mathbf{s}_i\|)^2, \quad (22)$$

where B_{ij} has been defined in (18) and all the other symbols have their usual meanings. In (22), we have transformed the k-community detection problem into a QUBO problem. An optimal solution, $\mathbf{S}$, corresponding to the maximum value of $\mathcal{M}^k$ will assign each node i to one of the k communities.

4.3   Choosing the Hyper-parameters

Our k-community detection algorithm has two hyperparameters, namely, the maximum number of communities that a user is searching for in the network (k) and the penalty coefficient (M). In the next two sections, we discuss the role of these two parameters, their features, and suggest a method for predetermining the appropriate values for these parameters.

Penalty Coefficient. In both the frustration and modularity approaches, we penalize configurations that try to simultaneously assign a node to multiple clusters (i.e., those configurations which violate the non-overlapping among clusters condition). We do this by adding to (subtracting from) the corresponding objective function of frustration (modularity) a term with a relatively large and positive coefficient M. The magnitude of the penalty coefficient depends on the other terms of the objective function.

Since, in our benchmarking examples, we deal with mostly small datasets, we limit our experimental setting to a fixed value for M (see Sect. 5). We expect that for larger datasets a fixed value for M will not result in satisfactory performance. Therefore, it becomes necessary to develop a procedure for choosing the appropriate penalty coefficient for each node, based on the structure of the SG, and we do so in what follows. Note that while we derive the method for the case of frustration, it can be applied to the modularity formulation in a similar way.

Thus far in our formulation (10), we have considered the penalty coefficient to have the same value M for each node. Given the topology of the network, different nodes will contribute differently to the objective function. Hence, it is reasonable to have different penalty values for each node. Let us first define the penalty vector $\mathbf{M}$ as

$$\mathbf{M} = [M_1, M_2, \cdots, M_n]. \tag{23}$$

We can then rewrite the second term on the right-hand side of (10) for the case that each penalty coefficient is different for each individual node. Let us call the penalty term $\mathscr{P}$ and write it as

$$\mathscr{P} = \sum_i M_i (1 - \|\mathbf{s}_i\|)^2, \tag{24}$$

where i iterates over all the nodes in the network and each term has a distinct penalty coefficient term M_i. Our goal here is to develop an approach to determining the value of each individual penalty term according to the topology of the SG.

To start, let us consider a feasible solution $\mathbf{s_j}$ which satisfies the j-th constraint in (24) (i.e., the penalty term evaluates to zero). When the constraint is not satisfied by node j, the penalty term will contribute a positive value to the total objective function. Changing the assignment $\mathbf{s_j}$ to one that violates the constraint for node j, there will be a drop in the first term of the right-hand side of (10) (we call this term the unconstrained frustration and denote it by $\mathcal{F}^k$). To find M_j for node j, we need to know by how much $\mathcal{F}^k$ will be lowered if we violate the j-th constraint imposed solely on node j. By setting M_j larger than this amount, we ensure that violating the constraint at node j will increase (rather than decrease) the overall objective function's value, thus ensuring that the optimal solution will satisfy the constraint.

To find an upper bound on the amount the objective function can change by violating a constraint, let us consider a node j. There are k variables encoding the communities of node j (see (3)). We check all the terms of $\mathcal{F}^k$ that involve node j and assume that, by violating the constraint, they will lower the $\mathcal{F}^k$ by the maximum possible value of one. Since there are $k \times d_j$ many such drops, we can choose the penalty term for the j-th node to be

$$M_j := 2d_j k \,, \tag{25}$$

where the factor 2 accounts for the terms resulting from the symmetry of the adjacency matrix of the SG. Using (25), we can guarantee that the optimal solution to the overall objective function $\mathcal{F}_{\mathscr{P}}^k$ will satisfy all constraints.

The Number of Communities, k. One of the features of our multi-community detection algorithm is that, given an upper bound by the user on the number of communities for which to search, the algorithm will assign each node to one of $k' \leq k$ communities, where k' is the optimal number of communities (provided that an optimal solution has been found by the QUBO solver for the corresponding k-community detection problem). While our numerical studies demonstrate this feature, we also provide the following argument to support our claim about this feature.

Let us assume that we are given an SG (e.g., $\mathbf{G}'$) for which we know that c communities exist. In other words, when we assign nodes to c non-overlapping communities, we get the optimal value for the overall frustration or modularity within $\mathbf{G}'$. Therefore, we call c the optimal number of communities in $\mathbf{G}'$. Now let us encode any given node i in $\mathbf{G}'$ into a k-dimensional ($k > c$) $\mathbf{s_i}$ vector. This means that either (10) or (22) will be formulated to find k communities within $\mathbf{G}'$. If optimizing the frustration or modularity yields more than c non-empty communities, then we have divided at least one of the c communities into sub-communities, increasing the total frustration or decreasing the modularity within the underlying SG. Likewise, when the algorithm returns a number of communities less then c, then at least two of the c communities have merged, again increasing (decreasing) the frustration (modularity) within the network. In both cases, neither the solution with more than, nor the solutions with fewer than, c communities is not actually optimal. Thus, even if the encoding size (which is equivalent to k) provided by the user is larger than the optimal number of communities, our algorithm returns the optimal number of communities. Note, however, that this argument is only valid if the underlying optimization algorithm returns the optimal solution to the QUBO problem.

5   Approach

This section discusses our approach to benchmarking our k-community detection algorithm. We give a detailed explanation of the datasets, the evaluation criteria for the performance of our algorithm on different datasets, the choice of the QUBO solvers and the optimization approach, and our experimental settings for optimization.

Benchmarking Datasets. We apply our multi-community detection algorithm on two synthesized datasets and one real-world example dataset [36]. We have

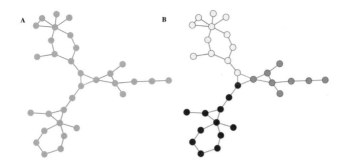

Fig. 2. A randomly generated signed graph ($\mathcal{D}_1$) comprising 32 nodes trivially divided into three clusters of 8, 12, and 12 nodes. Solid circles represent the users that are connected by purple (orange) solid lines, denoting a friendly (antagonistic) relation between them. (A) represents users as being part of one community (all users are represented in blue) which undergoes a graph clustering procedure such that (B) each user is assigned to one of three corresponding communities (each community is represented using green, black, or yellow).

limited the size of the benchmarking datasets to accommodate the size of the quantum annealer.

For the case of synthesized datasets, we consider two SGs of 32 and 64 nodes, where each graph can be trivially clustered into three communities such that the total frustration of the graph is zero. The first synthesized dataset has three clusters of sizes 8, 12, and 12 (Fig. 2A). The second graph consists of 64 nodes where three clusters of sizes 18, 22, and 24 form the entire graph (Fig. 3A). We employ the following procedure to synthesize the datasets. Given a specified sparsity (0.2), we generate three disjoint random clusters where all the nodes within one cluster are positively linked with probability 0.2. We then choose one node from each community and connect these nodes to each other with negative links, such that we end up with a synthesized dataset with a trivial clustering. For the case of the real-world example dataset, we consider a dataset [36] that describes the relation between sixteen tribal groups (represented using solid circles) of the Eastern Central Highlands of New Guinea. Based on previous studies, we know the ground truth (i.e., the optimal number of communities) for this dataset to be three (see Fig. 4A).

To more easily refer to each dataset in this work, we call the two synthesized datasets with 32 nodes and 64 nodes $\mathcal{D}_1$ and $\mathcal{D}_2$, respectively. We refer to the real-world example dataset with 16 nodes as $\mathcal{D}_3$.

QUBO Solvers. In (8) and (22), we have formulated the multi-community detection problems as QUBO problems. We use two QUBO solvers—PTICM and a quantum annealer (the D-Wave 2000Q). We have chosen PTICM because of its superior performance over a range of other QUBO solvers [24,25]. We also employ a quantum annealer to demonstrate that our algorithm can immediately benefit from advancements in the field of quantum computing.

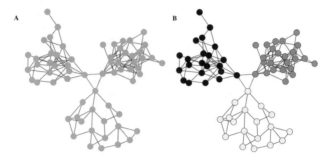

Fig. 3. A randomly generated signed graph ($\mathcal{D}_2$) comprising 64 nodes trivially divided into three clusters of 18, 22, and 24 nodes. Solid circles represent the users that are connected by purple (orange) solid lines, denoting a friendly (antagonistic) relation between them. (A) represents users as being part of one community (all users are represented in blue) which undergoes a graph clustering procedure such that (B) each user is assigned to one of three corresponding communities (each community is represented using green, black, or yellow).

Block Coordinate Descent. Given the large size of our dataset with respect to the size of the quantum annealer, we use the decomposition-based approach BCD on top of each QUBO solver to solve QUBO problems whose size exceeds the current capacity of the quantum annealer.

BCD works by iteratively solving subproblems while keeping the rest of the variables fixed. We have provided a schematic view of BCD in Fig. 5 that shows different steps of BCD at the i-th iteration. To summarize, the algorithm first decomposes the original QUBO problem into a few subproblems by dividing the original variables of the QUBO problem into disjoint subsets. It then uses a QUBO solver to solve each reduced QUBO problem (i.e., subproblem) and updates the solution for each subproblem. Finally, it combines all the subproblems to reconstruct the original QUBO problem with a newly obtained incumbent solution. Its performance on several QUBO problems is reported in [29].

Optimizer Settings. In addition to the hyperparameters of each QUBO solver, the performance of the optimization algorithm for solving the community detection problem is largely dependent on the structure of the graph and the formulation of the QUBO problem. In Sect. 4.3, we have proposed a formulation for choosing the penalty coefficient when the size of the graph is large. Here we have used a fixed penalty value for each of the community detection problems. Specifically, we have used a penalty value of 10 for the $\mathcal{D}_2$ dataset and 50 for the $\mathcal{D}_1$ and $\mathcal{D}_3$ datasets.

In order to collect statistics on the performance of each QUBO solver on the community detection problems, we run each solver 20 times. Each run includes 200 iterations of BCD (see Fig. 5). We use a different subspace size for the different problems. For $\mathcal{D}_1$ we choose the value $h = 4$ ($h = 2$) for the case of frustration (modularity). For $\mathcal{D}_2$, the subspace size is the same ($h = 4$) for both

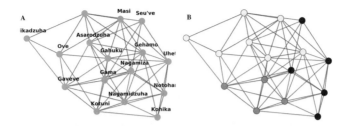

Fig. 4. An example community detection problem for a real-world example dataset [36]. The relation between tribal groups (represented using circles) of the Eastern Central Highlands of New Guinea is shown using solid purple (orange) links denoting a friendly (antagonistic) relation between groups. (A) represents tribal groups as one community (all tribes are represented in blue) which undergoes a community detection procedure such that (B) each tribe is assigned to one of three corresponding communities (each community is represented using green, black, or yellow).

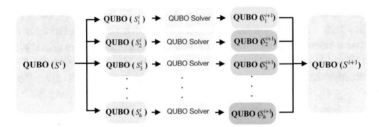

Fig. 5. Schematic view of a single iteration of block coordinate descent (BCD). The algorithm first decomposes the entire QUBO problem into h subproblems. Each individual QUBO problem in the reduced space is constructed by fixing the variables that are absent in that subspace. It then employs a QUBO solver algorithm to solve each reduced QUBO problem with S_h^i as the initial configuration (the best feasible solution up to the i-th step). At the next step, the algorithm returns the updated solution for each reduced QUBO problem and combines them to build the entire solution of the original QUBO problem such that the updated solution S^{i+1} is the new incumbent solution.

frustration and modularity. For the case of the real-world example dataset, we set the subspace size to $h = 4$ ($h = 5$) for the case of frustration (modularity) (see Sect. 5 for more details regarding the BCD and a definition of "subspace").

Benchmarking Criteria. To compare the performance of the two algorithms (i.e., modularity and frustration) we compare the total frustration, of the network after assigning each node to each cluster. We recognize that using frustration might not be a fair metric (especially for large datasets) in comparing the frustration- and modularity-based community detection algorithms. In general, the quality of the solution from each of these algorithms depends largely on the structure of the dataset, and there is no universal method for comparing the performance of different community detection algorithms. However, on small-sized datasets with a trivial structure, we can use frustration as a common means to compare the efficacy of two algorithms. To avoid confusion between the name of the frustration algorithm and our performance metric, we define a criterion of *badness*, $\mathcal{B}$, for each algorithm. The quantity $\mathcal{B}$ is a positive integer which has a lower bound of zero and can take any positive values depending on the structure of a graph. An assignment corresponding to the minimum value for $\mathcal{B}$ is called an optimal assignment, which correctly assigns each node to each community. We define $\mathcal{P}$ as the success probability of each QUBO solver's finding an optimal assignment over 20 runs.

Multi-cluster Encoding. As mentioned earlier, one of the advantages of our proposed multi-community detection algorithm is that it does not require a priori knowledge with respect to the number of communities. For each synthesized dataset, we encode each node into a four-dimensional one-hot encoding vector. This means that, although we know in advance that each of the synthesized datasets has three trivial communities, we still solve the problem as if four communities were present in the given network. A successful community detection algorithm should assign each node to one of the three clusters. In the case of the real-world example dataset, we perform the encoding over five clusters and expect the algorithm to assign each node to only three communities.

6   Results

In this section, we report the results of our experiments on the three datasets considered in this work. We begin with the smaller synthesized dataset, which has 32 nodes ($\mathcal{D}_1$). Figure 2B shows the outcome of an optimal assignment for the case of the $\mathcal{D}_1$ dataset for both the modularity and frustration methods. We also show the optimal assignment of the nodes for the $\mathcal{D}_2$ (Fig. 3B) and $\mathcal{D}_3$ datasets (Fig. 4) for both the modularity and frustration methods.

To show the convergence of different QUBO solvers on different problems, we have plotted the values of frustration or modularity versus the number of iterations. Figure 6 shows an example of such a plot for the case of the $\mathcal{D}_2$ dataset. Figure 6A (Fig. 6B) denotes the results for the case of frustration (modularity). As there are too many such plots to show, we summarize the results of our multi-community detection methods for the other datasets in Tables 1 and 2.

Specifically, we report the best, the mean, and the worst results for each of the frustration and modularity methods. We report the statistics of these quantities by considering the results at the 200th iteration over 20 runs. We also report the statistics (i.e., $\mathcal{B}_{\text{best}}$, $\mathcal{B}_{\text{mean}}$, and $\mathcal{B}_{\text{worst}}$) of the badness of each method for each solver as well as the success probability $\mathcal{P}$ of each solver to find the optimal assignment of the nodes to multiple clusters for both frustration and modularity.

Table 1. The performance of two QUBO solvers (the D-Wave 2000Q and PTICM) in detecting communities in three signed graphs $\mathcal{D}_1$ (a synthesized graph comprising 32 nodes), $\mathcal{D}_2$ (a synthesized graph comprising 64 nodes), and $\mathcal{D}_3$ (a real-world example graph comprising 16 nodes). Here we use frustration as a metric for finding communities. The best, mean, and worst quantities are taken from the last iteration of each of the 20 runs of each QUBO solver. $\mathcal{B}$ refers to the badness of the frustration-based algorithm (see Sect. 5), where a lower value of $\mathcal{B}$ denotes better performance of the frustration method over the modularity (see Table 2) method. $\mathcal{P}$ is the success probability of each QUBO solver finding the optimal solution for each $\mathcal{D}_i$ ($i \in \{1, 2, 3\}$) over the 20 runs for each QUBO solver.

		Frustration ($\mathcal{F}_{\mathscr{P}}^{k}$)						
		Best	Mean	Worst	$\mathcal{B}_{best}$	$\mathcal{B}_{mean}$	$\mathcal{B}_{worst}$	$\mathcal{P}$
$\mathcal{D}_1$	D-Wave Solver	-388	-382	-378	0.0	2.9	5.0	0.10
	PTICM	-388	-386	-382	0.0	0.7	3.0	0.45
$\mathcal{D}_2$	D-Wave Solver	-3470	-3448	-3404	0.0	11.0	33.0	0.05
	PTICM	-3470	-3459	-3432	0.0	5.4	19	0.05
$\mathcal{D}_3$	D-Wave Solver	-374	-374	-374	2.0	2.0	0.0	1.0
	PTICM	-374	-374	-374	2.0	2.0	0.0	1.0

7   Discussion

In this section, we discuss the performance of our algorithms on the three benchmarking datasets that we have considered in this work. We begin with Fig. 2A, which shows the original SG $\mathcal{D}_1$, where all nodes are assigned to one community. It is trivial for any successful community detection algorithm to discover the three individual communities within $\mathcal{D}_1$. We take a different approach from existing divisive-based multi-community detection algorithms to solve this problem. One of the advantages of our proposed method is that the algorithm requires only an upper-bound estimate of the number of communities from the user to search for the optimal number of communities. To demonstrate this, we initially assume that there exist four communities ($k = 4$) within the graph. The number of variables in the underlying QUBO algorithm increases linearly with k and is proportional to $n \times k$. Therefore, we have to solve a QUBO problem with 132 binary variables. Figure 2B shows that, despite the initial assumption that there are four communities within the graph, both frustration- and modularity-based methods report three communities as the optimal number of communities.

We have followed the same approach for the case of $\mathcal{D}_2$. We run the algorithm with the assumption that there are four communities ($k = 4$) within the graph. For this case, we have to solve a QUBO problem with $4 \times 64 = 256$ variables. Figure 3B shows the results of community detection for both the frustration and modularity methods. Both methods discover three communities despite the initial assumption of there being four communities. For the case of the real-world example dataset (Fig. 4A), we solve the problem such that we assume five communities within the graph. The corresponding QUBO problem size is

Table 2. The performance of two QUBO solvers (the D-Wave 2000Q and PTICM) in detecting communities in three signed graphs $\mathcal{D}_1$ (a synthesized graph comprising 32 nodes), $\mathcal{D}_2$ (a synthesized graph comprising 64 nodes), and $\mathcal{D}_3$ (a real-world example graph comprising 16 nodes). Here we use modularity as a metric for finding communities. The best, mean, and worst quantities are taken from the last iteration of each of the 20 runs of each QUBO solver. $\mathcal{B}$ is called the badness of the modularity-based algorithm, where a lower value of $\mathcal{B}$ denotes better performance of the modularity method over the frustration (see Table 1) method. $\mathcal{P}$ is the success probability of each QUBO solver finding the optimal solution for each $\mathcal{D}_i$ ($i \in \{1, 2, 3\}$) over the 20 runs for each QUBO solver.

		Modularity ($\mathcal{M}_{\mathscr{P}}^k$)						
		Best	Mean	Worst	$\mathcal{B}_{\text{best}}$	$\mathcal{B}_{\text{mean}}$	$\mathcal{B}_{\text{worst}}$	$\mathcal{P}$
$\mathcal{D}_1$	D-Wave Solver	-384	-383	-381	0.0	0.5	2.0	0.5
	PTICM	-384	-383	-381	0.0	0.5	2.0	0.50
$\mathcal{D}_2$	D-Wave Solver	-3457	-3443	-3410	0.0	7.5	25.0	0.05
	PTICM	-3457	-3444	-3421	0.0	6.8	19	0.05
$\mathcal{D}_3$	D-Wave Solver	-852	-852	-852	2.0	2.0	2.0	1.0
	PTICM	-852	-852	-852	2.0	2.0	2.0	1.0

$5 \times 16 = 80$. Both the modularity and frustration methods correctly assign each node to one of three communities.

To show that our method can benefit immediately from progress within quantum technology, we have considered the D-Wave 2000Q quantum annealer as one of the QUBO solvers. This quantum annealer can solve a community problem for a complete SG of size 64. Since the size of all the QUBO problems that we consider in this work exceeds the current capabilities of available quantum annealers, we employ the BCD algorithm to iteratively solve each subproblem of the original QUBO problem. In general, our community detection algorithm does not rely on the BCD algorithm and can be used without using BCD so long as the underlying computing device can handle the size of the corresponding QUBO problem.

We have summarized the performance of the two QUBO solvers on various benchmarking problems in Tables 1 and 2. Table 1 reports the results for the case of frustration ($\mathscr{F}_{\mathscr{P}}^k$) and Table 2 reports the results for the case of modularity ($\mathcal{M}_{\mathscr{P}}^k$). We begin with the results pertaining to the frustration method.

The first column of Table 1 shows that both the D-Wave solver and PTICM are able to find the optimal number of communities within all three datasets. This is also clear from the column that shows the $\mathcal{B}_{\text{best}}$ for different datasets. The badness measure for the first two datasets ($\mathcal{D}_1$ and $\mathcal{D}_2$) are zero, which corresponds to a perfect assignment of each node of these graphs to one of the three communities (see Figs. 2B and 3B). For the case of the real-world example dataset $\mathcal{D}_3$, the optimal assignment of the nodes to three individual communities

A   B

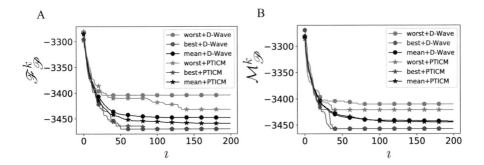

Fig. 6. D-Wave 2000Q (solid circle) and PTICM (solid star) results for the 64-node synthesized dataset. We show the worst (green line), best (blue line) and mean (black line) of the A) frustration ($\mathscr{F}^k$) and B) modularity measures vs. the number of iterations ($\imath$) of the BCD algorithm.

corresponds to a badness of 2.0, as the nontrivial structure of the network does not allow any assignment with $\mathcal{B} = 0$.

The results from the mean, worst, $\mathcal{B}_{\mathrm{mean}}$, and $\mathcal{B}_{\mathrm{worst}}$ columns confirm the superior performance of PTICM over the D-Wave solver for the case of frustration-based method. Over 20 runs, PTICM could find the optimal number of communities in 45% and 15% of the time for the case of $\mathcal{D}_1$ and $\mathcal{D}_2$, whereas these quantities are 10% and 5% for the D-Wave solver. Both algorithms perform the same on $\mathcal{D}_3$, with a success probability of $\mathcal{P} = 1.0$. We would like to stress that our results regarding the superiority of PTICM over the D-Wave solver are not conclusive, as we had resources insufficient for performing a proper hyper-parameter tuning on the D-Wave quantum annealer.

We now discuss Table 2, which includes the modularity ($\mathcal{M}^k_{\mathscr{P}}$) results. From the first column of the table, we can see that both the D-Wave and PTICM solvers find the optimal number of communities for all three datasets. Looking at $\mathcal{B}_{\mathrm{mean}}$, we can see that the performance of PTICM and the D-Wave solver is comparable on all three datasets. For each dataset, both solvers find the optimal number of communities with almost the same success probability.

8   Concluding Remarks and Future Work

In this work, we have devised a multi-community detection algorithm to find communities within signed graphs. We have tested our approach on three different datasets, including two randomly generated signed graphs and one real-world dataset. Our approach has two main advantages over existing algorithms. First, when searching for multiple communities within a graph, our algorithm preserves the global structure of the network as opposed to divisive community detection algorithms, which only consider the local structure. Second, our method does not require a priori knowledge of the number of communities. Given an upper bound on the number of communities, our algorithm will find the optimal number of communities, which will be equal to or less than the predefined upper bound chosen by the user.

Our main focus in this work has been to provide the formulation of a multi-community algorithm that uses frustration or modularity as a means to detect communities. The next step would be to test our method on larger datasets using available state-of-the-art classical optimizers such as the Digital Annealers [37].

References

1. Massa, P., Avesani, P.: Proceedings of the 20th National Conference on Artificial Intelligence - Volume 1, AAAI 2005, pp. 121–126. AAAI Press (2005)
2. Leskovec, J., Huttenlocher, D., Kleinberg, J.: Proceedings of the 19th International Conference on World Wide Web, WWW 2010, pp. 641–650. ACM, New York (2010)
3. Kunegis, J., Lommatzsch, A., Bauckhage, C.: Proceedings of the 18th International Conference on World Wide Web, WWW 2009, pp. 741–750. ACM, New York (2009)
4. Heider, F.: J. Psychol. 21(1), 107 (1946). PMID: 21010780
5. Bhagat, S., Cormode, G., Muthukrishnan, S.: Node Classification in Social Networks, pp. 115–148. Springer, Boston (2011). https://doi.org/10.1007/978-1-4419-8462-3_5
6. Liben-Nowell, D., Kleinberg, J.: Proceedings of the Twelfth International Conference on Information and Knowledge Management, CIKM 2003, pp. 556–559. ACM, New York (2003)
7. Aggarwal, C., Subbian, K.: ACM Comput. Surv. 47(1), 10:1 (2014)
8. Chiang, K.Y., Natarajan, N., Tewari, A., Dhillon, I.S.: Proceedings of the 20th ACM International Conference on Information and Knowledge Management, CIKM 2011, pp. 1157–1162. ACM, New York (2011)
9. Tang, J., Aggarwal, C., Liu, H.: Node Classification in Signed Social Networks, pp. 54–62 (2016)
10. Tang, J., Chang, Y., Aggarwal, C., Liu, H.: ACM Comput. Surv. 49(3), 42:1 (2016)
11. Fischer, K.H., Hertz, J.A.: Spin Glasses, vol. 1. Cambridge University Press, Cambridge (1993)
12. Srinivas, N., Deb, K.: Evol. Comput. 2(3), 221 (1994). https://doi.org/10.1162/evco.1994.2.3.221
13. Pizzuti, C.: 2009 21st IEEE International Conference on Tools with Artificial Intelligence, pp. 379–386 (2009). https://doi.org/10.1109/ICTAI.2009.58
14. Anchuri, P., Magdon-Ismail, M.: Proceedings of the 2012 International Conference on Advances in Social Networks Analysis and Mining (ASONAM 2012), ASONAM 2012, pp. 235–242. IEEE Computer Society, Washington, DC (2012)
15. Chen, J., Zhang, H., Guan, Z.H., Li, T.: Phys. A 391(4), 1848 (2012)
16. Salathé, M., Jones, J.H.: PLoS Comput. Biol. 6(4), 1 (2010)
17. Ferrara, E., Meo, P.D., Catanese, S., Fiumara, G.: Expert Syst. Appl. 41(13), 5733 (2014)
18. Waskiewicz, T.: Proceedings on the International Conference on Artificial Intelligence (ICAI). The Steering Committee of The World Congress in Computer Science, Computer Engineering and Applied Computing (WorldComp), p. 1 (2012)
19. Lin, S., Hu, Q., Wang, G., Yu, P.S.: Advances in Knowledge Discovery and Data Mining, pp. 82–95. Springer, Cham (2015). Cao, T., Lim, E.P., Zhou, Z.H., Ho, T.B., Cheung, D., Motoda, H. (eds.)

20. Amelio, A., Pizzuti, C.: 2013 IEEE/ACM International Conference on Advances in Social Networks Analysis and Mining (ASONAM 2013), pp. 95–99 (2013). https://doi.org/10.1145/2492517.2492641

21. Esmailian, P., Abtahi, S.E., Jalili, M.: Phys. Rev. E **90**, 042817 (2014)

22. MacKay, D.J.: Learning in Graphical Models, pp. 175–204. Springer, Dordrecht (1998)

23. Hukushima, K., Nemoto, K.: J. Phys. Soc. Jpn. **65**(6), 1604 (1996). https://doi.org/10.1143/jpsj.65.1604

24. Zhu, Z., Ochoa, A.J., Katzgraber, H.G.: Phys. Rev. Lett. **115**, 077201 (2015)

25. Zhu, Z., Fang, C., Katzgraber, H.G.: CoRR arXiv:1605.09399 (2016)

26. Johnson, M.W., Amin, M.H.S., Gildert, S., Lanting, T., Hamze, F., Dickson, N., Harris, R., Berkley, A.J., Johansson, J., Bunyk, P., Chapple, E.M., Enderud, C., Hilton, J.P., Karimi, K., Ladizinsky, E., Ladizinsky, N., Oh, T., Perminov, I., Rich, C., Thom, M.C., Tolkacheva, E., Truncik, C.J.S., Uchaikin, S., Wang, J., Wilson, B., Rose, G.: Nature **473**, 194 EP (2011)

27. Shaydulin, R., Ushijima-Mwesigwa, H., Safro, I., Mniszewski, S., Alexeev, Y.: arXiv preprint arXiv:1810.12484 (2018)

28. Zintchenko, I., Hastings, M.B., Troyer, M.: Phys. Rev. B **91**, 024201 (2015)

29. Rosenberg, G., Vazifeh, M., Woods, B., Haber, E.: Comput. Optim. Appl. **65**(3), 845 (2016). https://doi.org/10.1007/s10589-016-9844-y

30. Easley, D., Kleinberg, J.: Networks, Crowds, and Markets: Reasoning About a Highly Connected World. Cambridge University Press, Cambridge (2010)

31. Wasserman, S., Faust, K.: Social Network Analysis Methods and Applications, vol. 8 (1993)

32. Harary, F.: Michigan Math. J. **2**(2), 143 (1953)

33. Davis, J.A.: Hum. Relat. **20**(2), 181 (1967)

34. Newman, M.E.J., Girvan, M.: Phys. Rev. E **69**, 026113 (2004)

35. Newman, M.E.J.: Proc. Natl. Acad. Sci. **103**(23), 8577 (2006)

36. Read, K.E.: Southwestern J. Anthropol. **10**(1), 1 (1954)

37. Aramon, M., Rosenberg, G., Miyazawa, T., Tamura, H., Katzgraber, H.G.: arXiv preprint arXiv:1806.08815 (2018)

Simulation-Based Analysis of Equalization Algorithms on Active Balancing Battery Topologies for Electric Vehicles

Asadullah Khalid, Alexander Hernandez, Aditya Sundararajan, and Arif I. Sarwat[⊠]

Florida International University, Miami, FL, USA
asarwat@fiu.edu

Abstract. Determination of the cell electro-chemistry, topology, and application requirements are crucial to developing a battery management system for charge equalization in a series-connected stack of Lithium-ion (Li-ion) cells. The existing literature on topology categorization does not provide battery and battery model selection methodology for battery management system (BMS) development. To bridge this gap in the literature, this paper provides a unique simulation based analysis on the major steps required to build a BMS that include analysis of a variety of existing Lithium-ion cell electro-chemistries, equivalent models, equalization topologies and circuits. Equalization circuits and their variants are categorized based on components, topology, balancing time and configurations. Cell balancing simulations are then performed on a centralized and a distributed topology using an appropriate equivalent model identified by the analysis. In addition, the simulation also uses a unique cell equalization algorithm proposed in this paper. The results validate voltage and state of charge (SOC) equalization performance in terms of balancing time and energy efficiency. These factors play a crucial role in maintaining battery life and preventing thermal runaways in electric vehicles (EV) or energy storage systems (ESS).

Keywords: Equalization algorithm · Parameter identification · Electro-chemistry · Balancing time · Energy storage element · SOC convergence

1 Introduction

Electric vehicles (EVs) and energy storage systems (ESSs) use very large high-energy density Li-ion battery stacks consisting of multiple cells connected in series and monitored continuously [30]. Because of this, EV's and ESS face technical problems with respect to energy management throughout the system [37,68]. The energy does not get evenly distributed, resulting in charge mismatch in the stack. The charge mismatch results in decreased battery operational life and unexpected thermal failures [35]. A battery management system (BMS) is

© Springer Nature Switzerland AG 2020
K. Arai et al. (Eds.): FTC 2019, AISC 1069, pp. 708–728, 2020.
https://doi.org/10.1007/978-3-030-32520-6_52

expected to continuously monitor individual cell conditions, identify any over or under-voltage and fault scenarios, compensate for charge imbalances and be able to output, identify and estimate the state of charge (SOC) and state of health (SOH) information [34,36,69]. In order to model a physical battery management system, an equalization circuit or topology (centralized or distributed) and an equalization algorithm (that can be programmed into a micro controller) needs to be developed. With respect to BMS model simulation, in addition to a topology and algorithm, an appropriate battery model needs to be identified to be able to take precise electrochemical dynamics into consideration.

This paper compares different battery models based on their components, to identify the electrochemical feature that each of them would bring into BMS development. Most of the research is directed towards performance evaluation of equalization topologies alone [12,13], while largely ignoring the underlying control algorithm and the energy storage element involved. This paper also reviews and compares existing BMS topologies and provides in-depth explanation of their corresponding algorithms.

The key contributions of this paper are: (1) Li-ion chemistries, battery models and topologies comparison; (2) A unique algorithm to balance cells in variable topologies; and (3) Simulated validation of algorithm on the topologies.

The rest of the paper is organized as follows. Section 2 analyzes the electrochemistry to identify the right Li-ion chemistry, followed by electrical battery model and topology selection. Section 3 formulates the tested cell data on to the selected battery model followed by its simulation using the proposed algorithm to be fed into a battery model. Section 4 presents a summary of the post-simulation results, and finally Sect. 5 concludes the analysis and provides future direction for research. The terms 'cell' and 'battery' are interchangeably used in this paper.

2 Cell, Model and Topology Selection

2.1 Cell Selection

Li-ion is a highly researched cell chemistry because of its high specific energy, high cycle life, low self-discharge, and high nominal voltage. It supersedes other chemistries such as Lead-acid, Nickel Cadmium (Ni-Cd), and Nickel Metal Hydride (Ni-MH) in almost every category except over charge and over discharge situations [10]. This section compares and evaluates various commercially available Li-ion chemistries, by analyzing Lithium Manganese Oxide (LMO), Lithium Nickel Manganese Cobalt Oxide (NMC), Lithium Iron Phosphate (LFP), and Nickel Cobalt Aluminium (NCA) to identify the optimum chemistry for our application.

LMO offers high thermal stability due to addition of Manganese but also contributes to its low specific energy [17,31]. NMC provides high energy density but lower thermal stability [17,31,55]. LFP offers improved safety and thermal stability because of its structure, which also allows it to operate at higher temperatures [17,31] but it has a flat discharge curve. Comparison between LFP,

NMC, and NCA in [5] showed that an increasing SOC% increases the temperature of LFP by around 0.8 °C per SOC value. The corresponding values for NMC and NCA increase by 1.02 °C, and 1.74 °C per SOC%, respectively. The NCA lithium chemistry has twice the potential to catch fire, than LFP if overcharged. But NCA cells' very high specific energy and low thermal stability, sloping discharge curve characteristic, high voltage range and high electrical conductivity makes it an optimal choice for BMS [32,33,55].

2.2   Model Selection

In order to simulate the battery in a simulation environment based on the electrical test data obtained from the cell, model selection is required to have the highest accuracy for the SOC calculation. The evaluation of the models is performed in terms of dynamicity by analyzing the variable functions, battery types supported and simulation softwares on which they have been used. Rint or Simple model [6,18,23,57,64] and Linear model [25,26] do not allow modeling of the dynamic characteristics of the battery, thereby leading to higher calculation errors. SAFT RC or 2-capacitance model has higher accuracy but the discrepancy occurs due to the linear nature of the capacitors [23,56,73,79]. Thevenin models and their derivatives like basic thevenin [14,16,23,49,53,71,74,76], dual polarization or second order model [23,67,76,80], first order or modified thevenin model [11,18,63], resistive thevenin [18,20,24,26,57,59,76], reactive thevenin [18,20,57] and m-th order linear parameter varying (LPV) or electrical analogue model (Fig. 8) [43,46,76,78], have higher accuracy but their dynamicity or functions varies from one model to another. This also limits their applicability to a specific simulation software. m-th order LPV model constitutes all the components and inputs of the former models and support all types of lithium chemistry. Hence, This model can be simulated using MATLAB/Simulink software. In this paper, the value of m is taken as 2 for simulation.

2.3   Topology Selection

Classification of models based on the topology is necessary depending on the application's balancing time, complexity and energy efficiency requirement. The terms centralized and distributed are used broadly and can refer to either the number of converters used as discussed in [21,41,44,48,50,54], or [42,50,51,54,66] which categorize them based on the number of microcontroller units. Modular, a very broad term, is another criterion used to classify active balancing topologies with respect to converters [1,7,15,38,61], controller [52], and their ability to easily add and remove cells from the stack, being balanced [62,72]. The following section will use centralized and distributed to classify the number of converters used.

Distributed topologies, in our case, signifies a topology that has multiple converters for transferring charge. These topologies include, but are not limited to, Buck/Cuk based converters (Figs. 1 and 2) and variants of the multiple-winding transformer (Figs. 3 and 5) [52,60,62]. These topologies contain multiple

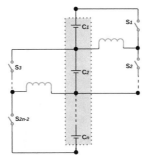

Fig. 1. Buck-boost-based equalizer circuit

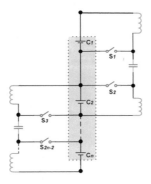

Fig. 2. Ćuk-based equalizer circuit

energy storage elements to transfer charge between cells. The Cuk/Buck based converters have every two cells shared an inductor, more in the case of the Cuk converter [61]. In this way the charge is transferred adjacently from cell to cell. On the other hand, the multi-winding transformer variants allow for charge transfer from cell to stack/stack to cell and cell-to-stack-to-cell. While the general statement of distributed topologies having better balancing times is true there are some exceptions to that. The increased availability of resources allows cells to operate independently of others during balancing. The problem with the design of the Buck/Cuk is that they can only transfer charge adjacently. This means the more cells that are added to the stack the slower the balancing speed. This is due to the fact that the top cell and bottom cell will become further apart and the distance charge has to travel is increased. Overall, distributed topologies have relatively better balancing times because the resource bottleneck is removed, which we will see as a problem in the following section in case of centralized topology. Lastly, the distributed topologies while faster, suffer from increased component use, increased size, and increased cost which are also important considerations to make when selecting the appropriate topology for the application. In the following sections, Multi Core Multi Winding topology (Table 1) will be used for simulation.

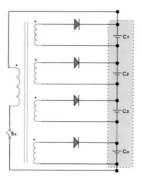

Fig. 3. Multi-windings transformer (single core) equalizer circuit.

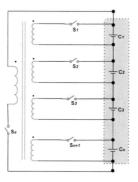

Fig. 4. Flyback converter-based equalizer circuit.

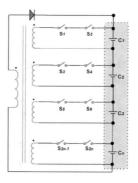

Fig. 5. Multi-winding transformer-based (single core) equalizer circuit.

Centralized topologies are defined by the use of only one converter for all cells in the stack. This can be seen in a topology like the flying capacitor (Fig. 6) [9] which is listed in Table 1. This topology is very simple, but it is slow at balancing a stack of cells. This is attributed to two main reasons. The first being the fact that only one cell can be either charged/discharged at any point in time.

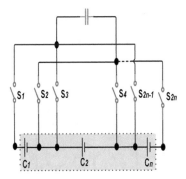

Fig. 6. Flying capacitor equalizer circuit.

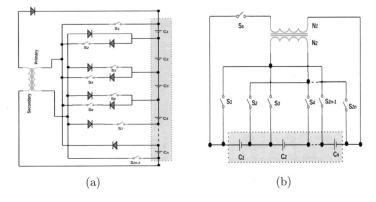

(a)   (b)

Fig. 7. (a) Flyback converter-based equalizer circuit, and (b) Single winding transformer equalizer circuit

Secondly, the topology uses a ceramic multi-layer capacitor for balancing which takes longer to equalize with a cell when the voltage difference between the cells is small. This means that closer the cells are to being balanced the higher would be the balancing time. The distributed analog to this model is the switched capacitor topology which has every pair of cells share a capacitor. This topology is still bottlenecked by the capacitor, but it is no longer hindered by the lack of available resources. The cells can now be balanced together, which greatly reduced balancing time when compared to the single winding topology. The capacitor in the flying capacitor topology can be swapped for a transformer that is not hindered by the voltage difference of the cells. This topology is listed as single-winding transformer (Fig. 7b) [9] in Table 1. The topology works in the same manner as the flying capacitor with the exception of the changed energy storage element. The distributed analog to this would be the multi-winding transformers discussed above. Again, the same comparison can be made where increased resources increase balancing time. However, a slow balancing time is not always true. Variants of the simpler initial models have been made and have improved

Table 1. List of features of general equalization topologies

Topology	Capacitor	Diode	Switch	Inductor	Transformer	Type (based on converter)	Time to balance	Energy storage element
Buck boost-based converter/ Single inductor [52,62]	/	/	$2n-2$	$n-1$	/	D	+	Shared inductor
Ćuk-based converter [61]	$n-1$	/	$2n-2$	$2n-2$	/	D	−	Inductor
Multi-winding transformer (single core) [60]	/	n	1	/	$1:n$	D	+++	Primary winding
Fly-back converter-based [60]	/	/	$n+1$	/	$1:n$	D	++	Primary winding (bottom balancing), Secondary windings (top balancing)
Multi-winding transformer-based [52,77]	/	1	$2n$	/	$1:n$	D	+++	Secondary windings
Multi core multi-winding [19]	/	/	$2n$	/	$n:n$	D	++	Primary winding (bottom balancing), Secondary windings (top balancing)
Flying capacitor [9]	1	/	$2n$	/	/	C	− −	Capacitor
Fly-back converter-based [28]	/	$2n-2$	$2n-2$	/	$1:1$	C	++	Primary winding
Single winding transformer [9]	/	/	$2n+1$	/	$1:1$	C	+++	Primary winding (bottom balancing), Secondary windings (top balancing)
Forward and flyback converter [22]	/	/	n	/	$1:n$	C	+++	Primary winding (bottom balancing), Secondary windings (top balancing)

Labels: Very Poor (− −), Poor(−), Good (+), Very Good (++), Excellent (+++), Not applicable (/), D: Distributed, C: Centralized

balancing time. The Forward and Flyback converter (Figs. 4 and 7a) in [22] is an example of this. Still, the model does require more components and its complexity is increased in order to provide better balancing time, which negate the main benefits of using a centralized topology. These include reduced complexity, reduced component use, reduced cost, and reduced size, which depending on the application can again be very important considerations. In the following sections, Forward Converter topology (Table 1) will be used for simulation.

3 Formulation and Simulation of Results

This section presents the steps for identifying the parameters for a battery model selected, based on the practical test results on NCA chemistry. This model is then fed into a battery management system topology, which is then simulated to operate in cell-to-cell, cell-to-stack and stack-to-cell charge transfer modes. Cell-to-stack and stack-to-cell are cumulatively written as cell-to-stack-to-cell in the following sections.

3.1 Cell Testing and Modeling

Hybrid pulse power characterization (HPPC) discharge and electrochemical impedance spectroscopy (EIS) [4] test performed on 3.6V 2.9Ah NCR18650PF, an NCA chemistry Li-ion battery (similar to the ones used in Tesla cars [2]) results for which were obtained from [40] have been used for building the battery model for simulation. The HPPC test conducted at $298.15K$ consists of 13 full pulse discharge cycles. Each full cycle is divided into 5 pulse discharge cycles of different discharge rates (I_L). Since the diffusion layer approximation accuracy (RC pair accuracy representing transient effects during charge and discharge [43,46,47,76,78]) increases with the increase in RC pairs, the number of RC pairs (m) selected here is 2. This consequently, increases the number of parameters required for computation. With $m = 2$, the required parameters/variables are $R_0(SOC)$, $R_{b_1}(SOC)$, $R_{b_2}(SOC)$, $\tau_1(SOC)$ ($= C_{i_1}R_{b_1}$), $\tau_2(SOC)$ ($= C_{i_2}R_{b_2}$), OCV(SOC), and $I_L(t)$. For the Electrical Analogue model used, a comprehensive description of steps for cell data extraction and formulation for feeding the battery model are listed below:

1. Identify the HPPC pulse discharge cycle periods corresponding to discharge rates (from voltage response against time data/plot). Each selected discharge pulse is divided into two time ranges, which represent dynamic responses (specifically τ_1 and τ_2). Identify SOC corresponding to time extremes (pulse rise and drop-in time) selected. This would give approximated $\tau_1(SOC)$ and $\tau_2(SOC)$ values for each discharge pulse [70]. The bandwidth values (frequency range) must be in the lower frequency range so as to obtain better resolution diffusion parameters [45].
2. Identify the voltages corresponding to the rest/wait periods between full pulse discharge cycles (from voltage response vs time data/plot). This is the OCV value for which the corresponding SOC value would approximately be same as the ones obtained in step (1). This would, in turn give OCV(SOC) [45].
3. Identify the voltages corresponding to the time range values obtained in step (1). Each voltage difference (absolute value) divided by the respective discharge rate would give the respective $R_{b_1}(SOC)$, and $R_{b_2}(SOC)$ values.

4. From EIS data, identify the output voltage corresponding to the OCV values obtained in step (b). These values must correspond to a higher frequency range. Each voltage difference between OCV and its corresponding output voltage divided by the respective discharge would give the corresponding R_0 values [27].

Above-mentioned steps would need to be repeated for different temperatures if thermal variation is to be analyzed. For modeling the 2nd Order Linear Parameter Varying Model, 13 full pulses are used to calculate the diffusion parameters. $0.5C$ and $6C$ are the rates (load currents) corresponding to which the parametric data is fed into the battery model. Frequency range for identifying τ values is less than $0.0005\,\mathrm{Hz}$ and for identifying R_0 is within 10–$60\,\mathrm{Hz}$. A snapshot of the values used for the parameters is shown in Table 2. These values were applied to all the 6 cells (in the stack) used for simulation, except for the SOC beginning values (B.V.), which were selected to indicate variability in capacities of the cells in the stack.

Table 2. Results showing: (a) Individual Cell Parameters, and (b) Parameters for Cells 1–6

(a) SOC B.V

	SOC B.V.
Cell 1	0.85903
Cell 2	0.6623
Cell 3	0.4791
Cell 4	0.3903
Cell 5	0.2123
Cell 6	0.09583

(b) Cells 1-6 Parameters

SOC	OCV	τ_1	τ_2	R_{b_1}	R_{b_2}	R_0
0.962321	4.102	2429.9	2421	0.013736501	0.005690015	0.000483939
0.912362	4.051	2430	2422	0.012242153	0.005230216	0.000442557
0.862372	3.992	2430	2430	0.012126632	0.007931162	0.002351762
0.762307	3.882	2430	2420	0.011839271	0.007298968	0.001083351
0.662303	3.797	2430	2430	0.011724327	0.004770192	0.001580485
0.562334	3.705	2430	2420	0.01138003	0.004655467	0.002286351
0.462483	3.622	2430	2430	0.010747806	0.006839513	0.003058814
0.362324	3.561	2430	2424	0.010862756	0.005172741	0.000223577
0.262472	3.498	2430	2420	0.010920231	0.005115266	0.000357494
0.2123	3.451	2430	2430	0.012242153	0.007816586	0.000209783
0.162303	3.391	2430	2420	0.013850798	0.006092052	3.27591E-05
0.127807	3.364	2430	2410	0.018909241	0.018334492	0.001711602
0.04861	3.225	2430	3550	0.420596834	0.353025539	0.011100998

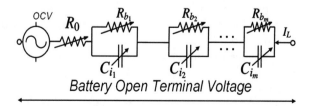

Fig. 8. The m-th order linear parameter varying (LPV) or electrical analogue model.

3.2   Centralized and Distributed Topologies Simulation

This section shows simulation results (using MATLAB/Simulink) on centralized and distributed topologies using a defined algorithm (Algorithm 1). Data from Table 2 is fed into the battery models, shown in Fig. 9a and b.

In the algorithm shown in this section, charge, represented by 'C', indicates the voltage of respective cells. Also in this section, although == (equal to) and != (not equal to) symbols are used to indicate cell voltage status, it is assumed that a delta (Δ V) range is defined (as deemed appropriate for the application) within which == (within delta range) and != (outside delta range) conditions are satisfied, as the voltages cannot ideally be precisely equal in practical applications because of variations in the cell chemistry. Further, the delta value range can vary within the algorithm. This range is recommended to be varied to optimize the algorithm. Algorithm 1 reads every cell (in the stack) voltage one by one to calculate the average. The algorithm checks every subsequent cell pair in the stack to identify whether the cell value is within a defined delta. Once the voltages of each cell are determined, every cell pair's deviation from the defined average is identified. If the cell with the higher voltage is deviating from the average, the cell with the higher voltage is discharged until the cell with the minimum voltage is deviating from the average, where the cell with the minimum voltage is charged using the energy stored in energy storage element. For every cell pair monitored, the cells are categorized into maximum and minimum cell voltage pairs, and the process repeats until the cell voltages in every pair are equal (or are within certain delta). This algorithm allows flow of charge from one cell to another and waits for a defined time before switching to a new maximum and minimum cell pair. The algorithm terminates when the deviation from average stack voltage drops below the defined delta value. For a cell-to-cell charge transfer, the β in line 8 of Algorithm 1 has been updated with C_{nmin}. Further, this algorithm has also been updated to include cell-to-stack and stack-to-cell(s) charge transfers as well. This update is performed by modifying β in line 8 with α and comparing $|C_{nmax} - \alpha|$ and $|C_{nmin} - \alpha|$ (as an additional condition on line 8 in Algorithm 1), where the higher value of the former would allow cell to stack charge transfer and higher value of the latter stack to cell charge transfer.

The centralized topology selected for simulation is the forward converter topology (Fig. 9a), in which six electrical analogue battery models are used for BMS simulation. Forward converter used in the topology is based on the bidirectional forward converter used in [29], where buck-boost converter allows cell-to-stack-to-cell charge transfer operation and the active clamp is controlled by the primary side switch. Algorithm 1 controls the switching and connects the higher voltage cell to a lower voltage cell via forward converter. In this topology, two pairs of switches, each pair corresponding to a cell with higher and lower voltage respectively are turned on such that the cells connected to forward converter could be charged or discharged. In this case, the primary winding of the forward converter is the energy storage element.

The distributed topology selected for simulation is the multiple transformer topology (Fig. 9b), in which the same set of six electrical analogue battery models

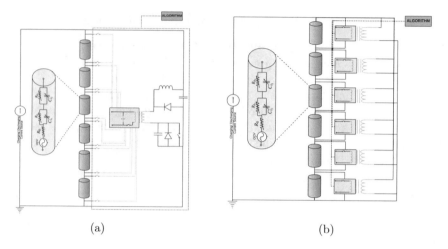

(a)   (b)

Fig. 9. Equalization scheme using (a) centralized forward converter, and (b) distributed multi core multi winding transformer

---

Algorithm 1. Proposed Centralized and Distributed BMS Simulation Algorithm for Cell-to-Cell, Cell-to-Stack-to-Cell Charge Transfers

---
1: **repeat**
2:　　*Monitor Cells*
3:　　**if** $C_{nmax} \mathrel{!}= C_{nmin}$ **then**
4:　　　　Find $\sum_{i=1}^{n} C_i, i \in \mathbb{Z}^+$
5:　　　　Calculate average, $\alpha = \frac{\sum_{i=1}^{n} C_i}{n}$
6:　　　　Calculate each consecutive cell voltage and set as $C_{nmax\&\&nmin}$
7:　　　　For each cell set as $C_{nmax\&\&nmin}$
8:　　　　**if** $|C_{nmax} - \beta| >$ delta **then**
9:　　　　　　Activate S_{nmax} or S_{nmax-P}
10:　　　　　Discharge C_{nmax} or C_{nmax-P}
11:　　　　　Allow energy storage element to charge
12:　　　　　Activate S_{nmin} or S_{nmin-P}
13:　　　　　Allow energy storage element to discharge
14:　　　　　Charge C_{nmin} or C_{nmin-P}
15:　　　　**end if**
16:　　**end if**
17: **until** $C_i == C_{i+1}, \forall i \in [1, n-1]$

---

are used for BMS simulation. Algorithm 1 controls the switching and connects the higher voltage cell to a lower voltage cell via a 1:2 transformer. In this topology, switches corresponding to a cell with higher and lower voltage respectively are turned on, such that the cells connected to transformer winding could be charged or discharged. In this case, the primary winding of the transformer is the energy storage element.

4   Simulation Results and Discussion

The results of the SOC convergence of the cell-to-cell and cell-to-stack-to-cell charge transfer centralized BMS topology are shown in Fig. 10a and b, respectively. Figure 11a and b show the SOC convergence results of the distributed cell-to-cell and cell-to-stack-to-cell charge transfer BMS topology respectively. The values corresponding to t_0, t_1, t_2, and t_3 for all simulated topologies are listed in Table 3 for reference.

In case of cell-to-cell charge transfer in centralized topology (Fig. 10a) at t_0, the cell with the maximum voltage gets discharged while the energy storage element gets charged. During this process, the energy storage element discharges the stored charge to the cell with the minimum voltage. Between t_0 and t_1, the voltage and SOC values begin to converge as the cells assigned $C_{nmax\&\&nmin}$ are being changed constantly, also resulting in change in average voltage values. At approximately 4951 s (t_2), the voltages of all the respective cells converge to 3.585V–3.60V range, hence dropping the average to a value below the delta value (≤ 0.05) which now roughly equals $|C_{nmax} - \alpha|$ at t_2. Hence, starting t_2, the algorithm would stay at the balanced state with voltage range within delta value, until load application or self discharge, at which point the algorithm would initiate again. Also, as shown in Fig. 10b, the corresponding SOC values also tend to converge at $40 \pm 0.5\%$ values with a difference of about 4.62%. The voltage and SOC difference values corresponding to the times are listed in Table 3a.

In case of cell-to-stack charge transfer in centralized topology (Fig. 10b) at t_0, maximum voltage cell (Cell 1) is being discharged to charge the stack with one cell (Cell 6) being charged by the stack (by stack to cell charge transfer). That means, the algorithm is trying to make $|C_{nmin} - \alpha|$ greater than $|C_{nmax} - \alpha|$ between t_0 and t_1. Hence more oscillations between cell-to-stack and stack-to-cell charge transfers can be noticed at t_1. At t_2 (and between t_1 and t_2), similar stack to cell discharges can again be noticed but with increased time as $|C_{nmax} - \alpha|$ and $|C_{nmin} - \alpha|$ values tend to get closer to delta, which results in repetitions of the loop in Algorithm 8. This is verified by the value at t_3 where voltage values are around 3.65 ± 0.02 V, hence converging at approximately 4000 s. This topology addresses the delay in SOC convergence seen in distributed (cell-to-cell) topology and converges to a SOC range of 46.7–49.6% at t_3 (SOC convergence following voltage convergence). In this case, although the algorithm converges the SOC values, the voltage values keep oscillating between $|C_{nmax} - \alpha|$ and $|C_{nmin} - \alpha|$, thereby entering and exiting the delta voltage value range for a short period of time. The voltage and SOC difference values corresponding to the times are listed in Table 3b.

In Fig. 11a, again at t_0, the cell with the maximum voltage gets discharged while the energy storage element gets charged. During this process, the energy storage element discharges the stored charge to the cell with the minimum voltage. Between t_0 and t_1, both voltage values and SOC tend to converge to a range of 3.61–3.64V range (within the delta range) and 36–50% around t_1. Although the voltage values converge faster in this topology, it still allows the SOC to

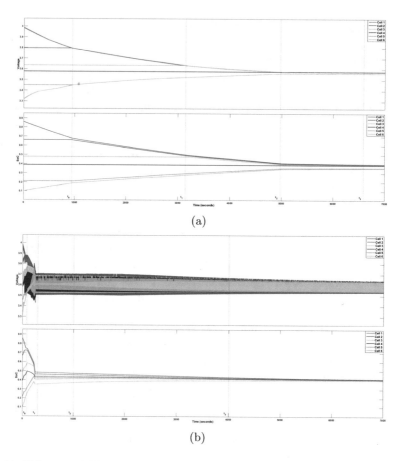

Fig. 10. Voltage and SOC vs time results for: (a) centralized cell equalization (cell-to-cell), (b) centralized cell equalization (cell-to-stack-to-cell).

converge around $50 \pm 0.5\%$ which happens around 4000 s. The voltage and SOC difference values corresponding to the times are listed in Table 3c.

In Fig. 11b, at t_0, multiple cells are being discharged to charge the stack (that is, $\{|C_{nmax} - \alpha| > |C_{nmin} - \alpha|\}$) with limited stack to cell charge transfers (that is, $\{|C_{nmax} - \alpha|$ becomes less than $|C_{nmin} - \alpha|$ for few cells in the stack$\}$). Between t_0 and t_1, the $|C_{nmin} - \alpha|$ becomes greater than $|C_{nmax} - \alpha|$, hence more stack to cell charge transfers can be noticed at t_1 (where voltage values range between 3.47–3.74V). At and around t_2, similar stack to cell discharges can again be noticed but with increased time as $|C_{nmax} - \alpha|$ and $|C_{nmin} - \alpha|$ values tend to get closer to delta, which results in repetitions of the loop in Algorithm 1. This is verified by the value at t_3 where voltage values are around 3.56 ± 0.02 V, hence converging at approximately 4000 s. This topology addresses the delay in SOC convergence seen in distributed (cell-to-cell) topology and converges to a SOC range of 33.9–37.5% at t_3 (SOC convergence following voltage convergence).

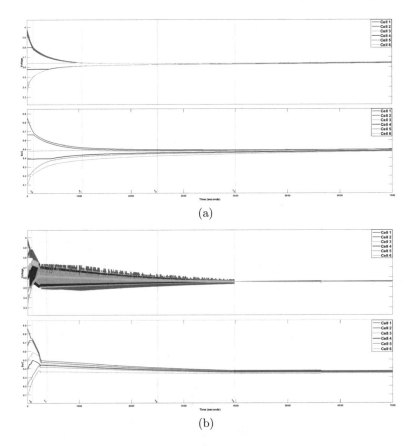

(a)

(b)

Fig. 11. Voltage and SOC vs time results for: (a) distributed cell equalization (cell-to-cell), (b) distributed cell equalization (cell-to-stack-to-cell).

The voltage and SOC difference values corresponding to the times are listed in Table 3d. The fluctuation in cell-to-stack-to-cell charge transfer case is highly noticeable because the stack is generating current to charge a single cell as opposed to the cell-to-cell charge transfer case where a single cell generates current to charge its counterpart.

The results from Fig. 10a, and b show that the cell-to-stack-to-cell balancing (alteration in algorithm and topology) reduces the time taken to balance the cells, and increases the SOC balancing range. The results from Fig. 11a and b also show that the balancing time gets reduced with cell-to-stack-to-cell balancing. The improved results in the distributed are also a result of the increased number of energy storage elements. Hence, a cell-to-stack-to-cell balancing algorithm is recommended for equalization algorithm design, provided design complexity is taken into consideration while developing of these topologies for the purpose of addressing balancing time limitation.

Table 3. Results showing Voltage and SOC values corresponding to time stamps t_0, t_1, t_2, and t_3 for:

(a) Centralized (cell-to-cell)

Time (in seconds)	t_0	t_1	t_2	t_3
	940	3150	4951.6	6500
Voltage Difference	0.346	0.111	0.015	0.0138
SOC difference	49.10%	21.90%	6.55%	4.62%

(b) Centralized (cell-to-stack and vice versa)

Time (in seconds)	t_0	t_1	t_2	t_3
	72.12	319	1000.5	4000
Voltage Difference	0.394	0.313	0.257	0.0288
SOC difference	54%	12.6%	9.75%	2.7%

(c) Distributed (cell-to-cell)

Time (in seconds)	t_0	t_1	t_2	t_3
	120	1100	2500	4000
Voltage Difference	0.295	0.0355	0.00949	0.00836
SOC difference	46.70%	14%	6.99%	4.52%

(d) Distributed (cell-to-stack and vice versa)

Time (in seconds)	t_0	t_1	t_2	t_3
	75	348	2500	3967
Voltage Difference	0.371	0.244	0.0914	0.0003
SOC difference	52%	12.5%	6.15%	3.62%

5   Conclusion and Future Directions of Research

This paper enlists the complete process of developing a battery management system from electro-chemistry, battery models to topology selection on a simulation platform. Also, a comparative analysis between centralized and distributed topologies are presented, which further re-enforce that the distributed topologies outperform the centralized topologies in terms of balancing time and energy efficiency. Table 3 lists the voltage and SOC difference values corresponding to the time steps discussed in the previous section and in Figs. 10a and b, and 11a and b. Results from Table 3 indicate that trade-off between design complexity and balancing time must be taken into consideration when developing a distributed balancing topology.

In practical application of BMS, measurement accuracy is another aspect which needs to be taken into account. In this paper, the data was obtained from testing performed on a Digatron Universal Battery Tester which has a measurement accuracy of ±1%. Galvanic isolation should also be placed between the batteries and the converter topology in practical applications. Practical designs must also consider parasitic elements during transient conditions as any drop in voltage could deceive the monitor to detect under-voltage condition [3].

Future works could also focus on analyzing battery monitoring techniques [75] carried out for each equalization algorithm listed in this paper. An analysis of influence of integration of advanced battery models [8], effect of battery degradation [58], implemented on battery management system using additional cell parameter modeling techniques [65] in a simulation environment can also be carried out. Future works could also focus on analyzing and performing optimization between the major performance indicators: cost, balancing time and energy efficiency by keeping SOC instead of voltage as the algorithm input variable [39]. Finally, the effect of topological changes (involving thermal variations on battery models), on performance indicators can be simulated.

Nomenclature

The following nomenclature is used in this paper.

n	Total number of cells in a stack, $n \in \mathbb{Z}^+$
C_{nmax}	Cell with the highest charge/voltage of cell with the highest charge among n cells
C_{nmin}	Cell with the lowest charge/voltage of cell with the lowest charge among n cells
C_{nmax-P}	Pair of cells connected to the switch S_{nmax-P}
C_{nmin-P}	Pair of cells connected to the switch S_{nmin-P}
$C_{nmax\&\&nmin}$	Set of two cells with highest and lowest voltages, respectively
C_i	i^{th} cell/voltage of the ith cell, $\forall i \in [1, n]$
S_{nmax}	Switch corresponding to the cell with the highest charge or voltage
S_{nmin}	Switch corresponding to the cell with the lowest charge or voltage
S_{nmax-P}	Pair of switches connected to the cells with the highest voltage in the stack
S_{nmin-P}	Pair of switches connected to the cells with lowest voltage in the stack
$S_{nmax-P\&\&nmin-P}$	Two pairs of switches corresponding to cells with highest and lowest voltages in the stack, respectively
S_s	Stack switch

Acknowledgments. The material published is a result of the research supported by the National Science Foundation under the Award number CNS-1553494.

References

1. Battery cell charging system having voltage threshold and bleeder current generating circuits, March 2002
2. Tesla model s 18650 cell test data, March 2015. https://teslamotorsclub.com/tmc/threads/teslamodel-s-18650-cell-test-data.45063/
3. Affanni, A., Bellini, A., Franceschini, G., Guglielmi, P., Tassoni, C.: Battery choice and management for new-generation electric vehicles. IEEE Trans. Industr. Electron. **52**(5), 1343–1349 (2005)
4. Andre, D., Meiler, M., Steiner, K., Walz, H., Soczka-Guth, T., Sauer, D.U.: Characterization of high-power lithium-ion batteries by electrochemical impedance spectroscopy. ii: Modelling. J. Power Sources **196**(12), 5349–5356 (2011). Selected papers presented at the 12th Ulm Electro Chemical Talks (UECT): 2015 Technologies on Batteries and Fuel Cells

5. Brand, M., Gläser, S., Geder, J., Menacher, S., Obpacher, S., Jossen, A., Quinger, D.: Electrical safety of commercial Li-ion cells based on NMC and NCA technology compared to LFP technology. In: 2013 World Electric Vehicle Symposium and Exhibition (EVS27), pp. 1–9, November 2013

6. Brando, G., Dannier, A., Spina, I., Piegari, L.: Comparison of accuracy of different LiFePO4 battery circuital models. In: 2014 International Symposium on Power Electronics, Electrical Drives, Automation and Motion, pp. 1092–1097, June 2014

7. Bui, T.M., Kim, C.-H., Kim, K.-H., Rhee, S.B.: A modular cell balancer based on multi-winding transformer and switched-capacitor circuits for a series-connected battery string in electric vehicles. Appl. Sci. 8(8), 1278 (2018)

8. Cacciato, M., Nobile, G., Scarcella, G., Scelba, G.: Real-time model-based estimation of SOC and SOH for energy storage systems. IEEE Trans. Power Electron. 32(1), 794–803 (2017)

9. Caspar, M., Eiler, T., Hohmann, S.: Systematic comparison of active balancing: a model-based quantitative analysis. IEEE Trans. Veh. Technol. 67(2), 920–934 (2018)

10. Chen, X., Shen, W., Vo, T.T., Cao, Z., Kapoor, A.: An overview of lithium-ion batteries for electric vehicles. In: 2012 10th International Power Energy Conference (IPEC), pp. 230–235, December 2012

11. Chiang, Y.-H., Sean, W.-Y., Ke, J.-C.: Online estimation of internal resistance and open-circuit voltage of lithium-ion batteries in electric vehicles. J. Power Sources 196(8), 3921–3932 (2011)

12. Daowd, M., Antoine, M., Omar, N., Lataire, P., Van Den Bossche, P., Van Mierlo, J.: Battery management system—balancing modularization based on a single switched capacitor and bi-directional DC/DC converter with the auxiliary battery. Energies 7(5), 2897–2937 (2014)

13. Daowd, M., Antoine, M., Omar, N., van den Bossche, P., van Mierlo, J.: Single switched capacitor battery balancing system enhancements. Energies 6(4), 2149–2174 (2013)

14. Daowd, M.A.A.H., Omar, N., Verbrugge, B., Van Den Bossche, P., Van Mierlo, J.: Battery Models Parameter Estimation based on Matlab/Simulink, November 2010

15. Du, J., Wang, Y., Tripathi, A., Lam, J.S.L.: Li-ion battery cell equalization by modules with chain structure switched capacitors. In: 2016 Asian Conference on Energy, Power and Transportation Electrification (ACEPT), pp. 1–6, October 2016

16. Dubarry, M., Vuillaume, N., Liaw, B.Y.: From single cell model to battery pack simulation for Li-ion batteries. J. Power Sources 186(2), 500–507 (2009)

17. Falconi, A.: Electrochemical Li-ion battery modeling for electric vehicles. Theses, Communaute Universite Grenoble ALPES, October 2017

18. Mousavi G., S.M., Nikdel, M.: Various battery models for various simulation studies and applications. Renew. Sustain. Energy Rev. 32, 477–485 (2014)

19. Gallardo-Lozano, J., Romero-Cadaval, E., Milanes-Montero, M.I., Guerrero-Martinez, M.A.: Battery equalization active methods. J. Power Sources 246, 934–949 (2014)

20. Gonzalez-Longatt, F.: Circuit based battery models: a review. In: Congreso Iberoamericano de estudiantes De Ingenieria Electrica, pp. 1–5 (2007)

21. Guo, Y., Lu, R., Wu, G., Zhu, C.: A high efficiency isolated bidirectional equalizer for lithium-ion battery string. In: 2012 IEEE Vehicle Power and Propulsion Conference, pp. 962–966, October 2012

22. Hannan, M.A., Hoque, M.M., Ker, P.J., Begum, R.A., Mohamed, A.: Charge equalization controller algorithm for series-connected lithium-ion battery storage systems: modeling and applications. Energies 10(9), 1390 (2017)

23. He, H., Xiong, R., Fan, J.: Evaluation of lithium-ion battery equivalent circuit models for state of charge estimation by an experimental approach. Energies **4**(4), 582–598 (2011)
24. Xiaosong, H., Li, S., Peng, H.: A comparative study of equivalent circuit models for Li-ion batteries. J. Power Sources **198**, 359–367 (2012)
25. Hussein, A.A.: Experimental modeling and analysis of lithium-ion battery temperature dependence. In: 2015 IEEE Applied Power Electronics Conference and Exposition (APEC), pp. 1084–1088, March 2015
26. Hussein, A.A., Batarseh, I.: An overview of generic battery models. In: 2011 IEEE Power and Energy Society General Meeting, pp. 1–6, July 2011
27. Hymel, S.: Measuring internal resistance of batteries, May 2013
28. Imtiaz, A.M., Khan, F.H.: "Time shared flyback converter" based regenerative cell balancing technique for series connected Li-ion battery strings. IEEE Trans. Power Electron. **28**(12), 5960–5975 (2013)
29. Texas Instruments. EMB1499Q bidirectional current DC-DC controller, September 2013
30. Islam, M., Omole, A., Islam, A., Domijan, A.: Dynamic capacity estimation for a typical grid-tied event programmable LI-FEPO4 battery. In: 2010 IEEE International Energy Conference, pp. 594–599, December 2010
31. Jeon, Y., Noh, H.K., Song, H.-K.: A lithium-ion battery using partially lithiated graphite anode and amphi-redox LiMn2O4 cathode. Sci. Rep. **7**(1), 14879 (2017)
32. Kam, K.C., Doeff, M.M.: Electrode materials for lithium ion batteries. Mater. Matters **7**, 56–60 (2012)
33. Karthigeyan, V., Aswin, M., Priyanka, L., Sailesh, K.N.D., Palanisamy, K.: A comparative study of lithium ion (LFP) to lead acid (VRLA) battery for use in telecom power system. In: 2017 International Conference on Computation of Power, Energy Information and Communication (ICCPEIC), pp. 742–748, March 2017
34. Khalid, A., Sundararajan, A., Acharya, I., Sarwat, A.I.: Prediction of Li-ion battery state of charge using multilayer perceptron and long short-term memory models. In: 2019 IEEE Transportation Electrification Conference (ITEC) (2019, in press)
35. Khalid, A., Sundararajan, A., Hernandez, A., Sarwat, A.: Facts approach to address cybersecurity issues in electric vehicle battery systems. In: IEEE Technology and Engineering Management Conference (TEMSCON) (2019, in press)
36. Khalid, A., Sundararajan, A., Sarwat, A.I.: A multi-step predictive model to estimate Li-ion state of charge for higher c-rates. In: 2019 IEEE International Conference on Environment and Electrical Engineering and 2019 IEEE Industrial and Commercial Power Systems Europe (EEEIC/I CPS Europe) (2019, in press)
37. Khalid, A.: Electricity usage monitoring using face recognition technique. Int. J. Emerg. Technol. Adv. Eng. **2**(10), 274–276 (2012)
38. Kim, C., Kim, M., Park, H., Moon, G.: A modularized two-stage charge equalizer with cell selection switches for series-connected lithium-ion battery string in an HEV. IEEE Trans. Power Electron. **27**(8), 3764–3774 (2012)
39. Kirchev, A.: Battery management and battery diagnostics. In: Moseley, P.T., Garche, J. (eds.) Electrochemical Energy Storage for Renewable Sources and Grid Balancing, chap. 20, pp. 411–435. Elsevier, Amsterdam (2015)
40. Kollmeyer, P.: Panasonic 18650PF Li-ion battery data (2018). https://data.mendeley.com/datasets/wykht8y7tg/1
41. Konishi, Y., Huang, Y.-S., Luor, T.-S.: Bridge battery voltage equalizer. US Patent US7612530B2 (2006)
42. Teja, G.K., Prabhaharan, S.R.S.: Smart battery management system with active cell balancing. Indian J. Sci. Technol. **8**, 1 (2015)

43. Kroeze, R.C., Krein, P.T.: Electrical battery model for use in dynamic electric vehicle simulations. In: 2008 IEEE Power Electronics Specialists Conference, pp. 1336–1342, June 2008

44. Kutkut, N.H., Wiegman, H.L.N., Divan, D.M., Novotny, D.W.: Charge equalization for an electric vehicle battery system. IEEE Trans. Aerosp. Electron. Syst. **34**(1), 235–246 (1998)

45. Li, J., Mazzola, M., Gafford, J., Younan, N.: A new parameter estimation algorithm for an electrical analogue battery model. In: 2012 Twenty-Seventh Annual IEEE Applied Power Electronics Conference and Exposition (APEC), pp. 427–433, February 2012

46. Li, J., Mazzola, M.S.: Accurate battery pack modeling for automotive applications. J. Power Sources **237**, 215–228 (2013)

47. Li, J., Mazzola, M.S., Gafford, J., Jia, B., Xin, M.: Bandwidth based electrical-analogue battery modeling for battery modules. J. Power Sources **218**, 331–340 (2012)

48. Li, Y., Han, Y.: A module-integrated distributed battery energy storage and management system. IEEE Trans. Power Electron. **31**(12), 8260–8270 (2016)

49. Liaw, B.Y., Jungst, R.G., Nagasubramanian, G., Case, H.L., Doughty, D.H.: Modeling capacity fade in lithium-ion cells. J. Power Sources **140**(1), 157–161 (2005)

50. Ling, R., Dan, Q., Zhang, J., Chen, G.: A distributed equalization control approach for series connected battery strings. In: The 26th Chinese Control and Decision Conference (2014 CCDC), pp. 5102–5106, May 2014

51. Liu, W., Song, Y., Liao, H., Li, H., Zhang, X., Jiao, Y., Peng, J., Huang, Z.: Distributed voltage equalization design for supercapacitors using state observer. IEEE Trans. Ind. Appl. **55**(1), 620–630 (2018)

52. Brandl, M., Gall, H., Wenger, M., Lorentz, V., Giegerich, M., Baronti, F., Fantechi, G., Fanucci, L., Roncella, R., Saletti, R., Saponara, S., Thaler, A., Cifrain, M., Prochazka, W.: Batteries and battery management systems for electric vehicles, pp. 971–976, March 2012

53. Mayer, S., Geddes, L.A., Bourland, J.D., Ogborn, L.: Faradic resistance of the electrode/electrolyte interface. Med. Biol. Eng. Comput. **30**(5), 538–542 (1992)

54. Narayanaswamy, S., Kauer, M., Steinhorst, S., Lukasiewycz, M., Chakraborty, S.: Modular active charge balancing for scalable battery packs. IEEE Trans. Very Large Scale Integr. (VLSI) Syst. **25**(3), 974–987 (2017)

55. Nitta, N., Wu, F., Lee, J.T., Yushin, G.: Li-ion battery materials: present and future. Mater. Today **15**(April), 252–264 (2015)

56. Omar, N., Widanage, D., Abdel Monem, M., Firouz, Y., Hegazy, O., Van den Bossche, P., Coosemans, T., Van Mierlo, J.: Optimization of an advanced battery model parameter minimization tool and development of a novel electrical model for lithium-ion batteries. Int. Trans. Electrical Energy Syst. **24**(12), 1747–1767 (2013)

57. Pang, S., Farrell, J., Du, J., Barth, M.: Battery state-of-charge estimation. In: Proceedings of the 2001 American Control Conference. (Cat. No. 01CH37148), vol. 2, pp. 1644–1649, June 2001

58. Perez, A., Moreno, R., Moreira, R., Orchard, M., Strbac, G.: Effect of battery degradation on multi-service portfolios of energy storage. IEEE Trans. Sustain. Energy **7**(4), 1718–1729 (2016)

59. Plett, G.L.: Extended Kalman filtering for battery management systems of LiPB-based HEV battery packs: part 2. Modeling and identification. J. Power Sources **134**(2), 262–276 (2004)

60. Qi, J., Lu, D.D.C.: Review of battery cell balancing techniques. In: 2014 Australasian Universities Power Engineering Conference (AUPEC), pp. 1–6, September 2014
61. Rehman, M.U.: Modular, Scalable Battery Systems with Integrated Cell Balancing and DC Bus Power Processing. Ph.D. thesis (2018)
62. Rui, L., Lizhi, W., Xueli, H., Qiang, D., Jie, Z.: A review of equalization topologies for lithium-ion battery packs. In: 2015 34th Chinese Control Conference (CCC), pp. 7922–7927, July 2015
63. Salameh, Z.M., Casacca, M.A., Lynch, W.A.: A mathematical model for lead-acid batteries. IEEE Trans. Energy Convers. **7**(1), 93–98 (1992)
64. Schonberger, J.: Modeling a lithium-ion cell using PLECS. In: Plexim GmbH, pp. 1–5 (2009)
65. Schweiger, H.G., Obeidi, O., Komesker, O., et al.: Comparison of several methods for determining the internal resistance of lithium ion cells. Sensors **10**(6), 5604–5625 (2010)
66. Steinhorst, S., Shao, Z., Chakraborty, S., Kauer, M., Li, S., Lukasiewycz, M., Narayanaswamy, S., Rafique, M.U., Wang, Q.: Distributed reconfigurable battery system management architectures. In: 2016 21st Asia and South Pacific Design Automation Conference (ASP-DAC), pp. 429–434, January 2016
67. Subburaj, A.S., Bayne, S.B.: Analysis of dual polarization battery model for grid applications. In: 2014 IEEE 36th International Telecommunications Energy Conference (INTELEC), pp. 1–7, September 2014
68. Sundararajan, A., Khan, T., Moghadasi, A., Sarwat, A.I.: Survey on synchrophasor data quality and cybersecurity challenges, and evaluation of their interdependencies. J. Mod. Power Syst. Clean Energy **7**(3), 449–467 (2018)
69. Sundararajan, A., Sarwat, A.I.: Roadmap to prepare distribution grid-tied photovoltaic site data for performance monitoring. In: 2017 International Conference on Big Data, IoT and Data Science (BID), pp. 110–115, December 2017
70. Thanagasundram, S., Arunachala, R., Makinejad, K., Teutsch, T., Jossen, A.: A cell level model for battery simulation, pp. 1–13, November 2012
71. Tsang, K.M., Chan, W.L., Wong, Y.K., Sun, L.: Lithium-ion battery models for computer simulation. In: 2010 IEEE International Conference on Automation and Logistics, pp. 98–102, August 2010
72. Uno, M., Kukita, A.: Bidirectional PWM converter integrating cell voltage equalizer using series-resonant voltage multiplier for series-connected energy storage cells. IEEE Trans. Power Electron. **30**(6), 3077–3090 (2015)
73. Johnson, V.H., Pesaran, A.A., Sack, T.: Temperature-dependent battery models for high-power lithium-ion batteries. In: 17th Annual Electric Vehicle Symposium, vol. 12, pp. 1–14 (2001)
74. Verbrugge, M.W., Conell, R.S.: Electrochemical and thermal characterization of battery modules commensurate with electric vehicle integration. J. Electrochem. Soc. **149**(1), A45–A53 (2002)
75. Waag, W., Fleischer, C., Sauer, D.U.: Critical review of the methods for monitoring of lithium-ion batteries in electric and hybrid vehicles. J. Power Sources **258**, 321–339 (2014)
76. Wehbe, J., Karami, N.: Battery equivalent circuits and brief summary of components value determination of lithium ion: a review. In: 2015 Third International Conference on Technological Advances in Electrical, Electronics and Computer Engineering (TAEECE), pp. 45–49, April 2015

77. Yun, J., Yeo, T., Park, J.: High efficiency active cell balancing circuit with soft-switching technique for series-connected battery string. In: 2013 Twenty-Eighth Annual IEEE Applied Power Electronics Conference and Exposition (APEC), pp. 3301–3304, March 2013
78. Zhang, C., Li, K., Mcloone, S., Yang, Z.: Battery modelling methods for electric vehicles - a review. In: 2014 European Control Conference (ECC), pp. 2673–2678, June 2014
79. Zhang, X., Zhang, W., Lei, G.: A review of Li-ion battery equivalent circuit models. Trans. Electr. Electron. Mater. **17**, 311–316 (2016)
80. Zhao, X., Cai, Y., Yang, L., Deng, Z., Qiang, J.: State of charge estimation based on a new dual-polarization-resistance model for electric vehicles. Energy **135**, 40–52 (2017)

Deep Learning-Based Document Modeling for Personality Detection from Turkish Texts

Tuncay Yılmaz, Abdullah Ergil, and Bahar İlgen[✉]

Istanbul Kültür University, Istanbul, Turkey
b.ilgen@iku.edu.tr

Abstract. The usage of social media is increasing exponentially since it has been the easiest and fastest way to share information between people or organizations. As a result of this broad usage and activity of people on social networks, considerable amount of data is generated continuously. The availability of user generated data makes it possible to analyze personality of people. Personality is the most distinctive feature for an individual. The results of these analyses can be utilized in several ways. They provide support for human resources recruitment units to consider suitable candidates. Similar products and services can be offered to people who share the similar personality characteristics. Personality traits help in diagnosis of certain mental illnesses. It is also helpful in forensics to use personality traits on suspects to clarify the forensic case. With the rapid dissemination of online documents in many different languages, the classification of these documents has become an important requirement. *Machine Learning* (ML) and *Natural Language Processing* (NLP) methods have been used to classify these digitized data. In this study, current ML techniques and methodologies have been used to classify text documents and analyze person characteristics from these datasets. As a result of classification, detailed information about the personality traits of the writer could be obtained. It was seen that the frequency-based analysis and the use of the emotional words at the word level are crucial in the textual personality analysis.

Keywords: Big five personality traits · Deep neural network · Natural Language Processing · Text mining · RNN · LSTM

1 Introduction

As a result of the increasing use and dissemination of electronic data in our daily lives, *Natural Language Processing* (NLP) studies for the classification of these data has become more and more important. Website news, biographies, comments, summaries, microblogs, e-mails, databases and different forms of electronic data are all unlabeled or semi-labeled documents. Using these textual resources, it is crucial to accurately classify and analyze news sources. Hence, the main purpose of text mining studies includes to extract valuable information from electronic data, to perform classification and summarization tasks on the user generated texts. Today, social media provides richer and more valuable information, since its' usage constitutes a large part of internet usage recently. Almost half of the world's population uses the internet. As a result,

© Springer Nature Switzerland AG 2020
K. Arai et al. (Eds.): FTC 2019, AISC 1069, pp. 729–736, 2020.
https://doi.org/10.1007/978-3-030-32520-6_53

there are millions of people who use several forms of social networks regularly. Therefore, the use of social media is an inevitable part of the internet. According to the researches, people login to their social media accounts 4–5 times a day. They stay at least 10 min every time they become online [1].

The most distinctive feature of an individual that makes a person unique is his/her personality. Personality is a phenomenon composed of a combination of emotions, behaviors and thoughts of a person [2]. Every person's personality plays a very important role in the progress of that person's life. Personality determines the friends people have, the choices they make and even their state of health. Analysis of the features that plays such an important role in our lives is of course very important and necessary. People actually describe their personalities through the social media. They present their personality traits to the knowledge of other people with or without awareness. Users constantly share clues about their selves through the online media in forms of texts, pictures and/or video. They make comments under specific contents, write blogs, post messages etc. In all of these, textual datasets are one of the most valuable form that leads researchers to get useful and accurate results in personality analysis.

In this study, we used widely accepted model of Big Five [3] to determine and detect the personality traits from texts. As one of the most accepted model, it explains and summarizes human personality using five dimensions. These include the personality dimensions given in Table 1.

Table 1. Big five personality dimensions

Five factors	Description
Extraversion	Degree of activeness, sociability, and talkativeness
Agreeableness	Personality traits such as cooperativeness, kindness and trust
Conscientiousness	Refers to the degree of achievement orientation. Being hardworking, organize and reliable is linked with high score
Neuroticism	Indicates the emotional stability degree. People who have low degree of neuroticism are calm, secure and confident whereas the opposite anxious, insecure and moody
Openness	Related with creative, cultural and intellectual interest

In the scope of this study, we used the labeled textual datasets to train our datasets. During the training phase, we used *Long Short-Term Memory* (LSTM) [4], which is the special kind of the *Recurrent Neural Networks* (RNNs). LSTM networks are capable of learning long-term dependencies. They produce satisfactory results on large variety of problems.

The rest of the paper is organized as follows: Sect. 2 summarizes the related work in the research area. In Sects. 3 and 4, we describe the methodology we used in this study and our experimental results respectively. Finally, Sect. 5 draws conclusion and future work.

2   Related Work

NLP methods and ML algorithms are used in sequential order for the analysis and classification of texts. Text mining is used to pre-process, classify, summarize and make inferences from texts [5]. Making discoveries to produce electronic data from various sources, classifying them and ensuring their integration is very important for the research area.

The importance given to personality analysis has increased in recent years. Accurate personality analyses have potential to help humanity in many areas. Experts in the area utilize mobile social networks and social internet sites in different languages frequently for these personality analyses. Some of the areas that benefit from the results of personality analyses include the recommendation systems, emotion analysis and crime fighting. Recently, human resources officials do not talk with people during interviews but rather make hires based on their sharing and writing on social media accounts [6]. On the other hand, e-commerce authorities use the analysis of personality to provide more attractive products to customers. In addition to these, personality analyses help people to take precautions for the certain illnesses. For example, it is known that there is a relation between anger and heart diseases. The health authorities can extensively consider this link between the disease and personality trait.

In [7], relationship between personality traits and software quality has been investigated. The software metrics used in the study include; depth of inheritance, coupling between objects, cyclomatic complexity and maintenance index. They conducted experiments on software developers working in an educational institution. They used the big-five data of the developers together with the software they produced. Then the links between given software metrics and personality traits have been observed. Their study shows that personality analyses can also be used to predict performance of software developers. In [8], a heterogeneous information ensemble framework was proposed for the prediction of user personality. The work integrates the heterogeneous information such as emoticons, avatars, self-language usage etc.

People can act to deceive other people in their relations. Taking such actions out of the texts written by them can provide a significant advantage in many areas. Some of the areas that will benefit from such research are military, legal and public areas. If we determine those words which are deceiving the people, those clues could be used automatically in many areas by software [9]. The proposed study examines and elaborates the methods to prevent the bad situations that may occur by analyzing the personality from the texts.

In [10], a corpus of 1.2M English tweets has been annotated and used in the experiments. In the study, Myers-Briggs personality type and gender information has been covered. Their experiments have shown that the certain personality dimensions such as Introvert-Extrovert and Feeling-Thinking could be distinguished successively. Another research on personality analysis [11] provided methods to detect personality of a person using ontology-based personality detection approach. In the examined study, a real-world application has been developed by providing questionnaires.

3   Methodology

In the scope of this project, the first step includes building and collecting our dataset for big-five personality analysis. Because of lack of Turkish textual datasets, a dataset in proper form had to be translated into Turkish. In addition to this, the agglutinative nature and existence of free word order of the language, Turkish has rich set of inflectional suffixes. Analyzing Turkish words was one of the most important steps in the study. We utilized Zemberek NLP library for the analysis of Turkish words. Because this library was developed using Java programming language, we used Java Virtual Machine (JVM) and functions in the library. The project has been developed using Python programming language. During the application development phase, Anaconda development platform has been preferred. The main reason to select Anaconda environment was the useful tools that are provided to develop data science programs. It also includes libraries that are useful and necessary while developing deep learning projects. Another reason to prefer Python programming language during the project and library development phase was to take the advantage of using popular and powerful libraries for deep learning written in Python programming language. In this project, all libraries used are open source and written using Python language except the NLP related classes of Zemberek library. The project can run on Python console and other python development environments.

Another important tool for project is Java Virtual Machine. Oracle's jdk (Java Development Kit) and jre (Java Runtime Environment) is installed on device to handle Turkish texts. Thanks to Java Virtual Machine, the project runs java-based program background. In this way, project utilizes its functions and gets the result. JPype library is needed to connect Java and Python in project to each other. It's a Python library that is customized to establish a bridge between Java program on Java Virtual Machine and Python project.

Libraries for reading and editing text data from excel and csv file have been used in the project. Numpy library helps to store data in multidimensional arrays. Pandas library takes turn to provide high-performance data structures for Python programming language. It basically converts the data into the table which is called data frame in ML terminology. JPype library's startJVM function takes target Java program's jar (Java Archive) path and provides accessibility to its functions. Regular Expression (re) library is needed to remove unnecessary and redundant characters from the text. Trstop library is used for eliminating Turkish stop words from text. It compares all words with its own list at background.

Second step was reading data from excel/csv file and clean it up. During the reading phase, pandas require document's file path and its' character type. Latin5 encoding type covers Turkish language. We had five categories and needed five models. In order to train five models, we need to handle their labels separately. Using the cleaned data classes both the labels are highlighted, and the text is cleared. Firstly, Pandas reads all data and create data frame. Then pandas.concat function declares necessary columns for the model to be trained. Dropna function delete row of data from data table if there is no data on one of declared column.

To analyze words, we firstly converted them to Turkish. Zemberek use ascii toleranced root picker to convert non-Turkish characters to Turkish characters. For example, 'gozlukcu' to 'gözlükçü' (i.e., optician in English). Then the converted word is checked to decide whether it is Turkish or not. If the returned result is Turkish, it will be ready for morphological analysis phase.

Finding word roots is important to do morphological analysis. Zemberek put letters of words on tree, iterate letter by letter and collect all encountered roots. Then, in analysis phase, all roots candidates are tried, and the result is returned.

Text to be trained must enter the deep learning algorithm in a shape. The shape is determined as the length of longest sentence. The unique words that have passed tokenizing process are transferred to the array. The array has elements as many as the number of words of the longest sentence. In pad sequence process, empty cells in this array is filled with zero.

The training layers consist of LSTM units. LSTM is an improved version of RNNs (see Fig. 1) [12]. LSTM units are typically composed of the following components; a cell, an input gate, an output gate and a forget gate. Each cell keeps values over arbitrary time intervals. There are also three gates which regulate the information flow into and out of the cell. A LSTM cell initially takes an input and stores this information during a period of time. This will be equivalent with applying identity function (f(x) = x) to input since its' derivate is constant and gradient does not vanish when the network is trained with backpropagation.

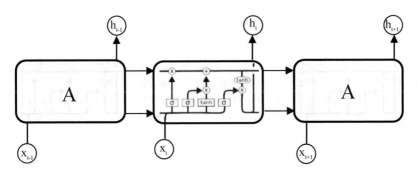

Fig. 1. LSTM architecture

The first step in LSTM consists of deciding to the information that will be eliminated from the cell state. The origin of this decision is sigmoid layer which is called the "forget gate layer". It looks at h_{t-1} and x_t and generates a number between 1 and 0 for the numbers in the cell state C_{t-1}. The former represents "completely keep this" while the latter represents "completely get rid of this" [12].

$$f_t = \sigma\left(W_f \cdot [h_{t-1}, x_t] + b_f\right) \quad (1)$$

$$i_t = \sigma(W_i \cdot [h_{t-1}, x_t] + b_i) \quad (2)$$

$$\widetilde{C}_t = tanh(W_C.[h_{t-1}, x_t] + b_C) \qquad (3)$$

In the following step the old cell state C_{t-1} is updated into the new cell state C_t. The action is determined during the previous steps and applied in the next one. The old state is multiplied by f_t and, $i_t * \widetilde{C}_t$ is added in (4). This is obtained as the new candidate value, scaled by how much it is decided to update each state value.

$$C_t = f_t * C_{t-1} + i_t * \widetilde{C}_t \qquad (4)$$

During the final part of training, "Fit function" takes turn and runs. Tensorflow keras fit function runs compiled model and applies machine learning tasks in this section. Fit function needs arguments to run. These are training dataset (X_train), test data (Y_train), epoch number, batch size, and validation data that include training and test data's results (X_test, Y_test). Epoch number determines the iteration count. The program runs itself up to the epoch number. This number may cause overfitting. Batch size determines how much data will be taken for an iteration while training.

At the end of the training session, trained model must be saved to use test session. Model's properties should be saved in *Javascript Object Notation* (JSON) format and model should be saved in h5 (Hierarchical Data Format) data format. There are other types but h5 is the most stable type for deep learning models. Also cleaned data must be saved to retokenize the words in test program.

4 Experimental Results

Our personality detection system was run on the Anaconda Platform which is a Pyhton IDE (Integrated Development Environment) on the Windows Operating System. Java Virtual Machine is built on Windows operating system in order to run Java codes in Python IDE. A dataset named *ocean.csv* which has 10000 different texts from social media were translated and used for training and testing purposes. Figure 2 shows the form of contents in the original dataset. Each line of the dataset includes the text and personality tags (*i.e., opn, con, ext, agr, neu*) in respective order.

```
text,opn,con,ext,agr,neu
likes the sound of thunder.,1,0,0,0,1
is so sleepy it's not even funny that's she can't get to sleep.,1,0,0,0,1
"is sore and wants the knot of muscles at the base of her neck to stop hurting. On the other hand, YAY I'M I
likes how the day sounds in this new song.,1,0,0,0,1
is home. <3,1,0,0,0,1
www.thejokerblogs.com,1,0,0,0,1
"saw a nun zombie, and liked it. Also, *PROPNAME* + Tentacle!Man + Psychic Powers = GREAT Party.",1,0,0,0,1
is in Kentucky. 421 miles into her 1100 mile journey home.,1,0,0,0,1
was about to finish a digital painting before her tablet went haywire. Is now contemplating the many ways sh
technology.,1,0,0,0,1
is celebrating her new haircut by listening to swinger music and generally looking like a doofus.,1,0,0,0,1
has a crush on the Green Lantern.,1,0,0,0,1
has magic on the brain.,1,0,0,0,1
"saw Transformers, Up, and Year One this week. Good movie overload. :D",1,0,0,0,1
Who wants to meet up on schedule pick-up day at Oviedo?,1,0,0,0,1
"desires the thrill of inspiration. Also, money.",1,0,0,0,1
is going to bed at 9:30! Yeah!,1,0,0,0,1
"is reading, admiring her permit, and occasionally glancing at her ner McDonald's uniform.",1,0,0,0,1
```

Fig. 2. Sample lines from Big Five Model Dataset

LSTM method and NLP libraries have been used together to create a program that can analyze the personality of an author from the text. In our RNN models in which it is applied using LSTM, we found accuracy 55% of EXT personality, 60% of NEU personality, 53% of AGR personality, 54% of CON personality and 68% of OPN personality characteristics. In our study, we found that the accuracy rate increased as the number of epochs increased, but the problem of overfitting occurred as a result. In order to solve the problem of overfitting, we have tried to find the right number of epochs to find the best solution. The following Table 2 summarizes training results of models.

Table 2. Training results of models.

Model name	Training accuracy	Test accuracy
EXT	0.8622	0.5499
NEU	0.8429	0.5970
AGR	0.8649	0.5321
CON	0.8470	0.5422
OPN	0.8991	0.6795

5 Conclusions and Future Work

Personality analyses on the texts reflect the characteristics of people in very detailed form. On the other hand, the usage of social networking has become increasingly widespread recently. As an expected result, personality analyses using these social networks are getting more attention in the research field. These studies are beneficial in many areas of our lives. People transfer their feelings to the texts that they generate regularly.

In the scope of this study, personality detection analyses on annotated datasets have been investigated using LSTM model. LSTM networks provide an improvement over RNN by deciding selectively remembering or forgetting things. Big-five datasets and texts collected from social media were used as dataset during the experiments. It has been shown that LSTM networks work well on the personality detection problem. The accuracy rates for each personality trait has been obtained in acceptable ranges. It is expected that the results will be improved by using more comprehensive personality datasets in near future. The results of personality analysis will provide support in many areas such as crime identification, recruitment, personalized marketing. In addition, these analyses can easily be expanded to investigate the additional links between personality traits and specific domain-related observation.

As a future work, we will expand our search by using larger and more comprehensive datasets to investigate the additional links between personality traits and several domains such as health, education and performance of individual on certain subject.

References

1. Ong, V., Rahmanto, A.D., Williem, W., Suhartono, D.: Exploring personality prediction from text on social media: a literature review. Internetworking Indonesia **9**(1), 65–70 (2017)
2. Majumder, N., Poria, S., Gelbukh, A., Cambria, E.: Deep learning-based document modeling for personality detection from text. IEEE Intell. Syst. **32**(2), 74–79 (2017)
3. John, O.P., Donahue, E.M., Kentle, R.L.: The big five inventory—versions 4a and 54 (1991)
4. Hochreiter, S., Schmidhuber, J.: Long short-term memory. Neural Comput. **9**(8), 1735–1780 (1997)
5. Khan, A., Baharudin, B., Lee, L.H., Khan, K.: A review of machine learning algorithms for text-documents classification. J. Adv. Inf. Technol. **1**(1), 4–20 (2010)
6. Agarwal, B.: Personality detection from text: a review. Int. J. Comput. Syst. **1**(1), 1–4 (2014)
7. Barroso, A.S., da Silva, J.S.M., Souza, T.D., Bryanne, S.D.A., Soares, M.S., do Nascimento, R.P.: Relationship between personality traits and software quality-big five model vs. object-oriented software metrics. In: ICEIS, no. 3, pp. 63–74 (2017)
8. Wei, H., et al.: Beyond the words: predicting user personality from heterogeneous information. In: Proceedings of the Tenth ACM International Conference on Web Search and Data Mining, pp. 305–314. ACM 2017
9. Zhou, L., Twitchell, D.P., Qin, T., Burgoon, J.K., Nunamaker, J.F.: An exploratory study into deception detection in text-based computer-mediated communication. In: Proceedings of the 36th Annual Hawaii International Conference on System Sciences, p. 10. IEEE (2003)
10. Plank, B., Hovy, D.: Personality traits on twitter—or—how to get 1,500 personality tests in a week. In: Proceedings of the 6th Workshop on Computational Approaches to Subjectivity, Sentiment and Social Media Analysis, pp. 92–98 (2015)
11. Sewwandi, D., Perera, K., Sandaruwan, S., Lakchani, O., Nugaliyadde, A., Thelijjagoda, S.: Linguistic features based personality recognition using social media data. In: 6th National Conference on Technology and Management (NCTM), pp. 63–68. IEEE (2017)
12. LSTM Networks. http://colah.github.io/posts/2015-08-Understanding-LSTMs/. Accessed 11 Aug 2018

Prediction-Based Reversible Watermarking for Safe Transfer of CT Scan Medical Images

Nisar Ahmed Memon$^{(\boxtimes)}$ and Shakeel Ahmed

College of Computer Science and Information Technology (CCSIT), King Faisal University, Al-Ahsa, Kingdom of Saudi Arabia
{nmemon, shakeel}@kfu.edu.sa

Abstract. Due to rapid development in the field of computer networks, exchange of medical records over the Internet has become common practice now-a-days. Medical records, which contain patient information and medical images, are exchanged among medical personnel for number of reasons. Discussing diagnostic and therapeutic measures is one of the main reasons of this exchange. Since patients are highly sensitive about their medical information; therefore these images require strict security. To achieve this stringent requirement digital image watermarking provides a good solution. In this paper, we present a prediction-based reversible medical image watermarking (MIW) system, which is comprised of two phases: embedding phase and extraction phase. In embedding phase, the proposed system, first divides the cover image in region of interest (ROI) and region of non-interest (RONI). Later it embeds a fragile watermark (*FW*) in ROI for achieving integrity control of ROI and a robust watermark (*RW*) in RONI for achieving security and authenticity of medical image. In extraction phase, the implanted watermarks are extracted first and then are compared with their reference watermarks to verify the integrity, security and confidentiality of the received image. We have used CT scan medical images for simulations. Experimental results show that proposed system provides better security and confidentiality for safe transfer of CT scan medical images. In addition, proposed system also outperforms the other reversible MIW techniques, which are currently reported in literature.

Keywords: Prediction-based reversible watermarking · Fragile watermarking · Robust watermarking · Authenticity · Confidentiality · LSB technique

1 Introduction

Transfer of medical information on electronic media, like the Internet have become a common practice now-a-days due to the fast expansion and rapid development in the fields of telecommunication and information technology [1]. Doctors and medical practitioners send and receive medical images and information related to patients from one hospital to another hospital located in different locations around the globe for number of reasons [2]. For example teleconferences among doctors, interdisciplinary exchange between radiologists for consultative purpose or for discussing the diagnostic and therapeutic measures. However, this transfer of medical information causes two main limitations: (i) Confidentiality: A proof that patient information is secure and has

© Springer Nature Switzerland AG 2020
K. Arai et al. (Eds.): FTC 2019, AISC 1069, pp. 737–749, 2020.
https://doi.org/10.1007/978-3-030-32520-6_54

not been accessed by any unauthorized user; and (ii) Integrity: A proof that patient information has not been tampered by invader [3].

In order to provide the facility for sharing of medical images and remote handling of patient record, MIW guarantees attractive properties [4–7]. In MIW techniques hidden information also known as watermark is embedded in the medical image in such a manner that it is attached to its original content, which is identified and recovered later [8]. A large number of watermarking techniques has been reported in literature for medical images. Some recently developed MIW techniques are described over here.

A reversible block based MIW method has been developed by Liew et al. [9]. The method first divides input image into ROI and RONI regions, which are then subdivided into non-overlapping blocks of sizes 8×8 and 6×6, respectively. The restoration information of each ROI block is embedded into its mapped block. Later, the method reported in [10] is used for tamper detection inside ROI and for recovering original ROI. In order to make the scheme reversible, the original bits stored inside RONI are used to replace the LSBs of pixels inside ROI. The main limitations of this system are: (1) It will not be possible to recover original ROI if both original block and its mapped block are tampered. (2) No authentication data is used by the method to check directly the tampered areas in ROI.

Tjokorda et al. [11] has reported another reversible MIW method in which ROI size is considered larger than RONI size. Before dividing the input image into ROI and RONI regions, first the original LSBs of the input image are stored in separate store and then are set to zero. After then, ROI and RONI regions are further partitioned into blocks of sizes 6×6 and 6×1, respectively. Each ROI block is mapped to another ROI block to store the recovery information. Using Run Length Encoding (RLE) technique, the originally stored LSBs are compressed and are implanted into 2 LSBs of 6×1 blocks in RONI. In the extraction phase, the watermarked medical image is segmented into ROI and RONI as done in embedding procedure. Then, tamper detection process is done inside ROI and original ROI is recovered by applying the method as proposed in [10]. The original medical image is obtained by extracting the original LSBs from RONI and restoring to their positions. This method also has same drawbacks as reported in [9] given above.

Al-Qershi et al. in [12] have presented another MIW method. In embedding phase, first input image is divided into ROI and RONI regions. Later, both the patient information and hash value of ROI are implanted into ROI. Then by using the technique of Tian, compressed form of ROI, average values of blocks inside ROI, embedding map of ROI, are all implanted into RONI. At extraction phase, information related to ROI is extracted from a secret area and is employed to identify ROI and RONI regions. Two main limitations of the technique are (1) extracting the data from RONI without knowing embedding map of RONI; and (2) using compressed form of ROI as recovery data for the ROI.

In [13], a reversible MIW technique has been developed by Deng et al. which employs quadtree composition. Input image having high level of homogeneity is segmented into blocks based on quadtree decomposition. Restoring information for each block is calculated using linear interpolation. Both restoring and quadtree information are implanted as first and second layer, respectively, in the image using integer wavelet transform. One main limitation of this method is that if single block is modified during exchange of image, it is not possible to recover the input image to its pristine state.

Thus important objective of this study is to present a watermarking system which can overcome the limitations of the techniques mentioned above. In addition, it can facilitate the doctors and the medical practitioners so that they can exchange the medical images without the fear of loss of medical information. Rest of the paper is organized as follows. Section 2 contains description of the tools and techniques used in the proposed system. Section 3 contains the detailed description of the system model being used for the sake of embedding and extracting the watermark into the host medical image in a step-by-step way. Section 4 contains the simulation results after experiments of the proposed system and Sect. 5 concludes the paper.

2 Prediction-Based Reversible Watermarking

We have used the prediction-based reversible watermarking which does not require a location map for reversibility. Thus, proposed system does not embed the substantial overhead information, which is usually required in reversible watermarking techniques. The proposed system not only increases the embedding capacity for embedding more patient information in the input image but also at the same time provides the method for conveying the patient information losslessly. The number of authors have proposed prediction-based reversible watermarking techniques [14–16]. In proposed system the prediction-based watermarking is employed as follows:

2.1 Embedding Procedure

1. The host image is first segmented into ROI and RONI regions.
2. Excluding the first row and first column, the RONI part of input image is scanned in raster order and pixel, P, is predicted as follows:

$$P'(i, j) = \left\lfloor \frac{P(i - 1, j) + P(i, j - 1)}{2} \right\rfloor \tag{1}$$

where $\lfloor . \rfloor$ denotes floor function. The prediction error, d, of each pixel is found by Eq. 2:

$$d = |P(i, j) - P'(i, j)| \tag{2}$$

3. Based on the value of T, d categorized in two cases:

$d \leq T$ and $d > T$, where first condition carry one bit of watermark information b and second condition do not. Here T is some predefined threshold.
Case 1: If $d \leq T$ then watermarked pixel is calculated by Eq. 3:

$$P^w(i, j) = \begin{cases} P'(i, j) + 2 \times d + b, & if \ P'(i, j) \leq P(i, j) \\ P'(i, j) - 2 \times d - b, & otherwise. \end{cases} \tag{3}$$

where $b \in \{0, 1\}$ denotes one bit of hidden information.

Case 2: If $d > T$ then P^w is computed by shifting the pixel $P(i, j)$ by Δ with no data implanted in it, by Eq. 4:

$$P^w(i, j) = \begin{cases} P'(i, j) + \Delta, & \text{if } P'(i, j) \leq P(i, j) \\ P'(i, j) - \Delta, & \text{otherwise.} \end{cases} \tag{4}$$

where $\Delta = T + 1$

4. The parameter, T will be transmitted to the decoder for extraction of data.

2.2   Extraction and Restoration Procedure

1. Divide the watermarked image into WROI and WRONI regions.
2. Excluding the first row and first column, the WRONI is scanned in raster scan order. Each pixel is predicted by the following Eq. 5:

$$P'(i, j) = \left\lfloor \frac{P^w(i - 1, j) + P^w(i, j - 1)}{2} \right\rfloor \tag{5}$$

3. The prediction error d^w of watermarked pixel is calculated using Eq. 6:

$$d^w = |P^w(i, j) - P'(i, j)| \tag{6}$$

4. Based on value of T, d^w is divided in two categories:

Case 1: If $d^w \leq 2T + 1$ then embedded watermark information b is extracted by Eq. 7 and the $P(x, y)$ is restored by Eq. 8:

$$b = d^w - 2\left\lfloor \frac{d^w}{2} \right\rfloor \tag{7}$$

$$P(i, j) = \begin{cases} P'(i, j) + \left\lfloor \frac{d^w}{2} \right\rfloor, & \text{if } P'(i, j) \leq P^w(i, j) \\ P'(i, j) - \left\lfloor \frac{d^w}{2} \right\rfloor, & \text{otherwise.} \end{cases} \tag{8}$$

Case 2: If $d^w > 2T + 1$, then nothing will be retrieved and pixel is restored by Eq. 9:

$$P(i, j) = \begin{cases} P'(i, j) - \Delta, & \text{if } P'(i, j) \leq P^w(i, j) \\ P'(i, j) + \Delta, & \text{otherwise.} \end{cases} \tag{9}$$

2.3   Underflow and Overflow Problem

In 8-bit gray scale images, the minimum value that can be stored is 0 and maximum value that can be stored is 255. The condition of underflow or overflow occurs when gray value goes less than 0 or becomes greater than 255. As the payload is embedded in LSB of each pixel of RONI which may cause underflow or overflow. In order to avoid this underflow or overflow, the extreme values in the input medical image are narrow down by the value of Δ. Thus histogram of input image which is originally is in range of [0 255] becomes [Δ 255$-\Delta$].

2.4   Generation of Watermarks

The watermarks used in proposed system are generated by following the procedure as described below:

The Fragile Watermark (*FW*): The fragile watermark is composed of alphabetic character "x" and is depicted in Fig. 1.

The Robust Watermark (*RW*): The robust watermark is a combination of 4 different watermarks and each is described below:

Patient Record (PR): *PR* represents the particulars (Name, age, sex, etc.) and history of disease related to the patient. This patient related data is converted in binary form as described in [17].

Doctor's ID (DI): *DI* is the unique string of 16 character used by doctor to identify that who is the creator or producer of the image.

Hospital Logo (HL): *HL* is logo which serves for the ownership of image. We have used the Logo as depicted in Fig. 2. The size of *HL* depends on the user. However, in our experiments the size 32 × 32 bits is used for logo.

LSB Information (LI): The *LI* is the collections of bits related to first bit-plane of ROI. The size of *LI* is actually ROI oriented. Higher the ROI, higher will be the size of *LI*.

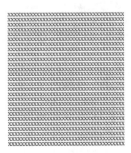

Fig. 1. The binary pattern embedded in ROI

Fig. 2. Logo used for hospital.

2.5   Segmentation of Medical Image

Different doctors divide the medical image into ROI and RONI regions according to their own choice. Some use geometrical shapes, like, square, ellipse or rectangle for defining ROI, other use free polygon. Since proposed system focus on CT Scan images in which lung parenchyma is usually the region of interest for a doctor. Therefore, for dividing the input CT Scan image in ROI and RONI regions, we have selected the segmentation algorithm as defined in [17].

3   The Proposed Watermarking System

A MIW system is proposed, which is based on image segmentation, prediction-based reversible watermarking technique and least significant bit substitution method. The system contains two phases: (i) watermark embedding, and (ii) watermark extraction and image recovery. In watermark embedding phase two separate watermarks namely fragile watermark and robust watermark are embedded in cover image after dividing the cover image in ROI and RONI regions. The *FW* is a binary pattern, which is casted in ROI by simple LSB substitution method, while *RW* is a composite watermark and is implanted in RONI by prediction-based reversible watermarking method. The purpose of *FW* is to identify the tampered areas if the tampering is done by invader during transmission of the image. Thus main purpose of *FW* is to get integrity control of ROI, while the main purpose of *RW* is to verify the authenticity and ownership issues. At extraction side, after dividing the watermarked image into watermarked region of interest (WROI) and watermarked region of non-interest (WRONI), the implanted watermarks are extracted. Each pixel in WRONI is restored to its original value after extracting the *RW* and it is brought back to its pristine state. After extraction of watermark bits from ROI same binary pattern is generated for visual inspection to verify the integrity of ROI while watermark information extracted from WRONI is divided in 4 different watermarks according to their predefined lengths and are compared with their corresponding counterparts to check the authenticity and copyright protection of the received image. After verification, *LI* information is set back to 1st bit-plane to restore ROI and then ROI and RONI are combined together to obtain the original image. As described in [18], two broad categories of MIW techniques are reported in literature: (1) techniques that emphasis on electronic patient data hiding, and (2) techniques that emphasis on tamper localization and authentication. Our proposed system represents the second category of MIW techniques.

3.1   Watermark Embedding Phase

In this phase, the cover image is first divided in ROI and RONI regions by following the algorithm as described in Sect. 2.5 above. After division, watermark casting process is done separately in these regions. Each step of watermark casting process in this phase is described as given below:

1. Segment the cover image, I, into ROI and RONI regions.
2. Get LSBs of first bit-plane of ROI and store them separately.
3. Produce the watermarks FW and RW.
4. Embed FW into ROI by replacing each LSB of ROI pixel to obtain WROI.
5. Apply prediction-based reversible watermarking on RONI to embed the RW to obtain WRONI.
6. To get watermarked image, I^w, combine both WROI and WRONI.

3.2   Watermark Extraction and Image Restoration Phase

The main steps of this phase are defined as under:

1. Segment the watermarked image, I^w, into WROI and WRONI regions.
2. Apply prediction-based reversible watermarking for extracting the watermark information (RW') and restore the WRONI to its original state, i.e. RONI.
3. Separate all watermarks from RW' to their predefined lengths.
4. Now compare these watermarks with their original watermarks for verification.
5. Replace the LSBs of WROI with extracted LI' to get original ROI.
6. Rejoin ROI and RONI to obtain the original image.

4   Experimental Results and Discussions

A large number of experiments were carried out on CT Scan images to test the validity of the proposed system. For simulations the original image was resized to 256×256 pixels. The results of experiment performed on some medical image database containing 60 CT Scan slices has been described as given below.

First the input image is divided in ROI and RONI regions by applying the algorithm as described in Sect. 2.5 above. The average segmentation time for each slice was about 12.05 s. Figure 3(a) depicts the input image, while Fig. 3(b) and (c) show the ROI and RONI regions, respectively. After then, watermarks are generated as described in Sect. 2.4 above and are embedded in input CT scan image. FW is inserted in the ROI by replacing the LSBs of ROI. After segmentation process, a binary mask has been produced for differentiating ROI and RONI areas. In binary mask all pixels belong to ROI are set as 1 (white area), while all pixels belonging to RONI are set as zero (black area). The binary mask is shown in Fig. 4.

The RW is inserted in the RONI by considering the same binary mask. All pixel values in cover image corresponding to black pixel values in binary mask are selected for inserting the RW in RONI. The prediction-based reversible watermarking has been employed for this purpose. We have used T = 2 for our simulations. Figure 5(a) to (c) shows the residual images of ROI, RONI and original input image, respectively, after watermarking.

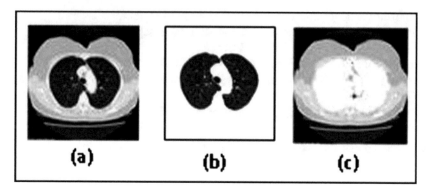

Fig. 3. (a) Original image (b) ROI (c) RONI

Fig. 4. Binary mask

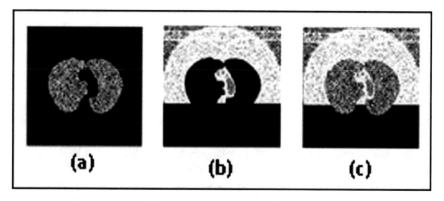

Fig. 5. (a) Residual image of ROI (b) Residual image of RONI (c) Residual image of original and watermarked image.

Table 1 shows the simulation results for CT Scan image database related of model patient. In Table 1, the results for every 5th slice from 1-60 slices are presented. Columns 1 to 6 in Table 1, show the Slice number, bits embedded as fragile water-mark, bits embedded as robust watermark, the total number of bits embedded in input image, total payload and the peak signal to noise ratio (PSNR) for every 5th slice, respectively. The PSNR is calculated using the Eqs. 10 and 11.

Table 1. Simulation results after embedding the watermark in cover image

Slice #	FW (bits)	RW (bits)	Total watermark (bits)	Payload (bpp)	PSNR (dB)
1	0	6144	6144	0.0937	44.2262
5	0	6144	6144	0.0937	44.2262
10	7598	13742	21340	0.3256	42.4991
15	13022	19166	32188	0.4911	41.7261
20	15040	21184	36224	0.5527	41.3044
25	15135	21279	36414	0.5556	41.1856
30	14203	20347	34550	0.5272	41.3250
35	12830	18974	31804	0.4853	41.6786
40	12103	18247	30350	0.4631	41.7562
45	7857	14001	21858	0.3335	42.9095
50	1737	7881	9618	0.1467	43.7137
55	0	6144	6144	0.0937	44.2457
60	0	6144	6144	0.0937	44.2457

In Eq. 10, R and MSE represents the maximum gray scale value in input image and mean square error, respectively. While in Eq. 11, I is the input medical image and I^w is the watermarked image.

$$PSNR = 10 \log_{10} \frac{R^2}{MSE} \qquad (10)$$

$$MSE = \frac{\sum_{M,N} [I(m,n) - I^w(m,n)]^2}{M \times N} \qquad (11)$$

In columns 2–3 of Table 1, the given size of FW and RW is actually depends on the size of ROI. This is due to the structure of lung parenchyma in human body. It has been observed that the size of ROI in the beginning and last slices is always smaller than the slices represent the middle part of lung parenchyma [17]. Thus, starting and last slices have less number of pixels in ROI, whereas in middle slices there is larger number of pixels in the ROI. The bar graph depicted in Fig. 6 shows such behavior for all 60 slices. The bar graph visualizes that staring slices (slice# 1–5) and last slices (slice# 53–60) show zero number of pixels in ROI whereas, in middle slices (Slice# 6–52) there is varying size of ROI. Due to this varying size behavior of ROI, less number watermark bits are embedded in starting and end slices, and large number of watermark information is embedded in middle slices.

This fact can also be observed from Fig. 7, which shows the payload v/s PSNR graph. The graph reveals that for slices having lower ROI, less payload is embedded

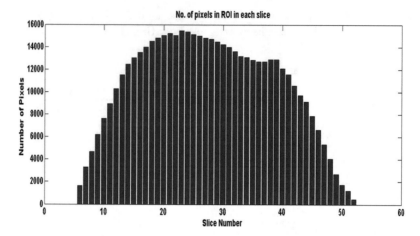

Fig. 6. Number of pixels in ROI in each slice

and thus having high value of PSNR. Whereas, the slices having higher ROI area, high payload has been embedded in those slices and having lower value of PSNR.

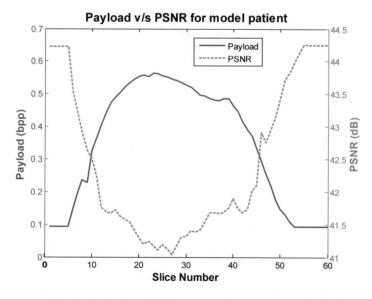

Fig. 7. Payload v/s PSNR for each slice of model patient

At extraction and restoration phase, first watermarked image was divided in WROI and WRONI regions. Then *FW* was extracted from WROI. Both patterns were found same and there was no any difference in original binary pattern (*LI*) and extracted binary pattern (*LI'*). Thus it was concluded that the content of ROI was intact and no attack was conducted on watermarked image during transmission. The extracted binary pattern is shown in Fig. 8. After that prediction-based reversible watermarking was used to recover the *RW'* from WRONI. During extraction of *RW'* each pixel was also restored to its original position which gives back original RONI. Further *RW'* was subdivided into *PR', DI', HL'* and *LI'* according to their predefined lengths. These extracted watermarks are then compared with their original counterparts by using the Normalized Hamming Distance (NHD) [3] measure. The NHD defines that Lower the value of distance measure, higher the percentage of accuracy between the embedded and extracted watermark. The distance of 0 defines the 100% accurqacy. In our case, distance of zero was found between the original (*PR, DI* and *HL* and *LI*) and extracted (*PR', DI'* and *HL'* and *LI'*) watermarks. Thus authenticity and confidentiality of received image was ensured. For restoring the image, first *LI'* information extracted from WRONI was set back into WROI by replacing its first biplane to get the original ROI. Finally restored ROI was combined with restored RONI. Thus input image was completely recovered to its pristine state.

At th end, we have compared the proposed method with other reversible watermarking techniques recently reported in literature as mentioned in Sect. 1 above. The comparison is shown in Table 2 which reveals that our proposed system outperforms and removes the deficiencies of these techniques as well.

Fig. 8. The extracted binary pattern

Table 2. Comparison of proposed method with other techniques.

Name of technique	ROI based	Embedding distortion inside ROI	Spotting tampers inside ROI	Accurate identification of tampered blocks	Recovery of tampered blocks or image
Liew et al. [9]	Yes	Yes	No	No	With average intensity of blocks
Tlokrda et al. [11]	Yes	Yes	No	No	With average intensity of blocks
Al-Qershi et al. [12]	Yes	Yes	No	No	With compressed form of ROI
Deng et al. [13]	No	–	No	No	With linear interpolation of pixel of blocks
Proposed	Yes	No	Yes	Yes	Pixels brought from RONI

5 Conclusions

A medical image watermarking system is presented which facilitates the medical doctor to transmit CT Scan images from one place to another without any fear of loss of information. By using the proposed system, the integrity, authenticity and confidentiality of CT scan image databases can be ensured during transmission. The integrity of medical image has been ensured by fragile watermarking while authenticity and confidentiality is ensured by robust watermarking. The computational cost of the proposed system is also very low. Thus proposed system leads to very efficient system which provides complete security and protection of CT scan images at very low cost. Thus proposed system can be used by low budget hospitals. The visual quality of watermarked image is also reasonable as witnessed by minimum value of PSNR of 41.18 dB as given in simulations above, which is greater than the benchmark 38.00 dB. Number of non-malicious and malicious attacks can be performed on the watermarked images. The proposed system is vulnerable to these attacks. Thus future directions can be developing a system which contains watermark that can withstand against non-malicious or friendly attacks and resist the malicious attacks.

Acknowledgment. The authors are very grateful to College of Computer Science and Information Technology (CCSIT), King Faisal University, Saudi Arabia for providing the resources to carry out this research.

References

1. Jamali, M., Samavi, S., Karimi, N., Soroushmehr, S.M.R., Ward, K. Najarian, K.: Robust watermarking in non-ROI of medical images based on DCT-DWT. In: 38th IEEE Annual International Conference on Engineering in Medicine and Biology Society (EMBC), Orlando, FL, USA, pp. 1200–1203 (2016)
2. Abbasi, F., Memon, N.A.: Reversible watermarking for security of medical image databases. In: 21st Saudi Computer Society National Computer Conference (CNN), Riyadh, Kingdom of Saudi Arabia (2018)
3. Memon, N.A., Chaudhary, A., Ahmad, M., Keerio, Z.A.: Hybrid watermarking of medical images for ROI authentication and recovery. Int. J. Comput. Math. **88**(10), 2057–2071 (2011)
4. Rahman, A., Sultan, K., Aldhafferi, N., Alqahtani, A., Mahmud, M.: Reversible and fragile watermarking for medical images. Comput. Math. Methods Med. **2018**, 7 (2018)
5. Qasim, A.F., Aspin, R., Meziane, F.: ROI-based reversible watermarking scheme for ensuring the integrity and authenticity of DICOM MR images. Multimedia Tools Appl. **78**, 16433–16463 (2018)
6. Mousavi, S.M., Naghsh, A., Abu-Bakar, S.A.R.: Watermarking techniques used in medical images: a survey. J. Digit. Imaging **27**, 714–729 (2014)
7. Parah, S.A., Ahad, F., Shaikh, J.A., Bhat, G.M.: Hiding clinical information in medical images: a new high capacity and reversible data hiding technique. J. Biomed. Inform. **66** (2017), 214–230 (2017)
8. Alsaade, F., Bhuiyan, A.: Digital watermarking for medical diagnosis. J. Appl. Sci. **15**(8), 1112–1119 (2015)

9. Liew, S.-C., Zain, J.M.: Reversible medical image watermarking for tamper detection and recovery with Run Length Encoding compression. In: International Conference on Computer Science and Information Technology (ICCSIT 2010), pp. 417–420 (2010)

10. Zain, J.M., Fauzi, A.R.M.: Medical image watermarking with tamper detection and recovery. In: Proceedings of the 28th Annual International Conference of the IEEE Engineering in Medicine and Biology Society (EMBS 2006), pp. 3270–3273 (2006)

11. Tjokorda Agung, B.W., Permana, F.P.: Medical image watermarking with tamper detection and recovery using reversible watermarking with LSB modification and run length encoding (RLE) compression. In: Proceedings of the IEEE International Conference on Communication, Networks and Satellite (ComNetSat 2012), Bali, Indonesia, pp. 167–171 (2012)

12. Al-Qershi, O.M., Khoo, B.E.: Authentication and data hiding using a reversible ROI-based watermarking scheme for DICOM images. In: Proceedings of International Conference on Medical Systems Engineering (ICEMSE 2009), pp. 829–834 (2009)

13. Deng, X., Chen, Z., Zeng, F., Zhang, Y., Mao, Y.: Authentication and recovery of medical diagnostic image using dual reversible digital watermarking. J. Nanosci. Nanotechnol. 13(3), 2099–2107 (2013)

14. Tseng, H.W., Hsieh, C.P.: Prediction-based reversible data-hiding. J. Inform. Sci. 179(14), 2460–2469 (2010)

15. Abokhdair, N.O., Manaf, A.A.: A prediction-based reversible watermarking for MRI images. Int. J. Comput. Electr. Autom. Control Inf. Eng. 7(2), 328–331 (2013)

16. Lee, C.-F., Chen, H.-L., Tso, H.-K.: Embedding capacity rasing in reversible data hiding based on prediction of difference expansion. J. Syst. Softw. 83, 1864–1872 (2010)

17. Memon, N.A., Keerio, Z.A., Abbasi, F.: Dual watermarking of CT scan medical images for content authentication and copyright protection. In: Shaikh, F.K., Chaudhary, B.S. (eds.) Communication Technologies, Information Security and Sustainable Development, IMTIC 2013. CCIS, vol. 14, pp. 73–183. Springer, Cham (2014)

18. Navas, K.A., Sasikumar, M.: Survey of medical image watermarking algorithms. In: Proceedings of the 4th International Conference on Sciences of Electronic Technologies of Informations and Telecommunications, pp. 1–6 (2007)

Diagnosis of Celiac Disease and Environmental Enteropathy on Biopsy Images Using Color Balancing on Convolutional Neural Networks

Kamran Kowsari[1], Rasoul Sali[1], Marium N. Khan[3], William Adorno[1], S. Asad Ali[4], Sean R. Moore[3], Beatrice C. Amadi[5], Paul Kelly[5,6], Sana Syed[2,4(✉)], and Donald E. Brown[1,2(✉)]

[1] Department of Systems and Information Engineering, University of Virginia, Charlottesville, VA, USA
[2] School of Data Science, University of Virginia, Charlottesville, VA, USA
{sana.syed,deb}@virginia.edu
[3] Department of Pediatrics, University of Virginia, Charlottesville, VA, USA
[4] Department of Pediatrics and Child Health, Aga Khan University, Karachi, Pakistan
[5] Tropical Gastroenterology and Nutrition Group, University of Zambia School of Medicine, Lusaka, Zambia
[6] Blizard Institute, Barts and The London School of Medicine, Queen Mary University of London, London, UK

Abstract. Celiac Disease (CD) and Environmental Enteropathy (EE) are common causes of malnutrition and adversely impact normal childhood development. CD is an autoimmune disorder that is prevalent worldwide and is caused by an increased sensitivity to gluten. Gluten exposure destructs the small intestinal epithelial barrier, resulting in nutrient mal-absorption and childhood under-nutrition. EE also results in barrier dysfunction but is thought to be caused by an increased vulnerability to infections. EE has been implicated as the predominant cause of under-nutrition, oral vaccine failure, and impaired cognitive development in low-and-middle-income countries. Both conditions require a tissue biopsy for diagnosis, and a major challenge of interpreting clinical biopsy images to differentiate between these gastrointestinal diseases is striking histopathologic overlap between them. In the current study, we propose a convolutional neural network (CNN) to classify duodenal biopsy images from subjects with CD, EE, and healthy controls. We evaluated the performance of our proposed model using a large cohort containing 1000 biopsy images. Our evaluations show that the proposed model achieves an area under ROC of 0.99, 1.00, and 0.97 for CD, EE, and healthy controls, respectively. These results demonstrate the discriminative power of the proposed model in duodenal biopsies classification.

Keywords: Convolutional neural networks · Medical imaging · Celiac Disease · Environmental Enteropathy

© Springer Nature Switzerland AG 2020
K. Arai et al. (Eds.): FTC 2019, AISC 1069, pp. 750–765, 2020.
https://doi.org/10.1007/978-3-030-32520-6_55

1 Introduction and Related Works

Under-nutrition is the underlying cause of approximately 45% of the 5 million under 5-year-old childhood deaths annually in low and middle-income countries (LMICs) [1] and is a major cause of mortality in this population. Linear growth failure (or stunting) is a major complication of under-nutrition, and is associated with irreversible physical and cognitive deficits, with profound developmental implications [32]. A common cause of stunting in LMICs is EE, for which there are no universally accepted, clear diagnostic algorithms or non-invasive biomarkers for accurate diagnosis [32], making this a critical priority [28]. EE has been described to be caused by chronic exposure to enteropathogens which results in a vicious cycle of constant mucosal inflammation, villous blunting, and a damaged epithelium [32]. These deficiencies contribute to a markedly reduced nutrient absorption and thus under-nutrition and stunting [32]. Interestingly, CD, a common cause of stunting in the United States, with an estimated 1% prevalence, is an autoimmune disorder caused by a gluten sensitivity [15] and has many shared histological features with EE (such as increased inflammatory cells and villous blunting) [32]. This resemblance has led to the major challenge of differentiating clinical biopsy images for these similar but distinct diseases. Therefore, there is a major clinical interest towards developing new, innovative methods to automate and enhance the detection of morphological features of EE versus CD, and to differentiate between diseased and healthy small intestinal tissue [4].

The overview of the methodology used is shown in Fig. 1.

In this paper, we propose a CNN-based model for classification of biopsy images. In recent years, Deep Learning architectures have received great attention after achieving state-of-the-art results in a wide variety of fundamental tasks such classification [13,18–20,24,29,35] or other medical domains [12,36]. CNNs in particular have proven to be very effective in medical image processing. CNNs preserve local image relations, while reducing dimensionality and for this reason are the most popular machine learning algorithm in image recognition and visual learning tasks [16]. CNNs have been widely used for classification and segmentation in various types of medical applications such as histopathological images of breast tissues, lung images, MRI images, medical X-Ray images, etc. [11,24]. Researchers produced advanced results on duodenal biopsies classification using CNNs [3], but those models are only robust to a single type of image stain or color distribution. Many researchers apply a stain normalization technique as part of the image pre-processing stage to both the training and validation datasets [27]. In this paper, varying levels of color balancing were applied during image pre-processing in order to account for multiple stain variations.

The rest of this paper is organized as follows: In Sect. 2, we describe the different data sets used in this work, as well as, the required pre-processing steps. The architecture of the model is explained in Sect. 4. Empirical results are elaborated in Sect. 5. Finally, Sect. 6 concludes the paper along with outlining future directions.

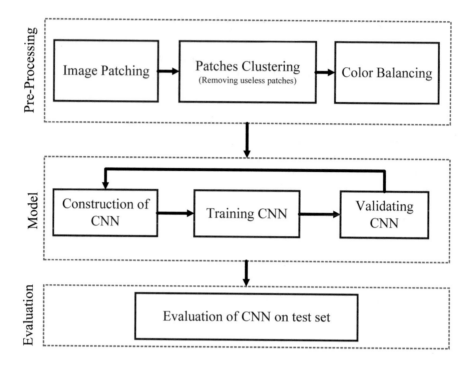

Fig. 1. Overview of methodology

2   Data Source

For this project, 121 Hematoxylin and Eosin (H&E) stained duodenal biopsy glass slides were retrieved from 102 patients. The slides were converted into 3118 whole slide images, and labeled as either EE, CD, or normal. The biopsy slides for EE patients were from the Aga Khan University Hospital (AKUH) in Karachi, Pakistan ($n = 29$ slides from 10 patients) and the University of Zambia Medical Center in Lusaka, Zambia ($n = 16$). The slides for CD patients ($n = 34$) and normal ($n = 42$) were retrieved from archives at the University of Virginia (UVa). The CD and normal slides were converted into whole slide images at 40x magnification using the Leica SCN 400 slide scanner (Meyer Instruments, Houston, TX) at UVa, and the digitized EE slides were of 20x magnification and shared via the Environmental Enteric Dysfunction Biopsy Investigators (EEDBI) Consortium shared WUPAX server. Characteristics of our patient population are as follows: the median ($Q1, Q3$) age of our entire study population was 31 (20.25, 75.5) months, and we had a roughly equal distribution of males (52%, $n = 53$) and females (48%, $n = 49$). The majority of our study population were histologically normal controls (41.2%), followed by CD patients (33.3%), and EE patients (25.5%).

3   Pre-processing

In this section, we cover all of the pre-processing steps which include image patching, image clustering, and color balancing. The biopsy images are unstructured (varying image sizes) and too large to process with deep neural networks; thus, requiring that images are split into multiple smaller images. After executing the split, some of the images do not contain much useful information. For instance, some only contain the mostly blank border region of the original image. In the image clustering section, the process to select useful images is described. Finally, color balancing is used to correct for varying color stains which is a common issue in histological image processing.

3.1   Image Patching

Although effectiveness of CNNs in image classification has been shown in various studies in different domains, training on high resolution Whole Slide Tissue Images (WSI) is not commonly preferred due to a high computational cost. However, applying CNNs on WSI enables losing a large amount of discriminative information due to extensive downsampling [14]. Due to a cellular level difference between Celiac, Environmental Entropathy and normal cases, a trained classifier on image patches is likely to perform as well as or even better than a trained WSI-level classifier. Many researchers in pathology image analysis have considered classification or feature extraction on image patches [14]. In this project, after generating patches from each images, labels were applied to each patch according to its associated original image. A CNN was trained to generate predictions on each individual patch.

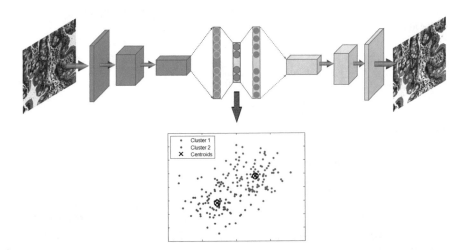

Fig. 2. Structure of clustering model with autoencoder and K-means combination

Table 1. The clustering results for all patches into two clusters

	Total	Cluster 1	Cluster 2
Celiac Disease (CD)	16, 832	7, 742 (46%)	9, 090 (54%)
Normal	15, 983	8, 953 (56%)	7, 030 (44%)
Environmental Enteropathy (EE)	22, 625	2, 034 (9%)	20, 591 (91%)
Total	55, 440	18, 729 (34%)	36, 711 (66%)

3.2   Clustering

In this study, after image patching, some of created patches do not contain any useful information regarding biopsies and should be removed from the data. These patches have been created from mostly background parts of WSIs. A two-step clustering process was applied to identify the unimportant patches. For the first step, a convolutional autoencoder was used to learn embedded features of each patch and in the second step we used k-means to cluster embedded features into two clusters: useful and not useful. In Fig. 2, the pipeline of our clustering technique is shown which contains both the autoencoder and k-mean clustering.

An autoencoder is a type of neural network that is designed to match the model's inputs to the outputs [10]. The autoencoder has achieved great success as a dimensionality reduction method via the powerful reprehensibility of neural networks [33]. The first version of autoencoder was introduced by *DE. Rumelhart et al.* [30] in 1985. The main idea is that one hidden layer between input and output layers has much fewer units [23] and can be used to reduce the dimensions of a feature space. For medical images which typically contain many features, using an autoencoder can help allow for faster, more efficient data processing.

A CNN-based autoencoder can be divided into two main steps [25]: encoding and decoding.

$$O_m(i,j) = a\left(\sum_{d=1}^{D} \sum_{u=-2k-1}^{2k+1} \sum_{v=-2k-1}^{2k+1} F_{m_d}^{(1)}(u,v) I_d(i-u, j-v) \right)$$
$$m = 1, \cdots, n \tag{1}$$

Where $F \in \{F_1^{(1)}, F_2^{(1)}, \ldots, F_n^{(1)},\}$ is a convolutional filter, with convolution among an input volume defined by $I = \{I_1, \cdots, I_D\}$ which it learns to represent the input by combining non-linear functions:

$$z_m = O_m = a(I * F_m^{(1)} + b_m^{(1)}) \quad m = 1, \cdots, m \tag{2}$$

where $b_m^{(1)}$ is the bias, and the number of zeros we want to pad the input with is such that: $\dim(I) = \dim(\text{decode}(\text{encode}(I)))$ Finally, the encoding convolution is equal to:

$$O_w = O_h = (I_w + 2(2k+1) - 2) - (2k+1) + 1$$
$$= I_w + (2k+1) - 1 \tag{3}$$

The decoding convolution step produces n feature maps $z_{m=1,...,n}$. The reconstructed results $\hat{I}$ is the result of the convolution between the volume of feature maps $Z = \{z_{i=1}\}^n$ and this convolutional filters volume $F^{(2)}$ [7,9].

$$\tilde{I} = a(Z * F_m^{(2)} + b^{(2)}) \tag{4}$$

$$O_w = O_h = (I_w + (2k+1) - 1) - (2k+1) + 1 = I_w = I_h \tag{5}$$

Where Eq. 5 shows the decoding convolution with I dimensions. The input's dimensions are equal to the output's dimensions.

Results of patch clustering has been summarized in Table 1 and Fig. 3. Obviously, patches in cluster 1, which were deemed useful, are used for the analysis in this paper.

3.3   Color Balancing

The concept of color balancing for this paper is to convert all images to the same color space to account for variations in H&E staining. The images can be represented with the illuminant spectral power distribution $I(\lambda)$, the surface spectral reflectance $S(\lambda)$, and the sensor spectral sensitivities $C(\lambda)$ [5,6]. Using this notation [6], the sensor responses at the pixel with coordinates (x, y) can be thus described as:

$$p(x, y) = \int_w I(x, y, \lambda)S(x, y, \lambda)C(\lambda)d\lambda \tag{6}$$

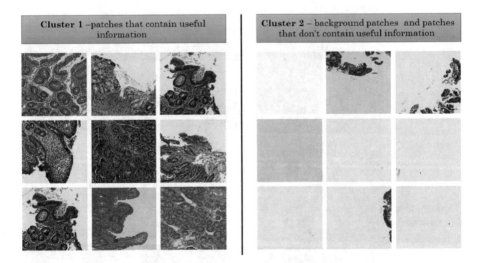

Fig. 3. Some samples of clustering results - cluster 1 includes patches with useful information and cluster 2 includes patches without useful information (mostly created from background parts of WSIs)

where w is the wavelength range of the visible light spectrum, ρ and $C(\lambda)$ are three-component vectors.

$$\begin{bmatrix} R \\ G \\ B \end{bmatrix}_{out} = \left(\alpha \begin{bmatrix} a_{11} & a_{12} & a_{13} \\ a_{21} & a_{22} & a_{23} \\ a_{31} & a_{32} & a_{33} \end{bmatrix} \times \begin{bmatrix} r_{awb} & 0 & 0 \\ 0 & g_{awb} & 0 \\ 0 & 0 & b_{awb} \end{bmatrix} \begin{bmatrix} R \\ G \\ B \end{bmatrix}_{in} \right)^{\gamma} \qquad (7)$$

where RGB_{in} is raw images from biopsy and RGB_{out} is results for CNN input. In the following, a more compact version of Eq. 7 is used:

$$RGB_{out} = (\alpha A I_w . RGB_{in})^{\gamma} \qquad (8)$$

where α is exposure compensation gain, I_w refers the diagonal matrix for the illuminant compensation and A indicates the color matrix transformation.

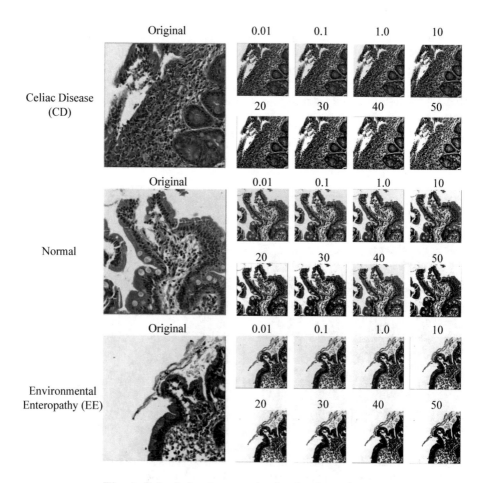

Fig. 4. Color balancing samples for the three classes

Figure 4 shows the results of color balancing for three classes (Celiac Disease (CD), Normal and Environmental Enteropathy (EE)) with different color balancing percentages between 0.01 and 50.

4   Method

In this section, we describe Convolutional Neural Networks (CNN) including the convolutional layers, pooling layers, activation functions, and optimizer. Then, we discuss our network architecture for diagnosis of Celiac Disease and Environmental Enteropathy. As shown in Fig. 5, the input layers starts with image patches (1000 × 1000) and is connected to the convolutional layer (*Conv* 1). Conv 1 is connected to the pooling layer (*MaxPooling*), and then connected to *Conv* 2. Finally, the last convolutional layer (*Conv* 3) is flattened and connected to a fully connected perception layer. The output layer contains three nodes which each node represent one class.

4.1   Convolutional Layer

CNN is a deep learning architecture that can be employed for hierarchical image classification. Originally, CNNs were built for image processing with an architecture similar to the visual cortex. CNNs have been used effectively for medical image processing. In a basic CNN used for image processing, an image tensor is convolved with a set of kernels of size $d \times d$. These convolution layers are called feature maps and can be stacked to provide multiple filters on the input. The element (feature) of input and output matrices can be different [22]. The process to compute a single output matrix is defined as follows:

$$A_j = f \left(\sum_{i=1}^{N} I_i * K_{i,j} + B_j \right) \tag{9}$$

Each input matrix $I-i$ is convolved with a corresponding kernel matrix $K_{i,j}$, and summed with a bias value B_j at each element. Finally, a non-linear activation function (see Sect. 4.3) is applied to each element [22].

In general, during the back propagation step of a CNN, the weights and biases are adjusted to create effective feature detection filters. The filters in the convolution layer are applied across all three 'channels' or Σ (size of the color space) [13].

4.2   Pooling Layer

To reduce the computational complexity, CNNs utilize the concept of pooling to reduce the size of the output from one layer to the next in the network. Different pooling techniques are used to reduce outputs while preserving important features [31]. The most common pooling method is max pooling where the maximum element is selected in the pooling window.

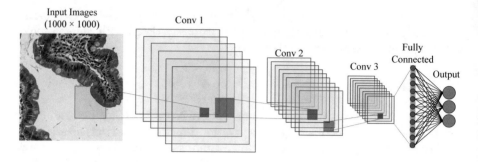

Fig. 5. Structure of convolutional neural net using multiple 2D feature detectors and 2D max-pooling

In order to feed the pooled output from stacked featured maps to the next layer, the maps are flattened into one column. The final layers in a CNN are typically fully connected [19].

4.3 Neuron Activation

The implementation of CNN is a discriminative trained model that uses standard back-propagation algorithm using a sigmoid (Eq. 10), (Rectified Linear Units (ReLU) [26] (Eq. 11) as activation function. The output layer for multiclass classification includes a *Softmax* function (as shown in Eq. 12).

$$f(x) = \frac{1}{1 + e^{-x}} \in (0, 1) \tag{10}$$

$$f(x) = \max(0, x) \tag{11}$$

$$\sigma(z)_j = \frac{e^{z_j}}{\sum_{k=1}^{K} e^{z_k}} \tag{12}$$

$$\forall \, j \in \{1, \dots, K\}$$

4.4 Optimizor

For this CNN architecture, the *Adam* optimizor [17] which is a stochastic gradient optimizer that uses only the average of the first two moments of gradient (v and m, shown in Eqs. 13, 14, 15 and 16). It can handle non-stationary of the objective function as in RMSProp, while overcoming the sparse gradient issue limitation of RMSProp [17].

$$\theta \leftarrow \theta - \frac{\alpha}{\sqrt{\hat{v}} + \epsilon} \hat{m} \tag{13}$$

$$g_{i,t} = \nabla_\theta J(\theta_i, x_i, y_i) \tag{14}$$

$$m_t = \beta_1 m_{t-1} + (1 - \beta_1) g_{i,t} \tag{15}$$

$$m_t = \beta_2 v_{t-1} + (1 - \beta_2) g_{i,t}^2 \tag{16}$$

where m_t is the first moment and v_t indicates second moment that both are estimated. $\hat{m}_t = \frac{m_t}{1-\beta_1^t}$ and $\hat{v}_t = \frac{v_t}{1-\beta_2^t}$.

4.5 Network Architecture

As shown in Table 2 and Fig. 6, our CNN architecture consists of three convolution layer each followed by a pooling layer. This model receives RGB image patches with dimensions of (1000×1000) as input. The first convolutional layer has 32 filters with kernel size of $(3,3)$. Then we have Pooling layer with size of $(5,5)$ which reduce the feature maps from (1000×1000) to (200×200). The second convolutional layers with 32 filters with kernel size of $(3,3)$. Then Pooling layer (MaxPooling 2D) with size of $(5,5)$ reduces the feature maps from (200×200) to (40×40). The third convolutional layer has 64 filters with kernel size of $(3,3)$, and final pooling layer (MaxPooling 2D) is scaled down to (8×8). The feature maps as shown in Table 2 is flatten and connected to fully connected layer with 128 nodes. The output layer with three nodes to represent the three classes: (Environmental Enteropathy, Celiac Disease, and Normal).

The optimizer used is Adam (see Sect. 4.4) with a learning rate of 0.001, $\beta_1 = 0.9$, $\beta_2 = 0.999$ and the loss considered is sparse categorical crossentropy [8]. Also for all layers, we use Rectified linear unit (ReLU) as activation function except output layer which we use *Softmax* (see Sect. 4.3).

Table 2. CNN architecture for diagnosis of diseased duodenal on biopsy images

	Layer (type)	Output shape	Trainable parameters
1	Convolutional layer	$(1000, 1000, 32)$	869
2	Max pooling	$(200, 200, 32)$	0
3	Convolutional layer	$(200, 200, 32)$	9, 248
4	Max pooling	$(40, 40, 32)$	0
5	Convolutional layer	$(40, 40, 64)$	18, 496
6	Max pooling	$(8, 8, 64)$	0
8	dense	128	524, 416
10	Output	3	387

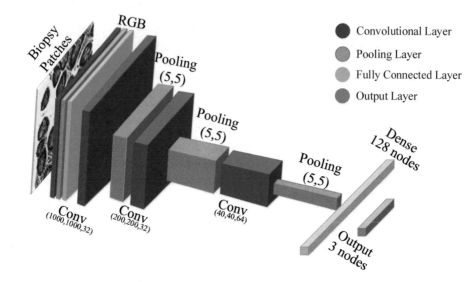

Fig. 6. Our convolutional neural networks' architecture

5 Empirical Results

5.1 Evaluation Setup

In the research community, comparable and shareable performance measures to evaluate algorithms are preferable. However, in reality such measures may only exist for a handful of methods. The major problem when evaluating image classification methods is the absence of standard data collection protocols. Even if a common collection method existed, simply choosing different training and test sets can introduce inconsistencies in model performance [34]. Another challenge with respect to method evaluation is being able to compare different performance measures used in separate experiments. Performance measures generally evaluate specific aspects of classification task performance, and thus do not always present identical information. In this section, we discuss evaluation metrics and performance measures and highlight ways in which the performance of classifiers can be compared.

Since the underlying mechanics of different evaluation metrics may vary, understanding what exactly each of these metrics represents and what kind of information they are trying to convey is crucial for comparability. Some examples of these metrics include recall, precision, accuracy, F-measure, micro-average, and macro-average. These metrics are based on a "confusion matrix" that comprises true positives (TP), false positives (FP), false negatives (FN) and true negatives (TN) [21]. The significance of these four elements may vary based on the classification application. The fraction of correct predictions over all predictions is called accuracy (Eq. 17). The proportion of correctly predicted positives to all positives is called precision, *i.e.* positive predictive value (Eq. 18).

$$accuracy = \frac{(TP + TN)}{(TP + FP + FN + TN)} \tag{17}$$

$$Precision = \frac{\sum_{l=1}^{L} TP_l}{\sum_{l=1}^{L} TP_l + FP_l} \tag{18}$$

$$Recall = \frac{\sum_{l=1}^{L} TP_l}{\sum_{l=1}^{L} TP_l + FN_l} \tag{19}$$

$$F1 - Score = \frac{\sum_{l=1}^{L} 2TP_l}{\sum_{l=1}^{L} 2TP_l + FP_l + FN_l} \tag{20}$$

5.2 Experimental Setup

The following results were obtained using a combination of central processing units (CPUs) and graphical processing units (GPUs). The processing was done on a *Xeon E*5 − 2640 (2.6 GHz) with 32 cores and 64 GB memory, and the GPU cards were two *Nvidia Titan Xp* and a *Nvidia Tesla K*20*c*. We implemented our approaches in Python using the Compute Unified Device Architecture (CUDA), which is a parallel computing platform and Application Programming Interface (API) model created by *Nvidia*. We also used Keras and TensorFlow libraries for creating the neural networks [2,8].

5.3 Experimental Results

In this section we show that CNN with color balancing can improve the robustness of medical image classification. The results for the model trained on 4 different color balancing values are shown in Table 3. The results shown in Table 4 are also based on the trained model using the same color balancing values. Although in Table 4, the test set is based on a different set of color balancing values: 0.5, 1.0, 1.5 and 2.0. By testing on a different set of color balancing, these results show that this technique can solve the issue of multiple stain variations during histological image analysis.

As shown in Table 3, the f1-score of three classes (Environmental Enteropathy (EE), Celiac Disease (CD), and Normal) are 0.98, 0.94, and 0.91 respectively. In Table 4, the f1-score is reduced, but not by a significant amount. The three classes (Environmental Enteropathy (EE), Celiac Disease (CD), and Normal) f1-scores are 0.94, 0.92, and 0.87 respectively. The result is very similar to *MA. Boni et al.* [3] which achieved 90.59% of accuracy in their mode, but without using the color balancing technique to allow differently stained images.

In Fig. 7, Receiver operating characteristics (ROC) curves are valuable graphical tools for evaluating classifiers. However, class imbalances (i.e. differences in prior class probabilities) can cause ROC curves to poorly represent the classifier performance. ROC curve plots true positive rate (TPR) and false positive rate (FPR). The ROC shows that AUC of Environmental Enteropathy (EE) is 1.00, Celiac Disease (CD) is 0.99, and Normal is 0.97.

Table 3. F1-score for train on a set with color balancing of 0.001, 0.01, 0.1, and 1.0. Then, we evaluate test set with same color balancing

	Precision	Recall	f1-score	Support
Celiac Disease (CD)	0.89	0.99	0.94	22,196
Normal	0.99	0.83	0.91	22,194
Environmental Enteropathy (EE)	0.96	1.00	0.98	22,198

Table 4. F1-score for train with color balancing of 0.001, 0.01, 0.1, and 1.0 and test with color balancing of 0.5, 1.0, 1.5 and 2.0

	Precision	Recall	f1-score	Support
Celiac Disease (CD)	0.90	0.94	0.92	22,196
Normal	0.96	0.80	0.87	22,194
Environmental Enteropathy (EE)	0.89	1.00	0.94	22,198

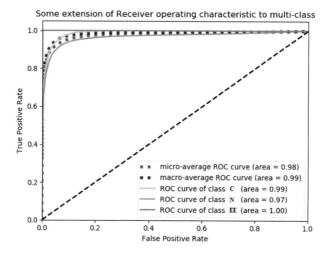

Fig. 7. Receiver operating characteristics (ROC) curves for three classes also the figure shows micro-average and macro-average of our classifier

As shown in Table 5, our model performs better compared to some other models in terms of accuracy. Among the compared models, only the fine-tuned ALEXNET [27] has considered the color staining problem. This model proposes a transfer learning based approach for the classification of stained histology images. They also applied stain normalization before using images for fine tuning the model.

Table 5. Comparison accuracy with different baseline methods

Method	Solve color staining problem	Model architecture	Accuracy
Shifting and reflections [3]	No	CNN	85.13%
Gamma [3]	No	CNN	90.59%
CLAHE [3]	No	CNN	86.79%
Gamma-CLAHE [3]	No	CNN	86.72%
Fine-tuned ALEXNET [27]	**Yes**	ALEXNET	89.95%
Ours	**Yes**	CNN	**93.39%**

6   Conclusion

In this paper, we proposed a data driven model for diagnosis of diseased duodenal architecture on biopsy images using color balancing on convolutional neural networks. Validation results of this model show that it can be utilized by pathologists in diagnostic operations regarding CD and EE. Furthermore, color consistency is an issue in digital histology images and different imaging systems reproduced the colors of a histological slide differently. Our results demonstrate that application of the color balancing technique can attenuate effect of this issue in image classification.

The methods described here can be improved in multiple ways. Additional training and testing with other color balancing techniques on data sets will continue to identify architectures that work best for these problems. Also, it is possible to extend the model to more than four different color balance percentages to capture more of the complexity in the medical image classification.

Acknowledgments. This research was supported by University of Virginia, Engineering in Medicine SEED Grant (*SS & DEB*), the University of Virginia Translational Health Research Institute of Virginia (*THRIV*) Mentored Career Development Award (*SS*), and the Bill and Melinda Gates Foundation (*AA, OPP*1138727; *SRM, OPP*1144149; *PK, OPP*1066118)

References

1. Who. children: reducing mortality. fact sheet 2017. http://www.who.int/mediacentre/factsheets/fs178/en/. Accessed 30 Jan 2019
2. Abadi, M., Agarwal, A., Barham, P., Brevdo, E., Chen, Z., Citro, C., Corrado, G.S., Davis, A., Dean, J., Devin, M., et al.: Tensorflow: large-scale machine learning on heterogeneous distributed systems. arXiv preprint arXiv:1603.04467 (2016)
3. Al Boni, M., Syed, S., Ali, A., Moore, S.R., Brown, D.E.: Duodenal biopsies classification and understanding using convolutional neural networks. American Medical Informatics Association (2019)
4. Bejnordi, B.E., Veta, M., Van Diest, P.J., Van Ginneken, B., Karssemeijer, N., Litjens, G., Van Der Laak, J.A., Hermsen, M., Manson, Q.F., Balkenhol, M., et al.: Diagnostic assessment of deep learning algorithms for detection of lymph node metastases in women with breast cancer. JAMA **318**(22), 2199–2210 (2017)

5. Bianco, S., Cusano, C., Napoletano, P., Schettini, R.: Improving CNN-based texture classification by color balancing. J. Imaging **3**(3), 33 (2017)
6. Bianco, S., Schettini, R.: Error-tolerant color rendering for digital cameras. J. Math. Imaging Vis. **50**(3), 235–245 (2014)
7. Chen, K., Seuret, M., Liwicki, M., Hennebert, J., Ingold, R.: Page segmentation of historical document images with convolutional autoencoders. In: 2015 13th International Conference on Document Analysis and Recognition (ICDAR), pp. 1011–1015. IEEE (2015)
8. Chollet, F., et al.: Keras: deep learning library for theano and tensorflow (2015). https://keras.io/
9. Geng, J., Fan, J., Wang, H., Ma, X., Li, B., Chen, F.: High-resolution sar image classification via deep convolutional autoencoders. IEEE Geosci. Remote Sens. Lett. **12**(11), 2351–2355 (2015)
10. Goodfellow, I., Bengio, Y., Courville, A., Bengio, Y.: Deep Learning, vol. 1. MIT Press, Cambridge (2016)
11. Gulshan, V., Peng, L., Coram, M., Stumpe, M.C., Wu, D., Narayanaswamy, A., Venugopalan, S., Widner, K., Madams, T., Cuadros, J., et al.: Development and validation of a deep learning algorithm for detection of diabetic retinopathy in retinal fundus photographs. JAMA **316**(22), 2402–2410 (2016)
12. Hegde, R.B., Prasad, K., Hebbar, H., Singh, B.M.K.: Comparison of traditional image processing and deep learning approaches for classification of white blood cells in peripheral blood smear images. Biocybern. Biomed. Eng. (2019)
13. Heidarysafa, M., Kowsari, K., Brown, D.E., Jafari Meimandi, K., Barnes, L.E.: An improvement of data classification using random multimodel deep learning (RMDL) **8**(4), 298–310 (2018). https://doi.org/10.18178/ijmlc.2018.8.4.703
14. Hou, L., Samaras, D., Kurc, T.M., Gao, Y., Davis, J.E., Saltz, J.H.: Patch-based convolutional neural network for whole slide tissue image classification. In: Proceedings of the IEEE Conference on Computer Vision and Pattern Recognition, pp. 2424–2433 (2016)
15. Husby, S., et al.: European society for pediatric gastroenterology, hepatology, and nutrition guidelines for the diagnosis of coeliac disease. J. Pediatr. Gastroenterol. Nutr. **54**(1), 136–160 (2012)
16. Ker, J., Wang, L., Rao, J., Lim, T.: Deep learning applications in medical image analysis. IEEE Access **6**, 9375–9389 (2018)
17. Kingma, D., Ba, J.: Adam: a method for stochastic optimization. arXiv preprint arXiv:1412.6980 (2014)
18. Kowsari, K., Brown, D.E., Heidarysafa, M., Meimandi, K.J., Gerber, M.S., Barnes, L.E.: HDLTex: hierarchical deep learning for text classification. In: 2017 16th IEEE International Conference on Machine Learning and Applications (ICMLA), pp. 364–371. IEEE (2017)
19. Kowsari, K., Heidarysafa, M., Brown, D.E., Meimandi, K.J., Barnes, L.E.: RMDL: random multimodel deep learning for classification. In: Proceedings of the 2nd International Conference on Information System and Data Mining, pp. 19–28. ACM (2018)
20. Kowsari, K., Jafari Meimandi, K., Heidarysafa, M., Mendu, S., Barnes, L., Brown, D.: Text classification algorithms: a survey. Information **10**(4) (2019). https://doi.org/10.3390/info10040150
21. Lever, J., Krzywinski, M., Altman, N.: Points of significance: classification evaluation (2016)

22. Li, Q., Cai, W., Wang, X., Zhou, Y., Feng, D.D., Chen, M.: Medical image classification with convolutional neural network. In: 2014 13th International Conference on Control Automation Robotics & Vision (ICARCV), pp. 844–848. IEEE (2014)
23. Liang, H., Sun, X., Sun, Y., Gao, Y.: Text feature extraction based on deep learning: a review. EURASIP J. Wirel. Commun. Networking **2017**(1), 211 (2017)
24. Litjens, G., Kooi, T., Bejnordi, B.E., Setio, A.A.A., Ciompi, F., Ghafoorian, M., Van Der Laak, J.A., Van Ginneken, B., Sánchez, C.I.: A survey on deep learning in medical image analysis. Med. Image Anal. **42**, 60–88 (2017)
25. Masci, J., Meier, U., Cireşan, D., Schmidhuber, J.: Stacked convolutional autoencoders for hierarchical feature extraction. In: International Conference on Artificial Neural Networks, pp. 52–59. Springer (2011)
26. Nair, V., Hinton, G.E.: Rectified linear units improve restricted Boltzmann machines. In: Proceedings of the 27th International Conference on Machine Learning (ICML 2010), pp. 807–814 (2010)
27. Nawaz, W., Ahmed, S., Tahir, A., Khan, H.A.: Classification of breast cancer histology images using ALEXNET. In: International Conference Image Analysis and Recognition, pp. 869–876. Springer (2018)
28. Naylor, C., Lu, M., Haque, R., Mondal, D., Buonomo, E., Nayak, U., Mychaleckyj, J.C., Kirkpatrick, B., Colgate, R., Carmolli, M., et al.: Environmental enteropathy, oral vaccine failure and growth faltering in infants in bangladesh. EBioMedicine **2**(11), 1759–1766 (2015)
29. Nobles, A.L., Glenn, J.J., Kowsari, K., Teachman, B.A., Barnes, L.E.: Identification of imminent suicide risk among young adults using text messages. In: Proceedings of the 2018 CHI Conference on Human Factors in Computing Systems, p. 413. ACM (2018)
30. Rumelhart, D.E., Hinton, G.E., Williams, R.J.: Learning internal representations by error propagation. Technical report, California Univ San Diego La Jolla Inst for Cognitive Science (1985)
31. Scherer, D., Müller, A., Behnke, S.: Evaluation of pooling operations in convolutional architectures for object recognition. In: Artificial Neural Networks-ICANN 2010, pp. 92–101 (2010)
32. Syed, S., Ali, A., Duggan, C.: Environmental enteric dysfunction in children: a review. J. Pediatr. Gastroenterol. Nutr. **63**(1), 6 (2016)
33. Wang, W., Huang, Y., Wang, Y., Wang, L.: Generalized autoencoder: a neural network framework for dimensionality reduction. In: Proceedings of the IEEE Conference on Computer Vision and Pattern Recognition Workshops, pp. 490–497 (2014)
34. Yang, Y.: An evaluation of statistical approaches to text categorization. Inf. Retrieval **1**(1–2), 69–90 (1999)
35. Zhai, S., Cheng, Y., Zhang, Z.M., Lu, W.: Doubly convolutional neural networks. In: Advances in Neural Information Processing Systems, pp. 1082–1090 (2016)
36. Zhang, J., Kowsari, K., Harrison, J.H., Lobo, J.M., Barnes, L.E.: Patient2Vec: a personalized interpretable deep representation of the longitudinal electronic health record. IEEE Access **6**, 65333–65346 (2018)

Diagnosis of Obstructive Sleep Apnea Using Logistic Regression and Artificial Neural Networks Models

Alaa Sheta[1]($\boxtimes$), Hamza Turabieh[2], Malik Braik[3], and Salim R. Surani[4]

[1] Computer Science Department, Southern Connecticut State University,
New Haven, CT 06515, USA
shetaa1@southernct.edu

[2] Department of Information Technology, Taif University, Taif, Saudi Arabia
h.turabieh@tu.edu.sa

[3] Department of Computer Science, Al-Balqa Applied University, Salt, Jordan
mbraik@bau.edu.jo

[4] Department of Medicine, Texas A&M University, Corpus Christi, TX, USA
srsurani@hotmail.com

Abstract. Regrettably, a large proportion of likely patients with sleep apnea are underdiagnosed. Obstructive sleep apnea (OSA) is one of the main causes of hypertension, type II diabetes, stroke, coronary artery disease, and heart failure. OSA affects not only adults but also children where it forms one of the sources of learning disabilities for children. This study aims to provide a classification model for one of the well-known sleep disorders known as OSA, which causes a serious malady that affects both men and women. OSA affects both genders with different scope. Men versus women diagnosed with OSA are about 8:1. In this research, logistic regression (LR) and artificial neural networks were applied successfully in several classification applications with promising results, particularly in the bio-statistics area. LR was used to derive a membership probability for a potential OSA system from a range of anthropometric features including weight, height, body mass index (BMI), hip, waist, age, neck circumference, modified Friedman, snoring, Epworth sleepiness scale (ESS), sex, and daytime sleepiness. We developed two models to predict OSA, one for men and one for women. The proposed sleep apnea diagnosis model has yielded accurate classification results and possibly a prototype software module that can be used at home. These findings shall reduce the patient's need to spend a night at a laboratory and make the study of sleep apnea to implement at home.

Keywords: Sleep apnea · Logistic regression · Artificial neural networks · Classification · Features selection

1 Introduction

Obstructive sleep apnea (OSA) is a long-term aerobic disease that takes place in about 10% of the overall population [1]. Thus, it has been recognized as het-

© Springer Nature Switzerland AG 2020
K. Arai et al. (Eds.): FTC 2019, AISC 1069, pp. 766–784, 2020.
https://doi.org/10.1007/978-3-030-32520-6_56

erogeneous and complexly upset. Apnea and lack of breathing (i.e., hypopnea) are two common respiratory events that occur most often for the time of sleep. Apnea could be defined as a lack of airflow in the nose and mouth for a period of more than 10 s [2]. A 50% dropping in airflow capacity or less for more than 10 s, a 3% reduction in oxygen saturation and arousal maturation are three factors that differentiate hypopnea [2]. Arousal is an abrupt turnover to a well-let sleep step or wakefulness through sleep [2]. In general, the root of OSA is the reiterative breakdown of the upper airway at bedtime, causing deteriorated gaseous interchanges and sleeps turmoil, often observed with a falloff in oxygen saturation. Studies reported in literature have demonstrated that OSA is associated with a rise in heart failure and the risk of strok [3]. As a consequence, OSA augments the peril of cardiovascular and cerebrovascular ailments [3]. Recent researches show that this heterogeneity occurs in the areas of the presenting symptoms [4], comorbid conditions [5] in addition to physiologic etiology [6]. Apart from the negative impact of the quality of life of the affected patients; severe snoring, night choking, frequent waking, sleep disturbance, insomnia, and severe sleepiness during the daytime are also a sign of OSA [7]. This pathological phenomenon has the largest prevalence of all sleep disorders, affecting about 2–5% of middle-aged women and 3–7% of the middle-aged men [7] in the population. Almost, it influences of about 2% to nearly 4% of the entire adult population [7].

1.1   OSA Diagnosis

OSA diagnosis events can be performed in full night polysomnography with particular equipments using a method referred to as the polysomnography (PSG) method [8] in a particular laboratory. Polysomnography represents a golden standard for OSA diagnoses. This method assists to detect several sleep-related diseases by registering in a lab environment. However, this method is highly expensive, needs a specific device and demands a long time to perform such an OSA diagnostic task [8]. The device that uses polysomnography is typically known as a PSG device. PSG represents a manner of observing the patients overnight using a signal acquisition device. This device can collect electrical signals from a varied set of electrodes and sensors connected to the body. Data are recorded in forms of time series signals such as chest-abdominal breathing movements, nasal air flow, respiratory movement in addition to oxygen saturation, when using this device with an electroencephalogram (EEG), electromyogram (EMG), electrooculogram and electrocardiogram (ECG).

A PSG device is typically used to monitor few body functions during sleep. Therefore, the patient needs to be associated with multiple probes (typically over 16 years) while staying overnight in a laboratory setting. Although PSG is broadly accepted as a systematic and reliable examination method, it still has some flaws [9].

- Firstly, PSG is, in many cases, inconvenient because the patient has to spend one night in a hospital or sleep center.

- Secondly, it is a costly process. For data processing, a technician has to monitor and manually "score" the outcomes of the measurements. Add to it, the hospital setup cost.
- Thirdly, the wait time is always a problem. In many cases, the patient has to wait for days or even months to do the test. Therefore, methods that provide a consistent diagnosis of sleep apnea based fewer measurements, and does not need any additional sleep laboratory facilities are urgently needed.

The data signals registered by the PSG device are regularly checked and examined by sleep experts and physicians, where they provide a report that analyzes and describes the signals of the PSG device. As a matter of fact, the physicians and sleep experts rely on statistical techniques and data analysis procedures in their decision-making process [10]. Even so, due to the complexity level and the amount of signals recorded, manual data analysis consumes much time and effort as well as large cost.

Many methods and applications have been suggested for the automatic detection of OSA events. Often, most automatic detection methods for OSA utilize signal processing techniques in the analysis of PSG records, where these methods extract statistical features to be used by a particular classifier. Without locating the features, such methods experience multiple problems with high dimensionality data or the data that has noise such as the PSG data. Specifically, a practical feature selection method is characterized by domain and it would be costly and time-consuming when implemented manually [11]. However, due to the high cost of sleep screening in the laboratory and the time spent to provide an in-depth study to the problem, over than 90% of patients suffered from sleep apnea are not cured. Therefore, alternative low cost methods are needed [12]. Accordingly, spolygraphic methods are increasingly utilized in OSA diagnostics [12].

Therefore, it is preferable to use the device in a laboratory environment or hospital by a qualified technician. Also, the device needs to use a lot of electrode usage [9]. These shortcomings delay the patient's sleep and attract the patient away from the normal sleeping environment [9]. It was found that sleeping environment can lessen the adequacy of the results. Photoplethysmography (PPG) could be defined as a noninvasive electro-optic method that provides information concerning the blood flow volume in the test area of the body near the skin. The PPG data records has recently inspected in literature and has plenty of information. The interrelation activities between apnea states and PPG have lately matured in the literature [13], but according to the knowledge, the number of publications related to detecting Apnea during the use of PPG signals is slight in literature. Numerous studies have been conducted on a very large number of sleep apnea cases based on the PPG signal observed in children between 2006 and 2014. These researchers concluded that the PPG signal is useful for OSA diagnosis [14]. Due to that oxygen saturation represents a parameter arrived at by the PPG signal, the respiration rate is typically realized by the PPG data [15]. The potentiality of detecting apnea state for patients during bedtime using the PPG signal was addressed in many studies [16]. For instance, Karmakar et al. [13] successfully presented a simple system to detect arousals using a PPG device.

The aim of this study is to present two classification models for the OSA problem using logistic regression (LR) technique and artificial neural networks (ANNs). The LR technique was applied to derive a membership probability of the OSA problem using a range of individual measurements features. These features are the weight, height, hip, waist, body mass index (BMI), age, neck circumference, modified Friedman, snoring, Epworth sleepiness scale (ESS), sex, and daytime sleepiness. As OSA causes an acute infirmity affecting both men and women at different scales, we have developed two models for such an interesting problem, one for men and the other for women. To reduce the dimensionality of the problem and select the most significant features, we apply a feature selection method as described later.

In Sect. 2, several methods of OSA are reviewed. Section 3 discusses the theory of logistic regression. The following Sect. 4 presents the artificial neural network concept and architecture. Section 5 then presents feature selection algorithms and its importance. Section 6 illustrates the proposed method of hybridizing feature selection with a machine learning classifier approach. Section 7 presents the medical data sets used in this paper. Performance criteria are presented in Sect. 8. The experimental setup and the evaluation results along with a comparison between the proposed models are presented in Sect. 9 with a summary of the research findings and concluding comments in Sect. 10.

2 Related Works

There are two common difficulties in many kinds of OSA detection methods that are attributed to the high-dimensionality of the features and decision-making process that could be handled as a black box conclusion. Currently, the core of most OSA detectors is time domain extraction, frequency domain and a variety of nonlinear features from many ECG-based signals. The extracted features are then used by classification methods to identify the presence of OSA. There are several types of feature selection methods that are widely used to reduce the dimensionality of the feature space. Such methods include statistical estimation [17], principal component analysis [18] and wrapper approach [19].

In a decision-making process, Song et al. [9] presented a guideline method to locate the features and identify OSA by associating the extracted features' values to predefined parameters, leading to a thorough description of the OSA detection results. As a result, a large number of artificial intelligence methods have been used in diagnosing OSA. In a study carried out in 1995, Siegwart et al. [20] identified OSA using the EEG signals (sleep stages) through the use of ANNs. In another study conducted in 2001, Nazeran et al. [20] used fuzzy inference algorithm to recognize OSA with a respiratory signal.

OSA types were categorized on their own using ANNs through studies presented in [21]. In 2008, OSA and narcolepsy identification was determined using EEG signals through the use of ANNS [22]. Shokrollahi et al. [23] has developed an OSA detection method using NN as a classification method performed based on the sound of snoring immediately taken from people; This is because the snoring sound occurs as a result of the tightness in the upper airway and the produced vacillation of the upper airway tissue. Hossein [24] proposed an NN-based features selection method for OSA identification where classification accuracy reached up to 70%. The NN was utilized for two key objectives: (1) selection of the optimum frequency bands that could be utilized to recognize the OSA at the point of time of features extraction, and (2) detection of OSA in the course of features matching process. A support vector machine (SVM) classifier was used for the first time by Khandoker et al. [25] to classify OSA patients. Khandoker et al. extracted the features using wavelet decomposition and RR intervals. Khandoker et al. proclaimed that they correctly detected about 90% of the subjects in the testing set.

Several methods were then presented to detect sleep apnea events from respiratory signals. Signal processing techniques were mostly applied as a pre-step to extract features, which are then used as inputs to the classification methods. For example, as presented in [26], the data was statistically analyzed using a wavelet transformation technique, where the extracted parameters of the wavelet features were extracted and used as inputs to SVM classifier. Some types of classifiers comprise ensemble methods [27] and feed-forward neural networks (FFNNs) [28]. The features of OSA are usually obtained from thoracic breathing signals or nasal, abdominal, chest and many other repository signals. In [29], the features were identified from blood oxygen saturation and airflow of fifteen records obtained from a PSG device. These features were then fed inputs to a time delay NNs. Nasal airflow signal was used in combination with adaptive fuzzy logic techniques to classify OSA events as normal or abnormal [30]. Most of the aforementioned OSA detection methods used ANN as a classifier and fragile feature extraction methods that often resulted in imperfect classification results. However, when reassessing all of the above studies, it was found that the classification process was implemented with only one structure. Due to such flaws, the ensemble classification method was reinforced in many research methods in the literature.

Although significant progress has been made in OSA detection and classification problems, performance is still not fully convinced in several cases. Many of the current methods have limited functionality resulting in relatively high false negative error rates. The better methods require training with a huge number of OSA cases to achieve a reasonable performance. So there is a room for further enhancements in both OSA detection and classification, particularly when a small dataset is available. The OSA classification method described here is designed to overcome these shortcomings. The originality and significance of this work are associated with the following items:

- We use a real data set that is gathered for both men and women participants. Therefore, to classify OSA efficiently, there is a demand for a better understanding of the functionality of this problem as well as its parameters.
- A secondary objective is to assess and quantify the performance of the developed classification models using an effective manner. In this work, we have explored a relative metric measure to give a precise judgment of the quality of the classification models.

- It is expected that the use of LR method for OSA classification for the first time will provide a possibility to achieve a high level of performance comparable to the most promising state-of-the-art classifiers.
- We have adopted a feature selection method to reduce the dimensionality of the problem as well as locating the most significant features.

The following sections introduce the logistic regression and artificial neural network models used for OSA classification:

3   Logistic Regression

Logistic regression is a supervised classification machine learning technique. On solving a classification problem using LR, we assume that we have a set of labeled examples characterized by a set of features (i.e., independent variables), where our target is to classify (i.e., label) these examples into groups. The LR model shall provide functionality for calculating the probability of determining whether a new example belongs to any of the previously defined classes. LR can be described mathematically as a possibility of occurrence of some events [31], in our case, the probability that an example belongs to a particular class appears as follows:

$$p = \frac{1}{1 + e^z} \tag{1}$$

where z is a linear equation of some features (i.e., predictor values or independent variables):

$$z = \theta_0 + \theta_1 f_1 + \theta_2 f_2 + \dots \theta_0 f_n \tag{2}$$

The probability p can take values in the range from 0 to 1 on an S-shaped curve, θ_0 is the intercept of the model and the other values $\theta_i, (i = 1, \dots, n)$ and the coefficients of the LR model. Our goal is to locate the optimal values of the parameters, $\theta's$ so that we achieve a high classification rate. In our case, the features f should be selected appropriately such that the presented model is accurate.

The classification problem can be described as follows: Given a set of classes (i.e., labels), in our case, Apnea (1) or No-Apnea (0), and a set of anthropometric features defined as $f = < f_1, f_2, \dots, f_n >$, we need to find the probability that a new example belongs to one of these two classes; $P(k|f)$, k is a given class with 0 or 1 (see Eq. 3).

$$P(k|f) = \frac{e^{\theta_0 + \sum_{i=1}^{n} \theta_i f_i}}{1 + e^{\theta_0 + \sum_{i=1}^{n} \theta_i f_i}} \tag{3}$$

The set of parameters θ is estimated from the given set of features $f_i, i = 1, \dots, n$ using the maximum likelihood (ML).

4   Artificial Neural Network

The ANN is defined as a computational method developed by the biological neural networks [32,33]. In this regards, ANN has a structure analogous to the biological nerve of human cells. Similar to the human cells, ANN possesses a combination of interrelated neurons in the form of layers, where each layer has several neurons. In other words, ANN is customarily known as an adaptive nervous system due to that it has a structure that can be described and modeled as a mathematical relationship of inputs and outputs.

4.1   Why ANN?

ANN has been broadly used in the past four decades due to its ability to provide high quality models for various nonlinear complex problems based on a simple learning algorithm. Moreover, ANN have been employed successfully as low-cost hardware for several applications. In general, several learning algorithms can be used to train ANN to mimic the human brain, that enables ANN to memorise a predetermined function, patterns, or behave as a decision maker based on a set of measurements. The ability to learn and generalize are two main advantages that make ANN an excellent beneficial approach in addressing a broad range of prediction and modeling problems. The following represent interesting characteristics for the ANNs:

- Learning process is a major process that affect the final performance of ANN, that is implemented based on a set of input training dataset. This data tunes the ANN weights based in predetermined learning algorithm. Once training process is converged successfully, the ANN model will be able to mimic the human brain once a new dataset (input) is presented.
- ANN has many eminent features such as: (1) great possibility of generalization, (2) a large capability to identify and describe the output from the input patterns of a system under consideration, (3) can effectively estimate the parameters of a particular model, (4) learn how to simulate the process model in order to estimate the future outcomes, and (5) address nonlinear, non-stationary as well as chaotic functions.

ANN has been used prosperously in addressing a large number of different prediction and identification problems and has confirmed its effectiveness in many applications [34–38].

4.2   ANN Architecture

McCulloch and Pitts had introduced ANN in 1943. The basic concept of ANN systems consists of several parallel working systems that contain many processing units, referred to as *neurons*, which are interconnected in a network and huddled to form a layer. In this context, a neuron has many inputs and a single output. Each input data has a value called weight. The neuron function as following:

the input data (signal) for every neuron is evaluated as the multiplicity of the equal weight, and the result of is summed and transferred using transfer function (i.e. sigmoid function). If the result of the aggregation process exceeds a certain limit, he output of the neuron will be activated; otherwise, the activation will not occur. Figure 1 shows a fully connected network, where all neurons in a single layer are linked to all neurons in the following layer. The first layer of this network is a hidden layer referred here to as h and the layer following this hidden layer is basically the output layer. The structure of the network in Fig. 1 composes of three inputs, two outputs in addition to two hidden units.

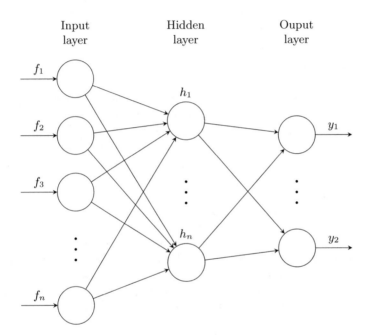

Fig. 1. Two output ANN for classification

4.3   Feedforward Neural Network

The feedforward-neural network (FF-NN) is usually known as a particular type of neural network, which is widely used to simulate nonlinear challenge problems. A multi-layer FFNN is named "multi-layer Perceptron (MLP)". MLP is a specific type of NN that is fully connected where all neurons in each layer are attached to all neurons in the next layer. As described above, the first and second layers are the hidden and output layers, respectively. The MLP model can be formulated as given in Eq. 4 [35, 39]:

$$\hat{y}_i(t) = g_i[f, \theta]$$

$$= \varphi_i \left[\sum_{j=1}^{n_h} W_{i,j} \phi_j \left(\sum_{l=1}^{n_f} w_{j,l} f_l + w_{j,0} \right) + W_{i,0} \right] \tag{4}$$

where $\hat{y}_i$ represents the output signal, g_i identifies the function that ANN achieves and θ stands for the parameter vector, which contains all the hyper-parameters of the NN or in other words the weights $w_{j,l}$, and the biases $W_{i,j}$. In general, the backpropagation (BP) learning algorithm is used to train the MLP model with a prime aim to match input and output signals. In reality, different input/output singles are used to achieve this match. The mean squared error (MSE) is the most familiar minimization metric measure.

5 Feature Selection

Features selection (FS) algorithms aim to find a minimal number of features that represent the original features. In general, there are two basic steps of the FS process (1) searching for minimal redacts, and (2) evaluation of the selected features. There are two methods for FS (1) *filter*, and (2) *wrapper*. The filtering method focuses on the relationship between features to determine the most valuable features, while the wrapper method uses a learning algorithm based on a validation method to evaluate the selected features. Filter methods are faster and less precise than wrapper methods [40].

In this study, we applied a wrapper approach known as minimal redundancy maximal relevance criterion (mRMR) [41] using a set of features S with features F_i, $(i = 1, 2, \ldots m)$, where m is the number of the selected features. The maximum relevance aims to locate the significant features so that the values of the mutual information between each feature and the proposed goal increase. Equation 5 shows the objective function for mutual information.

$$max \quad M(S, t) = \frac{1}{|S|} \sum_{F_i \in S} I(F_i, t) \tag{5}$$

where $M(S, t)$ indicates the average of the mutual information between each feature and target t, $I(i)$ denotes a measure of dependency the density of feature, x_i, and the density of the target, t. If two features have a powerful separability on the target t, it will not be acceptable to specify both features if they are highly interrelated. The key concept of minimum redundancy is to identify a particular number of features that are mutually different to a large extent. Equation 6 provides the minimization of minimum redundancy between two features.

$$min \quad R(S) = \frac{1}{|S|^2} \sum_{F_i, F_j \in S} I(F_i, F_j) \tag{6}$$

where $R(S, t)$ represents the average of the mutual information between pairs of selected features.

mRMR could be defined as the maximization of the mean of the respective function information between each feature F_i and target $M(S, t)$ and the minimization of the mutual information between pairs of features $R(S)$. Further information about mRMR can be found in [41].

6   Proposed Classification Methods

The proposed approach, based on feature selection and machine learning classifier, appears in Fig. 2. This approach consists of two main phases: (i) feature selection based on the mRMR algorithm; and (ii) building and evaluating a machine learning classifier using the LR method or ANN method. Feature selection phase begins after collecting data from patients. Section 7 presents the gathered dataset. mRMR algorithm will identify the most valuable features. A new subset of the original data set will be extracted to the next phase. Building and testing the classifier represent the second phase. The dataset is divided using a holdout approach (i.e., training (80%) and testing (20%) of the datasets). The training dataset is utilized to build the classifier and the testing dataset to evaluate the classifier. The evaluation process was developed based on the receiver operating characteristic - area under curve (ROC-AUC). Section 8 illustrates the AUC criteria.

The two main issues needed to be considered carefully when establishing a classification model for medical applications are the computational time and accuracy of prediction. The models developed for the OSA system under study indeed satisfy these two requirements.

7   Clinical Characteristics and Demographics

The medical study in this work included 1,418 subjects, where 659 subjects completed the data records and encompassed in the final analysis. A large number of patients were precluded from the analytical study which is approximately 94% of the total participant patients. This is attributed to the missing data related to the features, namely, ESS, comorbidities and snoring. Table 1 shows the basic medical attributes and demographics of the study presented in this work.

The participants mainly acknowledged themselves as Caucasian or Hispanic, where the mean BMI was calculated in the range of obesity. The baseline characteristics of participant patients lacking data were very similar to the patients involved in the final analysis of the results but were somewhat younger (53 ± 14 years versus 55 ± 14 years).

8   Performance Measure

Evaluating machine learning classifiers is a critical process for determining success or failure. Many criteria were used, such as accuracy, precision, recall, F-measure, as well as AUC. These criteria are influenced by the cut-off value

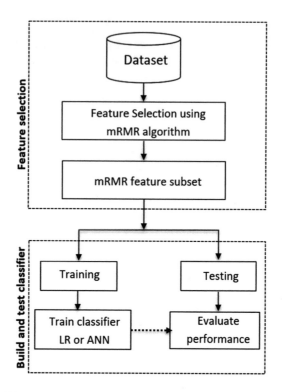

Fig. 2. A pictorial diagram of the proposed method.

on the expected probability of glitch instances except AUC values [42]. AUC criterion is not affected to the cut-off value, and AUC criterion is insensitive to changes in class distributions [43]. Many researchers prefer to use AUC criteria due to it is stable nature [44]. AUC criterion is defined in terms of the ratio between the TP rate against the FP rate. These parameters are described below:

1. TP (true positive): represents the number of correct classifications that take place when a patient has been classified.
2. TN (true negative): classifies the OSA cases that are not patients, and the test is negative.
3. FN (false negative): represents the patients in the database that are not classified.
4. FP (false positive): represents the number of wrong classifications obtained when the proposed classifier classifies an OSA case (patient) true case that is not an OSA.

A confusion matrix as shown in Table 2 is employed to evaluate the AUC criterion.

Sensitivity and specificity values are used to evaluate the AUC value from the confusion matrix. Sensitivity and specificity are evaluated, as shown in Eqs. 7 and 8, respectively.

Table 1. Clinical and demographic information of the study population Group Pilot Verification

Group	Male	Female
Number		
Age, mean (±s.d.)	56 ± 13	55 ± 14
Height, cm (±s.d.)	170 ± 11	171 ± 11
Weight, kg (±s.d.)	67 ± 14	69 ± 18
Body mass index (BMI), mean (±s.d.)	35 ± 7	35 ± 9
Male, no. (%)	86 (57%)	299 (59%)
Hispanic	75 (50%)	219 (43%)
Caucasian	75 (50%)	274 (54%)
Black	0 (0%)	15 (3%)

Table 2. The confusion matrix.

Actual class	Predicted class		
		Class = Yes	Class = No
	Class = Yes	True Positive (TP)	False Negative (FN)
	Class = No	False Positive (FP)	Ture Negative (TN)

$$Sensitivity = TP_{rate}$$
$$= \frac{TP}{P} \tag{7}$$

$$Specificity = TN_{rate}$$
$$= \frac{TN}{N} \tag{8}$$

where P represents the actual number of positive samples, and N represents the actual number of negative samples. Table 3 shows AUC rules to assess any classifier on the basis of AUC criterion [45].

9   Experimental Results

This section introduces the classification results of the proposed OSA prediction methods. The collected dataset is divided into two main clusters (i.e., male and female). For each cluster, we develop a classifier model using either LR or ANN. A feature selection algorithm based on an mRMR [41] has been applied to allocate the most significant features for OSA. A comparison has been made between the results of LR and ANN. To achieve this goal, a set of 31 experiments was conducted using MATLAB-R2015a environment. The following subsections present the obtained results.

Table 3. AUC values description [45].

AUC Range	Description
$0.00 \leq AUC < 0.50$	Bad classification
$0.50 \leq AUC < 0.60$	Poor classification
$0.60 \leq AUC < 0.70$	Fair classification
$0.70 \leq AUC < 0.80$	Acceptable classification
$0.80 \leq AUC < 0.90$	Excellent classification
$0.90 \leq AUC \leq 1.00$	Outstanding classification

9.1   Feature Selection Results

As aforementioned, the mRMR, as a wrapper feature selection method, was applied to the OSA prediction problem. Table 4 shows the most significant features obtained based-mRMR for male and female clusters. Male patients have different features compared to female patients. However, some features are shared between both male and female clusters, which are Height, Neck, M. Friedman, Epworth, and snoring features. Weight, waist, and Daytime sleepiness features are not selected for both clusters.

In a deeply consideration of the results presented in Table 4, the mRMR algorithm helps to understand the hidden information of the OSA problem. Removing such features will enhance the performance of the classifier and reduce the computational time for building a precise classifier.

The features labeled with "1" in Table 4 confirm that these features are significant to model the OSA problem based upon the collected datasets, while the features marked with "0" are not significant to the OSA model and might add a little contribution to the model along with sizeable computational burden.

9.2   Logistic Regression Model

The experimental results of the LR approach with and without using feature selection are shown in Table 5. The best, worst, and average AUC values are the three values reported in Table 5 for each evaluation experiment. The performance of LR approach with or without feature selection is capable of finding an outstanding classifier for male and female clusters (AUC ≥ 0.80) based on the best values of AUC. Furthermore, the average performance shows that the LR approach can locate an acceptable classifier for both clusters (AUC ≥ 0.70).

Figure 3 presents the boxplots diagram for evaluating the datasets with and without the presented feature selection method. For the male cluster based on the worst AUC values, the worst scenarios yields a good classifier (AUC ≥ 0.50). Moreover, from Fig. 3a, it is clear that the performance of LR over the male cluster is quite similar either with or without feature selection. This finding means that the feature selection approach selects the most valuable features from the male cluster without losing the final performance of LR. Figure 3b

Table 4. Selected features using the mRMR method for both clusters (1 identifies selected feature and 0 identifies unselected feature).

Feature	Male	Female
Height	1	1
Weight	0	0
Waist	0	0
Hip	0	1
BMI	1	0
Age	0	1
Neck	1	1
M. Friedman	1	1
ESS	1	1
Snoring	1	0
Daytime sleepiness	0	0

Table 5. Results of LR based on AUC values.

Cluster	Approach	Best	Worst	Average
Male	Without feature selection	0.854	0.561	0.705
	With feature selection	0.854	0.575	0.713
Female	Without feature selection	0.804	0.505	0.712
	With feature selection	0.851	0.539	0.703

explores the performance of the female cluster. The feature selection algorithm improves the overall performance of the LR approach.

9.3   ANN Model

The results of ANN with and without feature selection are presented in Table 6. The performance of ANN over male cluster with or without feature selection is reasonable on the basis of the best AUC values, and a fair classifier on the basis of the mean AUC values. However, the average performance of ANN over the female cluster was improved when looking at the average AUC values after applying the feature selection method.

Figure 4 shows the performance of both male and female clusters based on boxplots diagram. Figure 4a shows that ANN performance is more consistent with feature selection compared to the same approach without feature selection. However, ANN performance without feature selection over the female cluster is robust.

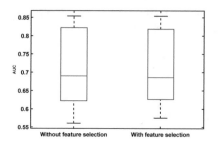 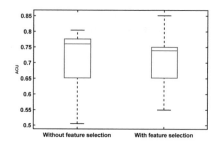

(a) Boxplots diagram for male cluster.   (b) Boxplots diagram for female cluster.

Fig. 3. Boxplots diagrams for LR classifier.

Table 6. Results of ANN based on AUC values.

Cluster	Approach	Best	Worst	Average
Male	Without feature selection	0.758	0.506	0.638
	With feature selection	0.761	0.527	0.636
Female	Without feature selection	0.839	0.443	0.619
	With feature selection	0.717	0.465	0.583

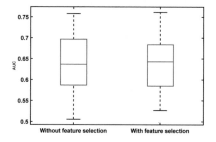 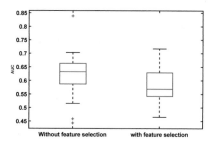

(a) Boxplots diagram for male cluster.   (b) Boxplots diagram for female cluster.

Fig. 4. Boxplots diagrams for ANN classifier.

9.4   Discussion

Based on the AUC results in Tables 5 and 6, the performance of LR and ANN varies when the feature selection algorithm is applied. Determining the best features (inputs) is essential to build a reasonable classifier. It is challenging to decide on the best features. As a result, the application of feature selection approaches based on machine learning will give further exploration about the dataset and what are the most valuable features for the classifier.

While the classifiers that do not utilize feature selection methods are simpler to implement, the application of feature selection will reduce the search space for the training process and accommodate the features that make patients suffer from the OSA problem. Identification of these features will save time and cost to collect such data from patients.

The results in Tables 5 and 6 reveal that LR and ANN methods form efficacious classifiers for the OSA problem with or without feature selection mechanism. They each reached an acceptable performance based on AUC values. However, the overall results illustrate that the proposed LR method as an OSA classifier can produce a more valuable and efficient classifier than what ANN can perform. To sum up, the sensible performance of the LR model demonstrates the capability of this method and establishes a classifier basis for this model to other classification problems.

10   Conclusion and Future Work

This paper has presented a model to predict obstructive sleep apnea (OSA) based on machine learning with feature selection method called minimal redundancy maximal relevance criterion (mRMA) algorithm. The presented mRMA algorithm identifies the most valuable features that reduce the dimensionality and redundancy of the dataset and further assist the presented machine learning classifiers in enhancing the prediction method. The logistic regression technique and artificial neural networks were used as two types of machine learning classifiers to develop two models in order to achieve the objective of this study. The obtained results show that mRMR is capable of reducing the dimensionality and redundancy of the collected data and boosting the average performance level of the classifiers. For future work, there is a plan to conduct different kinds of machine learning classifiers such as genetic programming (GP), k-nearest neighbor (KNN) and Naïve Bayes (NB) to build a computer model that can predict OSA. Further study is needed to attempt other feature selection methods. The extension of the proposed OSA models to more datasets would be valuable.

References

1. Dempsey, J.A., Veasey, S.C., Morgan, B.J., O'Donnell, C.P.: Pathophysiology of sleep apnea. Physiol. Rev. **90**(1), 47–112 (2010)
2. Berry, R.B., Budhiraja, R., Gottlieb, D.J., Gozal, D., Iber, C., Kapur, V.K., Marcus, C.L., Mehra, R., Parthasarathy, S., Quan, S.F., et al.: Rules for scoring respiratory events in sleep: update of the 2007 AASM manual for the scoring of sleep and associated events. J. Clin. Sleep Med. **8**(05), 597–619 (2012)
3. McNicholas, W., Bonsignore, M., M. C. of EU Cost Action B26, et al.: Sleep apnoea as an independent risk factor for cardiovascular disease: current evidence, basic mechanisms and research priorities. Europ. Respir. J. **29**(1), 156–178 (2007)
4. Ye, L., Pien, G.W., Ratcliffe, S.J., Björnsdottir, E., Arnardottir, E.S., Pack, A.I., Benediktsdottir, B., Gislason, T.: The different clinical faces of obstructive sleep apnoea: a cluster analysis. Eur. Respir. J. **44**(6), 1600–1607 (2014)
5. Eckert, D.J., White, D.P., Jordan, A.S., Malhotra, A., Wellman, A.: Defining phenotypic causes of obstructive sleep apnea. Identification of novel therapeutic targets. Am. J. Respir. Crit. Care Med. **188**(8), 996–1004 (2013)

6. Vavougios, G.D., Natsios, G., Pastaka, C., Zarogiannis, S.G., Gourgoulianis, K.I.: Phenotypes of comorbidity in osas patients: combining categorical principal component analysis with cluster analysis. J. Sleep Res. **25**(1), 31–38 (2016)

7. Fietze, I., Penzel, T., Alonderis, A., Barbe, F., Bonsignore, M., Calverly, P., De Backer, W., Diefenbach, K., Donic, V., Eijsvogel, M., et al.: Management of obstructive sleep apnea in Europe. Sleep Med. **12**(2), 190–197 (2011)

8. Borgström, A., Nerfeldt, P., Friberg, D.: Questionnaire OSA-18 has poor validity compared to polysomnography in pediatric obstructive sleep apnea. Int. J. Pediatr. Otorhinolaryngol. **77**(11), 1864–1868 (2013)

9. Song, C., Liu, K., Zhang, X., Chen, L., Xian, X.: An obstructive sleep apnea detection approach using a discriminative hidden markov model from ECG signals. IEEE Trans. Biomed. Eng. **63**(7), 1532–1542 (2016)

10. Karamanli, H., Yalcinoz, T., Yalcinoz, M.A., Yalcinoz, T.: A prediction model based on artificial neural networks for the diagnosis of obstructive sleep apnea. Sleep Breathing **20**(2), 509–514 (2016)

11. Längkvist, M., Karlsson, L., Loutfi, A.: A review of unsupervised feature learning and deep learning for time-series modeling. Pattern Recogn. Lett. **42**, 11–24 (2014)

12. de Chazal, P., Penzel, T., Heneghan, C.: Automated detection of obstructive sleep apnoea at different time scales using the electrocardiogram. Physiol. Meas. **25**(4), 967 (2004)

13. Karmakar, C., Khandoker, A., Penzel, T., Schöbel, C., Palaniswami, M.: Detection of respiratory arousals using photoplethysmography (PPG) signal in sleep apnea patients. IEEE J. Biomed. Health Inform. **18**(3), 1065–1073 (2014)

14. Lázaro, J., Gil, E., Vergara, J.M., Laguna, P.: OSAS detection in children by using PPG amplitude fluctuation decreases and pulse rate variability. In: 2012 IEEE Computing in Cardiology, pp. 185–188. IEEE (2012)

15. Shelley, K.H.: Photoplethysmography: beyond the calculation of arterial oxygen saturation and heart rate. Anesth. Analg. **105**(6), S31–S36 (2007)

16. Gaurav, G., Mohanasankar, S., Kumar, V.J.: Apnea sensing using photoplethysmography. In: 2013 Seventh International Conference on Sensing Technology (ICST), pp. 285–288. IEEE (2013)

17. Bsoul, M., Minn, H., Tamil, L.: Apnea medassist: real-time sleep apnea monitor using single-lead ECG. IEEE Trans. Inf. Technol. Biomed. **15**(3), 416–427 (2011)

18. Mendez, M.O., Bianchi, A.M., Matteucci, M., Cerutti, S., Penzel, T.: Sleep apnea screening by autoregressive models from a single ECG lead. IEEE Trans. Biomed. Eng. **56**(12), 2838–2850 (2009)

19. Isa, S.M., Fanany, M.I., Jatmiko, W., Arymurthy, A.M.: Sleep apnea detection from ECG signal: analysis on optimal features, principal components, and nonlinearity. In: 2011 5th International Conference on Bioinformatics and Biomedical Engineering, pp. 1–4. IEEE (2011)

20. Siegwart, D., Tarassenko, D., Roberts, S., Stradling, J., Partlett, J.: Sleep apnoea analysis from neural network post-processing (1995)

21. Tagluk, M.E., Akin, M., Sezgin, N.: Classification of sleep apnea by using wavelet transform and artificial neural networks. Expert Syst. Appl. **37**(2), 1600–1607 (2010)

22. Liu, D., Pang, Z., Lloyd, S.R.: A neural network method for detection of obstructive sleep apnea and narcolepsy based on pupil size and eeg. IEEE Trans. Neural Networks **19**(2), 308–318 (2008)

23. Shokrollahi, M., Saha, S., Hadi, P., Rudzicz, F., Yadollahi, A.: Snoring sound classification from respiratory signal. In: 38th Annual International Conference of the IEEE Engineering in Medicine and Biology Society (EMBC), pp. 3215–3218. IEEE (2016)

24. Cheng, M., Sori, W.J., Jiang, F., Khan, A., Liu, A.: Recurrent neural network based classification of ECG signal features for obstruction of sleep apnea detection. In: 2017 IEEE International Conference on Computational Science and Engineering (CSE) and IEEE International Conference on Embedded and Ubiquitous Computing (EUC), vol. 2, pp. 199–202. IEEE (2017)

25. Khandoker, A.H., Palaniswami, M., Karmakar, C.K.: Support vector machines for automated recognition of obstructive sleep apnea syndrome from ECG recordings. IEEE Trans. Inf. Technol. Biomed. **13**(1), 37–48 (2009)

26. Maali, Y., Al-Jumaily, A.: Hierarchical parallel PSO-SVM based subject-independent sleep apnea classification. In: International Conference on Neural Information Processing, pp. 500–507. Springer (2012)

27. Fontenla-Romero, O., Guijarro-Berdiñas, B., Alonso-Betanzos, A., Moret-Bonillo, V.: A new method for sleep apnea classification using wavelets and feedforward neural networks. Artif. Intell. Med. **34**(1), 65–76 (2005)

28. Avcı, C., Akbaş, A.: Sleep apnea classification based on respiration signals by using ensemble methods. Bio-Med. Mater. Eng. **26**(s1), S1703–S1710 (2015)

29. Tian, J., Liu, J.: Apnea detection based on time delay neural network. In: IEEE Engineering in Medicine and Biology 27th Annual Conference 2006, pp. 2571–2574. IEEE (2005)

30. Morsy, A.A., Al-Ashmouny, K.M.: Sleep apnea detection using an adaptive fuzzy logic based screening system. In: IEEE Engineering in Medicine and Biology 27th Annual Conference 2006, pp. 6124–6127. IEEE (2005)

31. Pradhan, B., Lee, S.: Delineation of landslide hazard areas using frequency ratio, logistic regression and artificial neural network model at Penang Island, Malaysia. Environ. Earth Sci. **60**, 1037–1054 (2009)

32. Sheta, A., Aljahdali, S., Braik, S.: Utilizing faults and time to finish estimating the number of software test workers using artificial neural networks and genetic programming. In: International Conference Europe Middle East & North Africa Information Systems and Technologies to Support Learning, pp. 613–624. Springer (2018)

33. Sheta, A., Braik, M., Al-Hiary, H.: Modeling the tennessee eastman chemical process reactor using bio-inspired feedforward neural network (BI-FF-NN). Int. J. Adv. Manuf. Technol., 1–22 (2019)

34. Sheta, A., Eghneem, A.: Training artificial neural networks using genetic algorithms to predict the price of the general index for amman stock exchange. In: Proceedings of the Midwest Artificial Intelligence and Cognitive Science Conference, DePaul University, Chicago, IL, USA, 21–22 April, pp. 7–13 (2007)

35. Sheta, A.F., Braik, M., Al-Hiary, H.: Identification and model predictive controller design of the tennessee eastman chemical process using ANN. In: IC-AI, pp. 25–31 (2009)

36. Sheta, A.F., Braik, M., Öznergiz, E., Ayesh, A., Masud, A.: Design and automation for manufacturing processes: an intelligent business modeling using adaptive neuro-fuzzy inference systems. In: Business Intelligence and Performance Management, pp. 191–208. Springer (2013)

37. Alkasassbeh, M., Alaa, H.F., Sheta, F., Turabieh, H.: Prediction of PM10 and TSP air pollution parameters using artificial neural network autoregressive, external input models: a case study in salt, jordan. Middle-East J. Sci. Res. **14**(7), 999–1009 (2013)
38. Kovač-Andrić, E., Sheta, A., Faris, H., Gajdošik, M.Š.: Forecasting ozone concentrations in the east of croatia using nonparametric neural network models. J. Earth Syst. Sci. **125**, 997–1006 (2016)
39. Norgaard, M., Ravn, O., Poulsen, N.K., Hansen, L.K.: Neural Networks for Modelling and Control of Dynamic Systems. Springer, London (2000)
40. Kudo, M., Sklansky, J.: Comparison of algorithms that select features for pattern classifiers. Pattern Recogn. **33**(1), 25–41 (2000)
41. Peng, H., Long, F., Ding, C.: Feature selection based on mutual information criteria of max-dependency, max-relevance, and min-redundancy. IEEE Trans. Pattern Anal. Mach. Intell. **27**(8), 1226–1238 (2005)
42. Zhang, F., Mockus, F., Keivanloo, I., Zou, Y.: Towards building a universal defect prediction model with rank transformed predictors. Empirical Softw. Eng. **21**(5), 2107–2145 (2016). https://doi.org/10.1007/s10664-015-9396-2
43. Fawcett, T.: Roc graphs: notes and practical considerations for researchers. Technical report (2004)
44. Turabieh, H., Mafarja, M., Li, X.: Iterated feature selection algorithms with layered recurrent neural network for software fault prediction. Expert Syst. Appl. **122**, 27–42 (2019). http://www.sciencedirect.com/science/article/pii/S0957417418308030
45. Hosmer, D.W., Lemeshow, S.: Applied Logistic Regression. Wiley Series in Probability and Statistics, 2nd edn. Wiley-Interscience Publication, New York (2000)

Building a Blockchain Application: A Show Case for Healthcare Providers and Insurance Companies

Kawther Saeedi, Arwa Wali, Dema Alahmadi, Amal Babour$^{(\boxtimes)}$,
Faten AlQahtani, Rawan AlQahtani, Raghad Khawaja,
and Zaina Rabah

Information Systems Department, King Abdulaziz University,
Jeddah, Kingdom of Saudi Arabia
{ksaeedi,ababor}@kau.edu.sa

Abstract. Blockchain is an evolving technology that provides trusted decentralized records of information through encrypted blocks of linked data. This technology promised to provide immutable and integral records shared among authorized parties. This infrastructure creates a vast range of opportunities and reforms wide range of business practices. However, this emerging technology is lacking application development experience in real life project. In this paper, building blockchain-based application is represented for two purposes. The first one is a show case of how blockchain application can reform current business practice. The case represents a blockchain application replacing common business practice of having a middleman (Third-party) delivering original bills between healthcare providers and insurance companies. A ClaimChain application is built to demonstrate the potential benefits of the blockchain application in comparison to conventional practice. The second purpose is a demonstration of design decisions made to build blockchain application with reference to blockchain application design approaches introduced by software architecture community. The paper ends with a summary of lessons learned and recommendations for development process of blockchain application.

Keywords: Blockchain · Software development · Healthcare · Insurance · Verification

1 Introduction

Blockchain revolution expands the world of technology with a new secured infrastructure; it transforms the businesses performance and created new business opportunities. Blockchain technology is a decentralized database that allows users to store unlimited data records that are linked though cryptographic techniques. This distributed and secure infrastructure promised to provide immutable, integral and transparent records. It enables new forms of agreements between business partners without central trusted body [1]. This characteristic encourages many engineers and developers to apply blockchain technology in solving many economic and financial transactions

© Springer Nature Switzerland AG 2020
K. Arai et al. (Eds.): FTC 2019, AISC 1069, pp. 785–801, 2020.
https://doi.org/10.1007/978-3-030-32520-6_57

problems. The cryptocurrency is widely known application of blockchain and the benefits of this technology are spread to all business domains.

However, building an application based on blockchain platform is a challenging task due to its complex architecture and limitation of research that outline experience and guidelines. This limitation is highlighted in [2–5] as a difficulty to use and understand blockchain; where different approaches to enhance the usability of blockchain are proposed. Thus, this paper contributes with an empirical experience for developing blockchain application. Our experience is captured throughout the development of blockchain application from scratch with a highlight on both technical and business design decisions made.

The paper is organized as follows. Section 2 provides an overview about blockchain platform and its applications. Section 3 presents our experience in building blockchain application to enhance claim process in healthcare insurance. Finally, we conclude in Sect. 4 with a discussion on challenges and future work.

2   Blockchain Concepts

2.1   Components of Blockchain Platform

In this section, we highlight the main components and the infrastructure required to build a blockchain platform. In principal, blockchain is a very recent technology, which can be considered as a container which stores data transactions in chained blocks. Each block in the sequence of blocks hold a complete list of data transactions records [6]. From Fig. 1, each block requires a pointer to the immediately previous block, which is a hash value, refers to the previous block so-called parent block. The *genesis* block, which is the first block in the block chain, doesn't have a parent block [7].

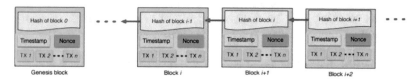

Fig. 1.   An example of blockchain. [8]

Each block consists of the block header and the block body. The block header includes some information such as parent's hash value that refers to the previous parent. It also includes the current timestamp in seconds. While the block body stores data about number of transactions and the transactions. Different size and number of transactions are based on the block size.

Another important concept is the digital signature. Each user manages two pair of keys private and public. The private key is used by the user to sign the transaction. Then, the digital signed transactions are spread over the entire network and can be accessed using public key by other users in the network.

Blockchain has the following key features [8]:

1. Decentralization: Blockchain technology guarantees transactions between peer-to-peer (P2P) without the need of authentication from third party or central agency. This improves the performance of central every from authenticating every transaction and reduce the cost.
2. Immutability: It is impossible to remove or modify any transaction registered on the blockchain. This feature can help to protect transaction from deletion and fraud [9].
3. Persistency: If blocks need to confirm and sign every transaction in the whole network, this makes any changes or misuse of transactions is impossible.
4. Anonymity: There is no central agency that maintains users' private information. In addition, users can produce several addresses to preserve their identity.
5. Auditability: Users can easily trace and check the validity of transactions as every transaction recorded within a timestamp. For the Time stamped reason, transactions are difficult to be tampered and altered. Thus, a transaction can be inserted to a block only when the entire network makes a consensus for a specific transaction to be included.

In summary, blockchain technology has a potential impact on the future internet systems; however, there is several technical challenges should be tackled. For example, scalability concern, the limited size of block affects the number of transactions per second. The increased in transactions size will affect the storage space needed, and this will decrease the growth of the blockchain network. Moreover, the algorithms for consensus such as Proof of Work (PoW) or proof of stake (PoS) have some drawbacks. For example, PoW consumes too much electricity energy to reach a consensus in the network.

2.2 Blockchain Related Applications

Blockchain has become the technology to transform businesses since the hype of the Internet. Governments, financial institutions, large and small corporations are transforming their processes adopting the blockchain. Furthermore, blockchain literature is rich with examples and success stories of how blockchain transforms businesses. However to support the proposed system, this section highlights processes of few applications in which a middlemen is replaced with blockchain technology.

The first example in the real estate field, Rentberry is a rental application and price negotiation platform brings tenants and landlords together. It is a decentralized web-based long term platform that allows both tenants and landlords to handle all rental tasks in one place making the rental process less costly and more secure. It automates the standard tasks such as searching for properties, making offers/bids, screening candidates, selecting the most suitable one, negotiating with the landlords, unlocking the security deposit, E-signing contracts, paying rents, submitting maintenance requests, allowing tenants and landlords to rate each other's, etc. The platform is for replacing the burden of all these standard tasks and making the whole process of rental through blockchain [10].

The Second example in the financial field, Sikoba is a global and decentralized money platform based on peer-to-peer IOU's and blockchain technology. It allows users who know and trust each other to grant each other credit lines and through it is using to 'credit conversion'; it makes transactions among users who do not know or do not trust each other are also possible. Both can pay each other using cryptocurrencies and fiat currency [11].

The Third example in the health field, Healthcare Data Gateway HGD is a smart application on smartphone enables patient to own, control and share their own data easily and securely on blockchain storage system without depending on a third party that may violate patient privacy. It provides a way to save and share healthcare data while keeping patients' privacy and avoiding scattering healthcare data in different systems [12]. Further, dHealthNetwork is a decentralized healthcare platform uses blockchain technology. It allows patients to easily access and shares their medical health information to experts from healthcare providers over the world and gets second opinion consultations in an affordable and a secure transparent platform [13].

3 Empirical Illustration of Blockchain Application Design

A common practice among business partners is to exchange original bills before closing a business deal. This practice is important as a second verification method to prevent fraud transactions. Most of the time a middle-man (agent/third party) takes care of transferring the bills based on approved schedule and budget. However, this process is inefficient as it causes delay in closing the deals, involves manual practices, which is prone to mistake, delay and privacy breach. ClaimChain application is built to eliminate the manual tasks of bill validation utilizing blockchain. The requirements are defined according to current practice in healthcare domain in Saudi Arabia. The following section illustrates the development methodology and its phases to build the application on blockchain.

3.1 Methodology

In order to build ClaimChain application, we are following the main four phases of software development life cycle (SDLC) analysis, design, implantation and testing [14]. We managed the development using scrum framework, which allows flexibility in requirements, minimizes risk and increases productivity. In the analysis phase, we first interviewed employees in insurance domain and receptionists in private healthcare providers. We extracted current business process and outline that using Business Process Model and Notation (BPMN). In this stage, we identified process weaknesses and potential benefits of adopting blockchain. In the design phase, we restructured the business process using blockchain, BPMN is also used to illustrate the new process. In this phase, the blockchain application design decisions [1] guide to design ClaimChain application. An open sources blockchain code is used at the implementation phase; we customize the code to fit the requirements. Finally, we perform unit testing, integration testing and user acceptance testing of our prototype. The user acceptance testing conducted with real employees in both healthcare and insurance domain.

3.2 Current Business Practice

This section illustrates the current process of transferring bills from healthcare provider to insurance company.

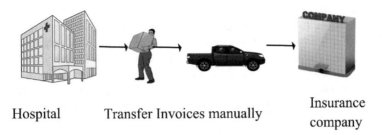

Hospital   Transfer Invoices manually   Insurance company

Fig. 2. Current practice transferring original bills from healthcare providers to insurance companies.

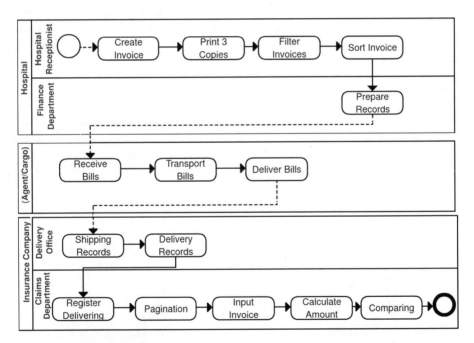

Fig. 3. Current business process of transferring original bills from healthcare providers to insurance companies.

The process starts once a patient with a valid insurance policy visits a healthcare provider. At the reception desk, patient issue her/his ID and insurance card. This proof of identity task is the first verification task for an insurance claim. The second verification task is verifying the validity of the insurance policy. A request for service approval is sent through policy number. If the insurance is valid the approval is received, based on the insurance policy the system issues a treatment bill. This bill includes the premium patient pay and the amount the insurance pay. Three copies of the bill are printed. The first one for the patent, the second one for finance and the third one to be send to the insurance company after approval from financial department in of healthcare provider.

A cargo service (middleman/agent) picks up pile of paper bills from the healthcare provider to the insurance company as shown in Fig. 2. Depending on agreement between insurance company and healthcare provider on how frequent cargo delivers the bills. In most cases, bills are delivered in a monthly basis. Once bills are received, a designated team needs to perform three main tasks. The first task unpacks the received bills, beside the physical effort required to accomplish this task, there is a small risk to pass pandemic to insurance office. The second task saves scanned copy of the bills then registers the amount of each bill into a balance sheet; even with the help of advance scanning machines this task is time consuming. The third task compares the balance sheet assembled by insurance team with the balance sheet sent by healthcare providers. These three tasks form the third validation task on an insurance claim. If the amount calculation is matching with the due amount, then payment is released without an issue.

On the other hand, if the amount does not match an investigation needs to take place. This process is captured after interviewing several employees working in different insurance companies; this is illustrated in BPMN diagram shown in Fig. 3.

3.3   Blockchain Application Design

This section highlights the business perspective of how blockchain will replace the middleman/agent transferring bills from hospitals to insurance companies in a secure and effective fashion as shown in Fig. 4. The application consists of three main components Bill Generator, Bill Retrieval and blockchain. Figure 5 illustrates the process with blockchain using BPMN.

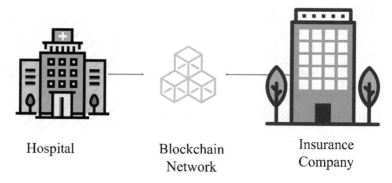

Hospital   Blockchain   Insurance
                  Network   Company

Fig. 4.   Blockchain network to store bills.

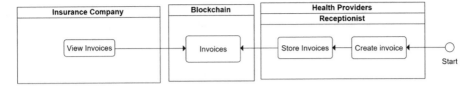

Fig. 5. ClaimChain business process.

The Bill Generator is a web application for hospitals allows authorized users to generate customer bills on the blockchain Network. The process at healthcare provider is the same as the conventional process. Also, the first and second verification tasks are the same, the reception desk verifies patient ID and insurance card then validity of the insurance policy is checked through insurance approval system. Approved bill issued by healthcare provider system is sent to the blockchain. A block is created includes bill information and joins the chain through mining process. The bills on the blockchain can be viewed by finance officers. Once a finance officer approves the bill the insurance company can retrieve the bill details through Bill Retrieval web application. This application provides access to bill information and generates reports to verify the balance sheet sent by healthcare providers. In this process, blockchain replaced the middleman/agent where bill information is encrypted, hashed and only accessed by the authorized insurance provider. Furthermore, the application generates the due amount without the need to manually revise all the bills.

3.4   Blockchain Data Structure

The structure of the blockchain is considered viable. The primary step of establishing the blockchain is storing the claim information (i.e. patient identity, service, bills, etc.) in blocks. The five Application Design Decisions highlighted in [1] helped us to design the blockchain structure of our application.

Application Design Decision 1: Where to keep the data and computation components, On-chain or off-chain. Although blockchain provides secure blocks of data, it requires high computational power to hash block information. This limits the storage capability on the blockchain. In ClaimChain application, the claim information of each patient is limited to codes, digits and brief description therefore all the detailed stored in the blockchain. For each day a blockchain is created includes block of bill information of the day. These blockchains are weekly built by healthcare employees and linked to the main chain. For both insurance companies and hospitals, only authorized individuals can access and modify the data in the blocks (see Fig. 6).

Fig. 6. Data structure in a blockchain.

Application Design Decision 2: This decision is related to access the scope of the blockchain (public, private or consortium/community) [15]. The use of ClaimChain application is limited to the insurance companies and health care providers who are our target users. To achieve higher privacy, each blockchain is private to a healthcare provider and insurance company. The right access only granted to healthcare provider to issue bills and retrieval access is granted for both parties. Figure 7 depicts the use of a private blockchain for each agreement between insurance and healthcare. The system structured to deal with M * N private blockchains, where M is the number of healthcare provider and N is the number of insurance company.

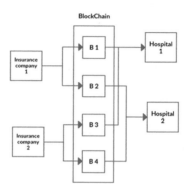

Fig. 7. Private blockchain for each agreement between insurance and healthcare provider.

Application Design Decision 3:   This decision is related to how many unique blockchain the system can maintain. Records can be separated according to parties or concerns. The previous section highlighted the use of a private blockchain for each agreement between a healthcare provider and an insurance company, therefore the application deals with M*N blockchains. Another separation decision is how and when insurance company accesses the bills: As soon as bill is approved? Every day? Every week? Or every month? On the daily practice a receptionist closes the day by handing the invoice copies and payment report to finance department. Therefore, the daily bills are linked together in a horizontal chain. Then a vertical chain links daily bills as shown in Fig. 8.

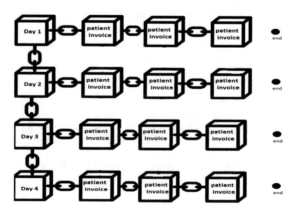

Fig. 8. Two-dimensional blockchain structure for daily bills.

Application Design Decision 4: This decision related where is the validation conditions of a block, is it external or internal. The external validation request addresses validation conditions manually or through external system. Whereas the internal validation request submits validation conditions to invoke smart contract functions of the blockchain. In ClaimChain application, healthcare providers are authorized to create the bills after verifying the identity of patient and validity of insurance policy. The first task is manual and second task conducted through communicating with insurance system. Therefore, validation of ClaimChain application is external.

Application Design Decision 5: This decision determines whether blockchain network participants have permission to manage the blockchain or only retrieve its information. Manage the blockchain means creation of block and optimizing the network rules. Permission less on blockchain allows network participants with sufficient computation power to change the records in the blockchain without proper consensus like the process on mining a Bitcoin whereas permission on blockchain participants and their rules are predefined. Participants with valid permission can submit transactions, mine block, create assets and manage permissions. The blockchain for ClaimChain is permissioned network where only healthcare providers create the blocks and insurance companies retrieve them.

3.5 Blockchain Application Implementation

To implement the prototype, we used python as a programming language, and JavaScript for designing the Graphical User Interface (GUI). Additionally, we used specific Node.js package manager (NPM) auditing commands to assure the required security and compliance. From the health provider side, the GUI allows to create invoices and displaying them by linking the python views page with the blockchain code. This linking helps storing the invoice data in a block in the blockchain and retrieving invoice information for displaying it to the user where the user id either the health provider or the insurance company.

Create an invoice: When ClaimChain application takes a user input, it makes a POST request to a connected node to add the transaction into the unconfirmed transactions pool. The transaction is then mined by the network, and finally it is fetched once we refresh the website.

To build the blockchain we used an open source Payton code that has the following blockchain functions:

- **Hash function:** We use the popular secure hash algorithm in python (sha256) from the library (hashlib) that hashes every block by 256 characters.
- **Proof of Work (PoW):** To ensure the validity of data, we call PoW function. We add additional constraint to this function to make our hash starts with two zeroes in order to determine a valid block.
- **Mining:** This process is used to add a block to the chain. Mining uses computing Proof of Work to confirm transactions. Once the constraints are satisfied, we can state that a block has been mined, and it is inserted into the blockchain.

- **Consensus:** When the chains of different participants in the network start to diverge, the consensus algorithm in ClaimChain is used to achieve agreement upon the longest valid chain in the blockchain; the longest chain is a good indicator of the most amount of work done in the system.

Retrieve Data from the Blockchain: A post method is fetched to get data from the node's chain endpoint, and the data, then, is locally parsed and stored.

3.6 Evaluation and Testing

Two types of testing and evaluation have been performed on the ClaimChain application: code testing and User Acceptance Testing (UAT). In this section, we briefly discuss the code testing and then describe the UAT in more detail.

3.6.1 Code Testing

Code Testing of the ClaimChain is completed using two types of measurements. The first type is a unit testing which is the initial level of software testing, which is performed on the source code mainly by the developers. Unit testing includes checking each individual unit/component of the software [9] to verify that the system is defect-free.

The second type is an integration testing which usually comes after the unit testing and is defined as a type of testing where software modules/components are logically integrated and tested as a group [9]. It concentrates essentially on the interfaces and the flow of data/information among modules. Regular integration testing executing is important in order to assure that the system is error-free. It is performed by checking the interfaces for each use-case with the intent of exposing errors. Operating both measurements of code testing will ensure that the ClaimChain is implemented correctly.

In order to perform integration testing, we tested the flow of the stored data in the blockchain, then retrieving data from the blockchain. The integration testing tests if the health provider can sign up and login correctly (Fig. 9(a)–(b)), if the program can transfer the health provider successfully to the home page (Fig. 10) for creating an invoice (Fig. 11) and sending it to the blockchain (Fig. 12) and if the health provider can show the invoices stored in the blockchain (Fig. 13).

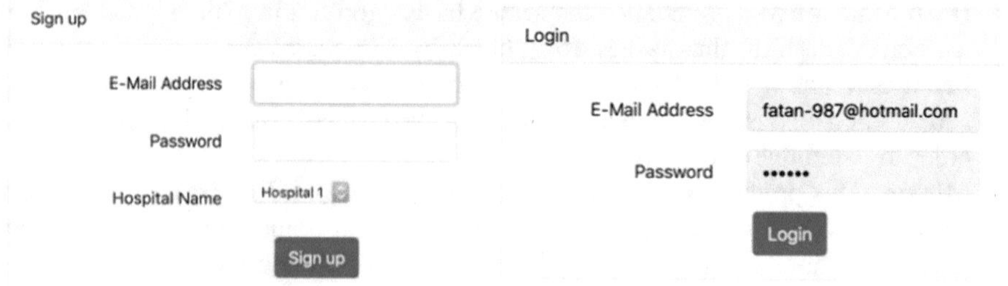

a. Sign Up Page   b. Login page

Fig. 9. Hospital sign up/login page.

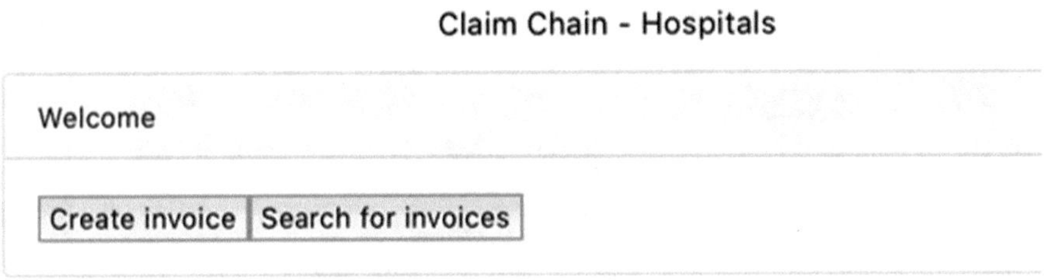

Claim Chain - Hospitals

Welcome

| Create invoice | Search for invoices |

Fig. 10. Hospitals home page.

Claim Chain - Hospitals

Create Invoice

Invoice Number

5460

Account Number

302414959

Medical Number

29880

Patient Name

Omer

Date of Birth

13/03/1996

Treatments taken

medicen2 ×   XYZ × |

Clinic

Hospital

Doctor Name

Ahmed

Insurance Percent

5

Amount: 300 SAR
Amount Covered by insurance: 15 SAR

Save Invoice

Invoice Number

5460   ✓

Account Number

302414959   ✓

Medical Number

29880   ✓

Patient Name

Omer

Date of Birth

13/03/1996

Treatments taken

medicen2 ×   XYZ ×

Clinic

Hospital   ✓

Doctor Name

Ahmed   ✓

Insurance Percent

5   ✓

Amount: 300 SAR
Amount Covered by insurance: 15 SAR

Save Invoice

> 127.0.0.1:5000 says
>
> Invoice saved successfully!
>
> OK

Fig. 11. Creating invoice page. 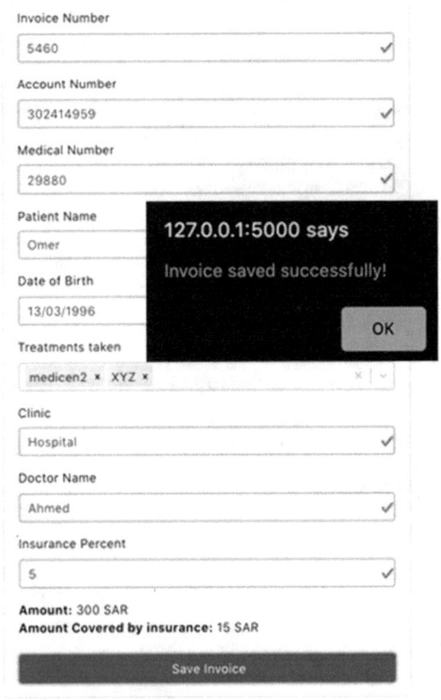   **Fig. 12.** Saving invoice.

After developing the blockchain technology and creating the interfaces for both hospital and insurance company, we connect the pages together and make the blockchain accessible for the insurance company as well.

Now the integration testing tests if the insurance company can sign up and login correctly (Fig. 14(a)–(b)), if the insurance company can search invoices by date or hospital name (Fig. 15) and display the invoice details (Fig. 16).

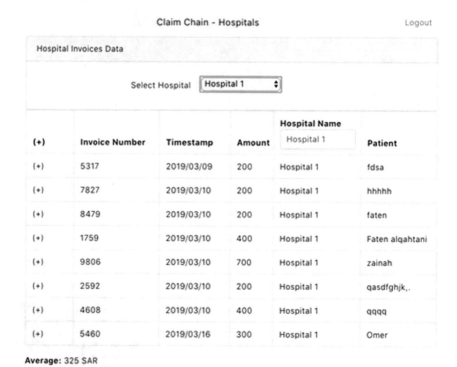

Fig. 13. Search for invoice by hospital name/date.

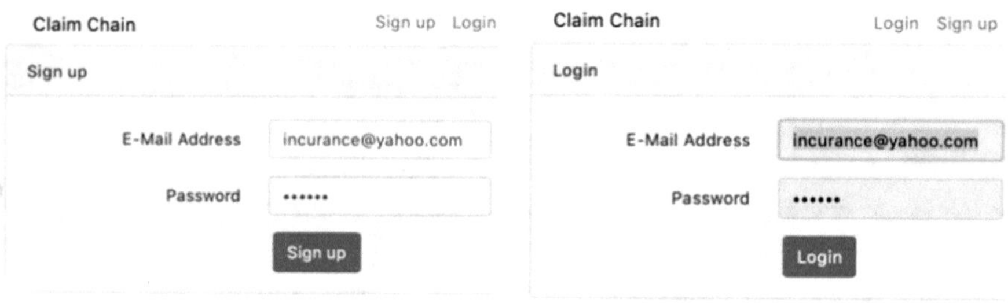

a. Sign Up Page   b. Login page

Fig. 14. Insurance company sign up/login page.

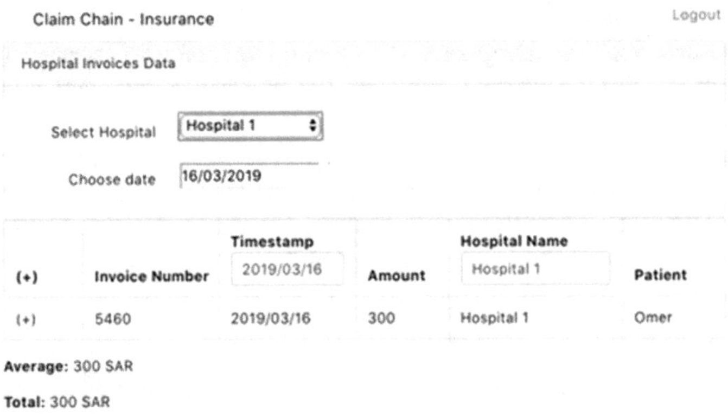

Claim Chain - Insurance   Logout

Hospital Invoices Data

Select Hospital | Hospital 1   ⬍ |

Choose date   16/03/2019

(+)	Invoice Number	Timestamp 2019/03/16	Amount	Hospital Name Hospital 1	Patient
(+)	5460	2019/03/16	300	Hospital 1	Omer

Average: 300 SAR

Total: 300 SAR

Fig. 15. Search for invoice by hospital name/date.

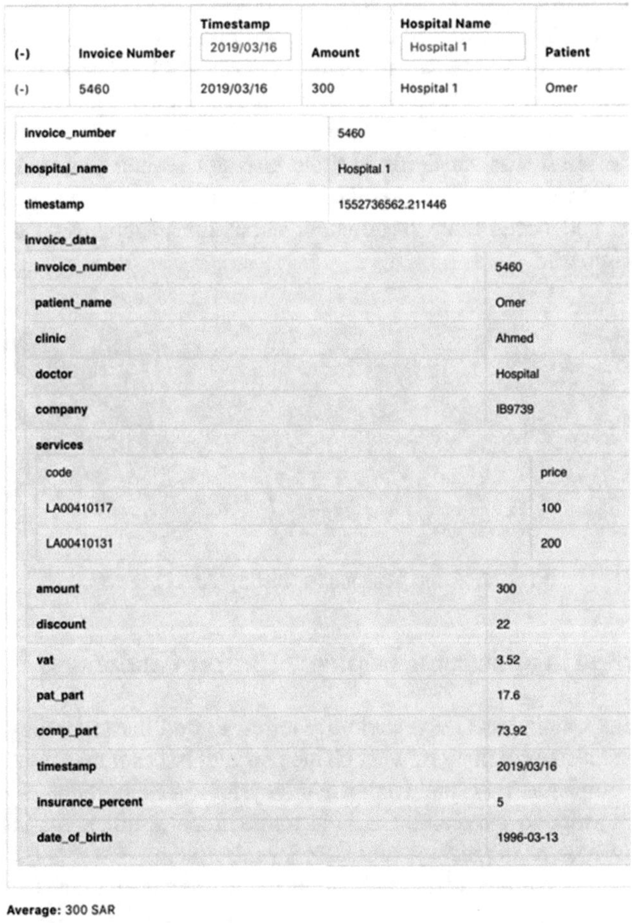

(-)	Invoice Number	Timestamp 2019/03/16	Amount	Hospital Name Hospital 1	Patient
(-)	5460	2019/03/16	300	Hospital 1	Omer

invoice_number	5460
hospital_name	Hospital 1
timestamp	1552736562.211446
invoice_data	
invoice_number	5460
patient_name	Omer
clinic	Ahmed
doctor	Hospital
company	IB9739
services	

code	price
LA00410117	100
LA00410131	200

amount	300
discount	22
vat	3.52
pat_part	17.6
comp_part	73.92
timestamp	2019/03/16
insurance_percent	5
date_of_birth	1996-03-13

Average: 300 SAR

Fig. 16. The invoice details.

3.6.2 User Acceptance Testing (UAT)

This section describes the acceptance testing and evaluation to measure how Claim-Chain project satisfies the user requirements. The first part of this section describes the acceptance test, followed by the System overall performance in the later section.

User acceptance testing (UAT), is a phase of software development in which the software is tested in the real world by the intended users [16, 17]. It includes a process for verifying that the presented solution stratifies the user needs [18]. Among different types of UAT, we chose the beta testing for the following reasons:

- Improving product quality via customer feedback.
- Reducing product failure risk via customer validation.
- Creating goodwill with customers and increases customer satisfaction.

Beta testing takes place in the user's environment and involves some extensive testing by a group of users who use the system in their environment. These beta testing then provide feedback, which in turn leads to improvements in the product [19]. It is done before the software is released to all customers.

The main criteria used in our UAT are, usefulness, saving time, accurately, and smoothness. Figure 17 shows some results of each UAT criteria for the users; the hospital, and the insurance company. Both found that our system is user friendly, clear and easy to understand, and learn it, and very helpful in getting automatic calculations. In general, the result was satisfying and we also got several feedbacks for possible improvement from the users, such as the fees part of the patient is not shown, needs more features, and needs more information about the patients (i.e., a label for the patient appointment).

Fig. 17. User Acceptance Testing (UAT) results.

4 Blockchain Application Implementation Challenges

Developing blockchain-based application requires a good understanding for this new technology, and also awareness of what challenges and issues it may entails. The major challenge of implementing the ClaimChain system was building the blockchain architecture in order to overcome the bills transactions problem for insurance companies and hospitals, as the main participants for this system. Goyal in [20] presents the 10 most challenges for any blockchain application implementation. In this light, we briefly explain the main challenges that we have encountered during Claimchain implementations:

- Data Privacy: In the proposed ClaimChain framework, the private patients information (i.e., name and address) and bills transactions will be available for all blockchain system participants. Therefore, convincing hospitals and insurance companies for existing of a reasonable fulfillment of data privacy in the blockchain technology was not an easy task.

- Transitioning difficulty: Insurance companies and hospitals in Saudi Arabia are using an old middleman transactions process for decades. Their staff and workers are well trained to use their old system structure to perform the required tasks. Leveraging the legacy process by adopting the blockchain technology, which needs a completely new data structures, will change the business strategy for the participants and may also get rid of some redundant jobs for the workers. In general, blockchain-based applications offer a solution that requires significant changes, or a complete replacement of existing systems [21].

- Insufficient blockchain literacy: The lack of needed information for explaining the blockchain technology, which are few and scattered in different websites and literature, makes developing a strong blockchain-based application difficult for us to solve the bill transactions problem. The blockchain is a new innovation and a wide specialization. It needs a considerable learning time to enhance our insight for programing this new ClaimChain system.

- Implementation and platform selection: There are numbers of open-source blockchain platforms exist. Each provides certain functionalities and solutions to the developers. With the lack of the experience and knowledge, selecting the most suitable blockchain platform is not easy and may come with a huge cost and time consuming. For instance, at the early stage of developing ClaimChain, we started with IBM Hyperledger fabric framework that was meant to make developing blockchain projects easier. However, during the installation, we encountered multiple compatibility and technical issues that made us, after a considerable amount of time, to switch to a python-based Flask framework. In most cases, the implementation methods are not clearly explained which also required effort and time for understanding a whole system configuration.

- Costs: In general, insurance companies and hospitals need to spend a lot of money for implementing and developing the blockchain-based applications, training employees, and setting up the infrastructure [22], which may hold them back from adopting such new technology in their organizations.

5   Conclusion

We expand the use of blockchain to the insurance sector by presenting a new system called ClaimChain. ClaimChain will allow hospitals to transfer bills information directly to the insurance companies without the need for the middleman. Insurance companies in Saudi Arabia, usually, use old manual methods for entering the hospitals' invoices data into their financial systems, which might cause information loss besides wasting the time and effort. Therefore, ClaimChain system can assist those companies not only for effectively transferring and securing their information but also minimizing the number of employees needed for checking and scanning patients' bills and insurance information. ClaimChain system also provides a paperless work environment for both, hospitals and insurance companies. In this paper, we highlighted our experience in building blockchain application. This is an empirical example that can guide other software developers on what decisions need to be considered when developing blockchain application. The paper also highlighted the challenges faced throughout development. With regards to future work, we need to implement more functions as recommend by target users after conducting user acceptant test. Another important work needs to be explored is the application integration with blockchain platforms. As we experienced difficulties to find support or tutorial guide how to integrate GUI with the available blockchain platform such as Hyperlager and Multichain.

References

1. Xu, X., Pautasso, C., Zhu, L., Gramoli, V., Ponomarev, A., Tran, A.B., Chen, S.: The blockchain as a software connector. In: WICSA (2016)
2. Beck, R., Avital, M., Rossi, M., Thatcher, J.B.: Blockchain technology in business and information systems research. Bus. Inf. Syst. Eng. **59**(6), 381–384 (2017)
3. Pillai, B., Muthukkumarasamy, V., Biswas, K.: Challenges in designing a blockchain platform (2017)
4. Mendling, J., Weber, I., Van Der Aalst, W., Brocke, J.V., Cabanillas, C., Daniel, F., Debois, S., Ciccio, C.D., Dumas, M., Dustdar, S., et al.: Blockchains for business process management-challenges and opportunities. ACM Trans. Manage. Inf. Syst. (TMIS) **9**(1), 4 (2018)
5. Xu, X., Lu, Q., Liu, Y., Zhu, L., Yao, H., Vasilakos, A.V.: Designing blockchain-based applications a case study for imported product traceability. Future Gener. Comput. Syst. **92**, 399–406 (2019)
6. Zheng, Z., et al.: Blockchain challenges and opportunities: a survey. In: Workshop Paper. Inderscience Publishers, Geneva (2016)
7. Buterin, V.: A next generation smart contract & decentralized application platform, February 2015. https://www.ethereum.org/pdfs/EthereumWhitePaper.pdf/
8. A Decentralized Global Home Rental Platform Whitepaper. 13 April 2019. https://cryptorating.eu/whitepapers/Rentberry/Rentberry_Whitepaper_en.pdf
9. What is Software Testing? Introduction, Definition, Basics & Types. https://www.guru99.com/software-testing-introduction-importance.html. Accessed 2 Apr 2019
10. Sikoba, a Decentralized P2P IOU Platform on blockchain, Launches Presale Ahead of Token ICO, 13 April 2019. https://www.prnewswire.com/newsreleases/sikoba-a-decentralized-p2p-iou-platform-on-blockchain-launches-presaleahead-of-token-ico-300446785.html
11. Yue, X., Wang, H., Jin, D., Li, M., Jiang, W.: Healthcare data gateways: found healthcare intelligence on blockchain with novel privacy risk control. J. Med. Syst. **40**(10), 218 (2016)
12. dHealthNetwork Transforming the Health Industry, 13 April 2019. https://www.dhealthnetwork.io/
13. Glaser, F.: Pervasive decentralization of digital infrastructures: a framework for blockchain enabled system and use case analysis. In: Proceedings of the 50th Hawaii International Conference on System Sciences (2017)
14. Sommerville, Software engineering. Dorling Kindersley, New Delhi (2011)

15. Integration Testing: What is, Types, Top Down & Bottom Up Example. https://www. guru99.com/integration-testing.html. Accessed 2 Apr 2019
16. What is user acceptance testing (UAT)? WhatIs.com. Accessed 20 Apr 2019
17. Davis, K.: User acceptance testing (UAT). https://searchsoftwarequality.techtarget.com/ definition/user-acceptance-testing-UAT. Accessed 30 Mar 2019
18. Cimperman, R.: UAT Defined: A Guide to Practical User Acceptance Testing. Addison Wesley Professional, Upper Saddle River (2006)
19. Alpha Testing Vs. Beta Testing: What's the Difference? https://www.guru99.com/alpha-beta-testing-demystified.html. Accessed 30 Mar 2019
20. Goyal, S.: Top 10 Enterprise blockchain Implementation Challenges, 5 March 2019. https:// 101blockchains.com/enterprise-blockchainimplementation-challenges/#prettyPhoto. Accessed 15 Apr 2019
21. blockchain Council|blockchain-council.org: Top 5 Challenges with Public blockchain, 3 July 2018.   https://www.blockchaincouncil.org/blockchain/top-5-challenges-with-pubilc-blockchain/. Accessed 13 Oct 2018
22. Southern Cross University Online (SCU Online): Can blockchain and its possibilities live up to the hype? 7 March 2018. https://online.scu.edu.au/blog/the-benefits-and-challenges-of-blockchain/. Accessed 15 Apr 2019

Using Formal Languages to Elicit Requirements for Healthcare Blockchain Applications

Bella Pavita[1] and Vinitha Hannah Subburaj[2(✉)] ⓘ

[1] School of Engineering, Computer Science and Mathematics, WTAMU,
Box 60767, Canyon, USA
bpavital@buffs.wtamu.edu
[2] West Texas A&M University, Canyon, TX 79016, USA
vsubburaj@wtamu.edu

Abstract. Specifying requirements for complex systems has an important role during software development. At times, the importance of the specification stage gets neglected by software engineers while developing software systems. Software engineering life cycle models are improved for better software quality and reliability. Even though powerful implementation strategies exist, if the requirements of a system to be developed are unclear and not specified properly, then the product can fail to satisfy user needs. Traditional methods use natural language for specifying the requirements of the software systems. The use of formal specifications to specify software systems remains a challenging research problem that needs to be addressed. In this research effort, several healthcare applications that are implemented using Blockchain are studied. Blockchain is a recent invention that uses the idea of decentralization to enforce security. Such applications come with complexity and specifying requirements for such systems using formal specification languages gets even more challenging. The challenges involved in writing formal specifications for such complex applications has been studied in this research effort. The Descartes specification language, a formal executable formal specification language, is used to specify the blockchain healthcare applications under study. Case study examples are used to illustrate the specification of Blockchain healthcare application requirements using the Descartes specification language.

Keywords: Blockchain applications · Security · Formal specifications

1 Introduction

Transactional inefficiency is a major concern in the healthcare industry. The healthcare system is currently losing $300 billion each year in poor data integration [1]. Researchers have discussed how blockchain may offer healthcare an effective and secure system to manage data transactions. Fundamentally, blockchain solves almost all transactional problems. Blockchain is a distributed digital ledger that keeps track of transactions from one party to another. Since it is distributed, its record is shared amongst each transactional entity [2]. Every transaction that is done must be matched

© Springer Nature Switzerland AG 2020
K. Arai et al. (Eds.): FTC 2019, AISC 1069, pp. 802–815, 2020.
https://doi.org/10.1007/978-3-030-32520-6_58

and verified by all parties. Moreover, previous transactions are immutable and protected by cryptography. These steps ensure the security and validity of the data transferred. Furthermore, the inefficiency of the need to reconcile several copies of the ledger will be removed because every ledger is in sync and in consensus with each other constantly. Today, Fortune 50 companies and countless organizations are developing blockchain technologies to solve similar problems. Therefore, there is a massive opportunity for a blockchain technology to revolutionize healthcare.

Some of the biggest areas that can be revolutionized to create a patient-focused system are health records and pharmaceutical logistics. First, health records are the backbone of every modern healthcare system. Healthcare data can be created, copied, and modified faster than ever before. Also, since every hospital and doctor's office has a way to store them, it is not always feasible for healthcare providers to reconcile this massive data. The goal of implementing blockchain is to give patients authority over their entire medical records and to provide a one-stop access to physicians while inherently bringing data security to the field. Blockchain can create a universal patient-identifier that aggregates data across multiple IT systems without having to move the data into a central location. Second, the pharmaceutical industry has the highest standards for product safety, security, and stability. Blockchain can monitor supply chain management reliably and transparently. This can greatly increase time efficiency and human error. The automation of blockchain architecture can also be used to record production costs, labor, and even waste in emissions in the supply chain. It can also authenticate of the origin of drugs to combat the counterfeit market.

These are just two possible solutions that blockchain can bring. With all these possibilities, blockchain engineers need a way to transform systems requirements into a much lower level of abstraction. The goal is to be able to maximize efficiency of interaction between user and engineers. In the software development cycle, the requirement elicitation phase is the most important phase to avoid excess maintenance and testing towards the end. With a wide gap of abstraction between the client's requirements and programming languages, the requirement's ability to be proven, tested, and compared is becoming very difficult. The engineers should be able to see the specifications and the source code together and be able to evaluate that they match each other. The objective of this research is to come up with the best formal language to elicit reliable specifications in building a healthcare blockchain application.

Formal specification language lies between natural language (English) and high-level language (programming languages). They are mathematically based techniques used to describe a system, analyze its behavior, and aid systems design by verifying key properties of interest through rigorous and effective reasoning tools [3]. It is important to understand that formal specifications describe what a system should do, not how the system should do it. Formal specification languages such as Z Notation, VDM, B Method, and Descartes have been used in the past to formally specify software systems [4, 5].

The remainder of this paper is structured as follows:

Section 2 discusses the existing work related to blockchain technology, the use of formal specifications to specify blockchain requirements in healthcare industry, and the Descartes specification language used to specify the blockchain healthcare applications. Section 3 discusses the functional and non-functional requirements of three blockchain healthcare applications, namely MedRec, NeuRon, and Nebula Genomics. Section 4 provides a case study example that illustrates the application of the Descartes specification language to formally specify the blockchain healthcare application MedRec. Section 5 summarizes the paper with a brief discussion of future work.

2 Background

2.1 Blockchain Technology

Blockchain technology has been gaining its popularity due to its nature to work in a trust-less environment. First introduced a decade ago by Satoshi Nakamoto, blockchain provides a globally-consistent append-only ledger that operates in a decentralized way [6]. These properties of blockchain are achieved by automated mechanisms such as *proof-of-work* mining scheme, smart contracts, and cryptographic hash function. Blockchain technology has been proposed by experts to solve data persistence, anonymity, fault-tolerance, auditability, and resilience problem in network security [7].

Other than cryptocurrencies, blockchain technology can be applied in other sectors that requires ultimate security. One example is to apply blockchain in cloud technology as an identification apparatus. Instead of storing a cryptocurrency's assets, the blockchain can store other digital assets such as Service Level Agreements (SLA), policies, certificates, and profile identities [8]. Another example would be managing an Electronic Health Record (EHR) system. Using the blockchain technology, the EHR system can be revolutionized; from being centralized and kept in a single medical institute, to being decentralized [9].

2.2 Formally Specifying Blockchain Applications

With all the possibilities that blockchain technology can bring, blockchain engineers need a way to transform systems requirements into a much lower level of abstraction. The goal is to be able to maximize efficiency of interaction between user and engineers. In the software development cycle, the requirement elicitation phase is the most important phase to avoid excess maintenance and testing towards the end [10]. With a wide gap of abstraction between the client's requirements and programming languages, the requirement's ability to be problem tested and compared is becoming very difficult. The engineers should be able to see the specifications and the source code together and be able to evaluate that they match each other.

Numerous researches had been conducted in writing formal specifications to be able to verify blockchain end-products. The first approach in modelling blockchain technology in formal language was done by using a functional programming language made for program verification called F*. First, the contracts were written in Solidity, and then compiled into an Ethereum Virtual Machine (EVM) byte-code. To ensure the verification was free of compiler-errors, the lower-level EVM code was also analyzed. The source codes of the contracts were obtained from [11]. The F* language was chosen because it has a rich type system that includes dependent types and monadic effects, which the researchers used to generate automated queries for an SMT solver to statically verify each properties of the contracts [6].

The second approach was to design and implement a new language for smart contracts called IELE, which addresses EVM's limitations. IELE is a language similar to Lower Level Virtual Machine (LLVM) that was specified formally and, in its implementation, generated a virtual machine. IELE was the first practical language that was designed and implemented as a formal specification with one of its biggest users, IOHK, a major blockchain company. The language was made to be both human readable and suitable as a compilation target for other high-level languages. Contracts in the IELE VM would be executed in the exact form as it was verified. This approach was enabled by a framework called the K Framework. The developers of this language also made a compiler to compile codes from Solidity to IELE [12].

The third approach was describing a formal verification protocol used by Authcoin, an authentication platform that uses a blockchain-based storage system. First, a model was created in Colored Petri Nets (CPNs), a graphical language used for constructing models of concurrent systems. This was done to eliminate eventual design flaws, missing specifications, and security issues. Then, a framework called CPN-Tools was used for designing, evaluating and verifying the CPN models [13].

2.3 The Descartes Specification Language

Urban [14] in 1977 developed a formal specification language named the Descartes specification language that uses Hoare tree structure to formally specify software systems. This specification language was designed in such a way that this can be used throughout the software engineering life cycle. Specification are written in such a way that the output data is a function of input data. Structuring methods namely direct product, discriminated union, and sequence are used to specify the systems. Direct product is used specify concatenated set of elements. If we have to specify a structure where one element is selected out of a set of elements, then discriminated union (+) is used. Zero or more repetition of a set of elements gets specified using the structuring method called sequence (*).

In the Descartes specification language, the Hoare tree has a sequence node and a sub node. The grammar that defines the specification of a literal, a node, a module in the Descartes specification language makes is easy for one specify any complex systems using this formal specification language. The original Descartes through powerful was used only to specify small less complex systems. During 2013, Subburaj [15] extended the original Descartes specification language to specify agent oriented systems. The extended formal specification language called as the Descartes – Agent used top-down design approach to specify large complex agent systems. Some of the new constructs that were added to the original Descartes specifications language were agent goal, attributes, roles, plans and communication protocol. Again, Descartes – Agent was taken into study in [16] and was extended to model the security requirements of Multi Agent Systems in.

3   Requirements of Healthcare Blockchain Applications

3.1   MedRec

MedRec is a decentralized medical record keeper designed to help patients move between providers. This is done by replacing centralized intermediaries by using decentralized framework such as Ethereum. Instead of a single node holding the data, the data is encoded in metadata and distributed across disparate providers. This project is being developed by the team at the MIT Media Lab [9].

3.1.1   Participants
The actors involved in the MedRec are defined as the following:

1. Patient – The owner of each medical records.
2. Provider – The modifier of each medical records.

3.1.2   Functional Requirements

1. Exchange value without deference to a middleman but using *smart contracts*. Enabling complex transactions to be ran on the blockchain, such as creating pointers to the specific locations of medical data and authentication.
2. Must create different types of contracts that can be identified to patients, providers, and more.
3. The only one who can add new information onto the blockchain are certified institutions. New data can only be added with the approval of the patient.
4. Everyone in the system can track a summary of their previous relationship with each other. This summary contains current and previous interactions with other nodes, a status variable that specifies when the relationship was initialized and change of relationships statuses by the patient.
5. There is an off-chain infrastructure that acts as Database Gatekeeper. This infrastructure runs a server that listens to query requests. These requests must be cryptographically signed by the creator. If the user can access the query, the gatekeeper returns the result to the client.
6. There is a local database at each patient nodes to act as a cache storage of the patient's information.
7. Must have a private peer-to-peer network of verified nodes acting as a Global Registrar. This maps all public identities to Ethereum addresses (this may contradict the decentralized aspect, but it is not unreasonable).
8. Must have a consensus mechanism to the main Ethereum blockchain.
9. Provider nodes are required to do mining mechanisms to sustain the distributed ledger. This requires spare computing power.

3.1.3   Non-functional Requirements

Accessibility

- The patient node must be light-weight and can be executed smoothly on a personal computer or a mobile device.
- There need to be a modular interoperability protocol to be able to interface with any provider backend and UI implementation.

Privacy

- Privacy of patient must be maintained. A patient's identity must not be able to be inferred from metadata of blockchain address.
- All nodes and miners must be a fully permissioned medical researchers.
- A full k-anonymity analysis of the query structure must be passed.
- Provider's identity must be distributed across the network instead of in a single provider address. This is done by using off-blockchain transaction with a verified provider at random to gather the information of the verified delegate.

Security

- Authorization data maintenance must be done in each node in the network.
- Storage of records must be persisted in both patient and relevant provider nodes.
- Single point of failure and tampering must be avoided because it may cause conflict and fail the consensus.
- Must be invulnerable to SQL injection attacks.

Scalability

- Only important changes done on medical record is recorded on the blockchain network, for example creation and modification.

3.2   NeuRon

NeuRoN is a decentralized artificial intelligence network designed to collectively training materials from patients. Using blockchain allows the AI to be able to authenticate where the data comes from. It also makes the data collection unbiased because it does not come from a control trial. The usage of blockchain also prevent breach of data. This project is being developed by Doc AI.

To start an epoch, first a research organization will broadcast their proposal. Users with appropriate attributes accept the training task and will be rewarded tokens for onboarding onto the platform and help train the model. After the proposal is completed, a new AI model is trained and distributed to the users who agreed to continue training the model. In the end, the synchronization phase will aggregate weights and learnings into a new model. This model is then returned to the users to predict their cases [17].

3.2.1 Participants

The actors involved in the NeuRoN learning phase are defined as the following:

1. Research organizations – Organizations who are interested in using NeuRoN architecture to train their models.
2. Users – People who are interested in interconnecting their devices and information to the network.
3. Neural Network Models – Neural models trained with the user's data.

3.2.2 Functional Requirements

1. Personal Portable Biology File
   a. A personalized biology profile that is global, portable, user-friendly and quantifiable.
   b. Has multi-signature capabilities.
   c. Displays biometrics information that users have entered.
   d. Lists health problems that occurred in medical record.
   e. Lists doctor's notes and procedures done.
   f. Lists allergies, microbiome, and immunizations of the patient.
   g. Lists contact persons for plan of care.
2. Onboarding Module with Computer Vision
   a. Selfie2BMI module uses Deep Neural Networks to predict anatomic features such as height, weight, BMI, age, and gender from face.
   b. Also monitors 23 facial attributes like skin, receding hairlines, wrinkles, teeth, etc.
3. Blood Test Decoder
   a. Deep conversational agent for users to discuss questions on 400+ blood biomarkers. This AI is designed to educate users using interactive content.
   b. Conversation can be personalized based on user's age, gender, and pre-conditions.
4. Genomics Test Decoder
   a. Deep conversational agent designed to provide genetic counseling after doing a genomics test.
   b. The AI records every visit and recommendation it has given.
5. Medicine Decoder
   a. Deep conversational agent designed to provide advice on medication dosage, side effects, and other personalized questions.
   b. It will be connected to the pharmacogenomic recommendation engine if genomic results are present.

3.2.3 Non-functional Requirements

Accessibility

- The patient node must be light-weight and can be executed on mobile phones and tablets. This includes execution of sensors that records patient's data.
- Distribute AI computation load to clients that are connected to a central server for batch training updates.
- Token compensation is made to operate as a free market. This is done to reflect the supply and demand of the data available to the network.

Privacy

- Users can accept or reject the model training offer.
- By decentralizing the data across all users, the data is cryptographically times-tamped and becomes immutable.
- To maintain Health Insurance Portability and Accountability Act (HIPAA) requirements, the data will be maintained on the edge device and not in a centralized server.

Security

- Must require a fee per attempted transaction in order to prevent malicious attacks such as DDoS of Sybil.
- The cost of participating in the network must increase in proportion to the demand on the mining network. This way, if an attacker initialized a high volume of transactions in a small amount of time, the cost should increase sharply.
- Tokens will be given back at the end of training phase or when the research organization concludes the learning.
- Use blockchain as a method to track and authenticate data sources in order to make great predictions.

3.3 Nebula Genomics

Nebula Genomics is a blockchain-enabled genomic data sharing and analysis platform. Since over the years the price of DNA sequencing has been decreased exponentially, it is predicted that within the next few years, genomic big data is predicted to outgrow text and video data. Nebula Genomics is attempting to pioneer DNA sequencing by making a stable and secure platform while removing middlemen. As more DNA sequences gathered and profiled, it will be much easier to interpret traits and diseases as compared to DNA genotyping. Nebula Genomics plans to facilitate the transactions using Nebula Tokens and build the stack on Blockstack platform and Ethereum-derived blockchain [18].

3.3.1 Participants

The actors involved in the Nebula Genomics model are defined as the following:

1. Individuals – genomic data owners.
2. Pharma companies – genomic data buyers.
3. Nebula sequencing facility – processes individual's genomic data.

3.3.2 Functional Requirements

1. The collection of high-quality trait information and the genomic data fragmentation problem will be solved by direct communication with data owners and the use of smart survey tool.
2. The Nebula smart contract-based survey tool enables data buyers to design multi-staged surveys that consist of interdependent questions and elicit true responses.
3. Data buyers can communicate with individuals to ask specific questions about the phenotype.
4. Data standardization helps data buyers curate vast datasets and reduce costs.
5. The use of smart contracts significantly accelerates data acquisition by automating the process of signing contracts, making payments, and transferring data.
6. Encoding scheme must support efficient computations, particularly for machine learning. This is done by representing the human genomes in hash arrays.
7. All nodes in the network can access the genome tile library containing individual's genome hash arrays.

3.3.3 Non-functional Requirements

Accessibility

- Decentralization of data storage allows for space-efficient data encoding. This may solve the problem of the growth of genomic data in the future.
- Decentralization removes limitation of network transfer speed in data sharing.
- Decentralization also solves the large amount processing and analysis times needed to handle genomic big data.

Privacy & Security

- Data owners will privately store their genomic data and have control in permissions.
- Shared data must be protected through zero-trust computing where the owners will be anonymous, and buyers will be required to be fully verify their identities.
- Shared data is encrypted and securely analyzed using Intel Software Guard Extensions (SGX) and homomorphic encryption.
- Data buyers shall never see genomic data in plaintext.
- All data transactions are immutable.

4   Formally Specifying MedRec Using the Descartes Specification Language

MedRec uses three kinds of smart contracts; registrar contract, summary contract, and patient-provider relationship contract. These contracts are used to locate and authenticate Electronic Health Record (EHR) locations. The registrar contract maps the user's IDs to their Ethereum addresses. This is where the regulation of new identities are encoded. The summary contract references the address of each identity string. It acts as a trail of breadcrumbs so each user can locate a summary of their previous relationships with others users. The patient-provider relationship contract connects the node that stores and manages medical records.

This architecture used in MedRec 2.0 involved two nodes, namely the provider and the patient node. Each node has an EHR manager, backend library, database gatekeeper, and its own Ethereum client that is connected to MedRec. Each Ethereum client contains the logic for creating and accessing summary and patient-provider relationship. As explained by [9] and referenced in the requirements document, patient approval is needed to be able to update the patient's EHR. The database gatekeeper executes a server listening to query requests which are cryptographically signed by the user. Also, there is a local database at each patient nodes to act as a cache storage of the patient's EHR. The read request is a consensus mechanism to the main Ethereum blockchain. From these parameters, we can produce a formal specification for updating patient's node as shown in Fig. 1.

The update patient node module accepts three inputs; patient node, provider node, and queries. Queries are stored in the database gatekeeper. The role of the database gatekeeper in the provider node is to propose queries to the patient node. Then, the database gatekeeper in the patient node lets the patient review the queries before the USER_PERMISSION node keeps track of the decisions. The return node evaluates the user permissions and calls the UPDATE node if necessary.

The role of the Ethereum client is to keep track of the location of each record. It also gives the provider node the ID of the patient according to the patient-provider relationship contract. Thus, accessibility is limited to only authorized providers. In this example, the relationship between the patient and the provider is assumed to be created prior to the update.

The provider node is proposing two update queries to the patient; to revise a previous record and to insert a new record. In the USER_PERMISSION node, the revision is permitted, but the insertion is not. Thus, the return node only calls the UPDATE node once for the revision query.

The Descartes specification for viewing patient's EHR is given in Fig. 2. Notice that there is an additional summary contract accessed on both nodes. The status contained in the summary contract is the indicator of whether the relationship has been approved by the patient. The patient is eligible for full control of acceptation, rejection, and deletion of the relationships.

```
UPDATE_(PATIENT_NODE)_FROM_(PROVIDER_NODE)_USING_(QUERIES)

PROVIDER_NODE
    EHR_MANAGER
    BACKEND_LIBRARY
    ETHEREUM_CLIENT
        READ_REQUEST(ETH_ID)
        GLOBAL_REGISTRAR(ETH_ID)
            ETH_ADDRESS
        PATIENT_PROVIDER_RELATIONSHIP(ETH_ID)
            PATIENT_ID
    DATABASE_GATEKEEPER
        QUERIES*

PATIENT_NODE
    EHR_MANAGER
    BACKEND_LIBRARY
    ETHEREUM_CLIENT
        READ_REQUEST(ETH_ID)
        GLOBAL_REGISTRAR(ETH_ID)
            ETH_ADDRESS
        PATIENT_PROVIDER_RELATIONSHIP(ETH_ID)
            PROVIDER_ID
    DATABASE_GATEKEEPER
        QUERIES*
    LOCAL_DATABASE
        RECORDS*
                record_id
                time
                place
                content

USER_PERMISSION(QUERIES)+*
    QUERIES*
        permitted
        not_permitted

UPDATE*
    QUERIES
    RECORDS|
        record_id
        time
        place
        content


return
            PROVIDER_NODE
            PROVIDER_ID
            PATIENT_NODE
            PATIENT_ID
            USER_PERMISSION(QUERIES)+
                PERMITTED
                        UPDATE
                NOT_PERMITTED
                        'Update not permitted'
```

Fig. 1. Update module of MedRec application specified using the Descartes specification language

```
VIEW_(PATIENT_NODE)_FROM_(PROVIDER_NODE)_USING_(QUERIES)
    PROVIDER_NODE
        EHR_MANAGER
        BACKEND_LIBRARY
        ETHEREUM_CLIENT
            READ_REQUEST(ETH_ID)
            GLOBAL_REGISTRAR(ETH_ID)
                ETH_ADDRESS
            PATIENT_PROVIDER_RELATIONSHIP(ETH_ID)
                PATIENT_ID
            SUMMARY_CONTRACT(ETH_ID)
                PATIENT_PROVIDER_RELATIONSHIPS*
                    STATUS
        DATABASE_GATEKEEPER
            QUERIES*

    PATIENT_NODE
        EHR_MANAGER
        BACKEND_LIBRARY
        ETHEREUM_CLIENT
            READ_REQUEST(ETH_ID)
            GLOBAL_REGISTRAR(ETH_ID)
                ETH_ADDRESS
            SUMMARY_CONTRACT(ETH_ID)
                PATIENT_PROVIDER_RELATIONSHIPS*
                    STATUS
        DATABASE_GATEKEEPER
            QUERIES*
        LOCAL_DATABASE
            RECORDS*
                record_id
                time
                place
                content

    USER_PERMISSION(QUERIES)+*
        QUERIES*
            permitted
            not_permitted

    VIEW*
        QUERIES
        RECORDS
            record_id
            time
            place
            content

    return
        PROVIDER_NODE
        PROVIDER_ID
        PATIENT_NODE
        PATIENT_ID
        USER_PERMISSION(QUERIES)+
            PERMITTED
                VIEW
            NOT_PERMITTED
                'View not permitted'
```

Fig. 2. View module of MedRec application specified using the Descartes specification language

Similar to the update patient node, the view patient node module accepts three inputs; patient node, provider node, and queries. Instead of appending or updating a list of records, the view patient node involves only select queries. Since a relationship must be initialized before viewing the patient node, the summary contract is needed.

5   Discussion of Research Findings, Conclusion, and Future Work

Requirements of three Blockchain healthcare applications namely MedRec, NeuRon, and Nebula Genomics were studied in this paper. The requirements stated in natural language for MedRec has been converted into formal specifications using the Descartes specification language. Some of the observations made through this research effort were: (1) The development of complex software systems like blockchain starts with the analysis of requirements to capture the functionalities of the system, followed by the design that elaborates the underlying feature then leading to the actual implementation. (2) Formal methods which are mathematical rules when applied to the software development process provide a way to validate the software systems under development. (3) Blockchain healthcare applications that gets specified by a formal language could be analyzed right from the initial stages of software development until the end to prevent unreliable software product. Key challenges observed while formally specifying requirements for Blockchain healthcare applications were: understanding the requirements of Blockchain applications in general and how they get used across healthcare industries; choice of formal constructs than can completely and correctly specify blockchain healthcare applications; identification of requirements early on during the development of healthcare applications and how to translate them into formal specifications; and lack of existing frameworks that define the underlying architecture of blockchain applications used in healthcare industry. The specification language used in this research effort can be used to specify software systems that have partial requirements. This feature allows for specifying blockchain requirements that are complex and coarse.

The challenging aspect of incorporating a formal executable specification language to specify security requirements for blockchain healthcare applications has been accomplished in this research effort. In this research effort, the MedRec application [9] was taken into study and was used to demonstrate the use of the Descartes specification formal specification language to specify blockchain healthcare applications. The blockchain healthcare specifications can be used as a model to be applied towards other similar applications.

As future work, a standardized blockchain framework for formally specifying any healthcare applications will be developed. Also, an automated software will be developed in the future to auto specify and validate specifications of blockchain healthcare applications written using a formal specification language. The overhead of manually verifying the correctness of formal specifications could be improved in the future research effort through software automation.

References

1. Manyika, J., Chui, M., Brown, B., Bughin, J., Dobbs, R., Roxburgh, C., Byers, A.H.: Big data: the next frontier for innovation, competition, and productivity (2011)
2. Iansiti, M., Lakhani, K.R.: The truth about blockchain. Harvard Bus. Rev. **95**(1), 118–127 (2017)
3. Hierons, R.M., Bogdanov, K., Bowen, J.P., Cleaveland, R., Derrick, J., Dick, J., Lüttgen, G.: Using formal specifications to support testing. ACM Comput. Surv. (CSUR) **41**(2), 9 (2009)
4. Pandey, Tulika, Srivastava, Saurabh: Comparative analysis of formal specification languages Z, VDM and B. Int. J. Current Eng. Technol. **5**(3), 2086–2091 (2015)
5. Subburaj, V.H., Urban, J.E.: Formal specification language and agent applications. In: Intelligent Agents in Data-intensive Computing, pp. 99–122. Springer, Cham (2016). https://doi.org/10.1007/978-3-319-23742-8_5
6. Bhargavan, K., Delignat-Lavaud, A., Fournet, C., Gollamudi, A., Gonthier, G., Kobeissi, N., Kulatova, N., Rastogi, A., Sibut-Pinote, T., Swamy, N., Zanella-Béguelin, S.: Formal verification of smart contracts: short paper. In: Proceedings of the 2016 ACM Workshop on Programming Languages and Analysis for Security, pp. 91–96. ACM, October 2016
7. Abdellatif, T., Brousmiche, K.L.: Formal verification of smart contracts based on users and blockchain behaviors models. In: 2018 9th IFIP International Conference on New Technologies, Mobility and Security (NTMS), pp. 1–5. IEEE, February 2018
8. Singh, I., Lee, S.W.: Comparative requirements analysis for the feasibility of blockchain for secure cloud. In: Asia Pacific Requirements Engineering Conference, pp. 57–72. Springer, Singapore, November 2017
9. Ekblaw, A., Azaria, A., Halamka, J.D., Lippman, A.: A case study for blockchain in healthcare: "MedRec" prototype for electronic health records and medical research data. In: Proceedings of IEEE Open & Big Data Conference, vol. 13, p. 13, August 2016
10. Eid, M.: Requirement Gathering Methods (2015)
11. Etherscan Homepage. http://etherscan.io
12. Kasampalis, T., Guth, D., Moore, B., Serbănută, T., Serbănută, V., Filaretti, D., Rosu, G., Johnson, R.: IELE: an intermediate-level blockchain language designed and implemented using formal semantics, pp. 1–25 (2018)
13. Leiding, B., Norta, A.: Mapping requirements specifications into a formalized blockchain-enabled authentication protocol for secured personal identity assurance. In: Dang, T.K., Wagner, R., Kung, J., Thoai, N., Takizawa, M., Neuhold, E.J. (eds.) Future Data and Security Engineering, pp. 181–196. Springer International Publishing (2017)
14. Urban, J.E.: A Specification Language and its Processor, Ph.D. Dissertation, Computer Science Department, University of Southwestern Louisiana, pp. 35–59 (1977)
15. Subburaj, V.H., Urban, J.E.: A formal specification language for modeling agent systems. In: 2013 Second International Conference on Informatics & Applications (ICIA), Lodz, pp. 300–305 (2013). https://doi.org/10.1109/icoia.2013.6650273
16. Subburaj, V.H., Urban, J.E.: Specifying security requirements in multi-agent systems using the descartes-agent specification language and AUML. In: Information Technology for Management: Emerging Research and Applications, pp. 93–111. Springer, Cham (2018)
17. Brouwer, W.D., Borda, M.: NeuRoN: decentralized artificial intelligence, distributing deep learning to the edge of the network, pp. 1–17 (2017)
18. Grishin, D., Obbad, K., Estep, P., Cifric, M., Zhao, Y., Church, G.: Nebula genomics: blockchain-enabled genomic data sharing and analysis platform, pp. 1–28 (2018)

Bridging the User Barriers of Home Telecare

Anita Woll[✉] and Jim Tørresen

Department of Informatics, University of Oslo,
Gaustadalleèn 23B, 0373 Oslo, Norway
anitwo@ifi.uio.no

Abstract. Scarce healthcare resources and a growing older population are pushing forward a need to transform elderly care arrangements into becoming more technology-enabled in the private homes. This paper present findings from an action research study where the participants tested telecare as a service for remote delivery of selected homecare services. The telecare solution was applied either by the participants' own television or using a modern tablet. However, several user barriers were experienced when the participants tested telecare in their everyday life. Thus, the telecare solution was rejected after end of pilot study. We found that the technical telecare solution was not enough supportive, especially during the days when the participants for various reasons not were able to perform the necessary preparation, the "invisible work" or behind the scene work ahead of the consultation. In previous research literature, the notion of the invisible work has mostly been used as theoretical lenses understanding formal care work. The move of assistive technologies into the home of the elderly people create a need to re-shape partakers in the elderly care work – whereas elderly care receivers also are active involved in the care arrangements. Moreover, the technology platform used for supporting elderly people should be design in a way that both support them in the visible work (the actual task the technology is set to support) and the invisible work (all the extra work to make it work) if we aim to better succeed in transforming pilot studies into sustainable care services.

Keywords: Home telecare · Elderly people · User barriers · Invisible work · Workarounds · Robotics

1 Introduction

1.1 The Increase in the Aging Population

The increase in the aging population is putting strains on today's health and care services. This resulting in a mismatch of available health care services and the demand for care services. Thus, the lack of staffing to assure human care services for all needing personal care and grooming in the future requires that the care work arrangements are transformed into more sustainable service deliveries. Consequently, elderly people who have the physical and cognitive capacity are motivated to self-care by the support of family, voluntary resources, and the use of assistive technology. Therefore, current health care services are increasingly technology-supported, e.g.; routine tasks are delegated to technology or parts of routine tasks are transform into technology –

© Springer Nature Switzerland AG 2020
K. Arai et al. (Eds.): FTC 2019, AISC 1069, pp. 816–835, 2020.
https://doi.org/10.1007/978-3-030-32520-6_59

supported services. The role that technology plays in elderly care is diverse, but the most essential role is to support users with safety and security measures. The prior main tool for that kind of support, the traditional safety alarm, also called the panic alarm is seen limited in function as it requires users to understand its usage regardless of cognitive and physical capabilities [1], and it is experienced in nursing home that approx. 80% for the residents do not understand its use. Therefore, it is required to explore upon alternative solutions as assistive technologies should be designed with different levels of automation to support shifting individual user needs and the diversity of user groups [2]. In our prior research is was recognized a need to separate assistive technology into the active and passive use of technology, whereas *active* telecare is considerate as a conscious motivated two-ways ICT-based communication between the care receiver and care provider [1] such as videoconferencing or using the personal safety alarm. While *passive* telecare is an unconsciously two-ways ICT-based communication; as sensors/monitoring equipment alert the care providers on behalf of the care receiver according to its set pre-programmed function, e.g., a fall detector, door sensor etc. [1].

In our previous research, it is also shown that elderly care is recognized as a collaborative effort including joint work contribution from elderly people (self-care), family and volunteers (informal care), and paid health care providers (formal care) [1]. Whereas, the use of technology is playing a role in the division of elderly care work, e.g., it can be a tool to delegate work responsibility back to the elderly people, with or without backup support from informal and formal care providers. However, there are still some user barriers that need to be bridge in order for assistive technology to fully reach its potential. A key constraint of elderly care in Norway is the primarily focus on solely formal care services, and the less attention it brings to technology-supported services beyond the use of the personal safety alarm. Moreover, technology-supported care services are often seen introduced too late, and then resulting in failing to support users in practice, which result in the technology being discarded after pilot projects. Some users experience gaps in the service delivery connected to the different housings including ordinary homes, care housing, and nursing homes. Moreover, the lack of infrastructure and standard for assistive technology is constraining for the users who wants to prolong independent living. Our previous findings indicate that the key constraints of using assistive technology are more pronounced than its potential benefits [1]. However, the identified key constraints are important knowledge in the processes of transforming elderly care, as the constraints can be solved by expanded use – or different use of technology. This paper is concerning the transformation of conventional home care practice into a new service delivery using assistive technology for home telecare. Moreover, how the lessons learned from this study can be used for future telecare studies.

1.2 Assistive Technology Used for Home Telecare

Assistive technology is in the Scandinavian countries named welfare technology. It relates to a broad range of technologies intended to support different user functions in everyday life: "The concept of welfare technology is primarily technological assistance that contributes to increased safety, security, social participation, mobility and physical and cultural activity, and strengthens the individual's ability to fend for themselves in the everyday lives despite illness and social, psychological or physical impaired functioning. Welfare technology can also act as technological support for relatives and otherwise help to improve the availability of resources and quality provision. Welfare technological solutions can, in some cases, prevent the need for services or institutionalization" [3, p. 99].

Home telecare can be one of several measures such as remote monitoring equipment (mobile and stationary) [4, 5]:

- Sensors (fall, wandering, logging activities etc.).
- Medical alert systems/digital medicine dispenser.
- Telecommunications, e.g., smartphone/tablet/PC etc.
- Video conferencing/telepresence.
- Digital/analog personal safety alarms.

The department of Health and Social care in UK [6] defines telecare as "...care provided at a distance using ICT. It has been described as the continuous, automatic and remote monitoring of real-time emergencies and lifestyle changes over time in order to manage the risks associated with independent living".

The move of technology into the home of the care receivers also has a vision of being cost-effective – scaling support into the home can transform services into being increased need-based and timely delivery of services, but it is seen that it is extremely challenging to implement and scale the use of assistive technology in practice [1, 7, 8]. In this paper, we will address the potential of home telecare based on related work and practices of using home telecare within elderly care. Moreover, we will identify barriers of its use by reporting on findings from own field studies and related research. The paper addresses the use of telecare on various platforms such as the television and the tablet for videoconferencing. Additionally, the paper will discuss the dilemma of finding the "right" platform for telecare in connection to choosing familiar, integrated or alien technology. These dilemmas must be viewed in the light of supporting the users' in a broader sense than merely with one function when operating various telecare solutions. Finally, we will make design considerations based on the identified user barriers. In which, we aim to make suggestion about the future organization of home telecare by bridging the user barriers by finding a technology platform that is more appropriate for its users regardless of their health conditions.

2   Related Researches

Greenhalgh et al. [8] have carried out a three-phased study of what is quality in assisted living technology including interviews, ethnographic case studies and co-design workshops. In the first phase, they were exploring the barriers of using telecare/telehealth by interviewing technology suppliers and service providers. Then they studied the actual users regarding their user needs, and lastly, they brought the all parties together for co-design workshops including users and their care providers as well as the technology suppliers and service providers. They study shows among other that technology development is crucial, however that the development should be grounded in the market and be user-driven to accomplish a greater attention to the performance referred to as "supporting technologies in use" rather than the technology itself. The authors also recognize a need for a shift in the design model approach whereas technology should be design in a manner that supports inter-operable components, which can be combined and used on multiple devices and platforms.

Proctor and co-authors [9] have explored upon the organization of work in a telecare call center. The telecare staffing is answering the calls and alarms received from various telecare technologies in order to assist elderly people living at private home (in opposite to people living in care housing and nursing homes with staffing in-house). Proctor et al. [9] report findings of a mismatch of the policy makers objectives of using such technologies and the practical use of such technologies. They argue so because the introduced technologies are not necessary supporting the elderly users actual need in daily life activities. The authors stress that despite the fact that remote care services have been debatable as the elderly care receivers have experienced less physical visits, the remote staff can also have a positive outcome as they experienced the staffing acting as the "glue" [9, p. 79] in the organization of care services, and by this are "providing the all-important link between otherwise fragmented services" [9, p. 79].

Farshchian et al. [10] report how the workers in a telecare call center are using various technologies for the delivery of remote care services to their end-users. The authors refer to the transfer value of the study of Roberts and Mort [11] who are stating that "[t]elecare systems introduce a paradox in that they introduce scale and a new form of distance into home care work, whilst simultaneously making care appear more immediate" [11, p. 142].

Proctor et al. [7] report findings from a study of the user experiences of the GPS technologies. The authors stated that the degree of articulation work beyond the formal care procedures was extensive, as well as the tacit knowledge within the care network are restricting the technologies scaling and sustainability – thus, the authors argues that methods must be found to capture the skills and "hidden work" knowledge of the staffing to better succeed implementing assistive technology that is sustainable solutions.

Fitzpatrick et al. [12] argue for the need of a more socio-technical approach when designing technologies supporting assistive living and telecare to better achieve designing solutions that supports the user's actual needs in their everyday life context. Moreover, Fitzpatrick et al. [12] emphasize the importance of designing for a modular infrastructure in the home. Whereas the infrastructure is established and introduced initially to support the user's well-being and social needs, then later can be expanded with services also supporting formal care services.

Aaløkke Ballegaard and co-authors [13] have studied how digital home-based services can support assisted living among elderly users. The authors state that the elderly people have shifting functional capacities, which require that technology used for supporting them need to adapt to their changing needs. They have observed that the use of assistive technologies regularly is first familiarized in follow-up services as support after a sudden decline of health, in so-called acute phases of an elderly person's lives. Aaløkke Ballegaard et al. [13] argue that technology should rather be introduced ahead of acute phases in order to prevent such situations. The authors also stress that designers of assistive technologies should consider designing technologies that can support the user's everyday life activities [13, p. 1808]. Moreover, the authors state that the use of technologies should be an integrated part of the home environment. This to avoid obviously visible technology to be seen by others approaching the home that could make the home dweller feeling of stigmatize.

Fitzpatrick and Ellingsen [14] are viewing the introduction of technology into the home as a "movement towards technology-enabled care at home with a greater focus on self-care." [14, p. 637]. The authors argue that the increased investigation of technology used for nursing or self-care can indicate a shift towards reduced human healthcare resources being physically present in the home, as well as an increased attention on the users and their experience of well-being and comfort.

Grönvall and Lundberg [15] report from a study where they have explored the complexity of introducing technology into the homes of elderly people – and in their work they have identified seven challenges that concern implementation of pervasive healthcare technologies in the homes of elderly people. The authors argue that the identified challenges go beyond application specific considerations, e.g., choosing the "right" sensor or developing an intuitive user-friendly interface. One (of several) relevant challenges in their research, is "appropriation", which they understand as aspects related to how technology becomes part of people's everyday lives whereas "new technology must be interpreted and ascribed meaning" [15, p. 28]. The authors stress that appropriation is time-consuming, an on-going process, and "it is through a dialogue between the user and a contextualized artifact that appropriation takes place" [15, p. 28]. Grönvoll and Lundberg also show awareness to the fact that introduction of healthcare technologies is appropriated in varying degrees into the users existing routines and daily lives.

Several relevant studies are about the use of a specific telecare solution such as how elderly users operates interactive digital interfaces [16–18], or how they use applications for social participation [19, 20], or self-monitoring and home-based technologies for rehabilitation [21, 22]. Moreover, other studies address the support of medication administration in the homes [23, 24] and how remote care technologies change the context of the private home and the traditional care work practices [25]. Among other, the study of Milligan, Roberts and Mort [25] show how the increased use of telecare technologies in the private home have implications for the context of the private space, and the users understanding of the care delivery. The authors state that "Telecare affects the nature of care interactions within the home; hence the widespread adoption of these technologies is likely to have a significant impact on the broader landscape of care." [25, p. 349]. Consequently, Milligan et al. [25] state that when the private spaces are transformed into an institutional context as a result of the expanded use of home telecare including all kinds of medical devices and public regulations – the transformation of the private space can conflict with the gain of staying in the home for these home dwellers. Milligan et al. [25] further pinpoint that designers have addressed these issues in recent time by developing and integrating telecare devices that match the layout of the home in order integrate assistive technologies in a subtle manner by providing "invisible" [25, p. 353] support in the home. For additional studies of contextual aspects concerning home telecare, see Joshi and Woll [26–28] and Woll [29, 30].

Greenhalgh et al. [31] raise concerns how the increased health issues of the elderly population are possible affecting their ability using telecare technologies. Moreover, how the technologies' material features are complicating the use among this user group. Greenhalgh et al. [31] refer to results from the EFORTT research team [31] who concluded that telecare is not a quick fix when supporting the increased number of elderly users. Telecare cannot replace traditional home care services as the technology itself cannot perform the care work. Thus, the authors argue for the importance of technology that has a function and then can be incorporated into the work practice of someone who can act and follow up during potential alerts.

Doyle, Bailey, and Scanaill [32] address who relative how few studies have examined the use of independent living technology in practice. The authors argue that technologies need to be moved out of the lab and into the homes of elderly people to be tested in real environments. This to assess the actual value of the design and to study the impact that the technology could have on their lives. Goodman-Deane and Lundell [33] also discuss how the design of technology must meet the actual user needs of the elderly people. They are doing so by highlighting the importance on capturing the user needs of "real older people, including "baby boomers" still in employment, frail older people with disabilities and the full range in between" [33, p. 3]. Compagna and Kohlbacher [34] have used the participatory technology development (pTD) approach when studying the use of care robots among elderly people. The authors report that such an approach has limitations, however, pTD can have potential to work if designers and developers move their work into real user environment and include the various end-users, e.g., the care workers and the elderly people.

Clemensen and Larsen [35] have examine the use of telemedicine for remote treatment of foot ulcer in the private home. The authors report findings of improved continuity of treatment as such consultation requires real-time collaboration with the home care nurse, patient, and the doctor. They also report findings of coordination issues, e.g., they experienced in several occasions that the physician had to wait for the home care nurse to arrive at the patient's home. Thus, the authors make design suggestions of an additional functionality of the telemedicine solution that can deal with the coordination problem of the technical solution.

Loe [36] has study the role that technology can play in supporting active ageing from the viewpoint of the oldest old, including elderly adapters of technology, reluctant users, and non-users. The author highlights the importance of capturing the oldest people's interpretation on technology – and she further argues that the policy makers, designers, and caregivers should listen carefully to the old people who already have adopted technology into their lives.

Blythe, Monk, and Doughty [37] have similar Loe [36] explored the user needs of the older people, and how the perspective of the older people can provide design implications for HCI. Blythe et al. [37] study is based on findings from structured interviews with health care professionals and older people. Blythe et al. [37] express concerns about technologies used for monitoring and the less attention that has been given to the social context of the home. Other HCI studies report findings from collaborative or interactive services where elderly people use the television from their home as a platform to receive telecare or similar services [38]. Several studies have made contributions that concern age-related challenges when designing for the elderly generation [39–41]. Others have provided new knowledge on how to develop interfaces usable for older people, e.g., Baunstrup and Larsen [42]. For instance, Baunstrup and Larsen point out that the television has evolved from a one-way monologue into a communications platform by offer increased dialogue-based services. They also argue that an iTV provides more "complex interaction paradigm" [42, p. 13].

3   Use of Home Telecare in Practice

This paper builds on fieldwork done as part of a two-year action research study in a care housing in Oslo, Norway (see Woll and Bratteteig [1] for more information about the housing's facilities, the role of the staffing and the additional technologies tested). The field work started at the same time as the residents moved into the new housing. Initially, several of the residents were able to self-care with the support they got from the in-house staffing, their relatives and the home care services. Though over time, we observed that many of the residents experienced increased and shifting health issues, which requiring them receiving increased home care assistance. Moreover, those with expansive need for care were given priority in the morning (e.g., to get support with the morning grooming). Thus, the residents with minor health problems expressed that they felt ignored and left waiting for the home care staff to arrive for their visit. Accordingly, one of the residents refused to continue receiving home care services because he experienced it stressful as the staff seldom arrived at the same time for the morning visit. Moreover, despite that he needed follow-up services after a temporary hospital admission, he found it easier to be independent. Moreover, another resident had concerns about the number of different nurses that came into her private home, which she found uncomfortable and intrusive. Thus, to address the various user issues concerned with active aging residents not being able to start their day early on, we initiated an action research study together with the district's home care service, aimed at using telecare as a means for videoconferencing of selected home care services.

3.1   Home Telecare Using the Television as the Platform for Care Services

An essential part of the action research study was the difficult choice of an appropriate technology platform for the telecare solution. Thus, we reviewed literature to get technology recommendations whereas findings shows (1) to use technology that fits into the elderly persons' daily life activities [43], (2) to build on familiar and existing technology in the home to avoid stigma [13], (3) to start with the use of ICT before acute illness [13], and (4) to mobilize physical visits in cases of uncertainty [11, p. 10]. Thus, we ended up using the television as the platform for telecare as all the residents at the housing had large television screens. The television is also representing a familiar technology that the participants can find understandable and easy to operate.

The television was also a preferred choice as we wanted a technology with fixed power supply to avoid re-charging any battery, which many elderly people find troublesome. By adding a wide-angle HD camera, we were able to capture both the living room and participants sitting in the television-chair/ comfy favorite chair. This in order to get the bigger picture of the home into the view, and not merely their faces – as the condition of a home can give additional information about a person's well-being. The study was done in collaboration with the home care staffing. The home care staffing had several pilot projects on their own concerning assistive technology as a result of scarce health care resources and continuous pressure of organizing services in a more cost-effective and sustainable manner. Thus, they were positive in testing and exploring new work arrangements to develop their services.

Initially, we started the study in the demo-apartment as the participants needed an introduction and training in how to operate the telecare solution. During the initial testing of the solution, all the participants handle the operations of the videoconferencing with ease. The participants were invited one-by-one at a set schedule and we did a number of test calls with the home care nurses after a fixed procedure. When the training was completed we moved the study from the demo-apartment into the homes of the participants. We then got a broader insight into the elderly care receivers' challenges of implementing technology support in their everyday life routines [27]. For example, over time the benefits of a stationary user interface in a fixed position were not flexible enough to handle situations when the participants were unexpected bedridden or bedridden for periods because of acute illness. Therefore, bedridden participants had to get traditional home care visits: they were not able to operate telecare sessions as the amount of "background work" or self-care abilities were decreased for a shorter or longer period. This "background work" was for example support in getting up from the bed and getting dressed. Consequently, when a nurse is arriving at a traditional home visit, s/he help with additional tasks depending on the care receivers' general condition or whereabout the person is in the morning routine, which are tasks not necessary a part of the nurse's work instructions. However, most of the participants recovered after some time and were able to return to receive home telecare sessions. This indicate that the balance of self-care and traditional formal care can vary over time, but that the balance goes both ways, and that people can recover and take up their self-care abilities. However, a lack in the technology support set-up was the need for additional context information about the participants condition when they did not respond to the telecare call – as many context-dependent user issues appeared during the test period of the home telecare solution. Thus, it was experienced lots of extra work to make sure that the participants were fine if they did not answer, e.g., after several calls' efforts were made with no success, we had to mobilize a physical home visit. The use of home telecare was highly affected by the users' context, and less concerned about their technical capabilities. However, some experienced the television as too complex to operate when additional devices to the platform were introduced. For example, to receive telecare calls they had to change the HDMI source from the set-up box to the camera, and afterwards back again to watch television – which was an operation that all participants struggled with and found difficult to learn. Additionally, some of the participants misplaced the remote controller for the camera and when the controller was «lost» they were not able to respond to the call. Moreover, the participants needed time to get used to the new practice of calls from the television as when the calls occurred when they were busy in their regular everyday life routine, they did not always understand the meaning of the "conference call sound". It was also evident that access to the television programs was their priority and it was experienced that they found the telecare call inconvenient if they were busy watching a program on the television. Some of the participants also had user preference that interfered with the telecare set-up, e.g., one participant had the television on constantly, but with the sound on mute. Thus, when he received telecare calls – he would not hear the nurse's voice as the television was muted. Additionally, another participant always unplugged the camera from day-to-day, which made the telecare set-up unfunctional. Telecare use is also vulnerable due to lack of control of the infrastructure, e.g., the network capacity in the private spaces and the camera connection to the television set. Moreover, home telecare such as video-conference requires a certain degree of functional capabilities and stable health condition to be experienced as appropriate and useful for the elderly participants.

The timing of the telecare call had also influence on the user's ability to incorporate ICT-supported care into their daily life activities. The timing issue was complicated; the participants had all agreed upon a suitable time for their call, so the timing problems seem less concerned about the time of the call than the unpredictability of the daily life activities that interfered with and prevented the participants from taking the call. We experienced that participants were unable to take the telecare call because they stayed in bed on "a bad day" with an unplanned set-back of the health condition, or they were eating breakfast and were unable to move from the kitchen area to the living room (in time) because of mobility issues, or they were absent from the home at the time the video consultation was supposed to take place etc. If they were not ready, in the sense that they were sitting in the television chair, they did not answer the call.

The participants also wanted to be "presentable" for the videoconferencing call as they were not ready before they were dressed and had comb their hair etc., see Fig. 1. However, when doing the physical visits their "presentable" appearance was not a topic of concern for the participants. Additionally, the participants' health conditions got worse as the study unfolded. A decline in health conditions is a reality for elderly people, and the changing user capabilities need to be considered when designing technical solutions for elderly care receivers. This finding is significant regarding actively vs. passively use of assistive technologies as part of the technology supported care delivery. Thus, to maintain telecare over time for these users, it would be a necessity to add additional solutions such as e.g., various types of sensors logging presence 24/7 as backup to support them in both "good" and "bad" days.

The lessons learned from the choice of television as the platform for telecare were as following:

- The stationary solution was not flexible enough for the user group.
- Maintenance issue of the telecare equipment – who is responsible when researchers leave?
- The television is first and foremost for television programs, as one participant stated "don't mess with my television".
- The television's user interface increases in complexity when adding additional devices.

The invisible work necessary ahead of being ready for a telecare call in the home was highly depended on the participants whereabouts in their daily routines and daily health conditions. The participants were experienced to need more care than what telecare could provide them regarding the invisible work, however the telecare study was set-up

Fig. 1. Illustrates home telecare using the television as the technology platform (Copyright A. Woll).

to perform merely the visible work e.g., the participants should take their medication. Thus, the technology support did not consider the invisible work, which was a prerequisite for doing the visible work, whereas physical home visits are seen more flexible in this matter.

3.2 Home Telecare Using the Tablet as the Platform for Care Services

During the action research study, we also tested the tablet as a platform for telecare, see Fig. 2. However, we observed that it was difficult for the residents to hold the tablet high enough for the telecare caller to see the participant's face – as the tablet was a mobile solution. The tablet was experienced as heavy for those with reduced strength in the arms. In addition, the tablet had to be put down on the table when the resident was to carry out a task with her/his hands (e.g., to take medication). Thus, then it also was impossible to observe if the participants took the medication or not – which was the main purpose of the call. Furthermore, the elderly participants often used a specific table for the tablet, "the tablet table", and they did not bring along their tablet when moving from place to place in the apartment, so in practical use, the tablet was experienced as just as fixed and stationary as the television. The tablet also requires recharge of battery, which was a drawback for a telecare solution require 24/7 operation. Thus, we experienced pros and cons with both the tablet and the television platform, thus when looking for other solutions we aimed at a technology being mobile and flexible in the sense of not being depending on the user to be mobile in order for the telecare set-up to work. There are several so-called robots for home telecare where a tablet/screen installation. The care provider controls the telepresence from remotely so the robot itself is not fully automated to work as a standalone installation. The benefit of this installation is that the robot is both mobile and stationary, while the staff can reach all the users from a single remote location. Moreover, such installation can be remotely controlled to locate the elderly person in the home.

Robots such as the giraffe, see Fig. 3, is flexible as it ideally can move from room to room in the opposite to the fix position of the television, and it is an adjustable installation in opposite to the hand-held tablet issue, as described above, as the giraffe can tilt and zoom in a person's face. But its weakness are doorsteps, carpets on the floor, furniture, staircase and closed doors as a closed bedroom door will keep the giraffe from entering to its user. Thus, we argue that robots such as the giraffe can be useful in up-to-date assistive living housing, however, we find them challenging to operate and function in houses with several floors or in older apartments.

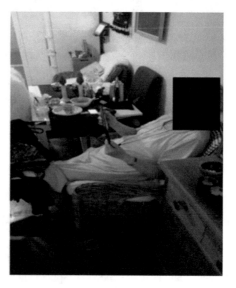

Fig. 2. Illustrates home telecare using the tablet as the technology platform (Copyright A. Woll).

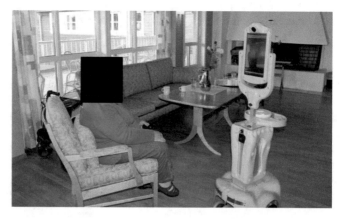

Fig. 3. Illustrates home telecare using the "robot" as the technology platform (Cesta et al. 2011).

In dementia care, where the personal safety alarm has out-played[1] its role because most of the users do not understand its purpose – a camera for automated-, as well as scheduled supervision with multiple functionalities has been an improved replacement for the personal alarm, see Fig. 4 illustration of such IR camera. In Oslo, Norway, a housing for persons with severe dementia have successfully implemented this kind of technology-support to increase their residents' independence, and the residents are then able to live outside institutional care. However, the technology-support including telecare, in-house staff during the day- and evening time, together with home care services at night shift have all in a collaboration supervised the residents' safety and security. The camera is installed outside reach of the residents to avoid them in un-mounting the camera. Additionally, the camera is out of sight as they are in the background of a large chandelier and an esthetician roof decoration to remove the residents' attention to the technology. The privacy of the resident is protected by the camera's attribute of IR-based depth sensor that provides two types of pictures, one anonymous where the resident's dressing, or non-dressing cannot be viewed (see Fig. 4 and picture number 2.), and one regular picture to look at the situation when alerts from the camera has been activated, e.g., the resident has fallen on the floor. Moreover, the camera stream is not stored and can only be viewed in a set time of seconds.

Fig. 4. Illustrates home telecare using camera as the technology platform (Copyright: Roommate).

The drawback of the camera is similar as the stationary television, it is only functional in a limited area of the apartment. Thus, in order to support the user in all part of the home – it will be necessity to install a camera in every room.

---

[1] The personal safety alarm is still useful as a "passive button" when the person with dementia is living in a housing with in-door positioning system or door entries detectors. However, it is challenges to make this user group to wear the alarm.

4   The "Invisible Work" of Technology-Supported Elderly Care

We recognize the notion of invisible work as introduced by Nardi and Engeström [44] as highly applicable to better understand the elderly care work as the home care nurses in general are performing several additional work tasks, often fluctuating work as things happens along their mobile "route" which make them have to re-arrange their set work flow, which are not strictly defined in their work instructions. The basis for our understanding of the invisible and visible work has arisen from the work of Nardi and Engeström's [44] introduction of the structure of *invisible work* referring to the incorporation of new technology that calls for a re-shaping of work practice. The authors state that while technology often replace a set function of the visible work whereas the visible work often is "mapped, flowcharted, quantified, measured" [44, p. 1] very carefully into sub-tasks – technology often is experienced failing to support or consider the practice of the invisible work, e.g., the background work or "behind the scene" work.

"When planning for restructuring or new technology, visible work is the focus of attention. It is the only work that is seen, so efforts to restructure center on how visible work can be manipulated, redrawn, reorganized, automated or supported with new technology. But a growing body of empirical evidence demonstrates that there is more to work than is captured in flow charts and conventional metrics" [44, p. 1].

The reason being when nurses enter the door to a home of a care receiver, they can never really know what is expecting them. The nurses can experience regular visits and then just carry out their work as planned – while other situations require for both workarounds and/or critical emergency situations. Elderly people as care receivers can have unstable and shifting user needs, as well as the various user group together have shifting and unstable user needs. For example, when the home care nurse arrives to their home care receivers, the care receiver may be in the middle of "something" that is required to be fixed for the nurse to do what s/he is planned to do of formal work tasks. Thus, some workaround tasks can be a necessity to carry out before they are able to do their planned and formal care work.

During the telecare study, we did not foresee the troublesome of the invisible work, as it was not an issue before it was not accomplished. However, the invisible «background work» of the elderly participants were a prerequisite for them to be ready at hand for telecare. The nature of the invisible work is it invisibility as most work is visible work, e.g., to make sure that a care receiver takes timely and accurate medication, see Fig. 5. The invisible work in our setting was for the participants to get up in the morning, get dressed, do the morning care and eating breakfast, turn on the television, swap to the HDMI camera source – and we experienced from time to time that all the necessary tasks ahead of telecare was challenging for them, especially over time when their health condition got worsen. Additionally, some of the participants also slept badly as they found it stressful if they over-sleep, and then were accidently running late for doing the telecare session.

Thus, the background work or behind the scene work is only seen when it is not carried out – and the structure/procedure falls apart. It was understood that the invisible work is often handled by the nurses doing physical visits – as they support the user in everyday life routines with various additional sub-tasks without them having awareness doing so. When the nurse is arriving at a home to assist a care provider with medication, s/he find it natural to assist the person with additional task, e.g., all from supporting a person in getting up from bed or doing parts of the morning grooming in the bathroom or provide a glass of water at the kitchen table – and then at the end of the visit make sure that the person gets his/hers medication, see Fig. 5. Furthermore, the nurse will not report on her informal/practical tasks, but merely on her/his formal tasks, when giving her daily report with exception if something of concern took place.

As telecare-supported services are moving into the home of elderly care receivers – where care receivers also are expected to contribute as an active self-caregiver– what happens then to the care arrangement both regarding visible and visible work of all the care contributors? The introduction of telecare in the home provides responsibilities to the care receiver that they may not be able to keep on daily basis. Thus, demands to be taken into the design considerations to be addressed in-situ with a more complete technical solution based on knowledge about how the work is performed in practice.

Moreover, the telecare support needs back-up solutions when the care receiver fails to do the self-care job, e.g., there should always be a mobilization plan.

Nardi and Engeström refers to four types of invisible work [44] that are referred to and listed below:

- Work carried out in invisible places, e.g., the authors refer to such work as the highly skilled behind-the-scenes work of reference librarian.

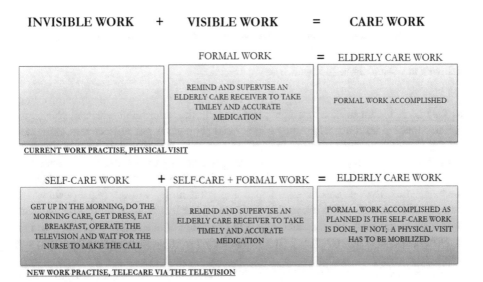

Fig. 5. Illustrates the type of work that needs to work in order to accomplish a successful active telecare arrangement.

- Work defined as routine or manual work that requires considerable problem solving and knowledge, e.g., the authors refer to such work as the work of telephone operators.
- Work done by invisible people such as domestics.
- Informal work processes that are not part of anybody's job description, but which are crucial for the collective functioning of the workplace, e.g., the authors refer to such work as regular but open-ended meetings without a specific agenda, informal conversations, gossip, humor, storytelling.

The authors emphasize the essential need to understand the nature and structure of invisible work when designing for new technology-supported work arrangements. Thus, when changing work practice it is easy to neglect the importance of invisible work as it is not part of the formal work instruction – but rather the work to make work actually work.

Nardi and Engeström [44] are also referring to the term invisible work as "second order work", which is the extra work to regular work tasks. They refer to this as special skills so-called "tacit knowledge", thus the staffing/users are essential when designing services or transforming work practices or implementing technology-supported measures.

Thus, when the second order work is delegated to the elderly care receivers, the services are getting more fragile as they are dependent on the elderly users daily general conditions in order to be performed – thus, the knowledge of the elderly people as partakers in the care work is of importance to be included in the design processes of assistive technology such as home telecare.

5 Discussion

"Seniors in general don't ask for help because they think they will inconvenience the health personnel"

(Bowes and McColgan [45, p. 33]).

The elderly people are a diverse group of users, all from those who never call for help as they do not want to be a "burden", and those in the other hand who call constantly for assistance – not only when they actual need support. A general concern expressed by the participants during the field work is that they feel that they have nothing to contribute with as they are not trained in using technology or that interacting with new modern technology makes them feel stupid when they are not able to operate the devices. Thus, they express concern in doing something "wrong" (e.g., pushing the "wrong" button, etc.) when interacting with unfamiliar technology. Thus, then we need to explain to the users that if technology is not working or is not able to operate with an easy intuitive, user interface, it is the technology that fails in poorly design, and not them.

In Norway, in the past decades, the acute emergency alarms have been prioritized, while the remote health care services such as telecare for elderly people have been left behind, as discussed by Farshchian et al. [10, p. 339]. However, this is under development, whereas more technology-enabled services beyond the personal safety alarm are established in the home. Also, hospital care is moving into the home, which require robust solutions for follow-up services and preferable a requirement of 24/7 operation. This calls for redundant solutions and back-up power on critical equipment, and organizational measures if/when technology fails when e.g., network infrastructure has accidental down-time. Moreover, it is essential that the technology introduced is useful for its users, in the sense that it is making sense for it set purpose and is operable. We acknowledge Greenhalgh et al. [8] approach about bringing all partakers together for co-design processes, the elderly users and the informal- and formal care providers as well as the technology suppliers. However, lesson learned is that technology-supported services are never plug and play, it is rather plug and play and evolve, thus, designing assistive technology require user-adjustments, often context-based adjustments over time as new restrictions or user needs came into the view in real life testing.

5.1 Familiar Integrated or Alien Technology

We learned during the telecare study that building services on the familiar technology in the home, the television, was not convenient in practice. Mostly, because the television is stationary, thus, not flexible in use when the participants were bedridden. Moreover, the users' main priority when using the television were to watch TV-programs. Therefore, we argue that for critical home telecare solutions that requires technology platform of 24/7 operation, it is more appropriate to choose a dedicated technology – even if it is viewed as an alien in the home, and it could potential be integrated in the home, but then in a manner where its more esthetically installed e.g., like the camera as mention for dementia care. The reason being is that a dedicated/specific device is set for its specific purpose, and then back-up solutions better can safeguard these to work. Additionally, it was found as a user restriction that the telecare solution via the television required users having the capacity of operating actively telecare. Over time, as the action research study unfold, we experienced that the participants health issues worsen – and that dealing with behind the scene work was found troublesome, see Fig. 6 showing various technologies set-up in the home.

However, isolated seen as experienced during initial trial in the demo-apartment, the telecare setup works smoothly using the television. Furthermore, it was learned that the daily-life activities were interrupting the telecare set-up and the use when the solution was implemented in the homes of the participants. Consequently, when moving telecare into the homes of the care receivers, the receivers suddenly were a partaker in the care work. Moreover, it was seen that the technical set-up was limited to the visible work of the participant, and not including the invisible work required ahead of operating home telecare. We experienced that a solution for telecare needs higher flexibility in its user support to assist users both in "good" and "bad" days, otherwise the solution will demand for lots of extra work the days when the users are experiencing declines in their general condition. The solution could also have some level of automation to better support the diversity of users having shifted and unstable health conditions.

We support Proctor et al. [7] findings concerning the occurrences of extensive articulation work beyond the formal care procedures, as well as the authors report about the tacit knowledge within the care network was hard to grasp. The authors stress that

Fig. 6. Illustrates technology as familiar, integrated or alien technology (Copyright 1: A. Woll, Copyright 2: Aaløkke Ballegard et al. 2006 p. 375, Copyright 2: unknown).

methods should be found for such hidden work. This is comparable to our research where we found the invisible work of the elderly care receivers in their self-care work – or rather when the self-care failed to work – their invisible work was suddenly appearing. We argue that technology can be used for discovering the invisible work, especially when the invisible work is not carried out using passive technology. Thus, in such situation the passive technology support can log presence in the home at room level and sensors can surveille their condition in order to mobilize required support (such as human support) timelier and more accurate. However, this call for a more advanced and comprehensive technology platform in the home. This is concurrent with Fitzpatrick et al. [12] who emphasize the importance of designing for a modular infrastructure in the home.

In our future planned research as part of the Multimodal Elderly Care Systems (MECS) we will study how the combination of intelligent sensors systems and mobile advanced robots can be introduced for such a platform to develop a smart robot infrastructure. Our lessons learned from prior studies show that the elderly care receivers demand for a technology-enabled care in the home that have the attributes of adding layers of support when they are experiencing declines in their health condition. Moreover, when the care receivers' health is stabilized, the level of technology-support can be at its lowest level of support – or according to the individual's need for support there and then. Moreover, the smart robot infrastructure must have coverage in the whole home – not merely in a room. This in order to safeguard the telecare solution to support the users in a more flexible and sustainable manner.

6   Conclusion

In this paper, we have presented lessons learned from an action research study where the television and tablet were used as the platform for home telecare of selected home care services. Moreover, we have identified barriers of its use by reporting on findings showing that when technology is enabled into the private homes of the elderly care receivers – the care receivers are delegated responsibilities and self-care work. However, when the care receivers' experiences decline in their health conditions – which many elderly care receivers do from time to time, their responsibilities are sometimes too much to handle, thus, they are not able to carry out the set self-care. The paper uses the notion of invisible work, the background work, as an approach to better understand the user situations when active telecare is not upkept. Moreover, we argue that when introducing technology-enabled care services in the home, the technology should be designed broader in scope, so the technical support itself can surveille the situation and alert when the care receiver need mobilizing of additional assistance. Such support requires a comprehensive infrastructure in the home, e.g., intelligent sensor systems and robotics to locate the user situation. In our future research we will explore how such an infrastructure can bridge the gaps experienced in prior home telecare solutions.

References

1. Woll, A., Bratteteig, T.: A trajectory for technology-supported elderly care work. In: Schmidt, K. (ed.) Computer Supported Cooperative Work (CSCW), vol. 28, no. 1–2, pp. 127–168. Springer, Heidelberg (2019)
2. Woll, A., Bratteteig, T.: Activity theory as a framework to analyze technology-mediated elderly care. In: Nardi, B. (ed.) Mind, Culture and Activity, vol. 25, no. 1, pp. 6–21. Taylor and Francis (2018)
3. NOU: Ministry of Education and Research - NOU 2011:11 – Innovation in care, Ministry of Health and Care Services, Oslo (2011)
4. Jerant, A.F., Azari, R., Nesbitt, T.S.: Reducing the cost of frequent hospital admissions for congestive heart failure: a randomized trial of a home telecare intervention. Med. Care. **39**(11), 1234–1245 (2001)
5. Barlow, J., Singh, D., Bayer, S., Curry, R.: A systematic review of the benefits of home telecare for frail elderly people and those with long-term conditions. J. Telemed. Telecare **13** (4), 172–179 (2007)
6. Integrating Community Equipment Services Telecare. Getting Started. Department of Health (2004). http://www.integratedcarenetwork.gov.uk/
7. Proctor, P., Wherton, P., Greenhalgh, J.: Hidden work and the challenges of scalability and sustainability in ambulatory assisted living. ACM Trans. Comput. Hum. Interact. **2**(25), 11–26 (2018)
8. Greenhalgh, T., Proctor, R., Sugarhood, P., Hinder, S., Rouncefield, M.: What is quality in assistive living technology. the ARCHIE framework for effective telehealth and telecare services. BMC Med. **13**, 1–15 (2015)
9. Procter, R., Wherton, J., Greenhalgh, T., Sugarhood, P., Rouncefield, M., Hinder, S.: Telecare call centre work and ageing in place. Comput. Support. Coop. Work (CSCW) **25**(1), 79–105 (2016)
10. Farshchian, A., Vilarinho, T., Mikalsen, M.: From episodes to continuity of care: a study of a call center for supporting independent living. Comput. Support. Coop. Work (CSCW) **26**, 309–343 (2017)
11. Roberts, C., Mort, M.: Reshaping what counts as care: older people, work and new technologies. ALTER – Eur. J. Disabil. Res. **3**(2), 138–158 (2011)
12. Fitzpatrick, G., Huldtgreen, A., Malmborg, L, Harley D., Ijsselsteijn, W.: Design for agency, adaptivity and reciprocity: reimagining AAL and telecare agendas. In: Computer Supported Cooperative Work, pp. 305–337 (2015)
13. Aaløkke Ballegaard, S., Bunde-Pedersen, J., Bardram, J.E.: Where to roberta? Reflecting on the role of technology in assisted living. In: Proceedings of NordiChi, pp. 373–376 (2006)

14. Fitzpatrick, G., Ellingsen, G.: A review of 25 years of CSCW research in healthcare: contributions, challenges and future agendas. Comput. Support. Coop. Work (CSCW) **22**(4–6), 609–665 (2013). J. Collaborative Comput. Work Pract.
15. Grönvall, E., Lundberg, S.: On challenges designing the home as a place for care. In: Holzinger, A., Ziefle, M., Röcker, C. (eds.) Pervasive Health. Human-Computer Interaction Series. Springer, London (2014)
16. Culén, A.L., Bratteteig, T.: Touch-screens and elderly users: a perfect match? In: Miller, L. et al. (eds.): ACHI 2013. Proceedings of the Sixth International Conference on Advances in Computer-Human Interactions, Nice, France, 24 February – 1 March 2013. IARIA XPS Press, pp. 460–465 (2013)
17. Häikiö, J., Wallin, A., Isomursu, M., Ailisto, H., Matinmikko, T., Huomo, T.: Touch-based user interface for elderly users. In: Proceedings of the 9th International Conference on Human Computer Interaction with Mobile Devices and Services, pp. 289–296. ACM (2007)
18. Heart, T., Kalderon, E.: Older adults: are they ready to adopt health-related ICT? Int. J. Med. Inform. **82**(11), 209–231 (2011)
19. Alaoui, M., Lewkow, M., Seffah, A.: Increasing elderly social relationships through TV-based services. In: Luo, G., Liu, J. (eds.) IHI 2012. Proceedings of the 2nd ACM SIGHIT International Health Informatics Symposium, Miami, Florida, USA, 28–30 January 2012, pp. 13–20. ACM Press, New York (2012)
20. Dewsbury, G., Rouncefield, M., Sommerville, I., Onditi, V., Bagnall, P.: Designing technology with older people. Univ. Access Inf. Soc. **6**(2), 207–217 (2007)
21. Grönvall, E., Verdezoto, N.: Beyond self-monitoring understanding non-functional aspects of home-based healthcare technology. In: Mattern, F., Santini, S. (eds.) UbiComp 2013. Proceedings of the 2013 ACM International Joint Conference on Pervasive and Ubiquitous Computing, Zurich, Switzerland, 8–12 September 2013, pp. 587–596. ACM, New York (2013)
22. Axelrod, L., Fitzpatrick, G., Burridge, J.H., Mawson, S.J., Smith, P.P., Rodden, T., Ricketts, I.W.: The reality of homes fit for heroes: design challenges for rehabilitation technology at home. J. Assistive Technol. **3**(2), 35–43 (2009)
23. Siek, K.A., Khan, D.U., Ross, S.E., Haverhals, L.M., Meyers, J., Cali, S.R.: Designing a personal health application for older adults to manage medications: a comprehensive case study. J. Med. Syst. **35**(5), 1099–1112 (2011)
24. Dalgaard, L.G., Grönvall, E., Verdezoto, N.: MediFrame: a tablet application to plan, inform, remind and sustain older adults medication intake. In: Yang, C.C., Ananiadou, S. (eds.): ICHI 2013. Proceedings of IEEE International Conference on Healthcare Informatics, pp. 36–45. IEEE (2013)
25. Milligan, C., Roberts, C., Mort, M.: Telecare and older people: who cares where? Soc. Sci. Med. **72**(3), 347–354 (2011)
26. Joshi, S.G., Woll, A.: A collaborative change experiment: telecare as a means for delivery of home care services. In: Design, User Experience, and Usability. User Experience Design for Everyday Life Applications and Services. Lecture Notes in Computer Science, vol. 8519, pp. 141–151 (2014)
27. Joshi, S.G., Woll, A.: A collaborative change experiment: diagnostic evaluation of telecare for elderly home dwellers. In: Digital Human Modeling. Applications in Health, Safety, Ergonomics and Risk Management: Ergonomics and Health, Lecture Notes in Computer Science, vol. 9185, pp. 423–434 (2015)

28. Joshi, S.G., Woll, A.: A collaborative change experiment: post-experiment evaluation of home telecare for elderly home dwellers. In: Ahram, T., Karwowski, W., Schmorrow, D. (eds.) AHFE 2015. Proceedings of the 6th International Conference on Applied Human Factors and Ergonomics (AHFE), Las Vegas, USA, 26–30 July 2015. Procedia Manufacturing, vol. 3, pp. 82–89 (2015)

29. Woll, A.: Is aging the new disease? In: Miller, L. et al. (ed.) ACHI 2016. Proceedings the 9th International Conference on Advances in Computer-Human Interactions, Venice, Italy, 24 – 28 April 2016. IARIA XPS Press, pp. 21–28 (2016)

30. Woll, A.: Use of welfare technology in elderly care. Ph.D. dissertation. University of Oslo. Oslo: Department of informatics, University of Oslo (2017). http://urn.nb.no/URN:NBN:no-58321

31. Greenhalgh, T., Wherton, J., Sugarhood, P., Hinder, S., Proctor, R., Stones, R.: What matters to older people with assisted living needs? A phenomenological analysis of the use and non-use of telehealth and telecare. Soc. Sci. Med. **93**, 86–94 (2013)

32. Doyle, J., Bailey, C., Scanaill, C.: Lessons learned in deploying independent living technologies to older adults' homes. Univ. Access Inf. Soc. **13**(2), 191–204 (2014)

33. Goodman-Deane, J., Lundell, J.: HCI and the older population. Interact. Comput. **17**(6), 613–620 (2005)

34. Compagna, D., Kohlbacher, F.: The limits of participatory technology development: the case of service robots in care facilities for older people. Technol. Forecast. Soc. Chang. **93**, 19–31 (2014)

35. Clemensen, J., Larsen, S.B.: Cooperation versus coordination: using real-time telemedicine at the home of diabetic foot ulcers. J. Telemed. Telecare **13**, 32–35 (2007)

36. Loe, M.: Comfort and medical ambivalence in old age. Technol. Forecast. Soc. Chang. **93**, 141–146 (2014)

37. Blythe, M.A., Monk, A.F., Doughty, K.: Socially dependable design: the challenge of ageing populations for HCI. Interact. Comput. **17**(6), 672–689 (2005)

38. Miyazaki, M., Sano, M., Mitsuya, S., Sumiyoshi, H., Naemura, M., Fujii, A.: Development and field trial of a social TV system for elderly people. In: Stephanidis, C., Antona, M. (eds.) UAHCI 2013, Part II. LNCS, vol. 8010, pp. 171–180. Springer, Heidelberg (2013)

39. Carmichael, A., Newell, A.F., Morgan, M.: The efficacy of narrative video for raising awareness in ICT designers about older users' requirements. Interact. Comput. **19**(5–6), 587–596 (2007)

40. O'Neill, S.A., et al.: Development of a technology adoption and usage prediction tool for assistive technology for people with dementia. Interact. Comput. **26**(2), 169–176 (2014)

41. Weiner, M.F., Rossetti, H.C., Harrah, K.: Videoconference diagnosis and management of Choctaw Indian dementia patients. Alzheimer's Dement. **7**(6), 562–566 (2011)

42. Baunstrup, M., Larsen, L.B.: Elderly's barriers and requirements for interactive TV. In: Stephanidis, C., Antona, M. (eds.) UAHCI 2013, Part II. LNCS, vol. 8010, pp. 13–22. Springer, Heidelberg (2013)

43. Aaløkke Ballegaard, S., Hansen, T., Kyng, M: Healthcare in everyday life: designing healthcare services for daily life. In: CHI 2008, Proceeding of the Twenty-Sixth Annual SIGCHI Conference on Human Factors in Computing Systems, pp. 1807–1816 (2008)

44. Nardi, B., Engeström, Y.: A web on the wind: the structure of invisible work. Comput. Support. Coop. Work **8**, 1–8 (1999)

45. Bowes, A., McColgan, G.: Telecare for older people: promoting independence, participation, and identity. Res. Aging **35**(1), 32–49 (2012)

An Intelligent System for Detecting a Person Sitting Position to Prevent Lumbar Diseases

Paul D. Rosero-Montalvo[1,2(✉)], Vivian López-Batista[2,3],
Vanessa E. Alvear Puertas[1], Edgar Maya-Olalla[1],
Mauricio Dominguez-Limaico[1], Marcelo Zambrano-Vizuete[1],
Ricardo P. Arciengas-Rocha[4], and Vanessa C. Erazo-Chamorro[4]

[1] University Técnica del Norte, Ibarra, Ecuador
pdrosero@utn.edu.ec
[2] University of Salamanca, Salamanca, Spain
[3] YachayTech, Urcuquí, Ecuador
[4] Instituto Tecnológico Superior 17 de Julio, Urcuquí, Ecuador

Abstract. The present system shows a position detection of a person in the sitting position by means of an accelerometer sensor and the implementation of data analysis stages of prototype selection and supervised classification. As a result, an optimal training matrix with a 100% classification performance is obtained. Finally, an interface is presented that shows the own decision of the system.

Keywords: Body position · Embedded system · Data analysis · Sensors

1 Introduction

The majority of activities of the human being at work is performed in the sitting position. This is because most of the work obligations are based on the monitoring of the functions performed by the machines [1]. On some occasions, people stay up to 10 h in this position. Some studies recommend changing activity every 30 min to avoid muscle atrophy. However, a large number of people, usually sit incorrectly while using their computers, this not only generates physical pain but also affect mental health [2]. On the one hand, they can cause exhaustion, distraction, as short-term affections. On the other hand, by remaining in one position, the body only consumes a third of the calories in relation to a person in constant movement. This causes long-term back pain, obesity and increases their mortality by up to 40% [3].

The position is taken at the time of sitting, if not adequate causes a series of diseases which can be avoided by this system that detects which is the correct position based on information collected, in Table 1 and Fig. 1 shows the positions most common that people adopt and the diseases they generate. Also, for classification purpose, each position has a label to identify them.

© Springer Nature Switzerland AG 2020
K. Arai et al. (Eds.): FTC 2019, AISC 1069, pp. 836–843, 2020.
https://doi.org/10.1007/978-3-030-32520-6_60

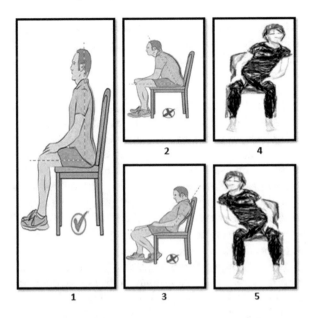

Fig. 1. Most common human posture in chair

Table 1. Diseases description in relation with each position

Label	Human posture	Diseases
1	Correct form	No diseases occur, helps improve circulation and relax muscles
2	Leaning forward	The natural lumbar curvature is lost, the pain in the back can become chronic. The disease is known as Cervicalgia
3	Bent back	The back is too stooped, causing frequent pain, interrupts circulation and inflames the nerves. Causing Kyphosis or Torticollis
4	Inclined to the right	Most of the weight is concentrated in the lower back, causing chronic lumbar pain. In addition, it affects internal organs that are in that part of the body
5	Inclined to the Left	As in the previous position, this causes chronic back pain

These health conditions can be avoided if there is constant monitoring of the position of the body in your workdays. For this, the embedded systems have the feature of adaptation. This implies that they can perform certain functions that emulate the human brain [4]. In this way, an embedded system can have the ability to make its own decisions and adapt to changes. This process is carried out by means of machine learning algorithms. The same ones that under the criterion of classification, the need for previous information to make a decision (training set). However, much of this information may be affected by errors in data acquisition. The sensors that are in charge of this process (convert a physical quantity into an electric one). In this sense, errors can be influenced by uncontrolled variables such as environmental factors or wear of the electronic elements.

One criterion for eliminating noise in the data is the prototype selection (PS). These algorithms are based on the fact that not all data present valuable information for the classifier. In this way, the data that show redundancy or reading errors can be eliminated. To achieve this, one of the algorithms used is from the nearest neighbor. The same one that performs a supervised classification in relation to the Euclidean distance between two points. As a result, a smaller training matrix is obtained that does not compromise the classification performance [5]. Finally, a system that has the ability to make its own decisions, must have an interface to the user. In this way, people can know this decision and extract knowledge from it.

Some works have developed some solutions to this problem. These systems present embedded circuits to detect lumbar or hip position However, there are some present disadvantages such as portability, visualization, among others [6–8]. For this reason, the present system shows a novel way of detecting a person posture in a chair for daily hours of work by means of accelerometer sensors. In this way, the system detects the angles of the column in which the person is working and alerts graphically the position and possible affection. To do this, it is necessary to present a stage of software filtering of data acquisition [9]. Subsequently, an analysis of data that allow the reduction of the training matrix and the comparison with the different criteria of classification algorithms to select the appropriate one. Finally, the system is connected via WiFi to an interface on the computer that shows in real time the position of the person [10].

The rest of the document is structured as follows. Section 2, the design of the electronic system and its conception blocks. Section 3 presents the data analysis for the correct classification of postures. The results are shown in Sect. 4. Finally, the conclusions and future works are shown in Sect. 5.

2   Electronic System Design

The MPU6050 sensor, combining an accelerometer and a 3-axis gyroscope, measures acceleration, inclination or vibration. The data is sent to the NODE MCU programmed in the Arduino environment. This module has been used for its integrated connection to WiFi. Figure 2 shows the connection between the sensor and the acquisition node.

The embedded (ES) system designed was tested in workers that spend more than 4 daily hours sitting in a chair. ES acquire data every five minutes to detect a new posture and the user must put ES in their chest with velcro straps. On this way, ES does not bother to different activities of the user.

The proposed data schema for the detection of posture was made in the following way: (i) The person locates the sensor node in the chest and performs his daily

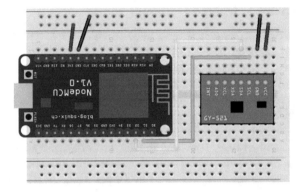

Fig. 2. Electronic system developed

tasks. An expert observes the experiment and detects the bad position of the person and assigns the label of Table 1. Subsequently, this data has stored in a matrix outside the system for analysis. This process is done with 10 people. (ii) The data analysis stage is performed and the reduced database is stored within the system. (iii) Subsequently, the classification algorithm is implemented and the selected label and the acquired data are sent in real time to the visualization stage that determines the person's position. This is shown in Fig. 3.

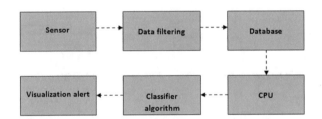

Fig. 3. Data analysis scheme

3 Data Analysis

3.1 Data Acquisition

The acquired data is stored in a matrix $Y \in \mathbb{R}^{m \times n}$, where: **m** is the number of samples and **n** represents the quantity of data acquired by the sensors and the camera. Meanwhile $L \in \mathbb{R}^{m \times 1}$ represents the vector of the labeling of the samples. In this case $m = 500$ and $n = 3$.

3.2   Data Analysis Scheme

In relation to PS algorithms, you can have three criteria. The first is to choose the points near the edge borders of each label to improve the classifier. The second approach is to choose points near the centroid of each subset. This can reduce the training matrix in a greater way. As a third criterion, a combination of the previous two can be performed. To choose the right one, you will perform performance tests with each of the classification algorithms [11].

The distance-based algorithm is considered k Nearest Neighbor (k-NN). According to the literature, the best results are obtained with $k = 3$ and with $k = 5$. A Bayesian classifier (criterion by probabilities), obtains the posterior probability of each class, C_i, using the Bayes rule, as the product of the probability *apriori* of the class by the conditional probability of attributes (E) of each class, divided by the probability of the attributes: $P(C_i|E) = P(C_i)P(E|C_i)/P(E)$. Finally, as a heuristic criterion, the classification tree algorithm is used. Since a classifier can be defined as a function $d(x)$ defined in the classification space $\mathbf{X}$ in $\mathbf{M}$ different subsets $A_1, A_2, ..., A_M$, being $\mathbf{X}$ the union of all of them for all x belonging to A_m to the predicted class C_m. At Fig. 4 shows data analysis scheme.

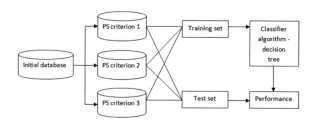

Fig. 4. Data analysis test with each PS and classifier algorithm

In Table 2, it shows the classification yields with the complete database. In relation to the algorithms k-NN, Bayesian classifier, and decision tree. For the use of the criteria should normalize the distances from the center to the rest of the data for each label. The first criterion uses the normalized distance closest to 1 (Table 3). For the second criterion, a criterion close to 0.2 is used and for the third criterion, standardized values between 0.5 and 0.8 are used (Table 4). As a result of this stage of PS. With the first criterion, 75% of the database was reduced, 50% with the second and 80% in the third (Table 5). With this, the matrices were determined $\mathbf{W} \in \mathbb{R}^{t \times n}$ with $\mathbf{t} = 125$, $\mathbf{Q} \in \mathbb{R}^{s \times n}$ with $\mathbf{s} = 250$ and $\mathbf{R} \in \mathbb{R}^{p \times n}$ with $\mathbf{p} = 100$. 20 for each position.

4   Results

Once the data analysis stage is completed, it is defined as the training base $\mathbf{R}$ since its average yield is greater than the rest of the matrices. It is also

Table 2. Full training set classification performance

| Algorithm | 1,2,3 | 3,5,6 | 2,8,1 | 3,6,1 | Average |
	Test 1	Test 2	Test 3	Test 4	
KNN = 1	1	1	1	1	1
KNN = 3	0,99	1	1	0,99	0,995
KNN = 5	0,97	1	1	0,99	0,99
Bayesian	0,98	0,97	0,98	0,97	0,975
Decision tree	1	1	1	1	1

Table 3. First criterion (training set W) classification performance

| Algorithm | 1,2,3 | 3,5,6 | 2,8,1 | 3,6,1 | Average |
	Test 1	Test 2	Test 3	Test 4	
KNN = 1	1	1	1	1	1
KNN = 3	1	1	1	1	1
KNN = 5	1	1	1	1	1
Bayesian	0,97	1	0,973	1	0,986
Decision tree	0,986	1	1	0,986	0,993

Table 4. Second criterion (training set Q) classification performance

| Algorithm | 1,2,3 | 3,5,6 | 2,8,1 | 3,6,1 | Average |
	Test 1	Test 2	Test 3	Test 4	
KNN = 1	1	1	1	1	1
KNN = 3	1	1	1	1	1
KNN = 5	1	1	1	1	1
Bayesian	1	1	0,983	1	0,996
Decision tree	1	1	1	1	1

Table 5. Third criterion (training set R) classification performance

| Algorithm | 1,2,3 | 3,5,6 | 2,8,1 | 3,6,1 | Average |
	Test 1	Test 2	Test 3	Test 4	
KNN = 1	1	1	1	1	1
KNN = 3	1	1	1	1	1
KNN = 5	1	1	1	1	1
Bayesian	1	1	1	1	1
Decision tree	1	1	1	1	1

implemented a decision tree and k-NN as classification algorithms. In this way, a double verification of the decision taken is made. Figures 5 and 6 shows the results of the interface performed upon receiving the system's decision. As a result of the implemented system, 100% accuracy was achieved in real conditions. In addition, it was possible to show that a person tends to change positions between 22 to 30 min. In this way, people choose 30% of the work day in a bad decision.

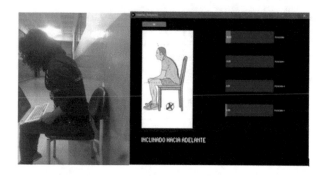

Fig. 5. Embedded systems tests

Fig. 6. Embedded systems tests

5   Conclusions and Future Works

The development of the system allowed for a scheme of an analysis of data suitable for the elimination of noise and improve the performance of the system. It was possible to demonstrate that the algorithms of lower computational cost presented a high accuracy in the decision making. Because of their k-NN development facility and decision tree, they can be implemented within an electronic system.

The detection of posture to remain long sitting allows to generate an alert to the person and can change their lifestyle and work. With this, new alternatives for active breaks in work can be presented with the aim of preventing occupational risks.

As future work, it is proposed to unite different systems that can send and share information of several people and look for trends in behavior.

References

1. Rabin, A., Portnoy, S., Kozol, Z.: The association between visual assessment of quality of movement and three-dimensional analysis of pelvis, hip, and knee kinematics during a lateral step down test. J. Strength Conditioning Res. **30**(11), 3204–3211 (2016)
2. Hughes, G.: A Review of recent perspectives on biomechanical risk factors associated with anterior cruciate ligament injury. Res. Sports Med. **22**(2), 193–212 (2014). https://www.tandfonline.com/doi/full/10.1080/15438627.2014.881821
3. Boissoneault, J., Sevel, L., Letzen, J., Robinson, M., Staud, R.: Biomarkers for musculoskeletal pain conditions: use of brain imaging and machine learning. Curr. Rheumatol. Rep. **19**(1), 5 (2017). http://link.springer.com/10.1007/s11926-017-0629-9
4. Cicchetti, D.V.: Guidelines, criteria, and rules of thumb for evaluating normed and standardized assessment instruments in psychology. Psychol. Assess. **6**(4), 284–290 (1994). http://doi.apa.org/getdoi.cfm?doi=10.1037/1040-3590.6.4.284
5. Hallgren, K.A.: Computing inter-rater reliability for observational data: an overview and tutorial. Tutorials Quant. Methods Psychol. **8**(1), 23–34 (2012). http://www.tqmp.org/RegularArticles/vol08-1/p023
6. Lin, J.F.S., Kulić, D.: Human pose recovery using wireless inertial measurement units. Physiol. Meas. **33**(12), 2099–2115 (2012). http://stacks.iop.org/0967-3334/33/i=12/a=2099?key=crossref.c3f211bf3d71a7a7f4d15b291c37b3f4
7. Das, D., Busetty, S.M., Bharti, V., Hegde, P.K.: Strength training: a fitness application for indoor based exercise recognition and comfort analysis. In: 2017 16th IEEE International Conference on Machine Learning and Applications (ICMLA), pp. 1126–1129. IEEE, December 2017. http://ieeexplore.ieee.org/document/8260796/
8. Nunez-Godoy, S., Alvear-Puertas, V., Realpe-Godoy, S., Pujota-Cuascota, E., Farinango-Endara, H., Navarrete-Insuasti, I., Vaca-Chapi, F., Rosero-Montalvo, P., Peluffo, D.H.: Human-sitting-pose detection using data classification and dimensionality reduction. In: 2016 IEEE Ecuador Technical Chapters Meeting (ETCM), pp. 1–5. IEEE, October 2016. http://ieeexplore.ieee.org/document/7750822/
9. Rosero-Montalvo, P., Peluffo-Ordonez, D.H., Umaquinga, A., Anaya, A., Serrano, J., Rosero, E., Vasquez, C., Suarez, L.: Prototype reduction algorithms comparison in nearest neighbor classification for sensor data: empirical study. In: 2017 IEEE Second Ecuador Technical Chapters Meeting (ETCM), pp. 1–5. IEEE, October 2017. http://ieeexplore.ieee.org/document/8247530/
10. Barshan, E., Ghodsi, A., Azimifar, Z., Zolghadri Jahromi, M.: Supervised principal component analysis: visualization, classification and regression on subspaces and submanifolds. Pattern Recogn. **44**(7), 1357–1371 (2011). http://linkinghub.elsevier.com/retrieve/pii/S0031320310005819
11. Kuncheva, L.I.: Editing for the k-nearest neighbors rule by a genetic algorithm. Pattern Recogn. Lett. **16**(8), 809–814 (1995). http://linkinghub.elsevier.com/retrieve/pii/016786559500047K

In Search of a Decision-Making Framework for Involving Users Who Have Learning Disabilities or Sensory Impairments in the Process of Designing Future Technologies

Jane Seale[1]([✉]), Helena Garcia Carrizosa[1], Jonathan Rix[1],
Kieron Sheehy[1], and Simon Hayhoe[2]

[1] Faculty of Wellness, Education and Language Studies, Open University,
Milton Keynes, UK
jane.seale@open.ac.uk
[2] Department of Education, University of Bath, Bath, UK

Abstract. A comprehensive literature review was undertaken in order to identify design approaches that have been employed with users who have learning disabilities or sensory impairment; the factors that influenced their choices and the extent to which the approaches and techniques adopted were successful. There was a huge variation across the corpus regarding whether a justification was offered for the choice of approach and the extent to which those justifications were supported by evidence. In addition there was a lack of comprehensive evaluation of the design approaches. Technology designers who intend working with users with learning disabilities or sensory impairments therefore currently have little to help them decide which design approach might be the most appropriate or effective.

Keywords: Disability · Design process · User participation · Decision-making

1 Introduction

This paper will present the results of a comprehensive literature review regarding methods for including adults with a diverse range of access preferences frequently associated with the labels of sensory impairment and learning disability in the design of technologies. The objective of this review is to identify if there is any consensus around which design approaches are appropriate and effective to use with these user groups and under what circumstances. The stimulus for the literature review presented in this paper is a Horizon 2020 funded project called ARCHES (Accessible Resources for Cultural Heritage EcoSystems) which involves museum education and technology partners across Europe [1]. The overarching aim of ARCHES is to create more inclusive cultural environments for adults who have a range of access preferences frequently associated with the labels of sensory impairments and learning disabilities [2]. One way in which the ARCHES project is attempting to achieve this aim is by

© Springer Nature Switzerland AG 2020
K. Arai et al. (Eds.): FTC 2019, AISC 1069, pp. 844–861, 2020.
https://doi.org/10.1007/978-3-030-32520-6_61

developing online resources, software applications and multisensory technologies to enable people with learning disabilities and sensory impairment to access museum learning opportunities. Participatory approaches were used to work collaboratively with over 100 participants from England (London), Spain (Madrid and Oviedo) and Austria (Vienna) along with 6 museums and 5 technology companies. Participants are taking a role in identifying existing useful technologies and resources that can promote inclusion; evaluating their experiences of activities and resources within museums; suggesting ways in which technologies might enhance their experiences or resources; evaluating test or beta-versions of technologies and analysing the processes and outcomes of the project as a whole. It was felt that it may be helpful to conduct a literature review in order to examine whether there is a consensus in the field regarding how best to include users with learning disability and sensory impairments in the design process and what factors influence the decisions that designers and developers make regarding their design practices. Such a review is needed because very little specific advice exists to guide interdisciplinary design teams about how best to include users with intellectual or sensory impairments in the process of designing technologies. This paper will begin by discussing what guidance currently exists to help designers decide whether to use these approaches with disabled users. An overview of the method used to undertake a literature review of studies that have involved users with learning disability or sensory impairments will then be provided. The results of the review will be presented and the extent to which analysis of the identified corpus of design literature enables a decision-making framework for choosing appropriate design approaches when designing with users who have learning disability or sensory impairments to be identified will be discussed. Finally, what implications and recommendations can be drawn from the review that can inform the design practices of future design projects focusing on learning disability or sensory impairment will be discussed.

2 Approaches to Including Users in the Research and Design of Technologies

Common approaches to including users in the research and design of technologies are User-centred Design (UCD), Participatory Design (PD) and Human-Centred Design (HCD). Broadly speaking, these approaches offer designers a framework which requires them to address a number of issues or premises relating to: Who, What, When, How & Why [3]. The 'What' relates to overarching focus or orientation of the approach [4, 5] and the underpinning values or principles [4, 6]. The 'How' relates to processes, tools and techniques [5]. The 'Why' relates to goals and motivations [7] (see Table 1). UCD, PD and HCD were not developed specifically with disabled users in mind. Some might argue that either Universal Design, Design for All, Accessible Design or Inclusive Design offer disability sensitive alternatives [8–10]. However, these offer design principles rather than design approaches and are therefore excluded from consideration in the paper. These offer designers a framework of rules, guidelines or standards that they are encouraged to comply with, they do not however elaborate on how exactly designers can enact these rules. Design approaches and associated techniques suggest specific actions, activities or processes. Consulting design principles may be an integral part of one or more of the stages within a design approach (see [23] for example) but the principles are just one aspect of a design approach. With this distinction in mind disability sensitive approaches to technology design have been sought elsewhere.

Table 1. A comparison of design approaches against a framework of design factors

Design factors		UCD	PD	HCD
WHO	Who are the actors in the design process	End-users and designers/developers	Designers, end-users, external stakeholders	Users and other stakeholders, designers
	How are the end-users of the artefact being conceptualised	User as Informants (providing feedback) User as subject	User as Partner, active or full participant, co-designer	Humans Active
WHAT	Design orientation- key focus, overarching characteristic	Usability	Collaboration	Empathy Meaning-making
	Working principles or values underpinning design approach	Improving the understanding of user and task requirements	Democracy Interactive two-way relationship between designer and user	Gaining a clear understanding of users, how they interact with their environment and their needs, desires, experiences & perspectives,
WHEN	Early-late in the process All or some stages in the process	Early in the development cycle (but not necessarily in the initial idea stage) Throughout	Throughout	Throughout
HOW	**Processes**	Iterative Design Empirical Measurement	Iterative	Reflective Evaluative Iterative
	Methods, Tools and techniques (that are unique to an approach or predominantly used)	Task analysis, needs analysis, Usability testing, heuristic evaluation, prototyping (lo-tech, rapid)	Ethnographic methods, Mock-ups, Games, role play, acting; Workshops; Diaries, scenarios	Consulting data-sets; ethnographic interviews and observations; focus groups, role-playing; think-aloud
WHY	Goals and/or motivations	A better (more usable) product	Better quality of life (through use of end-product) Better end-product	A better usable product Improved quality of life for users

2.1   Disability Sensitive Approaches to Technology Design

Some developers and researchers have offered alternative approaches to the standard UCD, PD and HCD approaches which they claim are more appropriate for working with disabled users. For example, Newell et al. argue that UCD methods provide little or no guidance about how to design for disabled people [11]. They also argue that traditional UCD is problematic when the user groups include some disabled users or is entirely composed of disabled users. This means there is a greater variety of user characteristics and functionality which may mean it is difficult to find designs that suit disabled and non-disabled users or disabled users with different kinds of needs. They suggest an extension to UCD that they call 'User-Sensitive Inclusive Design' which they argue requires designers to develop a strong empathy with their disabled user groups. They reject standard UCD methods such as usability tests and experiments where users are positioned as 'subjects'. They propose alternative methods such as ethnography, personas, scenarios and theatrical techniques involving professional actors as useful and appropriate techniques to use with disabled users. Newell et al. do not however explain why they have not positioned their alternative as PD or HCD (or something else) but choose instead to remain within the UCD paradigm.

Bühler offers an alternative design approach for those developers who were aiming for the empowered participation of disabled users in technology-focused research and development projects [12]. His framework, which he labels the 'FORTUNE concept' is underpinned by seven principles: partnership as a basis; users are members of user organisations (so that they advocate on behalf of whole user group as well for themselves individually) the accessibility of all relevant materials and premises are guaranteed; every partner guarantees confidentiality, respect and expertise; there is a detailed plan for the project including time and resource planning for user participation and partnership is implemented from the beginning of the project. Buhler does not explicitly position this approach as an extension of PD, but there are some elements in common such as conceptualising the user as partner and involving the user in all stages of the design process. Reflecting on the potential practical and philosophical validity of the Fortune principles, initial experience of working in the ARCHES project would suggest that it is important not to assume that members of a user group can effectively advocate for all members of the group. Some people can find it difficult to imagine how others in their group would respond and they may therefore need support to build this skill. Many disabled activists and researchers working in the field of critical disability studies would argue that empowerment for disabled people is not in the gift of non-disabled others, disabled people empower themselves by becoming agentic beings. Published in 2001, this approach appears to have had a limited influence on the field. A handful of studies that involve users with learning disability or sensory impairments have cited the FORTUNE concept as an example of a user participation framework or of PD, but they have not actually implemented it themselves [13–15].

2.2   Frameworks for Choosing the Most Appropriate Design Approaches

Given the limited influence of proposed disability-sensitive extensions or alternatives to the standard design approaches it seems then that designers who are new to the field and intend working with users with learning disability or sensory impairments have little to help them decide which design approach might be the most appropriate or effective. An inspection of the main similarities and differences between UCD, PD and HCD as summarised in Table 1 provide no obvious indications as to why designers who wish to work with users who have learning disability or sensory impairments would choose one approach over another. The focus on democracy within the PD approach could be attractive to those working with people with learning disability and who are familiar with participatory or inclusive research frameworks because of their emphasis on equal partnerships between participant and researcher and their positioning of people with learning disability as co-researchers.

Interestingly, drawing on participatory research literature, Draffan et al. propose a framework to enable assistive technology designers to decide the level of participation that disabled users will be afforded with each design project (from non-involvement through to participant initiated and directed). Their framework requires designers to consider the potential strengths of the user, the tasks required of the user, the resources required to enable participation (e.g. training) plus the expertise users bring with them, the environment in which they may be working and the tools they may need to support participation (e.g. communication aids). They argue that "careful analysis of all the components involved in the suggested framework can lead to better AT participatory design and research methodologies with potential users informing best practice" [16]. Whilst this framework might be helpful to PD designers, it does not help designers choose between UCD, PD and HCD, nor suggest any disability-specific adaptations of PD methods. However what this framework does offer is some series of questions (which may need to be extended further to generalise to UCD and HCD design projects) that designers can ask themselves in order to increase the chances of the employment of the design approach being successful. Questions relating to the user, what they will be asked to do, the environment in which they will be asked to design and the resources and time available to support participation in the design process.

Given the lack of broad frameworks that cover all three main design approaches it is important to interrogate the research and development literature in more detail in order to examine how designers decide which design approach to use with users who have learning disabilities or sensory impairments; what factors influence their choices and the extent to which the approaches adopted as a result are successful. In the following section the method used to undertake such a review and to answer the following questions will be outlined:

1. What design approaches are commonly used to include users with learning disability or sensory impairments in the design of technologies?
   a. What factors influence the choice of design approach?
   b. What justifications are given for the choices of design approach
   c. What factors influence the successful employment of the chosen design approach?
   d. What evaluative evidence is provided to demonstrate successful employment of the design approach with the intended user group?

3 Review Method

The literature review took place between October 2016 and March 2018. The SCOPUS database was searched as it includes a range of journals that reflect the multidisciplinary nature of research in the field of learning disability and technology design. In addition Scopus is the worlds' largest abstract and citation database of peer-reviewed literature containing 36,377 titles from approximately 11,678 publishers, of which 34,346 are peer-reviewed journals. A particular focus of the search was the design of technologies similar to those being developed within the ARCHES project. A range of keyword terms were used to search for outputs related to learning disability and sensory impairment in order to reflect the national and disciplinary differences in labels used to categorise this group of people. The parameters of the review include: the date range of the search was restricted to the last twelve years in anticipation that design approaches may be quite different for older technologies designed and evaluated prior to 2006; included where users with either learning disability or sensory impairments were included in the design process. Papers were excluded if the users were children below the age of eighteen or if the majority of the user group were classed autistic (which this paper is not defining as being as example of learning disability, but it is recognised that some authors do). The search produced 59 papers. A two-level filtering process reduced the number of papers down to a corpus of 32 [14, 17–47]. Once the 32 papers had been identified, they were each re-read and notes were made on anything within the paper that had implications for approaches to technology design. In the following sections an overview of the corpus of the 32 papers will be provided followed by an analysis of the decisions and evaluations made regarding design approaches.

4 Overview of the Corpus of Papers

In presenting the results of the literature review an overview of the corpus of papers found in the search will first be provided in order to offer a detailed context for the review findings; particularly in relation to access needs, age range, technologies and intended purpose of technology use. 18 papers involved users with sensory impairments as their primary user group. Of these, three involved blind users [20, 21, 32]; seven involved visually impaired users; three involved both blind and visually impaired users [17, 18, 20]; two involved deaf users [28, 30]; two involved deaf or hard of hearing users [24, 27] and one involved hard of hearing users [22]. 14 papers involved users with learning disability as their primary user group. Two papers involved users with both intellectual and sensory impairment [35, 39] and one paper also included users with complex communication needs [44]. None of the papers reported working solely with middle aged or older adults. The technologies being designed in the 32 papers were diverse and included haptic devices, games, robots, avatars, websites, interfaces and mobile applications. 18 papers reported focusing on designing new technologies. For example, one study worked with eight blind participants to create wearable controls for mobile devices [21]. The design of the technology was based on previous studies that should how hazardous it is for blind users to listen to their phone's guiding instructions whilst trying to move around the urban landscape. 15 papers reported focusing on re-designing existing technologies. For example one study involving users with sensory impairments focused on designing a tactile button interface that could control the native Voice-Over Gesture navigations of IOS devices [18]. There were nine intended purposes of the technologies that the projects were developing: communication, daily living, education, employment, health, accessing information, leisure, safety and travel. It is noticeable that the projects that focused on education involved just users with sensory impairments [18, 23, 27]. The projects that focused on health involved just users with learning disability [36, 44]. The average size of the user group was 11 (range 1 to 48).

5   Design Approaches Commonly Employed with Users Who Have Learning Disability or Sensory Impairments

When making decisions about how to categorise the design approach of each paper, any explicit claims the author made in the title, keywords, abstract or main body of the paper were taken into account. Where there was no explicit statement about the approach professional judgment was applied based on the nature of the design papers they cited in support of their work and/or how closely they fitted to the design characteristics outlined in Table 1.

Analysis revealed that two studies adopted HCD approaches; six used UCD, 12 employed PD approaches (including two that used co-design) and eight adopted a hybrid approach. Four papers adopted approaches other than UCD, PD, HCD or hybrid. Overall there was a clear preference for using PD and hybrid approaches with users with sensory impairments whilst approaches employed with users with learning disability was much more eclectic. In order to try and understand this pattern of adopted design approaches the justifications that designers gave for their choice of design approach and whether these were specifically linked to the difficulties and difficulties experienced by the intended user group will be examined.

5.1   Justifications for the Choice of Design Approach

There was a huge variation across the corpus regarding whether or not a justification was offered for the choice of approach, and the extent to which those justifications proffered were supported by evidence such as citing broad design literature or specific studies that have also employed the approach.

5.1.1   User-Centred Design

Two of the papers offered no definition of UCD or justification as to why UCD might be an appropriate approach to employ with the user group [30, 39]. One paper did not offer a definition of UCD, but did cite the ISO standard for UCD [48]. However, they did not engage in any justifying of the approach or make it clear how the approach they adopted with deaf participants mirrored the approach advocated by the ISO [28]. In a brief conference paper focusing on the design of hospital patient profiling software for people with complex communication needs and cognitive impairment, the researchers offered no definition of UCD but did state that little UCD work has been done with adults with complex communication needs who may also have cognitive impairments [44].

One project analysed the strengths and weaknesses of the UCD approach and argued that although it can better address the user needs and preferences it cannot analyse the user requirements and product function in detail, therefore requiring the involvement of additional usability experts [23]. In order to address the weakness of UCD therefore, they integrated the use of Universal Design principles and the GOMS (Goals, Operators, Methods and Selection) model into their approach. Whilst this SWOT analysis provides a rationale for the integration of UD and GOMS it does not provide a rationale as to why UCD is appropriate to use with visually impaired users or discuss the potential cons of using techniques such as UD and GOMS which do not require user involvement.

Hooper et al. describes a project in which they sought to design an online social platform that would facilitate inclusive research partnerships with people with learning disability [42]. The title of their paper includes reference to 'co-design'. Despite this they do not position the methods they used to design the platform as inclusive research methods, or PD, but rather UCD. They draw on a range of UCD studies and publications to position their work including Gould and Lewis [49]. They define UCD as involving:' the user of a product or service through all the stages of the design of that product or service'. Continuing their rather 'fluid positioning, Hooper et al. justify their use of UCD by arguing that it will result in more appropriate, acceptable designs. They also acknowledge that trying to support the 'more equitable involvement if users in pursuing this goal' is not without tensions and challenges.

5.1.2   Participatory Design

Eight papers offer no definition of their PD approach or if they did, they offered no rationale for why they were using it with their user groups beyond rather vague implications that PD enables user needs to be met (see for example Ferreira and Bonacin, p. 696). Just five papers offered some rationale. For example, Azenkot et al. sought to design specifications that detail how a building service robot could interact with and guide a blind person through a building in an effective and socially acceptable way [17]. Drawing on the work of Kensing and Blomberg and Sanders et al. [50, 51] they define PD as: "a method where a system is designed collaboratively by designers and target users". Their rationale for using PD appears to centre on the fact that PD has been used before with disabled people, although the one reference they cite in support of this, was for a project involving users with aphasia rather than blind people. Kim et al. have an explicit rationale for using PD with their disabled user group arguing that 'users with disabilities have specific needs and requirements for assistive technology applications that are hardly expected by designers without disabilities; thus, they should be involved throughout the entire design process' [14]. They also refer to the fact that the PD approach has been shown appropriate and effective for people with disabilities. They cite the work of Wu et al. [52] who used PD to design an orientation aid for amnesiacs and Wattenberg [53] who described the use of focus groups as an 'accessible research method' with visually impaired people. Tixier et al. justify the use of PD in general terms, rather than relating to specifically to why it is appropriate for use with disabled people [32]. They do however, state that few studies have focused on the use of PD in this field. In referring to the lack of studies, Tixier et al. do cite three papers, one that has used PD with users with learning disability and two that have used PD with users with visual impairment. Sahib et al. justify their use of PD with blind users, because they argue it is hard for sighted users to design for non-sighted users [29]. It is also noted that none of these papers made reference to Bühler's FORTUNE design framework.

5.1.3 Human-Centred Design

Chan and Siu cite just one HCD reference (the out of date ISO 1999 international standard for HCD processes) and they don't justify the use of HCD per se, but rather their use of 'user needs analysis [19]. Furthermore, their justification refers broadly to issues of diversity, rather than visual impairment. Dekelver et al. do not explicitly define HCD, but they indicate that there is a scarcity of literature documenting the use of a human-centred approach with people with learning disability [38]. Dekelver et al. do argue that a human-centred approach is an appropriate one to use because 'design must support the easiness of use'. But they do not make it clear why HCD would support easiness of use over and above other approaches such as UCD or PD.

5.1.4 Hybrid Approaches

Eight papers adopted a hybrid approach- combining two design approaches. Not all of them explicitly claimed that their approach was hybrid in nature, but when their description was interrogated and matched against the characteristics outlined in Table 1, it was concluded that there were elements of two approaches. Five papers combined PD with UCD [20, 25, 26, 31, 37] and three papers combined PD with HCD [24, 33, 46]. It is interesting to note that no studies combined UCD with HCD. Mi et al. (2014) describe a three phase project which was largely UCD in nature [25]. The first phase involved a comprehensive review of existing standards, guidelines and user requirements regarding mobile handheld device accessibility. The second phase included both heuristic evaluation and usability testing. The third phase configured the finalized design guidelines into a heuristic checklist for designing accessible smartphones, which could be generalized and applied to other mobile or touchscreen-based devices. However, in the first phase, the designers used PD to filter a set of preliminary user requirements. Da Silva et al. positioned their methodology as Design Science Research consisting of two 'steps': UCD and PD [37]. The first step used UCD to identify the system requirements which resulted in prototypes of augmentative communication screens. The second step employed the PD approach to enable users to choose the screens images and evaluate the system usability.

Three of the eight studies offered no definition, no references and no rationale for either the hybrid approach or why the hybrid approach might be appropriate to use with their disabled users [20, 26, 31]. Five studies offer some limited (typically implicit rather than explicit) rationale for adopting a hybrid approach- but not for why it would be appropriate with disabled users [23–25, 37, 46]. For example, Kawas et al. talk about the need for a 'holistic qualitative approach' view [24]. They do not however explicitly claim that their hybrid approach of HCD and PD would enable this or why such an approach is needed with users with sensory impairment. Furthermore, none of their 26 references relate to methods, instead they are all related to Automatic Speech

Recognition and captioning for deaf and hard of hearing people, which rather weakens any argument they are making about the validity or appropriateness of the method. Yuan et al. [33] employed what is considered by the authors of this paper to be a combination of PD and HCD. However their rationale for why their approach is appropriate to use with users who are visually impaired, focus more on the PD component than the HCD component. They argue: 'Such a PD process allows us to observe PVI's practices from a holistic perspective and to develop trust, which also benefits from a long-term engagement before we introduce design changes into these practices.' Yuan et al. do however cite a range of studies as support for their approach, including generic design papers [54] and those specifically describing design approaches with visually impaired users [55].

5.1.5   Approaches Other Than UCD, PD, HCD or Hybrid

Four papers reported using an approach other than UCD, PD, HCD or hybrid [34, 35, 40, 41]. All of them involved users with learning disability and adapted their approach in some way to cater for their needs (all except one do not specify how many users they involved). The only clue to how Brown et al. are positioning the design of their project is in a section heading title "User sensitive design' [35]. However in the text within the section Brown et al. do not define user sensitive design, nor do they cite the work of Newell et al. [11] regarding user sensitive design. Apart from occasionally using the language of inclusion with terms such as 'co-discovery', there is no other referral to the inclusive design literature or discourse. Three papers, all reporting on the same project (Sensory Objects Project) position their approach as inclusive research [34, 40, 41]. The researchers actually use the term 'inclusive design' to describe their approach, however the reference to researchers and co-researchers along with reference to the work of well-known participatory/inclusive research studies would suggest that they are sympathetic to inclusive research and perhaps see no difference between inclusive design and inclusive research [56]. This conflation of the two terms inclusive research and inclusive design may also reflect the multi-stakeholder nature of the project team.

5.2   Evaluations of Choice of Approach

Across the corpus, just eleven papers offered some evaluative reflections or comments on the perceived success or failure of their chosen design approach with the intended user group. These were spread evenly across the user groups (6 sensory impairment projects and 5 learning disability projects). Interestingly, there were no evaluations from studies that had employed UCD. Evaluations focused on seven areas: user needs, skills and difficulties; the experience of the process for the user, the quality of the end-design or product; the pragmatics of conducting the study; stakeholder needs and values and researcher skills, needs or difficulties.

5.2.1   User Related Evaluations

Chan and Siu argue that the iterative nature of the study enabled them to design a system based on the needs of visually impaired people [19]. This is however the extent of their evaluation of how successful or appropriate the use of HCD with their users group was. Sahib et al. provide a bit more information as to why involving blind users at an early stage allowed them to identify limitations with their own design ideas [29]. They share how 'participants would often question the practicality of the proposed interface features, requiring detailed explanations of how these interface components would be accessed in a realistically usable way with screen readers'. Similarly Xu et al. report: 'Without working with people with an intellectual disability, the team may not have realised how subtle changes to colour, icons, pictures and wording would have a large effect on how people with an intellectual disability understand and use Rove n Rave' [47]. However they also report that the biggest challenge for their team was the fact that users had such different reactions to one another despite all having the same 'label'. Allen et al. conclude that they have learnt not to underestimate their co-researchers interest and ability to use technology [34].

5.2.2   Design Process

Two studies that involved users with sensory impairments offered specific recommendations to other designers regarding the design process. One study that employed PD recommended that researchers introduce participants to the design at the early stages of the process, to spur creativity while providing some necessary constraints [17]. Another study that used PD made three recommendations. Firstly, to 'consider the whole process of an activity in design so as to identify actual needs and possible technology supports that take place at each stage and as a whole'. Secondly to shift the design focus away from steps (e.g. identify an item) towards activities (e.g. organising the pantry). Thirdly, not to get distracted and consumed by the 'mitigating deficits' of the users [33].

5.2.3   Experience of the Process for the User

Three studies report on the influence of their approach on the engagement and motivation levels of their users. Hollinworth et al. comment positively on the impact of using inclusive research methods with users with learning disability. They noted that their co-researchers were so highly engaged to the extent that they were keen to share the project with their peers [41]. They also suggest that they experienced an increase in confidence and empowerment, but present no explicit evidence for this claim. Usoro et al. claim that the use of a PD approach with young people with learning disability enhanced user engagement throughout the process [45]. Working with people with visual impairment, Yuan et al. claim that their hybrid of HCD and PD and in particular their detailed attention to shopping practices of the users led the users to trust the design team. They also comment on a growing willingness of the users to stand-up and testify about the project to external stakeholders [33].

5.2.4   Researcher Skills, Needs or Difficulties

Two projects reflected on their experiences regarding the nature and the level of skills that researchers require in order to successfully engage in design projects with people with learning disability. Dekelver et al. conclude that using HCD with intellectually impaired users requires sociological skills in order to fully understand the specific position of people with learning disability at home and in care and work placement centres [38]. Allen et al. conclude that they have learnt the importance of using all their senses in the development of museum interpretation in order to give more chance of engagement to people with different disabilities and interests [34].

5.2.5 Product Related

Two PD projects claim that using this approach resulted in working, usable technology [18, 36]. For example Buzzi et al. claim that allowing PD to drive the development of their learning platform resulted in feedback that led to making the games customizable in terms of discriminative stimuli, difficulty levels and reinforcement, as well as the creation of a game "engine" to easily set up new personalized exercises. They claim that these customization features not only meet the needs of the users, but broaden the appeal of the platform to a wider user group [36].

5.2.6 Study Pragmatics

Mi et al. conclude that one of the greatest challenges in conducting research with users with impairments is access to the participants themselves. An additional limitation they identified is the variability in the time each PD member spent learning how to use the prototype prior to evaluation [25]. They argue that 'factors such as work schedules and insufficient learning assistance may be potential threats to the study control, but also other factors, such as frustration with the new technology, may have negatively affected interest in phone exploration'. This problematizing of the user and not the technology is unhelpful and potentially inappropriate. It is unconvincing that a well-designed product would require a user to invest significant time to learn how to use it. Furthermore, from the experience of the authors, disabled people can be reluctant to take part in studies due to negative prior experiences, particularly if they felt that their participation was tokenistic and not taken seriously.

Another study involving users with sensory impairments that lasted for about a year also concluded that it was important to engage in PD for an extended period of time and that short-term engagements 'may not be sufficient for the designers to fully grasp users' needs and practices' [33]. However, they do not specify how they would define short-term engagement. Interestingly, an analysis of the duration of each of the studies in the corpus reveals that the duration of a study ranged from 1 day to 1095 days and the average (mean) duration for a study was longer for those involving users with learning disability (412 days) compared to those involving users with sensory impairments (320 days). When comparing average (mean) duration by approach, the shortest was UCD (26), followed by HCD (117 days), PD (122 days), Hybrid (488 days) and Other (1095 days). The figures for HCD probably does not reflect reality, given that HCD methods are meant to involve ethnographic studies of users' lives and experiences. However, it is possible the high figure for Hybrid studies reflects the fact that half of these studies included HCD as part of the 'mix'.

5.2.7 Stakeholder Needs or Values

Allen et al. report that the Visitor Experience Officer at the heritage site noted that the Sensory Objects workshop consultation process was important to the owners of site who wanted to make their exhibits more credible. It also fitted with their organisational philosophy [34].

6 Discussion

In the previous two sections the studies in the corpus have been analysed with respect to the decisions and evaluations made regarding the design approaches employed with users with learning disability or sensory impairment. In this section the findings will be summarised and the common factors that appear to influence design decisions and the common issues raised when evaluating the success of design projects will be highlighted.

The review reveals that UCD, PD and HCD were all employed within the corpus, but that PD was the most commonly used and HCD was the least commonly used. Given that HCD is quite a labour intensive method requiring a range of both computer science and social science skills [38] it is perhaps understandable why it might be the least used approach. On the other hand given that many of the intended purposes of the technologies being designed were to support disabled users undertake tasks and activities within their own environment (e.g. travel, leisure, employment and daily living skills such as shopping) and that HCD is a method that involves understanding how users interact with their environment it could also be surprising that more studies did not employ HCD.

6.1 Factors that Might Influence the Choice of Design Approach

When considering the factors that might influence the choice of design approach it was noted that PD was more commonly used with users with sensory impairments and that the choice of approach was more varied for studies involving users with learning disability. One reason why PD is more common approach to use might be that some designers may assume that people with learning disability do not have the mental capacity to engage in co-design activities. This needs further investigation however, and it is important to remember one of the conclusions from a study that did involve users with learning disability regarding not under-estimating the interest and ability of people with learning disability to use technology [34]. Those studies where approaches other than UCD, PD and HCD had been adopted with users with learning disability involved experienced multidisciplinary teams that had years of experience of involving people with learning disability, which perhaps gave them the knowledge and the confidence to find other creative design approaches [34, 35, 40, 41].

There was a huge variation across the corpus regarding whether or not a justification was offered for the choice of approach, whether the justification was related to the disabilities of the user groups and the extent to which those justifications proffered were supported by evidence. This makes it hard to discern whether there were any valid reasons for choosing one design approach over another when working with users with learning disability or sensory impairments. The tendency not to offer definitions of the approach being used made it difficult on many occasions to ascertain the overarching design orientation or the principles and values that the designers were using to underpin their design decisions. The tendency not to cite other studies that have been conducted with users with learning disability or sensory impairments could be argued to be due to a lack of studies in this area as some of the authors suggest [32, 38, 44]. Choosing to cite studies that did not involve users with intellectual or sensory impairments but did involve users with other impairments in order to support design choices may suggest that designers assume that disabled people are a homogenous group and that there is no need to consider their specific abilities and needs when considering which design approach to use [14].

6.2   Factors that Designers May Need to Take into Account in Order to Effectively Employ a Particular Approach

When considering the factors that designers may need to be taken into account in order to effectively employ a particular approach it is noted that UCD studies were on average the shortest in duration and that 'Other' design approaches (which typically involved employing elements of inclusive research, working with people with learning disability in particular) were the longest. It is also interesting to note that for most of the design approaches, the age of the user does not appear to be important, since it was most common for researchers not to report their age in their papers. Age was reported more commonly in the PD studies and a possible trend was observed in that more studies involved young adults than the other age groups. This may be because there was an assumption that younger adults are more frequent technology users and therefore could give more informed responses regarding the strengths and weaknesses of new technology designs. Similarly, if the potentially distorted figures for HCD and 'Other' are ignored, there is not a lot of difference between the average size of groups across the design approaches.

The lack of a comprehensive or detailed evaluation of the success or the failure of the chosen approaches with users with learning disability or sensory impairments makes it difficult to draw any confident conclusions regarding what factors influence the successful employment of a design approach with users with learning disability and sensory impairments. This lack of evaluation, particularly of any failures or weaknesses in the employment of their approach may be symptomatic of the researchers desire to show their work and product in a positive light in order to secure future funding. What little evaluation evidence has been identified suggests that:

- involving people with learning disability and sensory impairments in PD and HCD results in usable technologies [18, 19, 36];
- using PD, hybrid and 'other' approaches with users with learning disability and sensory impairments can lead to high levels of engagement and commitment [33, 41, 45]; and
- designers working with users with learning disability learn a lot about themselves and the needs of people with learning disability and sensory impairments when they adopt PD, HCD or 'Other' approaches to design [34, 37, 47].

Far more evidence is needed to support these tentative conclusions which requires future studies in this area to be far more evaluative than those in the corpus have been.

7 Conclusion

With regards to identifying a decision-making framework for deciding between design approaches the literature review has not revealed a clear framework. For example, whilst a pattern in favour of using PD with users with sensory impairments was noticed, the lack of evidence-based justifications for this or detailed evaluations of the success of the approach means that there is no clear reason behind such a decision. It is recommend therefore that future studies, irrespective of which approach they are employing make their decision-making process much more explicit and detailed. The review of the literature have revealed significant variation in the approaches used by designers and researchers along with large variation regarding whether or not a justification for the choice of design approach is offered. Where a justification is offered, there is huge variation in whether that justification is related to the needs of the intended user group or supported by evidence. In addition there is a lack of comprehensive and detailed evaluation of the design approaches employed within the corpus studies. Technology designers (and their partners from other disciplines) who are new to the field and intend working with users with intellectual or sensory impairments therefore currently have little to help them decide which design approach might be the most appropriate or effective. The value and effectiveness of future technologies will be severely limited unless more work is done to articulate and justify a meaningful decision-making framework. In terms of limitations of the study reported here it is possible that using databases other than SCOPUS might have revealed a different or larger set of studies and there is certainly value in future work updating the search to go beyond March 2018. Following this survey, the technology companies involved in the ARCHES company were interviewed about their experience and understanding of participatory approaches and analysis of these interviews suggest there is value in building on this to conduct a wider interview study where designers are questioned about their decision-making processes regarding their chosen design approaches in order to triangulate with the study presented here.

Acknowledgment. This work was performed within the framework of the H2020 project ARCHES (http://www.arches-project.eu), which has received funding from the European Union's Horizon 2020 research and innovation programme under grant agreement No. 693229.

References

1. https://www.arches-project.eu/
2. At the outset of the project a broad label was proposed: "People who experience differences and difficulties associated with perception, memory, cognition and communication". As the project progressed however, it became clear that not all the participants wished to be defined by this or any other label. There was a collective agreement therefore to subsequently refer to participants as having access preferences
3. Giacomin, J.: What is human centred design? Des. J. **17**, 606–623 (2014)
4. Ellis, R.D., Kurniawan, S.: Increasing the usability of online information for older users: a case study in participatory design. Int. J. Hum.-Comput. Interact. **12**, 263–276 (2000)

5. Blomberg, J.L., Henderson, A.: Reflections on participatory design: lessons from the trillium experience. In: Proceeding of Conference on Human Factors in Computing Systems, CHI 1990, Seattle, Washington, pp. 353–359 (1990)

6. van der Bijl-Brouwer, M., Dorst, K.: Advancing the strategic impact of human-centred design. Des. Stud. **53**, 1–23 (2017)

7. Muller, M.J., Haslwanter, J.H., Dayton, T.: Participatory practices in the software lifecycle. In: Helander, M., Landauer, T.K., Prabhu, P. (eds.) Handbook of Human-Computer Interaction, 2nd edn, pp. 255–297. Elsevier, Amsterdam (1997)

8. Clarkson, J.: Inclusive Design: Design for the Whole Population. Springer, London (2003)

9. Steinfeld, E., Maisel, J.: Universal Design: Creating Inclusive Environments. Wiley, New York (2012)

10. Klironomos, I., Antona, M., Basdekis, I., Stephanidis, C.: White paper: promoting design for all and e-accessibility in Europe. Univ. Access Inf. Soc. **5**, 105–119 (2006)

11. Newell, A.F., Gregor, P., Morgan, M., Pullin, G., Macaulay, C.: User-sensitive inclusive design. Univ. Access Inf. Soc. **10**, 235–243 (2011)

12. Bühler, C.: Empowered participation of users with disabilities in R&D projects. Int. J. Hum. Comput. Stud. **55**, 645–659 (2001)

13. Saridaki, M., Mourlas, C.: Integrating serious games in the educational experience of students with intellectual disabilities: towards a playful and integrative model. Int. J. Game-Based Learn. **3**, 10–20 (2013)

14. Kim, H.N., Smith-Jackson, T.L., Kleiner, B.M.: Accessible haptic user interface design approach for users with visual impairments. Univ. Access Inf. Soc. **13**, 415–437 (2014)

15. Millen, L., Cobb, S., Patel, H.: Participatory design approach with children with autism. Int. J. Disabil. Hum. Dev. **10**, 289–294 (2011)

16. Draffan, E.A., James, A., Wald, M., Idris, A.: Framework for selecting assistive technology user-participation methods. J. Assist. Technol. **10**, 92–101 (2016)

17. Azenkot, S., Feng, C., Cakmak, M.: Enabling building service robots to guide blind people: a participatory design approach. In: Proceedings of ACM/IEEE International Conference on Human-Robot Interaction, HRI 2016, Christchurch, New Zealand, pp. 3–10 (2016)

18. Batterman, J.M., Martin, V.F., Yeung, D., Walker, B.N.: Connected cane: tactile button input for controlling gestures of iOS voiceover embedded in a white cane. Assistive Technol. **30**, 91–99 (2018)

19. Chan, M.K., Siu, K.W.M.: Inclusivity: a study of Hong Kong museum environments. Int. J. Crit. Cult. Stud. **11**, 45–61 (2013)

20. Dietz, M., Garf, M.E., Damian, I., André, E.: Exploring eye-tracking-driven sonification for the visually impaired. In: Proceedings of the 7th Augmented Human International Conference, AH16, Article No 5, Geneva, Switzerland (2016)

21. Feng, C.: Designing wearable mobile device controllers for blind people: a co-design approach. In: Proceedings of the 18th International ACM SIGACCESS Conference on Computers and Accessibility, ASSETS 2016, Reno, USA, pp. 41–342 (2016)

22. Ferreira, M.A.M., Bonacin, R.: Analyzing barriers for people with hearing loss on the web: a semiotic study. In: Proceedings of International Conference on Universal Access in Human-Computer Interaction, UAHCI, Las Vegas, USA, pp. 694–703 (2013)

23. Huang, P.-H., Chiu, M.-C.: Integrating user centered design, universal design and goal, operation, method and selection rules to improve the usability of DAISY player for persons with visual impairments. Appl. Ergon. **52**, 29–42 (2016)

24. Kawas, S., Karalis, G., Wen, T., Ladner, R.E.: Improving real-time captioning experiences for deaf and hard of hearing students. In: Proceedings of the 18th International ACM SIGACCESS Conference on Computers and Accessibility, ASSETS 2016, Reno, USA, pp. 15–23 (2016)

25. Mi, N., Cavuoto, L.A., Benson, K., Smith-Jackson, T., Nussbaum, M.A.: A heuristic checklist for an accessible smartphone interface design. Univ. Access Inf. Soc. **13**, 351–365 (2014)

26. Parkinson, A., Tanaka, A.: The haptic wave: a device for feeling sound. In: Proceedings of Conference on Human Factors in Computing Systems, CHI 2016, San Jose, USA, pp. 3750–3753 (2016)

27. Peruma, A., El-Glaly, Y.N.: CollabAll: inclusive discussion support system for deaf and hearing students. In: Proceedings of the 19th International ACM SIGACCESS Conference on Computers and Accessibility, ASSETS 2017, New York, USA, pp. 315–316 (2017)

28. Rocha, T., Paredes, H., Soares, D., Fonseca, B., Barroso, J.: MyCarMobile: a travel assistance emergency mobile app for deaf people. In: Proceedings of IFIP Conference on Human Computer Interaction, INTERACT 2017, Mumbai, India, pp. 56–65 (2017)

29. Sahib, N.G., Stockman, T., Tombros, A., Metatla, O.: Participatory design with blind users: a scenario-based approach. In: Proceedings of IFIP Conference on Human Computer Interaction, INTERACT 2013, Cape Town, South Africa, pp. 685–701 (2013)

30. Smith, R.G., Nolan, B.: Emotional facial expressions in synthesised sign language avatars: a manual evaluation. Univ. Access Inf. Soc. **15**, 567–576 (2016)

31. Tanaka, A., Parkinson, A.: Haptic wave: a cross-modal interface for visually impaired audio producers. In: Proceedings of Conference on Human Factors in Computing Systems, CHI 2016, San Jose, USA, pp. 2150–2161 (2016)

32. Tixier, M., Lenay, C., Le Bihan, G., Gapenne, O., Aubert, D.: Designing interactive content with blind users for a perceptual supplementation system. In: Proceedings of the 7th International Conference on Tangible, Embedded and Embodied Interaction, Barcelona, Spain, pp. 229–236 (2013)

33. Yuan, C.W., Hanrahan, B.V., Lee, S., Rosson, M.B., Carroll, J.M.: Constructing a holistic view of shopping with people with visual impairment: a participatory design approach. Univ. Access Inf. Soc. **18**, 1–14 (2017)

34. Allen, K., Hollinworth, N., Minnion, A., Kwiatkowska, G., Lowe, T., Weldin, N., Hwang, F.: Interactive sensory objects for improving access to heritage. In: Proceedings of Conference on Human Factors in Computing Systems, CHI 2013, Paris, France, pp. 2899–2902 (2013)

35. Brown, D.J., McHugh, D., Standen, P., Evett, L., Shopland, N., Battersby, S.: Designing location-based learning experiences for people with intellectual disabilities and additional sensory impairments. Comput. Educ. **56**, 11–20 (2011)

36. Buzzi, M.C., Buzzi, M., Perrone, E., Rapisarda, B., Senette, C.: Learning games for the cognitively impaired people. In: Proceedings of 13th Web for All Conference, W4A 2016, Montreal, Canada, Article no 14 (2016)

37. da Silva, D.M.A., Berkenbrock, G.R., Berkenbrock, C.D.M.: An approach using the design science research for the development of a collaborative assistive system. In: Proceedings of CYTED-RITOS International Workshop on Groupware, CRIWG 2017, Saskatoon, Canada, pp. 180–195 (2017)

38. Dekelver, J., Daems, J., Solberg, S., Bosch, N., Van De Perre, L., De Vliegher, A.: Viamigo: a digital travel assistant for people with intellectual disabilities: modelling and design using contemporary intelligent technologies as a support for independent travelling of people with intellectual disabilities. In: Proceedings of 6th International Conference on Information, Intelligence, Systems and Applications, IISA 2015, Corfu, Greece, art. no. 7388014 (2015)

39. Hassell, J., James, A., Wright, M., Litterick, I.: Signing recognition and cloud bring advances for inclusion. J. Assist. Technol. **6**, 152–157 (2012)

40. Hollinworth, N., Allen, K., Kwiatkowska, G., Minnion, A., Hwang, F.: Interactive sensory objects for and by people with learning disabilities. SIGACCESS Newsl. **109**, 11–20 (2014)
41. Hollinworth, N., Allen, K., Hwang, F., Minnion, A., Kwiatkowska, G.: Interactive sensory objects for and by people with learning disabilities. Int. J. Inclusive Mus. **9**, 21–38 (2016)
42. Hooper, C.J., Nind, M., Parsons, S., Power, A., Collis, A.: Building a social machine: co-designing a TimeBank for inclusive research. In: Proceedings of the 2015 ACM Web Science Conference, Oxford, United Kingdom, Article Number 16 (2015)
43. Iversen, O.S., Leong, T.W.: Values-led participatory design - mediating the emergence of values. In: Proceedings of the 7th Nordic Conference on Human-Computer Interaction, NordiCHI 2012, Copenhagen, Denmark, pp. 468–477 (2012)
44. Prior, S.: HCI methods for including adults with disabilities in the design of CHAMPION. In: Proceedings of Conference on Human Factors in Computing Systems, CHI, 2010, Atlanta, USA, pp. 2891–2894 (2010)
45. Usoro, I., Connolly, T., Raman, S., French, T., Caulfield, S.: Using games based learning to support young people with learning disabilities stay safe online. In: Proceedings of the European Conference on Games-Based Learning, Paisley, Scotland, pp. 704–712 (2016)
46. Wilson, C., Sitbon, L., Brereton, M., Johnson, D., Koplick, S.: Put yourself in the picture': designing for futures with young adults with intellectual disability. In: Proceedings of the 28th Australian Computer-Human Interaction Conference, OzCHI 2016, Launceston, Australia, pp. 271–281 (2016)
47. Xu, Y., Zhang, J., Yagovkin, R., Maniero, S., Wangchunk, P., Koplick, S.: Rove n Rave™ development: a partnership between the university and the disability service provider to build a social website for people with an intellectual disability. In: Proceedings of the 26th Australian Computer-Human Interaction Conference, OzCHI 2014, Sydney, Australia, pp. 531–534 (2014)
48. ISO 13407. Human-Centred Design Processes for Interactive Systems (1999)
49. Gould, J., Lewis, C.: Designing for usability: key principles and what designers think. Commun. ACM **28**, 300–311 (1985)
50. Kensing, F., Blomberg, J.: Participatory design: issues and concerns. Comput. Support. Coop. Work **7**, 167–185 (1998)
51. Sanders, E.B.N., Brandt, E., Binder, T.: A framework for organizing the tools and techniques of participatory design. In: Proceedings of the 11th Biennial Participatory Design Conference, PDC 2010, New York, USA, pp. 195–198 (2010)
52. Wu, M., Baecker, R., Richards, B.: Participatory design of an orientation aid for amnesics. In: Proceedings of the SIGCHI Conference on Human Factors in Computing Systems, CHI 2005, Montreal, Canada, pp. 511–520 (2005)
53. Wattenberg, T.L.: Online focus groups used as an accessible participatory research method. In: Proceedings of the 7th International ACM SIGACCESS Conference on Computers and Accessibility, ASSETS 2005, Baltimore, USA, pp. 180–181 (2005)
54. Carroll, J.M., Chin, G., Rosson, M.B., Neale, D.C.: The development of cooperation: five years of participatory design in the virtual school. In: Proceedings of Designing Interactive Systems, DIS 2000, Newcastle Upon Tyne, UK, pp. 239–251 (2000)
55. Katz, B.F., Kammoun, S., Parseihian, G., Gutierrez, O., Brilhault, A., Auvray, M., Truillet, P., Denis, M., Thorpe, S., Jouffrais, C.: NAVIG: augmented reality guidance system for the visually impaired. Virtual Reality **16**, 253–269 (2012)
56. Walmsley, J., Johnson, K.: Inclusive Research with People with Learning Disabilities: Past, Present and Futures. Jessica Kingsley, London (2003)

Hybrid Cryptosystems for Protecting IoT Smart Devices with Comparative Analysis and Evaluation

Ahmed Ab. M. Ragab[1](✉), Ahmed Madani[1], A. M. Wahdan[2],
and Gamal M. I. Selim[1]

[1] Arab Academy for Science, Technology and Maritime Transport, Cairo, Egypt
a.abdelhamid92@gmail.com,
{madani82,dgamalselim}@aast.edu
[2] Department of Computer and Systems Engineering, Ain Shams University,
Cairo, Egypt
wahdan73@gmail.com

Abstract. There are limited numbers of reliable hybrid cryptosystems that can be used to protect IoT smart devices, specifically in smart cities, smart hospitals, smart homes, and industrial fields. Therefore, much related work has to be performed. The aim is to achieve more secure hybrid cryptosystem with high demanded performance. This paper investigates some recommended types of hybrid cryptosystems for protecting IoT smart devices from internet attacks. Then, a robust hybrid cryptosystem is proposed. These hybrid cryptosystems combine symmetric encryption algorithms such as TEA, XTEA, XXTEA, and asymmetric encryption algorithms such as RSA and ECC. They should have the capability to protect IoT smart devices from internet attacks. Since, they could efficiently achieve confidentiality, authenticity, integrity, and non-repudiation. Comparative analysis and evaluation are achieved. The analysis included the most important factors that have to be investigated in case of using lightweight ciphers to suit limited resources IoT smart devices. Among these factors are security level, memory size, power consumption, encryption time, decryption time, and throughput. Results show that the proposed hybrid cryptosystem that combined ECC and XXTEA gives better security and performance than RSA and XXTEA.

Keywords: Hybrid cryptosystems · Protecting IoT smart devices · Lightweight ciphers analysis · RSA · ECC · XXTEA

1 Introduction

There is extensive diversity of IoT devices, including sensor-enabled smart devices, and all types of wearables smart devices connect to the Internet. The cost of technology has decreased, in every area, we demand access to the internet which delivers a quantity of information in real time. IoTs are applied in several applications including smart homes, smart hospitals, smart industry, and smart cities. Moreover, some environments exist only on the internet, such as social networks, where all information is in the cloud.

© Springer Nature Switzerland AG 2020
K. Arai et al. (Eds.): FTC 2019, AISC 1069, pp. 862–876, 2020.
https://doi.org/10.1007/978-3-030-32520-6_62

Several advanced technologies, such as smart sensors, networks, wireless communications, data analysis techniques, and cloud computing, have been developed to realize the potential of IoT with different smart systems [1]. As technology rapidly rises at radio-frequency identification (RFID), Internet approaches have gained momentum in connecting everyday things to the internet and facilitating communication from machine to human and from machine to machine with the physical world [2]. The main advantages of IoT smart devices include enhanced data collection, provides real-world information and technology optimization. However, vulnerabilities that face IoT smart devices may include security, privacy, and complexity. IoT smart devices are vulnerable to security attacks [3]. So that protection schemes using encryption algorithms are used to protect IoT smart devices data transmission from different types of attacks.

Symmetric encryption technologies provide cost-effective means and efficiency to secure data without compromising security [4]; however, sharing a secret key is a problem. On the other hand, asymmetric technologies resolve the encryption key distribution problem; however, they are slow compared to symmetric encryption and consume more computer resources [5]. Hence, one of the best possible solutions for encryption is the complementary use of both symmetric and asymmetric encryption techniques [6]. In this paper, comparative analysis and evaluation of the hybrid cryptosystems based on the tiny family block ciphers algorithms (TEA, XTEA, XXTEA) and the asymmetric ciphers algorithms such as EEC and RSA are investigated. Then, a robust hybrid model combines two different cryptography including XXTEA and ECC to ensure the confidentiality, authenticity, integrity, and non-repudiation is proposed. The rest of the paper is as follows. Section 2 discusses related work. Section 3 explains the hybrid cryptosystem security features and services. Section 4 describes the proposed hybrid cryptosystem. Section 5 discusses performance analysis and evaluation and results. Section 6 is the conclusion.

2   Related Work

There are several hybrid encryption algorithms discussed in many kinds of literature based on different techniques. For example, a hybrid encryption method based on symmetric DES and asymmetric cryptography ECC to ensure the security of the database system in a mobile payment technology was applied in smart travel [7]. The drawback of this system is that it only achieves confidentiality and authenticity. To enhance security, a hybrid encryption technique described in [8] uses the Advance Encryption Standard (AES) and the asymmetric key to enabling strong security and low computational complexity through a combined encryption algorithm. It provides confidentiality, integrity, and non-repudiation on the data transmission for IoT to the database server. The drawback of this system is that AES occupy large size in ROM and RAM at processing [9]; also, the MD5 is susceptible to differential attack [10]. To increase efficiency and decrease memory consumption, there was a study to implement and observe parameters like time and memory for implementation of multi-level encryption using the Data Encryption Standard (DES [25]) and a modified version of the RSA Algorithm, the multi-prime RSA [11]. It consumes a larger size in memory to generate a longer key [10]. Another technique (HAN [15]) is used to enhance encryption's speed with less computational complexity. It is based on AES for data encryption and decryption and uses NTRU [16] for key encryption and decryption through the media. It applies new lattice basis reduction techniques to cryptanalyze NTRU to discover the original key to get original text [16].

In [19], a hybrid technique was used to combine the symmetric cryptographic algorithm (AES) and the asymmetric algorithm (RSA) and hashing function (MD5). These algorithms were combined to ensure confidentiality, authentication and data integrity. However, the drawbacks of AES are that it occupies large size in ROM and RAM at processing, also MD5 is susceptible to differential attack along with the large size of memory used to process RSA key. In [20] for securing medical data transmission in IoT environments, a healthcare security model was investigated. The model secure patients' data using hybrid encryption scheme created from AES and RSA algorithms. The drawbacks of AES are that it occupies large size in ROM and RAM at processing, also large size of memory used to process RSA key.

In [21] a hybrid model was constructed to ensure security and integrity assurance of the data during the transmission. The model is an implementation of two cryptographic algorithms including the SHA1, hash generation algorithm and AES for encryption and decryption of messages. The paper also discussed various other cryptographic algorithms and the reason why AES and SHA1 are preferred in an RFID system. The drawback of this system is that SHA1 is susceptible to collision attacks [22]. In [23] another context introduced a hybrid technique for cryptography using the symmetric algorithms AES and Blowfish. This combination gives high security, it also uses hashing Key based MD5 [24] that make hashing the key in the encryption process and make the same process in decryption. This design causes a CPU overload and memory consumption in the process of double encrypt of plaintext.

To increase the security of hybrid cryptosystem, another methodology was proposed using the lightweight algorithm XTEA for data encryption and ECC for key encryption and PBKDF2 for key generation in IoT and deployed on an Arduino kit for secure transmission between IoT WSN [17]. The drawback of this system is that XTEA got broken and weaknesses within PBKDF2 function due to its susceptibility to Brutal force attack [18].

A cyber security scheme for IoT [12] was proposed to facilitate an additional level of security based on the strength of symmetric and asymmetric algorithms such as AES and RSA for a closed system through tunneling technology to be applied in internet sensitive application and file storage. The drawbacks of this system are that AES consumes more memory and RSA needs more computational power to generate RSA key [26]. It was also shown that the lightweight block cipher XXTEA outperforms the performance of AES. However, XXTEA is susceptible to a related key attack [13]. To overcome this attack, XXTEA is joined with a chaotic system to produce a more secure algorithm [14, 27]. When the system sends data, it uses a chaotic algorithm to generate key during each transmission. This design only achieves confidentiality of data.

Besides the above, Table 1 shows a brief comparison of some additional work done and the types of cryptographic algorithms used. Table 2 illustrates the types of cryptographic algorithms used and security features achieved in IoT industrial systems.

To overcome most drawbacks of the described cryptographic models above, a new hybrid cryptosystem is proposed to protect IoT smart devices efficiently as described in the next sections.

Table 1. A brief comparison of some cryptographic algorithms used in some of the previous work.

Authors	References	Applications	Algorithms used
Ruan, Xiao, and Luo [7]	IEEE 3rd Int. Conf. on cloud computing and intelligent systems, 2014	Mobile payment system	DES, ECC
Xin [8]	IEEE Int. Conf. on cyber enabled distributed computing and knowledge discovery, 2015	IoT security transmission system	AES, ECC, MD5
Kaur, et al. [11]	Modern education and computer science (MECS) press, 2017	Security analysis	DES, RSA
Darwish, El-Gendy, Hasanien [12]	Springer International publishing AG, 2017	IoT security	ANSI X9.17, AES, RSA, SHA1
Bhasher and Rupa [28]	IEEE Intr. Conf. on innovations in power and advanced computing technology, 2017	Data transmission security	Chaos keys, XXTEA

Table 2. Types of security features used in IoT industrial systems.

Security features	Libelium IoT industry Hybrid cryptosystem [29]	Amazon IoT industry Hybrid cryptosystem [30]	ARM IoT industry Hybrid cryptosystem [31]
Authentication	RSA	RSA/ECDHE	RSA
Confidentiality	AES256	AES128	AES128
Integrity	MD5/SHA1	SHA256	SHA
Non-repudiation	RSA key signing message	ECDSA	Not defined

3   The Hybrid Cryptosystem Security Features and Services

The following are the security features that the proposed robust hybrid cryptosystem has to achieve.

- *Confidentiality.* This security service ensures that only those authorized have access to the content of the information. Hence, it prevents an unauthorized user to access the content of the protected information. It is sometimes referred to as secrecy.

- *Data integrity.* It provides a mean to detect whether data has been manipulated by an unauthorized party since the last time an authorized user created, stored or transmitted it. Data manipulation refers to operations such as insertion, deletion, or substitution.
- *Authentication.* Authentication is related to identification and it is often divided into two classes: data origin authentication and entity authentication.
  - Data origin authentication. It gives assurance that an entity is the original source of a message. Data origin authentication implicitly provides data integrity. Sometimes, it is referred to as message authentication.
  - Entity authentication. Entity authentication assures one entity about the identity of a second entity with which it is interacting. Usually, entity authentication implies data origin authentication.
- *Non-repudiation.* It is a security service that prevents an entity from denying a previous action or commitment. It is very useful in situations that can lead to disputes. When a dispute arises, a trusted third party is able to provide the evidence required to settle it.

These security features and services are provided by different symmetric and asymmetric cryptographic primitives that are implemented in the proposed hybrid cryptosystem proposed as explained next.

4   Proposed Hybrid Cryptosystem

Figure 1 shows the proposed hybrid cryptosystem. It is combined of three components including the XXTEA (symmetric block cipher) used to achieve confidentiality, the ECC (asymmetric cipher) used to achieve authenticity, integrity and non-repudiated based digital signature, and the scrypt (key derivation function) used to generate keys. The aim is to achieve the best performance with a high level of security. The functions of each component are described next.

4.1   The XXTEA Block Cipher

It is used to achieve confidentiality. The XXTEA [32, 33] is a block cipher comprising at least two 32-bit words, using a 128-bit key. It is designed to correct weaknesses in the original block TEA and XTEA. An XXTEA full cycle is n rounds, where n is the number of words in the block. The number of full cycles to enact over the block is given as $6 + 52/n$. The XXTEA memory consumptions and computational cycles (cost) are analyzed and compared with other block ciphers [16, 27, 40]. Tables 3(a) and 3(b) show the results. It is clear from these results that XXTEA outperforms AES128 and AES256 because it consumes less memory consumption and less computational cycles. For these reasons, the XXTEA is used in the proposed hybrid cryptosystem for data encryption and decryption to achieve confidentiality.

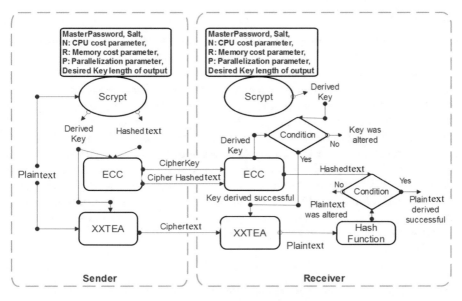

Fig. 1. The design of the proposed robust hybrid cryptosystem (RHCS).

Table 3a. Memory consumption (bytes).

IoT device controller	Mica2		Arduino pro	
Block cipher	RAM	ROM	RAM	ROM
XXTEA	**542**	**6312**	**226**	**4112**
Skipjack	3096	8658	398	4952
RC5	682	6110	350	3184
AES-128	1074	6296	814	3692
AES-256	1822	7932	1014	4190
CGEA	664	6268	548	3228

Table 3b. Computational cost (cycles)

IoT device controller	Mica2	Arduino pro
Block cipher	Cycles	Cycles
XXTEA	**24064**	**30464**
Skipjack	9820	12672
RC5	53014	61504
AES-128	37525	43200
AES-256	80344	88896
CGEA	67786	76212

4.2   The ECC

The elliptic curve cryptography (ECC) [34] is asymmetric cryptography engages the use of two keys. A private key used to decrypt messages and sign (create) signatures. And a public key used to encrypt messages and verify signatures. ECC is used to achieve authenticity, integrity, and non-repudiation with the help of some additional algorithms. ECDHE Digital signatures are *used* to verify the *authenticity* of messages and confirm that they have not been altered in transmission. To assure the integrity of the received data, a hashing algorithm is used. The Elliptic Curve Digital Signature Algorithm (ECDSA) algorithm uses ECC to provide a variant of the Digital Signature Algorithm (DSA). A pair of keys (public and private) are generated from an elliptic curve, and these can be used both for signing or verifying a message's signature. The ECDSA algorithm is used to achieve non-repudiation [30]. It ensures the sender and receiver from denying the sending or receiving of a message and the authenticity of their signature.

Generation of the public key (Q), Eq. (1).

$$Q = d * P \tag{1}$$

where: d is the random number chosen within the range (1 to n − 1); the private key, P is a point on the curve, and Q is the public key.

ECC-Encryption: Let "m" be the message that should be sent. this message has to be represented on the curve. Consider "m" has the point "M" on the curve "E". Randomly select "k" from [1 − (n − 1)]. We need to generate two cipher texts (C1 and C2); Eq. (2) and send them.

$$C_1 = k * P \text{ and } C_2 = M + k * Q \tag{2}$$

ECC-Decryption: We can get the original message using this function (3):

$$M = C_2 - d * C_1 \tag{3}$$

Several analyses are made for RSA and ECC to provide most efficient in IoT smart devices [27, 35–40]. Table 4 shows the recommended security bit level provided by NIST [36]. This means that, for the same level of security, significantly smaller key sizes can be used in ECC than RSA. For instance, to achieve 80 bits of security level, the RSA algorithm requires a key size of 1024 bits, while ECC needs a key size of 160 bits. Hence, RSA consumes longer encryption time and uses memory more than ECC. Performance evaluation of the proposed RHCS cryptosystem is discussed next.

Table 4. Comparing ECC and RSA security bit level.

Security bit level	RSA key size	ECC key size
80	1024	160
112	2048	224
128	3072	256
192	7680	384
256	15360	512

4.3   Scrypt

A hashing function is commonly used to check data integrity. Where a unique and fixed-length signature is created from hashing data. Scrypt is a key derivation function (KDF) that is used to generate multiple keys in each time data is being encrypted. It is used in the proposed hybrid cryptosystem because it is more robust and secure than PBKDF2 since it uses more parameters than PBKDF2 and required hardware platforms higher cost than PBKDF2 [41].

4.4   The Proposed Hybrid Cryptosystem Algorithm

Referring to Fig. 1, the following algorithm illustrates the steps of operations of the hybrid cryptosystem proposed.

1. The methodology of ECC, first Sender A, and Receiver B exchange their public key using ECC Diffie-Hellman key exchange algorithm.
2. Sender A wants to send an encrypted message to Receiver B.
3. Sender A constructs 128-bit hashed key for XXTEA through using of Scrypt Hashing Algorithm.
4. XXTEA encrypts the desired message using 128-bit hashed key and sends it to Receiver B.
5. Sender A sends 128-bit hashed key and hashed message after they are both encrypted by ECC.
6. Receiver B decrypts received the message sent using ECC to obtain 128-bit hashed key and hashed message.
7. After Receiver B receives an encrypted message, it decrypts it using 128-bit hashed key.
8. 128-bit hashed key derived from Receiver B with same parameters used at Sender A to compare it with received 128-bit hashed key as well as check received message sent from Sender A.

5   Results and Discussions

The hybrid cryptosystems are implemented by simulation using C software code that runs on intel i7 core processor Laptop with 4 GB RAM. For performance comparisons, the results are classified as follows.

5.1   The Hybrid Cryptosystem Based Tiny Family and RSA: Enc and Dec

Figures 2(a) and 2(b) show results of performance comparison between (TEA & RSA), (XTEA & RSA), and (XXTEA & RSA) for the relation between messages size (KB) and encryption time (sec) in case of messages encryption and decryption, respectively. Results show that the hybrid cryptosystem (TEA & RSA) consumes less encryption and decryption time while hybrid cryptosystem (XXTEA & RSA) consumes highest encryption and decryption time. This is because the TEA block cipher algorithm is much simpler than both XTEA and XXTEA, hence it consumes lesser encrypt and decrypts time.

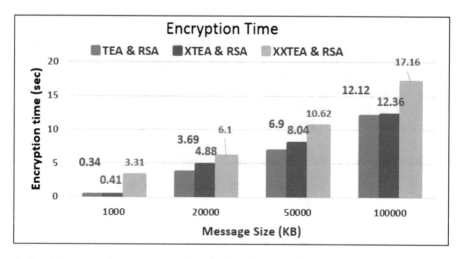

Fig. 2a. Encryption time vs message sizes for hybrid cryptosystems based tiny family and RSA.

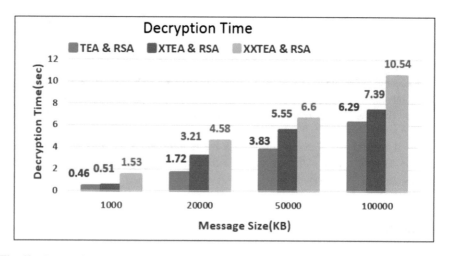

Fig. 2b. Decryption time vs message sizes for hybrid cryptosystems based tiny family and RSA.

5.2   The Hybrid Cryptosystem Based Tiny Family and ECC: Enc and Dec

Figures 3(a) and 3(b) show results of performance comparison between (TEA & ECC), (XTEA & ECC), and (XXTEA & ECC) for the relation between messages size (KB) and encryption time (sec) in case of messages encryption and decryption, respectively. Different message sizes are used to test the hybrid cryptosystems in case of information transfer between IoT devices and between cloud and IoT, ranged from Kbytes to Mbytes, as illustrated in the Figures. Results show that the hybrid cryptosystem (TEA & ECC) consumes less encryption and decryption time while the hybrid cryptosystem (XXTEA & ECC) consumes highest encryption and decryption time. The XTEA encryption time is longer than TEA because there is an additional 11 shift round in XTEA algorithm to make it irregular. Besides, XTEA uses whole 128 bit key in first 2 cycles while TEA uses 2 (32-bit key array) in the first 2 cycles. The XXTEA uses whole 128 bit key, input next and previous character in encryption and decryption. This causes little delay to increase security when encrypted data.

Although, the encryption and decryption time for XXTEA & ECC is longer than others. The hybrid cryptosystem based on XXTEA & ECC is recommended because XXTEA is more secure than both TEA and XTEA [42–44]. It uses a more involved round function which makes use of both the immediate neighbours in encrypting each word in the block. Besides, an XXTEA is more efficient than TEA and XTEA for encrypting longer messages [45]. This means that it can also be used as a general-purpose block cipher.

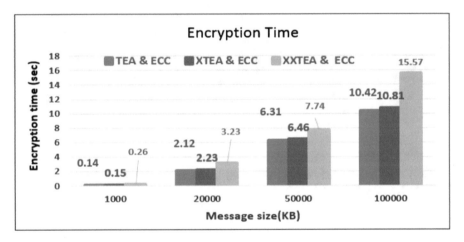

Fig. 3a. Encryption time vs message sizes for hybrid cryptosystems based tiny family and ECC.

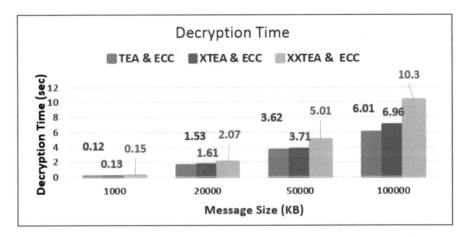

Fig. 3b. Decryption time vs message sizes for hybrid cryptosystems based tiny family and ECC.

5.3   The Throughput Performance Comparison and Evaluation

Figures 4(a) and 4(b) show results of performance comparison between the hybrid cryptosystems based (XXTEA & ECC), and (XXTEA & RSA) for the relation between messages size (Kb) and encryption and decryption throughput, respectively. Results show that the hybrid cryptosystem (XXTEA & ECC) encryption and decryption throughput is higher than encryption and decryption throughput of the hybrid cryptosystem based (XXTEA & RSA). This is because of the ECC algorithm is much simpler than RSA and consumes lesser execution time.

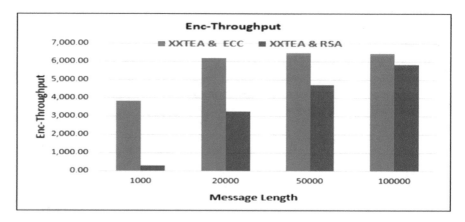

Fig. 4a. Comparing Enc-Throughput vs message sizes for hybrid cryptosystems (XXTEA & ECC) and XXTEA & RSA).

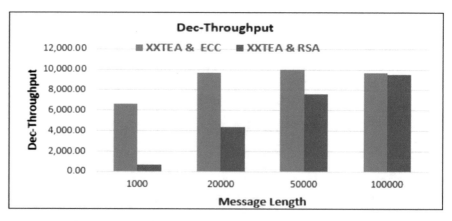

Fig. 4b. Comparing Dec-Throughput vs message sizes for hybrid cryptosystems (XXTEA & ECC) and (XXTEA & RSA).

Table 5 shows the encryption and decryption throughput of the hybrid cryptosystems, where:

The throughput $S = \sum$ Message Size$/\sum$ Total time taken.
The average throughput of encryption and decryption for
$(XXTEA + ECC) = 5.73 + 9 = 15\ Mb/s.$
The average throughput of encryption and decryption for
$(XXTEA + RSA) = 3.5 + 5.5 = 9\ Mb/s.$
Hence, the throughput efficiency $= [(15 - 9)/15] * 100 = 40\%.$

This result shows that the hybrid cryptosystem based (XXTEA + ECC) outperforms the hybrid cryptosystem based (XXTEA + RSA) by 40%.

Table 5. The average throughput of encryption and decryption of the hybrid cryptosystem proposed.

Enc- throughput (Kb/s)	Message size (Kb)				
Hybrid cryptosystem	1000	20000	50000	100000	Avg
XXTEA & ECC	3,846.15	6,191.95	6,459.95	6,422.61	5,730.17
XXTEA & RSA	302.11	3,278.69	4,708.10	5,827.51	3,529.10
Dec- throughput (Kb/s)	Message size (Kb)				
Hybrid cryptosystem	1000	20000	50000	100000	Avg
XXTEA & ECC	6,666.67	9,661.84	9,980.04	9,708.74	9,004.32
XXTEA & RSA	653.59	4,366.81	7,575.76	9,487.67	5,520.96

6 Conclusion

The comparative study, in this work, showed that the XXTEA lightweight block cipher is more suitable to be used in IoT smart devices for messages encryption since it requires less memory consumption and less computational cycles (cost). Besides, the ECC asymmetric cipher was used because it achieves a high level of bit security at smaller key sizes better than RSA. The ECC cipher was used to achieve authenticity, integrity, and non-repudiation. The XXTEA block cipher was used to achieve confidentiality. In addition, the scrypt hashing function is used to generate multiple keys in each time data is being encrypted and to check data integrity. Performance analysis and evaluation-based on simulation were performed. Results show that the hybrid cryptosystem based (XXTEA + ECC) out performs the hybrid cryptosystem based (XXTEA + RSA) with respect to encryption and decryption time as well as better throughput by 40%. The proposed hybrid cryptosystem combines ECC, XXTEA, and scrypt achieves the main demanded characteristics of cryptography, including confidentiality, authenticity, integrity, and non-repudiation. This helps to protect IoT smart devices from vulnerabilities related to security attacks. The hardware realization of the proposed hybrid cryptosystem will be tackled in future work.

References

1. Bhardwaj, S., Kole, A.: Review and study of the internet of things: it's the future. In: 2016 IEEE International Conference on Intelligent Control Power and Instrumentation (ICICPI)
2. Demblewski, M.: Security frameworks for machine-to-machine devices and networks. Ph.D. thesis, Nova Southeastern University (2015). https://nsuworks.nova.edu/cgi/viewcontent. cgi?article=1068&context=gscis_etd. Accessed 10 Feb 2019
3. Ali, B., Awad, A.: Cyber and physical security vulnerability assessment for IoT-based smart homes. Sensors **18**(3), 817 (2018)
4. Chandra, S., Paira, S., Alam, S., Sanya, G.: A comparative survey of symmetric and asymmetric key cryptography. In: International Conference on Electronics, Communication and Computational Engineering (ICECCE) (2014)
5. Kumar, Y., Munjal, R., Sharma, H.: Comparison of symmetric and asymmetric cryptography with existing vulnerabilities and countermeasures. IJCSMS Int. J. Comput. Sci. Manag. Stud. **11**(03), 60–63 (2011)
6. Henriques, M., Vernekar, N.: Using symmetric and asymmetric cryptography to secure communication between devices in IoT. In: 2017 IEEE International Conference on IoT and Application (ICIOT) (2017)
7. Ruan, C., Luo, J.: Design and implementation of a mobile payment system for intelligent travel. In: 2014 IEEE 3rd International Conference on Cloud Computing and Intelligence Systems (CCIS) (2014)
8. Xin, M.: A mixed encryption algorithm used in the internet of things security transmission system. In: 2015 IEEE International Conference on Cyber-Enabled Distributed Computing and Knowledge Discovery (CyberC) (2015)
9. Biswas, K., et al.: Performance evaluation of block ciphers for wireless sensor networks. In: Advanced Computing and Communication Technologies, pp. 443–452. Springer, Heidelberg (2016)
10. Ekera, M.: Differential Cryptanalysis of MD5. Master of Science Thesis Stockholm, Sweden (2009)
11. Kaur, S., et al.: Study of multi-level cryptography algorithm: multi-prime RSA and DES. Int. J. Comput. Netw. Inf. Secur. **9**(9), 22 (2017)
12. Darwish, A., El-Gendy, M.M., Hasanien, A.: A new hybrid cryptosystem for the internet of things applications. In: Multimedia Forensics and Security, pp. 365–380. Springer, Heidelberg (2017)
13. Yarrkov, A.: Cryptanalysis of XXTEA, 4 May 2010. https://eprint.iacr.org/2010/254.pdf. Accessed 10 Feb 2019
14. Bhaskar, C., Rupa, C.: An advanced symmetric block cipher based on chaotic systems. In: 2017 IEEE Innovations Power and Advanced Computing Technologies (i-PACT) (2017)
15. Yousefi, A., Jameii, S.: Improving the security of internet of things using encryption algorithms. In: 2017 IEEE International Conference on IoT and Application (ICIOT) (2017)
16. Singh, S., Padhye, S.: Cryptanalysis of NTRU with n Public Keys. In: 2017 ISEA Asia Security and Privacy (ISEASP) (2017)
17. Khomlyak, O.: An Investigation of Lightweight Cryptography and Using the Key Derivation Function for a Hybrid Scheme for Security in IoT (2017)
18. Hatzivasilis, G.: Password hashing status. Cryptography **1**(2), 1–31 (2017) MDPI Int. J.
19. Harini, A., et al.: A novel security mechanism using hybrid cryptography algorithms. In: 2017 IEEE International Conference on Electrical Instrumentation and Communication Engineering (ICEICE) (2017)

20. Elhoseny, M., et al.: Secure medical data transmission model for IoT-based healthcare systems. IEEE Access **6**, 20596–20608 (2018)
21. Njuki, S., et al.: An evaluation on securing cloud systems based on cryptographic key algorithms. In: Proceedings of the 2018 2nd International Conference on Algorithms, ACM Computing and Systems (2018)
22. Shoup, V.: Advances in Cryptology-CRYPTO. The 25th Annual International Cryptology Conference, Santa Barbara, California, USA, 14–18 August 2005 Proceedings, vol. 3621. Springer, Heidelberg (2005)
23. Abdelminaam, D.: Improving the security of cloud computing by building new hybrid cryptography algorithms. Int. J. Electron. Inf. Eng. **8**(1), 40–48 (2018)
24. Sagar, F.: Cryptographic Hashing Functions - MD5, September 2016. http://cs.indstate.edu/~fsagar/doc/paper.pdf. Accessed 20 Apr 2019
25. Habboush, A.: Multi-level encryption framework. (IJACSA) Int. J. Adv. Comput. Sci. Appl. **9**(4), 130–134 (2018)
26. Mahto, D., Khan, D., Yadav, D.: Security analysis of elliptic curve cryptography and RSA. In: Proceedings of the World Congress on Engineering 2016, WCE 2016, 29 June–1 July 2016, London, UK, vol. 1 (2016)
27. Ragab, A.: Robust hybrid cryptosystem for protecting smart devices in internet of things (IoT). Master thesis, record number 14120399, Department of Computer Engineering, Arab Academy for Science, Technology and Maritime Transport, Cairo, Egypt (2019)
28. Bhasher, U., Rupa, C.: An advanced symmetric block cipher based on chaotic systems. In: IEEE International Conference on Innovations in Power and Advanced Computing Technologies (2017)
29. Libelium-Techedge, Smart Industrial Protocols Solution Kit. https://www.the-iot-marketplace.com/libelium-techedge-smart-industrial-protocols-solution-kit. Accessed 10 Feb 2019
30. AWS IoT Device Defender, Security management for IoT devices. https://aws.amazon.com/iot-device-defender/?nc=sn&loc=2&dn=5. Accessed 20 Apr 2019
31. Security on Arm. https://developer.arm.com/technologies/security-on-arm. Accessed 20 Apr 2019
32. Wheeler, D., Needham, R.: Correction to XTEA. Computer Laboratory, Cambridge University, England (1998)
33. Andem, V.: A cryptanalysis of the tiny encryption algorithm. Master thesis, Department of Computer Science in the Graduate School of The University of Alabama (2003)
34. Miller, V.: Use of elliptic curves in cryptography. In: Conference on the Theory and Application of Cryptographic Techniques. Springer, Heidelberg (1985)
35. Kaur, S., Bharadwaj, P., Mankotia, S.: Study of multi-level cryptography algorithm: multi-prime RSA and DES. Int. J. Comput. Netw. Inf. Secur. **9**(9), 22 (2017)
36. Barker, E., Dang, Q.: Recommendation for Key Management. NIST Special Publication 800-57 Part-3 Revision 1, National Institute of Standards and Technology (NIST), January 2015
37. Percival, C.: Stronger Key Derivation via Sequential Memory-hard Functions, pp. 1–16 (2009)
38. Dunkelman, O., Preneel, B.: Improved meet-in-the-middle attacks on reduced-round DES. In: International Conference on Cryptology in India. Springer, Heidelberg (2007)
39. Stamp, M., Low, R.: Applied Cryptanalysis: Breaking Ciphers in the Real World. Wiley, Hoboken (2017)
40. Albela, M., Lamas, P., Caramés, T.: A practical evaluation on RSA and ECC-based cipher suites for IoT high-security energy-efficient fog and mist computing devices. Sensors **18**, 3868 (2018)

41. Ertaul, L., Kaur, M., Gudise, V.: Implementation and performance analysis of PBKDF2, bcrypt, scrypt algorithms. In: Proceedings of the International Conference on Wireless Networks (ICWN), Athens, pp. 66–72 (2016)

42. Rajesh, S., Paul, V., Menon, V., Khosravi, M.: A secure and efficient lightweight symmetric encryption scheme for transfer of text files between embedded IoT devices. Symmetry **11**, 293 (2019)

43. Ankit Shah, A., Engineer, M.: A survey of lightweight cryptographic algorithms for IoT-based applications. In: Advances in Intelligent Systems and Computing (2019). https://doi.org/10.1007/978-981-13-2414-7_27. Accessed 20 Apr 2019

44. Percival, C.: Stronger Key Derivation via Sequential Memory-Hard Functions. https://www.tarsnap.com/scrypt/scrypt.pdf. Accessed 20 Apr 2019

45. Sehrawat, D., Gill, N.S.: Lightweight block ciphers for IoT based applications: a review. Int. J. Appl. Eng. Res. **13**(5), 2258–2270 (2018)

Securing Combined Fog-to-Cloud Systems: Challenges and Directions

Sarang Kahvazadeh$^{(\boxtimes)}$, Xavi Masip-Bruin$^{(\boxtimes)}$,
Eva Marín-Tordera$^{(\boxtimes)}$, and Alejandro Gómez Cárdenas$^{(\boxtimes)}$

Advanced Network Architectures Lab (CRAAX),
Universitat Politècnica de Catalunya (UPC), Barcelona, Spain
{skahvaza,xmasip,eva,alejandg}@ac.upc.edu

Abstract. Nowadays, fog computing is emerged for providing computational power closer to the users. Fog computing brings real-time processing, low-latency, geo-distributed, etc. Although, fog computing do not come to compete cloud computing, it comes to collaborate. Recently, Fog-To-Cloud (F2C) continuum system is introduced to provide hierarchical computing system and facilitates fog-cloud collaboration. This F2C continuum system might encounter security issues and challenges due to their hierarchical and distributed nature. In this paper, we analyze attacks in different layer of F2C system and identify most potential security requirements and challenges for the F2C continuum system. Finally, we introduce the most remarkable efforts and trends for bringing secure F2C system.

Keywords: Security · Fog-to-cloud computing

1 Introduction: Combining Fog and Cloud

Nowadays, the wide and continuous deployment of smart devices at the edge, such as sensors, actuators, smartphones, tablets, etc. setting the roots for the emerging Internet of Things (IoT) [1] paradigm, along with the recent developments in network technologies, from the network core (e.g., elastic or flexible optical networks) up to the user edge (e.g., LoRa), with an eye on the promising 5G, all enabling fast users and (even more important) Machine-to-Machine (M2M) communication [2] are, all in all, paving the way towards an innovative but also unforeseen scenario where devices, users, services and also data will interact in a, more than probably disruptive way. Aligned to this evolution, cloud computing [3] has been widely adopted to address some of the key concerns related to data, including aggregation, processing and storage leveraging the large capacity massive data centers located at cloud are endowed with.

Unfortunately, the far distance from the datacenters at cloud to the users and/or devices requiring the cloud services, brings undesired concerns –in terms of for example scalability, security or quality of service (QoS), just to name a few–, that may become critical challenges for some applications, for instance applications demanding real-time processing. Recognized this weakness, a new computing paradigm, fog computing [4], recently came up, and intended to sort this weakness out by moving cloud capacities closer to the edge. To that end, nodes with some computing capacities

© Springer Nature Switzerland AG 2020
K. Arai et al. (Eds.): FTC 2019, AISC 1069, pp. 877–892, 2020.
https://doi.org/10.1007/978-3-030-32520-6_63

are distributed close to users and devices at the edge, referred to as fog nodes [5], thus facilitating real-time processing and low-latency guarantees. Interestingly, fog computing does not compete with cloud computing, rather they both complement each other to guarantee a service to be allocated to those resources best suiting its demands. Thus, a new scenario comes up, built upon considering the whole set of resources from the edge up to the cloud, coined as IoT continuum [6, 7].

A key challenge in this combined scenario refers to resources management assuming the specific requirements coming when considering the heterogeneous set of potential resources at the edge. Some ongoing efforts are already addressing this challenge, such as the OpenFog Consortium [8] or the EU mF2C project [9], the latter aimed at designing and developing the Fog-to-Cloud (F2C) concept proposed in [10], and both aimed at optimizing services execution and resource utilization. Recognized the potential benefits brought in by such optimization, a key and highly critical component refers to security provisioning. Indeed, there is no doubt that the edge context, putting together low control, low power, and generally speaking, considering a vast set of heterogeneous devices on-the-move, is highly vulnerable to attacks and security breaches that may put the whole system at risk. Therefore, it seems reasonable to devote some efforts to analyze and characterize what security provisioning means, what the challenges are and how could be addressed.

To that end, this paper illustrates the particular scenario built by the F2C approach, as a potential candidate to the final optimization objective, although certainly the analysis performed in the paper might be adopted for any solution aligned to a decentralized architecture for combined fog and cloud systems. Indeed, F2C is proposed as a hierarchical multi-layered control architecture designed to manage the whole stack of resources from the very edge up to the cloud in a coordinated way, putting together the advantages of both computing paradigms, i.e. proximity at the fog and high performance at the cloud.

From an organizational perspective, the combined F2C ecosystem is organized into areas, each including its whole set of resources (fog nodes and IoT devices) see Fig. 1, the exact scope of an area and the individual allocation of resources into each area are ongoing research topics, certainly affecting the scalability of the system, but out of the scope of this paper. One node at each area is selected to serve as the fog leader (policies yet to be defined), thus responsible for managing the devices inside the area as well as coordinating with higher hierarchical layers. The set of fog leaders are also connected through a higher layer, thus setting the envisioned multi-layered hierarchical control architecture (see Fig. 1 as a topological example, where fog leaders are connected through cloud).

However, as briefly introduced above, this highly distributed and heterogeneous scenario fuels many security gaps and weaknesses that must be properly addressed. Certainly, it is worth emphasizing that most of the IoT devices included in the different areas will have low computational power and consequently, will not be able to handle their own security, nor support any highly demanding external security solution either. In fact, the scenario is not promising, since neither cloud computing (as a centralized strategy located far from the devices to secure) nor fog computing (using fog node resources to support security functionalities may have a non-negligible impact on the individual fog node QoS performance) seem to be the proper solution to provide the expected security guarantees. Moreover, beyond specific security provisioning limitations in IoT devices, any combined F2C-like system might impose additional security requirements due to its distributed, decentralized and layered nature, thus making security provisioning even more challenging.

With the aim of illustrating the security concerns in combined F2C-like systems, this paper analyzes main security requirements and challenges for such scenario, with particular focus, for the sake of illustration, on the F2C approach.

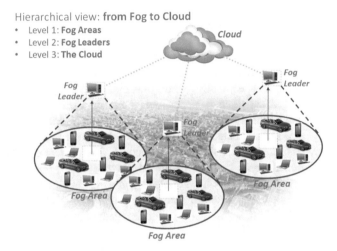

Fig. 1. Combined F2C system

This paper is organized as follows. In Sect. 2, we identify security requirements in combined F2C systems. Then, Sect. 3 discusses and analyzes security challenges and provides some directions to move forward and Sect. 4 illustrates some existing efforts and trend to provide F2C security. Finally, Sect. 5 concludes the paper.

2 Security Requirements for F2C-Like Systems

A mandatory step previous to a potentially successful design of a security solution for combined fog and cloud systems boils down identifying the particular set of security requirements demanded by such scenario. Contributions led by the OpenFog Consortium and the mF2C project, are currently trying to set the roots for such a definition. This section, focuses on the latter, considering the F2C architecture as an illustrative and sound approach for combined fog and cloud systems – although the security characteristics described in this paper should accommodate needs for any other F2C-like system. Moreover, it is worth highlighting that, from the design perspective, whatever the solution to be proposed is, it should benefit from recent technology innovations to improve or facilitate security provisioning. In this direction, just as an example, [11] briefly introduces potential benefits from adopting blockchain to security provisioning in fog systems.

Before digging into the set of security requirements, we first summarize the whole distributed security procedure as proposed for the F2C layered architecture, describing specific steps for each one of the two envisioned domains and then we introduce potential attacks a F2C-like system may deal with.

2.1   Security Process in F2C

As shown in Fig. 1, F2C proposes a distributed and layered architecture where devices closer to the edge are clustered into different Fog Areas, each managed by a Fog Leader, logically located at an upper layer. Certainly, Fog Leaders enable both area-to-area and area-to-cloud connectivity, hence consequently a secure communication on two domains, fog-cloud and device-fog, is a must to avoid any passive or active attack.

On the first domain, i.e. fog-cloud, some steps are needed to bring security in, as illustrated in Fig. 2. More specifically, we may consider that first, each fog leader must be securely discovered and then perform mutual authentication with cloud through some credentials and identities, in order to provide system and data integrity and confidentiality. Once authenticated, fog leaders either generate keys, or get keys from cloud if key generation is not enabled at fog leaders. Fog leaders can then use these keys to encrypt and decrypt the information flow with cloud, aimed at preventing attackers to eavesdropping, modifying, or deleting information between cloud and fogs. From a technology view, the network technologies supporting cloud and fog leader communications (e.g. wired, wireless, etc.) must be secured to avoid passive and active attacks, turning into secure channels (blue lines in Fig. 2(a)).

Once the fog leaders are properly installed and authenticated to the F2C cloud, they are authorized to provide computation, network, storage, and shareable computational environment to fog nodes and IoT devices at the edge of the network in a distributed way. Thus, the second domain, i.e. device-fog, as illustrated in Fig. 2(b), starts when a device reaches out to a fog area, demanding the device to be discovered by a fog leader in a secure way. Then, as done for the first domain, fog leaders and devices must be mutually authenticated through some credentials in order to guarantee data and system integrity and confidentiality at the edge of the network. Once authenticated, devices either generate keys or get keys from the fog leaders, in case they cannot generate keys, to be used for encryption and decryption. The information flow between devices (IoT devices and fog nodes) and fog leaders must be properly encrypted to prevent attackers from eavesdropping, modify, or delete the information. Similarly, all information must be conveyed through secure channels regardless the deployed technology (blue lines in Fig. 2(b)).

It is worth highlighting that in the first domain, cloud acts as access control providing the access to cloud data centers for the distributed fog leaders, according to their attributes and thus preventing any unauthorized access, while in the second domain, fog nodes are responsible for such task, preventing any unauthorized access to the F2C system at the edge of the network.

The F2C scenario described so far may be more complex if mobility is also considered. More specifically, whatever the security strategy proposed is, the effects of the handover process triggered by a device on the move leaving one area (i.e. a fog leader) and getting into another one must be considered. This shift may drive the need for both

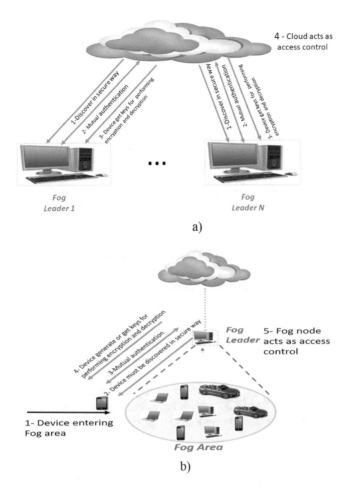

Fig. 2. Intra-layer security a) Fog-cloud layer security b) Device-fog layer security

areas to communicate each other to facilitate the handover process. Two options come up, one through cloud and the other one enabling horizontal communication among fog leaders. For the sake of processing time we may assume the latter to be the optimal one, hence demanding direct secure inter-communication between fog leaders.

2.2 Attacking a F2C System

Unfortunately, there are many potential security vulnerabilities in F2C-like systems, paving the way for attackers to launch attacks in different layers in the system. In this section, we identify most potential attacks to be faced by F2C-like systems as illustrated in Fig. 3, all grouped into three categories, as follows.

Man-in-the-Middle Attack: Attackers can take the network control between devices at different levels (IoT devices, fog nodes, fog leaders and cloud) to either eavesdrop communication, modify information or even to inject malicious information and code into the system. For example, attackers can obtain the identity of a F2C component and then impersonate it to be an eligible component. Due to the obtained identity, the attacker can impersonate a fog leader (malicious fog leader) thus getting devices and users information and locations. Also, an attacker can impersonate users and devices to take information or gain access to services it is not authorized to. In upper layers, i.e. fog-cloud communication, an attacker can impersonate fog leader or even cloud to launch a man-in-the-middle attack. In all these cases, attackers can launch the attack in passive (eavesdropping without changing information) and active (information modification, manipulation and malicious injection) ways. This type of attack affects the integrity and confidentiality of any F2C-like system (see Fig. 3A).

Denial of Service and Distributed Denial of Service (DoS and DDoS): In this case, attackers either launch multiple service requests to the fog leader or perform a jamming wireless communication between fog leader-devices to deplete the fog leader resources and consequently making it down. An attacker can use legitimate devices, such as IoT devices, fog devices or fog leaders to launch DoS and DDoS using their identities. DoS and DDoS attacks can also occur in upper layers such as fog leader-cloud. As a consequence, attackers successfully prevent legitimate users and devices from accessing services provided by a fog leader or even by cloud (see Fig. 3B). In short, this attack severely affects the availability of the F2C system.

Database Attacks: In a F2C system, databases may meet a hierarchical architecture, keeping for example one centralized at cloud and some other locally distributed at fog layers. If an attacker can access to these databases, it can modify, manipulate and even leak the data, what may have a high impact on the total system performance. Database attacks may be internal –coming from F2C service providers–, or external –legible and illegible users. This attack intensely affects the F2C integrity and confidentiality (see Fig. 3C).

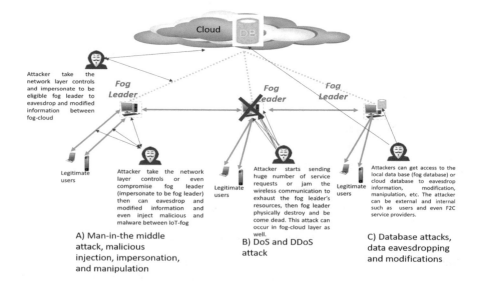

Fig. 3. Attacks in F2C systems

2.3   Security Requirements in F2C

After describing the main scenario characteristics for combined fog-cloud systems, the security strategy for F2C and the set of potential attacks a F2C system may suffer from, next, potential security requirements for combined fog-cloud systems are introduced in Table 1 (see also [12, 13] for more information), summarized into 9 main concepts, all in all driving a set of challenges, as introduced in next section, fueling novel research avenues. It is worth mentioning that despite the fact that the identified security requirements are not novel in their conception, what makes them specially challenging is the specific scenario security is to be designed for, i.e. F2C-like systems, where heterogeneity, volatility and dynamics are basic and undeniable characteristics.

Table 1. Combined F2C system security requirements

Security requirements in combined F2C system	Description
Authentication and authorization at the whole set of layers	Authentication must be done for all participant components in F2C systems to provide integrity and secure communication. A hierarchical authentication may be considered, in short, cloud authenticates fog leaders, and fog leaders authenticate edge devices (fog nodes and IoT devices)
Appropriate key management strategy	F2C systems must include a well-defined key management strategy for keys distribution and update as well as for key revocation
Access control policies to reduce intrusions	Access control must be supported at cloud level and distributed access controls at fog layers
Providing confidentiality, integrity, and availability (the CIA triad) as a widely adopted criteria for security assessment	In a F2C system, user's information must stay private not to be disclosed to unauthorized users (confidentiality), information must be complete, trustworthy and authentic (integrity), and finally the whole system must work properly, reacting to any disruption, failure or attack (availability)
All network infrastructure must be secure	All components in a F2C system (users, devices, fog leaders, fog nodes, and cloud) must communicate through secure channels regardless the specific network technology used to connect (wired, Bluetooth, wireless, ZigBee, etc.)
All components must be trustable	In the proposed hierarchical approach, the set of distributed fog leaders act as a key architectural pillar enabling data aggregation, filtering, and storing closer to the users, hence making trustness mandatory for fog leaders
Data privacy is a must	Data processing, aggregation, communication and storage must be deployed not to disclose any private information, or produce data leakage, data eavesdropping, data modifications, etc. To that end, data must be encrypted, and data access must not be allowed to unauthorized users. Moreover, assuming mobility a key bastion in F2C systems other

(continued)

Table 1. (*continued*)

Security requirements in combined F2C system	Description
	particular privacy related issues come up, such as for example geo-location
Preventing fake services and resources	Fake scenarios are highly malicious in F2C systems, hence some actions must be taken to prevent that to happen, such as services and resources must be discovered and identified correctly and services and resources allocation must be done securely
Removing any potential mobility impact on security	Fog nodes and IoT devices might be on the move, thus demanding the design of secure procedures to handle mobility related issues, such as devices handover

3   Open Challenges

This section, driven by the set of security attacks and requirements identified in Sect. 2, is intended to go deep into the open challenges coming from the requirements above and necessary to provide security guarantees in a scenario combining fog and cloud computing. For the sake of understanding, the F2C approach is used to illustrate the specific details brought in when combining fog and cloud that make security provisioning even more challenging. In fact, the objective is not only to identify the challenges but also to highlight the open issues making them so challenging. To that end, challenges are split into 5 security concepts as shown in Table 2, aligned to and extending previous works in the area of fog computing.

3.1   Trust and Authentication

Trust: Traditional strategies applied to static or well defined scenarios do not match the particularities rolled in by the potential stack of heterogeneous and largely distributed resources considered in F2C systems. Working on that direction, the Open fog consortium [8] proposes to embed the hardware root of trust into the fog nodes to provide security at the IoT layer. Certainly, this is still an ongoing effort where many challenges remain yet unsolved, such a potentially unaffordable cost or to what extent an attack on a fog node may compromise the security for its IoT devices. In the same direction, a joint effort within the mF2C project is analyzing a blockchain-based solution to provide a novel distributed trust strategy particularly tailored to face the specific trust needs and conditions of combined fog and cloud systems.

Authentication: Authentication is a key security component in any ICT system, but it is particularly important in scenarios where mobility and heterogeneity are undeniable characteristics. It is widely accepted that the use of the conceptually centralized cloud for handling device's and user's authentication cannot be sufficient for a F2C system. The reason for that is twofold. First, the huge number of messages forwarded between a large set of unstoppable increasing devices at the edge and cloud. Second, if cloud is down, compromised or attacked, the whole F2C system will be compromised. Fortunately, a solution based on the proposed hierarchical layered architecture may be proposed. The main rationale for this proposal is that cloud can be used to authenticate fog leaders, and then, the authenticated fog leaders can handle IoT devices', fog nodes', and users' authentication in their area in a distributed fashion. This solution would reduce cloud dependencies, would facilitate redundancy at fog layers to deal with potential fog node breaches and promises to be scalable (per layer). However, the work is not completely done and there are some questions yet to be solved. Indeed, a key question refers to "how can these hierarchical authentication processes be done and which type of authentication for each layer can be used".

Key Management: Scalability is also a non-negligible drawback when considering a centralized key generator center (KGC) in combined fog and cloud systems. Indeed, beyond the effects of compromising the KGC, a large number of messages would be expected. Although a distributed key management strategy for generating, assigning, updating and revoking keys seems to be the proper approach, many open questions come up, mainly referring to:

- How can IoT devices (in a large number and with low computational power) get keys to encrypt data?
- Would it be possible fog layers to handle distributed key management for both IoT devices and users?
- Would it be reasonable to assume cloud to act as key manager for fog layers and so the latter as distributed key managers for their areas? And if so, what should be the proper key management algorithm (symmetric or asymmetric) for each F2C layer?

Identity Management: The main aim is to assign an ID to all devices, what is pretty challenging for devices with low computational power, since ID storage might be not affordable. Indeed, some questions come up:

- How can distributed IDs manager in fog layer be secured?
- Would ID fragmentation be a candidate solution to minimize scalability issues? (i.e., using different fragment's size for each layer [14])
- If so, what would the right fragmentation policy to support an optimal fragment storage?
- And what is the max/min fragment size allowed per layer?

3.2   Access Control and Detection

Access Control: Considering the large set of devices at the edge as well as the different characteristics shown by the whole set of resources in the continuum from the edge up to the cloud, the key question may be stated as "how to design a global access control, supporting the different systems characteristics and constraints". The hierarchical model envisioned by the F2C model addresses this challenge proposing a hierarchical access control, deploying distributed access controls located at fog layers responsible for controlling devices and users at the edge of the network, and a centralized access control located at cloud layer responsible for controlling the distinct fog leaders. This may be a tentative solution that is currently being developed within the mF2C project.

Intrusion Detection Mechanism: Deploying a centralized intrusion detection solution managing the envisioned huge number of participant devices in F2C systems brings many weaknesses, for example malicious activities or nodes might not be detected due to either the huge volume of traffic analysis, or the centralized approach it self. Indeed, whether a centralized intrusion detector collapse or it is compromised, the whole intrusion detection solution may fail, what would not be sane for the system.

Malicious IoT and Fog Device Detection: Recognized the fact that the near to the edge a system is the more is its level of vulnerability as well, the massive deployment number of devices at the edge of the network facilitates attackers to successfully launch attacks or faking devices to eavesdrop the system. Obviously, if a device gets compromised or attacked, it must be properly detected and rapidly revoked from the system. Thus, the challenging question is "how can malicious devices in different layers be detected and what strategy or algorithm can be used to detect the malicious device on real-time processing and revoke them?".

3.3 Privacy and Sharing

Privacy: Data anonymization and data privacy are crucial components to protect user's private information. However, in scenarios where data is a must, notably leveraging data collection and processing to offer innovative services, a key challenge may be stated as "which data anonymization can be applied to the combined F2C scenario to keep the suitable trade-off between privacy and data utilization?" A tentative solution suggested for F2C systems may rely on keeping data as close to the edge as possible. To that end, fog leaders closer to users will take over the data processing, analysis and storage thus removing the need to go to higher layers in the hierarchy, consequently reducing the privacy gap. Another recently relevant concern refers to mobility aspects. Indeed, there are many services and apps demanding user's location to be executed, but the system should include strategies for users not willing to disclose their location. Indeed, some negotiation may be deployed between security and privacy, normally the more the security levels the lower the privacy guarantees and viceversa. In fact, when an attack is detected, the system must be able to react and consequently should be able to find out attacker's location. In short, privacy must be analyzed at different layers, privacy concerns must be identified, and finally, data and location privacy must be applied appropriately without impacting the whole security.

Secure Sharing Computation and Environment: In the envisioned combined model, resource sharing is an instrumental concept to move research but also market opportunities forward. In this sharing scenario fog nodes may complement IoT devices with low computational power, with additional resources. Although conceptually, this looks to be a promising model, attackers may benefit from it and may fake themselves as legible devices (as IoT devices, as fog nodes, and as fog leaders) to launch passive and active attacks. Three main relevant challenges come up: first, "how can a fog node or fog leader share their resources in a secure way with low computational power devices?" Second "how can IoT devices trust fog nodes?" Third "how can IoT devices outsource their service execution to fog nodes they share resources with?" Trust establishment and the ability to distinguish between legible and illegible devices is paramount here. Therefore, threat models and security analysis for the hierarchical shareable F2C environment must be done in each layer at an early stage, and all needed security requirements such as authentication, privacy, etc. must be provided before any device share their computational power with the system.

3.4   End-to-End Security Solution

Network Security: Interestingly, when moving close to the edge many different network technologies may be deployed, such as wireless, wired, zigbee, bluetooth, etc., that along with the candidate technologies to connect the edge to the cloud build a highly diverse technological scenario. Then, the main aim is to develop a network security strategy ensuring end-to-end security guarantees regardless the network technologies in place. It must be also considered that network technologies are not working in an isolated paradigm; rather different security protocols for different technologies might impact each other. Certainly, network technologies diversity is not a new problem and as said above, connectivity to cloud is agnostic of the technology used to connect. However, the single point failure concern and the long distance between the edges to cloud do not make cloud approach sound for the envisioned scenario. Therefore, network security must be re-designed to handle all type of network technologies, to provide end-to-end security, and to avoid the negative impact of handling different technologies and protocols.

Quality of Service (QoS): Service execution in combined fog and cloud scenarios must be supported by an optimal resource allocation, regardless where the resource is, as long as it perfectly meets service requirements, all in all to provide the expected QoS. However, whatever component is added for security provisioning some resources will be consumed, thus affecting the delivered QoS. The main rational behind this assessment seats in the huge volume of computational crypto requirements needed when implementing security. The services to be delivered through a combined fog and cloud deployment must positively benefit from such a resources combination in terms of a much better QoS while keeping solid levels of security on the whole stack of resources. Thus, to make it happen some open issues must be solved first, such as "if the distributed fog leaders take over security provisioning at the edge devices would they still have room to meet the expected QoS? or in other words, "how do both security and QoS meet expected performance in combined F2C systems?".

Heterogeneity: An F2C system is expected to manage heterogeneity at distinct levels, from network technologies, hardware or infrastructure providers, the latter being particularly interesting in terms of interoperability and also on a business oriented perspective. In fact, the envisioned F2C scenario should deal with distinct cloud providers (as already managed by the cloud sector) and also with fog providers. Certainly, what a fog provider may be is yet requiring sometime in the market, but it reasonable to consider cities, communities, malls, that is, "groups" of users that may become a "fog" provider for other users. In this scenario when providers become volatile (beyond the resources), setting agreements and "reliable" connections become a very challenging task. Consequently, a strategy should be sought to analyze how the different security strategies can be applied by the different providers, how they all can be compatible with each other and how the agreements may be set among them.

Secure Visualization: Interestingly, in a combined F2C system, fog nodes and fog leaders provide virtualization closer to the end users. The fog nodes and leaders might host the virtualization environment in their hypervisor. It means that, should the hypervisor get attacked, then the whole fog's virtualization environment might be compromised. This recognized the high vulnerability inherent to virtualization environments, in terms of virtualization attacks, such as virtual machine shape, hypervisor attacks, etc. Therefore, virtualization must implement in a secure way in the hierarchical F2C system.

Monitoring: The vast amount of devices distributed at the edge of the network, makes a centralized monitoring at cloud not adequate enough for F2C systems. The challenging questions are "which monitoring strategy/ies must be deployed to correctly monitor the huge amount of distributed devices located at different stack layers?" and "how the huge traffic analysis should be managed to detect malicious activities?" To that end, the solution pushed for by the mF2C project considers a distributed monitoring strategy implemented at fog layers to detect any malicious or abnormal behavior at the edge of the network, combined with a centralized monitoring strategy located at cloud for fog layers.

Security Management: The main challenges and questions here are, "can cloud as a centralized concept be sufficient to act as a security manager for the whole hierarchical F2C system?" The answer to this question is no, cloud as a centralized point failed to provide security and prevent several appeared attacks in past decades, although another challenge is, "What security management strategy must be taken into account for the distributed fogs and IoT devices?" To tackle with all these challenges, a new security management strategy, such as distributed security manager at fog layers and centralized at cloud, can be a suitable for the current security management for the F2C system.

Centralized vs. Distributed Security Management: The key question refers to the capacity cloud as a centralized approach that may have to serve as the security provider for the whole hierarchical F2C system. In fact, the correct answer is not positive since many reported attacks already exploited the cloud security vulnerabilities. Beyond that, the distributed nature of the F2C scenario, enriched with aspects related to mobility, volatility and heterogeneity make a centralized approach not to be the proper solution. Then, a distributed solution should be designed to handle the hierarchical nature of the F2C system, simultaneously handling the required security for distributed devices at the edge.

Secure Devices Bootstrapping: All components in the combined F2C system must bootstrap in a secure way by getting public and private parameters. In this scenario, the traditional centralized cloud or centralized trusted authority usually used for boot-strapping cannot be affordable due to the huge number of devices in the F2C system. Then, main questions are: "which distributed component must take cloud responsibility for bootstrapping devices at the edge of the network?", "can we apply a strategy where the cloud bootstraps fog leaders and fog leaders bootstrap devices in their area in a distributed fashion?".

3.5   Mobility Support

Secure Mobility: Recognized the inherent mobility shown by devices at the edge, the main challenge arises when trying to handle secure handover and secure mobility. As discussed before, the centralized and remote cloud cannot handle secure mobility for distributed devices at the edge of the network, fog leaders closer to the users might do so instead. To that end, distributed fog leaders must have secure intercommunication among them to provide secure handover for devices on the move. Open challenges are, "how does a fog leader hand secure mobility and secure handover for devices on the move?" and "should a fog leader is also on the move, who is providing its secure mobility and secure handover".

Secure Devices Joining and Leaving: The centralized cloud providing secure joining and leaving for the the huge number of devices in different layers cannot be sufficient for F2C systems, due to scalability issues. A hierarchical strategy can be useful to be applied here, such as, cloud can manage secure joining and leaving of fog leaders, and in parallel, each fog leader can control secure joining and leaving of devices in its area. On the other hand, in the F2C system, fog leaders should have secure intercommu-nication among them to provide secure devices joining in another area in case of mobility.

Secure Discovery and Allocation: All resources, services, and devices must be dis-covered in a secure way. Services must be allocated to resources, previously authen-ticated. Hence, different challenges arise: "how can services and devices be discovered in an authenticated secured way in the hierarchical F2C system?", "how can services be allocated to the corresponding authenticated resources securely?", "are fog leaders getting responsibility for securely discovering devices and allocating services to authenticated resources in a distributed fashion?", considering the different technolo-gies such as Wi-Fi, zigbee, bluetooth, etc. "which strategy can be applied in the F2C system to provide secure discovery for all mentioned technologies?" According to these challenges, "can a strategy for resource and service discovery, as well as allo-cation in a secure hierarchical authenticated fashion be re-designed for the combined F2C system?" With this idea, fog leaders can get authorization to provide distributed secure resources discovery.

To tackle all the questions and challenges mentioned above, proper security threats analysis must be done for the F2C system. Our proposal is then to re-design a hier-archical distributed security architecture for the combined F2C system, able to provide all the precious identified security.

Table 2. Main security challenges for F2C-like scenarios

Security area	Challenge	Description
Trust & authentication	Trust	Authentication is mandatory to prevent unauthorized users to access the system. The authentication mechanism needs identity or certificate to be verified and give users authorization to be involved into the system. Trust can be established between components after their authentication. Trust is one of the key components for establishing security between distributed fog nodes. Then, keys for encryption and decryption process can be distributed for components. Both Keys and identities need to be generated as unique, updated, and revoked during attacks, therefore, in F2C system handling key and identity management are the bottleneck due to hierarchical nature and huge number of distributed low-computational IoT devices. For Authentication and establishing trust, the traditional cloud as centralize point cannot be sufficient in F2C system due to distrusted nature. Therefore, as the main challenge here, trust and authentication must be redesigned to be handled in F2C system in hierarchical and distrusted way
	Authentication	
	Key management	
	Identity management	
Access control & detection	Access control	Access control is used to put rules that who and what can access the resources. In the case, the unauthorized users access the resources, intrusion and malicious device detection is needed. In case of access control and intrusion detection, handling the huge number of distributed IoT devices and fog nodes is one the main challenge in F2C security. Therefore, a need rises to re-design access control and intrusion detection in distributed way to be handled in F2C system
	Intrusion detection mechanism	
	Malicious IoT and fog device detection	
Privacy & sharing	Privacy	Privacy means that all the user's private information should not be disclosed to the others. In F2C system, fog nodes in hierarchical way would share their resources to users and IoT devices with low-computational power to run services. In this case, one of critical issues is how to handle user's, IoT devices', and Fog device's privacy in hierarchical F2C system without disclosing any critical information about each one of them to each other or even others
	Secure sharing computation and environment	
End-to-end security	Network security	Providing secure end-to-end communication between all components in a F2C system is one the challengeable issue due to different network protocols, huge amount on distributed devices at the edge of the network, and hierarchical F2C architecture. To provide secure communications, initially each one of the participant devices in F2C system must bootstrap in secure way. Fog nodes can be host virtualization environment to run the services, therefore, secure virtualization is a must at fog layers. All the secure communications must be monitored to detect any abnormal or malicious activities. All fog and cloud providers must set agreement to provide secure communications between their components in F2C system. At the end, a most challengeable secure communication issue is to design a new distributed security architecture to handle end-to-end security with less impact on the Quality of service
	Quality of service	
	Heterogeneity	
	Secure visualization	
	Monitoring	
	Centralized vs distributed security management	
	Secure devices bootstrapping	
Mobility support	Secure mobility	In the F2C system, devices such as IoT devices, mobiles, cars, etc. are dynamic. The devices are on the move. All devices arrive to the fog nodes must be securely discovered. A device joining in F2C system for the first time and even the existing device join the fog area must be done in secure way. Then, a securely leaving fog area to join another area must be considering as well. The most challengeable secure mobility issues is using cloud as single point of failure and even bring scalability issues. In hierarchical F2C system, a new distributed security must be design to handle device discovery, joining and leaving, mobility, and handover in secure way
	Secure devices joining and leaving	
	Secure discovery and allocation	

4   Remarkable Current Efforts and Trends

Certainly, there are many contributions dealing with security provisioning at cloud and recently many contributions came up focusing on the fog arena as well. Undoubtedly, the scenario brought in by combining cloud and fog (also including devices at the edge) poses several challenges not well covered by current solutions for either fog or cloud. However, any potential solution in the area of security must not start from scratch but learnt for past efforts in similar fields, what obviously includes efforts in cloud and fog. It is not the aim of this paper to go deep into these efforts but rather to highlight initiatives working on such a combined scenario and also other notable efforts pretty close to the F2C scenario, as briefly illustrated next.

There are two main ongoing efforts very aligned to the combined fog and cloud concept. The OpenFog Consortium (OFC) [8] aims to provide security by embedding hardware Root of Trust (HWRT) on the fog nodes. The HWRT can be programmed either when configuring (at the factory) or when booting. According to OFC expectations this module guarantees security against data breaches, spoofing, and hacking by providing secure identification, secure key store, protected operations, secure boot, etc. Another ongoing initiative is the mF2C project [9], yet in an early stage, theta proposes the use of distributed smart gateways (following a software approach) to ease the distribution of credentials and certificates. Thus, when a device enters an mF2C area for the first time, the device uses the smart gateway to connect to the certificate authority (CA) and gets credentials and certificate. Then, the device can establish secure and authenticated communication with the fog leader in that area, hence protecting users against man-in-the-middle attack, spoofing, etc.

Among other references found in the literature not linked to wide ongoing projects or initiatives, we may emphasize the work in [15] proposing end-to-end security mobility-aware for IoT-cloud systems using fog computing. In this work, smart gateways (Fog layer) provide device-users-cloud authentication and authorization remotely by using certificate-based data transport layer security (DTLS) handshakes. The end-to-end security between end users and devices is provided based on session resumption, and finally a robust secure mobility is implemented by secure inter-connection between smart gateways (Fogs). The solution proposal provides confidentiality, integrity, mutual authentication, forward security, scalability and reliability.

Similarly, authors in [16] propose a new security architecture for combined F2C systems, leveraging the use of a centralized F2C controller deployed at cloud and several distributed security control-area-units (CAUs) deployed at the edge of the network (fog layers) in different areas. Authentication and authorization procedures are defined to improve security guarantees.

5   Conclusion

This paper aims at highlighting the need to devote more efforts to sort out the large set of security concerns inherent to scenarios combining IoT, fog and cloud paradigms. In fact, this combined scenario brings together the different security challenges inherent to any of each paradigm, setting a complex setting demanding a comprehensive solution for security provisioning to guarantee the expected benefits brought in to run apps and services.

The paper introduces security requirements tailored to combined F2C-like systems, also analyzing what most sensitive attacks may be and ends up listing the. Main security challenges a F2C-like system must deal with to become a reality.

Acknowledgments. This work is supported by the H2020 projects mF2C (730929). It is also supported by the Spanish Ministry of Economy and Competitiveness and the European Regional Development Fund both under contract RTI2018-094532-B-100.

References

1. Al-Fuqaha, A., et al.: Internet of things: a survey on enabling technologies, protocols, and applications. IEEE Commun. Surv. Tutor. **17**(4), 2347–2376 (2015)
2. Chen, K.-C., Lien, S.-Y.: Machine-to-Machine Communications: Technologies and Challenges. Ad Hoc Netw. **18**, 3–23 (2014)
3. Gonzalez-Martinez, J., et al.: Cloud computing and education: a state-of the-art survey. Comput. Educ. **80**, 132–151 (2014)
4. Bonomi, F., et al.: Fog computing: a platform for internet of things and analytics. In: Big Data and Internet of Things: A Roadmap for Smart Environments. Studies in Computational Intelligence, vol. 546 (2014)
5. Marín-Tordera, E., et al.: Do we all really know what a fog node is? Current trends towards an open definition. Comput. Commun. **109**, 117–130 (2017)
6. Gupta, H., et al.: SDFog: A Software Defined Computing Architecture for QoS Aware Service Orchestration over Edge Devices. https://arxiv.org/pdf/1609.01190.pdf. Accessed Sept 2018
7. Coughlin, T.: Convergence through the cloud-to-thing consortium. IEEE Consum. Electron. Mag. **6**(3), 14–17 (2017)
8. The OpenFog Consortium. https://www.openfogconsortium.org. Accessed Apr 2018
9. mF2C project. http://www.mf2c-project.eu. Accessed Apr 2018
10. Masip-Bruin, X., et al.: Foggy clouds and cloudy fogs: a real need for coordinated management of fog-to-cloud computing systems. IEEE Wirel. Commun. **23**(5), 120–128 (2016)
11. Irwan, S.: Redesigning security for fog computing with blockchain. https://www.openfogconsortium.org/redesigning-security-for-fog-computing-with-blockchain/. Accessed Nov 2018
12. Ni, J., et al.: Securing fog computing for internet of things applications: challenges and solutions. IEEE Commun. Surv. Tutor. **20**(1), 601–628 (2017)
13. Martin, B.A., et al.: OpenFog security requirements and approaches. In: Fog World Congress (2017)
14. Gomez, A., et al.: A resource identity management strategy for combined fog-to-cloud systems. In: IoT-SoS (2018)
15. Moosavi, S.R., et al.: End-to-end security scheme for mobility enabled healthcare internet of things. Future Gener. Comput. Syst. **64**, 108–124 (2016)
16. Kahvazadeh, S., et al.: Securing combined fog-to-cloud system through SDN approach. In: Crosscloud (2017)

Low Power Area Network Sensors for Smart Parking

Aida Araujo Ferreira[1]([✉]), Gilmar Gonçalves de Brito[1],
Meuse Nogueira de Oliveira Junior[1], Gabriel Vanderlei[1], Samuel Simoes[1],
Ian Karlo[1], Iona Maria Beltrao Rameh Barbosa[1],
and Ronaldo Ribeiro Barbosa de Aquino[2]

[1] Federal Institute of Pernambuco,
Av. Professor Luis Freire, 500 - Cidade Universitaria, Recife, PE, Brazil
`aidaferreira@recife.ifpe.edu.br`
[2] Federal University of Pernambuco,
Av. Prof. Moraes Rego, 1235 - Cidade Universitária, Recife, PE, Brazil
`http://portal.ifpe.edu.br/`

Abstract. The Internet of Things consists of installing sensors for everything and connecting them to the Internet for information exchange for localization, tracking, monitoring and management. Smart parking is a typical IoT application that can provide many services to its users and parking owners. In order to create technological solutions for smart parking, an intelligent platform, known as Techpark, was designed. This platform consists of a web-based parking management system, an application for map view availability of parking lots, wireless sensor network controllers and car presence sensors. The car presence sensor was developed using Low Power Area Network technology with a SoC CC430 and an HMC5883L magnetometer. Results achieved by the sensor were promising, with a range of about 100 m, using radio frequency communication in the 900 MHz band, and with sensor life, using a battery with two 8.6 AH cells, estimated to be 8 years.

Keywords: Low Power Area Network · Smart parking ·
Internet of Things · Car presence sensor

1 Introduction

With the emergence of the Internet of Things (IoT), the ideal of creating a smart city is becoming possible and feasible. The concept of intelligent cities tries to take advantage of the evolution of technologies in general, directing their applications to the improvement of urban infrastructure, thereby making metropolises more efficient in the use of resources and, at the same time, better to live. The extent of citizen agglomerations has reached such a high order that issues such as pollution, access to health, housing and physical distancing to places of human activities (education, work, leisure, etc.), among others, sacrifice the environment and the economy of populations [1].

© Springer Nature Switzerland AG 2020
K. Arai et al. (Eds.): FTC 2019, AISC 1069, pp. 893–908, 2020.
https://doi.org/10.1007/978-3-030-32520-6_64

It is tried to face the urban logistics attacking it in several fronts, because all are parcels of the same problem, and a world-wide question is of the management of the fleet of vehicles. Their traffic consumes a part of the improvement effort, but on the other side are the parking lots, and the ways to make their use and management more efficient are eagerly sought [2]. In today's cities, finding a parking spot available is always difficult for drivers, and it tends to become more difficult with the increasing number of private car users [3]. This can be seen as an opportunity for smart cities to take action to increase the efficiency of their parking facilities, leading to reduced search times, traffic jams and traffic accidents.

Problems related to parking and traffic congestion can be solved if drivers can be informed in advance about the availability of parking spaces around the intended destination. Recent advances in building low-cost, low-power embedded systems are helping developers create new applications for IoT [4]. Almost every cosmopolitan city in the world suffers from congestion problems, which causes frustration to drivers, especially when looking for a parking space. Therefore, interest in this domain has become timely for scientists and researchers. Solving such a problem or even trying to relieve it offers benefits such as reducing driver frustration and stress, saving time and fuel, reducing greenhouse gas emissions, which in turn will affect pollution levels. Current systems have created an environment in which most modern statistics indicate discoveries regarding fuel waste during the search for parking spaces [2].

Most current parking systems can not be considered as smart parking, as they do not provide advance information on the availability of places, nor precise information on the spatial location of places, nor the reservation of a specific parking space. Generally, these systems depend on how many cars have entered the parking area and calculate the difference between that number and the maximum number of parking spaces to estimate the number of available parking spaces [5].

Existing solutions for so-called smart parking are based on electronic sensor technologies. These technologies can solve part of the challenge, for example, showing the user the existence of vacancies with some advance. On the other hand, they do not provide tools that aid in the management and depend on physical structures that limit their applicability.

The management of a car parking has been addressed using two models: parking gate monitoring and parking lot monitoring. A park gate monitoring may provide some services for drivers, such as the capability of checking the availability of a free spot and reservation over the Internet. In a parking lot monitoring, each parking spot is equipped with a sensor (camera or presence sensor) to detect the car presence/absence with the objective of building an availability map that can be used for parking guidance, reservation and the other services. The lot management model is more expensive than the gate monitoring the but it provides many more services to the drivers and the parking lot owner. Smart parking management, using lot monitoring, was classified in [6] into multi-parking management, used to manage different parking lots in different indoor and outdoor areas, or mono-parking management, generally targeting indoor parking lots and focusing on a single parking lot's management.

1.1 Related Work

According to [5], most of the smart parking systems, proposed in academia, depends on the knowledge of real-time parking information, based on which the system makes and upgrades allocations for drivers. A centralized architecture and a set of peripheral devices deployed over the parking area are the base for smart parking services, which connects all sensors together and transmits sensing data to the gateway and then to the driver through GUI via the server. It is necessary too a set of link layer technologies to develop a system that can support a possibly large amount of traffic resulting from the aggregation of an extremely high number of smaller data flows. Most of layer technologies enabling the realization a smart parking system are constrained technologies. They are characterized by low energy consumption and relatively low transfer rates, typically smaller than 1 Mbit/s. The more prominent solutions in this category are IEEE 802.15.4, Bluetooth, IEEE 802.11 Low Power, Near Field Communication (NFC), Radio-Frequency Identification (RFID) and Low Rate Wireless Personal Area Network (LWPAN) [7].

RFID technology and embedded systems with the objective of helping a user to check the availability of a parking space over the Internet were proposed by [8–10]. Smart parking's RFID solutions make it possible to manage permit parking easily, especially in the prototype stages. The main mechanism of RFID technology depends on an electromagnetic field to identify and track tags attached to objects automatically. However, it is still costly, the system circuits are complex, and the system has many disadvantages and limitations. Sometimes the RFID tags do not function or the accurate reading rate is very low, and the reader itself is quite expensive [5].

In [11], a smart parking system, using an ultrasonic detector, was developed to detect the car's presence/absence in a park slot. For each individual car park, one sensor is fixed on the ceiling above each parking space. Ultrasonic sensors operate based on echolocation. The sensor transmits a sound, which hits a solid object (car or ground) and is reflected back to the sensor. In [12], a combination of magnetic and ultrasonic sensors was proposed for the accurate and reliable detection of vehicles in a parking lot, and a modified version of the min max algorithm, for the detection of vehicles using magnetometers, was described.

A system that uses image detection was proposed in [13, 14] and [15]. This field of study includes methods for acquiring, processing, and analysing images. It uses computers to emulate human vision, including learning and being able to make inferences and take actions based on visual inputs, also called computer vision. Those systems have scalability and reliability very high, but regarding accuracy, false detection may occur. Motivation for developing this system comes from the fact that minimum cost is involved because image processing techniques are used rather than sensor-based techniques.

A Wireless Sensor Network (WSN) based intelligent car parking system was proposed in [16]. In this system, low-cost wireless sensors are deployed into a car park field, with each parking lot equipped with one sensor node, which detects and monitors the occupation of the parking lot. The status of the parking field detected by sensor nodes is reported periodically to a database via the deployed wireless sensor network and its gateway. The database can be accessed by the upper layer management system to perform various management functions, such as finding vacant parking lots, auto-toll, security management, and statistic report. Authors implemented a prototype of car sensor using a processor and a radio chip, referred as Motes Processor Radio boards (MPR).

An implementation of an energy-efficient and cost-effective, wireless sensor networks based vehicle parking system for a multi-floor indoor parking facility has been introduced in [17]. The system monitors the availability of free parking slots and guides the vehicle to the nearest free slot. The amount of time the vehicle has been parked is monitored for billing purposes. In this system, when a user enters the parking facility, at the entrance, there will be a keypad. The driver enters his mobile number using the keypad and then the parking slot is associated with the user. On successful entry of the mobile number, information regarding ID of the nearest empty parking slot, time of entry and route direction will be displayed on the monitor. When the vehicle is parked in the designated slot, an indication is sent to the Central Supervisory Station which starts a timer for the corresponding slot.

1.2   Contribution and Outline

There are lots sensing technologies could be utilized at the sensor layer for embedded parking solutions, such as the Radio Frequency Identification (RFID) for car parking access control; laser, passive infrared, microwave radar, ultrasonic, passive acoustic array sensors, or Closed-Circuit Television (CCTV) with video image processing for detecting the status of the car parking lots; users action between others. The focus of this work lies on multi-parking management and targets the presentation of the economic and engineering solutions of a smart parking system using hybrid wireless/wired communication between sensors, coordinators and server (cloud) [6]. The evolution of car presence sensors prototypes is presented as well as vantages and disadvantages in each prototype.

This paper is organized as follows: Sect. 2 presents our proposed smart parking system, called Techpark and the evolution of cars sensor prototypes built is highlights. Section 3 describes experiments setup and Sect. 4 presents the car detection sensor choosed to be the car's presence sensor in our smart parking solution mainly due to its low power consumption and range of communication and finally, Sect. 5 presents ours conclusions and future works.

2   The Smart Parking System

2.1   Techpark Platform for Smart Parking System

An IoT platform can be defined as a set of software and/or hardware that runs
the key operations of an IoT solution. This means that a platform is expected to
provide the means to collect device data, send it via internet to a server in the
cloud, store, process, present, and manage such data in a way that provides useful
information to the users. Techpark platform was designed in order to create an
IoT platform applied to the scope of the smart parking. The Techpark platform
provides for the integration of remote reconnaissance of parking space status
with system coordination units, publishing and information management over
the internet, and a smartphone application. The Techpark platform consists of
a web-based parking management system (cloud), an application for map view
of the managed parking and the availability of parking slots, wireless sensor
network controllers and prototypes for automobile presence (techpark sensors).

Figure 1 presents system architecture for Techpark intelligent parking. In
this architecture the car presence sensors (SENSOR) communicates with the
sub-coordinator through radio frequency ISM (Industrial Scientific and Medical) band. The sub-coordinator receives data from the sensors, treats such a
data and send it through a wireless network ethernet. The coordinator module receives such a data and sends it to the main server (Fig. 2). The Techpark
application was developed using Spring Boot technology. This application is
responsible for managing the requests coming from the web and mobile clients
and for providing the data requested by these applications through web services.
It is also responsible for processing the messages received from the coordinators,
through the MQTT broker [18], and recording them in the postgres database.
The application front end was developed using IONIC technology.

The communication between the sub-coordinator, coordinators and server is
performed over Message Queue Telemetry Transport [18] protocol. The coordinator sends the status of each sensor (address, availability, temperature, battery
level and firmware version) to the application server, which is responsible for
processing this information, storing it in a database, and making it available to
users through a geographic information system (GIS). The GIS provides web-
based parking administration functions and is responsible for providing spatial
configuration consultation services on the availability of parking spaces that can
be consumed by applications on mobile devices or in web browsers.

2.2   Development of LPWAN Sensors to Detect Automobile
Presence

With the objective of developing a vehicle presence sensor, capable of operating
in indoor or open-air parking lots such as those of the major urban centers,
solutions have been developed that go from the detection system itself to the
application services for user use, through the creation of the communication
networks necessary for the integration and control of the entire system.

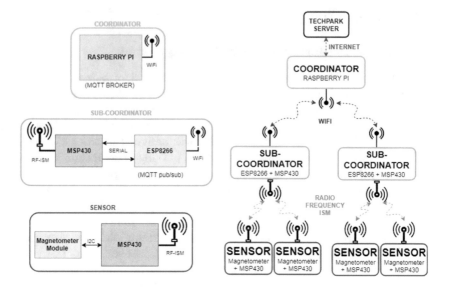

Fig. 1. Techpark architecture.

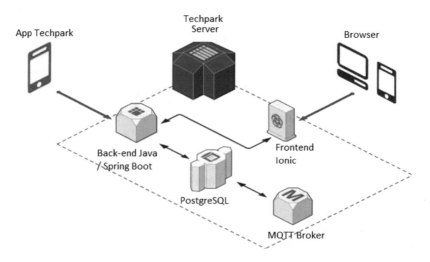

Fig. 2. Techpark server.

The principle of automobile presence detection, implemented in the Tech-Park sensor, is through the variation of the magnetic field of the environment caused by the body that is wanted to perceive. This is accomplished by reading a module using the HMC5883L Honeywell magnetometer. The development of the automobile presence sensor through the variation of the magnetic field was carried out in a gradual manner and resulted in eight versions. Versions 1 and 2 were built using the Arduino platform and radio into ISM band. Versions 3 and

4 were built using a limited implementation of TI CC430F6147 SoC. Versions 5 through 7 adopted the D1 mini module based on ESP8266 technology, as a concept test of using IEEE802 wireless network. On versions 8 the limitation of using TI CC430F6147 SoC was overcame so that it were proved as the best implementation for TechPark project. Versions 1 and 2 were built with the basic objective of the concept test and proof of the feasibility of detecting the presence of automobile through the variation of the magnetic field. From version 3 onwards the presence sensors were constructed and evaluated mainly in relation to the cost, the reach and the duration of the battery. Figure 3 shows the evolution of automobile presence sensor prototypes of automobile and Table 1 presents details about each prototype, including hardware and positive and negative aspects for each prototype.

Fig. 3. Evolution of automobile presence sensor prototype through the variation of magnetic field.

Among the eight versions of automobile presence sensors that were developed, version eight, which uses MSP 430 based SoC CC430, was selected as the TechPark sensor for its low power consumption, the main requirement for a sensor that should have a minimum of 3 years of service life without battery change. According to [19], the Texas Instruments MSP430 microcontrollers have a simple and yet powerful design. Some of the key aspects of the MSP430 architecture are: High performance operating at low operating voltage and low power consumption, has a reduced number of instructions (RISC architecture), ease of recording and debugging made possible by the use of the JTAG interface Test Action Group), which allows the designer to program and debug their software directly on the application board, without the need to use expensive equipment as emulators, also has several types of encapsulation and a large number of peripherals. Figure 4 shows the development platform EM430F6147RF900 from Texas Instruments.

Table 1. Evolution of sensors prototypes

Evolution of sensors prototypes

Version	Hardware	Positive	Negative
1	Arduino Pro Mini, HMC5883L Honeywell magnetometer, HMC5883L, an array of contacts, module Bluetooth HC-05	Low cost, low programming complexity and medium battery consumption	It needs an Android device for information gathering, short range, low battery consumption
2	Arduino Pro Mini, HMC5883L Honeywell magnetometer, WiFi NRF24l01	Low cost, low programming complexity and medium battery consumption	It needs an Android device for information gathering, short range, low battery consumption
3	MSP430cc, HMC5883L Honeywell magnetometer, radio communication	Medium processing power and low power consumption	Medium cost, programming with high difficulty, short range, low battery consumption, high range
4	MSP430cc, Arduino Pro Mini, HMC5883L Honeywell magnetometer, radio communication	Medium processing power and low power consumption	Medium cost, programming with high difficulty, short range, low battery consumption, high range
5	ESP8266, HMC5883L Honeywell magnetometer, WiFi	Low complexity to program, several available resources (peripherals, buses and protocols), scalability for the project	High range, high battery consumption
6	ESP12, HMC5883L Honeywell magnetometer, WiFi	Low cost, medium-high processing power, low complexity to program, several available resources (peripherals, buses and protocols), scalability for the project)	High range, high battery consumption
7	Atmega328P microcontroller, Module D1 Mini, HMC5883L Honeywell magnetometer, WiFi	Low cost, medium-high processing power, low complexity to program, several available resources (peripherals, buses and protocols), scalability for the project)	High range, medium-high battery consumption
8	CC430F6147 microcontroller, HMC5883L Honeywell magnetometer, ISM band	Low cost, medium-high processing power, low complexity to program, extremely low energy cost, easy scalability for industrial production)	High range, extremely low battery consumption

Among the several families of MSP430 microcontrollers, the CC430 SoC family was chosen to develop the TechPark sensor. The Texas Instruments CC430 family of ultra-low power, system-on-chip microcontrollers with integrated RF transceiver cores consists of multiple devices with different sets of peripherals targeted to a wide range of applications. The architecture, combined with seven low-power modes, is optimized to increase battery life in portable measurement applications. The device has a MSP430^{TM} 16-bit RISC CPU, 16-bit registers [20].

The CC430F614x series comprises the TechPark sensor unit and is a microcontroller that combines the CC1101 sub-1-GHz RF transceiver (ISM Band) with the MSP430 CPUXV2, up to 32 kB of programmable flash memory in the system, up to 4 kB of RAM, two 16-second timers 10-bit A/D converter with

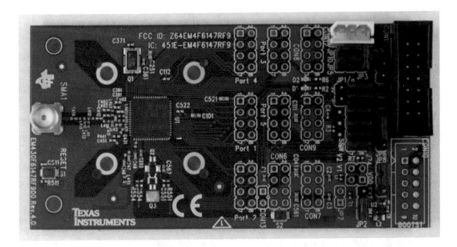

Fig. 4. Texas instruments EM430F6147RF900 development platform.

eight external inputs plus internal temperature and battery sensors, comparator, universal serial communication interfaces (USCIs), 128-bit AES security accelerator, hardware multiplier, DMA, clock module real-time alarm capabilities, LCD driver and up to 44 I/O pins [20].

The LPWAN sensor developed with a SoC CC430 and an HMC5883L magnetometer showed promising results with a range of about 100 m using radio frequency communication in the 900 MHz range and with sensor life using a battery of two cells 8.6 Ah, estimated at 8 years.

2.3 Sensor Network Structure

The TechPark sensor network coordinator serves as a link between the vehicle presence sensors and the server. A single switch must have the ability to communicate with multiple sensors and send their respective information to the rest of the system via Ethernet medium. In this way, a higher computational power than a simple microcontroller based board is required, so that you can efficiently meet the various needs. On the current project stage the coordinator is based on Raspberry platform, applying two networks conception, IEEE802 and ISM, for building the TechPark network. Such a conception is showed in Fig. 1. One central coordinator manages a IEEE802 wireless network supporting a set of sub-coordinators that manages its very ISM band network supporting the TechPark sensors.

The purpose of the central coordinator is to manage a large number of sub-coordinators spread over a specific geographic area, receiving their data and requesting their actions, and sending this data to a Spatial Data Bank. The sub-coordinators works on same manner but over an ISM band network. The way the TechPark network works allows the control panel to command all sensors to perform the self-calibration required for their operation and then start monitoring all parking spaces simultaneously.

The network topology used to enable communication between the sensors fixed in the parking slots and the controller was, in principle, the star topology. The designation star network means that all information must pass through an intelligent central station, which must connect each sensor in the network and distribute the traffic to a server. The central element that 'manages' the network data flow is virtually connected to each sensor Fig. 5.

Communication between subcoordinators and sensors is done by broadcast and the data flow mode is half-duplex. The data packet model is described in Fig. 6. The first two fields of the payload to be transmitted are composed by the MAC address of the controller and the sensor, respectively.

The MAC (Media Access Control) address is a physical and unique address that is associated with the communication interfaces used on network devices. The identification is recorded in hardware by network card manufacturers, becoming later part of equipment such as computers, routers, smartphones, tablets, printers among others [21].

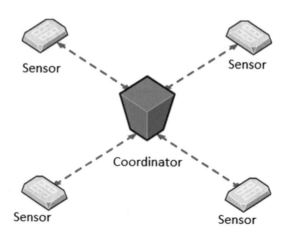

Fig. 5. Coordinator and its influence radius under the sensors placed in parking slots.

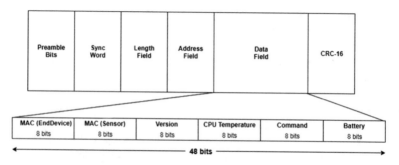

Fig. 6. Data packet layout.

3   Experiments

In current stage, this research has implemented a basic set of test toward elaborating advantage of using sub-1-GHz SoC platform compared to IEEE802.11 implementation, even for short range. In order to elucidate about energy consumption a power meter application was implemented, from now termed as Emeter (vide Fig. 7). The Emeter comprise two module: (i) the microcontrolled unit and (ii) the desktop application. The microcontrolled unit consist of a Mega2560 Arduino programmed for: (i) measuring device supply voltage, (ii) voltage measuring over shunt-like resistor put on device power supply branch (vide Fig. 8), and (iii) for processing the signal and send it for desktop application. The desktop application processes data from the microcontrolled unit rendering it by voltage, current and power graphics. In addition, the Emeter allows set up specific built experiment protocol, but it is not applied on this paper. Two scenarios was evaluated: wireless IEEE802.11 based net and sub-1-GHz SoC based net. Three set of measuring was performed, for no traffic consumption, data transmission consumption and data reception consumption.

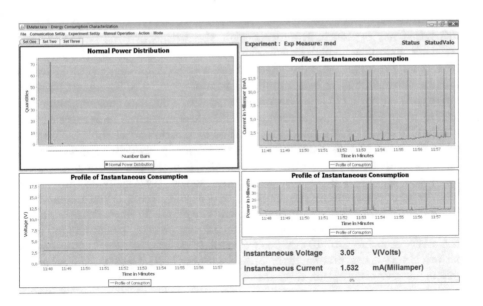

Fig. 7. Emeter

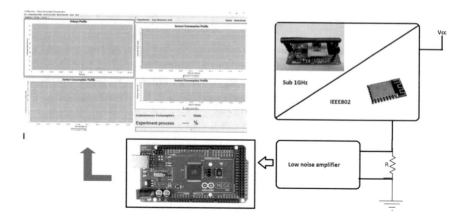

Fig. 8. Experiment diagram.

4   Results

As result of experiment realized, it was confirmed the advantage of sub-1-GHz SoC based net approach. Despite of high abstraction of wireless IEEE802.11 based net that allowing high level implementation and rapid integration on clouding servers, the spend of energy became the wireless approaches prohibitive for a sensor network. Figure 9 shows wire consumption pattern. There is a base line current of 12 mA, do not care the operation status. There are consumption peaks ranging up to 510 mW - associated to 102 mA peak current. The average current is about 40 mA, rippling according wireless connection reachability. The variable component of consumption represents the energy cost for support wireless IEEE802.11 device, being result of using TCP-IP protocols. There is no practical changes among different operation modes, so that practically nothing can be done for improving energy consumption by software. For a battery comprising two 8600 mAH cells this results in autonomy ranging of 18 days. On other hand, the sub-1-GHz SoC based net consumption pattern presents peaks ranging up to 52,8 mW - associated to 16 mA peak current (vide Fig. 10). Such consumption pattern is very acceptable since the protocol timing is configurable, impacting on system energy consumption rate. Currently, due to needing of polling for receiving data from net coordinator, the TechPark protocol have to operate with 0.0167 duty cycle at minimal. Such a condition imply average current of 266 uA. For a battery comprising two 8600 mAH cells this resulting in autonomy ranging of 8 years. If functionalities of remote firmware update and remote calibration was removed, the present sensor could remain operational during 20 years.

The CC430 SoC platform in conjunction with the HMC5883L magnetometer as a car detection sensor proved to be viable as a solution for intelligent parking using IoT mainly due to its low power consumption and range of communication. In order to improve energy saving, system scalability, space reduction and web integration in the system infrastructure, an ESP8266 + CC430 set was used

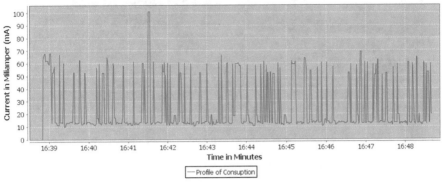

Fig. 9. Currents of IEEE802.11 based device.

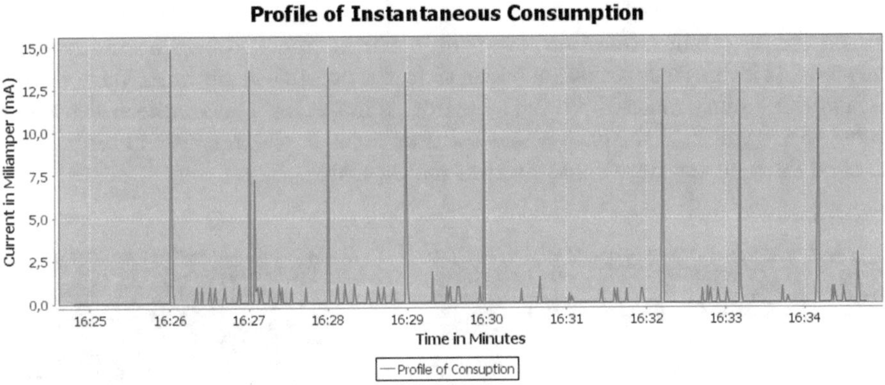

Fig. 10. sub-1-GHz SoC based enddevice.

to act as a subcoordinator for the sensors that will carry out the data management and will send the data to the Coordinator based in Raspberry Pi, already integrated into the Techpark platform.

The programming of the CC430 microcontroller was done in C language and all simulations were done by varying the power levels of the development platform and the temperature to prove the effectiveness of the SoC sensor detection. Both received packet data and packet data could be viewed using the IAR Embedded Workbench IDE, which we also used to program the microcontroller, and connecting the platform to the computer using the Olimex MSP430 USB programmer model MSP430-JTAG-TINY- V2. Figure 11 presents sensors prototypes built with CC430 microcontroller.

Establishment of radio communication between the sensor module and the sub-master's MSP was successful. Three sensors were tested and all were able to successfully send the transmission frame to the subcomputer. The architecture used was based on the Low Power Mode system to ensure compliance with the

Fig. 11. LPWAN sensor or sensor techpark.

low power function proposed by the project. The MSP's own operating mode, based on interruptions occurring in its low-power state, turns out to be the reason for its distinct learning line from other microcontrollers normally used. However, it is due to this characteristic that it is possible to obtain a lower energy expenditure, since most of the time the SoC is in the low consumption state only being active when, in the case of the TechPark sensor, the magnetometer is ready to send data or any data must be received or sent.

Fig. 12. LPWAN sensor and subcoordinator respectively.

The entire system was tested with up to three sensors and all data arrived properly in the subcoordinator, just as the subcoordinator sent the data properly to the Coordinator. These analyzes were performed with simulators developed for each platform (Fig. 12).

5   Conclusions

By requiring mostly low power projects, IoT applications use embedded computing systems that are efficient, reliable and communicate over the wireless network without sacrificing low power consumption. We found that the MSP430TM microcontrollers, especially the CC430 family, perfectly fit these requirements. The presence of transceivers in the development boards provided not only point-to-point communication but also using the star topology in an excellent way using only common batteries for powering the sensor modules. Therefore, the MSC430 SoC platform was adequate for the development of IoT applications.

The development difficulty presented in earlier versions of the project is due to the SoC architecture adapted to the low-power operation that makes the program better suited to scenarios where interruptions are the key to system operation. After overcoming the initial difficulties, the process of using the CC430 proved to be very advantageous in relation to the others available, especially when looking at the current scenario where the IoT concept is so promising. Aiming at the subcoordinator, the ESP8266 is an excellent platform in the IoT scenario, which, even with its high energy expenditure, when combined with microcontrollers that carry out the power management process is an excellent option in sending data via Wireless IEEE802 through the MQTT protocol.

The energy expenditure of the TechPark sensor must still be analyzed in a more analytical way, but due to the way the CC430 works in conjunction with the analyzed documentation and the experiments carried out in conjunction with the batteries, it was noticed that the platform has a great capacity of keep working with few amps.

Although preliminary results indicate the considerable efficiency and potentiality of the proposed cars presence sensor, experiments continuing be conducted in a controlled environment. Next steps are install sensors in the our institution's parking and collect their data (availability, temperature, battery level, humidity and firmware version) during 3 months to realize analyses from actual data.

Acknowledgments. Authors thanks to IFPE and FACEPE.

References

1. Kim, T.-H., Ramos, C., Mohammed, S.: Smart city and IoT. Future Gener. Comput. Syst. **76**, 159–162 (2017)
2. Faheem, Mahmud, S.A., Khan, G.M., Rahman, M., Zafar, H.: A survey of intelligent car parking system. J. Appl. Res. Technol. **11**(5), 714–726 (2013)
3. Long, A.R.: Urban Parking as Economic Solution. Technical report (2013)
4. Khanna, A.: IoT based Smart Parking System, November 2017
5. Fraifer, M., Fernström, M.: Investigation of smart parking systems and their technologies. In: Thirty Seventh International Conference on Information Systems. IoT Smart City Challenges Applications (ISCA 2016), Dublin, Ireland, pp. 1–14 (2016)
6. Bagula, A., Castelli, L., Zennaro, M.: On the design of smart parking networks in the smart cities: an optimal sensor placement model. Sensors **15**(7), 15443–15467 (2015)

7. Zanella, A., Bui, N., Castellani, A., Vangelista, L., Zorzi, M.: Internet of things for smart cities. IEEE Internet Things J. **1**(1), 22–32 (2014)
8. Li, X., Ranga, U.K.: Design and implementation of a digital parking lot management system. Technol. Interface J. (2009)
9. Jian, M.-S., Yang, K.S., Lee, C.-L.: Modular RFID parking management system based on existed gate system integration. WTOS **7**(6), 706–716 (2008)
10. Pala, Z., Inanc, N.: Smart parking applications using RFID technology. In: 2007 1st Annual RFID Eurasia, pp. 1–3, September 2007
11. Kianpisheh, A., Mustaffa, N., Limtrairut, P., Keikhosrokiani, P.: Smart parking system (SPS) architecture using ultrasonic detector. Int. J. Softw. Eng. Appl. **6**(3), 55–58 (2012)
12. Lee, S., Yoon, D., Ghosh, A.: Intelligent parking lot application using wireless sensor networks. In: 2008 International Symposium on Collaborative Technologies and Systems, pp. 48–57, May 2008
13. Yass, A.A., Yasin, N.M., Zaidan, B.B., Zeiden, A.A.: New design for intelligent parking system using the principles of management information system and image detection system. In: Proceedings of the 2009 International Conference on Computer Engineering and Applications (2009)
14. Bong, D.B.L., Ting, K.C., Lai, K.C.: Integrated approach in the design of car park occupancy information system (COINS). IAENG Int. J. Comput. Sci. **35**, 7–14 (2008)
15. Banerjee, S., Choudekar, P., Muju, M.K.: Real time car parking system using image processing. In: 2011 3rd International Conference on Electronics Computer Technology, vol. 2, pp. 99–103, April 2011
16. Tang, V.W.S., Zheng, Y., Cao, J.: An intelligent car park management system based on wireless sensor networks. In: Proceedings of the 1st International Symposium on Pervasive Computing and Applications, August 2006
17. Joseph, J., Patil, R.G., Narahari, S.K.K., Didagi, Y., Bapat, J., Das, D., et al.: Wireless sensor network based smart parking system. Sens. Transducers **162**(1), 5–10 (2014)
18. Singh, M., Rajan, M.A., Shivraj, V.L., Balamuralidhar, P.: Secure MQTT for internet of things (IoT). In: 2015 Fifth International Conference on Communication Systems and Network Technologies, pp. 746–751. IEEE, April 2015
19. Davies, J.H.: MSP430 Microcontroller Basics. Elsevier Science (2008)
20. EM430F6147RF900 CC430 wireless development tool — TI.com. http://www.ti.com/tool/EM430F6147RF900. Accessed 28 Dec 2018
21. Wikipedia contributors. MAC address, December 2018. https://en.wikipedia.org/w/index.php?title=MAC_address&oldid=874289123. Accessed 28 Dec 2018

Blockchain Technology Solutions for Supply Chains

Marius Bjerkenes and Moutaz Haddara[(✉)]

Kristiania University College, 0186 Oslo, Norway
mariusbjerkenes@gmail.com,
moutaz.haddara@kristiania.no

Abstract. Blockchain has become one of the most promising technologies in various industries. Within the supply chain management (SCM) domain, many researchers and practitioners argue that blockchain technologies can transform the supply chain for the better and shape its future. This review seeks to identify, summarize and analyze the extant literature in order to pinpoint the potential key benefits that the blockchain technologies can introduce into supply chains. Thus, this review aims at presenting the possible solutions that can address common supply chains issues when integrated with blockchain technologies. Six benefits are identified in literature; increased transparency, enhanced security, high veracity of data, cost-reductions, trust and better contract execution. In addition, several challenges were also discussed. Moreover, this paper may provide insights for those who seek pinpointing further research avenues in the blockchain within the SCM context.

Keywords: Blockchain · Supply Chain Management · Literature review

1 Introduction

Supply chains are no longer just traditional networks of companies. As digitalization and globalization have changed the industry, they have become large ecosystems with different products moving through multiple parties, trying to coordinate work together [1]. Shorter lifecycles combined with increased competition have made the need for supply chains to be more dynamic than ever before. Supply chain management (SCM) is the administration and management of upstream and downstream interactions and relationships between suppliers, partners and customers in order to deliver high customer value at the least cost possible [2].

In the top 2018 technologies for supply chains report by Gartner, several supply chain technology trends were identified. The top ten trends included technologies such as internet of things (IoT), artificial intelligence (AI) and advanced analytics [3]. However, one of the most interesting technologies mentioned in their report is the blockchain technology. The introduction of blockchain technology has opened for a substantial potential for companies entering the digital age. Blockchain is a decentralized transaction and data management technology, initially designed for the cryptocurrency Bitcoin [4]. The technology enables the data to be shared among and agreed upon a peer-to-peer network in a secure and trustworthy way. The data is linked by a

© Springer Nature Switzerland AG 2020

K. Arai et al. (Eds.): FTC 2019, AISC 1069, pp. 909–918, 2020.
https://doi.org/10.1007/978-3-030-32520-6_65

sequence of blocks with timestamps secured by cryptography and verified by other actors in the chain. When an item or data file is connected to the blockchain, it cannot be changed. Resulting in an immutable record of past activity. For supply chains with large supplier networks, this can be of great importance. The introduction of blockchain has generated exciting research areas, where the technology can provide guidance related to gaps that were identified earlier. Blockchain is relatively new for commercial use. Thus, there is a research gap in the area of blockchain in business contexts in general, and in SCM in specific. This paper aims to contribute to the knowledge that already exists regarding the use of blockchains for SCM, through investigating the following question: *what are the potential benefits of blockchain technologies integration in supply chains?*

The rest of the paper is organized as follows. Section 2 presents the research methodology adopted in this research. Section 3 provides an overview of the reviewed articles. The main findings are presented in Sect. 4. Section 5 provides a discussion of the findings, followed by conclusions and future research suggestions in Sect. 6.

2   Research Methodology

Literature reviews provide critical summaries and analyses of the existing body of knowledge and themes within a field [5]. Webster & Watson [6] define a systematic literature review as one that provides strong basis for advancing knowledge in an area. It assists theory development, fills a gap in areas where a plethora of research exist, and unearths areas where research is required [6]. Thus, this paper adopts a systematic review approach [6]. We first search for literature in our scope, then extract the relevant ones. After that the identified papers are assessed for quality to determine which articles to include. Lastly, the literature is analyzed and combined to pinpoint research gaps that can be of interest for future research. Blockchain's application in supply chain management has been valued as one of the most important trends for the future, and therefore an interesting research area for further research.

In order to identify key articles for this paper, a search strategy was set. This chosen strategy aided in making a more narrowed and focused search to locate relevant literature. This review covers articles published between 2015 and 2018, as there is minimal research before this time-period regarding non-financial applications. The primary search engines and databases used are Google scholar and Web of Science. Google scholar has a wider range of articles but has limited filtration tools that can be applied. Web of Science makes it easier to control the quality of the literature but has a low number of articles within this topic. Therefore, hybrid use of both search engines was applied. The following search procedures were applied to ensure a systematic methodology:

1. Search for literature through Google scholar with the given search terms in the search strategy. The combination of the keywords "supply chain management" and "blockchain" generated the best result to find relevant literature. Articles that seemed relevant where registered for the next screening process.

2. Same approach as in step one, but on Web of science. Resulted in fewer articles but has filtration tools that are useful in setting criteria for quality assessment, such as impact factors on outlets. However, high impact factor was not considered as a prerequisite for this paper, as literature was scarce.
3. The abstract of the articles found in step one and two were read by both authors to ensure that they were relevant for this review. Keywords helped to guarantee that the articles were within the researched domain.
4. We were planning to review papers with a minimum of 10 citations, however most of the articles were between 2017 and 2018; therefore, which did not have time for being cited. Given the small amount of literature available, we decided to include articles in any peer-reviewed outlet, regardless of their citation indexes, and the journal/conference's ranking.

3   Overview of Articles

Blockchain technology was initially introduced in 2008 in the cryptocurrency Bitcoin; however other areas of applications started to get research attention as late as in 2013 [7]. Swan [8] argues that a threefold classification can explain the literature on blockchain for various applications: The first stream of literature is called blockchain 1.0, which focuses on the contribution on financial applications with cryptocurrency. Blockchain 2.0 goes beyond the pure financial contracts and presents solutions such as smart contracts. Finally, blockchain 3.0 is the applications that go beyond the financial world, furthermore contributing to the implementation of this technology in other industries. Our stream of literature concentrates on the 3.0 phase. In total, 15 articles were identified and reviewed, shown in Table 1. Eight of the articles were journal articles, while the remaining seven are conference articles. The distribution of the publications per year is shown in Fig. 1.

Table 1. Overview of articles

Author(s)	Title	Publication channel
[9]	An agri-food supply chain traceability system for China based on RFID & blockchain technology	International Conference on Service Systems and Service Management (ICSSSM)
[10]	Will blockchain technology revolutionize excipient supply chain management?	Journal of Excipients and Food Chemicals
[11]	Blockchains everywhere - A use-case of blockchains in the pharma supply-chain	IFIP/IEEE Symposium on Integrated Network and Service
[12]	A Blockchain-Based Supply Chain Quality Management Framework	IEEE: International Conference on e-Business Engineering (ICEBE)
[13]	Digital Supply Chain Transformation toward Blockchain Integration	Hawaii International Conference on System Sciences

(continued)

Table 1. (*continued*)

Author(s)	Title	Publication channel
[14]	Information sharing for supply chain management based on blockchain technology	Conference on Business Informatics (CBI)
[15]	Configuring blockchain architectures for transaction information in blockchain consortiums: The case of accounting and supply chain systems	Intelligent Systems in Accounting, Finance and Management
[16]	Trace and track: Enhanced pharma supply chain infrastructure to prevent fraud	International Conference on Ubiquitous Communications and Network Computing
[17]	A distributed ledger for supply chain physical distribution visibility	MDPI: Information
[18]	Adaptable Blockchain-Based Systems: A Case Study for Product Traceability	IEEE Software
[18]	Research on agricultural supply chain system with double chain architecture based on blockchain technology	Future Generation Computer Systems
[19]	Blockchain's roles in meeting key supply chain management objectives	International Journal of Information Management
[20]	Infrastructural Grind: Introducing Blockchain Technology in the Shipping Domain	ACM International Conference on Supporting Group Work
[21]	The Supply Chain Has No Clothes: Technology Adoption of Blockchain for Supply Chain Transparency	MDPI: Logistics
[22]	How blockchain improves the supply chain: case study alimentary supply chain	Procedia Computer Science

Figure 1 below shows the number of publications per year. 13% of the literature was published in 2016, with a peak of 47% in 2017 and 40% in 2018, which exemplifies that this topic is a considerably new and upcoming area of research. The growth of papers within this field reflects the interest that blockchain has gotten the last few years. Applications within other areas than finance, known as the blockchain 3.0 phase may also contribute to the increase of literature within this field. Table 2 shows the articles that met the chosen requirements. The articles reflected upon several applications of blockchain in SCM, and therefore only the central insight from each article was included for further discussion.

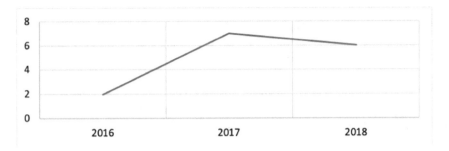

Fig. 1. Publication/year

4 Findings

As earlier mentioned, 15 articles were identified as relevant and address the integration of blockchain in SCM. The supply chain domain has a wide range of research areas; however, a small amount of these papers focused on the applications that blockchain can provide. The findings from the literature are organized based on theme or focus, as shown in Table 2. These themes are the main topics that the literature points at regarding competitive advantage areas that can be achieved in SCM with the integration of blockchain. Several papers have covered different themes. Thus, they may fall under several themes as presented in the table below.

Table 2. Area of focus in literature

Author(s)	Theme/Focus
[13, 14, 19]	Costs
[9, 10, 12, 13, 16, 18]	Security
[11, 20, 22]	Smart contracts
[17, 18, 20, 21, 23]	Transparency
[12, 14]	Trust
[14, 15, 17, 22, 23]	Veracity of data

In this section, the purpose is not to summarize article by article, but rather summarize the six main findings that the literature mentions as essential aspects.

The most discussed theme in literature is transparency. Author in [21] argue that technology before blockchain cannot meet the increasing requirements from customers regarding information about products. Furthermore, it accelerates transparency further upstream and downstream compared to what was possible before [20]. Integration of blockchain can help supply chains achieve a level of transparency which can enable customers gaining the information they require, as well as the company achieving across the supply chain. Wu et al. [17] found that blockchain will provide the visibility necessary for optimal decision-making given the data within the ledger, arguing that the trade-off between privacy and transparency is of crucial importance for the widespread adoption of the technology. Additionally, [18] further confirm that the technology will improve the transparency resulting in an overall efficient system. All in all, transparency is something that is of great value for supply chains. Not only does it make operations more efficient, but it also makes it easier to act fast to the challenges the supply chains encounter. It can help detect bottleneck that was hard to notice before as well as other operational problems. Literature confirms that blockchain will contribute to transparency, and therefore could generate competitive advantages for supply chains.

With transparency, the need for security becomes important for entities in the supply chains. As supply chains favor collaborative work, there may be some that act out of self-interest. Blockchain aims to combat that threat, and literature suggests that it succeeds in this given the securing of data and making them immutable of changes. Not only does the technology provide security for data, but it also makes physical products more secure concerning quality [16]. Apte & Petrovsky [10] argue that by removing intermediaries, it secures the data that is shared from A to B and additionally provides vital insight for every entity in the supply chain. Justified by the immutable technology giving access to non-falsifiable cloud-based transaction records. Verified information is a crucial insight from these articles. The core problem that these articles assess is how blockchain can help build a foundation based on consensus around how secure data is transferred, without worrying about the tampering of data. Author in [18] note that blockchain enables a consensus on security, validity, and integrity with highly decentralized decision-making power. The mathematical algorithm behind the technology guarantees the security of data for all entities in the supply chain, as well as providing privacy [13]. Chen et al. [12] found that blockchain contributes to security at a lower cost than traditional intermediaries, as this peer-to-peer network has no central system, but rather is put together by a decentralized system of blocks.

Two of the reviewed articles discuss the positive effects that blockchain can have on trust within the supply chain. Chen et al. [12] found that blockchain solves the issues related to distrust, given that information is immutable, as well as, traceable records are visible for the entities in the supply chain. Contrary to the traditionally centralized conduct, the decentralized structure that lies within blockchain contributes to mitigating the self-interest of supply chain entities, information asymmetry and the cost of quality assessment [12]. Author in [14] continues the notion that trust can be further ensured with blockchain: instead of relying on a third party, the information is secured through the cryptographic proof. Intermediaries can misuse data, and [14] argues that users should control and own their data, mitigating the threat of misconduct and stealing. An important aspect one should consider regarding blockchain integration is the potential cost reductions that occur. Articles that focused on this domain found that cost reduction is achieved due to the streamlining and efficiency that the blockchain technology holds. With the elimination of intermediaries and administrative work practices, transaction costs would shrink, and in some cases disappear [13]. Author in [19] performed a multiple case study, with companies that have used blockchain, where the research showed that there are zero to low marginal costs to generate blockchain code. Author in [13] also found that the streamlining nature of blockchain was positive on the bottom line. The main point to bring out of findings related to costs is that with the elimination of intermediaries and improved flow of information, transaction costs will drastically decrease.

One of the main reasons for the popularity of blockchain technology is the high veracity of data. Author in [22] found that with the integration of blockchain, shipments were tracked seamlessly with authentication of origin and destination. All the articles regarding the veracity of data found that this aspect is of crucial importance for entities in a supply chain. When the data is timestamped and immutable to later changes, it can create an environment for the supply chain that incorporates trust as well as higher responsiveness when that data is real-time and authenticated [17]. Information asymmetry is mitigated by such as system, which is a clear competitive advantage. The veracity of data contributes to other areas that were unthinkable with previous technology. An area that can change in tremendous ways is contracts, and blockchain introduces the use of smart contracts. Smart contracting is a revolutionizing new way of handling the contractual agreements and executions of the contracts. The articles reviewed argue that it helps in removing intermediaries in the exchange of goods or services, such as banks [11]. Just as regular contracts, the contract is defined given the agreements of the counterparts. However, with smart contracts, the obligations that are agreed upon are enforced automatically given the actions of the parties [20]. Instead of waiting for a signed contract, it is registered, and the money is transferred automatically, given that the conditions are met. Author in [11] argue that this technological development leads to less time spending on administrative activities, leading to more time allocated to their core competence. The articles found the use of smart contracts to be less time-consuming, more secure and had lower transaction costs.

5   Discussion

The reviewed articles are published in 15 different publication outlets. These range from journals that focus on logistics to others that research general information systems topics. Consequently, revealing that this area of research has reached other areas than the initial application of the technology. Blockchain has reached other areas than the initial intention of the technology, further revealed by the growth of research articles regarding this topic. However, one could argue that 15 articles on a three-year period are not a large sample. The extant research is limited but given the growth of articles in the last three years, as well as, the attention this technology has gotten, it seems that there will be more research conducted within this field in the future.

The decentralized approach that blockchain attains relies on the distributed network that validates the consensus for transactional records and prevents anyone for tampering with it [16]. Increased efficiency, cost-reductions and greater control of shared information are key arguments in moving from a centralized structure to a peer-to-peer network that blockchain enables [11]. Author in [21] argue that customers have become more empowered, and they demand more information about the products and services they purchase. After reviewing the literature, we have observed that transparency occurs with the integration of blockchain. Author in [16] found in their article that blockchain will make products such as pharmaceutical drugs to be safer for the end consumer. Justified by that the traceability of transactional records, as well as the tracing mechanisms, will show if the drugs are counterfeit, which can lead to catastrophically consequences for the end-consumer.

Furthermore, the imbalances caused by asymmetric information within supply chains will no longer be a problem. The flow of information upstream and downstream in the supply chains are validated, mitigating the risk of acting out of self-interest and unethical behavior [19]. When the information that is received is validated and correct, it will lead to less time spent on unnecessary activities- Furthermore, activities that were time-consuming before could now be executed in a matter of seconds. As transparency is achieved, common supply chain management challenges can be detected earlier and move more to a demand chain rather than a supply chain. Real-time data and information about all entities in the supply chain make it easier to act proactively for the market needs [20]. Bottlenecks will be discovered early, and the issue of overstock can be mitigated. Overall leading to more efficient decision-making and more time spent on core competencies [18]. The decentralized structure of blockchain together with permanent transaction records brings an increased net of security for supply chains that integrates blockchain, which is what articles within the trust domain found. The information that is available is immutable and verified, leading to a more transparent supply chain. Real-time data makes everyone in the supply chain able to extract the data they need and the possibility to make decisions based on that.

A common theme that the literature emphasizes is that contrary to before, the use of intermediaries will not be a necessity. Author in [16] argue that the presence of intermediaries may monopolize significant transactions, therefore if they are eliminated there are significant reductions in transactions costs. Furthermore, the dependency and vulnerability of third-parties are mitigated [14]. Leading to benefits of being independent of intermediaries and remain in power. The findings reveal that this is the prominent reason in consideration of leveraging blockchain technology. Literature shows that not only does the transaction costs shrink or disappear, but the supply chain also becomes less dependent on actors that exist outside of the supply chain. It is important to mention that blockchain does not add value purely by itself. It is a complementary attachment to existing activities and adds value when integrated. When integrated, it creates an opportunity to address existing challenges and problems to achieve improvements such as transparency, security, and cost-reductions [11]. With fewer intermediaries, transactions costs that were obligatory before becomes eradicated. The streamlining nature of blockchain mitigates the risk of asymmetric information leading to a more agile approach to future events. The veracity of data has been found in literature as an essential aspect of blockchain integration. As mathematical algorithms verify the information flow, it becomes impossible to provide false information to partners. Smart contracts introduce a new way of contracts, the automatic execution of agreements will contribute to lower transactions cost as well as eradicating time-consuming tasks [20]. Furthermore, the threat of default concerning contracts will be mitigated, given the design of smart contracts.

All in all, there are substantial possibilities in integrating blockchain. Literature has shown that the findings could generate considerable benefits for supply chains. There are opportunities that should be grasped by those that want to follow the technological developments and the market demands of the future. Literature found that one of the main reasons why companies integrate blockchain is to lower cost, increase security and improve overall performance [11, 12, 14]. These three factors summarize the benefits that are apparently the most important in blockchain integration, based on our review. There will be more time to be allocated to value-adding activities, the use of intermediaries will be eradicated, and the security will be improved.

6 Conclusion and Future Research Avenues

This paper contributes to both research and practice by providing a review of the literature that concerns the integration of blockchain in supply chains. For practice, it discloses the key benefits that supply chains can achieve. In addition, it provides insights into the possible applications of this technology. For research, this review could provide an overview of the focus of research within this domain, aiding researchers in identifying, interests, gaps, and future research avenues. However, this paper has its limitations. While the literature chosen for this paper focuses on the benefits achieved, one needs to understand the risks of blockchain. A limitation of this study lies in that there might exist more articles within this field that were not identified, which could have contributed to a broader review of this topic. Other articles could have identified different vital benefits, but given the range of the articles in this review, it seems like there is a collective agreement on the overall benefits.

During this research process, not a single article focusing on risks or disadvantages in blockchain integration was identified. Which could be a future research avenue. Furthermore, researching integration issues should be pursued. Those research areas should be researched further and can provide useful insights for those that consider blockchain as a new way of conducting business.

References

1. Brody, P. How blockchain is revolutionizing supply chain management. Digit. Mag. 2017
2. Christopher, M.: Logistics & Supply Chain Management. Pearson, UK (2016)
3. Titze, C., Searle, S., Klappich: The 2018 Top 8 Supply Chain Technology Trends You Can't Ignore. (2018). https://www.gartner.com/doc/3843663/top–supply-chain-technology. Accessed 9 Nov 2018
4. Nakamoto, S.: Bitcoin: A Peer-to-Peer Electronic Cash System (2008)
5. Blaxter, L.: How to Research. McGraw-Hill Education, UK (2010)
6. Webster, J., Watson, R.: Analyzing the past to prepare for the future: Writing a literature review. MIS Q. xiii–xxiii (2002)
7. White, G.R.: Future applications of blockchain in business and management: A delphi study. Strateg. Change 26(5), 439–451 (2017)
8. Swan, M.: Blueprint for a New Economy. O'Reilly Media Inc (2015)
9. Tian, F.: An agri-food supply chain traceability system for China based on RFID & blockchain technology. In: International Conference on Service Systems and Service Management (ICSSSM), pp. 1–6 (2016)
10. Apte, S., Petrovsky, N.: Will blockchain technology revolutionize excipient supply chain management? J. Excip. Food Chem. 7(3), 910 (2016)

11. Bocek, T., Rodrigues, B., Strasser, T., Stiller, B.: Blockchains everywhere-a use-case of blockchains in the pharma supply-chain. In: IFIP/IEEE Symposium on Integrated Network and Service, pp. 772–777 (2017)
12. Chen, S., Shi, R., Ren, Z., Yan, J., Shi, Y., Zhang, J.: A blockchain-based supply chain quality management framework. In: e-Business Engineering (ICEBE), 2017 IEEE 14th International Conference, pp. 172–176 (2017)
13. Korpela, K., Hallikas, J., Dahlberg, T.: Digital supply chain transformation toward blockchain integration. In: Proceedings of the 50th Hawaii International Conference on System Sciences (2017)
14. Nakasumi, M.: Information sharing for supply chain management based on block chain technology. In: Conference on Business Informatics (CBI), vol. 1, pp. 140–149 (2017)
15. O'Leary, D.E.: Configuring blockchain architectures for transaction information in blockchain consortiums: the case of accounting and supply chain systems. Intell. Syst. Acc. Financ. Manag. **24**(4), 138–147 (2017)
16. Alangot, B., Achuthan, K.: Trace and track: enhanced pharma supply chain infrastructure to prevent fraud. In: International Conference on Ubiquitous Communications and Network Computing, pp. 189–195 (2017)
17. Wu, H., Li, Z., King, B., Ben Miled, Z., Wassick, J., Tazelaar, J.: A distributed ledger for supply chain physical distribution visibility. Information **8**(4), 137 (2017)
18. Leng, K., Bi, Y., Jing, L., Fu, H., Nieuwenhuyse, V.: Research on agricultural supply chain system with double chain architecture based on blockchain technology. Future Gen. Comput. Syst. **86**, 641–649 (2018)
19. Kshetri, N.: Blockchain's roles in meeting key supply chain management objectives. Int. J. Inf. Manag. **39**, 80–89 (2018)
20. Jabbar, K., Bjørn, P.: Infrastructural grind: introducing blockchain technology in the shipping domain. In: Proceedings of the 2018 ACM Conference on Supporting Groupwork, pp. 297–308 (2018)
21. Fransisco, K., Swanson, D.: The supply chain has no clothes: technology adoption of blockchain for supply chain transparency. Logistics **2**(1), 2 (2018)
22. Casado-Vara, R., Prieto, J., De la Prieta, F., Corchado, J.: How blockchain improves the supply chain: case study alimentary supply chain. Procedia Comput. Sci. **134**, 393–398 (2018)
23. Lu, Q., Xu, X.: Adaptable blockchain-based systems: a case study for product traceability. IEEE Softw. **34**(6), 21–27 (2017)

An Optimal Design of Contactless Power Transfer System Applied for Electric Vehicles Using Electromagnetic Resonant Coupling

Junlong Duan$^{(\boxtimes)}$ and Weiji Wang

Department of Engineering and Design, University of Sussex,
Brighton BN1 9QT, UK
{J.Duan, W.J.Wang}@sussex.ac.uk

Abstract. Over the past decades, contactless power transfer (CPT) has been acquiring considerable attentions for investigations on wireless power transmitting (WPT) based electric vehicles (EV) charging solutions. This paper describes a geometrically improved CPT system using innovative H-shape ferrite core prototype and electromagnetically analytical methods. In order to address the key issues such as *system power transfer rating levels, maximization of system efficiency and charging distance of two coils*, the CPT prototype in this paper focuses on operating frequencies, coupling distances and electromagnetic performances. This H-shape CPT prototype has been modelled in 3D finite element method (FEM) environment, resulting in a maximum coil transmitting efficiency of 63%, an optimal system efficiency of over 40% and a maximum RMS real power of 20.39 kW on the load end, with an air gap of 30 mm. Moreover, the H-shape system with 20-mm air gap could be measured to output an RMS real power of 31.95 kW on the load of the CPT system, achieving a maximum coil transmitting efficiency and overall system efficiency of over 77% and 47%, respectively. Furthermore, from the perspective of electromagnetics, the proposed CPT coupling design in this paper tends to appear advantages on electromagnetic field performance by analyzing the generated parameters of flux linkage, flux line distributions, magnetic flux density and so on. In addition, the limitations and future works on the CPT technologies for EV have been discussed in this research paper.

Keywords: Contactless Power Transfer (CPT) · Inductive coupling · Wireless Power Transmitting (WPT) · Maxwell equations · Finite Element Method (FEM) · Ferrite cores · Electromagnetics · Electric Vehicles (EV) Charging Efficiency

1 Introduction

The fundamental concepts and principles of inductive energy transmitting have been proposed and developed by Nikola Tesla, Michael Faraday and Henry Poynting in the 19th century. Based on Maxwell's equations and formulations in 1862 and Poytning theorem in 1884 [1], Nikola Tesla highlighted the wireless power transfer (WPT) techniques by which modern contactless power transfer (CPT) technologies and

© Springer Nature Switzerland AG 2020
K. Arai et al. (Eds.): FTC 2019, AISC 1069, pp. 919–933, 2020.
https://doi.org/10.1007/978-3-030-32520-6_66

derivative research topics have been pervasively promoted and conducted especially over the past half century. With the development progress of electric vehicles (EV) and hybrid electric vehicles (HEV), the defects and limitations of plug-in charging solutions for EVs and HEVs have been pointed out, such as safety concerns, inconsistency of plug-in types by different charging stations and inflexibility dependent of weather conditions. In the meantime, the growing issues worldwide regarding global air pollutions, climate change and fossil energy reductions, the industries and research institutes have been focusing on the investigations of environmentally friendly transportation replacements and solutions. Therefore, CPT technologies tend to be more comprehensively optimistic charging options for EVs in the near future.

Numerous studies have paid attentions to the charging system efficiency, charging current issues, actual charging power to the load side by analyses and considerations of inductive charging system circuit improvements, supercapacitors as the system loads, electronic power control from the front supply end to the rear load end, and so on [2–5]. A rectangular coil based inductive coupling power transmitting system was built and evaluated in [6], which attempted to identify an optimal system configuration for the designed CPT application and obtained an efficiency of about 80% using a 2-kW inductive battery charging system. To improve inductive power transfer performance and produce higher coupling misalignment tolerance, conventional flat circular coils were adopted for the energy pickup CPT system designed in [7] and [8]. The experiments in [9] verified a coreless spiral coil IPT system using current-source-supply series-to-series (I-SS) and series-to-parallel (I-SP) compensation schemes, by which the 200-W IPT model achieved an efficiency of up to 88% excluding its inverter at a nonmagnetic coupling frequency, 100 kHz, as reported. Generally, the larger the effective coupling area is, the higher mutual inductance value is, a large circular coupler IPT system was built with a coil diameter of 700 mm and a coupling charging distance of 200 mm, which was reported to yield a transfer power of 2 kW [10, 11].

In order to achieve optimal inductive effectiveness, power transfer efficiency, real power ratings on the receiving side and electromagnetic field performance, an H-shape ferrite core coupler based CPT system illustrated in Fig. 1 has been analytically proposed in this paper and particularly implemented in a 3D finite-element method (FEM) platform.

2   System Structure

2.1   Systematic Layout

In terms of the actual performance of the electromagnetic field and the overall CPT prototype, the inductive coupling modular can play a vital role throughout the wireless energy transmitting processes due to the significant magnetic flux generations and dynamically inductive effectiveness in between the air gap and within the non-linear core materials. In order to form the flux line distributions generated by the system and facilitate the inductive coupling caused by the flux linkages, geometrically, the H-shape ferrite core is more able to utilize the total flux to produce optimal inductive coupling including electromagnetic resonant coupling at specific conditions. Besides, the high permeability and low core loss of ferrite materials could contribute to flux linkage, magnetic flux density and overall electromagnetic field performance of the CPT system [12]. Figure 1 provides an overview of the designed H-shape stationary CPT scheme configuration with four capacitance compensation topology options (SS, SP, PP, PS). In this paper, the series-to-series (SS) compensation is selectively adopted for the CPT system as shown in Fig. 2.

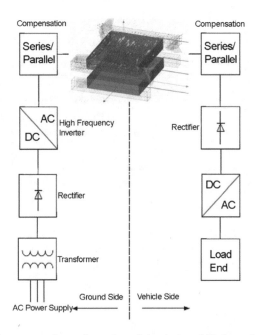

Fig. 1. The systematic configuration of the designed H-shape CPT system.

The circuit parameters of the proposed CPT system in Fig. 1 have been set up based on RLC circuit theories with initial system tests of this model and with calculations for resonant coupling conditions. The both side compensation capacitors are set with 150 nF in Fig. 2. The load is a resistor with a value of 50 Ω. The system power supply is voltage source with RMS value of 5 kV.

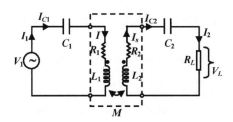

Fig. 2. The simplified equivalent system circuit with an SS compensation scheme.

2.2   Analytical Design and Modeling

Based on the characteristics of high permeability and the analytical geometry design considerations, the particularly shaped ferrite core could be able to form the magnetic flux paths mainly though the geometric directions as expected like the shape and to improve the generations of magnetic flux density and flux linkage in order for the system to produce higher performance. In this coupling modular built in Ansys 3D Maxwell environment, the Steel 1010 was employed as the ferrite core material as shown in Fig. 3.

Table 1. The geometric parameters of the designed H-shape IPT couplers.

Parameter	Value
Winding size	100 mm * 100 mm * 20 mm
Core size	150 mm * 150 mm * 20 mm
H-shaped core bar size of each side	150 mm * 15 mm * 20 mm
Primary winding number of turns	80
Secondary winding number of turns	80
Air gap of the CPT charging system	20 mm and 30 mm

Considering the benefits of skin effects reduction, proximity effects mitigation and high frequency application, Litz wire is selected for the windings for this model. A small sized magnetic coupling modular is modelled in this paper, the model design specifications are presented in Table 1.

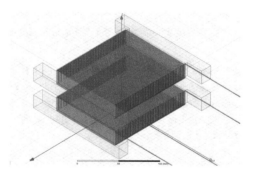

Fig. 3. The proposed H-shape couplers modelled in 3D FEM environment.

3   Electromagnetic Field and Numerical Representations

In order to analyze the inductive coupling performance and system outputs, the electric current density **J** induced on the secondary coil side of the CPT system is required to be determined. Therefore, the Ampere's law, the Gauss' law, the Faraday's law and the

B-H curve expressed by Eqs. (1), (2), (3) and (4) below, respectively, are required to be solved as constitutive equations:

$$\nabla \times \mathbf{H} = \mathbf{J} \tag{1}$$

$$\nabla \cdot \mathbf{B} = 0 \tag{2}$$

$$\nabla \times \mathbf{E} = -\frac{\partial \mathbf{B}}{\partial t} \tag{3}$$

$$\mathbf{B} = \mu \mathbf{H} \tag{4}$$

in which $\mathbf{H}$ is the magnetic field strength, $\mathbf{B}$ is the magnetic flux density, $\mathbf{E}$ is the electric field strength, and μ is the permeability non-linearly depending on local value of $\mathbf{B}$ in the B–H curve by $\mathbf{B} = f(\mathbf{H})$.

Constitutively with electric field $\mathbf{E} = f(\mathbf{J})$ by Eq. (5) and material equation of Pouillet's law, the 3D finite-element method (FEM) formulations are required to numerically solve Maxwell equations above and the relative material equations, using the methods of translating the differential equations into algebraic equations, discretizing the modelled 3D space by tetrahedrons and the methods of semi-iterative conjugate gradient for each magnetic vector potential $\mathbf{A}$ [13, 14].

$$\mathbf{E} = \rho \mathbf{J} \tag{5}$$

where ρ is the resistivity of the windings. Based on the equations above, Eq. (1) turns into:

$$\nabla \times \left(\frac{1}{\mu} \nabla \times \mathbf{A}\right) = \mathbf{J} \tag{6}$$

which is required to be numerically solved by 3D FEM computations in order to determine the magnetic field with values of magnetic flux ϕ.

To determine the self-inductance L produced by the electromagnetic field on the both sides of the CPT system, the relations between the flux linkage λ and coil current i_1 and i_2 satisfy:

$$\lambda_1 = N_1 \phi_1 = L_1 i_1 \tag{7}$$

$$\lambda_2 = N_2 \phi_2 = L_2 i_2 \tag{8}$$

$$L = \frac{\lambda}{i} \tag{9}$$

In addition, the approximate natural resonant frequency for the designed coupling system could be determined by the RLC circuit theories [15] to analyze and optimize the actual system performance by specifying operating frequencies.

To investigate the actual inductive coupling outcomes and electromagnetic field performance reflected by the numerical vectors and scalars above, the actual power generations given by Eq. (10) and the efficiencies given by Eqs. (11) and (12) from the front end to the load end of the CPT system are required to be analyzed and compared.

$$P_{RMS} = V_{RMS} \, I_{RMS} \, |\cos \varphi| \tag{10}$$

$$\eta_{coupler} = \frac{P_{RMS_{Secondary \, coil}}}{P_{RMS_{Primary \, coil}}} \tag{11}$$

$$\eta_{overal \, system} = \frac{P_{load}}{P_{RMS_{power \, supply}}} \tag{12}$$

4   Results and Analyses

4.1   The H-Shape Core CPT System with a 30-mm Air Gap

Electromagnetic Field. In terms of the actual performance of the electromagnetic field and the overall CPT, the coupling module of the system leads a significant role in the CPT prototype design and realization. As mentioned above, due to characteristics of high permeability and low core loss, the soft ferrite material deployed in the contactless power transfer model design is able to shape the magnetic flux lines enclosed through the both coils and air space in between the transmitter side and receiver side. Hence, the H-shaped ferrite core has been proposed in the paper to achieve expected optimizations regarding a CPT system performance.

Fig. 4. The electromagnetic field overlays for the designed H-shape couplers for the CPT system at the operating frequency of 8000 Hz, with an air gap of 30 mm.

Currents and Efficiencies. Theoretically, the currents tend to be maximum when the system is approximated to a status of resonance at specific operating frequencies and magnetic conditions after a basic impedance matching. Therefore, two resonant capacitors also as compensations are adopted in this CPT system, which has been used to identify the natural resonant frequencies and to optimize the real power transferred and overall efficiency produced. In this small sized CPT system scheme, the compensation capacitors C1 and C2 are 150 nF, the load is a resistor of 50 Ω, the system power supply is sinusoidal AC voltage source of RMS 5 kV and the number of turns is 80 for each Litz wire winding.

The designed modular coupler is shown in Fig. 4, illustrating an approximate magnetic resonant status the CPT prototype when the system is supplied with an operating frequency of 8000 Hz, which is specified by result analysis and calculations according to Eqs. (7), (8) and (9) in terms of actual generated RMS flux linkage and induced RMS currents from the coupling system. It can be seen from Fig. 4 that the magnetic field strength **H** and the magnetic flux density **B** could reach maximum values of 11.07 kA per meter and 0.468 T, respectively, when the system tends to be stable and magnetically coupling after about 10 periodic inductive coupling at the corresponding calculated resonant frequency.

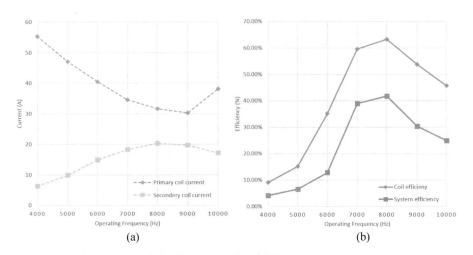

Fig. 5. Current (a), efficiency (b) versus system operating frequency for the H-shape CPT system with an air gap of 30 mm.

From Fig. 5(a) and (b) above, there a clear trend that the secondary coil current increases when the CPT model is supplied with the increasing operating frequency and the secondary coil current appears to reach its maximum value of 20.289 A at frequency of 8000 Hz. Meanwhile, it can be seen that the primary side current tends to decrease from about 55 A to a minimum value of 30.299 A from 4000 Hz to 9000 Hz. In regard to efficiency results, the overall system efficiency and the coupling coil efficiency roughly have the same trend in terms of the increase of operating frequency and both of them achieve their peak values, 41.78% and 63.23%, respectively, at 8000 Hz.

Powers in Terms of Frequency. As can be found from Fig. 6 below, the RMS real power transferred to the load gradually increases when the system operating frequency is being increased, which reaches the peak value of 20.398 kW on the load at the system operating frequency of 8000 Hz. Obviously, the secondary coil power shares a similar trend with the load power, and both the load end power and secondary coil power tend to decrease after achieving their maximum values at 8000 Hz in terms of increasing operating frequency. However, the primary coil real power drops to a bottom value of 14.835 kW at the operating frequency of 6000 Hz, after which the power produced on the primary coil turns to rise steadily up. In addition, by considering the trends of primary coil, secondary coil and load end power, it seems to be proved that the operating frequency range of 7000 to 8000 Hz is able to generate optimistic electromagnetic field phenomena which could provide the specific CPT system an optimal scope of high performance.

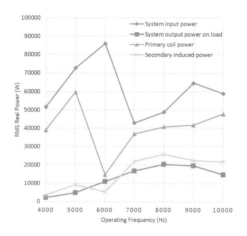

Fig. 6. Power versus system operating frequency.

Calculated Inductance and Natural Resonant Frequency. From the perspective of resonant coupling, it can be noticed, from Table 2 that both the calculated inductances of primary coil side and secondary coil side decrease dramatically when the supplied operating frequency is lower than 6000 Hz. After that, both the calculated inductances tend to be slightly changed and steadily reduced, which may mean that the calculated natural resonant frequency $f_{resonant} = \frac{1}{2\pi\sqrt{LC}}$ of both primary and secondary side has a tendency to be stable. Hence, in reality application, a possibility of being able to achieve a stable and long range of natural resonant frequency of a system could deliver a higher compatibility to the range of operating frequency settings, which means the designed CPT system seems to be much easier to achieve a satisfactory electromagnetic resonant coupling status by an acceptable long range of operating frequency power supply. Therefore, a system performance optimization for the 30-mm air gap CPT application can be realized in terms of RMS real power generation, system efficiency and operating frequency range according to results shown in Figs. 5, 6 and Table 2.

Table 2. Calculated inductance of the 30-mm air gap system.

Operating frequency (Hz)	Calculated inductance on primary coil (mH)	Calculated inductance on secondary coil (mH)
4000	22.265	17.386
5000	13.263	7.2216
6000	4.9298	5.8053
7000	5.3660	3.7830
8000	4.1205	2.9671
9000	3.7066	2.4288
10000	3.8396	1.9695

4.2   The H-Shape Core CPT System with a 20-mm Air Gap

Electromagnetic Field. It is well-known that the charging distance normally affects the inductive coupling performance in applications by flux line distributions and effects in between the intermediates. In this paper, in order to investigate the actual resulted relation between the air gap change and the overall system outputs, besides of 30-mm air gap CPT model, a 20-mm charging gap model has also been built and implemented.

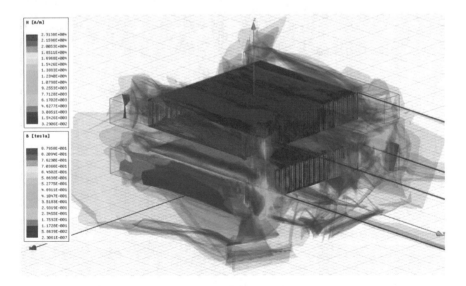

Fig. 7. The electromagnetic field overlays for the designed H-shape couplers for the CPT system at the operating frequency of 7000 Hz, with an air gap of 20 mm.

It can been noticed that the actual electromagnetic field performance has been significantly promoted in terms of the magnetic field strength **H** and the magnetic flux density **B** in Fig. 7 above. Compared with the generated values in Fig. 4 of the 30-mm air gap model, similarly when the modular magnetic field with its output waves tend to be stable after a few transient working cycles, the maximum magnetic flux density **B** has risen to a peak value of 0.87958 T at the H-shape edge sides in Fig. 7, which is much higher than 0.46838 T in the 30-mm distance model shown in Fig. 4. Moreover, the maximum value of the magnetic field strength **H** in the 20-mm distance model is also improved to 23.138 kA per meter, which is more than a double of the peak value of the previous model.

Along with the field overlay views presented in Fig. 8(a–c), it can be found that the magnetic flux density **B** in the primary core and winding seems much higher than that in Fig. 4. Furthermore, the magnetic field strength **H** values shown from Fig. 8(b) and (c) in both primary and secondary windings appear to be very strong especially at the bar sides of the H-shape cores towards each other. This effect reflects the effectiveness of using H-shape ferrite cores, manipulating a closer charging distance and approximating the calculated resonant frequency of the system towards improving the magnetic field outputs, which indirectly facilitates the electric power generations and efficiencies.

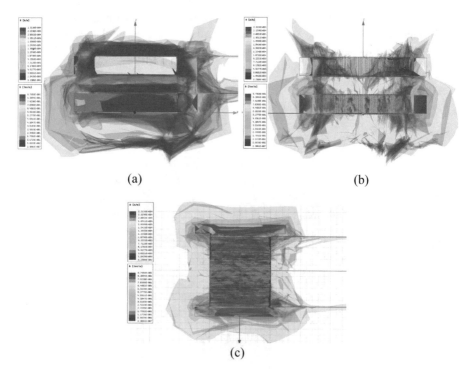

(a)   (b)

(c)

Fig. 8. The front (a), left (b) and top (c) views of the field for the designed H-shape couplers for the CPT system at the operating frequency of 7000 Hz, with an air gap of 20 mm.

Currents and Efficiencies. As shown in Fig. 9(a), the induced secondary coil current of the 20-mm distance system performs with higher overall values at each frequency point when compared to the 30-mm system. Between the frequency of 7000 to 10000 Hz, the induced current on the secondary side is mostly over 20 A, which is even higher than the peak current of 20.289 A at 8000 Hz in the previous system. Besides, the primary coil side current tends to be lower than 30 A in the 20-mm distance system after the operating frequency of 7000 Hz, which could reflect that more electric energy is utilized by magnetic coupling and transferred to the secondary coil thorough the electromagnetic field in between the 20-mm air gap.

Meanwhile, according to the resulted curves in Fig. 9(b), it could be found that the coil coupler transfer efficiency can be over 70% within a long range of operating frequency, which can a satisfactory response that the real power from the primary circuit has been utilized and optimized. As well, the overall system efficiency reaches its peak value of 47.75% at 7000 Hz and is able to be stabilized over 40% before the coupling reduction point at 10000 Hz.

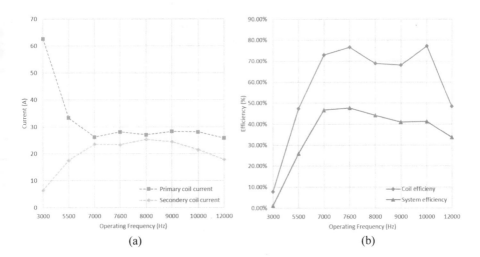

Fig. 9. Current (a), efficiency (b) versus system operating frequency for the H-shape CPT system with an air gap of 20 mm.

Powers in Terms of Frequency. It can be seen from Fig. 10 that both the secondary coil real power and system output power generations in the 20-mm distance system are very optimistic with much higher values than those of the previous 30-mm distance system. The system output power on the load reaches 31.9538 kW, which is higher than 20.398 kW in the 30-mm distance system and reflects much better power transfer performance. In reality application for EVs charging, a high real power transferred to a load could contribute to faster and more stable charging status. In addition, the system input power in Fig. 10 stays high, satisfactory and steady in terms of the increasing operating frequency within a long range since 5500 Hz, which could also reduce the front-end power supply design difficulties and function effectiveness of inverter and converter modules in real-world full power electronics system when compared with the previous 30-mm distance system.

Calculated Inductance and Natural Resonant Frequency. According to the results listed in Table 3, the calculated inductance on primary side tends to be more stable after the operating frequency point of 7000 Hz. However, the calculated inductance on secondary side continually decreases to slightly lower values compared to the primary side, which could mean that the resonant frequency of the secondary side $f_{resonant} = \frac{1}{2\pi\sqrt{LC}}$ continually increases while the resonant frequency of the primary side tends to be stable between 7000 to 7500 Hz. Comprehensively and expectedly based on Figs. 8, 9, 10 and Table 2, a quality system performance can be achieved and stayed when providing a 7000 to 7500 Hz operating frequency due to the significant magnetic coupling effects at this most approximate resonant frequency range.

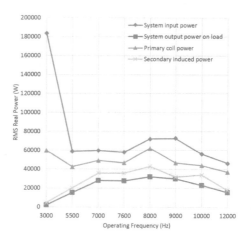

Fig. 10. Power versus system operating frequency.

Table 3. Calculated inductance of the 20-mm air gap system.

Operating frequency (Hz)	Calculated inductance on primary coil (mH)	Calculated inductance on secondary coil (mH)
3000	31.196	18.951
5500	7.3285	5.9119
7000	3.8317	2.9821
7600	3.6511	3.2778
8000	3.1277	2.9493
9000	2.8276	2.3850
10000	2.9384	2.0085
12000	3.0371	1.4537

5 Limitations and Discussion

In the 3D FEM model built in 3D Ansys Maxwell, regarding the boundary conditions, an open boundary is supposed to be established as a satisfactory approximation in order to be equivalent to real world CPT applications. However, there must be existing traces of flux escaping from the investigated model in real scenarios, for which about 15 times the coupler model has been implemented in this paper to sufficiently simulate the CPT charging cases. Besides, due to the flux line imperfection and element meshing limits of finite element methods, the boundary conditions for this designed prototype from an electromagnetic point of view could not be hundred percent flawless, which may affect the final computed results to some extent in this investigation.

In terms of materials, the air region employed in this model has a relative permeability of 1.0000004. The ferrite core material is a steel 1010 type with a nonlinear relative permeability and B–H curve. As for the losses, the core loss, eddy-current loss and iron hysteresis losses are not considered in this model due to the computationally intensive and numerical complications in the specific CPT system. Moreover, the main losses are contributed by the winding Litz wires with a copper relative permeability of 0.999991.

In addition, the misalignment tests and considerations have not been taken into account in this paper. In real application, depending on how the vehicle users park the cars, the actual lateral and longitudinal displacements would significantly affect the inductive coupling performance due to more imperfect electromagnetic field generations. In the experiments in this paper, the misalignment degree has been ideally set up.

Importantly, the cupper losses from the circuit and winding as the main energy loss in the entire CPT system are also required to be profoundly considered as a tradeoff criterion when the maximum RMS power transferred to the load is required. Besides, the actual phase differences and delays in the waves generated would be difficult to control for a very satisfactory power factor (PF) of every single module of the CPT system, which could be another subtopic and challenge towards overall CPT system enhancement purposes.

In the future investigations, all limitations mentioned above will be supposed to be addressed, along with a scaled-up larger coupler for the CPT system. It is believed that more accurate and profound results could be gained and larger air gap charging distance can be guaranteed once the coupler size is exactly suitable and compatible to be mounted on a real world vehicle chassis.

6 Conclusion

Based on the analysis on the designed H-shape core prototype proposed in this paper, the main focuses towards the maximization of system efficiency, system power transfer rating and air gaps of windings have been investigated and discussed. Two H-shape coupler systems in terms of variation of charging distances have been analytically modelled in 3D FEM and evaluated. A 30-mm distance system could produce maximum coil efficiency and overall system efficiency of over 63% and 40%, respectively, along with a maximum RMS power of 20.39 kW to the load end and when the CPT system is optimized to perform within the calculated natural resonant frequency range. Furthermore, a 20-mm distance CPT system has also been implemented and analyzed, resulting in a peak coil transmitting efficiency and system efficiency of 77% and 47%, respectively, with a maximized RMS power of 31.95 kW on the load, which shows the significant system performance improvements especially when the magnetic resonant coupling status tends to occur due to offering the system operating frequency with approximating the calculated natural resonant frequency.

The methodology on studying CPT topics in this paper has been presented from the perspective of electromagnetics. The actual electromagnetic indices and magnetic coupling parameters generated have been evaluated, which may offer some insight into the integrated system analysis methods of both electronic and magnetic field criteria towards an inductive coupling based contactless power transfer questions. In addition, further work is required in the future for practical implementation to address the main challenges on maximization of system efficiency, system power transfer rating and air gaps of coupling coils.

References

1. Duffin, W.J.: Electricity and Magnetism, 4th edn. McGraw-Hill, London (1990)
2. Fu, M., Ma, C., et al.: A cascaded boost-buck converter for high-efficiency wireless power transfer systems. IEEE Trans. Industr. Inf. **10**(3), 1972–1980 (2014)
3. Hu, A., et al.: Wireless power supply for ICP devices with hybrid supercapacitor and battery storage. IEEE J. Emerg. Sel. Top. Power Electron. **4**(1), 273–279 (2016)
4. Hiramatsu, T., et al.: Wireless charging power control for HESS through receiver side voltage control. In: 2015 IEEE Applied Power Electronics Conference and Exposition (APEC), pp. 1614–1619 (2015)
5. Mcdonough, M.: Integration of inductively coupled power transfer and hybrid energy storage system: a multiport power electronics interface for battery-powered electric vehicles. IEEE Trans. Power Electron. **30**(11), 6423–6433 (2015)
6. Sallan, J., et al.: Optimal design of ICPT systems applied to electric vehicle battery charge. IEEE Trans. Industr. Electron. **56**(6), 2140–2149 (2009)
7. Liu, X., et al.: Optimal design of a hybrid winding structure for planar contactless battery charging platform. IEEE Trans. Power Electron. **23**(1), 455–463 (2008)
8. Fernandez, C., et al.: Design issues of a core-less transformer for a contact-less application. In: APEC. Seventeenth Annual IEEE Applied Power Electronics Conference and Exposition (Cat. No. 02CH37335), vol. 1, pp. 339–345 (2002)
9. Sohn, Y., et al.: General unified analyses of two-capacitor inductive power transfer systems: equivalence of current-source SS and SP compensations. IEEE Trans. Power Electron. **30** (11), 6030–6045 (2015)
10. Raju, S., et al.: Modeling of mutual coupling between planar inductors in wireless power applications. IEEE Trans. Power Electron. **29**(1), 481–490 (2014)
11. Budhia, M., et al.: Development of a single-sided flux magnetic coupler for electric vehicle IPT charging systems. IEEE Trans. Industr. Electron. **60**(1), 318–328 (2013)
12. Duan, J., Wang, W.: Electromagnetic coupling optimization by coil design improvements for contactless power transfer of electric vehicles. In: Proceedings of the Future Technologies Conference (FTC) 2018. Advances in Intelligent Systems and Computing, vol 881, pp. 944–958. Springer, Cham (2018)

13. Sadiku, M.N.O.: Numerical Techniques in Electromagnetics. CRC Press, Boca Raton (1992)
14. Silvester, P., Ferrari, R.: Finite Elements for Electrical Engineers, 2nd edn. Cambridge University Press, Cambridge (1990)
15. Grover, F.: Inductance Calculations: Working Formulas and Tables. Dover, Mineola (1962)

PROBE: Preparing for Roads in Advance of Barriers and Errors

Mohammed Alharbi[1,2]([✉]) and Hassan A. Karimi[1]

[1] Geoinformatics Laboratory, School of Computing and Information,
University of Pittsburgh, Pittsburgh, PA 15260, USA
{m.harbi,hkarimi}@pitt.edu
[2] Taibah University, Medina, Saudi Arabia

Abstract. In this paper, we analyze the performance of sensors to find optimal navigation solutions for fully autonomous vehicles (AVs). We consider sensory and environmental uncertainties that may prevent AVs from achieving optimal navigation accuracy, which could in turn threaten the safety of road users. As the fusion of all sensors might not be able to resolve these problems, we propose a new approach called PROBE (preparing for roads in advance of barriers and errors) that consists of five algorithms: error analyzer, path finder, challenge detector, challenge analyzer, and path marker. The objective of these algorithms is to provide the highest level of navigation accuracy obtainable in navigation-challenging conditions. The final outcome of these algorithms is visualization of challenging conditions on a selected route. We experimented with PROBE using routes in different countries and the results show that PROBE performs well in detecting, analyzing, and marking navigation challenges in advance. PROBE's outcome can be used to decide on the appropriate sensors in advance of challenging conditions.

Keywords: Autonomous vehicles · Location estimation · Navigation

1 Introduction

Fully autonomous vehicles (AVs) will be on the road in the near future. By 2030, approximately 50 percent of all passenger vehicles will be available with Society of Automation Engineers (SAE) Levels 3 and 4 technologies (i.e., highly autonomous) where drivers are required to monitor and rectify driving functions of the vehicle under certain conditions [1,2]. Moreover, 15 percent of vehicles on the market will have SAE Level 5 technology (i.e., fully autonomous) where a driver is not necessary at all [1,2].

Autonomous driving is assumed to increase driver/passenger safety, efficiency, and comfort with regard to mobility experience. For example, while driving, people would be able to concentrate on other activities, such as reading [3,4]. Importantly, transport and traffic authorities envision that AVs will reduce traffic-related fatalities because over 90 percent of such deaths currently result

© Springer Nature Switzerland AG 2020
K. Arai et al. (Eds.): FTC 2019, AISC 1069, pp. 934–957, 2020.
https://doi.org/10.1007/978-3-030-32520-6_67

from human failure [3,5]. AVs also can improve the throughput of transportation systems [5]. Moreover, AVs could provide independent mobility alternatives and easier access to essential services for people who cannot drive by themselves for medical reasons or legal constraints (e.g., the absence of a driving license), thus helping to mitigate their social isolation [6].

Sensors are an essential requirement for autonomous driving. In order to monitor, analyze, and understand the vehicle's surroundings, AVs utilize multiple onboard sensors, including global positioning system (GPS), inertial measurement unit (IMU), cameras, radar, and light detection and ranging (LiDAR). The quality and functionality of these integrated technologies determine their costs, which range from extremely expensive to remarkably cheap. Thus, there is a tradeoff between the price of AVs and the quality of onboard sensors. Different tasks, such as navigation, path planning, object detection, and obstacle avoidance, involve various combinations of on-board sensors.

For AVs, navigation is one of the core functions whose aim is to pilot a car from its current location to a destination in real time on a path that maximizes safety and minimizes obstacles. The navigation process involves digital maps and sensor fusion to localize the vehicle and find a feasible trajectory [7–9]. Sensors and techniques such as GPS, cameras, LiDAR, odometry, and dead reckoning are used in the navigation process but vary with regard to accuracy, reliability, consistency, and availability. These variations occur due to two main error/noise sources. First, sensor-related limitations, which are caused mainly by the technologies chosen and are built into device specifications, create some level of noise. For example, a GPS has an error level for its range measurements, which refers to the closeness of the measurement to the true value. Sensor-related limitations also include a restricted coverage range for most sensors, including cameras, radar, and LiDAR. The second main error/noise source is situational whereby the environment drives noise for sensory readings. Common environment-related errors are caused by signal blockage and weather conditions. For instance, vision sensors, such as cameras, are sensitive to the brightness level, making this sensor unsuitable for driving at night. GPS signal also cannot pass through (line-of-sight) solid objects, so GPS receivers cannot function properly while driving through, for example, a tunnel. Currently, autonomous navigation is attracting significant research interest, and most recent research has focused on sensor fusion to perform accurate navigation [7–9]. However, although sensors can provide optimal navigation solutions in an integrated fashion, if one sensor or a combination of sensors is not working, then the potential outcome could be devastating.

In this study, we investigated the impact of sensors' functionality on AV navigation performance. We also examined whether or not, in the case of a malfunctioning subset of sensors for navigation, the remaining sensors would provide the required accuracy for reliable navigation. For instance, under dim lighting or glare conditions, can the navigation system in an AV achieve good performance without a vision sensor? For addressing such concerns, we propose a new methodology called PROBE (preparing for roads in advance of barriers and errors) that consists of five algorithms: error analyzer, path finder, challenge detector, challenge analyzer, and path marker. The PROBE methodology steps are as follows. First, we analyze sensory errors in real datasets by adapting an extended Kalman filter (EKF) to evaluate sensors individually and collectively for various scenarios. Second, we find an optimal path by using the multi-level Dijkstra's algorithm [10]. Third, we analyze the computed path to detect challenging conditions where navigation sensors might perform poorly. Fourth, we analyze the detected challenging conditions and determine the best combination of sensors that can provide the highest navigation accuracy. Fifth, we segment the path and mark each segment based on the degrees of navigation accuracy. The main contributions of this article are three of these algorithms. One algorithm checks the computed path to detect challenging conditions. The second algorithm analyzes each detected challenging condition for navigation accuracy. The third algorithm marks each segment of the computed path with predicted navigation accuracy.

This paper is organized as follows. Section 2 provides background information regarding AV navigation research, the EKF theory, and road networks. Section 3 investigates navigation challenging conditions. Section 4 presents our proposed PROBE methodology. Section 5 discusses the outcomes of sensory error analysis using an EKF and our algorithms. Section 6 presents conclusions and future research directions.

2 Background

An autonomous navigation system has two main components: perception and motion planning. The perception component allows a vehicle to understand its environment in order to perform tasks such as obstacle avoidance, lane keeping, localization, and mapping. Taking localization and mapping as examples of perception, these tasks can be performed independently with and without a mapped environment that contains landmarks to assist in the positioning estimation and a reference trajectory with no road infrastructure obstacles [7–9,11–14]. The motion planning component determines how a vehicle moves by dividing its movements into discrete motions that satisfy certain constraints. Motion planning includes two modules: lane changing and path planning. Lane-changing is complex and involves maneuvers in available lanes to meet several objectives, including following a route, merging onto a road, minimizing travel time and cost (i.e., fuel, work-related time/costs, etc.), and avoiding traffic congestion and accidents. The lane-changing module consists of three main groups. The first group considers highly autonomous vehicles [15–19] in which the driver is required to judge the situation and make a decision. The second group involves an approach without any human interaction (i.e., a stand-alone unit) [20–22]. The third group involves a cooperative lane-changing unit that makes all vehicles at a specific scene collectively perform a lane-changing maneuver [23,24]. The path planning module estimates a trajectory to a global objective and includes such factors as minimal fuel consumption, braking, turning, and travel time. The path planning module contains a global planner and a local planner. The global planner searches road network data to find the optimal path [25–33], whereas the local planner iteratively recomputes the path to avoid dynamic obstacles [34–38]. As the path planning module is responsible for finding vehicle trajectories, our work is concentrated on the global planning component (i.e., for planning in advance) and considers fully autonomous vehicles. Our objective is to find a suitable solution for navigating challenging conditions that affect the sensory information that is needed to navigate a selected route. To the best of our knowledge, the literature is devoid of studies about how to address this problem. In our work, first, we examine the sensor performance of AVs in challenging conditions. Then, we detect the challenging conditions within a selected route. Once the challenging conditions are detected, we analyze each to assess it for navigation accuracy. Finally, we mark the road segments based on navigation accuracy. To develop our algorithms, we used real data obtained from GPS, IMU, camera, and LiDAR on or in an AV that navigated different scenarios.

In order to fuse different sensory readings, we use a Kalman filter (KF), which is an estimation algorithm, taking a series of measurements with inaccurate, uncertain, and indirect readings and producing states of a system [39]. As such, a KF output is more accurate than outputs based on measurements alone. The noise in the measurement and process of the system are explicitly considered in a KF. However, the limitation of the KF is the linearity assumption in which the state transitions and measurements of a system are assumed to be linear; unfortunately, they are rarely linear in practice as most real-life systems are nonlinear. Thus, the KF has been extended (EKF) [40] to be applied to nonlinear systems. Specifically, the EKF linearizes the system model by assuming that the state transition and measurement are controlled by nonlinear functions instead of matrices. Overall, the EKF incorporates a previous system state to predict a new state. Then it compares the predicted state with the measured state and corrects the estimates by weighing the variance of the predicted state and the measured state.

A road network is a database of points (intersections) and edges (segments connecting intersections). Road network databases assist AV navigation and are used to compute optimal paths between pairs of locations. The main objectives of AVs are saving time and minimizing cost. In terms of saving time, the vehicle can optimize its travel time as well as the passengers' time. One of the major benefits of AVs is that they never give out and or need breaks. An AV could drop off passengers for an important activity and then drive to another location to pick up other passengers. Thus, a route would be optimized to save passengers' time. An AV also might optimize the travel time by using choosing a route with the least travel time and fewest number of stops. Moreover, minimizing costs is another route optimization goal. Costs include fuel, battery life, and computational resources. In order to optimize these resources, the AV must consider road infrastructures such as steep roads, which affect resource consumption.

3 Navigation Challenging Conditions

Whether driving on open highways or in close urban areas, AVs require the capability to find their trajectories rapidly, robustly, and accurately. However, navigating different environments may include situations where AVs' locating information is imperfect or unknown. These situations, referred to as "navigation challenging conditions" in this study, can confuse AVs with different locations found by each affected sensor. Such challenging conditions can be especially dangerous because the navigation sensors readings contradict each other and fail to estimate the actual position. As a result, the navigation function of the AV is impaired due to uncertain conditions, which could lead to significant negative consequences. For instance, an AV on a multilane road could collide with other vehicles if its navigation is relying on a sensor with low precision. Another critical scenario is an AV that is operating with a sensor that generates optimal readings only in a certain environment. As a result, if the AV uses the sensor readings in areas where the sensor operates poorly, the vehicle risks a potential collision.

Table 1. Sources of errors for navigation sensors

Sensor	Error type	Source
GPS	Sensory	High centimeter-level accuracy (low precision)
	Situational	Obstructions such as high-rise building, tunnel, trees
Camera	Sensory	Light sensitivity (measured by ISO[a])
		Resolution (dpi[b], or megapixels)
		Response time
		Degree-level horizontal field of view
	Situational	Variations in illumination conditions
		Weather conditions
		Reflective and transparent surfaces [41]
LiDAR	Sensory	Sparse point cloud with limited spatial resolution
		Range (m)
		Number of channels (points/sec)
		Meter-level accuracy
	Situational	Reflective and transparent surfaces [41]

[a]ISO is International Standardization Organization
[b]dpi is dots per inch

In terms of navigation sensors, the challenging conditions originate from the sensors' own imprecise readings and/or readings affected by challenging areas. Table 1 presents the two main types of error source, sensory and situational, that contribute to navigation uncertainties. These errors diminish the performance of the various types of sensors that affect AV navigation. To achieve optimal outcomes, these error sources must be minimized.

First, sensor limitations can produce noise that impacts navigation performance. For instance, GPS has an error level in range measurements and thus can only be accurate within its range. A GPS receiver with high-level accuracy leads to low precision. Another limitation for most sensors, including cameras, radar, and LiDAR, is coverage range. These sensors can sense the environment only within a certain range. In addition, a laser scan is obtainable only for a sparse set of points. Such points have an uneven distribution with a noticeable bias toward the near field of an AV. Therefore, the sensors collect more points for close objects than the ones located further away, which increases the possibility of errors. To improve AV navigation performance, this type of error must be considered and mitigated.

Second, situational conditions also can potentially manipulate sensory readings. Examples of the noise that the environment can generate include signal blockage and weather conditions that affect GPS signal and most sensors, respectively. For example, driving through a tunnel prevents GPS receivers from functioning properly because GPS signal cannot pass through solid objects. Variations in illumination conditions can also negatively affect visual sensors. Color and grayscale cameras are known to be sensitive to the brightness level, making them unreliable sources of information for driving at night. Due to the amount of available light during the day, a camera can produce pictures at high ISO levels without noise. At night, however, noise and dim lighting are major issues for a moving camera. To avoid malfunctioning AVs, these situational errors must be alleviated.

Table 2. Challenging conditions for navigation sensors

Challenge type	Condition	Impact	Influenced sensor
Road infrastructure	Steep inclines	Sensor deficiency	– LiDAR
			– Camera
	Tunnels		– GPS
Time	Night-time dim lighting or glare	Sensor deficiency	– Camera
	Daytime bright		
Climate	Fog	Sensor deficiency	– Camera
	Rain		– LiDAR
	Snow		
Environment	Urban area	Sensor deficiency	– GPS
	Rural areas		

Several different challenging conditions can impair sensor readings that degrade AV navigation. Table 2 shows that these conditions can be categorized into different types. First, road infrastructure, such as steep inclines, can affect the sensors' ability to recognize surroundings. The light conditions while driving also can be a challenge that compromises sensor performance. For instance, dim lighting, glare from headlights, or other lights at night as well as bright sunlight or surface glare during daylight can significantly increase camera malfunctions, potentially causing AVs to make poor decisions. Another challenge is climate conditions that cause vehicles' sensors to generate noise. Specifically, foggy or rainy weather can impede camera visibility and confuse LiDAR readings, which then diminishes AV performance. Finally, urban areas with high-rise buildings and dense traffic create challenges that affect GPS signal. Similarly, rural areas with dense vegetation or wooded areas, especially on curves, may obstruct GPS signal. When AVs incorporate the affected GPS readings in their decision-making component, they are more likely to make inaccurate choices.

In this study, we focus only on two navigation challenging conditions: tunnels and night-time. The dominant sensors for navigation under these conditions are GPS and camera. Tunnels create challenges for GPS by blocking signal, and night-time creates challenges for cameras by not obtaining clear images. Furthermore, we chose to investigate only these two conditions due to the availability of real data for experimentation. For that, we only investigate tunnel and night-time challenges.

4   PROBE

The PROBE methodology consists of an error analyzer that is based on an EKF, aimed to assess the performance of navigation sensors in terms of collected datasets, a path finder, a challenge detector, a challenge analyzer, and a path marker. Details of the proposed methodology are described in the following sections.

4.1   Error Analyzer

EKFs have been utilized to fuse navigation sensors and then assess the sensors' performance collectively and individually. Using an EKF, the aim is to determine which combinations of sensors are able to provide the highest level of accuracy. The error analyzer is comprised of three procedures: localization based on odometry and dead reckoning techniques, optimal estimation using an EKF and evaluation of the estimates.

Localization. To navigate autonomously, an AV needs to recognize its pose (position and orientation). The techniques needed to perform localization are divided into absolute positioning and relative positioning. Absolute positioning acquires the absolute position from external sources, including beacons and satellite signals (e.g., from GPS). Relative positioning relies on information gleaned by onboard sensors, including gyroscopes, accelerometers, and cameras. For absolute positioning, the position of the AV is determined externally, so its accuracy is independent, and error is not cumulative with regard to travel duration. For relative localization, however the integrated data are noisy, so the error is cumulative. Different techniques can minimize such error and estimate the relative position of AVs. Commonly used relative localization techniques are odometry and dead reckoning.

Odometry is used to estimate the change in the position of an AV over time using motion sensors. For this purpose, we utilized odometry for localizing AVs using visual and LiDAR sensors. ORB-SLAM2 was adopted for stereo visual odometry as it exhibits outstanding performance using a KITTI dataset [42]. For LiDAR odometry, we used the Laser Odometry and Mapping (LOAM) method [14], which ranks high in the KITTI benchmark. LOAM ranks as the third best method for odometry and the best method specifically for LiDAR odometry. Both odometry algorithms, i.e., ORB-SLAM2 and LOAM, can localize an object within 6 degrees of freedom (DoFs). In this study, we concentrated on 2-DoF, so the z-axis value and rotation angles are ignored for simplicity.

Dead reckoning is used to compute the dynamic object location after a time Δt based on its previous location and taking into consideration object features. We applied a dead reckoning algorithm to gyroscope and accelerometer measurements, which are included in IMUs. We modeled the object state using an RPB model, where each letter indicates a feature: rotating (R), constant position rate (P), and body axis coordinates (B) [43]. Given the heading direction (ψ) and velocity (v), the RPB formulas are as follows:

$$l_x(\Delta t) = l_x^0 + v_t \Delta t \sin(\theta) \tag{1a}$$

$$l_y(\Delta t) = l_y^0 + v_t \Delta t \cos(\theta) \tag{1b}$$

$$\theta = \psi_0 - \psi_t \tag{1c}$$

where, l_x and l_y indicate the current positions on Cartesian coordinates; l_x^0 and l_y^0 denote the previous positions on the Cartesian coordinates; v_t is the heading velocity at this time; and Δt is the time increment for a dead reckoning step.

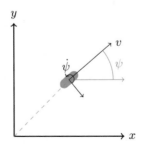

Fig. 1. Constant turn rate and velocity model.

Extended Kalman Filter. Given the localization information, an EKF is adopted for optimal estimation. Due to noisy data, the absolute position, odometry, and dead reckoning outcomes may not correlate. The EKF fuses these resultant positions in order to estimate the closest position to the actual one. We specify the AV state as the vehicle position (x, y), heading direction (ψ), forward velocity (v), and yaw rate sensor $(\dot{\psi})$ (see Fig. 1). The following vector state (x_t) represents the state at time t.

$$x_t = \begin{bmatrix} x & y & \psi & v & \dot{\psi} \end{bmatrix}^T \tag{2}$$

To describe the AV's motion, we investigated different motion models to improve the accuracy and robustness of the estimation. Each motion model is designed based on a certain assumption. The commonly used linear motion models have an assumption of a constant velocity (CV) or a constant acceleration (CA). Even though these models offer linearity, they do not take rotation angles into account. Also, Schubert et al. [44] found that linear models potentially are unable to generalize data, which results in a large positioning error. More complex models include the constant turn rate and velocity (CTRV) and constant turn rate and acceleration (CTRA) models that consider the heading direction (ψ) angle, which is the rotation around the z-axis. The CTRV model is less complicated than the CTRA model and provides reasonable estimations. On this ground, we adopted the CTRV model (see Fig. 1) for our study, so the state at step $(x_{t+\Delta t})$ can be computed using the following nonlinear state transition [44]:

$$x_{t+\Delta t} = \begin{bmatrix} x + \frac{v}{\dot{\psi}}\left(-\sin(\psi) + \sin\left(\Delta t \dot{\psi} + \psi\right)\right) \\ y + \frac{v}{\dot{\psi}}\left(\cos(\psi) - \cos\left(\Delta t \dot{\psi} + \psi\right)\right) \\ \Delta t \dot{\psi} + \psi \\ v \\ \dot{\psi} \end{bmatrix} \tag{3}$$

The state transition or update matrix (F) is computed by the Jacobian matrix of the dynamic matrix (Eq. 3) with respect to the state vector (Eq. 2). This matrix (Eq. 4) must be computed at every filter step because it consists of state variables.

$$F = \begin{bmatrix} 1 & 0 & \frac{v(-\cos(\psi)+\cos(T\dot{\psi}+\psi))}{\dot{\psi}} & \frac{-\sin(\psi)+\sin(T\dot{\psi}+\psi)}{\dot{\psi}} & \frac{Tv\cos(T\dot{\psi}+\psi)}{\dot{\psi}} - \frac{v(-\sin(\psi)+\sin(T\dot{\psi}+\psi))}{\dot{\psi}^2} \\ 0 & 1 & \frac{v(-\sin(\psi)+\sin(T\dot{\psi}+\psi))}{\dot{\psi}} & \frac{\cos(\psi)-\cos(T\dot{\psi}+\psi)}{\dot{\psi}} & \frac{Tv\sin(T\dot{\psi}+\psi)}{\dot{\psi}} - \frac{v(\cos(\psi)-\cos(T\dot{\psi}+\psi))}{\dot{\psi}^2} \\ 0 & 0 & 1 & 0 & T \\ 0 & 0 & 0 & 1 & 0 \\ 0 & 0 & 0 & 0 & 1 \end{bmatrix} \tag{4}$$

The remaining EKF configurations are as follows.
Uncertainty covariance matrix of state (P):

$$P = diag(1000, 1000, 1000, 1000, 1000) \tag{5}$$

Process noise covariance matrix (Q):

$$Q = diag\left(\sigma_x^2, \sigma_y^2, \sigma_\psi^2, \sigma_v^2, \sigma_{\dot{\psi}}^2\right) \tag{6}$$

Measurement vector (z):

$$z = \begin{bmatrix} x_{s1} & y_{s1} & v & \dot{\psi} & x_{s2} & y_{s2} & x_{s3} & y_{s3} & x_{s4} & y_{s4} \end{bmatrix}^T \tag{7}$$

where s1, ..., s4 are the navigation sensors.

Extraction measurement matrix (H):

$$H = \begin{bmatrix} 1 & 0 & 0 & 0 & 0 \\ 0 & 1 & 0 & 0 & 0 \\ 0 & 0 & 0 & 1 & 0 \\ 0 & 0 & 0 & 0 & 1 \\ 1 & 0 & 0 & 0 & 0 \\ 0 & 1 & 0 & 0 & 0 \\ 1 & 0 & 0 & 0 & 0 \\ 0 & 1 & 0 & 0 & 0 \\ 1 & 0 & 0 & 0 & 0 \\ 0 & 1 & 0 & 0 & 0 \end{bmatrix} \tag{8}$$

Measurement noise covariance matrix (R):

$$R = diag\left(\sigma^2_{x_{s1}}, \sigma^2_{y_{s1}}, \sigma^2_v, \sigma^2_{\dot\psi}, \sigma^2_{x_{s2}}, \sigma^2_{y_{s2}}, \sigma^2_{x_{s3}}, \sigma^2_{y_{s3}}, \sigma^2_{x_{s4}}, \sigma^2_{y_{s4}} \right) \tag{9}$$

where noise strength σ^2_i denotes the sensory noise variances that are determined experimentally or computed from the ground truth. We computed noise strength from the ground truth of selected sequences as

$$\sigma^2 = \frac{1}{2} \sum_{\tau \in T} (x^\tau_{gt} - x^\tau_s)^2 \tag{10}$$

where x^τ_{gt} denotes the reference location at time-stamp τ and x^τ_s denotes the sensory or estimated location at the same time τ.

Absolute Trajectory Error. To assess the EKF outcomes, we evaluated the global consistency of an estimated trajectory using the absolute trajectory error (ATE) method with respect to the transition [45]. The ATE finds the difference in absolute distance between the ground truth and estimated trajectories. As both trajectories can have different coordinate frames, we rotated and translated points in one frame to match their corresponding points in the other. This work can be achieved in closed form using the Umeyama alignment method [46]. Given two trajectories ($\mathcal{P}^{gt}$, $\mathcal{P}^{est}$), the Umeyama algorithm finds the rigid-body transformation B that minimizes the residual sum of squares between a point in the ground truth trajectory $B \times \mathcal{P}^{gt}_j$ and its corresponding point in the estimated trajectory $\mathcal{P}^{est}_j$. Given the aligned trajectories, the ATE at step j can be computed as

$$\mathcal{E}_j = (\mathcal{P}^{gt}_j)^{-1} B \mathcal{P}^{est}_j = \mathcal{P}^{est}_j \ominus \mathcal{P}^{gt}_j \in \mathbb{R}^{1 \times 2} \tag{11}$$

Algorithm 1: Challenge Detector Algorithm (Tunnel)

Input: Path data (P), current location (l), current time (t)
Output: updated path data (P), set of challenging locations (D)

$D \leftarrow []$; // initialize the challenging areas with an empty set
// tunnel checker
for *each road segment e in P* **do**
    $osmds \leftarrow$ query OSM to acquire all POIs within e;
    for *each d in osmds* **do**
        if *(d is a tunnel)* **then**
            append(D, [$d.id$, "Tunnel", $d.geometry$]);
            split road segment e where $d.geometry$ is and update P;
        end
    end
end
return P, D;

Then, the root mean squared error (RMSE) is evaluated for all steps as

$$RMSE(\mathcal{E}_{1:n}) = \sqrt{\frac{1}{n}\sum_{j=1}^{n}||\mathcal{E}_j||^2} \tag{12}$$

4.2   Path Finder

The purpose of the path finder algorithm is to find the route between two points. In this study, we adopted the Open Source Routing Machine (OSRM) system to solve the fastest path problem [47]. The assigned goal of the OSRM is to minimize the travel time. The OSRM provides the optimal path according to a certain configuration and relies on the multi-level Dijkstra pipeline [10]. We sought to find the lowest cost path P between source s and destination d on road network G. We configured the OSRM to compute the path based on the least travel time as an objective function by taking into account the length of each road segment $|e_i|$, and the speed limit v_{e_i} on each road segment, and the time for stops and turns. Giving source s and destination d, this algorithm computes the fastest route between s and d by using the travel time on each road segments.

4.3   Challenge Detector

After finding the fastest path, we used the navigation challenge detector algorithms, shown below as Algorithms 1 and 2 for tunnels and night-time, respectively. The purpose of these algorithms is to search for navigation challenges on the computed path so that we can analyze the sensors' performance with regard to each challenging condition and choose the sensors that collectively or individually provide superior accuracy. As discussed in Sect. 3, this step is significant

Algorithm 2: Challenge Detector Algorithm (Night-time)

Input: Path data (P), current location (l), current time (t)
Output: updated path data (P), set of challenging locations (D)

$D \leftarrow []$;
$e \leftarrow$ map the current location l to the nearest road segment $e \in P$;
$\alpha \leftarrow$ navigating rate in seconds;
$\beta \leftarrow$ night-time hours period;
while *navigating* **do**

    if *(hour(t) within β)* **then**

        split road segment e where l is located (e_1, e_2) and update P;
        $e \leftarrow e_2$;
        append(D, [$e.id$, "Night-time", $e.geometry$]);

    end

    meters_per_second $\leftarrow$ convert $e.speed$ to meter per second, if needed;
    $l \leftarrow l + \alpha \times$ meters_per_second;
    $e \leftarrow$ map the current location l to the nearest road segment $e \in P$;
    increment t by α;

end

return P, D;

---

because sensors cannot always perform well on their own and in all environments. In designing the detector, we focused on two navigation challenges: tunnels and night-time driving conditions. A tunnel is a challenge for autonomous navigation. For this, Algorithm 1 searches the road network (e.g., OpenStreetMap or OSM) database for the presence of tunnels on the segments of the selected route and splits the road segment where a tunnel is sited. This means that tunnels are identified as separate road segments. Another challenge is night-time conditions. Algorithm 2 navigates the computed path virtually to recognize this challenge. It starts from the current location and time and continues incrementally via displacement that is computed from a navigating rate of seconds and the road segment speed. The night-time period is stated in terms of hour, whereby if the algorithm reaches a point where the AV will presumably reach this point within that time, it would split the edge into two segments and mark the segment that includes this point as night-time. The algorithm terminates when it completes navigating the whole path.

4.4   Challenge Analyzer

The purpose of the challenge analyzer is to decide which combination of sensors should be considered and the accuracy of that combination. The analyzer takes the outcomes of the navigation challenge detector and the EKF as input. A route that is detected to have challenges is analyzed by using the EKF to compute position estimations for all possible combinations of sensors; with four sensors (GPS, dead reckoning, camera, and LiDAR), there are 15 combinations. The position estimation of each combination is recorded, and it is expected that

---

Algorithm 3: Challenge Analyzer Algorithm

Input: Path data (P), set of challenging locations (D)
Output: updated path data (P)

for *each road segment e in P* **do**
   if *(e.geometry in D* **then**
      challenge ← get *d.challenge* form *D* where
      (*d.geometry == e.geometry*);
      e.inf_sensor ← *e.inf_sensor* − sensor_not_working(*challenge*);
   end
end
for *each e in P* **do**
   combin ← get sensor combinations that do not have *e.inf_sensor*;
   e.accuracy ← min(*get_performance(combin)*);
end

return P;

---

the highest accuracy is achieved when all four sensors are considered and the environment has no, or least, effect on sensors performance. For each navigation challenging condition, we eliminate the affected sensors for computing position estimation while driving through the condition. The challenge analyzer makes a distinction, by taking into account the combination of sensors, between those segments which are not challenging and those which are challenging on a route. Then, for each segment, challenging or not, we use the EKF results to decide the highest accuracy level or least RMSE. The logic of the challenge analyzer is detailed in Algorithm 3. An example situation is when the challenge detector has detected a tunnel on a route. In this case, Algorithm 3 finds the combination of sensors that provides the highest position estimation (accuracy) while passing through the tunnel.

4.5   Path Marker

The final step is to mark the route that corresponds to the detected challenges. The path marker aims to mark and compress the resultant data obtained from the navigation challenge decoder and analyzer. The path marker provides a simple yet thorough representation of the navigation-challenging conditions on a route. This representation will assist AVs in perceiving the risk of sensory uncertainties as well as areas of challenging conditions and then making the optimal navigating decision. Specifically, AVs can avoid areas where their sensors are expected to yield poor accuracy. This is accomplished by marking these navigation challenges on the computed route and assisting AVs in avoiding those road segments with poor localization accuracy. The path marker functions are as follows:

1. Acquire the updated path data P from the navigation challenge analyzer.
2. Acquire the set of challenging locations D.

3. Acquire the road network G from OSM.
4. Plot the road network G.
5. Generate a color map of road segment accuracy.
6. Plot the road segments with the designated color.

5   Experimentation

We tested the PROBE algorithms using a simulated environment. We designed all trials to search for navigation challenges addressed by our detector algorithm and performed tests using routes in different countries. For the purpose of evaluating our algorithms, we utilized two datasets: OSM for road network data and KITTI for AV data. In this section, details about our experiments are given, including dataset selection, sensor error analysis using an EKF, and overall PROBE algorithms experiments.

Table 3. LiDAR performance for position estimation

DATASET	RMSE	MIN	MAX
04	6.4050	1.6873	15.1343
06	40.9363	5.5591	84.4246
09	74.9452	1.8311	154.3155
10	27.0849	1.7619	67.2701

5.1   Datasets

We adopted OSM for acquiring the road network data for this work. OSM is a collaborative project that provides a rich source of global road network infrastructure data [48] and is currently the most commonly used world mapping software that allows free access without restrictions. Using these data, we found the fastest paths for our experiments and where tunnels are located.

In addition, our experiments relied on the KITTI vision benchmark dataset [49]. These data were collected outdoors from a moving vehicle in various urban areas in Karlsruhe, Germany. The dataset provides stereo images together with Velodyne laser scans (i.e., GPS and IMU data) in an inertial navigation system. It provides a large amount of sensory data with a highly accurate reference trajectory for several tasks. The data include realistic conditions such as shadows, glare, specular reflections, changes in camera gain, etc. which confuse the positioning estimation. The data are registered and synchronized at 10 Hz. Specifically, we used the odometry dataset sequences 04, 06, 09, and 10. Each sequence represents a certain scenario. Sequence 04 provides a scenario where an AV drives in a straight multilane highway with trees and their shadows in some areas. Sequence 06 has various road plans and environments, including single-lane roads with dense trees on both sides, bidirectional traffic, curves, and residential areas with parked cars. Sequence 09 has single-lane roads and U-turn features. Sequence 10 is in a rural area with single-lane and narrow roads, dense trees and tight curves.

5.2   Error Analysis

We began our tests by evaluating sensor performance. Table 3 and Fig. 2 present the sensor fusion results for the four sequences (04, 06, 09, and 10) obtained from the KITTI dataset. In each of the four sequences, we fused the sensory and estimated positioning data obtained from GPS, dead reckoning, LiDAR, and a stereo camera into an EKF in order to estimate the actual position. In scenario 04, the GPS, LiDAR, and dead reckoning data exhibited the best performance collectively, whereas the camera yielded moderate error. Moreover, sensor fusion of all the sensors resulted in performance that was close to optimal performance. In terms of individual performance, the GPS was best. It is worthwhile to note that fusing dead reckoning data always improved the accuracy. In scenario 06, the best accuracy was achieved by the GPS and camera. The use of the camera

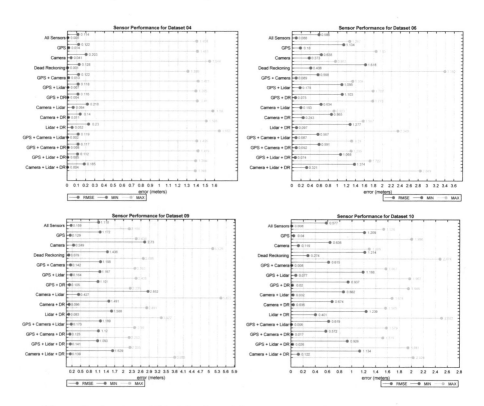

Fig. 2. Evaluation of EKF for four different datasets for sensor performance.

had a significant impact on improving accuracy. In scenario 09, we relied on the GPS and dead reckoning for the best accuracy and avoided using the camera and LiDAR. In scenario 10, the combination of GPS, camera, and LiDAR achieved optimal accuracy. Most sensors were unable to provide reasonable accuracy on their own.

Overall, it became noticeable that LiDAR did not fit into the navigation context if there were no additional sensors to correct its readings. Table 3 shows the LiDAR performance for all four scenarios. A minimum of around 6 meters of RMSE was achieved in scenario 04. This error decreased to as low as 0.11 if the sensor was fused with another. Moreover, a subset of sensors could exhibit different performance for different environments or situations. As shown in Fig. 2, no single subset of sensors was reliable. Furthermore, fusing more sensors might not necessarily result in optimal performance. That is, fusing all available sensors comes close to optimal performance, but none is superior. However, a stand-alone sensor may provide reasonable functionality. Figure 3 through Fig. 6 present the four different time series of the ATE for each of the four scenarios. It is worth noting that some sensors, such as the camera in scenarios 06 and 10, exhibited outstanding performance compared to the others. This finding confirms our claim that certain sensors have shortcomings because of their specifications or the surrounding environment (Figs. 3, 4, 5, and 6).

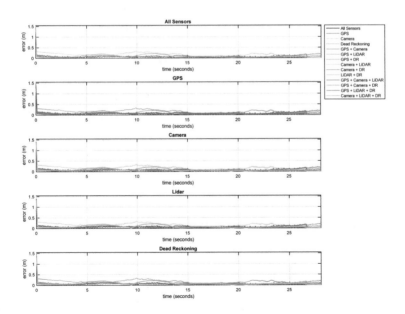

Fig. 3. Time series: ATE (sequence 04).

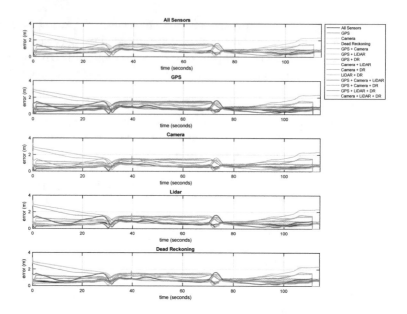

Fig. 4. Time series: ATE (sequence 06).

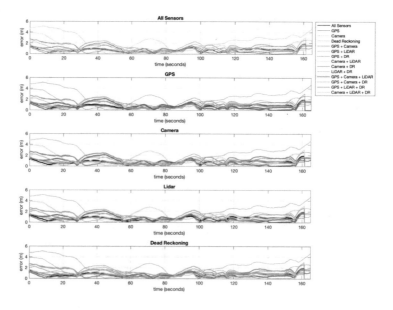

Fig. 5. Time series: ATE (sequence 09).

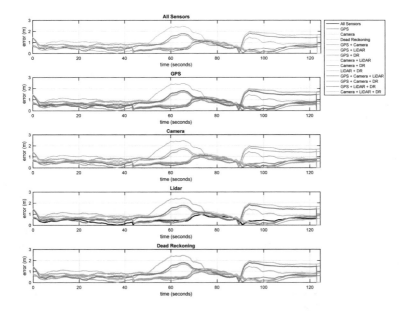

Fig. 6. Time series: ATE (sequence 10).

5.3 PROBE Algorithms

After analyzing sensor error, we ran the remaining PROBE algorithms, i.e., the path finder, challenge detector, challenge analyzer, and path marker, to assess their functionalities. For the path finder, we took advantage of the OSRM, a routing engine that employs the underlying OSM data to compile a road network graph with weights as a travel time for finding the fastest route. The OSRM was installed in our machine and configured to produce a route that has the least travel time for each (origin, destination) pair. For each journey, the challenge detector took the computed route and searched for navigation challenges (i.e., tunnels and night-time). Each challenge is independent from the other and perceived by a separate algorithm. The tunnel and night-time challenge detectors were run in parallel. For detecting tunnels and their corresponding geometry, we used OSM road network data. Given a computed path P, we utilized OSM to search each e_i on this path to collect points of interest (POIs) that fall on e_i. We treated the POIs as metadata of road segments. For the night-time conditions, we virtually navigated the route and recognized night-time if the navigation time was between 5 p.m. and 5 a.m. The challenge detectors passed the geometries of each challenging condition and their types to the challenge analyzer. We then applied the challenge analyzer to determine the highest level of accuracy of the sensors for these challenges. The sensors' accuracy data were acquired by the EKF. Then, we marked the route with these challenges and their accuracy.

Currently, to the best of the authors' knowledge, there are no comparable approaches for evaluating our approach. For this, we tested the PROBE algorithms using routes in different countries, including the United States, the United Kingdom, France, and Norway, with different time settings. Figure 7 through Fig. 10 present the results for each of the countries, respectively, where A denotes day-time, B indicates a time before sunset, and C implies night-time. The colors on these maps show the amount of error in meters. The purple color labels the base lines where all the sensors are available with normal performance and the other colors represent the highest accuracy that can be achieved when the affected sensor is absent. As shown in the Figure As, while driving in the morning, the only challenge that might be experienced is the tunnel that blocks the GPS signal. Our algorithms successfully marked the tunnel in red according to the best accuracy that could be obtained from the remaining subsets of sensors. In this case, AVs might consider alternative routes if possible. Moreover, while AVs traveled before sunset (shown in Figure Bs), they reached some road segments at which the camera was assumed not to be working properly after 5 p.m. Our algorithms marked the night-time challenges with cyan based on sensor accuracy. In this case, the AVs do not have an option to avoid night-time, but they could consider roads that offer adequate lighting or better illumination. Lastly, night-time had the highest potential risk, as AVs encountered difficulties in understanding and processing the whole route, seen in Figure Cs. Therefore, switching to an alternative might not be helpful if the current road has insufficient lighting. For all cases, using the PROBE algorithms, the AVs would be capable of making accurate decisions and, thus, the effects of these navigation-challenging conditions are manageable (Figs. 7, 8, 9, and 10).

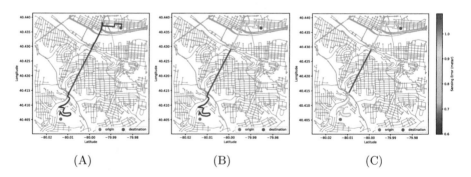

| (A) | (B) | (C) |

Fig. 7. Visualization of a path generated for a trip in the United States: the AV drives from an origin (bottom point) to a destination (upper point) at different times: (A) daytime, (B) a few minutes before sunset, and (C) night-time.

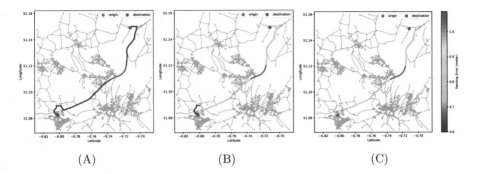

(A)   (B)   (C)

Fig. 8. Visualization of a path generated for a trip in the United Kingdom: the AV drives from an origin (bottom point) to a destination (upper point) at different times: (A) daytime, (B) a few minutes before sunset, and (C) night-time.

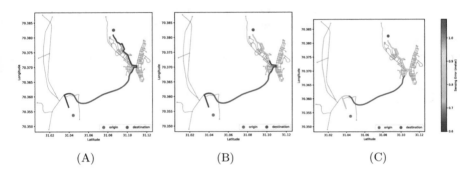

(A)   (B)   (C)

Fig. 9. Visualization of a path generated for a trip in Norway: the AV drives from an origin (bottom point) to a destination (upper point) at different times: (A) daytime, (B) a few minutes before sunset, and (C) night-time.

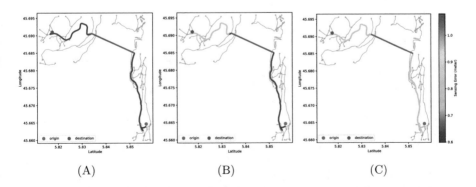

(A)   (B)   (C)

Fig. 10. Visualization of a path generated for a trip in France: the AV drives from an origin (bottom point) to a destination (upper point) at different times: (A) daytime, (B) a few minutes before sunset, and (C) night-time.

6   Conclusions and Research Directions

Navigation sensor performance is unreliable and inconsistent under all circumstances. In this paper, we demonstrate that inconsistency is problematic at best, and its seriousness is directly related to navigation-challenging conditions. To mitigate these circumstances, we introduce our PROBE approach that considers sensor failure at certain locations and for certain situations. The PROBE algorithms help AVs to predict and plan for navigation uncertainties in challenging conditions. The outcomes of the algorithms can be used to adjust the appropriate sensors in advance of challenging conditions.

In future work, we intend to extend the PROBE algorithms to take into account more challenges, as the current work considers only tunnels and nighttime conditions. To improve the algorithms, we need to consider other navigation challenges, such as high-rise buildings, trees and vegetation, glare, weather conditions, etc. Moreover, the algorithms presented in this paper do not attempt to solve routing problems in terms of sensor failure. A future research direction might be routing optimization with two objectives: safety and cost. This paper focused on navigation sensors, including GPS, IMU, camera, and LiDAR. Further work will include radar sensors. Furthermore, additional sensors will contribute to this topic of research because each sensor has its own deficiencies that differ from the shortcomings of other sensors. Another possible research direction is to design a vehicle-to-vehicle framework that can predict and perceive challenging locations/conditions.

References

1. Gao, P., Kaas, H.-W., Mohr, D., Wee, D.: Automotive revolution-perspective towards 2030. How the convergence of disruptive technology-driven trends could transform the auto industry. Technical report, Advanced Industries, McKinsey & Company (2016)
2. National Highway Traffic Safety Administration. Federal automated vehicles policy: accelerating the next revolution in roadway safety. US Department of Transportation (2016). https://www.nhtsa.gov/sites/nhtsa.dot.gov/files/federal_automated_vehicles_policy.pdf
3. Fraedrich, E., Beiker, S., Lenz, B.: Transition pathways to fully automated driving and its implications for the sociotechnical system of automobility. Eur. J. Futures Res. 3(1), 11 (2015). https://doi.org/10.1007/s40309-015-0067-8. ISSN: 2195-2248
4. Silberg, G., Wallace, R., Matuszak, G., Plessers, J., Brower, C., Subramanian, D.: Self-driving cars: the next revolution. White paper, KPMG LLP & Center of Automotive Research, p. 36 (2012)
5. Fagnant, D.J., Kockelman, K.: Preparing a nation for autonomous vehicles: opportunities, barriers and policy recommendations. Transp. Res. Part A Policy Pract. 77, 167-181 (2015). https://doi.org/10.1016/.tra.2015.04.003. http://www.sciencedirect.com/science/article/pii/S0965856415000804, ISSN: 0965-8564
6. Anderson, J.M., Nidhi, K., Stanley, K.D., Sorensen, P., Samaras, C., Oluwatola, O.A.: Autonomous Vehicle Technology: A Guide for Policymakers. Rand Corporation (2014). ISBN: 0833084372

7. Häne, C., et al.: 3D visual perception for self-driving cars using a multi-camera system: calibration, mapping, localization, and obstacle detection. Image Vis. Comput. **68**, 14–27 (2017). ISSN: 0262-8856

8. Li, J., et al.: Real-time self-driving car navigation and obstacle avoidance using mobile 3D laser scanner and GNSS. Multimedia Tools Appl. **76**(21), 23017–23039 (2017). ISSN: 1380-7501

9. Sukkarieh, S., Nebot, E.M., Durrant-Whyte, H.F.: A high integrity IMU/GPS navigation loop for autonomous land vehicle applications. IEEE Trans. Robot. Autom. **15**(3), 572–578 (1999). ISSN: 1042-296X

10. Delling, D., Goldberg, A.V., Pajor, T., Werneck, R.F.: Customizable route planning in road networks. Transp. Sci. **51**(2), 566–591 (2015)

11. Kim, H., Liu, B., Goh, C.Y., Lee, S., Myung, H.: Robust vehicle localization using entropy-weighted particle filter-based data fusion of vertical and road intensity information for a large scale urban area. IEEE Robot. Autom. Lett. **2**(3), 1518–1524 (2017). ISSN: 2377-3766

12. Ort, T., Paull, L., Rus, D.: Autonomous vehicle navigation in rural environments without detailed prior maps. In: 2018 IEEE International Conference on Robotics and Automation (ICRA), pp. 2040–2047. IEEE (2018)

13. Paton, M., Pomerleau, F., MacTavish, K., Ostafew, C.J., Barfoot, T.D.: Expanding the limits of vision-based localization for long-term route-following autonomy. J. Field Robot. **34**(1), 98–122 (2017). ISSN: 1556-4959

14. Zhang, J., Singh, S.: LOAM: lidar odometry and mapping in real-time. In: Robotics: Science and Systems, vol. 2, p. 9 (2014)

15. Díaz Alonso, J., Ros Vidal, E., Rotter, A., Muhlenberg, M.: Lane-change decision aid system based on motion-driven vehicle tracking. IEEE Trans. Veh. Technol. **57**(5), 2736–2746, 2008. https://doi.org/10.1109/TVT.2008.917220. http://ieeexplore.ieee.org/document/4439300/, ISSN: 0018-9545

16. Habenicht, S., Winner, H., Bone, S., Sasse, F., Korzenietz, P.: A maneuver-based lane change assistance system. In: Intelligent Vehicles Symposium (IV), 2011, pp. 375–380. IEEE (2011). ISBN: 1457708914

17. Hou, Y., Edara, P., Sun, C.: Modeling mandatory lane changing using bayes classifier and decision trees. IEEE Trans. Intell. Transp. Syst. **15**(2), 647–655 (2014). ISSN: 1524-9050

18. Hou, Y., Edara, P., Sun, C.: Situation assessment and decision making for lane change assistance using ensemble learning methods. Expert Syst. Appl. **42**(8), 3875–3882 (2015). ISSN: 0957-4174

19. Zheng, J., Suzuki, K., Fujita, M.: Predicting driver's lane-changing decisions using a neural network model. Simul. Model. Pract. Theory **42**, 73–83 (2014). https://doi.org/10.1016/j.simpat.2013.12.007. ISSN: 1569190X

20. Chen, C., Seff, A., Kornhauser, A., Xiao, J.: DeepDriving: Learning affordance for direct perception in autonomous driving. In: Proceedings of the IEEE International Conference on Computer Vision, pp. 2722–2730. IEEE (2015). https://doi.org/10.1109/ICCV.2015.312. ISBN: 9781467383912

21. Hatipoglu, C., Ozguner, U., Redmill, K.A.: Automated lane change controller design. IEEE Trans. Intell. Transp. Syst. **4**(1), 13–22 (2003). ISSN: 1524-9050

22. Ulbrich, S., Maurer, M.: Towards tactical lane change behavior planning for automated vehicles. In: 2015 IEEE 18th International Conference on Intelligent Transportation Systems (ITSC), pp. 989–995. IEEE (2015). ISBN: 1467365963

23. Nie, J., Zhang, J., Ding, W., Wan, X., Chen, X., Ran, B.: Decentralized cooperative lane-changing decision-making for connected autonomous vehicles. IEEE Access **4**, 9413–9420 (2016). https://doi.org/10.1109/ACCESS.2017.2649567. ISSN: 21693536

24. Wang, M., Hoogendoorn, S.P., Daamen, W., van Arem, B., Happee, R.: Game theoretic approach for predictive lane-changing and car-following control. Transp. Res. Part C Emerg. Technol. **58**, 73–92 (2015). https://doi.org/10.1016/j.trc.2015.07.009, http://www.sciencedirect.com/science/article/pii/S0968090X15002491, ISSN 0968-090X

25. Bhattacharya, P., Gavrilova, M.L.: Voronoi diagram in optimal path planning. In: 4th International Symposium on Voronoi Diagrams in Science and Engineering (ISVD 2007) (ISVD), pp. 38–47 (2007). https://doi.org/10.1109/ISVD.2007.43. ISBN: 0769528694

26. Canny, J.: A new algebraic method for robot motion planning and real geometry. In: 28th Annual Symposium on Foundations of Computer Science (SFCS 1987), pp. 39–48. IEEE (1987). https://doi.org/10.1109/SFCS.1987.1. ISBN: 0-8186-0807-2

27. Dechter, R., Pearl, J.: Generalized best-first search strategies and the optimality of a*. J. ACM, **32**(3), 505–536 (1985). https://doi.org/10.1145/3828.3830. http://doi.acm.org/10.1145/3828.3830, ISSN: 0004-5411

28. Gelperin, David: On the optimality of a*. Artif. Intell. **8**(1), 69–76 (1977). https://doi.org/10.1016/0004-3702(77)90005-4. ISSN: 0004-3702

29. Lavalle, S.M.: Rapidly-exploring random trees: a new tool for path planning. Technical report, Iowa State University (1998)

30. Petricek, T., Svoboda, T.: Point cloud registration from local feature correspondences-evaluation on challenging datasets. PLoS ONE **12**(11), e0187943 (2017). https://doi.org/10.1371/journal.pone.0187943. http://www.ncbi.nlm.nih.gov/pmc/articles/PMC5685596/, ISSN: 1932-6203

31. Seda, M.: Roadmap methods vs. cell decomposition in robot motion planning. In: Proceedings of the 6th WSEAS International Conference on Signal Processing, Robotics and Automation, ISPRA 2007, pp. 127–132. World Scientific and Engineering Academy and Society (WSEAS), Stevens Point, Wisconsin, USA (2007). http://dl.acm.org/citation.cfm?id=1355681.1355703. ISBN: 978-960-8457-59-1

32. Vadakkepat, P., Tan, K.C., Ming-Liang, W.: Evolutionary artificial potential fields and their application in real time robot path planning. In: Proceedings of the 2000 Congress on Evolutionary Computation, CEC00 (Cat. No.00TH8512), vol. 1, pp. 256–263. IEEE, July 2000. https://doi.org/10.1109/CEC.2000.870304

33. Yang, S.X., Luo, C.: A neural network approach to complete coverage path planning. IEEE Trans. Syst. Man Cybern. **34**(1), 718–724 (2004). https://doi.org/10.1109/TSMCB.2003.811769. ISSN 1083-4419

34. Gim, S., Adouane, L., Lee, S., Dérutin, J.P.: Clothoids composition method for smooth path generation of car-like vehicle navigation. J. Intell. Robot. Syst. Theory Appl. **88**(1), 129–146 (2017). https://doi.org/10.1007/s10846-017-0531-8. ISSN: 15730409

35. Piazzi, A., Bianco, C.G.L., Bertozzi, M., Fascioli, A., Broggi, A.: Quintic g_2-splines for the iterative steering of vision-based autonomous vehicles. IEEE Trans. Intell. Transp. Syst. **3**(1), 27–36 (2002). https://doi.org/10.1109/6979.994793. ISSN: 15249050

36. Rastelli, J.P., Lattarulo, R., Nashashibi, F.: Dynamic trajectory generation using continuous-curvature algorithms for door to door assistance vehicles. In: 2014 IEEE Intelligent Vehicles Symposium Proceedings, pp. 510–515. IEEE, June 2014. https://doi.org/10.1109/IVS.2014.6856526

37. Reeds, J.A., Shepp, L.A.: Optimal paths for a car that goes both forwards and backwards. Pacific J. Math. **145**(2), 367–393 (1990). http://projecteuclid.org/euclid. pjm/1102645450, ISSN: 0030-8730

38. Rosmann, C., Feiten, W., Wosch, T., Hoffmann, F., Bertram, T.: Efficient trajectory optimization using a sparse model. In: 2013 European Conference on Mobile Robots, pp. 138–143. IEEE (2013). https://doi.org/10.1109/ECMR.2013.6698833. ISBN: 9781479902637

39. Kalman, R.E.: A new approach to linear filtering and prediction problems. J. Basic Eng. **82**, 35–45 (1960). https://doi.org/10.1115/1.3662552. ISSN: 00219223

40. Smith, G.L., Schmidt, S.F., McGee, L.A.: Application of statistical filter theory to the optimal estimation of position and velocity on board a circumlunar vehicle. National Aeronautics and Space Administration (1962)

41. Janai, J., Güney, F., Behl, A., Geiger, A.: Computer vision for autonomous vehicles: problems, datasets and state-of-the-art. CoRR, abs/1704.05519: 67 (2017). http:// arxiv.org/abs/1704.05519

42. Mur-Artal, R., Tardós, J.D.: ORB-SLAM2: an open-source slam system for monocular, stereo, and RGB-D cameras. IEEE Trans. Robot. **33**(5), 1255–1262 (2017). ISSN: 1552-3098

43. IEEE Computer Society: IEEE standard for distributed interactive simulation - application protocols. In: IEEE Std 1278.1-1995, p. i (1996). https://doi.org/10. 1109/IEEESTD.1996.80831

44. Schubert, R., Richter, E., Wanielik, G.: Comparison and evaluation of advanced motion models for vehicle tracking. In: 2008 11th International Conference on Information Fusion, pp. 1–6. IEEE (2008). ISBN: 3800730928

45. Sturm, J., Engelhard, N., Endres, F., Burgard, W., Cremers, D.: A benchmark for the evaluation of RGB-D slam systems. In: 2012 IEEE/RSJ International Conference on Intelligent Robots and Systems (IROS), pp. 573–580. IEEE (2012). ISBN: 1467317365

46. Umeyama, S.: Least-squares estimation of transformation parameters between two point patterns. IEEE Trans. Pattern Anal. Mach. Intell. **13**, 376–380 (1991). ISSN: 0162-8828

47. Luxen, D., Vetter, C.: Real-time routing with openstreetmap data. In: Proceedings of the 19th ACM SIGSPATIAL International Conference on Advances in Geographic Information Systems, GIS 2011, pp. 513– 516, ACM, New York (2011). https://doi.org/10.1145/2093973.2094062, http://doi.acm.org/10.1145/2093973.2094062, ISBN: 978-1-4503-1031-4

48. OpenStreetMap contributors. Planet dump (2017). https://planet.osm.org, https://www.openstreetmap.org

49. Geiger, A., Lenz, P., Stiller, C., Urtasun, R.: Vision meets robotics: the kitti dataset. Int. J. Robot. Res. **32**, 1231–1237 (2013). https://doi.org/10.1177/ 0278364913491297. ISSN: 02783649

Industrial Cloud Automation for Interconnected Factories

Application to Remote Control of Water Stations

Lamine Chalal[(✉)], Allal Saadane, and Ahmed Rhiat

Icam of Lille, 6 Rue Auber, BP 10079, 59016 Lille Cedex, France
{lamine.chalal,allal.saadane,ahmed.rhiat}@icam.fr

Abstract. In the next industrial revolution, called Industry 4.0, manufacturers are focusing on client-specific production and added-value products. For this purpose, factories need new key automation technologies with small extra engineering effort. To make factories smarter, a lot of researches in the literature deal with the digital technologies potential such as Cloud computing and communication networks. That said, there is still lack of concrete applications. This paper is aimed to present our work on implementing and evaluating Proficloud as a new technology suitable to connect distributed factories for remote control and management purpose. For concrete application, we have designed a drinking water distribution system emulator as demonstrator. Proficloud was chosen for its ability to interconnect plants and facilities over long distances. This is done by combining the internet backbone and Profinet, the well-known industrial protocol. In this work, a literature review of modern remote control as well as a methodology to deploy Proficloud technology is presented. The main objective is to give a first evaluation of Proficloud performances based on experimental results, with a focus on latency issues.

Keywords: Automation · Industry 4.0 · Remote and distributed control · Supervisory control · Teleoperation · Profinet · Water station

1 Introduction

Advanced digital technologies are initiating a new industrial era, which will entirely change the way machines operate and communicate with each other and with human operators. This is called the industry of the future (France), industry 4.0 (Germany) or smart manufacturing (USA). It will transform isolated or remote factories to globally connected installations [1]. Thanks to Internet Of Things (IOT), physical devices, sensors, robots, drives and so would be able to communicate each other via networks. Some studies forecast more than 25 billion connected objects (IOT) by 2025 [2].

Industrial processes, data collection (i.e. from IOT) by a centralized system can participate to improve production management, optimize operations, prevent breakdowns and increase the life-cycle of machines. To do this, the cloud can be used to collect data, process it and act accordingly. Indeed, Cloud Computing technologies are maturing and the cost of cloud services is decreasing. Some technologies can extend local communication protocol to the cloud. This is the case of Proficloud, which can

© Springer Nature Switzerland AG 2020
K. Arai et al. (Eds.): FTC 2019, AISC 1069, pp. 958–972, 2020.
https://doi.org/10.1007/978-3-030-32520-6_68

connect directly local high-speed industrial networks (such as PROFINET) throw Internet [3]. As part of INCASE[1] project (Industry 4.0 via Networked Control Applications and Sustainable Engineering), a large interregional interconnected Proficloud pilot is developed, tested and evaluated.

In order to design a pilot as a practical application, we have chosen to emulate drinking water plants, given their distributed nature. The pilot includes remote pump and purification stations. It also includes an automated water treatment thanks to a gantry crane and logistics. These parts are distributed in a Franco-Belgian cross-border region. Each partner has developed its local demonstrator and the connection of the whole system is done by the ProfiCloud technology. Involved partners are Icam, Univ. Lille and KUL (Gent).

In this paper, we focus on local demonstrator developed at Icam of Lille (France) as part of the interregional pilot [4].

The dispersion of installations throughout the territory implies specific provisions, in particular inter-site communication. ICT innovations regularly influence the technological choices of remote control applications. Nowadays, communication for remote control and monitoring of water stations are mainly provided by PSTN (Public Switched Telephone Network). We can also find GSM data, GPRS, radio, satellite, ADSL, Power line communication etc. as shown in Fig. 1 [5].

PLC (Programmable Logic Controller) manufacturers provide adapted solutions to all types of pumping stations. The common trend is that all these systems are increasingly communicating both with each other and with central supervision system.

Figure 1 depicts components and general configuration of a SCADA (Supervisory Control And Data Acquisition) system. The control center hosts Master Terminal Unit (MTU) and communication routers. The MTU allows remote control of autonomous installations, which are dispersed over a large geographical area. It also includes the

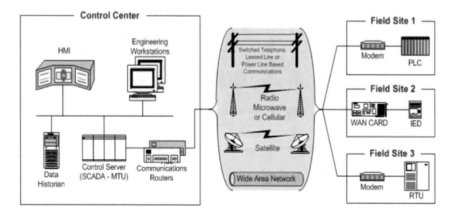

Fig. 1. SCADA system general layout [6]

[1] This work has received funding from the European Union's INTERREG V (2014-2020) program under grant agreement 2S01-049 (https://www.incase2seas.eu).

Human Machine Interface (HMI), which allows engineers to collect data from remote processes and generate command actions. The field site called RTU (Remote Terminal Unit) allows local automation and remote control.

Author in [7] describe some systems, which can be of varying degree of complexity depending in the process and the specific implementation. An essential feature of real-time remote control systems is that reliability depends not only in the accuracy of the processing performed, but also on the latency of data communication.

Figure 2 shows the CIM concept (Computer-integrated manufacturing) which describes data flow in a factory. We can notice that as the pyramid is narrowing, the amount of information to transmit is increasing. At the top of the pyramid, the amount of information is important, but the data transfer rate is lower and therefore real-time constraints are less important.

The main aim of this paper is to present the Proficloud technology and evaluate it for remote control of industrial installations. The application is to control and monitor decentralized pump stations, which uses Profinet backbone as a communication network.

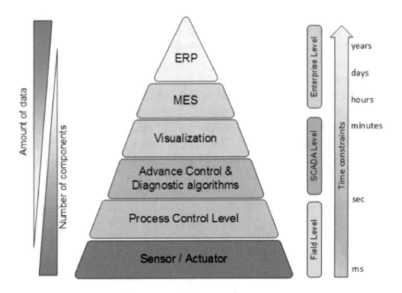

Fig. 2. Automation pyramid [8]

2 State of the Art of Communication Technologies

2.1 Standard Communication Technologies

Mainly, there are two means of communication between MTU and various RTUs, namely wired and wireless communications. Wireless communications have some advantages compared to wired technologies, for example, low deployment costs. However, depending on the nature of the transmission mode, the signal can be attenuated for long distances. On the other hand, wired solutions have few interference problems [5].

Two levels of communication are used in the case of water supply systems. The first flow concerns data communication with local RTU. Indeed, in case of complex telemetry applications (difficult access, hazardous locations, etc.), battery-powered wireless sensors can be used. For example, see [9], transmission is carried out by 2.4 GHz radio waves over short distances (1 km). As suggested in [10], communications can also be accomplished through other wireless communications, such as Zig-Bee, 6LowPAN, Z-wave, and so on.

For the second flow of information, between the different RTUs and the MTU, cellular or Internet technologies can be used. However, some limiting factors must be taken into account when deploying them, such as operational costs, the rural/urban environment, etc. Table 1 presents an overview of communication technologies.

Existing cellular networks can be a good option for communication between RTU and MTU. This avoids operating costs and additional time to set up a dedicated communication infrastructure.

In some critical applications, it is necessary to have continuity of communications service. Indeed, cellular network services are shared with customers, which can lead to network congestion and a general network slowdown. In this case, the optical fiber or private telephone line can be used.

Table 1. Example of communication technologies

Technology	Spectrum	Data rate	Distance	Application
GPRS	900–1800 MHz	14.4 Kbps	1–10 km	Home area network
Bluetooth	2.4 GHz	721 Kbps	1–100 m	Home area network
3G Cellular	1.6–2.5 GHz	2 Mbps	1–10 km	Field area network
PLC	1.8–86.0 MHz	1.8–86.0 MHz	1–3 km	Home area network
Radio	2.4 GHz	50 kbps–100 kbps	1 km	Home area network
LORA	150 MHz–1 GHz	0.3–50 Kb/s	<30 km	Field area network

2.2 Supervision Based on Cloud Computing

Another way to make these systems more communicative is to use the Cloud. To reduce costs and improve efficiency, web-based software services, more commonly called "Cloud Computing" are used in almost all types of companies.

Since systems often require new technological resources to optimize their operations, cloud-computing offers water utilities a viable solution without significant investment [11].

According to NIST (National Institute of Standards and Technology), Cloud-computing is a model for enabling ubiquitous, convenient, on-demand network access to a shared pool of configurable computing resources (e.g., networks, servers, storage, applications, and services) that can be rapidly provisioned and released with minimal management effort or service provider interaction [12]. These Cloud-computing services have several major advantages:

- Quick to deploy;
- Very low investment;
- Experts maintain the system so that utilities can focus on their core competencies;
- Software updates automatically implemented;
- Elasticity or rapid expansion to meet changing utility needs;
- Subscription rates offer a pay-per-use option.

In this way, Xylem has developed an architecture where intelligence is integrated into the pumps and connected directly to the Cloud without going through PLCs [13]. The manufacturer Crouzet develops small PLCs connected to Ethernet and which can regularly upload data logs [14]. The Siemens RTU 3030C offers the same data transmission capabilities [15]. Some current solutions available on the market for urban, industrial and agricultural water management are listed below:

- IBM Intelligent Water [16]
- EcoStructure for Water and Wastewater [16]
- Siemens – Smart Water Platform [11]
- TakaDu [17].

Recording, digitizing and linking the physical world to the virtual one are the core of Industry 4.0. For this purpose, PLCs should be deployed to the cloud by extending a variety of Fieldbus protocols. This is why alliances are emerging between different PLC manufacturers and cloud specialists. For example, the alliance between Beckhoff and Microsoft, which will enable optimal integration of automation systems into information systems, particularly the Cloud [18].

Nowadays, there is a wide range of commercial brands available in the industrial market that can be used to implement Cloud technologies; these include Siemens, Allen Bradley, Mitsubishi, Omron, Schneider, GE Fanuc, Beckhoff, Moeller, Hitachi, ABB, Phoenix Contact, and Bosch- Rexroth.

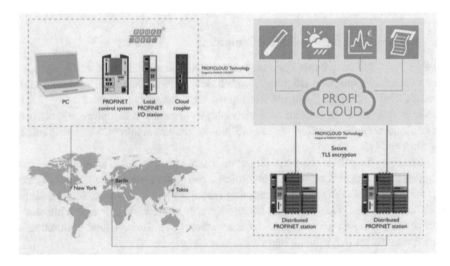

Fig. 3. Overview of the Proficloud technology [3]

In this work, the ProfiCloud technology was chosen. This technology is aiming to interconnect smart factories and facilities over long distance as shown in Fig. 3.

From Fig. 3, we can see with Proficloud, data can be exchanged worldwide in secure way (Transport Layer Security encryption). Indeed, Proficloud can be at the end, implemented by using public internet providers without additional infrastructure investment.

Furthermore, Proficloud webservers and third-party applications can be easily connected and integrated into the system.

However, to use Proficloud and Proficloud webservices cloud credit are required (paid service). Another aspect to be considered is internet network congestion or a possible loss of connection. Indeed, it can be problematic for remote control. It should be noted that proficloud is a proprietary protocol developed by Phoenix Contat.

3   Proficloud Technology

Proficloud technology, as shown in Fig. 3, allows Profinet users to use modern cloud computing services and to develop centralized supervision applications for remote facilities.

Indeed, by using this technology, the local Profinet network can support cloud functions. With Proficloud, users can import valuable data from Profinet systems to cloud servers in order to optimize management, to process data, to perform predictive maintenance, etc. Proficloud implementation has at least one cloud coupler and one cloud device.

In Fig. 3, we can see how Profinet devices are connected to Proficloud. The cloud coupler has two Ethernet ports, one for the local network connection (Profinet) and one for connecting to the Internet. The Coupler automatically sets up a connection to the Proficloud so that it is operational after a short period. This is accomplished by pre-configuring (UUID identification) the cloud devices on the Proficloud web service www.proficloud.net. Connecting devices to the Internet presents a risk for unauthorized access. Proficloud considers this by only allowing outbound traffic and Data transmission is encrypted with TLS technology, which is also used, for example, in banking solutions [3].

A cloud device is a Profinet PLC that can be connected to an open internet connexion (DMZ). It receives an IP address by means of DHCP. Then the network can be configured by using PC Worx software. Right clicking the PROFINET icon allows easily scanning the network for reachable devices. As shown in Fig. 4, Proficloud components are seen as Profinet virtual device on the network.

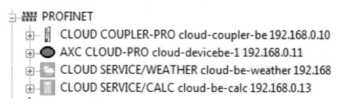

Fig. 4. Bus structure under Pc Worx

Once the network is well configured and prepared, the PLC programmer can develop control algorithms on PCworx environment. The remote PLC is seen as a remote Profinet I/O device. Via exchange tables, the local PLC and the remote one can communicates via the Proficloud.

The main features of ProfiCloud technology as illustrated in Fig. 5 are:

- Smart, reliable and secure data exchange between machines;
- Worldwide networking of manufacturing systems;
- Integration of 3rd party applications;
- Design and running of web-based services and applications.

An example of a service provided by the Proficloud is the weather service. This is possible by introducing the longitude and latitude of the location to be monitored. Each device, real or virtual (such as weather services) consumes credits.

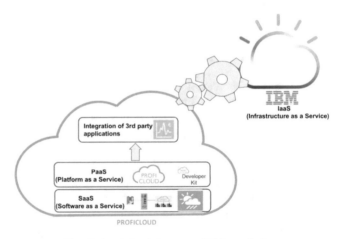

Fig. 5. Proficloud Internet of Things platform

4   Implementation of Industrial Cloud for Supervision of Water

4.1   System Design and Experimental Setup

Icam demonstrator represents a local site of production (pumping), storage and distribution of water drinking. The following diagram illustrates its functional configuration, which includes a pumping source, a storage area and the distribution network to the users (Fig. 6).

In order to emulate such system at Icam, we use three tanks setup representing all parts of our water station system (Fig. 7). The experimental hydraulic setup is based on:

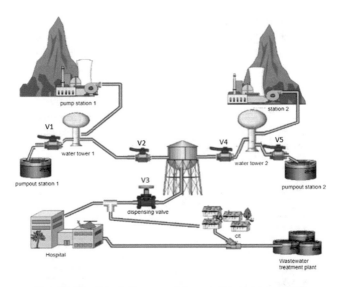

Fig. 6. Functional diagram of the emulated water system

- The three tanks emulating local storages reservoir (left and right tanks) and distribution reservoir (middle tank). Each tank include level sensor.
- Lower reservoir emulating underground water and consumption/leakage.
- Two pumps feeding two tanks (left and right).
- Two valves (V2 and V4) supplying the distribution reservoir (middle tank);
- A proportional valve emulating consumer demands (valve V3).
- Two valves for draining use (V1 and V5).
- The control system is built on:
  - AXC Cloud Pro (industrial PLC from Phoenix Contact).
  - I/O devices from Phoenix Contact.

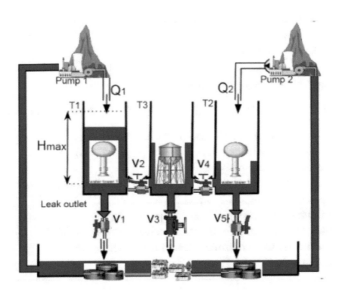

Fig. 7. Proficloud application at Icam: physical water system

In order to control this system, a controller is implemented on AXC 1050 PLC at remote side. This PLC is programmed to send commands via an analogue card throw the Proficloud coupler to both pumps (output device in Fig. 8). At the local side (the physical emulator), the servo-amplifier thus send the supply voltage to DC motors, which drives the pumps. A relay system has been added and wired in the control cabinet in order to be able to control the electro valves of the three-tank system in both directions of rotation. This allows simulating leaks, clogs, increased or decreased water consumption. The relay commands are carried out in On/Off mode with a 24 V DC digital card (On/off device in Fig. 8).

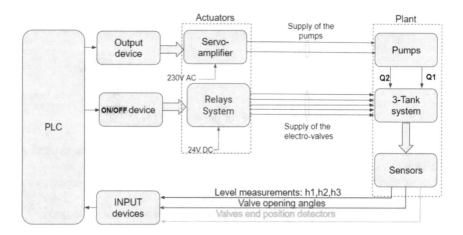

Fig. 8. Control system architecture

The test bed is also instrumented in order to have all relevant information to optimize control system. There are three types of measurements:

- Tank level measurements: h1, h2 and h3.
- Measurement of valve opening angles.
- Detection of the valves limits.

4.2   Proficloud Network and Features

Figures 9 and 10 show the topology of Icam ProfiCloud demonstrator. It is based on:

- Experimental setup described in the previous section.
- IP camera allowing real time video streaming of the setup.
- A remote control system composed of:
  - AXC 1050 (industrial PLC) running the remote controller program;
  - A computer including PC Worx and Visu+ softwares for programming AX1050 and designing of a HMI,
  - A Cloud Coupler device (from Phoenix Contact) allowing the connection between the remote controller (running on AXC 1050) and the experimental setup using AXC Cloud Pro.

For local test and evaluation of our platform, we installed the remote control system and the experimental setup in 2 separate rooms, each one connected using a different network and both connected to the cloud by Proficloud technology. In this way, the local Profinet networks are extended into a larger one using public network (Internet).

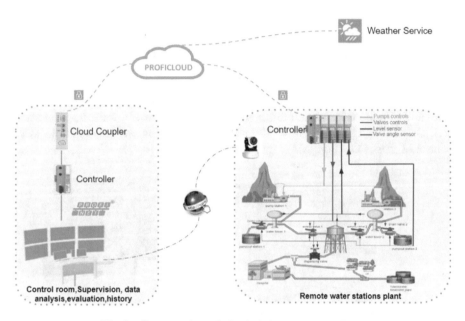

Fig. 9. Shows a view of physical demonstrator at Icam

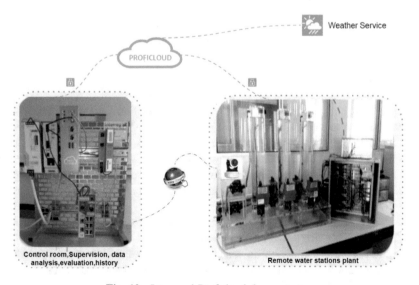

Fig. 10. Icam real Proficloud demonstrator

4.3   Automation and Supervision System

We consider TANK 3 as the distribution to the customer. Valve 3 (Valve3) thus represents customer demand. It is possible to vary this request from zero to 100%. Depending on the difference between the measured level and the desired level, the flow rate of the pumps (pump1, pump2) will be more or less important. Note that it is possible to optimize the system's response time by adjusting the gain. Valves 2 and 4 play the role of the distribution valves. They are opened automatically when the desired level is higher than the actual level and close again when the set value is reached. Valves 1 and 5 are not affected by the regulation. These are drain valves that can be operated manually. They can be used for maintenance operations. An HMI interface is developed for remote use of the platform (Fig. 11) It offers two features: on the one hand, it allows the visualization of the process data (tank levels, water demand status, etc.). On the other hand, it allows the real time interaction with the control devices (control of valves and pumps).

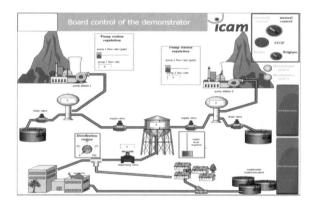

Fig. 11. View of the developed HMI

5   Evaluation

In order to test and validate the Proficloud implementation for remote water station management, we placed the remote control in a room; the coupler is connected to a public internet address and the process in another room where the Cloud Pro is also connected to another public address. The connection via the Proficloud was successful.

The experimentation tests are carried from the main screen of SCADA application. Figure 12 shows the level set point for the distributed tank (Red line). The blue line indicate the measured level.

As can be seen, the set point is well respected between 0 and 12 min approximately. During this time, the consumption valve is 10% open. At $t = 12$ min the consumption valve is 75% open. In order to reduce the gap between the set point and the measurements a proportional controller has been implemented. Thus the system reacts to keep the deviation to a minimum by starting the pump. At $t = 42$ min, the

Measured level (Cm)

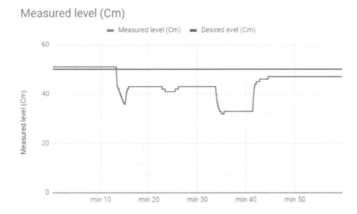

Fig. 12. Remote control analysis: proof of concept

opening of the consumption valve is reduced from 100% to 10%). It can be seen that the gap between the set point and the measurement is narrowing. In order to evaluate the delay times induced by the network, we performed several tests by controlling pump 1. Maximum of the control signal on pump 1 (100% of its rated power) and observed the water level in the tank. As soon as the water level is stabilized, a new control to the pump is sent. The visu+ software is used to send remote commands via Proficloud and to collect the measurements as shown in Fig. 13.

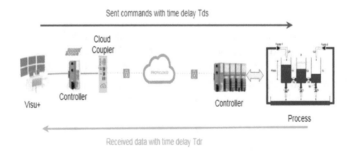

Fig. 13. Remote control of three-tank system using the Proficloud

Figure 14 shows the time delay duration (between 16 h:41 min:47 and 16 h:41 min:51). The time required to send order and receive the measurement is about 4 s. It is the time delay induced par Proficloud technology.

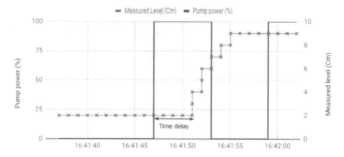

Fig. 14. Time delay during remote control via the ProfiCloud

The following table summarizes analysis of delays during local control. Table 2 shows that latency is quite different for both tests in two different days. Based on the measurements Proficloud induced, in worst case, a delay time of 3.7 s.

Table 2. Analysis of time delays in case of remote control

Time delays (Tds + Tdr)	Max	Min	Mean value	Standard deviation
Test on 12-12-2018	3.68 s	2,44 s	2.89 s	0,45
Test on 10-01-2019	3.69 s	2.04 s	2.82 s	0,55

Since the deviation is quite low, we can consider that the average of 2.8 s is representative. It is the time induced by ProfiCloud, Visu+ and PLCs as well as their I/O device to send the pump commands and receive the level measurement data. Since we control a very slow dynamic system, this latency time do not affect the stability of the system. Hence, good results obtained during the tests.

6 Conclusion

Analysis of current trends in remote control and management system has shown that Cloud Computing technology is a major challenge for intelligent remote factories. In this study, we managed to interconnect PLC with a CLOUD platform. First, we emulated pumping stations system by a physical test bench. The choice of pumping stations was motivated its distributed aspect over large areas. It is an ideal candidate to implement the Proficloud technology.

The main results of our work show the real potential of Proficloud technology for this kind of systems. Indeed Proficloud Technology is very suitable for slow dynamic system. Time delays induced by the Proficloud do not exceed 3 s (round-trip). Hence, the good results obtained during tests. However, this is not the case of rapid systems for example mobile robots. Indeed, time delays fluctuation can deteriorate system stability.

Faster connections would require what is now sometimes called the "Tactile Internet", requiring e.g. new Ethernet/internet technologies (such as TSN: Time Sensitive Networking) and 5G wireless connection.

As a future work, we plan to compare our results to other technologies and apply predictive control by using Proficloud weather service to better water management as done in reference [19].

References

1. Julien, N., Martin, E.: L'usine du futur: Stratégies et déploiement. Dunod, Mayenne (2018)
2. Manyika, J., Chui, M., Bisson, P., Woetzel, J., Dobbs, R., Bughin, J., Aharon, D.: Unlocking the potential of the Internet of Things, June 2015. https://www.mckinsey.com/business-functions/digital-mckinsey/our-insights/the-internet-of-things-the-value-of-digitizing-the-physical-world. Accessed 01 decembre 2018
3. Weßelmann, M.: Taking profinet networks into the cloud, Industrial Ethernet Book, no. 86, February 2018
4. Chalal, L., Saadane, A., Rhiat, A., Paty, T.: Contrôle à distance et supervision via Proficloud- Application à un systeme de distribution d'eau. L'EAU, L'INDUSTRIE, LES NUISANCES **421**, 57–60 (2019)
5. Gungor, V.C., Sahin, D., Kocak, T., Ergüt, S., Buccella, C., Cecati, C., Hancke, G.P.: Smart grid technologies: communication technologies and standards. IEEE Trans. Ind. Inform. 529–539 (2011)
6. Stouffer, K., Falco, J., Scarfone, K.: Guide to Industrial Control Systems (ICS) Security. Technical report, Gaithersburg, MD, United States (2015)
7. Branislav, A., Šagi, M.: Proposal of a modern SCADA system. In: Proceedings of 19th Telecommunications forum TELFOR, Serbia, Belgrade (2011)
8. Bajer, M.: Dataflow in modern industrial automation systems. Theory and practice. Int. J. Appl. Control Electr. Electron. Eng. **2**(4) (2014)
9. Schneider Electric: Battery-powered wireless sensors: Accutech BR20, February 2018. https://www.schneider-electric.be/fr/product-range-presentation/61237-accutech/#tabs-top. Accessed 12 decembre 2018
10. Wenpeng, L., Duncan, S., Sol, L.: Smart grid communication network capacity planning for power utilities. In: IEEE/PES Transmission and Distribution Conference and Exposition (T&D), New Orleans, LA, USA (2010)
11. Guey Ler, L.: Flood resilience and smart water management: implementation strategies for smart cities. PhD thesis, University of Côte d'Azur (2018)
12. Mel, P., Grance, T.: The NIST Definition of Cloud. NIST Special Publication 800-145, Gaithersburg, USA (2011)
13. Abelin, S.M.: Intelligent monitoring & control capability. World Pumps **17**(2), 30–32 (2017)
14. Niedermaier, M., Malchow, J.-O., Fischer, F., Marzin, D., Merli, D., Roth, V., Von Bodisco, A.: You snooze, you lose: measuring {PLC} cycle times under attacks. In: 12th USENIX Workshop on Offensive Technologies, Baltimore, MD (2018)
15. Fuster Losa, M.: Proyecto de puesta en marcha de unidades de telecontrol compactas y autónomas energéticamente en el contexto de la industria 4.0. Universitat Politècnica de València, València (2018)
16. Merchant, A., Mohan Kumar, M., Ravindra, P., Vyas, P., Manohar, U.: Analytics driven water management system for Bangalore city. Procedia Eng. **70**, 1137–1146 (2014)

17. Armon, A., Gutner, S., Rosenberg, A., Scolnicov, H.: Algorithmic network monitoring for a modern water utility: a case study in Jerusalem. Water Sci. Technol. **63**(2), 233–239 (2011)
18. Elkaseer, A., Salama, M., Ali, H., Scholz, S.: Approaches to a practical implementation of Industry 4.0. In: The Eleventh International Conference on Advances in Computer-Human Interactions, Rome, Italy (2018)
19. Chalal, L., Dieulot, J.-Y., Dauphin-Tanguy, G., Colas, F.: Supervisory predictive control of a hybrid solar panels, microturbine and battery power generation plant. In: IFAC Power Plants and Power Systems Control 2012, Toulouse, France (2012)

Multi-Layer Perceptron Artificial Neural Network Based IoT Botnet Traffic Classification

Yousra Javed[✉] and Navid Rajabi

Illinois State University, Normal, USA
{yjaved,nrajabi}@ilstu.edu

Abstract. Internet of Things (IoT) is becoming an integral part of our homes today. Internet-connected devices, such as smart speakers, smart bulbs, and security cameras are improving our convenience and security. With the growth in smart environments, there is an increasing concern over the security and privacy issues related to IoT devices. The issue of the IoT security has received considerable attention due to (1) the intrinsic technological constraints of IoT devices (computing and storage limitations) and (2) its prevalence in people's life's, in close proximity. IoT devices can be easily compromised (much easier than PCs and/or smart phones) and can be utilized for generating botnet attacks. In this paper, we propose an Artificial Intelligence (AI) based solution for malicious traffic detection. We explore the accuracy of Multi-Layer Perceptron (MLP) Artificial Neural Network (ANN) learning algorithm in detecting botnet traffic from IoT devices infected by two major IoT botnets, namely, Mirai and Bashlite (also known as Gafgyt). After tuning and optimization, the MLP-ANN algorithm achieved an accuracy rate of 100% in the testing phase of IoT botnet traffic classification.

Keywords: Multi-Layer Perceptron Artificial Neural Networks · IoT security · Botnets · Mirai · Bashlite

1 Introduction

Recent developments in high-speed communication networks (3G, 4G, LTE, 5G) and cutting-edge technologies, such as Software-Defined Networking (SDN) and Network Function Virtualization (NFV) have made the concept of smart cities more feasible. Consequently, these advancements are constantly trying to offer services with higher Quality of Service (QoS) and Quality of Experience (QoE). This has therefore increased the demand of smart world implementation among people and vendors.

With the growth in the demand of smart cities, there is an increasing concern among the researchers, vendors, and customers over the resulting security and privacy issues. Recent trends have led to a proliferation of studies in Internet of Things (IoT) security and possible solutions to overcome this obstacle.

© Springer Nature Switzerland AG 2020
K. Arai et al. (Eds.): FTC 2019, AISC 1069, pp. 973–984, 2020.
https://doi.org/10.1007/978-3-030-32520-6_69

Many factors compromise the security of IoT. For example, there are major impediments in applying the conventional network security best practices to IoT networks. These include: (1) the lack of sufficient power among the major IoT equipment like sensors, and actuators that restricts the time complexity of the underlying algorithms, and (2) the lack of storage among IoT devices that constrains the storage complexity. Human weaknesses is another factor. Many IoT device owners do not change the factory default or hard-coded usernames and passwords.

Recently, two famous botnets exploited the weak security on IoT devices and utilized them to run distributed denial of service (DDoS) attacks. These botnets are known as *Mirai* and *Bashlite* (also known as Gafgyt) [1,6].

According to Symantec, Mirai appeared in several attacks in 2016 [12]. These include the attack on French hosting company OVH, reaching at 1 Tbps traffic, the massive DDoS attacks targeting the website of journalist Brian Krebs (which reached 620 Gbps), the attacks targeting DNS provider Dyn (resulting the inaccessibility of several high-profile websites such as Twitter, GitHub, Netflix, etc.), the Liberia Lonestar attacks, and the Deutsche Telekom CWMP exploit.

Mirai operates by continuously scanning for IoT devices that are accessible over the Internet and are protected by factory default or hard-coded usernames and passwords. It then infects these devices with malware that forces them to report to a central control server, turning them into a bot that can be used in DDoS attacks. Mirai targets are usually Routers, DVRs, CCTV cameras, and any other 'smart', Internet-connected appliances.

In this paper, we propose an Artificial Intelligence (AI) based solution for malicious traffic detection and explore the Multi-Layer Perceptron (MLP) Artificial Neural Network (ANN) learning algorithm [3] for malicious traffic detection. Several applications of MLP have been investigated in the atmospheric sciences [2] and in Internet traffic prediction [11]. However, to the best of our knowledge, the accuracy of MLP-ANN learning algorithm in detecting IoT botnet traffic has not been investigated. We tested the MLP-ANN model on dataset on two IoT devices infected by Mirai and Bashlite. After tuning and optimization, the MLP-ANN algorithm achieved an accuracy rate of 100% in the testing phase of IoT botnet traffic classification.

We chose Deep Learning over Machine Learning due to the strong correlation between the size of the dataset and the performance of the model in Deep Learning (see Fig. 1). Another superiority of Deep Learning approaches over Machine Learning is that there is no need for separate feature extraction phase before the classification phase in Deep Learning. This leads to less human resource dependency and more automated environment.

The remainder of this paper is organized as follows. Section 2 describes the Multi-Layer Perceptron Artificial Neural Network learning algorithm in detail. Section 3 discusses the related work. Section 4 explains our proposed model's specifications. Section 5 describes the dataset used for testing the performance of our proposed model. Section 6 discusses the performance of our tuned and

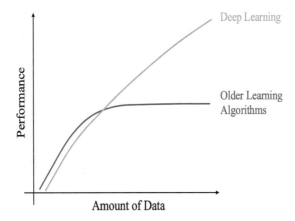

Fig. 1. Deep Learning vs Machine Learning performance difference as dataset gets larger [9]

optimized MLP ANN algorithm on this dataset. Finally, Sect. 7 concludes this paper with our main findings.

2   Multi-Layer Perceptron Artifical Neural Networks

MLP-ANN [3] is one of the most common architectures for neural networks which have one input layer (input features), one output layer (classes or labels), and one or more hidden layers in between. MLP-ANN is a fully-connected graph of these layers in which each hidden layer may contain an arbitrary number of neurons. MLP-ANN can be used for classification purposes when the dataset is labeled (also known as Supervised Learning). The MLP model needs to be trained using the training dataset (a portion of the whole dataset) to learn the associations between the input features and output labels. The accuracy of the model is measured by testing the model using the test dataset (the remaining data set which has not been seen by the model yet).

MLP-ANN learning algorithm comprises of two major functions for performing a classification task: (1) Forward-Propagation and (2) Back-Propagation:

Forward-Propagation is the process of feeding the data to the input layer of the MLP-ANN. In this way, the input layer can propagate the data through the next layers (hidden layers). Then, each neuron multiplies the input data by a weight term and adds a bias term to it. So, each neuron (expect the input layer neurons) calculates the weighted sum z as follows:

$$z_j^{[i]} = {w_j^{[i]}}^T x + b_j^{[i]}$$

Where w, x, b and z represent the initial weights matrix of the previous layer neurons, the input values vector of the previous layer neurons, the bias vector of the previous layer, and the output respectively. Also, i is the layer number and j is the neuron number within the layer. After calculating z, the output of z is passed to an activation function to squeeze the real numbers into a specific range.

These steps continue all the way through the neural network to reach the output layer. Then, the predicted label is compared with the actual label to obtain the corresponding loss and realizing the appropriateness of the weights and the biases using a Loss Function such as Cross-entropy:

$$L(z, y) = -[ylog(z) + (1 - y)log(1 - z)]$$

Back-Propagation is a method for updating the weights, considering the predicted output, desired output and their difference. The partial derivative with respect to w is computed by using chain rule as follows:

$$\frac{\partial L(z, y)}{\partial w} = \frac{\partial L(z, y)}{\partial a} \times \frac{\partial a}{\partial z} \times \frac{\partial z}{\partial w}$$

Where a is the output of any activation function that gets z and squeezes it into a desired range. Then, the computed gradient is deducted from weights w to achieve a new w as follows:

$$w \leftarrow w - \eta \frac{\partial L(z, y)}{\partial w}$$

Where η is the learning rate. These partial derivative calculations are done by an optimizer (e.g. Stochastic Gradient Descent) as a convex optimizer. If we define the loss function as a 2D diagram in which y-axis represents the loss (error) value and x-axis represents weights w, the optimization process helps to find the global minimum of the convex in an iterative manner, starting with an initial point. The ultimate goal is to find an optimum weight w in which the output of the loss function reaches a value near zero. This means if we choose that specific weight w, the generated output label and the desired output label will be at the closest possible distance. A similar process is done for biases b and these to values are propagated backward to the earlier layers neurons for justification (up to the first hidden layer). Then, each neuron performs the calculations based on the updated w and b and follows the feed-forward propagation process again. This iterative process will be done for the training phase until the model reaches it's highest possible efficiency.

3   Related Work

A considerable amount of literature has been published on IoT Botnet detection and Intrusion Detection Systems. These studies can be classified as follows: (1) The first category represents comprehensive surveys that have investigated the nature and behavior of Mirai malware [1]. (2) The second category contains the packet inspection approaches [14]. However, on an enterprise scale, these lightweight approaches tend to be less realistic considering the complicated nature and behavior of modern botnets. (3) The third category includes the Host-based intrusion detection systems (HIDS) [13].

On an enterprise scale with fragile IoT devices, it would more efficient to use a Network-based intrusion detection system (NIDS). The detection time could be minimized using high speed techniques of Machine Learning and Deep Learning to instruct the system to act more efficiently even in situations that are totally unprecedented and unseen by the model [8].

Recently, there have been research efforts on using Deep Learning based approaches for IoT Botnet detection. One of them is a Recurrent Neural Networks (RNN) based approach for learning traffic behavior as time passes (similar to Time-Series analysis approach) [15]. The second one is a Deep Autoencoder (DAE) approach known as N-BaIoT [7].

DAE is a type of Deep Learning model that can be defined as an encoder plus a decoder and has a symmetrical architecture. Each hidden layer includes less number of neurons compared to the immediate previous layer until the middle layer (which can be defined as the encoding phase). In the next layer, the exact reverse scenario happens from the middle layer to the output layer (which can be determined as the decoding phase). In DAE, input features are given to the first layer (input layer) and the model generates an encoded (compressed) version of the input data at the middle of the neural network. Then, the model tries to regenerate the exact same input data as an output. So, the accuracy of the model is examined by the similarity between the actual input data and the regenerated version of the input data. In this way, the model demonstrates that it has learned the existing dependencies and connections within the dataset.

In N-BaIoT, three Machine Learning-based approaches were used in addition to DAE, namely, Local Outlier Factor (LOF), One-Class SVM, and Isolation Forest. However, DAE model achieved the best True Postive Rate (TPR), False Positive Rate (FPR) and detection time among the other techniques [7].

MLP has shown less complexity and less Normalized Root Mean Squared Error compared to Stacked Autoencoders (SAE) in the internet traffic prediction use case [11]. On the other hand, since the architecture of DAE/SAE is based on the dimensionality reduction, it is more powerful in detecting common (frequent) patterns than uncommon patterns [7]. This can be construed as a downside for DAE/SAE since it is more logical to invest in building/training deep learning models that can be leveraged for transfer learning in the future. So, it may seem that these two models achieve the same level of performance (accuracy rate) in this dataset size, but it is completely probable that trivial/uncommon/outlier traffic records in small datasets can form a frequent malicious pattern when we scale up the dataset. This can make a huge difference between MLP ANN and DAE/SAE performance level. We came up with the idea to transform this problem to a supervised learning problem and test the feasibility of classifying the output label of this dataset by designing, tuning and optimizing an MLP ANN model.

4 Proposed Model's Specifications

4.1 Architecture

The proposed model is a 4-Layer Perceptron ANN, consisting of one input layer with 115 neurons (since we have 115 input features in our data set), followed by three hidden layers with 114 neurons each, and an output layer with two neurons (since we have two labels as output classes in which 0 represents benign traffic and 1 represents malicious traffic). We can simply define the architecture as {115, 114, 114, 114, 2} and use it as a representation convention to show the MLP-ANN architecture in this paper.

The motivation for this architecture is the "accuracy" and "complexity" trade-off in Deep Neural Networks (DNN) design. By definition, if the model doesn't contain any hidden layer, it is a Single-Layer Perceptron (SLP). In this case, the input layer is directly connected to the output layer without any hidden layer in between and the model won't achieve an acceptable level of performance in practice. Consequently, DNNs have been implemented to increase the performance by increasing the number of hidden layers and neurons. However, the computational complexity of MLP-ANN should be taken into account to decrease both the training and testing times, and to achieve a faster convergence. According to Scikit-learn Neural Networks documentation, the computational complexity (during the Back-propagation process) of a generic DNN or MLP is in order of $O(n.m.h^k.o.i)$ [5], where n is the number of training samples, m is the number of features, h is the number neurons within a layer, k is the number of hidden layers, o is the number of output neurons, and i is the number of iterations. Therefore, each of the parameters needs to be selected with scrutiny to achieve a descent level of efficiency, especially the number of neurons and hidden layers which grow exponentially.

4.2 Data Pre-processing and Normalization

The original dataset is not labeled and the owner of the dataset used unsupervised technique, known as Deep Autoencoders for classification. Therefore, we labeled the dataset by assigning 0 to all the benign traffic instances and 1 to all the malicious traffic. In addition, since the original dataset separated the traffic by the type of the attack and its protocol, we merged all the malicious traffic into a single dataset including the benign traffic. Then, we used the "Shuffle" feature to make sure all the data records are randomly distributed.

According the Scikit-learn documentation (Section 1.17.8. Tips on Practical Use), MLP-ANN is sensitive to feature scaling [5]. Therefore, we considered their recommendation and used the StandardScaler method to scale our data.

4.3 Optimization and Tuning

In this paper, we used Sigmoid (also known as logistic) activation function (σ) to pass the z value into the logistic function to obtain output values between 0 and 1 by considering 0.5 as a distinguishing threshold:

$$\sigma(z) = \frac{1}{1 + e^{-z}}$$

The logic behind this activation function selection is that we have two output classes (labels) in this problem. Consequently, the binary classification approach should be followed.

In terms of the loss function, we used Cross-entropy which is a popular loss function for ANNs. Cross-entropy (C) can be defined as follows [10]:

$$C = -\frac{1}{n} \sum_{x} [y ln a + (1 - y) ln(1 - a)]$$

In which:

$$a = \sigma(z) = \sigma(\sum_{j} w_j x_j + b)$$

For optimization, we used Adaptive Moment estimation (Adam) [4]. Adam is the optimized version of the Stochastic Gradient Descent. The Scikit-learn documentation states that Adam is very robust for large dataset [5]. Adam consists of four parameters that should be tuned, known as α, $\beta 1$, $\beta 2$ and ϵ that represent the learning rate, the exponential decay rate for the first moment estimates, the exponential decay rate for the second-moment estimates and a very small number to prevent any division by zero respectively. We tuned our model with the best practices suggested by Adam authors ($\alpha = 0.001$, $\beta 1 = 0.9$, $\beta 2 = 0.999$ and $\epsilon = $ 1e-08) as well as the Nesterovs Momentum, which is one the best practices asserted by Scikit-Learn. In Adam, weights w and biases b will be updated using the following formulas:

$$w \leftarrow w - \alpha \frac{V_{dw}}{\sqrt{S_{dw}} + \epsilon}$$

$$b \leftarrow b - \alpha \frac{V_{db}}{\sqrt{S_{db}} + \epsilon}$$

Another reason for replacing SGD with Adam is that Adam combines the best properties of two famous optimizers (AdaGrad and RMSProp) to come up with an optimization algorithm that can work efficiently in the noisy problems.

For the hyperparameters optimization, we have started with an initial architecture as {115, 115, 115, 115, 2} since we wanted to achieve the best possible performance with the least computational complexity. Considering the fact that we have 115 input features and do not want to take all of them into account to increase the performance, we didn't have the possibility of decreasing the number of neurons drastically.

On the other hand, we were trying to train an MLP-ANN model with the lowest complexity. Choosing 3 hidden layers as a trade-off for training achieved close to 100% accuracy in the testing phase. Therefore, we tried to change the number of neurons of the hidden layers for the final tuning and achieved the exact 100% accuracy using the {115, 114, 114, 114, 2} architecture.

Another challenge in Machine Learning and Deep Learning is over-fitting (also known as selection bias). Over-fitting is a phenomena in which the model is over-trained using too much training data. Consequently, the model works well in the training phase, but unsatisfactory in the testing phase. We took two major steps to overcome this challenge: (1) we shuffled our dataset using a random function to avoid selection bias as much as possible, and (2) we used 33/67 ratio instead of 30/70 ratio when splitting the dataset into testing/training portions to be as close as possible to the N-BaIoT [7] approach.

5   Dataset

The dataset we used in our experiment contains 10 classes of attacks, captured from 9 IoT devices infected with Mirai and Bashlite (Gafgyt) [7]. These devices are as follows:

1. Danmini Doorbell
2. Ennio Doorbell
3. Ecobee Thermostat
4. Philips B120N/10 Baby Monitor
5. Provision PT-737E Security Camera
6. Provision PT-838 Security Camera
7. SimpleHome XCS7-1002-WHT Security Camera
8. SimpleHome XCS7-1003-WHT Security Camera
9. Samsung SNH 1011 N Webcam

This dataset has two major advantages over the other datasets: (1) It is Network-based, instead of being Host-based. (2) It has been captured using real devices, instead of simulated network traffic. In this dataset, Mirai attack consists of the following malicious traffic:

Scan represents the automatic scanning of vulnerable devices
Ack represents Ack flooding
Syn represents Syn flooding
UDP represents UDP flooding
UDPplain represents UDP flooding with fewer options, optimized for higher PPS

The Bashlite (Gafgyt) attack includes the following malicious traffic:

Scan indicates scanning the network for vulnerable devices
Junk indicates sending spam data
UDP indicates UDP flooding
TCP indicates TCP flooding
COMBO indicates sending spam data and opening a connection to a specified IP address and port

The dataset contains 115 statistical features of the raw network traffic data, representing the behavior of IoT devices in the network (associated with the source and destination IP and MAC addresses). These statistical features include mean, variance, number, magnitude, radius, co-variance, and correlation coefficient etc. extracted from packet streams through the network. Our analysis incorporates all the above-mentioned attack classes (10 classes) plus the benign class (which forms 11 classes in total).

6   Analysis and Results

We used the Scikit-learn MLPClassifier as the model for training, testing, and optimization. We leveraged a subset of the dataset in our analysis by focusing on the traffic on two of the nine devices included in the N-BaIoT dataset [7]. These devices are Philips B120N/10 Baby Monitor and Provision PT-838 Security Camera. We focused on these two devices for the following reasons:

1. Firstly, the major targets of Mirai and BASHLITE botnets are CCTVs and Security Cameras.
2. Secondly, these two devices are the top two in terms of the traffic volume. So, analyzing these two can give a realistic perspective on the whole dataset.
3. Thirdly, the Philips B120N/10 Baby Monitor is the most complicated and sensitive device among the others since it can be considered as a package of IoT devices, such as webcam, temperature sensing, air quality sensing, etc.
4. Finally, we had computational and storage constraints for running our experiment (Core i5 CPU and 8 GB RAM).

Table 1. Classification performance of our MLP-ANN model on botnet traffic from two IoT devices

Measurements	Security camera	Baby monitor
Precision	1.00	1.00
Recall	1.00	1.00
f1-score	1.00	1.00
Support (0)	32444	57858
Support (1)	243731	304706
Micro avg	1.00	1.00
Macro avg	1.00	1.00
Weighted avg	1.00	1.00
Training instances	560716	736113
Testing instances	276175	362564
Total instances	836891	1098677
Test set score	1.00	1.00

Table 1 represents the statistical performance of our MLP-ANN model: If we define Confusion Matrix (C) as follows:

$$C = \begin{bmatrix} TP & FP \\ FN & TN \end{bmatrix}$$

The Confusion Matrix of the Provision Security Camera is as follows:

$$C_{Provision} = \begin{bmatrix} 32444 & 0 \\ 0 & 243731 \end{bmatrix}$$

The model achieved a 100% accuracy in True Positive Rate (TPR) since the sum of TP + TN is exactly equal to the number of total testing data instances:

$$TP + TN = 32444 + 243731 = 276175$$

The Confusion Matrix of the Philips Baby Monitor is as follows:

$$C_{Philips} = \begin{bmatrix} 57858 & 0 \\ 0 & 304706 \end{bmatrix}$$

Once again the model achieved a 100% accuracy rate in TPR:

$$TP + TN = 57858 + 304706 = 362564$$

Figures 2 and 3 demonstrate the loss function during the training phase for Provision Security Camera and Philips Baby Monitor, respectively.

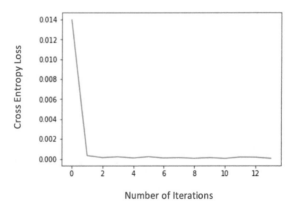

Fig. 2. Provision Security Camera's loss function during the training phase

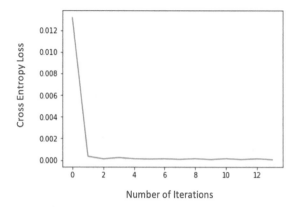

Fig. 3. Philips Baby Monitor's loss function during the training phase

7   Conclusion

In this paper, we proposed an MLP-ANN model for IoT botnet classification. We tested our model on UCI's N-BaIoT dataset for two devices and achieved the same TPR and accuracy rate (100%) as N-BaIoT [7]. However, our model uses a supervised learning approach unlike N-BaIoT's unsupervised approach. In addition, we used a subset of their dataset to prove that MLP-ANN can achieve the same accuracy level even with limited resources. The performance comparison between MLP-ANN and Stacked Autoencoders (SAE) [11] shows that both deep learning models achieved a similar level of performance. However, MLP-ANN has shown less Normalized Root Mean Squared Error (NRMSE) and less complexity over SAE in Internet traffic prediction. Another advantage of MLP ANN over DAE/SAE is the fact that MLP ANN can be more sensitive to every single record/pattern than DAE/SAE, regardless of being frequent or not, which can be beneficial in the future to focus on training a model which is sensitive from the beginning.

It is worth mentioning that MLP-ANN model can become more powerful by being trained on more data from real IoT devices. An advantage of Deep Learning is that the trained models can be useful in the future using the Transfer Learning concept. In addition, hybrid models are also recommended to intensify the reliability and robustness of the classifier. In this way, a combination of robust techniques can be utilized for traffic classification, instead of relying on a single technique.

References

1. Antonakakis, M., April, T., Bailey, M., Bernhard, M., Bursztein, E., Cochran, J., Durumeric, Z., Halderman, J.A., Invernizzi, L., Kallitsis, M., et al.: Understanding the Mirai botnet. In: USENIX Security Symposium, pp. 1092–1110 (2017)

2. Gardner, M.W., Dorling, S.R.: Artificial neural networks (the multilayer perceptron)–a review of applications in the atmospheric sciences. Atmos. Environ. **32**(14–15), 2627–2636 (1998)
3. Jain, A.K., Mohiuddin, K.M.: Artificial neural networks: a tutorial. Computer **29**(3), 31–44 (1996)
4. Kingma, D.P., Ba, J.: Adam: a method for stochastic optimization. arXiv preprint arXiv:1412.6980 (2014)
5. Scikit Learn: Neural network models documentation (supervised learning). https:// scikit-learn.org/stable/modules/neural_networks_supervised.html
6. Marzano, A., Alexander, D., Fonseca, O., Fazzion, E., Hoepers, C., Steding-Jessen, K., Chaves, M.H., Cunha, Í., Guedes, D., Meira, W.: The evolution of Bashlite and Mirai IoT botnets. In: 2018 IEEE Symposium on Computers and Communications (ISCC), pp. 00813–00818. IEEE (2018)
7. Meidan, Y., et al.: N-BaIoT–network-based detection of IoT botnet attacks using deep autoencoders. IEEE Pervasive Comput. **17**(3), 12–22 (2018)
8. Midi, D., Rullo, A., Mudgerikar, A., Bertino, E.: Kalis—a system for knowledge-driven adaptable intrusion detection for the Internet of Things. In: 2017 IEEE 37th International Conference on Distributed Computing Systems (ICDCS), pp. 656–666. IEEE (2017)
9. Ng, A.: What data scientists should know about deep learning. https://goo.gl/ nJqHsP (2015)
10. Nielsen, M.A.: Neural Networks and Deep Learning, vol. 25. Determination Press, USA (2015)
11. Oliveira, T.P., Barbar, J.S., Soares, A.S.: Multilayer perceptron and stacked autoencoder for internet traffic prediction. In: IFIP International Conference on Network and Parallel Computing, pp. 61–71. Springer (2014)
12. Symantec Security Response: Mirai: what you need to know about the botnet behind recent major DDoS attacks. https://www.symantec.com/connect/blogs/ mirai-what-you-need-know-about-botnet-behind-recent-major-ddos-attacks (2016)
13. Sedjelmaci, H., Senouci, S.M., Al-Bahri, M.: A lightweight anomaly detection technique for low-resource IoT devices: a game-theoretic methodology. In: 2016 IEEE International Conference on Communications (ICC), pp. 1–6. IEEE (2016)
14. Summerville, D.H., Zach, K.M., Chen, Y.: Ultra-lightweight deep packet anomaly detection for internet of things devices. In: 2015 IEEE 34th International Performance Computing and Communications Conference (IPCCC), pp. 1–8. IEEE (2015)
15. Tuor, A., Kaplan, S., Hutchinson, B., Nichols, N., Robinson, S.: Deep learning for unsupervised insider threat detection in structured cybersecurity data streams. arXiv preprint arXiv:1710.00811 (2017)

Project BUMP: Developing Communication Tools for the Older Adult Population

Claudia B. Rebola[✉] and Shi He

University of Cincinnati, Cincinnati, OH 45221, USA
rebolacb@ucmail.uc.edu, hesh@mail.uc.edu

Abstract. This paper describes the process of designing and developing robotic technologies as prosthetics for the home space to respond to the needs and advance the quality of life of older adults. The trend demographics report a significant increase of the number of older adults in the next decade in the United States, which would generate the most significant social transformations of the 21st century. Supporting older adult to age in place gracefully is a critical need. Even though there are a number of technologies that have been designed to support older adult, there are still challenges in their implementation and use, especially in products related to enabling communication. This project aims at exploring opportunities to expand the capabilities of existing homes and assist older adults in activities of daily life in their home. The project BUMP portrays a supportive human environment and how to extend older adults' home in other's homes. The significance of this project is to share pragmatic examples on how to better design technologies that are more natural, embedded and embodied communication tool for the older adult population.

Keywords: Environment · Robotics · Older adults · Well-being

1 Introduction

Trend demographics report a significant increase of the number of older adults in the next decades [1, 2]. This will lead to the most significant social transformations of the twenty-first century. In addition, indicators of well-being of older adults shows that the majority of older adults isolated from society. Older Americans spend approximately one third of their time in leisure activities (approximately 7 h) [3]. Of those hours, they spend approximately 4 h watching TV, 13 min of recreation/exercise and 42 min socializing/communicating. Socialization and communication are a critical component in human life to live meaningfully with a high quality of life [4–6].

According to market research, within the areas of digital transformation and IoT technology, there has been a growing interest in designing for living spaces that enable people to interact with human-made environments to access digital life and welfare [7]. However, there are still challenges to designing and developing a human-centered enabling and supportive environments for older adults. Current market solutions that connect via Internet with smart phones and other IoT devices for older adults are not friendly and need very high learn costs [8]. They pose a series of challenges for smart devices design and development.

© Springer Nature Switzerland AG 2020
K. Arai et al. (Eds.): FTC 2019, AISC 1069, pp. 985–990, 2020.
https://doi.org/10.1007/978-3-030-32520-6_70

As the aging population continues to grow, predictions estimate that there are more people who can expect to live into their 60 s and beyond in the twenty-one century. The opportunities for technology developments and consequences for health and health systems are profound [9]. Even though there have been innovations addressing older adults' needs, technologies for the population have not kept up with this demographic trend to help improve their lifestyles and aging in place. Although there are many IoT technologies in the market, older adults are still challenged by adopting technologies. Due to the fact that most hardware and software in the market do not cater older adults' needs, they are not familiar with the changes the technology made [10, 11].

2 Design Principle

In this paper, BUMP is presented as a design of a new communication to bridge unmet participation [12]. The goal of the project is to extend older adult's home spaces that enables them to connect with families and the environment in more natural ways. This project's contribution is to explore the possibilities to improve the environment, making the environment smarter and more supportive of its inhabitants. In addition, the goal is to advance technologies that can improve the older adults' quality of life.

A research needs to be proposed that through design methodology to develop IoT interventions for the home by tightening the relationships between the individual, the objects in the home, and the environment to care for the personal health. Literature review especially the research of older adults' key behavior that staying at home and watching television for a long time demonstrates that the problems such as long-time sedentary lifestyle and isolation from families and friends and further societies. Therefore, the older adults need to pay attention to their physical health, social health and mental health, which are largely influenced by their daily life. In order to bring well-being to older adults, it is necessary to improve older adults' daily life at home, so that they are able to move more frequently, connect easily with their families and friends, and realize that they are still a part of society and begin to self-motivate. Therefore, those problems and behaviors result in a list of opportunities in the following implications:

2.1 Physical Health

The design should encourage older adults to be active at home. Due to the fact that older adults spend most of spare time sitting on the sofa and watching TV, being active is necessary. The design should release right and positive triggers for older adults to move on their own. Moreover, it should be clear, simple, and easy to understand that does not need complicated operations but interpret human's behaviors and create experiences naturally.

2.2 Social Health

Design a way of communication to connect older adults with families, friends and care givers. Older adults' mobility difficulties and public space barriers isolate them from their families, friends, and care givers and further the society. The design should not only create a cellphone-like way to communicate, but should it extend older adults' senses through linking outside spaces with their limited indoor spaces.

2.3   Mental Health

Help older adults to focus on the self and mind. Older adults' long-time isolated lifestyle is the one of the main issues causing their mental health. The design should always consider the mental influences on older adults including whether it inspires them to be curious to unfamiliar stuffs, how enhance the desires of focusing on themselves, and motivate them to be willing to communicate with others.

3   Bump

The BUMP – Bridging Unmet Modes of Participation is a micro-interactive installation. People in two different sites will see each other through BUMP the stream video connection when they show up in front of the screens at the same time (see Fig. 1). This enables older adults to communicate with families and friends more easily.

The digital prototype is designed and rendered using Rhinoceros and Keyshot. In order to make the electronic experience carried on the screen more elegant and accessible for older adults, the stiffness and sense of technology brought by the screen itself should be avoided. By placing the large screen vertically and adding a frame to the screen, it becomes more like an ornament in the home environment.

The BUMP feature requires a motion sensor to pick up the person's dynamic signal, as well as a webcam to capture the video. People in both places should see each other at the same time. Therefore, we use Arduino to install PIR sensor, and a small computer to run the program and Arduino (see Fig. 2).

According to the first iteration on the paper BUMP- Bridging Unmet Modes of Participation [12], the delay dramatically impact user's feelings. Because the network supported by the server is too slow. Then we try to use Skype which has a better server as the second iteration to support stream video, and python to turn on/off screens when the data input comes from motion sensor at two places. However, the signal transmission from each side generates errors frequently. Because in order to avoid error messages, the computer should confirm inputs from motion sensor for more than 3 times. The signal reading speed greatly reduces the possibilities that people will see each other through BUMP which does not meet the design purpose of BUMP, and the program is error-prone.

Therefore, in this project, the streaming platform Kafka [13] is adapted to build stream videos. When people's actions trigger the PIR sensor, the data will command the webcam to capture the people and the Kafka will help send videos to the server, and the computer fetches the videos which uploaded by another computer (see Fig. 3).

Fig. 1. Using BUMP

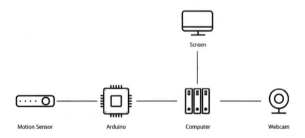

Fig. 2. Installation diagram

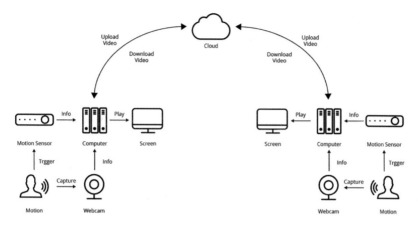

Fig. 3. Stream video service schematic diagram

4 Conclusion

The project BUMP, a new communication way which is triggered by participants' actions instead of traditionally pressing a button on a phone screen. When people in two different places appear at the same time, they will see each other through the screens (see Fig. 1).

This concept for home space is designed to extend older adults' living spaces. Bringing some surprises named as BUMP to their routine is a great way to create connections with other people such as care-givers, friends or families and build new modern relationships with them. Meanwhile, the wider extensive uses of the BUMP could be exploited in not merely short communication between families or friends but medical use when facing emergencies or knowing each other's situations at home.

However, the limitation of this design is that due to the short delay, the user's immediate experience is dramatically impacted. Also, the location of this device at home needs to be further researched. It is necessary to know older adults' daily activities routes and use BUMP in the right place where it protects privacy, avoids scaring and reveals the intimacies to motivate older adults to move properly.

The next step for this project is exploring other important usages in different spaces. Several additional usages are possible: (1) Working spaces; (2) Public spaces. In addition to exploring usages, user behaviors that how people respond to and understand such a responsive spatial installation and the new way communication will be assessed.

References

1. WHO | World report on ageing and health. http://www.who.int/ageing/events/world-report-2015-launch/en/. Accessed 22 Mar 2017

2. United Nations: World Population Ageing. (United Nations), 114 (2013). http://doi. org/ST/ESA/SER.A/348. https://www.google.com/search?client=safari&rls=en&q=United +Nations.+(2013).+World+Population+Ageing.+(+United+Nations),+114.+. http://doi. org/ST/ESA/SER.A/348&ie=UTF-8&oe=UTF-8. Accessed 22 Mar 2017
3. F.I.F.A. Statistics: Older Americans 2016 Key Indicators of Well-Being. CreateSpace Independent Publishing Platform (2016)
4. Kasem Cares: Socialization Leads to a Better Quality of Life for Seniors. https://www. kasemcares.org/socialization_leads_to_a_better_quality_of_life_for_seniors
5. Cornwell, E.Y., Waite, L.J.: Social disconnectedness, perceived isolation, and health among older adults. J. Health Soc. Behav. **50**(1), 31–48 (2009)
6. National Institute of Mental Health: Older Adults and Depression (2014). http://www.nimh. nih.gov/health/publications/older-adults-and-depression/index.shtml
7. K.H.-P. Care, Schlesinger, M., Golinkoff, R.M., Esther: The New Humanism: Technology should enhance, not replace, human interactions. Brookings, 11 June 2018
8. Center for Technology and Aging: Technologies to Help Older Adults Maintain Independence: Advancing Technology Adoption (2009)
9. Jones, B., Winegarden, C.R., Rogers, W.A.: Supporting Healthy Aging with New Technologies. Interactions **16** (2009)
10. Fisk, A.D., Rogers, W.A., Charness, N., Czaja, S.J., Sharit, J.: Designing for Older Adults: Principles and Creative Human Factors Approaches, 2nd edn. CRC Press, Boca Raton (2009)
11. Rebola, C.B.: Designed Technologies for Healthy Aging. Morgan & Claypool Publishers, San Rafael (2015)
12. Chu, C., Rebola, C.B., Kao, J.: BUMP: bridging unmet modes of participation. In: Proceedings of the 2015 British HCI Conference, New York, NY, USA, pp. 261–262 (2015)
13. Apache Kafka: Apache Kafka. https://kafka.apache.org/. Accessed 30 Mar 2019

Author Index

© Springer Nature Switzerland AG 2020
K. Arai et al. (Eds.): FTC 2019, AISC 1069, pp. 991–993, 2020.
https://doi.org/10.1007/978-3-030-32520-6

Printed in the United States
By Bookmasters